Springer Aerospace Technology

Series Editors

Sergio De Rosa, DII, University of Naples Federico II, NAPOLI, Italy

Yao Zheng, School of Aeronautics and Astronautics, Zhejiang University, Hangzhou, Zhejiang, China

The series explores the technology and the science related to the aircraft and spacecraft including concept, design, assembly, control and maintenance. The topics cover aircraft, missiles, space vehicles, aircraft engines and propulsion units. The volumes of the series present the fundamentals, the applications and the advances in all the fields related to aerospace engineering, including:

- structural analysis,
- aerodynamics,
- aeroelasticity,
- aeroacoustics,
- flight mechanics and dynamics
- orbital maneuvers,
- avionics,
- systems design,
- materials technology,
- launch technology,
- payload and satellite technology,
- space industry, medicine and biology.

The series' scope includes monographs, professional books, advanced textbooks, as well as selected contributions from specialized conferences and workshops.

The volumes of the series are single-blind peer-reviewed.

To submit a proposal or request further information, please contact:

Mr. Pierpaolo Riva at pierpaolo.riva@springer.com (Europe and Americas) Mr. Mengchu Huang at mengchu.huang@springer.com (China)

More information about this series at http://www.springer.com/series/8613

Alessandro de Iaco Veris

Fundamental Concepts
of Liquid-Propellant Rocket
Engines

 Springer

Alessandro de Iaco Veris
Rome, Italy

ISSN 1869-1730   ISSN 1869-1749   (electronic)
Springer Aerospace Technology
ISBN 978-3-030-54706-6   ISBN 978-3-030-54704-2   (eBook)
https://doi.org/10.1007/978-3-030-54704-2

Cover illustration: This cover image is taken from NASA website (https://www.nasa.gov/audience/foreducators/rocketry/imagegallery/rpd_spaceX.jpg.html).

This Springer imprint is published by the registered company Springer Nature Switzerland AG
The registered company address is: Gewerbestrasse 11, 6330 Cham, Switzerland

Preface

The present book is conceived for undergraduate or graduate aerospace engineers and in particular for those who design liquid-propellant engines for rocket propulsion.

This book gives several data on existing engines, typical values of design parameters, and worked examples of application of the concepts discussed with their numerical results. This enables the reader to apply the concepts introduced to cases of practical interest.

This book uses only the metric system (SI).

Rome, Italy

Alessandro de Iaco Veris

October 2019

Contents

Chapter 1
Fundamental Concepts on Liquid-Propellant Rocket Engines

1.1 The Generation of Thrust

A rocket is a vehicle which moves through space by ejecting a propellant gas at high velocity oppositely to the direction in which the vehicle is desired to move. High-thrust propulsion systems for rocket engines use chemical propellants, which are burned in the combustion chamber of a rocket to generate thrust. According to Newton's third law of motion (to every action there is always opposed an equal reaction), the continuous ejection of a stream of hot gases in one direction causes a steady motion of the rocket in the opposite direction.

A couple of propellants comprises a fuel and an oxidiser. A fuel is a substance, such as liquid hydrogen (H_2) or a mixture of hydrocarbons, which burns in the presence of oxygen to form the hot gases which are accelerated and then ejected at high speed through a nozzle. An oxidiser is either liquid oxygen (O_2) or a substance releasing oxygen, such as hydrogen peroxide (H_2O_2), nitrous oxide (N_2O), nitronium perchlorate (NO_2ClO_4), or another substance, such as fluorine (F_2), having a strong tendency to accept electrons. The oxidiser combines with the fuel in a proper mixture ratio, as will be shown in Sect. 1.2.

Four categories of high-thrust rocket motors may be identified, according to the physical state of the propellants carried within such rockets. These categories are solid-propellant motors, liquid-propellant engines, gaseous-propellant engines and hybrid-propellant engines, the last of them being those which use propellants stored in at least two of the three (solid, liquid, and gaseous) physical states of matter. The following figure, due to the courtesy of NASA-JPL [1], shows a scheme of a pump-fed liquid bi-propellant rocket engine.

A. de Iaco Veris, *Fundamental Concepts of Liquid-Propellant Rocket Engines*, Springer Aerospace Technology, https://doi.org/10.1007/978-3-030-54704-2_1

Fuel   Oxidizer   Pumps   Combustion   Nozzle
                                                 Chamber
                                                                   Exhaust

A_e V_e
$\dot{m}$ P_e

$\dot{m}$ = Mass flow rate
V_e = Exit Velocity P_o
P_e = Exit Pressure
P_o = Outside Pressure   Throat
A_e = Nozzle Exit Area

$$\text{Thrust} = F = \dot{m} V_e + (p_e - p_o) A_e$$

A rocket carries a combination of propellants (a fuel and an oxidiser) within itself in order to be capable of operating inside or outside the atmosphere of the Earth. In the rocket engine illustrated above, the magnitude F of the thrust vector $\mathbf{F}$ depends on the propellant mass flow rate $\dot{m} \equiv dm/dt$ through the engine, on the magnitude v_e of the exit velocity vector $\mathbf{v}_e$ of the exhaust gas, and on the pressure p_e of the gas at the exit plane of the nozzle.

The principle of functioning of a converging-diverging nozzle (also known as a de Laval nozzle, after the name of its inventor, Gustaf de Laval) is illustrated in the following figure, due to the courtesy of NASA-JPL [1].

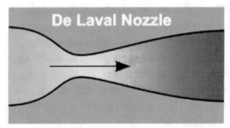

Flow velocity increases from green to red.

As has been shown by Crown [2], a compressible fluid at virtually zero velocity in the combustion chamber is accelerated through the converging portion of the nozzle to sonic speed ($M = 1$) in the throat where, if the converging portion is properly designed, the flow is uniform and parallel. The fluid is then expanded in the diverging portion of the nozzle until the desired Mach number $M > 1$ is reached in the test section, where the flow is again uniform and parallel.

In other words, the cross-section (perpendicular to the gas flow) of the nozzle immediately downstream of the combustion chamber decreases, in order for the subsonic flow ($M < 1$) to increase its speed up to the speed of sound ($M = 1$).

The sonic flow condition is reached in the throat, where the cross-sectional area A of the nozzle reaches its minimum value A_t. In order for the flow to increase further its speed in supersonic conditions ($M > 1$), the exhaust gases must expand downstream of the throat, and therefore the cross-sectional area A of the nozzle must increase ($A > A_t$). This is because, in the converging portion of the nozzle, the maximum uniform velocity which can be reached by the flow in any section is the velocity corresponding to the local velocity of sound. Further increases in velocity can only be obtained by expanding the gas in the subsequent diverging portion of the nozzle.

For optimum performance in terms of thrust, the gas pressure, p_e, at the exit plane of the nozzle should be exactly equal to the pressure, p_0, due to the environment around it. In the vacuum of space ($p_0 = 0$), this is impossible. The bigger the exit area, A_e, of the nozzle, the closer the rocket gets to the optimum thrust. However, at some point, the additional thrust gained is not worth the added mass which is necessary to make the exit area bigger. In conditions of optimum expansion ratio, the rocket motor produces the maximum thrust. A nozzle, which ends before the exhaust gas reaches the pressure of the outside environment, is called an under-expanded nozzle. In these conditions, the rocket does not get all the thrust available from its engine, because the expansion ratio is too low, and therefore the exhaust gas pressure is greater than the environmental pressure. By contrast, a nozzle, whose cross-sectional area at the exit plane is too large, is called an over-expanded nozzle. This occurs when the expansion ratio A_e/A_t is too high, and the pressure of the exhaust gas is less than the environmental pressure. Since the cross-sectional area at the exit plane A_e is too large, then the exhaust gas completely expands before reaching the exit plane of the nozzle. The gas flow downstream of the exit plane, as a function of the ambient pressure, is shown in the following figure, due to the courtesy of NASA [3].

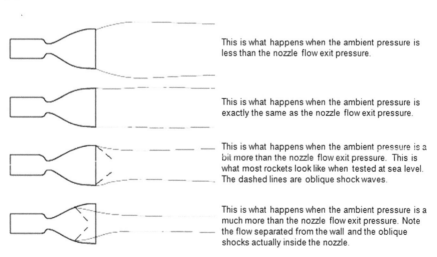

This is what happens when the ambient pressure is less than the nozzle flow exit pressure.

This is what happens when the ambient pressure is exactly the same as the nozzle flow exit pressure.

This is what happens when the ambient pressure is a bit more than the nozzle flow exit pressure. This is what most rockets look like when tested at sea level. The dashed lines are oblique shock waves.

This is what happens when the ambient pressure is a much more than the nozzle flow exit pressure. Note the flow separated from the wall and the oblique shocks actually inside the nozzle.

In the atmosphere, it is difficult to get the optimum expansion ratio, because the air pressure changes with temperature and altitude. Since a given expansion ratio results in the optimum expansion only at a specific altitude, then the design expansion ratio of the nozzle must be selected in such a way as to give the best average performance during powered flight.

With reference to the following figure, due to the courtesy of NASA [4], the thrust imparted to the spacecraft by the gas flow can be expressed as follows

$$F = \dot{m} v_e$$

where $\dot{m}$ is the mass flow rate of the exhaust gas, and v_e is its velocity at the exit plane of the nozzle. Therefore, for a constant mass flow rate $\dot{m}$ or weight flow rate $\dot{W} = g_0 \dot{m}$ of exhaust gas, the thrust F is proportional to the velocity v_e of the gas at the exit plane.

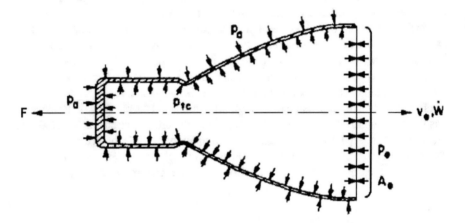

A spacecraft is subject not only to the thrust due to its engine, but also to the pressure, p_0, due to the environment around it. At high altitudes, p_0 can be assumed to be equal to zero. In these conditions, the net force acting on the gas contained in the thrust chamber (comprising the combustion chamber and the nozzle) is the sum of the reactions coming from the walls and of the reaction of the absolute pressure of gas at the exit plane. These two reaction forces are opposed, as shown in the preceding figure.

Therefore, according to the second principle of dynamics, the net force acting on the gas must be equal to the momentum flux out of the thrust chamber, which is

$$\int_{A_{tc}}^{0} p_{tc} dA - A_e p_e = \dot{m} v_e$$

where A_{tc} and p_{tc} are respectively the area and the pressure of the thrust chamber, A_e, v_e, and p_e are respectively the area, the gas velocity, and the static absolute pressure of the gas at the exit plane of the nozzle, and $\dot{m}$ is the steady mass flow rate of the gas. The integral, that is, the first term on the left-hand side of the preceding equation, represent the integral of the pressure forces (resultant) acting on the thrust chamber (combustion chamber and nozzle) and projected onto a plane normal to the axis of symmetry of the thrust chamber, as shown in the preceding figure. This integral is just the force, F, which acts on the thrust chamber (and, therefore, on the spacecraft). Consequently, the preceding equation may also be written as follows

$$F = \dot{m} v_e + A_e p_e$$

This equality holds under the following hypotheses:

- the injection flow velocity of the propellants is negligible;
- the flow is constant and does not change with time during the burn;
- the products of combustion are in chemical equilibrium after the burn;
- the flow of gas through the exit plane of the nozzle is one-dimensional, that is, all the molecules of the ejected gas move on straight lines which are parallel to the axis of symmetry of the nozzle;
- the flow is compressible, that is, the gas density is subject to significant changes;
- the flow is isentropic, that is, frictionless and adiabatic (without heat loss), and therefore depends only on the cross-sectional areas of the nozzle; and
- the gas flowing in the nozzle is a perfect gas, so that the well-known equation of state $p = \rho R T$ involving pressure p, density ρ, and temperature T can be used, where R is the constant of the specific gas (the universal gas constant R^* = 8314.460 N m kmol^{-1} K^{-1} divided by the average molar mass $\mathcal{M}$ (kg/kmol) of the exhaust gas; for example, if the exhaust gas were water, H_2O, then R = 8314.460/18 = 461.9 N m kg^{-1} K^{-1}).

In the thrust equation written above, the quantity $\dot{m}$ is called the momentum thrust, and the quantity $A_e p_e$ is called the pressure thrust. The latter term indicates that some (but not all) of the total forces due to the pressure p_{tc} in the thrust chamber contribute to the kinetic energy possessed by the exhaust gases flowing in the nozzle. In other words, a part of the gas pressure generated by the release of chemical energy is used to increase the gas momentum, whereas another part is not. The greater the part of gas pressure used for this purpose, the greater the efficiency of the nozzle.

Let us assume now that the rocket is fired at an ambient pressure p_0 greater than zero, that is, at low-altitudes. In these conditions, the pressure forces due to the environment and acting on the outside of the thrust chamber walls have no effect on the gas inside. However, such pressure forces subtract a part ($A_e p_0$) from the total pressure thrust ($A_e p_e$). Since the exhaust gas flows at supersonic velocity through the exit plane of area A_e, then the ambient pressure p_0 cannot gain access to it. Therefore, the ambient pressure generates a net unbalanced force onto the projected thrust chamber area, in the direction opposed to the thrust, of magnitude $A_e p_0$. This fact is taken into account by re-writing the equation of thrust as follows

$$F = \dot{m} v_e + A_e(p_e - p_0)$$

which is just the equation given in [1]. The same equation may also be written as follows

$$F = \dot{m}v_e + A_e(p_e - p_0) = \dot{m}c$$

where

$$c = v_e + \frac{A_e(p_e - p_0)}{\dot{m}}$$

is called the effective exhaust velocity, whose value is the same as that of the exhaust gas velocity v_e at the exit plane of the nozzle only when the gas pressure p_e at the same plane is equal to the ambient pressure p_0. As has been shown above, the presence of the second addend on the right-hand side of the preceding equation indicates that the optimum value of v_e has not been reached.

As an application of these concepts to a practical case, taken from [4], it is required to compute the exhaust gas velocity v_e at the exit plane of the nozzle, the thrust in space, and the effective exhaust velocities at sea level (c) and in space (c_s), for a rocket engine having a thrust $F = 4.448 \times 10^5$ N at sea level, a mass flow rate $\dot{m} = 167.5$ kg/s due to propellant consumption, an area $A_e = 0.4905$ m^2 at the exit plane of the nozzle, and an absolute pressure $p_e = 7.377 \times 10^4$ N/m^2 at the same plane. The atmospheric pressure p_0 at sea level has the standard value $101325 \approx 1.013 \times 10^5$ N/m^2.

In order to compute the value of v_e, we use the equation of thrust written above solved for v_e, as follows

$$v_e = \frac{F - A_e(p_e - p_0)}{\dot{m}} = \frac{4.448 \times 10^5 - 0.4905 \times (0.7377 - 1.013) \times 10^5}{167.5}$$
$$= 2736 \text{ m/s}$$

Since the pressure p_e of the exhaust gas at the exit plane of the nozzle is smaller than the atmospheric pressure p_0 at sea level, then the nozzle of the rocket engine considered in the present example is too long for sea level conditions.

During the ascent of the rocket, the value of the atmospheric pressure decreases continuously. When the rocket reaches a certain altitude h^* above the sea level, the atmospheric pressure at that altitude and the exhaust gas pressure p_e at the exit plane of the nozzle have the same value (in the present case, this value is 7.377×10^4 N/m^2), and therefore the value of the component $A_e(p_e - p_0)$ of the thrust becomes equal to zero. This event represents an ideal expansion of the gas flow. At an altitude h greater than h^*, the value of the atmospheric pressure p_0 becomes less than 7.377×10^4 N/m^2, and is equal to zero in space. When this happens, the value of the term $A_e p_0$, which appears on the right-hand side of the equation of thrust, is also equal to zero. The two equations which express the thrust F at sea level and the thrust F_s in space are respectively

$$F = \dot{m}v_e + A_e(p_e - p_0)$$

$$F_s = \dot{m}v_e + A_e p_e$$

By subtracting F from F_s and solving for F_s, there results

$$F_s = F + A_e p_0 = 4.448 \times 10^5 + 0.4905 \times 1.013 \times 10^5 = 4.945 \times 10^5 \text{N}$$

The effective exhaust velocity at sea level results from the preceding equation

$$c = v_e + \frac{A_e(p_e - p_0)}{\dot{m}} = 2736 + \frac{0.4905 \times (0.7377 - 1.013) \times 10^5}{167.5} = 2655 \text{ m/s}$$

The effective exhaust velocity in space (where $p_0 = 0$) is

$$c_s = v_e + \frac{A_e p_e}{\dot{m}} = 2736 + \frac{0.4905 \times 0.7377 \times 10^5}{167.5} = 2952 \text{ m/s}$$

1.2 The Gas Flow Through the Combustion Chamber and the Nozzle

Under the seven hypotheses indicated in Sect. 1.1, it is possible to compute the exhaust gas velocity, v_e, at the exit plane of the nozzle as a function of the physical properties of the gas. Three of the four fundamental equations used for this purpose express the conservation of energy, mass, and momentum, and the fourth is the equation of state of perfect gases. They are shown below.

The energy equation (also known as the Bernoulli equation) states the principle of conservation of energy. In the hypotheses of Sect. 1.1, let us consider an adiabatic flow between two cross-sections (1 and 2) of a nozzle perpendicular to its axis of symmetry, as shown in the following figure, due to the courtesy of NASA [5].

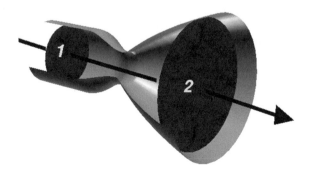

The energy equation for an adiabatic flow between the two cross-sections shown above can be written as follows [6]

$$h_1 - h_2 = c_p(T_1 - T_2) = \frac{1}{2}\frac{v_2^2 - v_1^2}{J}$$

where J is the mechanical equivalent of heat ($J = 1$ because the SI is used in this book, that is, because heat is measured in joules), $h = u + p/\rho$ is the enthalpy per unit mass of the fluid, u is the internal energy per unit mass of the fluid, c_p is the specific heat of the fluid at constant pressure, and T is the temperature of the fluid. The enthalpy measures the total energy of a thermodynamic system, and includes the internal energy of the gas and the amount of energy required to make room for it by displacing its environment and establishing its density and its pressure.

The energy equation states that the energy per unit mass available for heat transfer, $h_1 - h_2$, is converted into kinetic energy per unit mass, $\frac{1}{2}(v_2^2 - v_1^2)$, of the flow. In other words, an amount of energy in form of heat possessed by the gas is used to increase the velocity of the flow. The term $c_p(T_1 - T_2)$ indicates the decrease in temperature resulting from the energy conversion. The specific heat at constant pressure, c_p, has been assumed constant, and depends on the composition of the gas resulting from the combustion. In particular, when the cross-section 1 is located at the exit of the combustion chamber (subscript c), as shown in the preceding figure, then there results $v_1 \approx 0$, $T_1 = (T_c)_{ns}$, $v_2 \equiv v$, and $T_2 \equiv T_i$. Therefore, for an isentropic flow ($p/\rho^\gamma = $ constant), the energy equation can be written as follows

$$(T_c)_{ns} = T_i\left(1 + \frac{\gamma - 1}{2}M_i^2\right)$$

$$(p_c)_{ns} = p_i\left(1 + \frac{\gamma - 1}{2}M_i^2\right)^{\frac{\gamma}{\gamma-1}}$$

$$(\rho_c)_{ns} = \rho_i\left(1 + \frac{\gamma - 1}{2}M_i^2\right)^{\frac{1}{\gamma-1}}$$

where $M_i = v_i/a_i$ is the Mach number at the nozzle inlet, $\gamma \equiv c_p/c_v$ is the specific heat ratio of the gas, R is the specific constant of the gas, and $a_i = (\gamma R T_i)^{1/2}$ is the sonic velocity of the flow at the nozzle inlet.

The second of the four fundamental equations mentioned above states the principle of mass conservation. Since the flow is constant, then the rate $\dot{m}$ at which a quantity of mass m passes through a cross-section A of the nozzle must be independent of the position of this cross-section along the axis of symmetry of the nozzle, that is,

$$\dot{m} = \rho A v = \text{constant}$$

where ρ, A, and v are respectively the local density of the gas, the local cross-sectional area of the nozzle, and the local velocity of the flow.

Since the mass flow rate through the nozzle must be constant in all sections of the nozzle, the continuity equation may also be written as follows

$$\rho A v = \rho_t A_t v_t$$

where the subscript t indicates the throat of the nozzle, where the Mach number is equal to unity. By using the equations written above, it is possible to express the area ratio between any cross-sectional area A_x (where $M = M_x$) and the throat cross-sectional area A_t (where $M_t = 1$) of the nozzle as follows [2]:

$$\left(\frac{A_x}{A_t}\right)^2 = \frac{1}{M_x^2}\left[\left(\frac{2}{\gamma+1}\right)\left(1 + \frac{\gamma-1}{2}M_x^2\right)\right]^{\frac{\gamma+1}{\gamma-1}}$$

Let 1 and 2 be two cross-sections of the nozzle at two arbitrary points along its axis of symmetry. With reference to these sections, the equation of energy can be written as follows

$$h_1 - h_2 = \frac{1}{2}\left(v_2^2 - v_1^2\right)$$

The preceding equation, solved for v_2, yields

$$v_2 = \left[2(h_1 - h_2) + v_1^2\right]^{\frac{1}{2}}$$

which can also be written as follows [6]:

$$v_2 = \left\{\frac{2\gamma}{\gamma-1}RT_1\left[1 - \left(\frac{p_2}{p_1}\right)^{\frac{\gamma-1}{\gamma}}\right] + v_1^2\right\}^{\frac{1}{2}}$$

By choosing the cross-section 2 as the exit plane of the nozzle (subscript e) and the cross-section 1 as the exit plane of the combustion chamber (subscript c), the second addend ($v_1^2 \equiv v_c^2$) within curly brackets becomes negligible in comparison with the first. Therefore, the velocity of the exhaust gas at the exit plane of the nozzle is expressible as follows

$$v_e = \left\{\frac{2\gamma}{\gamma-1}RT_c\left[1 - \left(\frac{p_e}{p_c}\right)^{\frac{\gamma-1}{\gamma}}\right]\right\}^{\frac{1}{2}}$$

This is because the cross-section of the combustion chamber is larger than that of the throat. Therefore, the flow velocity, v_c, at the exit plane of the combustion chamber can be neglected in the expression given above. In addition, the temperature, T_c, of the combustion chamber and of the nozzle inlet differs very little, in isentropic conditions, from the stagnation temperature.

The maximum theoretical value of the exhaust gas velocity, v_e, at the exit plane of the nozzle is reached when the pressure value, p_e, at the same plane is zero (infinite expansion). Setting $p_e = 0$ in the preceding equation yields

$$(v_e)_{MAX} = \left(\frac{2\gamma}{\gamma - 1} RT_c \right)^{\frac{1}{2}}$$

The third of the fundamental equations mentioned above expresses the principle of conservation of momentum in a steady one-dimensional flow. With reference to the following figure, let us consider an infinitesimal particle of gas moving along a streamline, s, in a steady, one-dimensional flow, that is, in a flow whose velocity, v, is independent of time ($dv/dt = 0$).

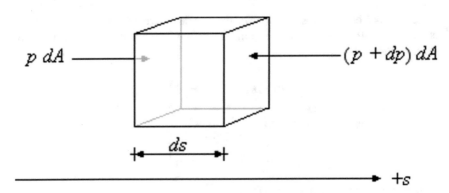

We consider here, of all forces which may act on the gaseous particle, only those due to pressure imbalances. Therefore, gravitational, magnetic, and viscous forces are neglected. By applying the second law of dynamics to the motion of the infinitesimal particle of gas along the streamline, the sum of the forces, $\sum F$, in the stream direction, s, is equal to the mass, m, of the particle multiplied by the rate of change, dv/dt, of its velocity, that is,

$$\sum F = m \frac{dv}{dt} = pdA - (p + dp)dA$$

Since the mass, m, of the infinitesimal particle is equal to its density, ρ, times the infinitesimal distance, ds, times the infinitesimal area, dA, then there results

$$m = \rho\, ds\, dA$$

In addition, since the velocity vector of the particle is always tangent to the streamline, then the rate of change of velocity with time is expressible as follows

$$\frac{dv}{dt} = \left(\frac{dv}{ds} \right) \left(\frac{ds}{dt} \right) = v \frac{dv}{ds}$$

By substituting the expressions of mass and acceleration in the equation written above, there results

$$pd A - (p + dp)d A = (\rho ds d A)\left(v\frac{dv}{ds}\right)$$

that is,

$$-dp = \rho v\, dv$$

which is the momentum equation. The minus sign in front of dp is due to the fact that the gas particle moves along the streamline, s, from a region of high pressure to a region of low pressure; therefore, when the velocity of the particle increases with s, then its pressure decreases. In other words, to each increase in the flow velocity of a particle there corresponds a decrease in its pressure. The preceding equation, also known as the Euler equation, is a particular case of the general Navier-Stokes equations [7]. Since the flow is not only steady but also isentropic ($p/\rho^\gamma = $ constant), then the following equality holds

$$\frac{dp}{p} = \gamma\frac{d\rho}{\rho}$$

The two equations written above, combined together, yield

$$vdv = -\frac{dp}{\rho} = -\left(\frac{dp}{d\rho}\right)\left(\frac{d\rho}{\rho}\right) = -a^2\frac{d\rho}{\rho}$$

where

$$a = \left(\frac{dp}{d\rho}\right)^{\frac{1}{2}}$$

is the velocity of sound in the flowing gas.

Since $M = v/a$ is the Mach number, then by multiplying and dividing the quantity on the left-hand side of the preceding equation by v there results

$$v^2\frac{dv}{v} = -a^2\frac{d\rho}{\rho}$$

that is,

$$M^2\frac{dv}{v} = -\frac{d\rho}{\rho}$$

By combining the preceding equation with the equation ($\rho A v = $ constant) which expresses the principle of mass conservation, there results

$$(1 - M^2)\frac{dv}{v} = \frac{dA}{A}$$

For subsonic flow ($M < 1$), a decrease in the cross-sectional area ($dA < 0$) of the nozzle causes the flow velocity to increase ($dv > 0$). This is what happens in the convergent portion of the nozzle, downstream of the combustion chamber, until the sonic condition ($M = 1$) is reached in the throat. By contrast, for supersonic flow, an increase in the cross-sectional area ($dA > 0$) of the nozzle causes the flow velocity to increase ($dv > 0$). This is what happens in the divergent portion of the nozzle, downstream of the throat. The equations written above (energy, continuity, momentum, and state) make it possible to express the propellant mass flow rate and the thrust in terms of the pressure of the fluid in the combustion chamber (subscript c) and the area of the nozzle throat (subscript t), as will be shown below. The energy equation for the combustion chamber (where $v_c \approx 0$) is

$$\frac{\rho_c}{\rho} = \left(1 + \frac{\gamma - 1}{2}M^2\right)^{\frac{1}{\gamma - 1}}$$

When M^2 is much smaller than unity, then the preceding equation yields $\rho \approx \rho_c$, and consequently

$$\frac{p_c}{p} = 1 + \frac{1}{2}M^2 = 1 + \frac{v^2}{2RT} = 1 + \frac{\rho v^2}{2p}$$

Hence

$$p_c = p + \frac{1}{2}\rho v^2$$

The mass flow rate per unit area is

$$\frac{\dot{m}}{A} = \rho v$$

Since $M = v/a$, $a = (\gamma RT)^{\frac{1}{2}}$, and $T_c/T = 1 + \frac{1}{2}(\gamma - 1)M^2$, then

$$v = M\left[\frac{\gamma RT_c}{1 + \frac{\gamma - 1}{2}M^2}\right]^{\frac{1}{2}}$$

and therefore

$$\frac{\dot{m}}{A} = \rho v = \frac{p_c \gamma^{\frac{1}{2}} M}{(RT_c)^{\frac{1}{2}}} \left[\frac{1}{1 + \frac{\gamma-1}{2}M^2}\right]^{\frac{\gamma+1}{2(\gamma-1)}}$$

The preceding equation, written for the throat (where $A = A_t$ and $M = 1$), yields

$$\frac{\dot{m}}{A_t} = \frac{p_c \gamma^{\frac{1}{2}}}{(RT_c)^{\frac{1}{2}}} \left[\frac{2}{\gamma+1}\right]^{\frac{\gamma+1}{2(\gamma-1)}}$$

that is,

$$\dot{m} = \Gamma(\gamma)\frac{A_t p_c}{(RT_c)^{\frac{1}{2}}}$$

where

$$\Gamma(\gamma) = \gamma^{\frac{1}{2}}\left(\frac{2}{\gamma+1}\right)^{\frac{\gamma+1}{2(\gamma-1)}}$$

is the Vandenkerkhove function. By introducing the expressions of $\dot{m}_n$ and v_e derived above in the equation of thrust

$$F = \dot{m}v_e + A_e(p_e - p_0)$$

it is possible to re-write this equation as follows [8]:

$$F = A_t p_c \left\{\frac{2\gamma^2}{\gamma-1}\left(\frac{2}{\gamma+1}\right)^{\frac{\gamma+1}{\gamma-1}}\left[1 - \left(\frac{p_e}{p_c}\right)^{\frac{\gamma-1}{\gamma}}\right]\right\}^{\frac{1}{2}} + A_e(p_e - p_0)$$

which expresses the thrust in terms of the pressure of the fluid in the combustion chamber and of the area of the nozzle throat. In other words, when the pressure thrust is zero, which occurs when $p_e = p_0$, then the thrust is directly proportional to the throat area, A_t, and nearly directly proportional to the pressure, p_c, of the fluid within the combustion chamber.

The chemical reaction, which takes place in the combustion chamber when the fuel combines with the oxidiser, is an oxidation. When the fuel is hydrogen (H_2) and the oxidiser is oxygen (O_2), the product of the combustion is water (H_2O) in its gaseous state, due to of the high temperature generated when hydrogen burns in the presence of oxygen, according to the well-known reaction

$$2H_2 + O_2 \rightarrow 2H_2O + \text{heat}$$

This means that, in each second, two kilomoles of hydrogen (that is, $2 \times 2 \times 1$ = 4 kg of hydrogen) react with one kilomole of oxygen (that is, $2 \times 16 = 32$ kg of oxygen) to form two kilomoles of water (that is, $2 \times (2 + 16) = 36$ kg of water). The corresponding stoichiometric mixture ratio (o/f) of the oxidiser (o) to the fuel (f) is

$$\frac{o}{f} \equiv \frac{\dot{m}_o}{\dot{m}_f} = \frac{32}{4} = 8$$

In other words, 8 kg of oxygen are necessary to burn 1 kg of hydrogen in 1 s. When used for rockets, hydrogen and oxygen are most often stored in their liquid state of aggregation, at very low temperatures (cryogenic propellants). This is because they take less space in the liquid state than in the gaseous state at normal temperature and pressure conditions. Hydrogen remains liquid without evaporating at atmospheric pressure (101325 N/m^2) at a temperature not higher than 20 K, and oxygen does the same at a temperature not higher than 90 K [9].

Another cryogenic fuel used for space propulsion is liquid methane (CH_4), which requires a storage at a temperature not higher than 111 K [10]. The related chemical reaction with liquid oxygen is [10]:

$$CH_4 + 2O_2 \rightarrow CO_2 + 2H_2O + \text{ heat}$$

In each second, one kilomole of methane (that is, $12 + 4 \times 1 = 16$ kg of methane) reacts with two kilomoles of oxygen (that is, $2 \times 2 \times 16 = 64$ kg of oxygen) to form one kilomole of carbon dioxide (that is, $12 + 2 \times 16 = 44$ kg of carbon dioxide) and two kilomoles of water (that is, $2 \times (2 + 16) = 36$ kg of water). The corresponding stoichiometric mixture ratio (o/f) between the oxidiser (o) and the fuel (f) is

$$\frac{o}{f} \equiv \frac{\dot{m}_o}{\dot{m}_f} = \frac{64}{16} = 4$$

A cryogenic oxidiser which can be used either alone or in combination with liquid oxygen is liquid fluorine (F_2), which requires a storage at a temperature not higher than 85 K. Fluorine is extremely toxic and reacts violently with most substances [10]. Cryogenic propellants are difficult to store over long periods of time, due to the low temperatures required by them. Therefore, they are not used in military rockets, which must be taken in a state of readiness for launch over times of several months. In addition, liquid hydrogen requires a storage volume many times greater than that required by other fuels, because of its very low density (70.85 kg/m^3). On the other hand, the higher performance made available by cryogenic propellants makes them particularly attractive when constraints of storage and reaction time are not critical. For example, according to Aerojet Rocketdyne [11], the RS-25 engines aboard the Space Shuttle orbiter use liquid hydrogen and liquid oxygen and have a specific impulse in space (see Sect. 1.3) of 451 s. The RL10 engines on the Centaur, the US first liquid-hydrogen/liquid-oxygen rocket stage, have a specific impulse in space of 433 s [12]. The J-2 engines used on the Saturn V second and third stages and on the second stage of the Saturn 1B also burn the liquid-hydrogen/liquid-oxygen combination. They have a specific impulse in space of 424 s [13]. By comparison, the liquid-oxygen/kerosene combination used in the cluster of five F-1 engines in the Saturn V first stage has a specific impulse in space of 304 s [14]. In a few words, liquid hydrogen, in comparison with other fuels (such as kerosene), yields more power per unit volume of storage.

When the fuel is a hydrocarbon (for example, kerosene), the oxidation reaction may be represented [10] as follows

$$2C_{12}H_{26} + 25O_2 \rightarrow 24CO + 26H_2O + \ \text{heat}$$

In each second, two kilomoles of dodecane (that is, $2 \times 12 \times 12 + 2 \times 26 = 340$ kg of dodecane) react with twenty-five kilomoles of oxygen (that is, $25 \times 2 \times 16 = 800$ kg of oxygen) to form twenty-four kilomoles of carbon monoxide (that is, $24 \times (12 + 16) = 672$ kg of carbon monoxide) and twenty-six kilomoles of water (that is, $26 \times (2 + 16) = 468$ kg of water). The corresponding stoichiometric mixture ratio (o/f) between the oxidiser (o) and the fuel (f) is

$$\frac{o}{f} \equiv \frac{\dot{m}_o}{\dot{m}_f} = \frac{800}{340} \approx 2.35$$

It is to be noted that kerosene is a mixture of hydrocarbons. The chemical composition of kerosene depends on its source. It usually consists of about ten different hydrocarbons, each of them containing from 10 to 16 carbon atoms per molecule; the constituents include n-dodecane, alkyl benzenes, and naphthalene and its derivatives. Kerosene is usually represented by the single compound n-dodecane. In particular, RP-1 is a special type of highly-refined kerosene which conforms to Military Specification MIL-R-25576 and is used as a fuel for rocket engines [10]. For example, RP-1 is used in the first-stage boosters of the Delta and Atlas-Centaur rockets. It also powered the first stages of the Saturn 1B and Saturn V [10]. Hydrocarbon-based fuels do not pose the severe constraints of storage which are peculiar to cryogenic fuels. On the other hand, as has been shown above, the specific impulse yielded by them is considerably less than that made available by cryogenic fuels.

Another class of substances used for space propulsion includes the hypergolic propellants, that is, specific couples of fuels and oxidisers which ignite spontaneously and violently when one gets in touch with the other, without the need for any ignition source. Such propellants remain liquid at normal temperatures. These properties make them particularly suitable for propulsion and hydraulic power systems carried aboard spacecraft, and in particular for those which are meant to be used for orbital manoeuvring. On the other hand, hypergolic propellants are extremely toxic and/or corrosive and must be handled with the highest care. The most common fuels used in hypergolic propellants are hydrazine (N_2H_4), monomethyl hydrazine ($CH_3(NH)NH_2$), unsymmetrical dimethyl hydrazine ($H_2NN(CH_3)_2$), and Aerozine 50, the last being an equal mixture of hydrazine and unsymmetrical dimethyl hydrazine.

According to Nufer [15], the oxidisers used in combination with the fuels named above are usually nitrogen tetroxide (N_2O_4) and various blends of nitrogen tetroxide with nitric oxide (NO). Hypergolic propellants are used in the core liquid-propellant stages of the Titan family of launch vehicles, and on the Delta launcher. The Space Shuttle orbiter uses hypergolic propellants in its orbital manoeuvring subsystem for orbital insertion, main orbital manoeuvres, and de-orbit. It also uses hypergols in its reaction control system for attitude control [10]. The efficiency (in terms of specific impulse) reached in the Space Shuttle by combining monomethyl hydrazine with nitrogen tetroxide ranges from 260 to 280 s in the reaction control system, and to 313 s in the orbital manoeuvring subsystem, the higher efficiency of the latter being attributed by NASA [10] to higher expansion ratios in the nozzles and higher pressures in the combustion chambers.

In a liquid-propellant rocket, the fuel and the oxidiser are stored in two separate containers, and are sent to the combustion chamber by means of two separate systems of pipes, valves, and pumps. As the sequel will show, a liquid-propellant engine is more complex than a solid-propellant motor, but has several advantages over the latter. This is because the former offers the possibility of controlling the flow of propellant towards the combustion chamber, thereby permitting of throttling, stopping, or re-starting the engine.

The liquid propellants are introduced into the combustion chamber at the injecting plane with a small axial velocity, v_{inj}, which is assumed to be equal to zero in the present calculation. The combustion proceeds throughout the length of the combustion chamber, and is presumed to be complete at the inlet plane of the nozzle. The density of the gas decreases from one section of the combustion chamber to the other, due to the heat of combustion released between these sections. Since the mass flow rate remains constant, the gas accelerates toward the inlet section of the nozzle, and its pressure decreases.

In practice, the gas flow within the combustion chamber is not entirely isentropic, that is, the expansion process within the combustion chamber is neither fully irreversible nor fully adiabatic. The stagnation temperature, T_s, remains nearly constant, but the stagnation pressure, p_s, decreases. This causes an energy loss, which depends on the specific heat ratio, γ, of the gas, and also on the nozzle contraction area ratio, A_c/A_t, where A_c is the area of the cross-section of the combustion chamber, and A_t is the area of the cross-section of the throat. The energy loss occurs where the acceleration of gases is affected by expansion due to heat release, as is the case in the combustion chamber, rather than by a change of area, as is the case in the nozzle. The greater the acceleration of the gas flow given by the nozzle, the more efficient the process is.

The following figure, adapted from [16], shows the static pressure p, the temperature T, and the velocity v of a gas flowing from the inlet plane to the exit plane of a converging-diverging nozzle.

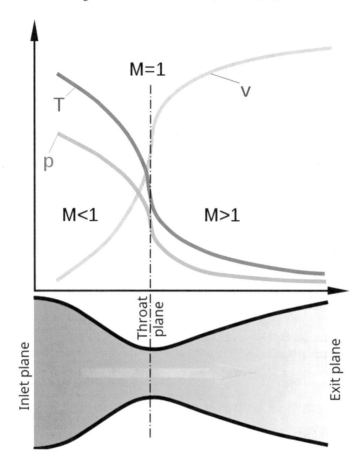

Neglecting the velocity of the gas flow at the injecting plane ($v_{inj} = 0$) and assuming the total pressure of the combustion chamber at the injecting plane $(p_c)_{inj}$ to be equal to the pressure of the injector p_{inj} (that is, assuming $(p_c)_{inj} = p_{inj}$), the total pressure ratio $(p_c)_{inj}/(p_c)_{ns}$ can be expressed as follows [4]:

$$\frac{(p_c)_{inj}}{(p_c)_{ns}} = \frac{1 + \gamma M_i^2}{\left(1 + \frac{\gamma-1}{2} M_i^2\right)^{\frac{\gamma}{\gamma-1}}}$$

where $(p_c)_{ns}$ is the total pressure in the combustion chamber at the nozzle inlet, M_i is the Mach number at the nozzle inlet plane, and $\gamma \equiv c_p/c_v$ is the specific heat ratio of the gas. The preceding equation results from the equation of energy written for the converging portion of nozzle, which goes from the inlet plane (where the Mach number of the gas flow is M_i) to the throat plane (where the Mach number of the gas flow is $M_t = 1$). In case of the static pressure ratio p_{inj}/p_i, the preceding equation can be simplified as follows

$$\frac{p_{inj}}{p_i} = 1 + \gamma M_i^2$$

In order to reduce the energy loss mentioned above, a small value (that is, considerably smaller than unity) is desirable for M_i. A typical value of 0.31 is indicated by Huzel and Huang [4] for a thrust chamber having a contraction area ratio, A_c/A_t, equal to 2 and a gas having a specific heat ratio, γ, equal to 1.2.

In this manner, a combustion chamber performs efficiently its function of converting a propellant into a gas of high temperature and pressure through combustion. This gas is then accelerated within the nozzle, as will be shown below. The nozzles used in rocket engines are of the converging-diverging (de Laval) type, as has been shown in Sect. 1.1. In a nozzle of this type, the area of the cross-section decreases and reaches its minimum value at the throat. Downstream of the throat, this area increases to the exit plane of the nozzle. The velocity of the gas flow increases, reaches the sonic value at the throat, and then increases further to supersonic values in the diverging portion of the nozzle.

As has been shown in Sect. 1.1, the gas flow through the nozzle is assumed to be an isentropic expansion. The total temperature and the total pressure of the gas are also assumed to remain constant throughout the nozzle. In these conditions, the pressure ratio, $p_t/(p_c)_{ns}$, between the static pressure at the throat and the total pressure in the combustion chamber at the nozzle inlet is the critical pressure ratio, which depends only on the specific heat ratio, γ, as follows

$$\frac{p_t}{(p_c)_{ns}} = \left(\frac{2}{\gamma + 1}\right)^{\frac{\gamma}{\gamma-1}}$$

which results from the energy equation. The sonic velocity ($M_t = 1$) of the gas flow is reached in the throat independently of whether the value of the ambient pressure (p_0) at the exit plane of the nozzle be, or be not, less than the value of the gas pressure at the throat.

In case of an ideal expansion ($p_0 = p_e$), the pressure of the gas flowing in the diverging portion of the nozzle continues to decrease, as shown in the preceding figure. Otherwise, an increase in pressure occurs in the diverging portion of the nozzle. This increase in pressure may take place either by isentropic subsonic deceleration of the gas flow, or by non-isentropic discontinuities (called shock waves), or by a combination of both of these manners.

An example of shock waves in a gas flow is shown in the following figure, which is due to the courtesy of NASA [17].

The blue cones of light, which appear below the three main engines of the Space Shuttle, are known as shock (or Mach) diamonds. They are due to the formation of shock waves in the exhaust plume of the three main engines. The exhaust gas (H_2O) flowing from the engines reaches the speed of Mach 10, and the increase in pressure results in the shock diamonds.

Lower values of pressure than that of the ambient pressure (p_0) can be reached in the diverging portion of a nozzle, where the exhaust gas flows at supersonic speed. The higher ambient pressure cannot advance upstream of the nozzle, because the gas flows at supersonic speed. However, along the nozzle walls, due to the friction, there may be a boundary layer of gas moving at low speeds. Within this boundary layer, the gas moves at subsonic speeds, and therefore the ambient pressure can advance upstream for a distance. Therefore, the low-pressure central flow is forced away from the nozzle walls.

The following example shows a calculation relating to an ideal liquid-propellant rocket engine. The following data are known (from [4]): propellant mass flow rate in the combustion chamber $\dot{m}_{tc} = 163.6$ kg/s, total pressure in the combustion chamber at the nozzle inlet $(p_c)_{ns} = p_i[1 + \frac{1}{2}(\gamma - 1)M_i^2]^{\gamma/\gamma - 1} = 6.895 \times 10^6$ N/m^2, total temperature of the combustion chamber $(T_c)_{ns} = T_i[1 + \frac{1}{2}(\gamma - 1)M_i^2]$ $= 3633$ K (T_i and M_i being respectively the flow temperature and the Mach number at the nozzle inlet), molar mass of the combustion products $\mathcal{M} = 22.67$ kg/kmol, specific heat ratio of the combusted gas $\gamma = 1.20$, and expansion area ratio of the nozzle $A_e/A_t = 12$. The Mach number M_{inj} at the injection plane is assumed to be equal to zero, and the Mach number M_i at the inlet plane of the nozzle is assumed to be equal to 0.4. In practice, the design values used for M_i range from 0.15 to 0.45. We want to compute the static pressures p_{inj}, p_t, p_x (at a cross-section x of area $A_x = 4A_t$), and p_e of the gas. As has been shown above, the total pressure ratio $(p_c)_{inj}/(p_c)_{ns}$ can be expressed by means of the following equation

$$\frac{(p_c)_{inj}}{(p_c)_{ns}} = \frac{1 + \gamma M_i^2}{\left(1 + \frac{\gamma - 1}{2} M_i^2\right)^{\frac{\gamma}{\gamma - 1}}}$$

In the present case, there results

$$\frac{(p_c)_{inj}}{(p_c)_{ns}} = \frac{1 + 1.20 \times 0.4^2}{\left(1 + \frac{1.20 - 1}{2} \times 0.4^2\right)^{\frac{1.20}{1.20 - 1}}} = 1.084$$

Since $(p_c)_{ns} = 6.895 \times 10^6$ N/m^2, then the total pressure at the injection plane is

$$(p_c)_{inj} = 6.895 \times 10^6 \times 1.084 = 7.474 \times 10^6 \text{N/m}^2$$

Since M_{inj} has been assumed to be equal to zero, then the static pressure at the injecting plane is

$$p_{inj} = (p_c)_{inj} = 7.474 \times 10^6 \text{N/m}^2$$

At the inlet plane of the nozzle, the Mach number is assumed to be $M_i = 0.4$. Since the static pressure ratio p_{inj}/p_i is expressed by the preceding equation

$$\frac{p_{inj}}{p_i} = 1 + \gamma M_i^2$$

then the static pressure at the inlet plane of the nozzle is

$$p_i = \frac{p_{inj}}{1 + \gamma M_i^2} = \frac{7.474 \times 10^6}{1 + 1.20 \times 0.4^2} = 6.270 \times 10^6 \text{N/m}^2$$

The static pressure of the gas at the throat results from the preceding equation

$$\frac{p_t}{(p_c)_{ns}} = \left(\frac{2}{\gamma + 1}\right)^{\frac{\gamma}{\gamma - 1}}$$

In the present case, since $(p_c)_{ns} = 6.895 \times 10^6$ N/m^2 and $\gamma = 1.20$, then

$$p_t = 6.895 \times 10^6 \times \left(\frac{2}{1.2 + 1}\right)^{\frac{1.2}{1.2 - 1}} = 3.892 \times 10^6 \text{N/m}^2$$

The static pressure p_x at the cross-section x of area $A_x = 4A_t$ results from the following equation given in [4]:

$$\frac{A_x}{A_t} = \frac{\left(\frac{2}{\gamma+1}\right)^{\frac{1}{\gamma-1}}\left(\frac{(p_c)_{ns}}{p_x}\right)^{\frac{1}{\gamma}}}{\left\{\frac{\gamma+1}{\gamma-1}\left[1 - \left(\frac{p_x}{(p_c)_{ns}}\right)^{\frac{\gamma-1}{\gamma}}\right]\right\}^{\frac{1}{2}}}$$

The unknown value of p_x can be computed numerically, as will be shown below. For convenience, we set $z = p_x/(p_c)_{ns}$, where z is an auxiliary variable.

Müller's method of root finding uses a quadratic interpolation and consequently requires the knowledge of three points in the vicinity of the unknown value of the root z to be found. This method is briefly described below.

Let z_2 and z_1 be the endpoints of an interval $z_2 \le z \le z_1$ containing the unknown value, z, of the root of a given equation. Let $f(z)$ be the function which is required to be equal to zero in the point z. The existence of at least one real root of $f(z) = 0$ lying between z_2 and z_1 is assured if $f_2 \equiv f(z_2)$ and $f_1 \equiv f(z_1)$ have opposite signs. Following Gerald and Wheatley [18], the computation is performed as follows:

- take a third point z_0 placed between z_2 and z_1 and compute $f_0 \equiv f(z_0)$;
- set $h_1 = z_1 - z_0$, $h_2 = z_0 - z_2$, and $k = h_2/h_1$;
- compute the following three coefficients

$$A = \frac{kf_1 - f_0(1+k) + f_2}{kh_1^2(1+k)}$$

$$B = \frac{f_1 - f_0 - Ah_1^2}{h_1}$$

$$C = f_0$$

of the interpolating parabola $f(z) = A(z - z_0)^2 + B(z - z_0) + C$;

- compute the estimated root of $f(z) = 0$ as follows

$$z = z_0 - \frac{2C}{B \pm \left(B^2 - 4AC\right)^{\frac{1}{2}}}$$

where the sign plus or the sign minus is chosen so that the denominator should have the maximum absolute value (that is, if $B > 0$, choose plus; if $B < 0$, choose minus; if $B = 0$, choose either) and compute $f \equiv f(z)$;

- check the computed value of z to determine which set of three points should be used in the next iteration (if z is greater than z_0, take z_0, z_1, and the root z for the next iteration; if z is less than z_0, take z_0, z_2, and the root z for the next iteration); and

- reset the subscripts 0, 1, and 2 so that z_0 should be placed between z_2 and z_1, and repeat the cycle while the computed value of z does not satisfy the condition $f(z)$ = 0 to some acceptable degree of tolerance.

In the present case ($A_x/A_t = 4$ and $\gamma = 1.2$), we define a function $f(z)$ such that

$$f(z) \equiv 4^2 - \frac{\left(\frac{2}{1.2+1}\right)^{\frac{2}{1.2-1}} \left(\frac{1}{z}\right)^{\frac{2}{1.2}}}{\left(\frac{1.2+1}{1.2-1}\right)\left(1 - z^{\frac{1.2-1}{1.2}}\right)}$$

Since $p_t = 3.892 \times 10^6$ N/m^2 and $(p_c)_{ns} = 6.895 \times 10^6$ N/m^2, then the unknown value of the pressure ratio $z \equiv p_x/(p_c)_{ns}$ is less than $p_t/(p_c)_{ns} = 0.5645$. Therefore, we search the unknown value of z by trying values which are progressively smaller than 0.5645. By so doing, we find the following interval

$$0.04 \leq z \leq 0.05$$

in which the function $f(z)$ changes sign, as will be shown below. At the upper end ($z_1 = 0.05$) of the interval indicated above, there results

$$f_1 \equiv f(0.05) = 2.859$$

At the lower end ($z_2 = 0.04$) of the same interval, there results

$$f_2 \equiv f(0.04) = -2.044$$

Since the product $f_1 f_2$ is less than zero, then the function $f(z)$ defined above has at least one zero in the interval $0.04 \leq z \leq 0.05$. We choose arbitrarily another point $z_0 = 0.045$ between 0.04 and 0.05, and compute

$$f_0 \equiv f(0.045) = 0.7464$$

Starting from the values f_2, f_0, and f_1 obtained above, we compute

$$h_1 = z_1 - z_0 = 0.050 - 0.045 = 0.005$$
$$h_2 = z_0 - z_2 = 0.045 - 0.040 = 0.005$$
$$k = h_2/h_1 = 0.005/0.005 = 1$$

and then the three coefficients (A, B, and C) of the quadratic polynomial which interpolates the three points (z_2, f_2), (z_0, f_0), and (z_1, f_1), as follows

$$A = \frac{kf_1 - f_0(1+k) + f_2}{kh_1^2(1+k)} = \frac{1 \times 2.859 - 0.7464 \times (1+1) - 2.044}{1 \times 0.005^2 \times (1+1)}$$
$$= -1.355 \times 10^4$$

$$B = \frac{f_1 - f_0 - Ah_1^2}{h_1} = \frac{2.859 - 0.7464 + 1.355 \times 10^4 \times 0.005^2}{0.005} = 490.3$$

$$C = f_0 = 0.7464$$

The estimated root z of $f(z) = 0$ is computed as follows

$$z = z_0 - \frac{2C}{B + (B^2 - 4AC)^{\frac{1}{2}}}$$

$$= 0.045 - \frac{2 \times 0.7404}{490.3 + (490.3^2 - 4 \times 1.355 \times 10^4 \times 0.7464)^{\frac{1}{2}}} = 0.04354$$

By applying repeatedly Müller's method, we find, with four significant figures, $z = 0.04351$. Therefore, remembering the definition of the auxiliary variable z, we have

$$p_x = z(p_c)_{ns} = 0.04351 \times 6.895 \times 10^6 = 3.0 \times 10^5 \text{N/m}^2$$

The static pressure p_e at the exit plane of the nozzle such that $A_e = 12A_t$ results from the same equation given above

$$\frac{A_x}{A_t} = \frac{\left(\frac{2}{\gamma+1}\right)^{\frac{1}{\gamma-1}} \left(\frac{(p_c)_{ns}}{p_x}\right)^{\frac{1}{\gamma}}}{\left\{\frac{\gamma+1}{\gamma-1}\left[1 - \left(\frac{p_x}{(p_c)_{ns}}\right)^{\frac{\gamma-1}{\gamma}}\right]\right\}^{\frac{1}{2}}}$$

In order to compute p_e, the preceding equation is re-written by replacing p_x with p_e, and A_x with A_e, as follows

$$\frac{A_e}{A_t} = \frac{\left(\frac{2}{\gamma+1}\right)^{\frac{1}{\gamma-1}} \left(\frac{(p_c)_{ns}}{p_e}\right)^{\frac{1}{\gamma}}}{\left\{\frac{\gamma+1}{\gamma-1}\left[1 - \left(\frac{p_e}{(p_c)_{ns}}\right)^{\frac{\gamma-1}{\gamma}}\right]\right\}^{\frac{1}{2}}}$$

In the present case ($A_e/A_t = 12$ and $\gamma = 1.2$), we define a function $f(z)$ such that

$$f(z) \equiv 12^2 - \frac{\left(\frac{2}{1.2+1}\right)^{\frac{2}{1.2-1}} \left(\frac{1}{z}\right)^{\frac{2}{1.2}}}{\left(\frac{1.2+1}{1.2-1}\right)\left(1 - z^{\frac{1.2-1}{1.2}}\right)}$$

where the auxiliary variable z is defined as follows $z \equiv p_e/(p_c)_{ns}$.

By applying repeatedly Müller's method, we find, with four significant figures, $z = 0.009859$. Therefore, according to the definition of z, we have

$$p_e = z(p_c)_{ns} = 0.009859 \times 6.895 \times 10^6 = 6.798 \times 10^4 \text{N/m}^2$$

Now, we want to compute the flow temperatures at the injection plane (T_{inj}), at the nozzle inlet (T_i), at the throat (T_t), at the cross-section x defined above (T_x), and at the exit plane of the nozzle (T_e).

Since the Mach number at the injection plane, M_{inj}, has been assumed to be equal to zero, then the flow temperature at the injection plane of the nozzle, T_{inj}, is equal to the total temperature of the combustion chamber ($T_c)_{ns} = 3633$ K.

By definition, the following equation holds between total temperature of the combustion chamber ($T_c)_{ns}$ and the temperature at the nozzle inlet T_i:

$$T_i = \frac{(T_c)_{ns}}{1 + \frac{1}{2}(\gamma - 1)M_i^2}$$

Since the Mach number of the flow at the inlet plane of the nozzle, M_i, has been assumed equal to 0.4, then there results

$$T_i = \frac{3633}{1 + 0.5 \times (1.2 - 1) \times 0.4^2} = 3576 \text{ K}$$

In case of an isentropic flow ($p/\rho^\gamma = $ constant), the energy equation, written for any two points 1 and 2 placed along the axis of symmetry of the nozzle, yields

$$\frac{T_1}{T_2} = \left(\frac{p_1}{p_2}\right)^{\frac{\gamma-1}{\gamma}}$$

Since ($T_c)_{ns} = 3633$ K and ($p_c)_{ns} = 6.895 \times 10^6$ N/m^2 are respectively the total temperature and the total pressure of the combustion chamber at the inlet plane of the nozzle, then the preceding equation, written for respectively the throat, the section x, and the exit plane of the nozzle, yields

$$T_t = (T_c)_{ns}\left[\frac{p_t}{(p_c)_{ns}}\right]^{\frac{\gamma-1}{\gamma}} = 3633 \times \left(\frac{3.892 \times 10^6}{6.895 \times 10^6}\right)^{\frac{1.2-1}{1.2}} = 3303 \text{ K}$$

$$T_x = (T_c)_{ns}\left[\frac{p_x}{(p_c)_{ns}}\right]^{\frac{\gamma-1}{\gamma}} = 3633 \times \left(\frac{3.0 \times 10^5}{6.895 \times 10^6}\right)^{\frac{1.2-1}{1.2}} = 2155 \text{ K}$$

$$T_e = (T_c)_{ns}\left[\frac{p_e}{(p_c)_{ns}}\right]^{\frac{\gamma-1}{\gamma}} = 3633 \times \left(\frac{6.798 \times 10^4}{6.895 \times 10^6}\right)^{\frac{1.2-1}{1.2}} = 1682 \text{ K}$$

Now, we want to compute the densities of gas at the injection plane (ρ_{inj}), at the nozzle inlet (ρ_i), at the throat (ρ_t), at the cross-section x defined above (ρ_x), and at the exit plane of the nozzle (ρ_e). The law of perfect gases states that

$$p = \rho R T$$

Since the molar mass of the combustion products is $\mathcal{M} = 22.67$ kg/kmol, and the universal gas constant is $R^* = 8314.460$ N m kmol^{-1} K^{-1}, then the densities of gas at the sections of interest are

$$\rho_{inj} = \frac{\mathcal{M} p_{inj}}{R^* T_{inj}} = \frac{22.67 \times 7.474 \times 10^6}{8314.46 \times 3633} = 5.609 \text{ kg/m}^3$$

$$\rho_i = \frac{\mathcal{M} p_i}{R^* T_i} = \frac{22.67 \times 6.270 \times 10^6}{8314.46 \times 3576} = 4.781 \text{ kg/m}^3$$

$$\rho_t = \frac{\mathcal{M} p_t}{R^* T_t} = \frac{22.67 \times 3.892 \times 10^6}{8314.46 \times 3303} = 3.213 \text{ kg/m}^3$$

$$\rho_x = \frac{\mathcal{M} p_x}{R^* T_x} = \frac{22.67 \times 3.0 \times 10^5}{8314.46 \times 2155} = 0.3796 \text{ kg/m}^3$$

$$\rho_e = \frac{\mathcal{M} p_e}{R^* T_e} = \frac{22.67 \times 6.798 \times 10^4}{8314.46 \times 1682} = 0.1102 \text{ kg/m}^3$$

We also want to compute the flow velocities at the nozzle inlet (v_i), at the throat (v_t), at the cross-section x defined above (v_x), and at the exit plane of the nozzle (v_e). The sonic velocity at the nozzle inlet is expressed by the following equation

$$a_i = (\gamma R T_i)^{\frac{1}{2}}$$

The flow velocity at the nozzle inlet, where $M_i = 0.4$ and $T_i = 3576$ K, is

$$v_i = M_i a_i = 0.4 \times (1.2 \times 366.8 \times 3576)^{\frac{1}{2}} = 501.8 \text{ m/s}$$

The flow velocity at the throat, where $M_t = 1$ and $T_t = 3303$ K, is

$$v_t = M_t a_t = 1 \times (1.2 \times 366.8 \times 3303)^{\frac{1}{2}} = 1206 \text{ m/s}$$

The flow velocity at the cross-section x, where $A_x/A_t = 4$ and $p_x = 3.0 \times 10^5$ N/m^2, is expressed by the following equation of [4]:

$$v_x = \left\{ \frac{2\gamma}{\gamma - 1} R(T_c)_{ns} \left[1 - \left(\frac{p_x}{(p_c)_{ns}} \right)^{\frac{\gamma-1}{\gamma}} \right] \right\}^{\frac{1}{2}}$$

where $(T_c)_{ns} = 3633$ K and $(p_c)_{ns} = 6.895 \times 10^6$ N/m^2 are respectively the total temperature and the total pressure of the combustion chamber at the nozzle inlet. After substituting these values into the preceding equation, there results

$$v_x = \left\{ \frac{2 \times 1.2}{1.2 - 1} \times 366.8 \times 3633 \times \left[1 - \left(\frac{3.0 \times 10^5}{6.895 \times 10^6} \right)^{\frac{1.2-1}{1.2}} \right] \right\}^{\frac{1}{2}} = 2551 \text{ m/s}$$

Likewise, the flow velocity at the exit plane of the nozzle, where the static pressure $p_e = 6.798 \times 10^4$ N/m^2, is expressed by the following equation

$$v_e = \left\{ \frac{2\gamma}{\gamma - 1} R(T_c)_{ns} \left[1 - \left(\frac{p_e}{(p_c)_{ns}} \right)^{\frac{\gamma-1}{\gamma}} \right] \right\}^{\frac{1}{2}}$$

After substituting this value into the preceding equation, there results

$$v_e = \left\{ \frac{2 \times 1.2}{1.2 - 1} \times 366.8 \times 3633 \times \left[1 - \left(\frac{6.798 \times 10^4}{6.895 \times 10^6} \right)^{\frac{1.2-1}{1.2}} \right] \right\}^{\frac{1}{2}} = 2930 \text{ m/s}$$

Now, we want to compute the Mach numbers of the gas flow at the cross-section x (M_x) and at the exit plane of the nozzle (M_e).

The sonic velocity of the gas flow at the cross-section x of the nozzle is

$$a_x = (\gamma R T_x)^{\frac{1}{2}} = (1.2 \times 366.8 \times 2155)^{\frac{1}{2}} = 973.8 \text{ m/s}$$

Therefore, the Mach number of the gas flow at the cross-section x of the nozzle is

$$M_x = \frac{v_x}{a_x} = \frac{2551}{973.8} = 2.619$$

Likewise, the sonic velocity of the gas flow at the exit plane of the nozzle is

$$a_e = (\gamma R T_e)^{\frac{1}{2}} = (1.2 \times 366.8 \times 1682)^{\frac{1}{2}} = 860.5 \text{ m/s}$$

Therefore, the Mach number of the gas flow at the exit plane of the nozzle is

$$M_e = \frac{v_e}{a_e} = \frac{2930}{860.5} = 3.405$$

Finally, we want to compute the areas of the cross-sections of the gas flow at the inlet plane of the nozzle (A_i), at the combustion chamber (A_c), at the throat (A_t), at the section x of the nozzle (A_x), and at the exit plane of the nozzle (A_e).

As has been shown in Sect. 1.2, owing to the principle of mass conservation, the rate at which a quantity of mass passes through a cross-section of the nozzle is independent of the position of this cross-section along the axis of symmetry of the nozzle, that is,

$$\dot{m} = \rho A v = \text{constant}$$

In the present case, we know the mass flow rate of the propellant in the combustion chamber to be $\dot{m}_{tc} = 163.6$ kg/s, which value is also constant at any cross-section of the nozzle. Therefore, the areas of the cross-sections of the gas flow at the inlet plane of the nozzle (A_i), at the combustion chamber (A_c), and at the throat (A_t) are

$$A_i = \frac{\dot{m}_{tc}}{v_i \rho_i} = \frac{163.6}{501.8 \times 4.781} = 0.06819 \text{ m}^2$$

$$A_c = A_i = 0.06819 \text{ m}^2$$

$$A_t = \frac{\dot{m}_{tc}}{v_t \rho_t} = \frac{163.6}{1206 \times 3.213} = 0.04222 \text{ m}^2$$

By definition, the areas of the cross-sections of the gas flow at the section x of the nozzle (A_x) and at the exit plane of the nozzle (A_e) are

$$A_x = 4A_t = 4 \times 0.04222 = 0.1689 \text{ m}^2$$

$$A_e = 12A_t = 12 \times 0.04222 = 0.5066 \text{ m}^2$$

1.3 Performance Indicators

The performance of a rocket engine, independently of whether its propellant may be liquid or solid, is measured by a quantity, which is called specific impulse (I_s). This quantity is defined as the ratio of the thrust F (N) imparted to the rocket to the weight flow rate $\dot{W}$ (N/s) of the propellant on the surface of the Earth and at the sea level, as follows

$$I_s = \frac{F}{\dot{W}} = \frac{F}{\dot{m} g_0}$$

This quantity is measured in seconds. The following table, due to the courtesy of the Government of the United States ([19], page 44), gives the specific impulse of some typical chemical propellants for rockets.

In case of liquid-propellant engines, it is important to specify whether a given specific impulse takes account of the thrust chamber only, in which case it is denoted here by $(I_s)_{tc}$, or of the overall engine, in which case it is denoted here by $(I_s)_{oa}$. This distinction is necessary, because, when a rocket engine is fed by turbo-pumps, then its specific impulse relating to the overall engine may include the power required by its turbines, by its Vernier rockets (by the way, a Vernier rocket is a small additional engine, which is placed at the bottom of the main rocket for the purpose of generating a control torque), and by its devices used for attitude control. The propellant necessary to fed the devices indicated above may come from one or more tanks carried on board a rocket. The amount by which $(I_s)_{tc}$ is greater than $(I_s)_{oa}$ is usually 1 or 2%.

TABLE 1.—*Specific impulse of some typical chemical propellants* [1]

Propellant combinations:	Isp range (sec)
Monopropellants (liquid):	
Low-energy monopropellants	160 to 190.
Hydrazine	
Ethylene oxide	
Hydrogen peroxide	
High-energy monopropellants:	
Nitromethane	190 to 230.
Bipropellants (liquid):	
Low-energy bipropellants	200 to 230.
Perchloryl fluoride—Available fuel	
Analine—Acid	
JP-4—Acid	
Hydrogen peroxide—JP-4	
Medium-energy bipropellants	230 to 260.
Hydrazine—Acid	
Ammonia—Nitrogen tetroxide	
High-energy bipropellants	250 to 270.
Liquid oxygen—JP-4	
Liquid oxygen—Alcohol	
Hydrazine—Chlorine trifluoride	
Very high-energy bipropellants	270 to 330.
Liquid oxygen and fluorine—JP-4	
Liquid oxygen and ozone—JP-4	
Liquid oxygen—Hydrazine	
Super high-energy bipropellants	300 to 385.
Fluorine—Hydrogen	
Fluorine—Ammonia	
Ozone—Hydrogen	
Fluorine—Diborane	
Oxidizer-binder combinations (solid):	
Potassium perchlorate:	
Thiokol or asphalt	170 to 210.
Ammonium perchlorate:	
Thiokol	170 to 210.
Rubber	170 to 210.
Polyurethane	210 to 250.
Nitropolymer	210 to 250.
Ammonium nitrate:	
Polyester	170 to 210.
Rubber	170 to 210.
Nitropolymer	210 to 250.
Double base	170 to 250.
Boron metal components and oxidant	200 to 250.
Lithium metal components and oxidant	200 to 250.
Aluminum metal components and oxidant	200 to 250.
Magnesium metal components and oxidant	200 to 250.
Perfluoro-type propellants	250 and above.

[1] Some Considerations Pertaining to Space Navigation, Aerojet-General Corp., Special Rept. No. 1450, May 1958.

Another performance indicator is the propellant mass fraction, R_p, of the whole vehicle which is propelled by a rocket engine. The propellant mass fraction is defined as the ratio of the mass of the usable propellant (m_p) to the initial mass of the rocket (m_0), as follows

$$R_p = \frac{m_p}{m_0}$$

where the initial mass of the rocket (m_0) includes the masses due to the engine at burnout, the structure, the guidance system, the propellant, and the payload.

Let $\dot{m}_e > 0$ be the rate at which the exhaust gas flows through the exit plane of the nozzle. Due to this flow, the mass m of the rocket decreases in time at the rate dm/dt. According to the principle of mass conservation, this decrease in mass must be equal to the exhaust gas flowing through the nozzle, as follows

$$\frac{dm}{dt} = -\dot{m}_e$$

Assuming $\dot{m}_e = $ constant, the mass m of the rocket decreases in time as follows

$$m = m_0 - \dot{m}_e t$$

Let m and v be respectively the total mass and the velocity of the rocket with respect to the Earth at a given time t. At the same time t, the linear momentum of the rocket is mv. After an infinitesimal interval of time dt, due to the flow of propellant, the mass m of the rocket decreases by dm, and the velocity v of the rocket increases by dv. This is because a positive mass $-dm$ has been expelled from the rocket at a velocity $-v_e$ (with respect to the rocket) and $v - v_e$ (with respect to the Earth). Therefore, at the time $t + dt$, the mass of the rocket becomes $m + dm$ (with $dm < 0$), and its velocity becomes $v + dv$ (with $dv > 0$).

The total linear momentum of the system (comprising the rocket and the exhaust gas) at the time $t + dt$ results from the linear momentum due to the rocket plus the linear momentum due to the exhaust gas, as follows

$$(m + dm)(v + dv) + (-dm)(v - v_e)$$

The principle of conservation of linear momentum states that the linear momentum of the system (gas and rocket) at the time t is equal to the linear momentum of the system at the time $t + dt$, as follows

$$mv = (m + dm)(v + dv) + (-dm)(v - v_e)$$

This is because the acceleration of the ejected mass of gas and the corresponding reaction (thrust) acting on the rocket (which are two forces equal in magnitude and oppositely directed) are the sole forces considered as acting on the system. Aerodynamic forces (lift and drag) and gravitational attraction are neglected.

After executing the operations on the right-hand side of the preceding equation, neglecting the second-order differential $dm\,dv$, and cancelling the two terms mv, there results

$$0 = m\,dv + v_e dm$$

which, solved for dv, yields

$$dv = -v_e \frac{dm}{m}$$

The preceding differential equation, integrated over the time interval going from $t = 0$ to the time of burnout t_{bo}, yields

$$\Delta v = -v_e \ln\left(\frac{m_{bo}}{m_0}\right) = v_e \ln\left(\frac{m_0}{m_{bo}}\right)$$

where $m_{bo} = m_0 - m_p$ is the mass of the rocket at burnout, m_0 is the initial mass of the rocket, and m_p is the mass of propellant. The preceding equation was derived in 1903 by the Russian scientist Konstantin Eduardovich Tsiolkovsky, and is known as the rocket equation.

In the ideal case, in which the ambient pressure p_0 is equal to the pressure p_e of the exhaust gas at the exit plane of the nozzle, the equation of thrust

$$F = \dot{m}v_e + A_e(p_e - p_0)$$

reduces to

$$F = \dot{m}v_e$$

Remembering the definition of specific impulse

$$I_s = \frac{F}{W} = \frac{F}{\dot{m}g_0}$$

and substituting $F = \dot{m}v_e$ into the preceding equation, the equation of rocket may also be written as follows

$$\Delta v = -I_s g_0 \ln\left(\frac{m_{bo}}{m_0}\right) = I_s g_0 \ln\left(\frac{m_0}{m_{bo}}\right)$$

In addition, since $m_{bo} = m_0 - m_p$, and $R_p = m_p/m_0$, then the rocket equation may also be written as follows

$$\Delta v = I_s \, g_0 \, \ln \left(\frac{m_0}{m_0 - m_p} \right) = I_s \, g_0 \, \ln \left(\frac{1}{1 - R_p} \right)$$

As has been shown above, the preceding equation holds in the ideal case, in which gravitational and aerodynamic forces are neglected. When these forces and the losses due to power requirements are taken into account, the preceding equation may be written as follows

$$\Delta v = C_{vc} (I_s)_{oa} \, g_0 \, \ln \left(\frac{1}{1 - R_p} \right)$$

where C_{vc} is a coefficient which takes account of the gravitational and aerodynamic forces, and $(I_s)_{oa}$ is the specific impulse of the overall engine.

In particular, when the initial velocity v_0 is equal to zero and the final velocity is the velocity of the rocket at burnout v_{bo}, then the preceding equation may be written as follows

$$v_{bo} = C_{vc} (I_s)_{oa} \, g_0 \, \ln \left(\frac{1}{1 - R_p} \right)$$

The performance of a thrust camber only is measured by the corresponding specific impulse $(I_s)_{tc}$. This performance depends on the propellant (fuel and oxidiser) combination, the combustion efficiency in the combustion chamber, and the expansion of the exhaust gas in the nozzle. When the thrust and the weight flow rate of the thrust chamber are known for a given rocket engine, then the specific impulse relating to the thrust chamber can be determined as follows

$$(I_s)_{tc} = \frac{F}{\dot{W}_{tc}}$$

As an option, when we know the effective exhaust velocity (c), defined in Sect. 1.1, then the specific impulse relating to the thrust chamber can be determined as follows

$$(I_s)_{tc} = \frac{c}{g_0}$$

The effective exhaust velocity (c) may also be defined as the product of two quantities, namely c^* and C_F, as follows

$$c = c^* C_F$$

where c^* is the characteristic velocity, which depends on the combustion performance of the propellant, and C_F is a dimensionless quantity, called the thrust coefficient, which measures the performance of the gas expansion through the nozzle. By using this definition of c, the specific impulse relating to the thrust chamber can be determined as follows

$$(I_s)_{tc} = \frac{c^* C_F}{g_0}$$

Of the performance indicators defined above, the specific impulse (I_s) and the propellant mass fraction (R_p) are of great importance to the designer of a rocket vehicle as a whole. On the other hand, the characteristic velocity (c^*) and the thrust coefficient (C_F) are of great importance to the designer of a rocket engine, as will be shown below.

The characteristic velocity (c^*) of a rocket engine, in which the sonic velocity of the gas flow is reached at the throat, measures the energy possessed by the propellant and the quality level characterising the injector and the combustion chamber. It may be expressed by the following equation [4]:

$$c^* = \frac{(p_c)_{ns} A_t}{\dot{m}_{tc}} = \frac{(p_c)_{ns} A_t g_0}{\dot{W}_{tc}}$$

where $(p_c)_{ns} = p_i [1 + \frac{1}{2}(\gamma - 1)M_i^2]^{\gamma/(\gamma-1)}$ is the total pressure at the nozzle inlet, p_i is the static pressure at the nozzle inlet, $\gamma \equiv c_p/c_v$ is the specific heat ratio of the combusted gas, M_i is the Mach number at the nozzle inlet, A_t is the area of the cross-section at the throat, $\dot{m}_{tc}$ and $\dot{W}_{tc}$ are respectively the mass flow rate and the weight flow rate of the propellant at the thrust chamber, and $g_0 = 9.80665$ m/s^2 is the acceleration of gravity at sea level. This equation indicates the mass or the weight flow rate of propellant which must be burned to maintain the required total pressure $(p_c)_{ns}$ at the nozzle inlet. A smaller value of propellant consumption $\dot{m}_{tc}$ or $\dot{W}_{tc}$ corresponds to a higher value of c^*.

Another expression of the characteristic velocity c^* is given in [4] as follows

$$c^* = \frac{\left[\gamma R(T_c)_{ns}\right]^{\frac{1}{2}}}{\gamma\left[\left(\frac{2}{\gamma+1}\right)^{\frac{\gamma+1}{\gamma-1}}\right]^{\frac{1}{2}}}$$

where $(T_c)_{ns} = T_i [1 + \frac{1}{2}(\gamma - 1)M_i^2]$ is the total temperature of the combustion chamber at the nozzle inlet, M_i is the Mach number at the nozzle inlet, $\gamma \equiv c_p/c_v$ is the specific heat ratio of the combusted gas, and R is the constant of the specific gas (that is, the universal gas constant $R^* = 8314.460$ N m kmol^{-1} K^{-1} divided by the average molar mass $\mathcal{M}$ of the combusted gas). The preceding equation shows that the characteristic velocity c^* depends on the properties of the combusted gas at the nozzle inlet, such properties being the specific heat ratio γ, the constant R of the specific gas, and the total temperature of the combustion chamber at the nozzle inlet $(T_c)_{ns}$.

The thrust coefficient C_F depends on the expansion of the exhaust gas and also on the design of the nozzle. By combining the three following equations derived above

$$(I_s)_{tc} = \frac{F}{\dot{W}_{tc}}$$

$$(I_s)_{tc} = \frac{c^* C_F}{g_0}$$

$$c^* = \frac{(p_c)_{ns} A_t \, g_0}{\dot{W}_{tc}}$$

the thrust coefficient can be expressed as follows

$$C_F = \frac{F}{A_t (p_c)_{ns}}$$

The thrust coefficient, put in this form, is the ratio of the actual thrust F acting on the rocket to the force $A_t \, (p_c)_{ns}$ which would act on the rocket if there were no expansion of the exhaust gas downstream of the throat, that is, if the total pressure on the combustion chamber acted only on the area of the throat.

The following equation, given in [4], expresses the theoretical value of the thrust coefficient at any altitude:

$$C_F = \left\{ \frac{2\gamma^2}{\gamma - 1} \left(\frac{2}{\gamma + 1} \right)^{\frac{\gamma+1}{\gamma-1}} \left[1 - \left(\frac{p_e}{(p_c)_{ns}} \right)^{\frac{\gamma-1}{\gamma}} \right] \right\}^{\frac{1}{2}} + \frac{A_e}{A_t} \left[\frac{p_e - p_0}{(p_c)_{ns}} \right]$$

where $\gamma \equiv c_p/c_v$ is the specific heat ratio of the combusted gas, p_e is the pressure of the exhaust gas at the exit plane of the nozzle, $(p_c)_{ns}$ is the total pressure in the combustion chamber at the nozzle inlet, p_0 is the ambient pressure, and A_e/A_t is the ratio of the area of the cross-section at the exit plane of the nozzle to the area of the cross-section at the throat. As has been shown in Sect. 1.1, when the ambient pressure increases from zero (in space) to $p_0 > 0$, then the thrust generated by a rocket engine decreases from F_s (in space) to F by an amount $A_e p_0$. Such is also the case with the thrust coefficient of a rocket engine, which decreases from $(C_F)_s$ (in space) to C_F by an amount $(A_e/A_t) \, p_0/(p_c)_{ns}$, as shown by the preceding equation. The same equation may also be rewritten as follows

$$C_F = (C_F)_s - \frac{A_e}{A_t} \left[\frac{p_0}{(p_c)_{ns}} \right]$$

In other words, higher values of thrust F and thrust coefficient C_F correspond to lower values of ambient pressure p_0.

For a given value p_0 of ambient pressure, the optimum value of thrust is reached when the expansion area ratio A_e/A_t of the nozzle is such that $p_e = p_0$. Let us consider a region in the divergent portion of a nozzle where $p_e > p_0$. In this region, the thrust of a rocket engine increases, and reaches its maximum value at the plane where $p_e = p_0$. When the divergent portion of a nozzle extends downstream of this plane, then the pressure p_e of the exhaust gas at the exit plane becomes less than the ambient pressure p_0, causing a decrease in the value of thrust. The value of the expansion area ratio A_e/A_t of the nozzle corresponding to the condition $p_e = p_0$ is called the optimum expansion area ratio.

Since the value of the ambient pressure decreases with altitude, no value of A_e/A_t is optimum at all altitudes, and therefore the design value of A_e/A_t is to be found on the basis of a compromise. This compromise is not necessary in case of upper stages of rockets, which operate at ambient pressures of zero or near zero. In case of rockets operating in space ($p_0 = 0$), a value of expansion area ratio greater than 25 has been found to be scarcely useful [4]. Therefore, in order to reduce weight, this value is not exceeded.

The specific heat ratio $\gamma \equiv c_p/c_v$ is an indicator of the quantity of energy stored in the molecules of exhaust gas. In particular, a small value of γ indicates a high capacity of storing energy, which results in a high performance of a rocket engine. The equations given above, which express c^* and C_F as functions of γ, indicate high values of c^* and C_F for low values of γ. These equations are re-written below for convenience of the reader.

$$c^* = \frac{\left[\gamma R(T_c)_{ns}\right]^{\frac{1}{2}}}{\gamma\left[\left(\frac{2}{\gamma+1}\right)^{\frac{\gamma+1}{\gamma-1}}\right]^{\frac{1}{2}}}$$

$$C_F = \left\{\frac{2\gamma^2}{\gamma-1}\left(\frac{2}{\gamma+1}\right)^{\frac{\gamma+1}{\gamma-1}}\left[1-\left(\frac{p_e}{(p_c)_{ns}}\right)^{\frac{\gamma-1}{\gamma}}\right]\right\}^{\frac{1}{2}} + \frac{A_e}{A_t}\left[\frac{p_e - p_0}{(p_c)_{ns}}\right]$$

In order for the specific heat ratio γ of the exhaust gas to have a small value, it is necessary to choose accurately the propellant to be used.

The effect of the constant $R = R^*/M$ of the specific exhaust gas on the performance of a rocket engine is shown by the first of the two preceding equations, which expresses the characteristic velocity c^* as a function of R. Since c^* is proportional to the square root of R, then the value of c^* increases for increasing values of R.

The effect of the total pressure in the combustion chamber at the nozzle inlet, $(p_c)_{ns}$, is shown by the second of two equations written above, which expresses the thrust coefficient C_F as a function of $(p_c)_{ns}$. It is to be noted that the total pressure $(p_c)_{ns}$ appears above in the two pressure ratios $p_e/(p_c)_{ns}$ and $p_0/(p_c)_{ns}$.

The following equation of Sect. 1.2

$$\frac{A_e}{A_t} = \frac{\left(\frac{2}{\gamma+1}\right)^{\frac{1}{\gamma-1}}\left(\frac{(p_c)_{ns}}{p_e}\right)^{\frac{1}{\gamma}}}{\left\{\frac{\gamma+1}{\gamma-1}\left[1-\left(\frac{p_e}{(p_c)_{ns}}\right)^{\frac{\gamma-1}{\gamma}}\right]\right\}^{\frac{1}{2}}}$$

shows that, for given values of A_e/A_t and γ, the ratio $p_e/(p_c)_{ns}$ has a singular value, corresponding to the condition $1 - (p_e/(p_c)_{ns})^{(\gamma-1)/\gamma} = 0$. The preceding equation

$$C_F = \left\{ \frac{2\gamma^2}{\gamma-1} \left(\frac{2}{\gamma+1} \right)^{\frac{\gamma+1}{\gamma-1}} \left[1 - \left(\frac{p_e}{(p_c)_{ns}} \right)^{\frac{\gamma-1}{\gamma}} \right] \right\}^{\frac{1}{2}} + \frac{A_e}{A_t} \left[\frac{p_e - p_0}{(p_c)_{ns}} \right]$$

shows that the ratio $p_e/(p_c)_{ns}$ affects the thrust coefficient C_F only through of the term $-p_0/(p_c)_{ns}$. When $(p_c)_{ns}$ increases, then $-p_0/(p_c)_{ns}$ decreases, and therefore C_F increases. This effect is particularly important for high values of the ambient pressure p_0. The preceding equation

$$C_F = \frac{F}{A_t (p_c)_{ns}}$$

shows that, for a given area of the cross-section of the nozzle at the throat A_t, the thrust F is proportional to both the total pressure in the combustion chamber at the nozzle inlet $(p_c)_{ns}$ and the thrust coefficient C_F. Therefore, when the total pressure $(p_c)_{ns}$ increases in a given combustion chamber, then the thrust F also increases. Increasing values of $(p_c)_{ns}$ also cause the values of the total temperature in the combustion chamber $(T_c)_{ns}$ to increase, and the values of γ and R to decrease. According to Huzel and Huang [4], these effects are slight, especially when the value of the total pressure $(p_c)_{ns}$ is greater than 2.068×10^6 N/m².

The performance indicators considered so far refer to an ideal rocket engine, which differs from a real rocket engine for the following reasons. A real engine is subject to friction, heat transfer, effects due to non-perfect gases, misalignment of the gas flow with respect to the axis of symmetry of the nozzle, non-uniform reacting substances and flow distribution, and changes in the composition of the exhaust gas. A change in gas composition, in turn, is due to non-uniform values of the quantities γ, $\mathcal{M}$, and R along the axis of symmetry of the nozzle.

The actual quantities result from multiplying the corresponding theoretical quantities by correction factors.

In particular, the correction factor η_F, whose values range from 0.92 to 1.00 [4], makes it possible to compute the actual values of thrust $(\overline{F})$ and thrust coefficient $(\overline{C}_F)$ from the corresponding theoretical values $(F$ and $C_F)$, as follows

$$\overline{F} = \eta_E F$$
$$\overline{C}_F = \eta_F C_F$$

The correction factor η_v, whose values range from 0.85 to 0.98 [4], makes it possible to compute the actual values of exhaust velocity $(\overline{v})$ and specific impulse $(\overline{I}_s)$ from the corresponding theoretical values $(v$ and $I_s)$, as follows

$$\overline{v} = \eta_x v$$
$$\overline{I}_s = \eta_x I_s$$

The correction factor η_{c*}, whose values range from 0.87 to 1.03 [4], makes it possible to compute the actual values of characteristic velocity ($\overline{c}^*$) from the corresponding theoretical values (c^*), as follows

$$\overline{c}^* = \eta_{c*}\, c^*$$

The correction factor η_W, whose values range from 0.98 to 1.15 [4], makes it possible to compute the actual values of weight flow rate ($\overline{W}$) from the corresponding theoretical values ($\dot{W}$), as follows

$$\overline{\dot{W}} = \eta_W\, \dot{W}$$

According to Huzel and Huang [4], the following relations exist for the correction factors defined above

$$\eta_v = \eta_{c*}\eta_F$$

$$\eta_v = \frac{1}{\eta_W}$$

Ranges of the actual values of the quantities relating to liquid-propellant rocket engines are given in the following table (from [4]).

Gas temperature, T (K)	2220–3890
Nozzle stagnation pressure, $(p_c)_{ns}$ (N/m^2)	68950–1.723 × 10^7
Molar mass, $\mathcal{M}$ (kg/kmol)	2–30
Specific gas constant, R (N m kg^{-1} K^{-1})	143.815–2157.23
Mach number of gas flow, M	0–4.5
Specific heat ratio, γ	1.13–1.66
Nozzle expansion area ratio, A_e/A_t	3.5–100
Nozzle contraction area ratio, A_c/A_t	1.3–6.0
Thrust coefficient, C_F	1.3–2.0
Characteristic velocity, c^* (m/s)	914–2440
Effective exhaust velocity, c (m/s)	1220–3660
Specific impulse in vacuo, I_s (s)	150–480

An example of application is given below. The rocket motor of Sect. 1.2 has the following data: propellant weight flow rate in the combustion chamber $\dot{W}_{tc} = 9.807 \times 163.6 = 1604$ N/s, total pressure in the combustion chamber at the nozzle inlet $(p_c)_{ns} = p_i[1 + \frac{1}{2}(\gamma - 1)M_i^2]^{\gamma/(\gamma-1)} = 6.895 \times 10^6$ N/m^2, total temperature of the combustion chamber $(T_c)_{ns} = T_i[1 + \frac{1}{2}(\gamma - 1)M_i^2] = 3633$ K (T_i and M_i being respectively the flow temperature and the Mach number at the nozzle inlet), molar mass of the combustion products $\mathcal{M} = 22.67$ kg/kmol, specific ratio of the gas $\gamma = 1.20$, and expansion area ratio of the nozzle $A_e/A_t = 12$. The Mach number at the injection plane, M_{inj}, is assumed to be equal to zero, and the Mach number at the inlet plane of the nozzle, M_i, is assumed to be equal to 0.4. The correction factors for respectively the characteristic velocity and the coefficient of thrust are $\eta_{c*} = 0.97$ and $\eta_F = 0.983$.

It is required to compute the theoretical value of the characteristic velocity, the theoretical values of the thrust coefficient at sea level and in space, the theoretical values of the specific impulse (due to the thrust chamber only) at sea level and in space, the actual value of the characteristic velocity assuming a correction factor $\eta_{c*} = 0.97$, the actual values of the thrust coefficient at sea level and in space assuming a correction factor $\eta_F = 0.983$, the actual values of the specific impulse (due to the thrust chamber only) at sea level and in space, the correction factor for the specific impulse (due to the thrust chamber only), the actual values of the thrust at sea level and in space, and the actual values of the areas of the cross-sections at the throat and at the exit section of the nozzle.

The theoretical value of the characteristic velocity, $c*$, results from

$$c* = \frac{\left[\gamma R(T_c)_{ns}\right]^{\frac{1}{2}}}{\gamma\left[\left(\frac{2}{\gamma+1}\right)^{\frac{\gamma+1}{\gamma-1}}\right]^{\frac{1}{2}}} = \frac{(1.2 \times 366.8 \times 3633)^{\frac{1}{2}}}{1.2 \times \left[\left(\frac{2}{1.2+1}\right)^{\frac{1.2+1}{1.2-1}}\right]^{\frac{1}{2}}} = 1780 \text{ m/s}$$

where the value of R has been determined in Sect. 1.2 as follows

$$R = \frac{R^*}{\mathcal{M}} = \frac{8314.460}{22.67} = 366.8 \text{ N m kg}^{-1}\text{K}^{-1}$$

The theoretical value of the thrust coefficient at sea level, C_F, results from

$$C_F = \left\{\frac{2\gamma^2}{\gamma-1}\left(\frac{2}{\gamma+1}\right)^{\frac{\gamma+1}{\gamma-1}}\left[1 - \left(\frac{p_e}{(p_c)_{ns}}\right)^{\frac{\gamma-1}{\gamma}}\right]\right\}^{\frac{1}{2}} + \frac{A_e}{A_t}\left[\frac{p_e - p_0}{(p_c)_{ns}}\right]$$

Remembering that p_e has been determined in Sect. 1.2 as follows

$$p_e = z(p_c)_{ns} = 0.009859 \times 6.895 \times 10^6 = 6.798 \times 10^4 \text{N/m}^2$$

and substituting $\gamma = 1.2$, $(p_c)_{ns} = 6.895 \times 10^6$ N/m², $A_e/A_t = 12$, and $p_0 = 1.013 \times 10^5$ N/m² into the expression of C_F, we find

$$C_F = 1.588$$

The theoretical value of the thrust coefficient in space, $(C_F)_s$, results from

$$(C_F)_s = C_F + \frac{A_e}{A_t}\left[\frac{p_0}{(p_c)_{ns}}\right] = 1.588 + 12 \times \frac{1.013 \times 10^5}{6.895 \times 10^6} = 1.764$$

Since the theoretical values of the characteristic velocity and the thrust coefficient (c^* and C_F) have been computed above, the theoretical value of the specific impulse at sea level relating to the thrust chamber only, $(I_s)_{tc}$, results from

$$(I_s)_{tc} = \frac{c^* C_F}{g_0} = \frac{1780 \times 1.588}{9.807} = 288.2 \text{ s}$$

Likewise, the theoretical value of the specific impulse in space relating to the thrust chamber only, $(I_s)_{tcs}$, results from

$$(I_s)_{tcs} = \frac{c^* (C_F)_s}{g_0} = \frac{1780 \times 1.764}{9.807} = 320.2 \text{ s}$$

Since the correction factor for the characteristic velocity is $\eta_{c^*} = 0.97$, then the actual value of the characteristic velocity results from

$$\overline{c}^* = \eta_{c^*} c^* = 0.97 \times 1780 = 1727 \text{ m/s}$$

Since the correction factor for the coefficient of thrust is $\eta_F = 0.983$, then the actual value of the coefficient of thrust at sea level results from

$$\overline{C}_F = \eta_E C_F = 0.983 \times 1.588 = 1.561$$

Likewise, the actual value of the coefficient of thrust in space results from

$$(\overline{C}_F)_s = \eta_E (C_F)_s = 0.983 \times 1.764 = 1.734$$

The actual value of the specific impulse at sea level relating to the thrust chamber only results from

$$(\overline{I}_s)_{tc} = \frac{\overline{c}^* \overline{C}_F}{g_0} = \frac{1727 \times 1.561}{9.807} = 274.9 \text{ s}$$

Likewise, the actual value of the specific impulse in space relating to the thrust chamber only results from

$$(\overline{I}_s)_{tcs} = \frac{\overline{c}^* (\overline{C}_F)_s}{g_0} = \frac{1727 \times 1.764}{9.807} = 310.6 \text{ s}$$

The correction factor η_v for the actual values of exhaust velocity and specific impulse results from

$$\eta_v = \frac{(\overline{I}_s)_{tc}}{(I_s)_{tc}} = \frac{274.9}{288.2} = 0.9539$$

The actual value of the thrust at sea level results from

$$\overline{F} = (\overline{I}_s)_{tc} \dot{W}_{tc} = 274.9 \times 1604 = 4.409 \times 10^5 \mathrm{N}$$

Likewise, the actual value of the thrust in space results from

$$(\overline{F})_s = (\overline{I}_s)_{tcs} \dot{W}_{tc} = 310.6 \times 1604 = 4.982 \times 10^5 \mathrm{N}$$

The actual values of the areas of the cross-sections of respectively the throat $(\overline{A}_t)$ and the exit plane of the nozzle $(\overline{A}_e)$ result from

$$\overline{A}_t = \frac{\overline{F}}{\overline{C}_F (p_c)_{ns}} = \frac{4.409 \times 10^5}{1.561 \times 6.895 \times 10^6} = 0.04096 \ \mathrm{m}^2$$

$$\overline{A}_e = 12\overline{A}_t = 12 \times 0.04096 = 0.4916 \ \mathrm{m}^2$$

1.4 Liquid Propellants for High-Thrust Rocket Engines

Some fundamental concepts on solid, liquid, and gaseous propellants for high-thrust rockets have been given in Sects. 1.1 and 1.2. The present paragraph deals specifically with liquid propellants. They may be classified into the following categories:

- mono-propellants;
- bi-propellants;
- cryogenic propellants; and
- storable propellants.

A mono-propellant is either a mixture of a fuel with an oxidiser, or a single substance which decomposes into a hot gas in a thrust chamber, in the presence of an appropriate catalyst. This chemical decomposition releases thermal energy, which is contained within the chemical bonds of the molecules involved in the reaction. For example, hydrogen peroxide (H_2O_2) decomposes into hot water vapour (H_2O) and gaseous oxygen (O_2) when made to pass through a platinum catalyst mesh, according to the following reaction:

$$2H_2O_2 \rightarrow 2H_2O + O_2$$

The gases resulting from this reaction are ejected through a nozzle to generate thrust. A gaseous mono-propellant, which can be used instead of hydrogen peroxide, is nitrous oxide (N_2O). This substance, when heated or passed over a catalyst bed, decomposes exothermically into gaseous nitrogen (N_2) and gaseous oxygen (O_2), according to the following reaction:

$$2N_2O \rightarrow 2N_2 + O_2$$

A scheme, which illustrates a rocket engine fed by nitrous oxide, is shown in the following figure, (re-drawn from [20]). Further information on this type of mono-propellant can be found, for example, in [20].

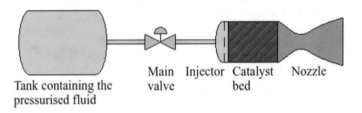

Tank containing the   Main   Injector Catalyst   Nozzle
pressurised fluid   valve   bed

The most widely used mono-propellant is hydrazine (anhydrous N_2H_4, also written H_2NNH_2), which is a strong reducer, that is, a substance which donates electrons to another substance in a redox chemical reaction. In a mono-propellant engine, hydrazine decomposes (in the presence of a catalyst such as iridium metal supported by high-surface-area alumina, or carbon nanofibres, or molybdenum nitride on alumina) into gaseous nitrogen (N_2) and gaseous hydrogen (H_2), or into ammonia (NH_3) and gaseous nitrogen (N_2), according to the following exothermic reactions [21]:

$$H_2NNH_2 \rightarrow N_2 + 2H_2$$
$$3H_2NNH_2 \rightarrow 4NH_3 + N_2$$

A scheme illustrating a mono-propellant rocket engine fed by hydrazine is shown in the following figure, which is due to the courtesy of NASA [22].

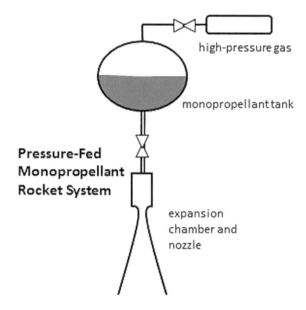

high-pressure gas

monopropellant tank

Pressure-Fed Monopropellant Rocket System

expansion chamber and nozzle

With reference to the preceding figure, the working principle of a hydrazine mono-propellant engine can be described as follows. An electric solenoid valve, placed between the hydrazine tank and the thrust chamber, opens in order to allow liquid hydrazine, in a pulsed or continuous flow, to reach the injector. This hydrazine enters the thrust chamber as a spray and gets in touch with the catalyst bed. This bed consists of alumina (Al_2O_3) pellets impregnated with iridium. The catalyst bed and the hot gases leaving the catalyst particles cause the liquid hydrazine to vaporise, and its temperature to rise to a point where the chemical reaction of dissociation is self-sustaining. Then, the gases resulting from hydrazine decomposition leave the catalyst bed and exit from the chamber through a nozzle of high expansion ratio to produce thrust.

Mono-propellant hydrazine engines produce a specific impulse of about 230–240 s. Mono-propellant engines using hydrogen peroxide and nitrous oxide have a lower performance than that obtained with hydrazine, since their specific impulses are respectively about 150 and 170 s [10].

Since hydrazine is corrosive, highly toxic and carcinogenic, and is also dangerous to handle and store, then new substances, called green propellants, are being studied and tested for its replacement. A green propellant is an aqueous solution of a high-energy oxidiser (such as hydroxyl ammonium nitrate, ammonium dinitramide, and others) and a fuel (such as methanol, ethanol, glycerol, and others). Hydroxyl ammonium nitrate, also known as HAN, whose chemical formula is NH_3OHNO_3, is a salt derived from hydroxyl amine (NH_2OH) and nitric acid (HNO_3), which can be used as a solution in mono-propellants, or also as a solid oxidiser in bi-propellants. The Air Force Research Laboratory at Edwards Air Force Base in California, USA, has developed a hydroxyl ammonium nitrate-based propellant known as AF-M315E. This propellant is less toxic and easier to handle than hydrazine, and has a specific impulse $I_s = 257$ s, which is about 12% greater than the specific impulse of hydrazine, the latter being $I_s = 230$ s. It requires a catalyst bed preheating at a temperature exceeding 558 K to be ready for general operation [23].

Ammonium dinitramide, also known as ADN, whose chemical formula is $NH_4N(NO_2)_2$, is the ammonium (NH_4^+) salt of the dinitraminic acid ($HN(NO_2)_2$), and was invented in the 1970s in the former Soviet Union and independently invented again in 1989 in the United States by SRI International. Gaseous ammonium dinitramide decomposes under heat into ammonia (NH_3), nitrous oxide (N_2O), and nitric acid (HNO_3), according to the following two-branch reaction [24]:

$$NH_4N(NO_2)_2 \rightarrow NH_3 + HN(NO_2)_2$$
$$\rightarrow NH_3 + N_2O + HNO_3$$

The Swedish company EURENCO Bofors produces a liquid mono-propellant, called LMP-103S, as a substitute for hydrazine by dissolving 65% ammonium dinitramide in 35% water solution of methanol (CH_3OH) and ammonia (NH_3). LMP-103S has 6% higher specific impulse and 30% higher impulse density (see below) than hydrazine mono-propellant [25]. LMP-103S has been tested on the PRISMA (Prototype Research Instruments and Space Mission technology Advancement, COSPAR designation 2010-028B and 2010-028F) mission in 2010. Rocket engines using mono-propellants are simple, require only one tank for the propellant, and can be readily turned on and off. They are mainly used to perform such functions as orbit maintenance and attitude control of satellites.

A liquid bi-propellant comprises two substances, namely a fuel and an oxidiser, which are held in separate tanks. They are not mixed before being injected into the combustion chamber of a rocket engine. They may be fed to the combustion chamber either by pumps or by pressure in the tanks, as shown in the following figure, due to the courtesy of NASA [26]. In a pump-fed engine (left), the liquid fuel (H_2) picks up heat, as it circulates through the coolant jacket of the thrust chamber, becomes gaseous, and drives a turbine. The gas exhausted by the turbine is then injected into the combustion chamber. This arrangement is called the expander cycle. Control in this cycle is achieved by using: (a) the turbine valve to regulate the thrust, and (b) the mixture valve to maintain the desired mixture ratio. In a pressure-fed engine (right), the pressure in the propellant tanks is sufficient to force the two propellants (H_2 and O_2) through the injector into the combustion chamber. The fuel (H_2) first circulates through the coolant jacket, and is then delivered to the combustion chamber. In a pressure-fed engine, the pressures in the supply tanks can be regulated to yield the desired pressure and mixture ratio in the combustion chamber [26].

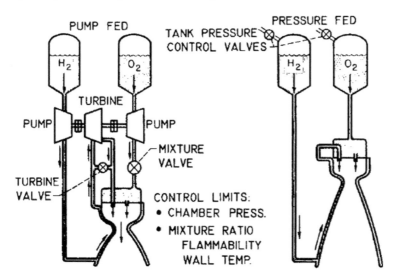

Liquid-propellant rocket engines using bi-propellants have higher performance than those using mono-propellants, as shown in the general table of Sect. 1.3. In addition, the former are often easier to operate than the latter, particularly in consideration of the risks connected with the use of hydrazine.

When the fuel and the oxidiser react spontaneously, as is the case with the hypergolic substances described in Sect. 1.2, then no ignition device is needed in a liquid-propellant engine. Otherwise, an ignition device is necessary in the combustion chamber of a bi-propellant rocket engine. Some of these devices are briefly described below. One of them is a spark plug igniter. A spark plug igniter is a small combustion chamber having two spark plugs. It provides the flame to ignite the propellant (fuel and oxidiser) in the main thrust chamber. When the engine is started, the spark exciters energise the spark plugs as the oxidiser and the fuel flow to the spark igniter. As the gaseous oxidiser and fuel enter this small combustion chamber, they mix and are touched off by an electrical spark. The pilot flame provided by the spark igniter lights off the main combustion chamber. A spark plug igniter is capable of multiple re-ignitions.

The Rocketdyne J-2 rocket engine had a particular spark plug igniter unit, called augmented spark igniter or spark torch igniter, which formed an integral part of the thrust chamber injector, as shown in the following figure, which is due to the courtesy of the United States Air Force [27].

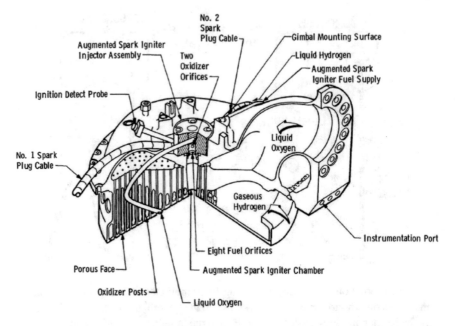

The J-2 rocket engine was used for the Saturn IB and Saturn V launch vehicles for the NASA Apollo programme. The main engines of the Space Shuttle also had an augmented spark igniter, whose chamber was located in the centre of the injector. The dual-redundant igniter was used during the engine start sequence to initiate combustion. The igniters were turned off after approximately three seconds because the combustion process is self-sustaining [28].

The following figure, due to the courtesy of the United States Air Force [27], shows the position of the augmented spark igniter in the J-2 rocket engine.

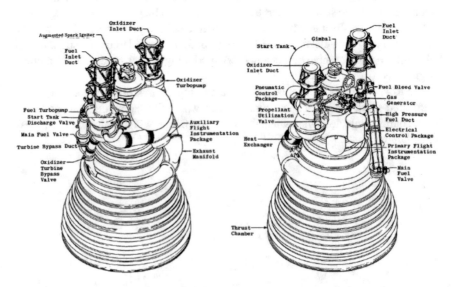

Another ignition device is a pyrotechnic igniter, which is shown in the following figure, due to the courtesy of NASA [29].

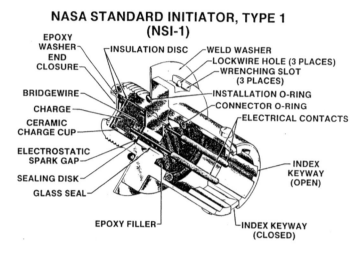

NASA STANDARD INITIATOR, TYPE 1 (NSI-1)

A pyrotechnic igniter is set off by an electric current, which heats a bridge wire enclosed in a clump of easily ignited materials. Each of these materials, in turn, comprises an oxidiser (usually potassium perchlorate or potassium nitrate) and a fuel (usually titanium, or titanium hydride, or zirconium, or zirconium hydride, or boron). Such materials, when ignited by the hot wire, generate sparks and hot gases due to the chemical reaction between the oxidiser and the fuel. These, in turn, ignite the mixture of gaseous propellant in the combustion chamber. Pyrotechnic igniters are safe and reliable. They are one-shot devices, and therefore cannot be used to re-start a rocket engine [30].

Still another ignition method used for liquid propellants consists in injecting some spontaneously ignitable fluid, called a pyrophoric fluid, ahead of the propellant. Pyrophoric substances (for example, ferrous sulphide and many reactive metals) have the property of igniting spontaneously, that is, without an external source of heat, upon exposure to air, moisture in the air, oxygen, or water. Such substances may be solid, liquid, or gaseous. Most of them are metals and react spontaneously with oxygen only when they are in a very finely divided state. Pyrophoric substances are a special class of hypergolic substances (described in Sect. 1.2), because the oxidising agent for the former class is restricted to atmospheric oxygen [31]. They can be handled safely in atmospheres of argon or, with a few exceptions, nitrogen. A list of pyrophoric materials can be found in [32]. These three methods of ignition (electric spark, chemical injection, and pyrotechnic) are shown together in the following figure, due to the courtesy of NASA [33].

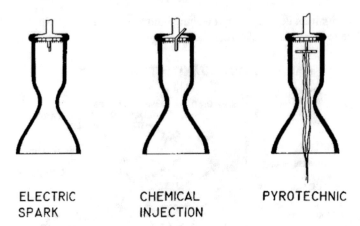

| ELECTRIC | CHEMICAL | PYROTECHNIC |
| SPARK | INJECTION | |

Finally, liquid propellants can also be ignited by using a small combustor, wherein pyrotechnic devices or electric spark plugs ignite the propellant. The hot gas produced, in turn, ignites the propellant in the main combustion chamber.

Fundamental concepts on cryogenic propellants have been given in Sect. 1.2. The scope of cryogenics has been defined by NASA ([34], page 1) as follows: "Cryogenics is the discipline that involves the properties and use of materials at extremely low temperatures; it includes the production, storage, and use of cryogenic fluids. A gas is considered to be cryogenic if it can be changed to a liquid by the removal of heat and by subsequent temperature reduction to a very low value. The temperature range that is of interest in cryogenics is not defined precisely; however, most researchers consider a gas to be cryogenic if it can be liquefied at or below −240 °F. The most common cryogenic fluids are air, argon, helium, hydrogen, methane, neon, nitrogen, and oxygen. Other gases that are being used in space probes and high-energy liquid-propellant rockets include fluorine and nitrogen trifluoride.". In accordance with the preceding definition, by cryogenic propellants we mean gases which can be liquefied at or below 122 K. The most common cryogenic propellants for rocket applications are liquid oxygen (O_2), liquid hydrogen (H_2), liquid fluorine (F_2), and oxygen difluoride (OF_2), or mixtures of some of these substances.

These propellants should be stored and handled in such a way as to reduce the increase in their temperature of storage to a minimum. This temperature depends on the particular propellant and also on the time of storage. The thermal insulation techniques used for reservoirs should reduce the increase in temperature. The gases developed from the evaporating liquids should have a venting system, in order to allow them to escape. Accurate control should be exerted on the environmental humidity. In addition, the design adopted and the materials selected for rocket engines using cryogenic propellants should take account of the properties of such propellants. These disadvantages are offset by high values of specific impulse, as has been shown in Sect. 1.2.

In contrast with cryogenic liquid propellants, storable liquid propellants are stable over a wide range of temperature and pressure, and are also scarcely reactive with the materials used for their reservoirs. Therefore, they can be stored for periods of a year or more. These properties make them readily available and reliable. They are used in military vehicles and in upper stages of space vehicles.

For example, a liquid storable fuel is Aerozine 50 (see Sect. 1.2), which is a 50/50 mixture by weight of hydrazine (H_2NNH_2) and unsymmetrical dimethyl hydrazine ($H_2NN(CH_3)_2$). Aerozine 50 is used with nitrogen tetroxide (N_2O_4) as the oxidiser, with which it is hypergolic. This combination of fuel and oxidiser is storable at room temperature, and therefore a loaded missile can be stored for several years without the maintenance requirements associated with cryogenic propellants. This fuel was developed in 1950s by Aerojet General Corporation for the Titan II ICBM rocket engines, and has been used in Aerojet's Titan liquid rocket engine [35] and Delta II stage 2 engine [36]. Aerozine 50 is mainly used for interplanetary probes and spacecraft propulsion, and is also used for intercontinental ballistic missiles, which require long-term storage and launch on short notice. Another liquid storable fuel, also used with nitrogen tetroxide (N_2O_4) as the oxidiser, is UH 24, which is a mixture of 75% unsymmetrical dimethyl hydrazine ($H_2NN(CH_3)_2$) and 25% hydrazine hydrate ($H_2NNH_2 \cdot H_2O$). This combination is also hypergolic, and has been used in the Ariane rocket versions 2 through 4 [37], and also in the Indian Geosynchronous Satellite Launch Vehicle Mark III [38].

The liquid propellants indicated above are sometimes mixed with additive substances, which are meant to either facilitate ignition, or stabilise combustion, or reduce corrosive effects, or lower freezing point, or improve cooling properties of propellants. Some attempts have been made to promote hypergolicity in propellants by adding substances in the cryogenic oxidiser or in the fuel. An account on these attempts has been given by Clark [39].

In particular, it has been proposed to add aluminium tri-ethyl ($Al_2(C_2H_5)_6$) to the fuel, or ozone fluoride (O_3F_2) to the oxidiser. According to Dickinson [40], using hypergolic additives to fuels is not a good practice, because it increases the danger of inadvertent ignition on exposure of the fuel to air. The additive might be injected as a secondary stream in the fuel ahead of the combustion chamber, but this would imply an added complication in the design of a rocket engine. It seems better to mix additives to the oxidiser, because the danger is confined to direct mixing of fluid with oxidiser, or to direct spilling of the oxidiser on combustible materials.

As has been shown in Sect. 1.2, there is a mixture ratio (o/f), called the stoichiometric mixture ratio, between the oxidiser (o) and the fuel (f). The stoichiometric mixture ratio for a rocket engine is the ideal ratio of oxidiser to fuel that burns all fuel with no excess oxidiser. In practice, there is another mixture ratio, called the optimum mixture ratio, between the oxidiser and the fuel which leads to the maximum value of performance. The optimum mixture ratio is less than the stoichiometric mixture ratio, because the temperature of the flame reaches its maximum value when the quantity of fuel exceeds the value defined by the stoichiometric mixture ratio. This happens because a combustion richer in fuel generates an exhaust gas having a lower molar mass. The optimum mixture ratio varies slightly with the pressure in the combustion chamber. In practice, the actual value chosen for the mixture ratio may differ from the optimum value, because the temperature of the combustion chamber must be kept within the limits imposed by the materials, or because of the coolant flow, or because the combustion must be made stable.

The performance of a liquid propellant depends not only on its specific impulse, I_s, but also on its density, because a low-density propellant implies large storage tanks, which in turn imply a high mass of a rocket engine. By contrast, a high-density propellant can be stored in small tanks and requires small pumps to be fed to the combustion chamber. This indicator of performance of a propellant combination is taken into account by means of the so-called impulse density, I_d, which is defined as follows [6]:

$$I_d = I_s \delta_{av}$$

where δ_{av} is a dimensionless quantity, called average specific gravity of a propellant combination (oxidiser and fuel), defined as follows

$$\delta_{av} = \frac{\delta_o \delta_f (1 + r)}{r \delta_f + \delta_o}$$

$r \equiv o/f$ is the oxidiser-to-fuel weight mixture ratio, δ_o is the specific gravity (ratio of the density of a given substance to the density of water at 277 K and atmospheric pressure) of the oxidiser, and δ_f is the specific gravity of the fuel. Since δ_{av} is a dimensionless quantity, then the impulse density, I_d, of a propellant combination is measured in seconds, as is the case with the specific impulse, I_s. Specific gravities of various propellants as a function of temperature are given in [6], page 243. Values of I_s and I_d for various propellants are given in [4], pages 25, 26, and 27. Further performance data, in terms of I_s and I_d, concerning liquid hydrocarbons and aluminium-hydrocarbon fuels, are given in [41], pages 5, 6, and 8.

A propellant combination for a specific application is chosen taking into account the advantages and the disadvantages of each combination. It is often necessary to adopt a compromise. Some criteria to be considered in a choice have been identified by Huzel and Huang [4]. They are:

- high release of energy per unit mass of propellant, combined with low molar mass of the gases resulting from combustion or decomposition, in order to have high values of specific impulse;
- easiness of igniting the propellant;
- stability of combustion;
- high density of the propellant or high impulse density, in order to reduce the volume and the weight of the tanks and the pumps;
- capability possessed by the propellant of cooling effectively the thrust chamber;
- low vapour pressure of the propellant a temperature of about 344 K, for low weight of tanks and pumps;
- low freezing point of the propellant, possibly less than 219 K, for easy engine operation at low temperatures;

- absence of corrosiveness and compatibility of the propellant with the materials used for the engine;
- high boiling point, possibly above 344 K, for storable propellants;
- low viscosity, possibly less than 0.01 Ns/m^2, to reduce pressure drops through the feed system and the injector;
- high thermal and shock stability, to reduce risks of explosion and fire;
- low toxicity of the propellant itself and its reaction products;
- low cost; and
- ready availability.

1.5 Combustion of Propellants in Steady State

The combustion process which occurs in steady-state conditions in a liquid-propellant rocket engine can be described by the following sequence of events.

The propellants are injected from a distributing manifold into the combustion chamber of the engine through orifices. The propellants are in the form of liquid jets. These jets are made to break up into small droplets, which then vaporise at high temperature in the combustion chamber. The breakup of the jets is often achieved by causing two or three of like (or sometimes unlike) substances to impinge. The impinging jets produce thin liquid sheets or fans which disintegrate rapidly. In bi-propellant engines, the reactive vapours containing fuel and oxidiser get mixed one with the other. The mixed vapours interact, and the hot gases resulting from the combustion process flow out of the combustion chamber toward the throat of the nozzle. Due to various reasons, the actual combustion process in steady state may be more complex than the ideal process described above, as shown in the following figure, due to the courtesy of NASA [42].

A reason is the internal shape of some injectors, which are designed so that the liquid propellants flowing through them are partially or totally mixed, or broken up into droplets, or vaporised before entering the combustion chamber. This effect can also be obtained by injecting controlled streams of gases into the injector passages, or by using other means.

Another reason is the occurrence of combustion reactions in condensed phase with certain propellants, before their mixture or vaporisation.

Still another reason is the occurrence of heterogeneous reactions in the absence of mixing in liquid phase, as is the case with some combinations of hypergolic propellants.

Still another reason is an incomplete mixture of fuel with oxidiser in gaseous phase. This happens near the walls of the combustion chamber, where fuel-rich streams are intentionally sprayed for purposes of cooling. This also happens because of condensed phases which may be present in the mixture of the combustion products, when the propellants used contain either metals or carbon. This also happens because part of combustion products may recirculate back to the vicinity of the injector plate between spray fans, instead of going directly toward the nozzle, thereby causing a fraction of gas to remain in the combustion chamber for a long time.

Still another reason is the pressure level in the combustion chamber, which may increase above the critical point of one or both of the propellants, thereby causing the sharp distinction between liquid and gas to disappear.

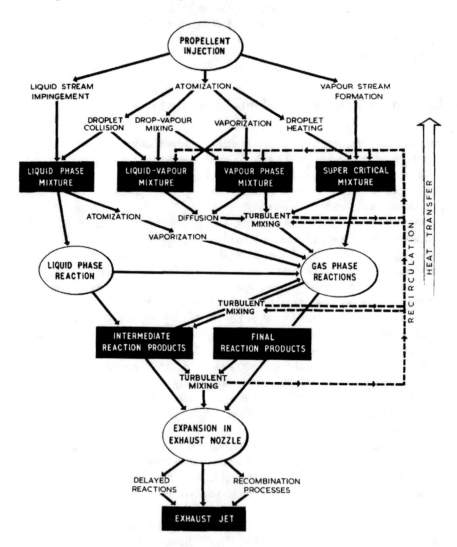

Due to one or more of these reasons, the actual combustion process may differ from the simple process indicated above. Analytical models of the combustion phenomena are described at length in [42, 43].

The phenomena which occur in a combustion chamber of cylindrical shape can be described by dividing the combustion chamber into a series of adjacent cylindrical segments, stacked onto one another in the axial direction from the injection plane to the inlet plane of the nozzle. The thickness of each segment, the phenomena which actually occur into it, and the shape of its surfaces of separation from the adjacent segments depend on factors such as the specific combination of propellants used, the operating conditions (pressure, temperature, mixture ratio, et c.), and the type of injectors. The surfaces which separate two adjacent cylindrical segments are not planar, but have undulations which vary with time.

The combustion phenomena in a rocket engine depend on the combination of propellants used. For example, when the fuel is liquid hydrogen which has been used, before injection, to cool the walls of the combustion chamber, then this fuel is in the gaseous state at a temperature varying from 60 to 500 K [6]. In such conditions, there are neither droplets of liquid hydrogen nor evaporation.

When the liquid propellants used are hypergolic, then an initial chemical reaction occurs in liquid phase at the moment in which a droplet of fuel impinges on a droplet of oxidiser. Therefore, care must be taken by the designer to avoid local explosions and excessive releases of energy which could generate shock waves.

The cylindrical segment immediately adjacent to the injector plate is known as the injection-and-breakup zone, where two different liquid propellants enter through the orifices of the injector. This happens with either storable propellants or combinations of liquid oxygen with hydrocarbons. The injection velocities range from about 7 to 60 m/s [6]. The type of injector used (including the pattern, the size, the number, and the distribution of the orifices) has a strong influence on the behaviour and in particular on the stability of the combustion in this zone. The same influence have the pressure drop, the configuration of the manifold, or the roughness of the walls of the injection ducts. In this zone, the individual jets or streams of propellants break up into droplets by impingement of one jet with another or with a surface, or by inherent instability of liquid sprays, or by interactions of liquids with gases having different velocities and temperatures. These droplets are heated by radiation coming from the next zone (called rapid combustion zone) and also by convection of gas in the injection-and-breakup zone. This heat causes the droplets to evaporate and create regions rich in either fuel vapour or oxidiser vapour. The injection-and-breakup zone is heterogeneous, because it contains liquid propellants, vaporised propellants, and hot gases resulting from combustion products. Since the propellants in liquid phase are located at discrete places, then large variations are present in mass flow rates, mixture ratios, sizes and distributions of droplets, and properties of resulting gases. The chemical reactions occurring in the injection-and-breakup zone generate heat at low rates, and hot gases may recirculate back to the injector plate. Such gases create vortices and turbulent motions which contribute to the initial evaporation of the liquids injected. The processes described above occur in a different manner when one of the propellants is in the gaseous state. Such is the case with gaseous hydrogen coming in contact with liquid oxygen. Gaseous hydrogen has no droplets and does not evaporate. Since the injection velocity of gaseous hydrogen is much higher (above 120 m/s, according to [6]) than that of liquid oxygen, then shear forces act on the liquid jets, and such forces foster the formation and the evaporation of droplets. This case requires the use of an injector of a particular type, which differs from the injectors used for two propellants entering the combustion chamber in the liquid state. To this regard, the following figure, due to the courtesy of NASA [42], shows an injector assembly of coaxial-tube injection elements like that used in the J-2 rocket engine. In this assembly, which is typical of an O_2/H_2 concentric orifice injector, the oxygen tubes are recessed and hydrogen enters the injection elements through inlet holes or slots in their sleeves.

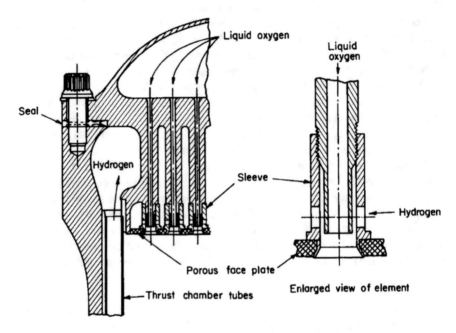

The cylindrical sector contiguous to the injection-and-breakup zone in the direction of the nozzle is called rapid combustion zone. In this zone, rapid and intense chemical reactions occur at increasing high temperatures. Droplets of propellants which may be left are vaporised by convective heat. Consequently, fuel-rich and fuel-lean gases are mixed. The oxidation of the fuel occurs rapidly in this zone, and generates heat at higher rates than those of the preceding zone, with consequent decrease in density and increase in velocity of the burning gas. The chemical composition and the mixture ratio become more uniform as the burning gas moves through this zone. The axial component of the velocity vector of a molecule of combusted gas moving along a stream line becomes gradually greater than its transverse component.

The cylindrical sector contiguous to the rapid combustion zone in the axial direction is called stream tube combustion zone. In this zone, oxidation reactions continue to occur, but at a lower rate than that of the previous zone. The gas mixture approaches an equilibrium composition. The axial component of the velocity vector (ranging from 200 to 600 m/s, according to [6]) of a gas molecule is much higher than its transverse component, and therefore there is little turbulent mixing between gaseous layers. The residence time of a gaseous molecule in the stream tube combustion zone is little in comparison with its residence time in the previous zones. The shape of the stream lines, the inviscid flow, and the tendency of the combustion products toward chemical equilibrium persist in the stream tube combustion zone.

The burning process which actually takes place in a combustion chamber is to be considered as a series of events which occur rather simultaneously than sequentially. Due to turbulence phenomena occurring in various degrees in all of the three zones indicated above, the flame front in a combustion chamber is not a planar surface.

The time spent by single molecules of reactants or combustion products in the combustion chamber of a liquid-propellant rocket engine is of the order of magnitude of 10 ms. The heat release per unit volume is about 3.7×10^5 J/m^3, which value is much higher than in a turbojet. In addition, the higher temperatures reached in the combustion chamber of a rocket engine cause chemical reactions to occur at much higher rates than in a turbojet [6].

1.6 Combustion of Propellants in Unsteady State

Transients can be induced in the process of combustion either intentionally, as is the case with engine start-up or shut-down, or because of the insurgence of undesired phenomena of instability.

As will be shown at length in Chap. 2, the combustion chamber of a liquid-propellant rocket engine is normally designed to operate in steady state or in conditions which vary slowly with time. However, an unstable behaviour may occur there as a result of small perturbations, which cause self-sustaining oscillations of pressure in the burning gas. The frequencies of these oscillations vary in a wide range. Specifically, frequencies going from less than 100 Hz to over 15,000 Hz have been measured in the combustion chamber at amplitudes going from 10 to 1000% of the pressure in steady state [42].

According to Culick and Kuentzmann [43], the energy taken by these oscillations is only a small part of the chemical energy made available by the propellants. Therefore, except in very severe instances, the mean thrust or the steady power of the engine is not affected by the oscillations. However, serious problems may arise because of structural vibrations excited by the oscillating pressure in the combustion chamber, and also because of oscillations of the thrust generated by the engine. In addition, the heat transfer rates between the hot gas and the internal surface of the combustion chamber can be highly increased, with consequent erosion of the chamber wall.

The unstable motion of the hot gas is self-excited as a result of an interaction of the combustion process with the structural modes of the rocket vehicle. This instability arises because a very small part of the chemical energy contained in the propellants is sufficient to produce large unsteady motions, and also because the processes tending to attenuate such motions are weak, unless appropriate steps are taken.

An example of these self-sustained oscillations is provided by the so-called pogo instability (see Chap. 2, Sect. 2.9, and Chap. 7, Sect. 7.6), which occurs in the feed lines of large liquid-propellant rocket vehicles, such as space launch vehicles or ballistic missiles. This particular instability is due to a feedback interaction between the propulsion system and the structure of a rocket stage, and occurs principally in the first longitudinal mode of the structure of the vehicle during operation of its first stage.

The most destructive type of combustion instability is characterised by oscillations of high frequency, and is also known as acoustic instability. The frequencies of these oscillations are equal to or greater than 1000 Hz [4].

The term acoustic instability is due to an observed correspondence, in both frequency and phase, between the pressure oscillations observed experimentally in a combustion chamber and the oscillations calculated for the acoustic resonance of the chamber. High-frequency instability includes both longitudinal and transverse modes, and the latter include, in turn, radial and tangential modes.

In general terms, a source of oscillating energy is necessary to sustain instability. In the particular case of high-frequency instability, the source of oscillating energy is the combustion of the propellants used in a rocket engine, and depends weakly on the feed system of the engine. The oscillating energy must, to sustain instability, be properly phased in time with the oscillating pressure, as will be shown at length in Chap. 2, Sect. 2.9.

Sustaining mechanisms which have been proposed for high-frequency instability include loss of ignition, sensitive chemical preparation time, physical time delays, detonation processes, pressure or temperature sensitive chemical kinetics, the burst of droplets heated beyond their critical temperature and pressure, and the shattering and mixing of the streams, fans, or drops by the motion of gas particles [42].

Instability in a combustion chamber may start either spontaneously or as a result of some artificial perturbation. Acoustic instability may have a threshold amplitude, above which a perturbation is sustained and below which the same perturbation is damped. A rocket engine is said to be inherently stable, when it can absorb large perturbations and yet return to its operation in steady state. The degree of inherent stability of a rocket engine can be measured by rating devices which provide artificial perturbations to its combustion chamber. Such perturbations involve operating conditions, such as mixture ratio of the propellants, pressure in the combustion chamber, temperature of the fuel, and so on.

An instability of the spontaneous type requires no initial perturbation to start, and grows out of the noise inherent to the combustion process. An instability of this type can be expected to occur just after an engine has reached its normal operating conditions, because no perturbation is required for its occurrence. Variations in test conditions and the closeness to a stability boundary may sometimes delay the occurrence of a spontaneous instability.

Instability in a combustion chamber may also be induced by natural or artificial perturbations which may occur in the combustion process. Such perturbations are also known as spikes or pops, where a spike indicates a significant overpressure in the combustion chamber upon ignition of the engine, and a pop indicates a similar overpressure occurring spontaneously during engine operation at nominal pressure in the combustion chamber. For example, at high altitudes, some combinations of hypergolic bi-propellants may start with an extremely high spike of pressure in the combustion chamber. This spike has been attributed to the explosion or deflagration of the propellants collected in the combustion chamber or accumulated on its walls during the period of ignition delay. This spike of pressure may be of sufficient magnitude to either cause destructive failure in the combustion chamber or adversely affect the guidance sensor systems. In addition, for small rocket engines used for intermittent operation, the resultant thrust or the total impulse is seriously altered from that of a smooth start. For large engines, the ignition spike can trigger resonance in the combustion process, with consequent hardware destruction [42].

In most cases, the combustion process in a rocket engine shows a non-linear behaviour, which requires some kind of trigger, be it natural or artificial, to cause pressure oscillations of high frequency and amplitude in the combustion chamber. Consequently, it is necessary to determine the types and the magnitudes of the triggers which may occur in an engine during flight. This done, it is also necessary to evaluate the stability of that engine with artificial triggers of the assumed types. However, as has been shown above, a pressure oscillation may in some cases grow out of combustion noise, in the absence of any observable trigger.

The dynamic stability of a rocket engine is concerned with the responses of the engine to transients occurring during its operation. A rocket engine must, to be dynamically stable, return to its normal operating conditions after transients of any type which may occur to it. In other words, in order for a rocket engine to be dynamically stable, the transients resulting from any type of operating conditions must die out, or the amplitudes of the subsequent sustained oscillations of pressure must be sufficiently small. In order to evaluate the dynamic stability of a rocket engine, the engine is driven by any means into pressure oscillations of high amplitude. If these oscillations subsequently decay to those proper to steady-state conditions, then there is sufficient assurance that no oscillations of high amplitude exist within the range of the given perturbations. A rocket engine, which has shown a dynamically stable behaviour within the range of the operating conditions expected in flight, should also remain stable in actual flight.

Another method which may be used to gain confidence in the stability of a rocket engine consists in conducting a large number of tests and flight. A confidence gained in this way is called statistical stability. This confidence indicates only that instability has rarely or never occurred in the operating condition tested, but not that instability can never occur.

As will be shown at length in Chap. 2, Sect. 2.9, the problem of assuring a stable combustion in an existing rocket engine (without changing the dimensions of the combustion chamber, the hydraulic resistances of the propellant feed system, the type of injector, the heat flux in the combustion chamber, and the engine performance) can be solved by using baffles, which are simple damping devices. Baffles can be mounted on existing injectors, in order to solve the stability problem with minimum effort and time. The early injectors had baffles consisting of an even number of blades extending radially from a central hub. Later on, baffles have been used in conjunction with injectors having larger orifices in order to gain stability by modifying the combustion process [42].

Attempts made so far to induce artificial pressure perturbations in the combustion chamber of a rocket engine have used pulses of very short duration and sufficient amplitude to excite the acoustic modes of the chamber. Sometimes, these pulses have been provided by explosive devices. Three types of techniques have been used to test the stability of combustion in a rocket engine. They are the inert gas (nitrogen or helium) pulse, the pulse gun, and the non-directional bomb.

The first technique acts as a velocity perturbation to the combustion process, whereas the pulse gun and the non-directional bomb generate perturbations in both pressure and velocity by using explosives. Therefore, the resulting perturbation is of the triggered (or induced) type. A gas pulse has been used in engines burning liquid oxygen and RP-1. For such engines, this technique has been found effective in producing perturbations triggering sustained instability. A gas pulse has also been used in engines burning nitrogen tetroxide and Aerozine 50 with negligible effects on the combustion.

A pulse gun is a device which resembles a gun. It consists of a breech into which an explosive charge is placed, usually in a cartridge case, a firing mechanism, a barrel, and often a diaphragm to protect the explosive charge from the environment of the combustion chamber of the engine. The barrel of a pulse gun is usually attached to the wall of the combustion chamber of the engine, in order for the pulse to be fired in the tangential direction or in the radial direction. A pulse gun is acted upon by a command given to its firing mechanism, which is usually a mechanical detonator acting on the main charge. The explosive charge of a pulse gun contains a mass of gun powder ranging from 194.4 to 259.2 mg, but may also contain a mass as high as 6480 mg of high explosive. A typical pulse gun is shown in the following figure, due to the courtesy of NASA [42].

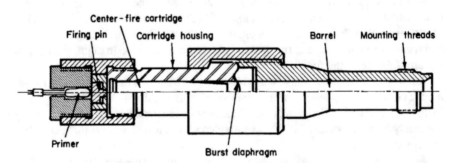

A non-directional bomb mounted on the wall of a combustion chamber is shown in the following figure, also due to the courtesy of NASA [42].

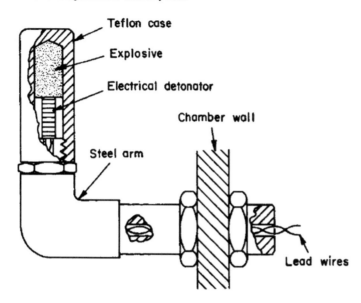

It consists of three principal parts, which are an explosive charge (for example, RDX), a detonator which may be commanded either by the rocket gases or by an electric signal, and a case which insulates the explosive charge and the detonator from the environment of the rocket engine and also contains the explosive charge. Unlike an inert gas pulse and a pulse gun, a non-directional bomb is usually mounted inside the combustion chamber and is not restricted to a location on the wall. Pulse guns and non-directional bombs can induce pressure perturbations ranging from 10 to 500% of the normal value in the combustion chamber. In the development of liquid-propellant rocket engines, non-directional bombs are the type of device most frequently used to induce pressure perturbations in the combustion chamber.

1.7 Principal Components of a Liquid-Propellant Rocket Engine

A liquid-propellant rocket vehicle is composed of the following essential parts:

- propulsion system;
- structure of the vehicle:
- guidance system;
- payload; and
- accessories.

This book deals only with the propulsion system and its parts. The parts of the propulsion system considered here include only those which are strictly necessary to generate thrust and keep it in the desired direction. Therefore, the tanks and their accessory parts are included in the number. The following figure, due to the courtesy of NASA [44], shows a scheme of the propulsion system of a typical liquid-bi-propellant engine fed by turbo-pumps.

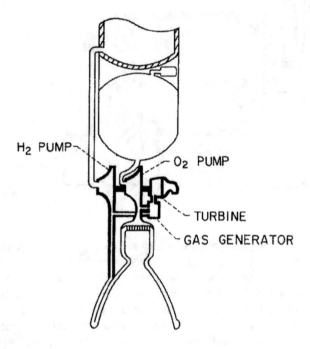

The following figure, also due to the courtesy of NASA [45], shows the principal components of the F-1 engine, used in the first stage of the Saturn V.

F-I ENGINE MAJOR COMPONENTS

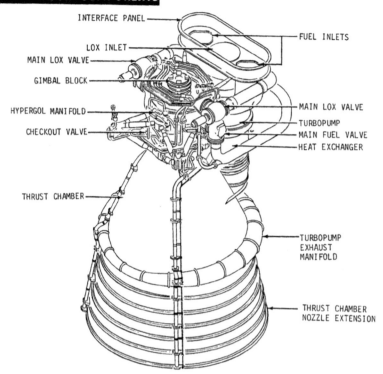

The propulsion system of a liquid-propellant rocket vehicle is composed of the following principal parts:

- a thrust chamber assembly;
- a feed system for the propellants, including a gas pressurising system;
- valves and other systems of propellant control;
- tanks containing the fuel and the oxidiser; and
- interconnecting components and structures.

The design of each of these parts will be shown in a specific chapter.

References

1. NASA, Jet Propulsion Laboratory, California Institute of Technology, Basics of space flight. http://www2.jpl.nasa.gov/basics/index.php
2. Crown JC (1948, June) Supersonic nozzle design, NACA Technical Note No. 1651, NACA, Washington, D.C., 35pp. https://ntrs.nasa.gov/archive/nasa/casi.ntrs.nasa.gov/19930082268. pdf

3. Greene W (2012) J-2X progress: the next phase for E10001, NASA, 3 Feb 2012. https://blogs. nasa.gov/J2X/tag/convergent-divergent-nozzle/
4. Huzel DK, Huang DH (1967) Design of liquid propellant rocket engines. NASA SP-125, NASA, Washington, D.C., 472pp. https://ntrs.nasa.gov/archive/nasa/casi.ntrs.nasa.gov/197100 19929.pdf
5. NASA, Glenn Research Centre, Bernoulli Equation. https://www.grc.nasa.gov/www/k-12/air plane/bern.html
6. Sutton GP, Biblarz O (2001) Rocket propulsion elements, 7th edn. Wiley, New York. ISBN 0-471-32642-9
7. Liepmann HF, Puckett AE (1947) Introduction to aerodynamics of a compressible fluid. Wiley, New York
8. Hill PG, Peterson CR (1992) Mechanics and thermodynamics of propulsion, 2nd edn. Addison-Wesley, Reading, Massachusetts. ISBN 0-201-14659-2
9. Braeunig RA Rocket and space technology—rocket propellants. http://braeunig.us/space/pro pel.htm
10. Braeunig RA Rocket and space technology—rocket propulsion. http://braeunig.us/space/pro puls.htm
11. Aerojet Rocketdyne, RS-25 Engine. https://www.rocket.com/rs-25-engine
12. Aerojet Rocketdyne, RL10 engine. http://www.rocket.com/rl10-engine
13. NASA (1968, December) Saturn V news reference, J-2 engine fact sheet. https://www.nasa. gov/centers/marshall/pdf/499245main_J2_Engine_fs.pdf
14. Astronautix. http://www.astronautix.com/f/f-1.html
15. Nufer BM (2009, June) A summary of NASA and USAF hypergolic propellant related spill and fires. NASA/TP-2009-214769, 112pp. https://ntrs.nasa.gov/archive/nasa/casi.ntrs.nasa. gov/20090029348.pdf
16. IOK (Own work) [Public domain], via Wikimedia Commons. https://upload.wikimedia.org/ wikipedia/commons/4/4c/Nozzle_de_Laval_diagram.svg
17. NASA/Tom Farrar, Scott Haun, Raphael Hernandez. https://www.nasa.gov/mission_pages/shu ttle/flyout/multimedia/discovery/2007-10-23.html)
18. Gerald CF, Wheatley PO (1984) Applied numerical analysis. Addison-Wesley, Reading. ISBN 0-201-11577-8
19. Buchheim RW et al (1959) Space handbook: astronautics and its applications. United States Government, Printing Office, Washington, D.C. http://history.nasa.gov/conghand/propelnt.htm
20. Lohner KA, Scherson YD, Lariviere BW, Cantwell BJ, Kenny TW (2008) Nitrous oxide mono-propellant gas generator development. Stanford University, 13pp. http://web.stanford.edu/~can twell/Recent_publications/Lohner_Scherson_JANNAF_2008.pdf
21. Kitson BA, Oliaee SN (2016) Selective, catalytic decomposition of hydrazine, 29 Apr 2016. Honours Research Projects, 286, 25pp. http://ideaexchange.uakron.edu/honors_research_pro jects/286
22. Greene WD (2013) Inside the LEO doghouse, start me up! NASA, 19 Dec 2013. https://blogs. nasa.gov/J2X/2013/12/19/inside-the-leo-doghouse-start-me-up/
23. Spores RA, Masse R, Kimbrel S, McLean C (2013) GPIM AF-M315E propulsion system. In: 50th AIAA/ASME/SAE/ASEE joint propulsion conference & exhibit, 28–30 July 2013, Cleveland, Ohio, USA, 12pp. https://ntrs.nasa.gov/archive/nasa/casi.ntrs.nasa.gov/20140012587. pdf
24. Park J, Chakraborty D, Lin MC (1998) Thermal decomposition of gaseous ammonium dinitramide at low pressure: kinetic modelling of product formation with ab initio MO/cVRRKM calculations. In: Twenty seventh symposium (international) on combustion/The Combustion Institute, pp 2351–2357. http://www.chemistry.emory.edu/faculty/lin/refs/adn.pdf
25. Sjöberg P, Skifs H, Thormählen P, Anflo K (2009) A stable liquid mono-propellant based on ADN. In: Insensitive munitions and energetic materials technology symposium, Tucson, USA, 11–14 May 2009. http://www.dtic.mil/ndia/2009/insensitive/8Asjoberg.pdf

26. Sanders JC, Wenzel LM (1962) Dynamics and control of chemical rockets. In: Proceedings of the NASA-university conference on the science and technology of space exploration, vol 2. NASA SP-11, Chicago, Illinois, USA, 1–3 November 1962, pp 53–56. http://www.dtic.mil/dtic/tr/fulltext/u2/a396443.pdf
27. Dougherty NS Jr, Rafferty CA (1969, February) Altitude developmental testing of the J-2 rocket engine. AEDC-TR-68-266, 261pp. http://www.dtic.mil/get-tr-doc/pdf?AD=AD0848188
28. NASA, Human space flight, Space Shuttle main engines. https://spaceflight.nasa.gov/shuttle/reference/shutref/orbiter/prop/engines.html
29. Varghese PL (1988, January) Investigation of energy transfer in a NASA standard initiator. NASA-CR-184673, 31pp. https://ntrs.nasa.gov/archive/nasa/casi.ntrs.nasa.gov/19890005798.pdf
30. Turner MJL (2006) Rocket and spacecraft propulsion, 2nd edn. Springer, Berlin. ISBN 3-540-22190-5
31. Anonymous (1994, December) DOE Handbook, Primer on spontaneous heating and pyrophoricity. DOE-HDBK-1081-94, U.S. Department of Energy, Washington, D.C., U.S.A. http://www.hss.energy.gov/nuclearsafety/ns/techstds/standard/hdbk1081/hdbk1081.pdf
32. Anonymous, Pyrophoric materials, handling Pyrophoric and other air/water reactive materials. University of Illinois. https://www.drs.illinois.edu/SafetyLibrary/PyrophoricMaterials
33. Jonash ER, Tomazic WA (1962) Current research and development on thrust chambers. In: Proceedings of the NASA-university conference on the science and technology of space exploration, vol 2. NASA SP-11, Chicago, Illinois, 1–3 Nov 1962, pp 43–52. http://www.dtic.mil/dtic/tr/fulltext/u2/a396443.pdf
34. Davis ML, Allgeier RK Jr, Rogers TG, Rysavy G (1970) The development of cryogenic storage systems for space flight. NASA SP-247, 132pp. https://ntrs.nasa.gov/search.jsp?R=19710021434
35. Aerojet Rocketdyne, Titan liquid rocket engine. http://www.rocket.com/titan-liquid-rocket-engine
36. Aerojet Rocketdyne, Delta II stage 2 engine. http://www.rocket.com/delta-ii-stage-2-engine
37. Technical University of Delft. https://blackboard.tudelft.nl/bbcswebdav/users/bzandbergen/LVC/Launch%20Vehicle%20Catalogue/Fiches/ARIANE_42_L.pdf
38. Spaceflight101, India's most-powerful rocket successfully reaches orbit. https://spaceflight101.com/gslv-mk3-d1/gslv-mk-iii-first-orbital-launch-success/
39. Clark JD (1972) Ignition!: an informal history of liquid rocket propellants. Rutgers University Press, New Brunswick, New Jersey. ISBN 0-8135-0725-1
40. Dickinson LA (1964) Technical problems in the production of solid and liquid propellants, Stanford Research Institute. In: Casci C (ed) Fuels and new propellants, pp 265–280, Proceedings of the conference held in Milan by Federazione Associazioni Scientifiche e Tecniche and sponsored by Consiglio Nazionale delle Ricerche, Elsevier. ISBN 978-1-4831-9829-3
41. Rapp DC (1990) High energy-density rocket fuel performance. NASA-CR-185279, 1st July 1990. https://ntrs.nasa.gov/archive/nasa/casi.ntrs.nasa.gov/19900019426.pdf
42. Harrje DT, Reardon FH (eds) (1972, January) Liquid propellant rocket combustion instability. NASA SP-194, 657pp. https://ntrs.nasa.gov/archive/nasa/casi.ntrs.nasa.gov/19720026079.pdf
43. Culick FE, Kuentzmann P (2006, December) Unsteady motions in combustion chambers for propulsion systems. NATO, RTO AGARDograph AG-AVT-039, 663pp. ISBN 978-92-837-0059-3. https://apps.dtic.mil/dtic/tr/fulltext/u2/a466461.pdf
44. Hartmann MJ, Ball CL (1962) New problems encountered with pumps and turbines. In: Proceedings of the NASA-university conference on the science and technology of space exploration, vol 2, NASA SP-11, Chicago, Illinois, USA, 1–3 Nov 1962, pp 23–35. http://www.dtic.mil/dtic/tr/fulltext/u2/a396443.pdf
45. Anonymous (1968, November) Saturn V flight manual SA 503, Technical Manual MSFC-MAN-503, NASA TM X-72151, 243pp. https://ntrs.nasa.gov/archive/nasa/casi.ntrs.nasa.gov/19750063889.pdf

Chapter 2
The Thrust Chamber Assembly

2.1 The Principal Components of a Thrust Chamber

The thrust chamber of a rocket engine comprises essentially a combustion chamber and a nozzle. This is the part of the engine in which the chemical energy of the propellants is converted into the kinetic energy of the combusted gas.

The following figure, due to the courtesy of NASA [1], is a simple scheme of the phenomena which take place in and near each element of an injector.

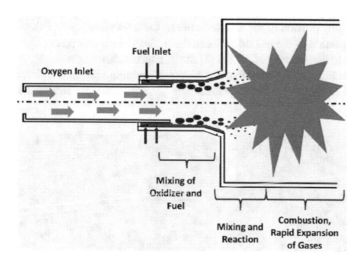

For a liquid bi-propellant rocket engine, the combustion process has been shown in Chap. 1, Sect. 1.5. It is summarised below for convenience of the reader.

A. de Iaco Veris, *Fundamental Concepts of Liquid-Propellant Rocket Engines*,
Springer Aerospace Technology,
https://doi.org/10.1007/978-3-030-54704-2_2

(1)   Firstly, the fuel and the oxidiser are injected, in the proper mixture ratio, into
      the combustion chamber through the elements of the injector, as shown in the
      preceding figure. The injection velocity ranges from about 7 to 60 m/s [2]. The
      droplets of the two propellants may either mix together in form of a spray or go
      separately into the combustion chamber.
(2)   Then, the high temperature existing in the combustion chamber causes the
      droplets to vaporise.
(3)   Then, the vaporised combination of fuel and oxidiser is further heated and burns
      at its stoichiometric mixture ratio, causing a continuous increase in mass flow
      rate in the combustion chamber. The combustion is aided by the high velocity
      of the molecules within the combustion chamber, and takes place completely
      upstream of the throat plane. Care must be taken by the designer in order for the
      combustion process to be stable, that is, free from shocks and detonation waves
      in the combustion front.
(4)   Finally, the combusted gas flows in the converging portion of the nozzle. The
      velocity of the gas increases in subsonic conditions ($M < 1$) and reaches the sonic
      value ($M = 1$) at the throat plane of the nozzle. The subsequent passage through
      the diverging portion of the nozzle causes the velocity of the gas molecules to
      increase further in supersonic conditions ($M > 1$). The combusted gas is then
      ejected from the thrust chamber through the exit plane of the nozzle.

The principal components of a thrust chamber are the injector including the propellant inlets and distributing manifolds, the ignition device (which is necessary in case
of a rocket engine burning non-hypergolic propellants), the combustion chamber,
the converging portion of the nozzle between the inlet plane and the throat, and the
diverging portion of the nozzle between the throat and the exit plane. The following
figure, due to the courtesy of NASA [3] shows a thrust chamber assembly comprising
the injector (left) and the combustion chamber including the converging portion of
the nozzle (right), for a rocket engine whose propellants are methane and liquid
oxygen.

The combustion chamber illustrated in the preceding figure is a body of tubular shape. The cross-section narrows in the converging portion of the nozzle and reaches its minimum value at the throat. The diverging portion of the nozzle is usually bell-shaped, in order to allow the combusted gas to expand towards the exit section.

The injector has the purpose of distributing the propellants into the combustion chamber at the proper mixture ratio, pressure, and spray pattern, in order to initiate and sustain a stable combustion. A cutaway view of the injector used in the thrust chamber of a rocket engine burning liquid hydrogen and liquid oxygen (with gaseous fluorine for ignition only) is shown in the following figure, due to the courtesy of NASA [4].

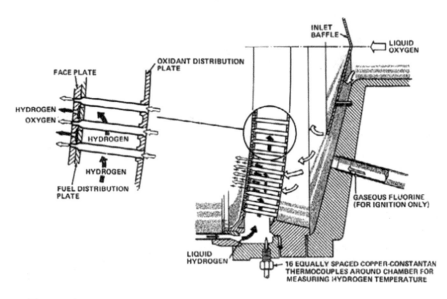

The particular injector shown above is of the shower-head type (described in Sect. 2.6), where both the fuel and the oxidiser injection holes are drilled at angles which allow the convergent streams of the propellants to meet at a common point placed at a given distance from the injector face.

Generally speaking, an injector is a round plate, having a honeycomb structure, with circular and radial inner passages, leading to drilled orifices. It is usually made of steel with nickel-plated surfaces, and is held in position at the fuel manifold below the liquid oxygen dome by means of high-strength bolts [5].

The following figure, due to the courtesy of NASA [6], shows the injector plate used for the Rocketdyne H-1 engine, which was used for the S-I and S-IB first stages of the Saturn I and Saturn IB rockets.

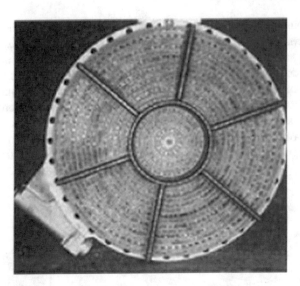

The seals between the injector and the body of the thrust chamber are of the O-ring type, made of rubber selected for compatibility with the fuel. The H-1 engine cited above burns liquid oxygen and RP-1 (kerosene). A threaded hole is provided in the centre of the injector face to permit the installation of the igniter. The fuel and the oxidiser are kept separate by the distribution system. The injection orifices may be arranged at angles such that the impingement angles should be either equal (uni-planar impingement) or different (multi-planar impingement) for the two propellants.

The following figure, due to the courtesy of NASA [6], shows the liquid oxygen dome, which is bolted in position above the injector. This figure relates to the same H-1 engine indicated above.

The liquid oxygen dome provides the inlet to the liquid oxygen, and is also the attachment interface between the thrust chamber and the vehicle. It is a single-piece, aluminium alloy die forging. The liquid oxygen dome and the injector have flanges which are sealed by a spirally wound gasket made of stainless steel strips with non-asbestos fillers (aramid or graphite fibres). The type of gasket used depends on the range of temperature in working conditions. For thrust-vector-controlled engines, as is the case, for example, with the Rocketdyne J-2 engine, the liquid oxygen dome also serves as a mount for the gimbal bearing. The thrust vector control of rocket engines will be considered in Sect. 2.2.

Finally, the pyrotechnic igniter of an engine is fired electrically, and is fixed to the injector surface by means of a threaded joint. It is designed for a single start, and must therefore be replaced after each firing. Igniters of this and other types will be described and illustrated at length in Sect. 2.8.

2.2 The Design of a Thrust Chamber for Thrust Vector Control

The present paragraph concerns the methods used to control the direction of the thrust vector in a rocket or spacecraft. These methods are considered here in view of their bearing on the design of a thrust chamber, as will be shown below. By thrust vector control we mean the ability of a space vehicle to deflect the direction of the thrust away (that is, at an angle, θ, other than zero) from the longitudinal axis of the vehicle. The following figure, due to the courtesy of NASA [7], shows the principal axes of inertia and the rotations in roll, pitch and yaw for the Space Shuttle.

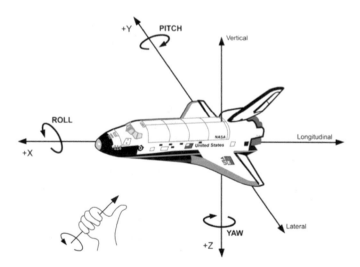

The necessity of this control arises from several causes. First, an intentional change of the direction of the flight path followed by the centre of mass of the vehicle. Second, an intentional change of attitude (or rotation) of the vehicle about one or more of the principal axes of inertia passing through its centre of mass. Third, a correction of either the flight path or the attitude which becomes necessary in order for the vehicle to maintain the desired trajectory and orientation. Fourth, a correction of the misalignment between the thrust vector and the gravity force vector. As to the last issue, the thrust vector is applied to the nozzle of the vehicle, whereas the gravity force vector is applied to its centre of gravity, as shown in the following figure, due to the courtesy of NASA [8].

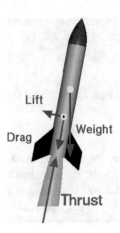

In order for the vehicle not to be subject to unwanted torques, the directions of the thrust and weight vectors must be aligned at π radians (180°). A rocket is also subject, when flying through the atmosphere, to aerodynamic forces (lift and drag), which are applied to its centre of pressure, not to its centre of gravity, as shown in the preceding figure.

These forces produce moments about the principal axes of inertia of the vehicle. These moments, in turn, cause the vehicle to rotate about its centre of gravity. For a stable flight, the centre of gravity of a rocket must be above its centre of pressure. The longitudinal axis of the rocket, or the line joining the tip of the nose with the centre of the exit section of the nozzle, is called the roll axis, and a motion of the rocket about the roll axis is called a rolling motion. The centre of gravity of the rocket lies along the roll axis. The pitch and yaw axes are mutually perpendicular and form a plane passing through the centre of gravity and perpendicular to the roll axis. Pitch moments tend to either lower or raise the nose of the rocket. Yaw moments cause the nose to move from side to side.

Both mechanical and aerodynamic methods can be used to re-direct the rocket thrust and provide the necessary steering forces. Some of such methods use static fins, movable fins, jet vanes, jetevators, canards, gimballed (that is, swivelled) nozzles, Vernier rockets, fuel injectors, and attitude-control rockets, as shown in the following figure, due to the courtesy of NASA [9].

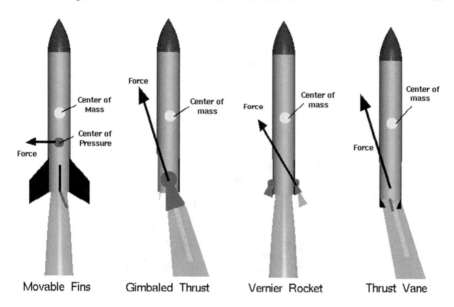

Movable Fins   Gimbaled Thrust   Vernier Rocket   Thrust Vane

As long as a rocket travels in the atmosphere, static fins at the tail of the rocket can generate aerodynamic forces which act at the centre of pressure and generate moments about the centre of gravity. These moments oppose the deflection of the rocket in the directions of the pitch and yaw axes. Of course, the static fins must be so sized as to generate the amount of aerodynamic forces required to counteract the deviations. On the other hand, the fins generate further drag, in the direction opposed to the rocket velocity vector. In spite of that, aerodynamic fins, be they fixed or movable, are very effective for controlling a vehicle flying through the atmosphere. They continue to be used in weather rockets, anti-aircraft missiles, and air-to surface missiles.

Generally speaking, a thrust vector control system used in a rocket may be either passive or active. A passive control system is a fixed device meant to stabilise a rocket by its very presence on the outside of the rocket. An example of such a device is a cluster of fins mounted around the lower end of a rocket, near the nozzle. The purpose of the fins is to keep the centre of pressure below the centre of gravity of the rocket.

In order to overcome the disadvantages (higher drag and mass) of the fins, active control systems have been developed. They are described below.

Jet vanes and movable fins are planar surfaces, which are used in the jet stream of a rocket to deflect the exhaust gases. They are shown in the following figure, due to the courtesy of NASA [10].

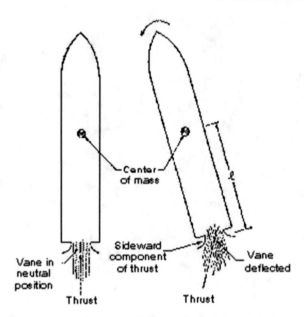

Thrust vanes have been employed on the V-2 and Redstone missiles to deflect the exhaust gas jet of the main propulsion motors by carbon vanes.

Jetevators, which are shown in the following figure, due to the courtesy of Lockheed [11], are control surfaces which can be moved into or against the jet stream of a rocket, in order to change the direction of the jet flow. The jetevator invented by Dr. Willy Fiedler is a solid ring with a spherical inside surface which is hinged over the rocket nozzle [12, page 52]. This device, which was used in the A1 version of the Polaris missile and then replaced by rotatable nozzles in the A2 version, deflects the flow when turned into the exhaust stream. It has the advantage of not causing propulsion losses when in the neutral position, since, unlike vanes, it does not interfere with the exhaust flow [10]. The same device has also been described by Edwards and Parker as a semi-spherical shell hinged to a rocket nozzle and rotated, at the command of a sensing unit, into the exhaust flow to produce a control force. Jetevators have been most frequently applied to control the direction of thrust of solid-propellant rockets [10].

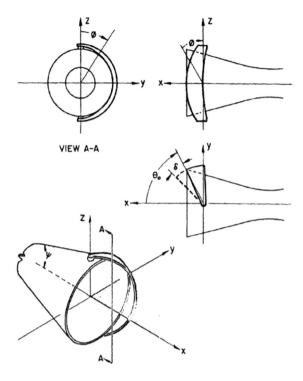

Canards are fore-plane surfaces mounted on the front end of a rocket or aircraft. The following figure, due to the courtesy of NASA [13], shows such surfaces.

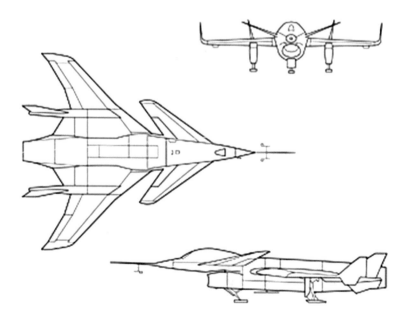

Movable fins and canards are quite similar to each other in appearance. The only real difference is their location on the rocket, because canards are mounted on the front end, whereas the movable fins are at the rear end. In flight, both of them tilt as rudders to deflect the air flow and cause the rocket to correct its course, in the event of unwanted directional changes being detected by motion sensors [14].

Another method for changing the direction of the exhaust gases, and hence the direction of the rocket, is to gimbal the nozzle. For this purpose, the engine can be mounted in a two-axis ring-suspension system, as shown in the following figure, due to the courtesy of NASA [15]. This figure, which relates to the engine gimbal system of the Viking Orbiter 1975, shows the two actuators mounted along directions parallel to, but not coincident with, respectively, the pitch axis (Y) and the yaw axis (Z) of the vehicle. This is because the origin of the roll, pitch, and yaw axes is the centre of mass of the vehicle.

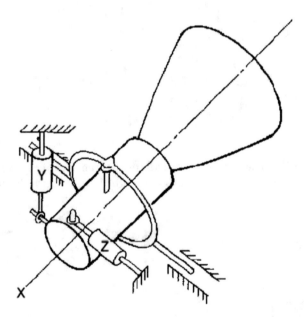

The Viking 1 orbiter (NSSDCA/COSPAR ID: 1975-075A) was launched from Cape Canaveral on the 20th of August 1975. The propulsion of the Viking 1 orbiter was furnished by a bi-propellant (monomethyl hydrazine, CH_3NHNH_2, and nitrogen tetroxide, N_2O_4) liquid-fuelled rocket engine which could be tilted up to $\pi/20$ rad (9°). Attitude control during engine burns was provided in roll (X-axis) by the attitude control system (using 12 small compressed-nitrogen gas jets located at the solar panel tips), and in pitch and yaw by an autopilot which commanded the engine gimbal system shown above [16].

Another type of gimbal system, relating to the Rocketdyne J-2 engine, is shown at the top of the following figure, due to the courtesy of NASA [17].

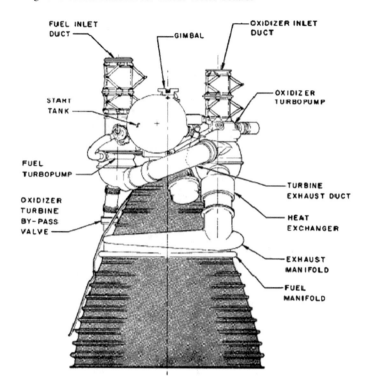

This type of gimbal system has a highly loaded (1.38×10^8 N/m^2) universal joint which is a spherical, socket-type bearing with a Teflon®/fibreglass composition coating which provides a dry, low friction bearing surface. It also has a lateral adjustment device for aligning the thrust chamber with the vehicle. This gimbal system transmits the thrust from the injector assembly to the thrust structure of the vehicle, and provides a pivot bearing for deflection of the thrust vector. As shown in the preceding figure, the gimbal is mounted on the top of the injector and liquid oxygen dome assembly [17]. This gimbal bearing system has been used in the Saturn V rocket and in the Space Shuttle [14]. In a gimballed thrust control system, the nozzle of the rocket can be swivelled from side to side. As the nozzle direction is deflected away from the axis of symmetry of the rocket, so does the direction of the thrust change with respect to the centre of gravity of the vehicle.

In a Vernier thrust control system, small rockets are mounted on the outside of the main thruster. In case of need, such rockets are fired in the proper direction, in order to produce the desired course change [14]. The following figure, due to the courtesy of NASA [18], shows a Vernier rocket mounted on the outside of the Atlas rocket.

Attitude-control rockets, shown below, are used to trim misalignments of the translational thruster, the basic attitude control with respect to the body principal axes being provided by a separate set of thrusters.

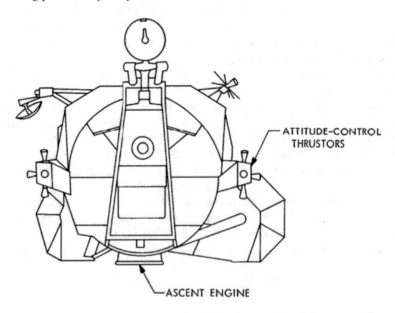

By so doing, the main engine is used only for the motion of the centre of mass of the vehicle. Attitude-control rockets are used in the coasting phase of flight. For this purpose, clusters of small rockets are mounted all around a space vehicle. By firing these small rockets in the proper combination, it is possible to turn the vehicle in the desired direction. After the vehicle has been aimed as required, the main engine is fired, in order to send the vehicle in the new direction. For example, the ascent stage of the Apollo Lunar Module used a fixed, high-thrust engine for translation and a series of small liquid-propellant rockets for attitude control, as shown in the preceding figure, which is due to the courtesy of NASA [19].

The methods described above use, all of them, mechanical means to control the direction of the thrust vector. They require actuating components which must work efficiently in the high-temperature environment of the rocket exhaust and are invariably associated with a loss of axial thrust when performing manoeuvres of thrust vector control [20].

Other methods use fluidic means to perform the same function. These methods use a static nozzle and the injection (or the removal) of a secondary flow into (or from) the region between the primary flow, which generates the thrust, and the nozzle. They do not require any kinematic structure and mechanical actuators. There is a variety of methods based on fluidic thrust vectoring, which differ one from the other by the way in which the secondary flow is used for thrust vector control. They may be classified as follows.

- Co-flow for fluidic thrust vectoring, which is based on the Coanda effect. The Coanda effect (so named after the Romanian engineer Henri-Marie Coanda) is the phenomenon in which a stream of fluid, ejected at high speed from a slot and coming in contact with a convexly curved surface, adheres to that surface along its curvature rather than continue to travel in a straight line.

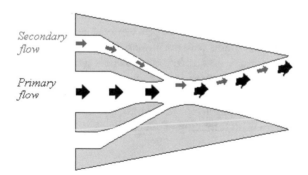

In other words, the stream is deflected from the axis of flow and follows the slope (or curvature) of a divergent wall whilst increasing in velocity and in mass by entraining additional fluid [21]. This happens because of the increase in the flow velocity over the curved surface, which causes pressure to decrease. The low pressure causes not only the injected co-flow but also the primary flow to be deflected off the nozzle axis toward the divergent wall, as shown in the preceding figure, where the black arrows represent the primary flow, and the red arrows represent the injected co-flow. Of all the fluidic methods for controlling the thrust vector, the co-flow method has been found [22] to be the one which produces the smallest deflection angle.

- Counter-flow for fluidic thrust vectoring, which is also based on the Coanda effect to deflect the thrust vector, as shown in the following figure.

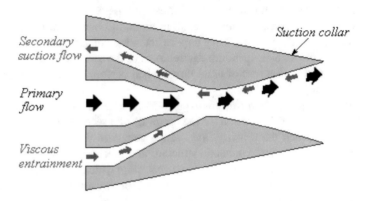

Strykowski et al [23] have shown that the thrust due to a jet stream can be continuously deflected to at least 0.2793 rad (16°) by creating a secondary counter-flowing stream between the primary jet and an adjacent curved surface. For this purpose, suction is applied asymmetrically between the trailing edge of a primary nozzle and an aft suction collar. This creates a low-pressure region along the suction collar, and causes the primary jet to turn [24). The results found by Strykowski et al. [23] at Mach 2 show that the thrust loss is less than 4% and the required mass flow rates are less than approximately 2% of the primary jet. A co-flow, due to the viscous entrainment generated by the primary flow, takes place at the wall of the suction collar, and interferes with the primary flow. By activating asymmetrically this co-flow, the primary flow is deflected towards the side on which the suction flow is applied, due to the pressure drop which causes the thrust to turn [22].

- Throat-shifting for fluidic thrust vectoring, which is performed by injecting secondary flow at or just upstream of the throat, as shown in the following figure. By so doing, there is no formation of shock waves.

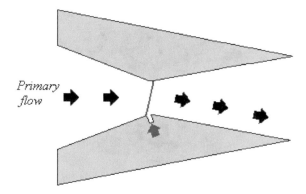

This injection skews the sonic plane and deflects the flow. Another version of this method uses variable recessed cavities, which deflect the primary flow by means of vortices in the cavities, as shown in the following figure.

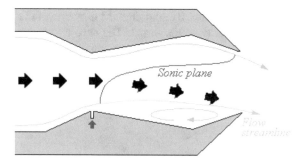

The recessed cavity portion of the nozzle shown above is located between the upstream minimum area and the downstream minimum area of the nozzle itself. The fluidic injection is performed at the upstream minimum area. A simulation study carried out by Deere et al. [25] has shown that substantial thrust-vector angles are obtainable without large penalties in thrust efficiency. This version is a combined method, because the recessed-cavity technique is used in addition to the throat-shifting technique in order to obtain greater performance.

- Shock vector control for fluidic thrust vectoring, which is performed by injecting secondary flow downstream of the throat, as shown in the following figure.

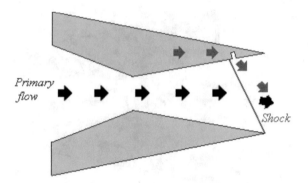

The secondary injected flow acts as a compression ramp in the direction of the primary supersonic flow. This compression induces an oblique shock wave in the diverging portion of the nozzle at some angle with respect to the direction of the primary flow. The primary flow, when interacting with the oblique shock wave, turns away from the axis of symmetry of the nozzle. This turn changes the direction of the thrust vector. By so doing, the direction of the primary flow does not change in the vicinity of the throat. The shock vector control may also be combined with the throat shifting, by using two ports instead of one to inject the secondary flow into the nozzle, as shown in the following figure.

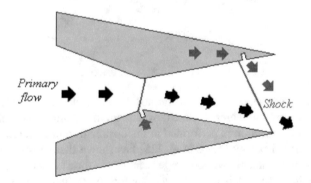

The aerospace vehicles which use fluidic instead of mechanical thrust vectoring methods have the advantages of being lightweight, free from moving parts, less expensive, easy to integrate, and less detectable by radars, the last advantage being of particular interest in military applications. On the other hand, they must be designed from the outset as such, and existing vehicles currently in use cannot be retrofitted. In addition, their capability of directional change is often held to be lower than that of vehicles using mechanical thrust vectoring. On the last issue, Strykowski et alii [23] have obtained a thrust vector angle of 0.2793 rad (16°) by using the counter-flow technique described above, and Wing and Giuliano [26] have obtained a value of up to $\pi/10$ rad (18°) for this angle, by using the shock vector control technique. Such values are by no means small in comparison with those cited by Sutton and Biblarz [2, page 611], Table 16.1 for mechanical thrust vectoring systems. They are lower only than the value of $\pi/9$ rad ($20°$), relating to the movable nozzle (rotary ball with gas seal). Another disadvantage of the fluidic thrust vectoring methods has been found in their need for a source of secondary flow. In the event of that source being the same as that of the primary flow which generates the thrust, it has been argued, the performance of the vehicle would decrease, at least in the phase of thrust vector control. As to the loss in performance, Strykowski et al. [23] have found, at Mach 2, a thrust loss lower than 4%, with the required mass flow rates being lower than approximately 2% of the primary jet. Consequently, the only strong disadvantage of the fluidic thrust vectoring methods seems to us to be the deficiency or insufficiency of data on their behaviour gathered by testing them in flight.

2.3   Performance of a Thrust Chamber

The performance of the thrust chamber of a rocket engine is measured by some indicators, which have been discussed at length in Chap. 1, Sect. 1.3. For convenience of the reader, a list of the principal performance indicators is also given below.

(1)   Specific impulse $(I_s)_{tc}$ of the thrust chamber, which is measured in seconds and is defined by

$$(I_s)_{tc} = \frac{F}{\dot{W}_{tc}} = \frac{c^* C_F}{g_0}$$

where F (N) is the thrust, $\dot{W}_{tc} = g_0 \dot{m}_{tc}$ (N/s) is the weight flow rate of the propellant at the thrust chamber, c^* (m/s) is the characteristic velocity, C_F is the dimensionless thrust coefficient, and $g_0 = 9.80665$ m/s^2 is the acceleration of gravity of the Earth at the sea level.

(2)   Characteristic velocity c^*, which is measured in m/s and is defined by

$$c^* = \frac{\left[\gamma R (T_c)_{ns}\right]^{\frac{1}{2}}}{\gamma \left[\left(\frac{2}{\gamma+1}\right)^{\frac{\gamma+1}{\gamma-1}}\right]^{\frac{1}{2}}}$$

where $(T_c)_{ns} = T_i [1 + \frac{1}{2}(\gamma - 1)M_i^2]$ is the total temperature (K) of the combustion chamber at the nozzle inlet, M_i is the Mach number at the nozzle inlet, $\gamma \equiv c_p/c_v$ is the specific heat ratio of the combusted gas, and R is the constant of the specific gas, that is, the universal gas constant $R^* = 8314.460$ N m kmol^{-1} K^{-1} divided by the average molar mass $\mathcal{M}$ (kg/kmol) of the combusted gas. When the propellant combination and its mixture ratio have been chosen, then the specific heat ratio γ

and the constant $R = R^*/\mathcal{M}$ of the combusted gas are also determined within known limits. In these limits, the characteristic velocity depends on the total temperature $(T_c)_{ns}$ of the gas in the combustion chamber at the nozzle inlet.

(3) Thrust coefficient C_F, which is dimensionless and is defined by

$$C_F = \left\{ \frac{2\gamma^2}{\gamma - 1} \left(\frac{2}{\gamma + 1} \right)^{\frac{\gamma+1}{\gamma-1}} \left[1 - \left(\frac{p_e}{(p_c)_{ns}} \right)^{\frac{\gamma-1}{\gamma}} \right] \right\}^{\frac{1}{2}} + \frac{A_e}{A_t} \left[\frac{p_e - p_0}{(p_c)_{ns}} \right]$$

where $\gamma \equiv c_p/c_v$ is the specific heat ratio of the combusted gas, p_e (N/m^2) is the pressure of the exhaust gas at the exit plane of the nozzle, $(p_c)_{ns}$ (N/m^2) is the total pressure in the combustion chamber at the nozzle inlet, p_0 (N/m^2) is the ambient pressure, and $\varepsilon \equiv A_e/A_t$ is the ratio of the area of the cross-section at the exit plane of the nozzle to the area of the cross-section at the throat. When the performance of the combustion process has been determined, then the value of the first addend on the right-hand side of the preceding equation is known. The value of the second addend depends on the geometric characteristics of the diverging portion of the nozzle, namely, on the expansion ratio $\varepsilon \equiv A_e/A_t$, which determines the pressure ratios $p_e/(p_c)_{ns}$ and $p_0/(p_c)_{ns}$.

As an example, it is required to determine the specific impulse of the thrust chamber, the characteristic velocity, and the thrust coefficient for a rocket, whose first stage has the properties indicated below. The propellant combination is liquid oxygen with RP-1 (kerosene), the oxidiser-to-fuel mixture ratio at the thrust chamber is $o/f = 2.35$, the total absolute pressure in the combustion chamber at the nozzle inlet is $(p_c)_{ns} = 6.895 \times 10^6$ N/m^2, the total temperature of the combustion chamber at the nozzle inlet is $(T_c)_{ns} = 3589$ K, the molar mass of the combusted gas is $\mathcal{M} = 22.5$ kg/kmol, the specific heat ratio of the combusted gas is $\gamma \equiv c_p/c_v = 1.222$, and the expansion area ratio of the nozzle is $A_e/A_t = 14$. The atmospheric pressure and the acceleration of gravity at sea level have the standard values $p_0 = 101325$ N/m^2 and $g_0 = 9.80665$ m/s^2.

The constant of the specific gas results from

$$R = \frac{R^*}{\mathcal{M}} = \frac{8314.460}{22.5} = 369.5 \, \text{N m K}^{-1} \, \text{kg}^{-1}$$

The static pressure p_e (N/m^2) at the exit section of the nozzle results from the following equation of Chap. 1, Sect. 1.2:

$$\frac{A_e}{A_t} = \frac{\left(\frac{2}{\gamma+1} \right)^{\frac{1}{\gamma-1}} \left(\frac{(p_c)_{ns}}{p_e} \right)^{\frac{1}{\gamma}}}{\left\{ \frac{\gamma+1}{\gamma-1} \left[1 - \left(\frac{p_e}{(p_c)_{ns}} \right)^{\frac{\gamma-1}{\gamma}} \right] \right\}^{\frac{1}{2}}}$$

where $A_e/A_t = 14$, $\gamma = 1.222$, and $(p_c)_{ns} = 6.895 \times 10^6$ N/m². As has been shown in Chap. 1, Sect. 1.2, the value of p_e can be determined numerically by defining an auxiliary variable $z = p_e/(p_c)_{ns}$ and a function $f(z)$ such that

$$f(z) \equiv 14^2 - \frac{\left(\frac{2}{1.222+1}\right)^{\frac{2}{1.222-1}} \left(\frac{1}{z}\right)^{\frac{2}{1.222}}}{\frac{1.222+1}{1.222-1}\left(1 - z^{\frac{1.222-1}{1.222}}\right)}$$

We search a zero of the function $f(z)$ in the interval $0.007 \leq z \leq 0.008$, because $f(z)$ changes sign in this interval. By using the Müller method shown in Chap. 1, Sect. 1.2, we find, with four significant figures, $z = 0.007538$. Hence, we have $p_e = z\,(p_c)_{ns} = 0.007538 \times 6.895 \times 10^6 = 5.197 \times 10^4$ N/m².

The theoretical value c^* of the characteristic velocity results from

$$c^* = \frac{[\gamma R(T_c)_{ns}]^{\frac{1}{2}}}{\gamma \left[\left(\frac{2}{\gamma+1}\right)^{\frac{\gamma+1}{\gamma-1}}\right]^{\frac{1}{2}}} = \frac{[1.222 \times 369.5 \times 3589]^{\frac{1}{2}}}{1.222 \times \left[\left(\frac{2}{1.222+1}\right)^{\frac{1.222+1}{1.222-1}}\right]^{\frac{1}{2}}} = 1764 \text{ m/s}$$

By introducing a correction factor $\eta_{c*} = 0.975$, the design value $\bar{c}^*$. of the characteristic velocity results from the theoretical value c^* as follows

$$\bar{c}^* = \eta_{c*} c^* = 0.975 \times 1764 = 1720 \text{ m/s}$$

The theoretical value C_F of the thrust coefficient at sea level results from

$$C_F = \left\{ \frac{2\gamma^2}{\gamma - 1}\left(\frac{2}{\gamma + 1}\right)^{\frac{\gamma+1}{\gamma-1}} \left[1 - \left(\frac{p_e}{(p_c)_{ns}}\right)^{\frac{\gamma-1}{\gamma}}\right] \right\}^{\frac{1}{2}} + \frac{A_e}{A_t}\left[\frac{p_e - p_0}{(p_c)_{ns}}\right]$$

After substituting $\gamma = 1.222$, $p_e = 5.197 \times 10^4$ N/m², $(p_c)_{ns} = 6.895 \times 10^6$ N/m², $p_0 = 1.013 \times 10^5$ N/m², and $A_e/A_t = 14$ into the preceding equation, we find $C_F = 1.561$.

By introducing a correction factor $\eta_F = 0.98$, the design value $\overline{C}_F$. of the thrust coefficient at sea level can be computed from the corresponding theoretical value C_F as follows

$$\overline{C}_F = \eta_F C_F = 0.98 \times 1.561 = 1.530$$

By using the design values computed above of, respectively, the characteristic velocity and the thrust coefficient at the sea level, the design value of the specific impulse of the thrust chamber at sea level results from

$$(\bar{I}_s)_{tc} = \frac{\bar{c}^* \overline{C}_F}{g_0} = \frac{1720 \times 1.530}{9.807} = 268.3 \text{ s}$$

In the following example, we want to determine the specific impulse of the thrust chamber, the characteristic velocity, and the thrust coefficient for a rocket, whose second stage burns a combination of liquid oxygen with liquid hydrogen, and has the properties specified below. The oxidiser-to-fuel mixture ratio at the thrust chamber is $o/f = 5.22$, the total absolute pressure and the total temperature in the combustion chamber at the nozzle inlet are respectively $(p_c)_{ns} = 5.516 \times 10^6$ N/m² and $(T_c)_{ns} = 3356$ K, the molar mass of the combusted gas is $\mathcal{M} = 12$ kg/kmol, the specific heat ratio of the combusted gas is $\gamma \equiv c_p/c_v = 1.213$, and the expansion area ratio of the nozzle is $A_e/A_t = 40$. The constant of the specific gas results from

$$ R = \frac{R^*}{\mathcal{M}} = \frac{8314.460}{12} = 692.9 \, \text{N m K}^{-1} \, \text{kg}^{-1} $$

The static pressure p_e (N/m²) at the exit section of the nozzle results from the following equation of Chap. 1, Sect. 1.2:

$$ \frac{A_e}{A_t} = \frac{\left(\frac{2}{\gamma+1}\right)^{\frac{1}{\gamma-1}} \left(\frac{(p_c)_{ns}}{p_e}\right)^{\frac{1}{\gamma}}}{\left\{\frac{\gamma+1}{\gamma-1}\left[1 - \left(\frac{p_e}{(p_c)_{ns}}\right)^{\frac{\gamma-1}{\gamma}}\right]\right\}^{\frac{1}{2}}} $$

where $A_e/A_t = 40$, $\gamma = 1.213$, and $(p_c)_{ns} = 5.516 \times 10^6$ N/m². By solving numerically the preceding equation for p_e as has been shown in the preceding example, we find $p_e = 1.094 \times 10^4$ N/m².

The theoretical value c^* of the characteristic velocity results from

$$ c^* = \frac{\left[\gamma R(T_c)_{ns}\right]^{\frac{1}{2}}}{\gamma\left[\left(\frac{2}{\gamma+1}\right)^{\frac{\gamma+1}{\gamma-1}}\right]^{\frac{1}{2}}} = \frac{[1.213 \times 692.9 \times 3356]^{\frac{1}{2}}}{1.213 \times \left[\left(\frac{2}{1.213+1}\right)^{\frac{1.213+1}{1.213-1}}\right]^{\frac{1}{2}}} = 2342 \, \text{m/s} $$

By introducing a correction factor $\eta_{c*} = 0.975$, the design value $\bar{c}^*$ of the characteristic velocity results from the theoretical value c^* as follows

$$ \bar{c}^* = \eta_{c*}c^* = 0.975 \times 2342 = 2284 \, \text{m/s} $$

The theoretical value C_F of the thrust coefficient in vacuo ($p_0 = 0$) results from

$$ C_F = \left\{\frac{2\gamma^2}{\gamma-1}\left(\frac{2}{\gamma+1}\right)^{\frac{\gamma+1}{\gamma-1}}\left[1 - \left(\frac{p_e}{(p_c)_{ns}}\right)^{\frac{\gamma-1}{\gamma}}\right]\right\}^{\frac{1}{2}} + \frac{A_e}{A_t}\left[\frac{p_e}{(p_c)_{ns}}\right] $$

After substituting $\gamma = 1.213$, $p_e = 1.094 \times 10^4$ N/m², $(p_c)_{ns} = 5.516 \times 10^6$ N/m², and $A_e/A_t = 40$ into the preceding equation, we find $C_F = 1.871$.

A value of 1.01 can be taken for the correction factor η_F of C_F. Therefore, the design value $\overline{C}_F$ of the thrust coefficient in vacuo can be computed from the corresponding theoretical value C_F as follows

$$\overline{C}_F = \eta_F C_F = 1.01 \times 1.871 = 1.890$$

By using the design values computed above of, respectively, the characteristic velocity and the thrust coefficient in vacuo, the design value of the specific impulse of the thrust chamber in vacuo results from

$$\left(\overline{I}_s\right)_{tc} = \frac{\overline{c}^* \overline{C}_F}{g_0} = \frac{2284 \times 1.890}{9.807} = 440.2 \text{ s}$$

2.4 Configuration and Design of a Thrust Chamber

In the design of a combustion chamber, account must be taken of the so-called stay time t_s (measured in seconds), which is the time required by the fuel to mix with the oxidiser and burn completely before the combusted gas is expelled through the nozzle.

The volume V_c (m^3) occupied by a combustion chamber is also important for the combustion efficiency. This volume depends on the mass flow rate $\dot{m}_{tc}$(kg/s) of the propellant components, on their average density ρ (kg/m^3), and on the stay time t_s (s) defined above, as follows

$$V_c = \frac{\dot{m}_{tc} t_s}{\rho}$$

Of course, the same equality also holds when the mass flow rate and the average density are replaced by, respectively, the weight flow rate (N/s) of the propellant components and their average specific weight (N/m^3).

Another important quantity in the design of a combustion chamber is its characteristic length L^* (m), which is defined as the ratio of the volume V_c (m^3) of the combustion chamber to the area A_t (m^2) of the cross section of the nozzle at the throat, as follows

$$L^* = \frac{V_c}{A_t} = \frac{\dot{m}_{tc} t_s}{\rho A_t}$$

As shown by the preceding equation, the characteristic length of a combustion chamber depends on the stay time of the propellant in the combustion chamber.

The characteristic velocity c^* of a rocket engine increases with the characteristic length L^* of its combustion chamber, and approaches asymptotically a maximum value. This value has been determined experimentally for each combination of propellants. Consequently, it is not advantageous to increase L^* beyond the value corresponding to the maximum value of c^*, because a higher characteristic length implies higher volume and mass, a higher surface to be cooled, and higher losses due to friction at the chamber walls.

Recommended values of the characteristic length L^* (m) of a combustion chamber for various combinations of propellants are given in the following table, which is due to the courtesy of NASA [5].

Propellant combination	L^* (m)
Chlorine trifluoride/hydrazine-base fuel	0.76–0.90
Liquid fluorine/hydrazine	0.61–0.71
Liquid fluorine/liquid hydrogen (GH$_2$ injection)	0.56–0.66
Liquid fluorine/liquid hydrogen (LH$_2$ injection)	0.64–0.76
Hydrogen peroxide/RP-1 (including catalyst bed)	1.6–1.8
Nitric acid/hydrazine-base fuel	0.76–0.90
Nitrogen tetroxide/hydrazine-base fuel	0.76–0.90
Liquid oxygen/ammonia	0.76–1.0
Liquid oxygen/liquid hydrogen (GH$_2$ injection)	0.56–0.71
Liquid oxygen/liquid hydrogen (LH$_2$ injection)	0.76–1.0
Liquid oxygen/RP-1	1.0–1.3

After selecting a combination of propellants, the cross section area A_t (m^2) of the nozzle at the throat, and the minimum value of the characteristic length L^* (m), the volume V_c (m^3) of a combustion chamber can be determined as follows

$$V_c = L^* A_t$$

The stay time t_s (s) of the propellant in the combustion chamber depends on the volume but not on the shape of the combustion chamber. This shape is chosen according to the following criteria.

A cylindrical combustion chamber having a small cross section and a high length is subject to high losses due to the friction of the combusted gas with the chamber walls. In addition, it is difficult to place the necessary number of orifices in a planar surface of small cross section. On the other hand, a cylindrical combustion chamber having a large cross section and a small length leaves a space sufficient for propellant mixing but insufficient for a complete combustion. Further considerations to be taken into account in choosing a shape concern heat transfer, stability of combustion, weight, and easiness of manufacturing.

The following figure, due to the courtesy of NASA [27], shows the two principal shapes, namely, spherical (or near-spherical) and cylindrical, which may be chosen for a combustion chamber. The spherical shape (left) was chosen for the combustion chamber of the V-2 missile, whereas the cylindrical shape (right) was chosen for the combustion chamber of the Navaho missile.

A spherical combustion chamber has, in comparison with a cylindrical one of the same volume, a smaller mass and a smaller surface to be cooled. In addition, for the same pressure and for the same strength of the materials used, the walls of a spherical combustion chamber can be less thick than those of a cylindrical combustion chamber. On the other hand, a spherical combustion chamber is more difficult to manufacture and offers a lower performance than is the case with a cylindrical combustion chamber. Therefore, a cylindrical combustion chamber is considered firstly, and other shapes will be discussed successively.

Let the cross-sectional area $A_t = \pi R_t^2$ of the throat and the characteristic length L^* of the combustion chamber be known for a cylindrical combustion chamber. The value to be given to the contraction area ratio $\varepsilon_c \equiv A_c/A_t$, where $A_c = \pi R_c^2$ is the cross-sectional area of the combustion chamber, depends on various factors connected with the performance of the process of combustion. However, on the basis of the experience gained so far, Huzel and Huang [5] suggest the following values for the contraction area ratio: 2–5 in case of pressurised-gas low-thrust engines, and 1.3–2.5 in case of turbo-pump high-thrust engines.

A simple scheme of a rocket engine is illustrated in the following figure, which shows, from left to right, the injector, the cylindrical combustion chamber, and the converging-diverging nozzle.

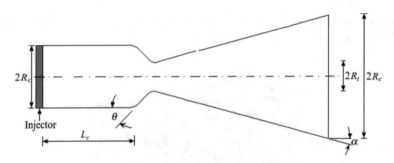

The length L_c of the combustion chamber is measured from the internal face of the injector to the inlet plane of the nozzle, as shown in the preceding figure.

Let V_c and A_{tot} be respectively the volume (measured from the injector face to the throat plane of the nozzle) and the total area, minus the injector face, of the combustion chamber. Huzel and Huang [5] suggest to use the following approximate formulae to determine the volume V_c and the total area A_{tot} of the combustion chamber as functions of the quantities L_c, A_t, A_c/A_t, and θ:

$$V_c = A_t L_c \left(\frac{A_c}{A_t} \right) + \frac{A_t}{3} \left(\frac{A_t}{\pi} \right)^{\frac{1}{2}} (\cot \theta) \left[\left(\frac{A_c}{A_t} \right)^{\frac{1}{3}} - 1 \right]$$

$$A_{tot} = 2L_c \left[\pi \left(\frac{A_c}{A_t} \right) A_t \right]^{\frac{1}{2}} + (\csc \theta) \left[\left(\frac{A_c}{A_t} \right) - 1 \right] A_t$$

As has been shown in Chap. 1, Sect. 1.1, for the optimum performance of a rocket engine in terms of thrust, the gas pressure p_e at the exit plane of the nozzle should be exactly equal to the pressure p_0 due to the environment around it. Since a given expansion ratio $\varepsilon \equiv A_e/A_t$ results in the optimum expansion only at a specific altitude, then the design expansion ratio of a nozzle must be selected in such a way as to give the best average performance during powered flight.

The supersonic expansion of the combusted gas occurs in the diverging portion of the nozzle. The two principal types (bell-shaped and conical) of the diverging portion of a nozzle are shown in the following figure, due to the courtesy of NASA [28].

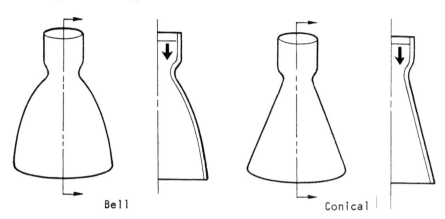

Bell   Conical

A bell-shaped nozzle (left) has the advantages, over a conical nozzle (right), of better performance and shorter length. The former has a radial-flow section in the initial divergent region. This fact generates a uniform flow directed along the symmetry axis at the exit cross-section of the nozzle. In addition, the gradual change of the wall angle with respect to the axis prevents oblique shocks.

Some of the methods (in particular, those due to Prandtl-Busemann, Puckett, and Foelsch) to determine a nozzle contour have been described by Crown [29]. Rao [30] has applied the calculus of variations to determine the shape of a nozzle contour leading to the maximum thrust. For engineering purposes, near-optimum parabolic contours are suitable for many applications [28].

According to Huzel and Huang [5], the optimum shape of a nozzle having a given expansion ratio is chosen on the basis of the following considerations:

- uniform parallel axial flow of the combusted gas at the exit section of the nozzle for the maximum magnitude of the momentum vector;
- minimum losses due to separation and turbulence within the nozzle;
- shortest possible length of the nozzle for minimum requirements of space envelope, weight, losses due to friction at the walls, and cooling; and
- easiness of manufacturing.

In practice, the cone and the bell are the most frequently used of all shapes for the diverging portion of a rocket nozzle.

The advantages of a conical nozzle over a bell-shaped nozzle are easiness of manufacturing, and capability of increasing or decreasing its exit section, without the necessity of re-designing the full surface of the nozzle. The diverging portion of a conical nozzle is shown in the following figure.

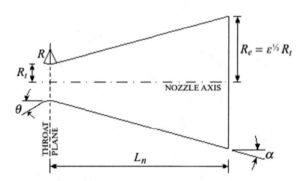

On both sides of the throat plane, the contour of the nozzle is a circular arc, whose radius R ranges from 0.5 to 1.5 times the radius R_t of the throat [5]. In the converging portion (not shown in the preceding figure) of the nozzle, the angle θ of semi-aperture ranges from $\pi/9$ rad to $\pi/4$ rad (from 20 to 45°). The semi-aperture angle α of the diverging portion of the nozzle varies from about $\pi/15$ rad to $\pi/10$ rad (from 12 to 18°) [5]. The length L_n of the diverging portion of a conical nozzle is expressed by the following equation of [5]:

$$L_n = \frac{R_t\left[\left(\frac{A_e}{A_t}\right)^{\frac{1}{2}} - 1\right] + R(\sec \alpha - 1)}{\tan \alpha}$$

In practice, $\pi/12$ rad (15°) is the most frequently chosen value for the divergence angle α of a conical nozzle, for reasons of weight, length, and performance. In case of a conical nozzle, the velocity vector v_e of the exhaust gas at the exit plane of the nozzle forms the divergence angle α with respect to the axis of symmetry of the nozzle. Therefore, not all the gas momentum is in the axial direction, there being a non-axial component of the exhaust gas velocity vector. This loss of axial momentum in a conical nozzle is taken into account by means of the factor

$$\lambda = \frac{1}{2}(1 + \cos \alpha)$$

where λ is the ratio between the momentum of the gases in a nozzle with a semi-aperture angle α and the momentum of an ideal nozzle with all gases flowing in the axial direction [2]. With this factor, the thrust of a rocket engine having a conical nozzle is [31]:

$$F = \lambda \dot{m} v_e + A_e(p_e - p_0)$$

In the ideal case of all the gas momentum being in the axial direction at the exit plane of a nozzle, the factor λ is equal to unity. In the case of a conical nozzle having a semi-aperture angle $\alpha = \pi/12$ rad $= 15°$, there results $\lambda = 0.983$, and therefore the velocity of the exhaust gas at the exit plane of a conical nozzle is in magnitude 98.3% of the velocity of the gas for an ideal nozzle.

For the purposes of reducing the length L_n of the divergent portion of a nozzle and increasing the performance, a bell-shaped nozzle has been studied. This type of nozzle offers the advantages of decreasing the portion of nozzle where the gas flow expands, and changing gradually the local angle of semi-aperture, so as to have a uniform, nearly axially directed flow at the exit plane. In addition, since the slope of the wall contour changes continuously, then the formation of oblique shocks is avoided. The optimum contour for a bell-shaped nozzle which leads to the maximum magnitude of thrust has been determined in 1958 by Rao [30], by using the method of characteristics combined with the calculus of variations. Several computer programmes have subsequently been developed (see, for example, 32) in order to determine the optimum contour by applying the method described by Rao [30]. In practice, an equivalent conical nozzle having a semi-aperture angle $\alpha = \pi/12 \approx 0.2618$ rad $= 15°$ is used as a standard to specify the length of a bell-shaped nozzle. For example, a bell-shaped nozzle is said to have a fractional length $L_f = n\%$ when its true length L_n, measured from the throat plane to the exit plane, is

$$L_n = \frac{n}{100} \frac{R_t}{\tan 0.2618} \left[\varepsilon^{\frac{1}{2}} - 1 + 1.5 \left(\frac{1}{\cos 0.2618} - 1 \right) \right]$$

that is, when the true length L_n of that nozzle is $n/100$ times the length of a conical nozzle having a semi-aperture angle $\alpha = \pi/12$ rad $= 15°$, the same radius R_t at the throat plane, and the same expansion ratio $\varepsilon \equiv A_e/A_t$. It has been proved experimentally that a bell-shaped nozzle whose fractional length L_f is greater than about 85% does not offer substantial advantages for the purpose of increasing the nozzle correction factor λ, because an increase in length implies an increase in mass. This fact is shown in the following figure, due to the courtesy of NASA [5].

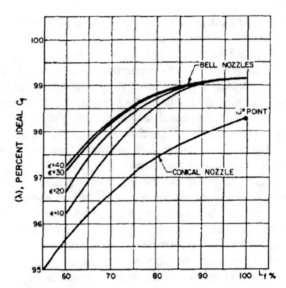

FRACTIONAL NOZZLE LENGTH (L$_f$) BASED ON A 15° HALF ANGLE
CONICAL NOZZLE WITH ANY AREA RATIO ϵ

Rao [33] and other authors have found a convenient way to design a near-optimum bell contour by using a parabolic approximation to the optimum contour. This parabolic approximation is shown in the following figure, due to the courtesy of NASA [5]. In the converging portion of the nozzle, immediately upstream of the throat plane, the nozzle contour is a circular arc, whose radius is equal to 1.5 times R_t, where R_t is the radius of the throat. This arc terminates at the point T, where T is the point in which this arc intersects the throat plane. The angle which subtends this circular arc is to be chosen by the designer. In the diverging portion of the nozzle, immediately downstream of the throat plane, the nozzle contour is also a circular arc, whose radius is equal to 0.382 times R_t. This circular arc goes from the point T to the point of inflection N, where N is the point in which the parabolic segment begins. The parabolic segment goes from to the point of inflection N to the point E, where E is the point in which the parabolic segment intersects the exit plane of the nozzle.

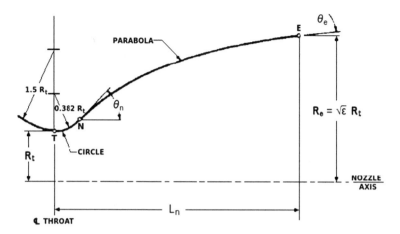

In order to design a specific bell-shaped nozzle, the following data must be known: the radius R_t of the throat, the true length L_n of the nozzle (measured from the throat plane to the exit plane along the axis of symmetry of the nozzle), the expansion area ratio $\varepsilon \equiv A_e/A_t$, the angle θ_n which the tangent in N to the parabola forms with the axis of symmetry, and the angle θ_e which the tangent in E to the parabola forms with the axis of symmetry.

Of course, the true length L_n of a bell-shaped nozzle may be expressed in terms of the fractional length L_f, with reference to an equivalent conical nozzle having a semi-aperture angle $\pi/12$ radians (15°).

The values of the angles θ_n and θ_e are given in, respectively, the upper part and the lower part of the following figure, due to the courtesy of NASA [5], as functions of the expansion area ratio $\varepsilon \equiv A_e/A_t$. The values of the angles θ_n and θ_e are those of the optimal (maximum thrust) nozzle determined by Rao [30].

This figure shows that the value of the angle θ_e at the exit plane of the nozzle is never equal to zero. The same figure can also be found in [34, 35]. In particular, Newlands [35] has extrapolated the curves found by Rao [30] to values greater than 50 of the expansion area ratio $\varepsilon \equiv A_e/A_t$. Experience has shown that these curves are substantially the same for all the values of the specific heat ratio $\gamma \equiv c_p/c_v$ which are of practical interest.

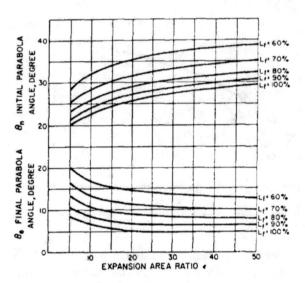

In order to draw a parabola from the point of inflection N to the point of exit E, Newlands [35] has suggested to use an ancient geometrical method, which is briefly described below.

With reference to the following figure, adapted from [5], a straight line is drawn at an angle θ_n from N, and then another straight line is drawn back at an angle θ_e from E. Let Q be the point of intersection of these straight lines. Next, both of these lines are divided into an equal number of segments, which is four in the following figure. The terminal points of these segments are labelled a, b, c and e, f, g. A straight line is drawn from point a to point e, then another from b to f, and another from c to g. These straight lines form a mesh, whose edge gives the parabola outline. This parabola is also tangent to the straight lines QN and QE. By using many more divisions, for example by means of a CAD package, there results a sharper contour. A series of straight-line segments results from removing most of the mesh. By joining the midpoint of each segment with a smooth curve, such as a CAD spline, there results the nozzle contour.

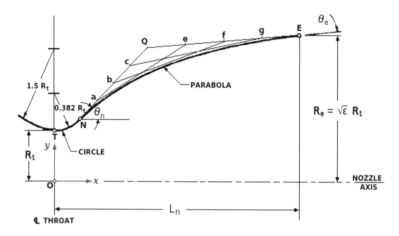

For the same purpose, an analytical method based on Bézier curves has also been suggested by Newlands [35]. Without loss of generality, this method is described below by means of a numerical example. With reference to the preceding figure, we consider a nozzle having a radius $R_t = 13.8$ cm measured in the plane of the throat, an expansion area ratio $\varepsilon \equiv A_e/A_t = 70$, and a fractional length $L_f = 80\%$ of a $\pi/12$-rad $= 15°$ equivalent conical nozzle. We want to determine the equation of the parabolic segment of nozzle going from the inflexion point N to the exit point E, where the axes x and y and their origin O are shown in the preceding figure. For this value of fractional length, the Rao curves give the angles $\theta_n = 0.5760$ rad $= 33°$ and $\theta_e = 0.1222$ rad $= 7°$ which the tangents to the contour of the nozzle in the points respectively N and E form with the axis of symmetry (x-axis) of the nozzle. These wall angles result from extrapolating the graphs given above. By substituting $n = 80$, $R_t = 13.8$ cm, and $\varepsilon = 70$ in the following formula, the true length of the nozzle results

$$L_n = \frac{n}{100}\frac{R_t}{\tan 0.2618}\left[\varepsilon^{\frac{1}{2}} - 1 + 1.5\left(\frac{1}{\cos 0.2618} - 1\right)\right] = 305.70\text{cm}$$

We take 0.5760 rad for the value of the angle α subtending the circular arc TN of radius $0.382\,R_t$, downstream of the throat. Therefore, the co-ordinates of N are

$$x_N = 0.382 R_t \sin \alpha = 0.382 \times 13.8 \times \sin 0.5760 = 2.8711\text{cm}$$
$$y_N = R_t[1 + 0.382(1 - \cos \alpha)] = 13.8 \times [1 + 0.382 \times (1 - \cos 0.576)]$$
$$= 14.650\text{cm}$$

The co-ordinates of the point E laying on the exit plane of the nozzle are

$$x_E = L_n = 305.70\text{ cm}$$
$$y_E = R_E = \varepsilon^{\frac{1}{2}} R_t = 70^{\frac{1}{2}} \times 13.8 = 115.46\text{ cm}$$

From the inflection point N, we produce the straight line NQ, which forms the angle θ_n with the x-axis. As is well known, the equation of NQ is

$$y - y_N = (x - x_N) \tan \theta_n$$

By substituting $y_N = 14.650$ cm, $x_N = 2.8711$ cm, and $\theta_n = 0.5760$ rad into the preceding equation, there results

$$y = 0.64941x + 12.785$$

Likewise, from the exit point E, we produce the straight line QE, which forms the angle θ_e with the x-axis. The equation of QE is

$$y - y_E = (x - x_E) \tan \theta_e$$

By substituting $y_E = 115.46$ cm, $x_E = 305.70$ cm, and $\theta_e = 0.1222$ rad into the preceding equation, there results

$$y = 0.12278x + 77.925$$

The straight line NQ (whose equation is $y = 0.64941\,x + 12.785$) intersects the straight line QE (whose equation is $y = 0.12278\,x + 77.925$) in the point Q, whose co-ordinates are easily found to be

$$x_Q = 123.69 \text{ cm}$$
$$y_Q = 93.112 \text{ cm}$$

Since the co-ordinates of the points N, Q, and E are known, then the parabolic segment of nozzle going from N to E can be expressed by means of a Bézier quadratic curve having the following parametric equations

$$x(t) = (1-t)^2 x_N + 2(1-t)t x_Q + t^2 x_E \quad (0 \le t \le 1)$$
$$y(t) = (1-t)^2 y_N + 2(1-t)t y_Q + t^2 y_E \quad (0 \le t \le 1)$$

The preceding equations specify the co-ordinates x and y of any point P of this curve corresponding to a value of t taken between 0 and 1. For example, the co-ordinates of three points P_1, P_2, and P_3 (other than N and E) of the curve can be determined by assigning the values 0.25, 0.50, and 0.75 to the parameter t.

In the present case, after substituting the values of (x_N, y_N), (x_Q, y_Q), and (x_E, y_E) computed above into the preceding equations, there results

$$x(t) = 2.8711(1-t)^2 + 247.38(1-t)t + 305.70t^2 \quad (0 \le t \le 1)$$
$$y(t) = 14.650(1-t)^2 + 186.22(1-t)t + 115.46t^2 \quad (0 \le t \le 1)$$

The preceding equations express analytically the segment of nozzle going from the inflection point N to the exit point E in the reference system Oxy shown in the preceding figure. As has been shown above, the coefficients of these equations have been determined in such a way as to satisfy the four boundary conditions, which specify the co-ordinates of the endpoints N and E and the angles θ_n and θ_e which the tangents to the parabola in these points form with the x-axis.

In the general case, there may be no set of three coefficients a, b, and c whose values satisfy exactly the boundary conditions specified above, in order to express the contour of a nozzle by means of the equation $y(x) = ax^2 + bx + c$. Such is the case with the numerical example given above, since $x_Q \neq (x_N + x_E)/2$. This is because a parabola can be expressed by the equation $y = ax^2 + bx + c$ only when its axis of symmetry is parallel to the y-axis.

When the requirements posed to the designer of a bell nozzle are: (a) the continuity of its contour; (b) the continuity of its slope; and (c) the respect of the four boundary conditions specified above, independently of whether a parabola or any other smooth curve may be chosen, then the contour can be expressed by a cubic equation, as follows

$$y(x) = ax^3 + bx^2 + cx + d$$

where the unknown values of the coefficients a, b, c, and d result from solving the following system of four linear equations

$$y_N = ax_N^3 + bx_N^2 + cx_N + d$$
$$y_E = ax_E^3 + bx_E^2 + cx_E + d$$
$$\tan\theta_n = 3ax_N^2 + 2bx_N + c$$
$$\tan\theta_e = 3ax_E^2 + 2bx_E + c$$

In the numerical example given above, after substituting the known values of x_N, y_N, x_E, y_E, θ_n and θ_e in the four preceding equations, we find the following values: $a = 1.1602 \times 10^{-6}$, $b = -0.0014065$, $c = 0.65746$, and $d = 12.774$.

The exhaust gas generated within the combustion chamber need not flow along the axis of symmetry of the nozzle. Such is the case with the two annular nozzles illustrated in the following figure, due to the courtesy of NASA [28].

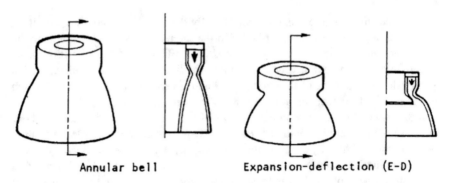

Annular bell   Expansion-deflection (E-D)

Annular (or plug or altitude-compensating) nozzles are so called because the propellant is combusted into a ring, also called an annulus, which is located around the base of the nozzle. They are shown in the following figure, due to the courtesy of NASA [5], which also illustrates a conical nozzle and a bell-shaped nozzle.

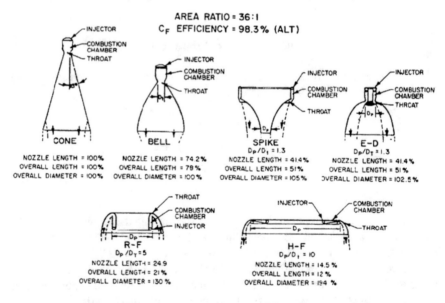

A central body, also called a plug, keeps the gas flow away from a central portion of the nozzle. There are two principal types of annular nozzles. They are the radial in-flow type (spike nozzle) and the radial out-flow type. The latter type, in turn, includes expansion-deflection or E-D, reverse-flow or R-F, and horizontal-flow or H-F nozzles. The nozzles shown above have, all of them, the same level of thrust, the same expansion area ratio, and the same theoretical thrust coefficient C_F (see Chap. 1, Sect. 1.3). Annular nozzles are shorter than conical or bell-shaped nozzles, and are therefore advantageous over the latter in terms of length and mass of the whole vehicle.

For an annular nozzle, the expansion area ratio ε is defined as the ratio of the projected area of the contoured nozzle wall to the area of the throat (A_t). The projected area of the contoured nozzle wall, in turn, is the area of the nozzle at the exit plane (A_e) minus the projected area of the central body (A_p). Therefore

$$\varepsilon \equiv \frac{A_e - A_p}{A_t}$$

The preceding figure indicates the values of the annular diameter ratio, D_p/D_t, where D_p is the diameter of the central body, and D_t is the diameter of the throat of an equivalent nozzle having a circular cross section. The value of annular diameter ratio makes it possible to compare an annular nozzle with a conventional (conical or bell-shaped) nozzle.

Annular nozzles are not subject to losses which affect nozzles of the bell or conical type. This is because, in nozzles of the latter type, the exhaust gas may expand to pressures which are considerably smaller than the ambient pressure before the gas flow detaches from the nozzle wall. This over-expansion results in thrust losses at high altitudes. The property of annular nozzles which avoids an over-expansion of the exhaust gas flow is shown in the following figure, due to the courtesy of NASA [5], for an E-D (expansion-deflection) nozzle.

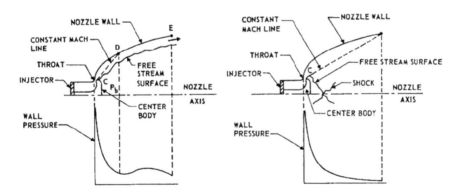

This figure illustrates an E-D nozzle operating at low altitudes (left) and at high altitudes (right). In this figure, p_b indicates the pressure at the back face of the central body of the nozzle. The value of p_b depends on the value of the ambient pressure p_0, and is usually smaller than that of p_0. The point C indicates the shoulder of the central body. The dashed line CD is a line of constant Mach number. Downstream of the throat, the exhaust gas expands unaffected around the shoulder C of the central body as long as its pressure at the wall is greater than p_b. After the gas expands from the throat to the point D along the line CD, the flow downstream of D depends on the nozzle contour DE, and also on the pressure p_b, whose value affects the free stream surface of the boundary of the inner jet. The gas flow is deflected by the contour of the curved wall, and therefore is subject to compression, which increases the pressure at the wall. This increase in wall pressure at low altitudes is shown on the left-hand side of the preceding figure. At such altitudes, the exhaust gas remains attached to the wall, as is the case with a conical or bell-shaped nozzle.

By contrast, at high altitudes, the value of p_b becomes so low, that the gas flow deviates towards the central body. This deviation is shown on the right-hand side of the preceding figure. In such conditions, a shock wave may occur, or the gas flow may continue unaffected up to the exit plane of the nozzle, depending on the flow conditions. This decrease in wall pressure at high altitudes is shown on the right-hand side of the same figure.

The following figure, due to the courtesy of NASA [36], illustrates a plug (or truncated aerodynamic spike) nozzle, which is an annular nozzle discharging exhaust gas having a radial inward component of velocity. This figure also shows the better performance of this aerospike nozzle in comparison with a bell nozzle in terms of specific impulse.

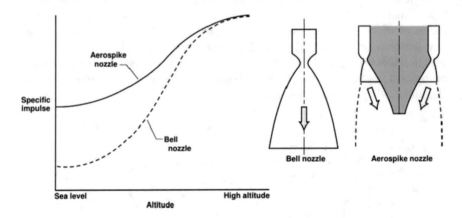

The advantage of the aerospike nozzle over the bell-shaped nozzle resides in its ability to adjust with altitude changes to the static pressure of the free stream. This results in a higher specific impulse than that of a bell-shaped nozzle at low altitudes, as shown in the preceding figure. This altitude compensation is due to the shape of the aerospike nozzle, which has a central ramp terminating either in a plug base or in a spike in the centre, and is open to the atmosphere on the sides.

An aerospike nozzle having a central ramp terminating in a plug base is shown in the following figure, which is due to the courtesy of NASA [36].

The central spike need not be a solid wall, because it can be aerodynamically formed by injecting gases from the engine base, as will be shown below. The exhaust gas is free to expand on the open sides of the nozzle, and to adjust its static pressure with the static pressure of the ambient, which decreases with altitude. Thus, a nozzle of very high area ratio (of high performance in vacuo) can also operate efficiently and safely at sea level.

By contrast, a bell-shaped nozzle can be designed to be optimum (that is, to operate at its maximum level of efficiency) at only one altitude. By maximum level of efficiency, we mean the optimal expansion of the combusted gas, which results in the maximum thrust. For example, the initial stage of the Saturn rocket, which carried the Apollo astronauts to the Moon, had a narrow bell-shaped nozzle to produce an ideal straight-edged column of exhaust gas at sea level. However, the command module, which orbited around the Moon, had a much wider bell-shaped nozzle, which was better suited than the nozzle of the initial stage to the expansion of the combusted gas in the vacuum of space.

As shown above, in an aerospike nozzle, the static pressure (p_e) of the exhaust gas is the same as the static pressure (p_0) of the ambient, and therefore the term $A_e(p_e - p_0)$ in the equation of thrust (see Chap. 1, Sect. 1.1):

$$F = \dot{m}v_e + A_e(p_e - p_0)$$

reduces to zero. In addition, an aerospike engine may also be made of individual thruster segments, which can be turned on or off to provide thrust vector control in order to steer the vehicle [37], instead of using one of the techniques described in Sect. 2.2.

An aerospike rocket engine was developed by Rockwell Aerospace/Rocketdyne Division of Canoga Park, California. In the 1960s, Rocketdyne developed a rounded aerospike engine, whose thrust chamber consisted of a truncated annular spike nozzle, of the radial in-flow type, and a number of discrete combustion chambers arranged around the periphery of the nozzle, so as to discharge their gases along the surface of the nozzle. Subsequently, in 1972, this design gave rise to a linear aerospike engine, where the gases are discharged along the surface of a rectangular wedge rather than around a round spike-shape. The following figure, due to the courtesy of NASA [38], illustrates (left) a normal rocket engine having a bell-shaped nozzle, and (right) the linear aerospike rocket engine designed for the Lockheed-Martin X-33 wing-shaped vehicle, which was a technology demonstrator (cancelled on the 1st of March 2001) for NASA's "next-generation" of space launch vehicles [39].

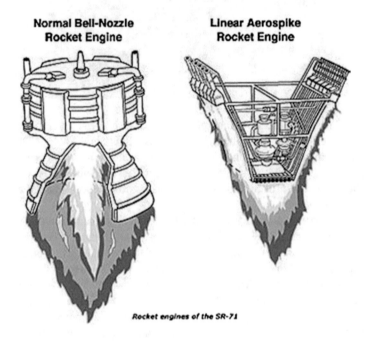

Normal Bell-Nozzle Rocket Engine   **Linear Aerospike Rocket Engine**

Rocket engines of the SR-71

The performance of an aerospike nozzle depends on various factors, such as the shape of the nozzle, the amount of the secondary flow (about 1% of the primary flow), the manner of introducing the secondary flow, and the energy ratio of the secondary flow to the primary.

The working principle of an aerospike nozzle can be explained as follows. The following figure, due to the courtesy of NASA [5], illustrates the gas flow, under altitude conditions (that is, high value of the pressure ratio p_c/p_0), of a rocket engine having a toroidal combustion chamber, an annular throat, and a truncated spike nozzle ending in a circular base.

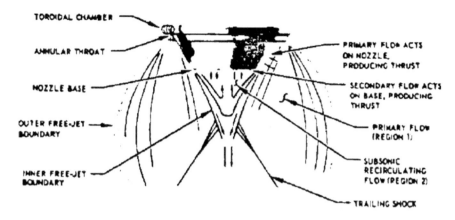

The primary flow of the exhaust gas expands along the wall of the nozzle, in the Region 1 of the preceding figure, and generates thrust. The primary flow continues to expand beyond the base of the nozzle, and gives rise to a subsonic recirculating flow in the Region 2 of the same figure. The pressure acting on the base generates additional thrust. By adding a small amount of secondary flow to the recirculating flow at the base of the nuzzle, the pressure acting on the base is increased further. There is a limit to this increase in efficiency, which determines the optimum amount of secondary flow to be added for each given shape of the nozzle. The outer surface of the primary flow is the boundary of a free stream, and therefore depends on the ambient pressure. This fact provides the property of altitude compensation to the aerospike nozzle illustrated above.

The performance of the gas expansion in nozzles of various types can be expressed by means of a graph of the thrust coefficient C_F (defined in Chap. 1, Sect. 1.3) versus the pressure ratio p_c/p_0 of the static pressure in the combustion chamber p_c to the ambient pressure p_0. In the following figure, due to the courtesy of NASA [5], we consider the thrust coefficient C_F for: (a) an ideal nozzle of variable expansion area ratio having the optimum expansion for any value of the pressure ratio p_c/p_0, (b) an aerospike nozzle of high area ratio, and (c) a bell-shaped nozzle of high area ratio. This graph shows that the dashed line of the aerospike nozzle approximates the solid line of the ideal nozzle much better than does the dash-and-dot line of the bell-shaped nozzle.

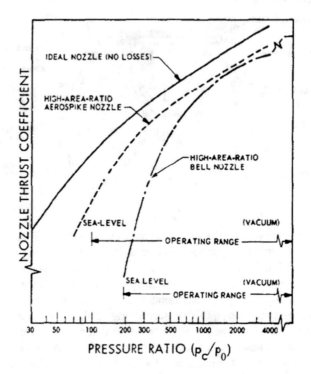

The same graph also shows that, for all the three types of nozzle indicated above, the value of the thrust coefficient C_F increases with the value of the pressure ratio p_c/p_0.

The choice of an aerospike nozzle instead of a traditional nozzle has important consequences on the design of a rocket engine, particularly in case of multistage rockets. Huzel and Huang [5] have identified four advantages and three disadvantages of aerospike nozzles, in the following order:

- shorter length, and therefore reduced mass, for the same performance;
- better performance at sea level;
- possibility of using the stagnant region in the centre of the nozzle to install gas generators, turbo-pumps, tanks, auxiliary equipment, and turbine gas discharges;
- possibility of propelling a rocket vehicle by means of a cluster of engines around a contoured plug instead of a large single engine, as shown in the following figure, due to the courtesy of NASA [40];
- higher cooling requirements;
- heavier structures in some applications; and
- increased difficulty of manufacturing.

The following figure shows a cluster of bell nozzles of high expansion area ratio which have been scarfed for the external expansion of their exhaust gases.

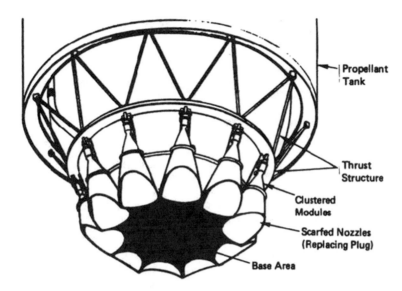

By the way, according to the definition given by Shyne and Keith [41], a scarfed nozzle is a two-dimensional asymmetric nozzle whose lower end (the cowl) is terminated at the point where the last characteristic which emanates from the upper nozzle wall (the ramp) intersects the cowl.

According to Aukerman [36], there are two distinct applications for aerospike nozzles:

- the first stage of a rocket, where the external expansion of the exhaust gas makes it possible to increase the effective expansion area ratio of the engine, due to the property of altitude compensation; and
- the stages operating at high altitudes or in space, where the nozzle can be truncated to a very short length, with minimum detriment of performance.

According to O'Leary and Beck [42], during the 1960s, Pratt & Whitney Rocketdyne tested numerous aerospike engines, ranging in size from subscale, cold-flow models to a 1.112×10^6 N thrust oxygen/hydrogen engine, which was tested at a test stand in Nevada.

A numerical example of design of a thrust chamber is given below. Let us consider a rocket whose first stage burns a combination of liquid oxygen with RP-1 (kerosene), and has the data given in Sect. 2.3. Let $F_{tc} = 3.323 \times 10^6$ N be the desired thrust of the first stage at sea level.

As has been shown in Sect. 2.3, the design value of the thrust coefficient at sea level is $\bar{C}_F = 1.530$, the total absolute pressure in the combustion chamber at the nozzle inlet is $(p_c)_{ns} = 6.895 \times 10^6$ N/m^2, and the expansion area ratio of the nozzle is $\varepsilon \equiv A_e/A_t = 14$. After substituting these values in the following equation of Chap. 1, Sect. 1.3

$$\bar{C}_F = \frac{F}{A_t (p_c)_{ns}}$$

and solving this equation for the area A_t of the cross section of the nozzle at the throat plane, there results

$$A_t = \frac{F}{\bar{C}_F (p_c)_{ns}} = \frac{3.323 \times 10^6}{1.530 \times 6.895 \times 10^6} = 0.3150 \, \text{m}^2$$

The radius R_t of the cross section of the nozzle at the throat plane is

$$R_t = \left(\frac{A_t}{\pi}\right)^{\frac{1}{2}} = \left(\frac{0.3150}{3.1416}\right)^{\frac{1}{2}} = 0.3166 \, \text{m}$$

Since the expansion area ratio of the nozzle is $\varepsilon \equiv A_e/A_t = 14$, then the radius of the cross section of the nozzle at the exit plane is

$$R_e = \varepsilon^{\frac{1}{2}} R_t = 14^{\frac{1}{2}} \times 0.3166 = 1.185 \, \text{m}$$

The table of Sect. 2.4 gives recommended values of the characteristic length L^* of a combustion chamber for various combinations of propellants. In case of the combination of liquid oxygen with RP-1, this table indicates a value of L^* falling in the interval from 1.0 m to 1.3 m. We choose $L^* = 1.143$ m. This makes it possible to determine the volume V_c of the combustion chamber by means of the following equation of Sect. 2.4

$$V_c = L^* A_t$$

After substituting $L^* = 1.143$ m and $A_t = 0.3150$ m^2 in the preceding equation, we find $V_c = 0.3600$ m^3.

As has been shown in Sect. 2.4, the value of the angle θ of semi-aperture in the converging portion of the nozzle ranges from 0.3491 rad to 0.7854 rad. We take $\theta = 0.3491$ rad. We also take 1.6 for the contraction area ratio $\varepsilon_c \equiv A_c/A_t$, where $A_c = \pi R_c^2$ is the cross-sectional area of the combustion chamber.

As has also been shown in Sect. 2.4, the contour of the nozzle upstream of the throat plane is a circular arc, whose radius R ranges from 0.5 to 1.5 times the radius R_t of the throat. We take

$$R = 1.5 R_t = 1.5 \times 0.3166 = 0.4749 \, \text{m}$$

Since 1.6 is the value chosen for the contraction area ratio A_c/A_t, then the radius R_c of the cross-section of the cylindrical combustion chamber is

$$R_c = \left(\frac{A_c}{A_t}\right)^{\frac{1}{2}} R_t = 1.6^{\frac{1}{2}} \times 0.3166 = 0.4005 \, \text{m}$$

Since the converging portion of the nozzle is a cone, whose semi-aperture angle is $\theta = 0.3491$ rad, then the length L_{conv} of the converging portion of the nozzle results from the following formula of Sect. 2.4

$$L_{conv} = \frac{R_t\left[\left(\frac{A_c}{A_t}\right)^{\frac{1}{2}} - 1\right] + R\left(\frac{1}{\cos\theta} - 1\right)}{\tan\theta}$$

After substituting $R_t = 0.3166$ m, $A_c/A_t = 1.6$, $R = 0.4749$ m, and $\theta = 0.3491$ rad in the preceding equation, we find $L_{conv} = 0.3142$ m.

Since the volume of a frustum of a right circular cone is $\frac{1}{3}\pi h(R^2 + Rr + r^2)$, where R and r are the radii of the two circular bases and h is the height of the frustum, then the approximate volume V_{conv} of the converging portion of the nozzle is

$$V_{conv} = \frac{1}{3}\times 3.1416\times 0.3142\times (0.4005^2 + 0.4005\times 0.3166 + 0.3166^2) = 0.1275\,\text{m}^3$$

This value is approximate, because the rounding at the throat plane and the rounding at the inlet plane of the nozzle are not taken into account.

The volume V_{cyl} of the cylindrical combustion chamber, whose length goes from the internal face of the injector to the inlet plane of the nozzle, is

$$V_{cyl} = V_c - V_{conv} = 0.3600 - 0.1275 = 0.2325\,\text{m}^3$$

The length L_{cyl} of this cylindrical combustion chamber is

$$L_{cyl} = \frac{V_{cyl}}{1.6A_t} = \frac{0.2325}{1.6\times 0.3150} = 0.4613\,\text{m}$$

The total distance L_{total} from the internal face of the injector to the plane of the throat is

$$L_{total} = L_{cyl} + L_{conv} = 0.4613 + 0.3142 = 0.7755\,\text{m}$$

For the diverging portion of the nozzle, we choose a fractional length $L_f = 80\%$ of a $\pi/12$ rad equivalent conical nozzle. Therefore, the true length L_n of the diverging portion of the nozzle is

$$L_n = \frac{n}{100}\frac{R_t}{\tan 0.2618}\left[\varepsilon^{\frac{1}{2}} - 1 + 1.5\left(\frac{1}{\cos 0.2618} - 1\right)\right]$$

After substituting $n = 80$, $R_t = 0.3166$ m and $\varepsilon \equiv A_e/A_t = 14$ in the preceding equation, we find $L_n = 2.642$ m. Let us consider again the following figure, adapted from [5].

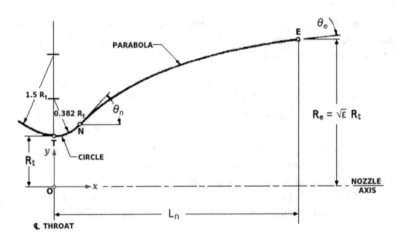

We take the optimum values of the wall angles θ_n and θ_e from the Rao curves for $L_n = 80\%$ and $\varepsilon = 14$. These curves indicate $\theta_n = 0.4782$ rad and $\theta_e = 0.1710$ rad.

Then, we take $\alpha = \theta_n = 0.4782$ rad for the value of the angle α which subtends the circular arc TN of radius $0.382\,R_t$, where N is the inflexion point of the nozzle contour. We use again the system of reference Oxy, whose origin O is on the axis of symmetry of the nozzle, in the plane of the throat, whose x-axis is the axis of symmetry of the nozzle pointing downstream, and whose y-axis is perpendicular to the axis of symmetry and points upward. The co-ordinates x and y of the points N and E in the system of reference defined above are

$$x_N = 0.382 R_t \sin\alpha = 0.382 \times 0.3166 \times \sin 0.4782 = 0.05566\,\text{m}$$
$$y_N = R_t[1 + 0.382(1 - \cos\alpha)] = 0.3166 \times [1 + 0.382 \times (1 - \cos 0.4782)]$$
$$\qquad = 0.3302\,\text{m}$$
$$x_E = L_n = 2.642\,\text{m}$$
$$y_E = R_E = \varepsilon^{\frac{1}{2}} R_t = 14^{\frac{1}{2}} \times 0.3166 = 1.185\,\text{m}$$

A parabola is chosen for the segment of nozzle going from the inflexion point N to the point E laying on the exit plane of the nozzle. In order to determine the co-ordinates x and y of each point of this parabola, we use again the analytical method suggested by Newlands [31], as will be shown below.

The equation of the straight line NQ tangent in N to the parabola is

$$y - y_N = (x - x_N)\tan\theta_n$$

In the present case, the equation of NQ is $y = 0.5183\,x + 0.3014$.
The equation of the straight line EQ tangent in E to the parabola is

$$y - y_E = (x - x_E)\tan\theta_e$$

In the present case, the equation of EQ is $y = 0.1727 x + 0.7288$.

The straight line NQ (whose equation is $y = 0.5183 x + 0.3014$) intersects the straight line EQ (whose equation is $y = 0.1727 x + 0.7288$) in the point Q, whose co-ordinates are easily found to be

$$x_Q = 1.237 \, \text{m}$$
$$y_Q = 0.9424 \, \text{m}$$

Since the co-ordinates of the points N, Q, and E are known, then the parabolic segment of nozzle going from N to E can be expressed by means of a Bézier quadratic curve having the following parametric equations

$$x(t) = (1 - t)^2 x_N + 2(1 - t)t x_Q + t^2 x_E \quad (0 \le t \le 1)$$
$$y(t) = (1 - t)^2 y_N + 2(1 - t)t y_Q + t^2 y_E \quad (0 \le t \le 1)$$

In the present case, after substituting the values of (x_N, y_N), (x_Q, y_Q), and (x_E, y_E) computed above into the preceding equations, there results

$$x(t) = 0.05566(1 - t)^2 + 2.474(1 - t)t + 2.642t^2 \quad (0 \le t \le 1)$$
$$y(t) = 0.3302(1 - t)^2 + 1.885(1 - t)t + 1.185t^2 \quad (0 \le t \le 1)$$

Another numerical example is given below. It concerns the design of a thrust chamber for a rocket whose second stage burns a combination of liquid oxygen with liquid hydrogen, and has the properties specified in Sect. 2.3, which are also given here for convenience. The oxidiser-to-fuel mixture ratio at the thrust chamber is $o/f = 5.22$, the total absolute pressure and the total temperature in the combustion chamber at the nozzle inlet are respectively $(p_c)_{ns} = 5.516 \times 10^6 \, \text{N/m}^2$ and $(T_c)_{ns} = 3356 \, \text{K}$, the molar mass of the combusted gas is $\mathcal{M} = 12 \, \text{kg/kmol}$, the specific heat ratio of the combusted gas is $\gamma = 1.213$, and the expansion area ratio of the nozzle is $A_e/A_t = 40$. Let $F_{tc} = 6.650 \times 10^5 \, \text{N}$ be the desired thrust of the second stage in vacuo.

As has been shown in Sect. 2.3, the design value of the thrust coefficient in vacuo is $\bar{C}_F = 1.890$. By substituting this value and $(p_c)_{ns} = 5.516 \times 10^6 \, \text{N/m}^2$ in the following equation of Chap. 1, Sect. 1.3

$$\bar{C}_F = \frac{F}{A_t (p_c)_{ns}}$$

and solving this equation for the area A_t of the cross section of the nozzle at the throat plane, there results

$$A_t = \frac{F}{\bar{C}_F (p_c)_{ns}} = \frac{6.650 \times 10^5}{1.890 \times 5.516 \times 10^6} = 0.06379 \, \text{m}^2$$

The radius R_t of the cross section of the nozzle at the throat plane is

$$R_t = \left(\frac{A_t}{\pi}\right)^{\frac{1}{2}} = \left(\frac{0.06379}{3.1416}\right)^{\frac{1}{2}} = 0.1425\,\text{m}$$

Since the expansion area ratio of the nozzle is $\varepsilon \equiv A_e/A_t = 40$, then the radius of the cross section of the nozzle at the exit plane is

$$R_e = \varepsilon^{\frac{1}{2}} R_t = 40^{\frac{1}{2}} \times 0.1425 = 0.9012\,\text{m}$$

By using the values given in the table of Sect. 2.4, we choose $L^* = 0.6604$ m for the characteristic length of the combustion chamber.

The volume V_c of the combustion chamber results from the following equation of Sect. 2.4

$$V_c = L^* A_t$$

After substituting $L^* = 0.6604$ m and $A_t = 0.06379$ m² in the preceding equation, we find $V_c = 0.6604 \times 0.06379 = 0.04213$ m³.

We take $\theta = 0.3491$ rad for the value of the angle of semi-aperture in the converging portion of the nozzle. We also take 1.6 for the contraction area ratio $\varepsilon_c \equiv A_c/A_t$, where $A_c = \pi R_c^2$ is the cross-sectional area of the combustion chamber.

The contour of the nozzle upstream of the throat plane is a circular arc, whose radius R ranges from 0.5 to 1.5 times the radius R_t of the throat. We take

$$R = 1.5 R_t = 1.5 \times 0.1425 = 0.2138\,\text{m}$$

Since 1.6 is the value chosen for the contraction area ratio A_c/A_t, then the radius R_c of the cross-section of the cylindrical combustion chamber is

$$R_c = \left(\frac{A_c}{A_t}\right)^{\frac{1}{2}} R_t = 1.6^{\frac{1}{2}} \times 0.1425 = 0.1802\,\text{m}$$

Since the converging portion of the nozzle is a cone, whose semi-aperture angle is $\theta = 0.3491$ rad, then the length L_{conv} of the converging portion of the nozzle results from

$$L_{conv} = \frac{R_t\left[\left(\frac{A_c}{A_t}\right)^{\frac{1}{2}} - 1\right] + R\left(\frac{1}{\cos\theta} - 1\right)}{\tan\theta}$$

After substituting $R_t = 0.1425$ m, $A_c/A_t = 1.6$, $R = 0.2138$ m, and $\theta = 0.3491$ rad in the preceding equation, we find $L_{conv} = 0.1414$ m.

The approximate volume V_{conv} of the converging portion of the nozzle is

$$V_{conv} = \frac{1}{3} \times 3.1416 \times 0.1414 \times \left(0.1802^2 + 0.1802 \times 0.1425 + 0.1425^2\right) = 0.01162\,\text{m}^3$$

The volume V_{cyl} of the cylindrical combustion chamber, whose length goes from the internal face of the injector to the inlet plane of the nozzle, is

$$V_{cyl} = V_c - V_{conv} = 0.04213 - 0.01162 = 0.03051\,\text{m}^3$$

The length L_{cyl} of the cylindrical combustion chamber is

$$L_{cyl} = \frac{V_{cyl}}{1.6A_t} = \frac{0.03051}{1.6 \times 0.06379} = 0.2989\,\text{m}$$

The total distance L_{total} from the internal face of the injector to the plane of the throat is

$$L_{total} = L_{cyl} + L_{conv} = 0.2989 + 0.1414 = 0.4403\,\text{m}$$

For the diverging portion of the nozzle, we choose a fractional length $L_f = 75\%$ of a $\pi/12$ rad equivalent conical nozzle. Therefore, the true length L_n of the diverging portion of the nozzle is

$$L_n = \frac{n}{100} \frac{R_t}{\tan 0.2618}\left[\varepsilon^{\frac{1}{2}} - 1 + 1.5\left(\frac{1}{\cos 0.2618} - 1\right)\right]$$

After substituting $n = 75$, $R_t = 0.1425$ m and $\varepsilon \equiv A_e/A_t = 40$ in the preceding equation, we find $L_n = 2.145$ m.

We take the optimum values of the wall angles θ_n and θ_e from the Rao curves for $L_n = 75\%$ and $\varepsilon = 40$. These curves indicate $\theta_n = 0.5760$ rad and $\theta_e = 0.1614$ rad.

We also take $\alpha = \theta_n = 0.5760$ rad for the value of the angle α which subtends the circular arc TN of radius $0.382\,R_t$, where N is the inflexion point of the nozzle contour. The co-ordinates x and y of the points N and E in the system of reference defined above are

$$x_N = 0.382R_t \sin \alpha = 0.382 \times 0.1425 \times \sin 0.5760 = 0.02965\,\text{m}$$
$$y_N = R_t[1 + 0.382(1 - \cos \alpha)] = 0.1425$$
$$\times [1 + 0.382 \times (1 - \cos 0.5760)] = 0.1513\,\text{m}$$
$$x_E = L_n = 2.145\,\text{m}$$
$$y_E = R_E = \varepsilon^{\frac{1}{2}} R_t = 40^{\frac{1}{2}} \times 0.1425 = 0.9013\,\text{m}$$

A parabola is chosen for the segment of nozzle going from the inflexion point N to the point E laying on the exit plane of the nozzle. We determine the co-ordinates x and y of each point of this parabola by using again the analytical method suggested by Newlands [31]. The equation of the straight line NQ tangent in N to the parabola is

$$y - y_N = (x - x_N) \tan \theta_n$$

In the present case, the equation of NQ is $y = 0.6495\, x + 0.1320$.
The equation of the straight line EQ tangent in E to the parabola is

$$y - y_E = (x - x_E) \tan \theta_e$$

In the present case, the equation of EQ is $y = 0.1628\, x + 0.5521$.
The straight line NQ (whose equation is $y = 0.6495\, x + 0.1320$) intersects the straight line EQ (whose equation is $y = 0.1628\, x + 0.5521$) in the point Q, whose co-ordinates are easily found to be

$$x_Q = 0.8632 \text{ m}$$
$$y_Q = 0.6926 \text{ m}$$

Since the co-ordinates of the points N, Q, and E are known, then the parabolic segment of nozzle going from N to E can be expressed by means of a Bézier quadratic curve having the following parametric equations

$$x(t) = (1 - t)^2 x_N + 2(1 - t)t x_Q + t^2 x_E \quad (0 \le t \le 1)$$
$$y(t) = (1 - t)^2 y_N + 2(1 - t)t y_Q + t^2 y_E \quad (0 \le t \le 1)$$

In the present case, after substituting the values of (x_N, y_N), (x_Q, y_Q), and (x_E, y_E) computed above into the preceding equations, there results

$$x(t) = 0.02965(1 - t)^2 + 1.726(1 - t)t + 2.145t^2 \quad (0 \le t \le 1)$$
$$y(t) = 0.1513(1 - t)^2 + 1.385(1 - t)t + 0.9013t^2 \quad (0 \le t \le 1)$$

2.5 Cooling of a Thrust Chamber

The walls of the thrust chamber of a liquid-propellant rocket engine must be protected from the high temperatures of combustion, which could melt the materials or seriously damage the walls. This is because the hot gases contained in a thrust chamber reach high temperatures, and also transfer high heat fluxes to the walls.

As to the values of temperatures and heat fluxes reached in a liquid-propellant rocket engine, the opinions of the authors are not unanimous. According to Huzel and Huang [5], the combustion temperatures range from 2500 to 3600 K and the heat fluxes range from 817 to 82000 kW/m². According to Wieseneck [43], the combustion temperatures in oxygen-hydrogen rocket engines (such as J-2, J-2S, and M-1) range from 3600 to 4700 K and the heat fluxes range from 28000 to

57000 kW/m^2. Wieseneck also cite the case of the engines of the Space Shuttle, where the maximum value of design for the heat flux was 118000 kW/m^2. According to Sutton and Biblarz [2], the combustion temperatures are well above the melting points of the materials of which the walls are made, and the heat fluxes range from 500 to 160000 kW/m^2. According to Turner [44], a typical temperature is 3000 K, but the melting point of most metals is below 2000 K.

There are several methods for maintaining the temperatures of the walls at levels of safety. Van Huff and Fairchild [45] cite the following methods:

- regenerative cooling, obtained by forcing one or both of the propellants to flow into tubes or channels running longitudinally along the outer surface of the wall to be cooled, before being discharged into a special gas generator or directly into the combustion chamber;
- transpiration cooling, obtained by cooling a porous inner wall by means of a cooling fluid which is forced to flow through the porous material;
- film cooling, obtained by maintaining a thin layer of cooling fluid over the inner surface of the wall; and
- coating, obtained by depositing a layer of low-conductivity material which acts as a thermal barrier on the inner side of the wall.

Huzel and Huang [5] add the following methods:

- dump cooling, obtained by feeding a small percentage of propellant (for example, hydrogen, in a liquid hydrogen-liquid oxygen engine) through passages in the wall of the thrust chamber for cooling, and subsequently dumping it overboard through openings at the rear end of the nozzle skirt;
- ablative cooling, obtained by intentional loss of the inner wall of the thrust chamber, whose material (usually fibre-reinforced organic material) is melted or vaporised away, in order to dissipate heat and save the material of which the outer wall is made; and
- radiation cooling, obtained by radiating heat away from the surface of the outer wall of the thrust chamber.

The choice of one of the methods indicated above depends on the design of the thrust chamber. Huzel and Huang [5] suggest some factors to be considered for a choice. They are indicated below.

(a) Propellants used, because their combustion products have properties (such as temperature, specific heat ratio, density, viscosity, etc.) which determine the heat fluxes and therefore the cooling requirements.
(b) Pressure in the combustion chamber, whose value also determines the heat flow rate. In case of high pressure in the combustion chamber, combined regenerative and film cooling methods are frequently used.
(c) Propellant feed system, which determines the values of pressure in the combustion chamber. For example, in case of a rocket engine fed by turbo-pumps, a high amount of pressure is available for cooling. This amount can be used for regenerative cooling, in order to force the cooling propellant to flow into the

cooling tubes before reaching the injector. By contrast, a rocket engine fed by pressurised gas is limited by a low amount of pressure, and therefore works at a low pressure in the combustion chamber. In the latter case, film, ablative, or radiation cooling methods are to be used.

(d) Shape of the combustion chamber, which determines the local mass flow rates of combusted gas and the areas of the wall to be cooled.

(e) Materials used in the construction of a thrust chamber. This is because strength and heat conductivity at high temperatures are desirable properties in materials used for regeneratively cooled thrust chambers. In case of film cooled thrust chambers, materials allowing higher working temperatures are necessary for the purpose of reducing the heat flow rates and therefore the flow rates of the film coolant. The possibility of using radiative cooling depends on the availability of refractory alloys capable of resisting temperatures of 1922 K and more. Likewise, the possibility of using ablative cooling depends on the availability of reinforced plastic materials.

As a general rule, the choice of cooling method has a deep influence on the design of a thrust chamber, and vice versa.

An analysis is made below of the heat transfer from the combustion products contained in a thrust chamber and the walls of that chamber. Due to the high velocity of the hot gases, the heat transfer occurs through convection, and therefore the heat propagates because of the motion of gaseous masses in the thrust chamber. The heat passes from the moving gas firstly to the stagnant boundary layer along the wall, and then to the wall itself.

Let q (measured in W/m^2) be the quantity of heat per unit time per unit surface transferred across the boundary layer. Let h_g (W m^{-2} K^{-1}) be the heat transfer coefficient on the hot gas side. Let T_{aw} (K) be the temperature of the hot gas at the adiabatic wall. The temperature T_{aw} is assumed to be equal to the total (or stagnation) temperature in the thrust chamber multiplied by a recovery factor (whose value ranges from 0.90 to 0.98) of the turbulent boundary layer. Let T_{wg} (K) be the local temperature of the wall on the hot gas side. The heat transfer through convection is governed by the following equation

$$q = h_g\left(T_{aw} - T_{wg}\right)$$

The value of the convective heat transfer coefficient h_g can be determined by using the Bartz equation [46, page 30, Eq. 50], as follows

$$h_g = \left[\frac{0.026}{D_t^{0.2}}\left(\frac{\mu^{0.2}c_p}{Pr^{0.6}}\right)_{ns}\left(\frac{(p_c)_{ns}}{\bar{c}^*}\right)^{0.8}\left(\frac{D_t}{\bar{R}}\right)^{0.1}\right]\left(\frac{A_t}{A}\right)^{0.9}\sigma$$

where D_t (m) is the diameter of the nozzle in the throat plane, μ (N s m^{-2}) is the coefficient of dynamic viscosity of the gas, c_p (J kg^{-1} K^{-1}) is the specific heat of the gas at constant pressure, $Pr = \mu c_p/k$ is the Prandtl number, k (W m^{-1} K^{-1}) is the thermal conductivity of the gas, $(p_c)_{ns}$ (N/m^2) is the total pressure of the combustion chamber at the inlet plane of the nozzle, $\bar{c}^*$ (m/s) is the design value of the characteristic velocity, $\overline{R}$ (m) is the mean radius of curvature of the throat in a plane which contains the axis of symmetry of the nozzle, A_t (m^2) is the area of the cross-section of the nozzle in the throat plane, and A (m^2) is the area of the cross-section under consideration along the axis of symmetry of the nozzle.

The dimensionless factor σ of correction is [46, page 30, Eq. 49]:

$$\sigma = \frac{1}{\left[\frac{1}{2}\frac{T_{wg}}{(T_c)_{ns}}\left(1 + \frac{\gamma-1}{2}M^2\right) + \frac{1}{2}\right]^{0.8-0.2\omega}\left[1 + \frac{\gamma-1}{2}M^2\right]^{0.2\omega}}$$

where $(T_c)_{ns} = T\,[1 + \frac{1}{2}(\gamma - 1)M^2]$ is the total temperature (K) of the combustion chamber at the nozzle inlet, M is the local Mach number, $\gamma \equiv c_p/c_v$ is the specific heat ratio of the combusted gas, T_{wg} (K) is the local temperature of the wall on the hot gas side, and ω is the exponent of the temperature dependence of viscosity given below. The value of ω is equal to 0.6 for diatomic gases [47].

According to Wang et al. [48, page 911, Eqs. 4 and 5], in case of data on Pr and μ (N s m^{-2}) not being available for particular mixtures of combusted gas, it is possible to determine approximate values by means of the following equations

$$Pr = \frac{4\gamma}{9\gamma - 5}$$

$$\mu = \kappa T^\omega = \left(1.184 \times 10^{-7}\right)\mathcal{M}^{0.5}T^{0.6}$$

where $\mathcal{M}$ (kg/kmol) is the molar mass, T (K) is the temperature of the mixture of combusted gas, and the exponent ω of T has been taken equal to 0.6.

The preceding equations give approximate values of the convective heat transfer coefficient h_g on the hot gas side. The calculated value of h_g may be lower than the real value, because a substantial part of combusted gases may transfer heat through radiation, or because a substantial part of gaseous molecules dissociate and then recombine near the wall of the thrust chamber, or because the gas flow is unstable. Conversely, the calculated value of h_g may be higher than the real value, because the chemical reactions occurring in the combustion process may be incomplete in the combustion chamber, or because the combusted gases may deposit solid particles, which in turn create insulating layers on the wall of the chamber. These solid particles are made of carbon, in case of a combination of liquid oxygen with RP-1 (kerosene). This carbon layer increases the thermal insulation of the wall on the hot gas side.

In order to take account of the solid deposit of carbon on the chamber walls in the computation of the heat transfer, the following equation may be used

$$q = h_{gc}\left(T_{aw} - T_{wg}\right)$$

where h_{gc} (W m^{-2} K^{-1}), which is the overall heat transfer coefficient on the hot gas side, has the following expression

$$h_{gc} = \frac{1}{\frac{1}{h_g} + R_d}$$

where R_d (m^2 K W^{-1}) is the thermal resistance caused by the solid deposit. This thermal resistance vanishes ($R_d = 0$) in the absence of deposit.

As an application, taken from [5], we want to compute the approximate value of the overall heat transfer coefficient h_{gc} on the hot gas side in the combustion chamber, at the throat plane, and at the exit plane, for a nozzle whose area expansion ratio is $\varepsilon \equiv A_e/A_t = 5$, for a regeneratively cooled thrust chamber of a rocket engine which burns a combination of liquid oxygen with RP-1 (kerosene), and has the data given in Sect. 2.3. These data are also given below for convenience: the oxidiser-to-fuel mixture ratio at the thrust chamber is $o/f = 2.35$, and the total absolute pressure in the combustion chamber at the nozzle inlet is $(p_c)_{ns} = 6.895 \times 10^6$ N/m^2. The chemical reactions in the combustion chamber are assumed to be homogeneous and complete.

As has been shown in Sect. 2.3, the total temperature of the combustion chamber at the nozzle inlet is $(T_c)_{ns} = 3589$ K, the molar mass of the combusted gas is $\mathcal{M} = 22.5$ kg/kmol, and the specific heat ratio of the combusted gas is $\gamma \equiv c_p/c_v = 1.222$.

The design value $(\overline{T}_c)_{ns}$ of the total temperature of the combustion chamber at the nozzle inlet is equal to the theoretical value $(T_c)_{ns}$ given above multiplied by the square of the correction factor η_{c*} of the characteristic velocity c^*. Since we have taken $\eta_{c*} = 0.975$ in Sect. 2.3, then

$$(\overline{T}_c)_{ns} = (T_c)_{ns}\eta_{c*}^2 = 3589 \times 0.975^2 = 3412 \text{ K}$$

As has been found in Sect. 2.3, the design value of the characteristic velocity is

$$\overline{c}^* = \eta_{c*} c^* = 0.975 \times 1764 = 1720 \text{ m/s}$$

As has been found in Sect. 2.4, the diameter of the cross section of the nozzle at the throat plane is

$$D_t = 2R_t = 2 \times 0.3166 \text{ m} = 0.6332 \text{ m}$$

The radius of curvature R_{us} of the nozzle contour upstream of the throat plane is

$$R_{us} = 1.5R_t = 1.5 \times 0.3166 \text{ m} = 0.4749 \text{ m}$$

The radius of curvature R_{ds} of the nozzle contour downstream of the throat plane is

$$R_{ds} = 0.382R_t = 0.382 \times 0.3166 \text{ m} = 0.1209 \text{ m}$$

The mean radius $\overline{R}$ of curvature of the nozzle contour at the throat is

$$\overline{R} = \frac{1}{2}(R_{us} + R_{ds}) = 0.5 \times (0.4749 + 0.1209) = 0.2979\,\text{m}$$

In Sect. 2.3, the value of the constant R of the specific gas has been found to be

$$R = \frac{R^*}{\mathcal{M}} = \frac{8314.460}{22.5} = 369.5\,\text{J K}^{-1}\,\text{kg}^{-1}$$

Since $\gamma \equiv c_p/c_v = 1.222$ and $c_p - c_v = R = 369.5\,\text{J K}^{-1}\,\text{kg}^{-1}$, then

$$c_p = \frac{\gamma}{\gamma - 1} R = \frac{1.222}{1.222 - 1} \times 369.5 = 2034\,\text{J K}^{-1}\,\text{kg}^{-1}$$

The Prandtl number is computed as follows

$$Pr = \frac{4\gamma}{9\gamma - 5} = \frac{4 \times 1.222}{9 \times 1.222 - 5} = 0.8149$$

The coefficient of dynamic viscosity of the gas is computed as follows

$$\mu = (1.184 \times 10^{-7})\mathcal{M}^{0.5}T^{0.6} = (1.184 \times 10^{-7}) \times (22.5)^{0.5} \times (3412)^{0.6}$$
$$= 7.400 \times 10^{-5}\,\text{N s m}^{-2}$$

By using the Bartz equation, the convective heat transfer coefficient is computed as follows

$$h_g = \left[\frac{0.026}{D_t^{0.2}} \left(\frac{\mu^{0.2}c_p}{Pr^{0.6}} \right)_{ns} \left(\frac{(p_c)_{ns}}{\bar{c}^*} \right)^{0.8} \left(\frac{D_t}{\overline{R}} \right)^{0.1} \right] \left(\frac{A_t}{A} \right)^{0.9} \sigma$$

After substituting $D_t = 0.6332$ m, $\mu = 7.400 \times 10^{-5}$ N s m^{-2}, $c_p = 2034$ J K^{-1} kg^{-1}, $Pr = 0.8149$, $(p_c)_{ns} = 6.895 \times 10^6$ N m^{-2}, $\bar{c}^* = 1720$ m s^{-1}, and $\overline{R} = 0.2979$ m in the preceding equation, we find

$$h_g = 8042 \left(\frac{A_t}{A} \right)^{0.9} \sigma \quad \text{W m}^{-2}\text{K}^{-1}$$

In order to determine the value of the correction factor σ at the exit plane of the nozzle, we compute the static pressure p_e of the combusted gas at the exit plane, as will be shown below. The static pressure p_e at the exit section of area $A_e = 5A_t$ results from the following equation

$$\frac{A_e}{A_t} = \frac{\left(\frac{2}{\gamma+1}\right)^{\frac{1}{\gamma-1}} \left(\frac{(p_c)_{ns}}{p_e}\right)^{\frac{1}{\gamma}}}{\left\{\frac{\gamma+1}{\gamma-1}\left[1 - \left(\frac{p_e}{(p_c)_{ns}}\right)^{\frac{\gamma-1}{\gamma}}\right]\right\}^{\frac{1}{2}}}$$

The unknown value of p_e can be computed numerically. For this purpose, we define $z \equiv p_e/(p_c)_{ns}$, where z is an auxiliary variable. In the present case ($A_e/A_t = 5$ and $\gamma = 1.222$), we define a function $f(z)$ such that

$$f(z) \equiv 5^2 - \frac{\left(\frac{2}{1.222+1}\right)^{\frac{2}{1.222-1}}\left(\frac{1}{z}\right)^{\frac{2}{1.222}}}{\frac{1.222+1}{1.222-1}\left(1 - z^{\frac{1.222-1}{1.222}}\right)}$$

We search the unknown value of z in the interval $0.025 \le z \le 0.035$, because the value of the function $f(z)$ changes sign in this interval.

By applying repeatedly Müller's method (see Chap. 1, Sect. 1.2), we find, with four significant figures, $z = 0.03044$. Therefore, remembering the definition of the auxiliary variable z, we have

$$p_e = z(p_c)_{ns} = 0.03044 \times 6.895 \times 10^6 = 2.099 \times 10^5 \, \text{N/m}^2$$

Now, we compute the flow temperature at the exit plane of the nozzle (T_e). In case of an isentropic flow ($p/\rho^\gamma = \text{constant}$), the energy equation, written for any two points 1 and 2 placed along the axis of symmetry of the nozzle, yields

$$\frac{T_1}{T_2} = \left(\frac{p_1}{p_2}\right)^{\frac{\gamma-1}{\gamma}}$$

Since $(\overline{T}_c)_{ns} = 3412 \, \text{K}$ and $(p_c)_{ns} = 6.895 \times 10^6 \, \text{N/m}^2$ are respectively the total temperature and the total pressure of the combustion chamber at the inlet plane of the nozzle, then the preceding equation, solved for the flow temperature T_e at the exit plane of the nozzle, yields

$$T_e = (\overline{T}_c)_{ns}\left[\frac{p_e}{(p_c)_{ns}}\right]^{\frac{\gamma-1}{\gamma}} = 3412 \times \left(\frac{2.099 \times 10^5}{6.895 \times 10^6}\right)^{\frac{1.222-1}{1.222}} = 1809 \, \text{K}$$

Now we compute the flow velocity v_e at the exit plane of the nozzle. The flow velocity at the exit plane of the nozzle, where the static pressure of the combusted gas is $p_e = 2.099 \times 10^5 \, \text{N/m}^2$, is expressed by the following equation

$$v_e = \left\{\frac{2\gamma}{\gamma-1} R(\overline{T}_c)_{ns}\left[1 - \left(\frac{p_e}{(p_c)_{ns}}\right)^{\frac{\gamma-1}{\gamma}}\right]\right\}^{\frac{1}{2}}$$

After substituting $p_e = 2.099 \times 10^5$ N/m^2 into the preceding equation, there results

$$v_e = \left\{ \frac{2 \times 1.222}{1.222 - 1} \times 369.5 \times 3412 \times \left[1 - \left(\frac{2.099 \times 10^5}{6.895 \times 10^6} \right)^{\frac{1.222-1}{1.222}} \right] \right\}^{\frac{1}{2}} = 2553 \text{ m/s}$$

Finally, we compute the Mach number M_e of the gas flow at the exit plane of the nozzle. The sonic velocity a_e of the gas flow at the exit plane of the nozzle is

$$a_e = (\gamma R T_e)^{\frac{1}{2}} = (1.222 \times 369.5 \times 1809)^{\frac{1}{2}} = 903.8 \text{ m/s}$$

Therefore, the Mach number of the gas flow at the exit plane of the nozzle is

$$M_e = \frac{v_e}{a_e} = \frac{2553}{903.8} = 2.825$$

Since the temperature T_{wg} of the carbon deposit approaches the temperature $(T_c)_{ns}$ of the combusted gas, then we take 0.8 as the value of the ratio $T_{wg}/(T_c)_{ns}$. We have also taken 0.4 as the value of the Mach number M at the inlet plane of the nozzle. After introducing these values in the following equation

$$\sigma = \frac{1}{\left[\frac{1}{2} \frac{T_{wg}}{(T_c)_{ns}} \left(1 + \frac{\gamma-1}{2} M^2 \right) + \frac{1}{2} \right]^{0.68} \left[1 + \frac{\gamma-1}{2} M^2 \right]^{0.12}}$$

where the exponent ω has been set equal to 0.6, we find $\sigma = 1.066$ at the inlet plane of the nozzle.

We take 1.6 (see Sect. 2.4) for the value of the contraction area ratio A_c/A_t of the nozzle, and find at the inlet plane of the nozzle

$$\left(\frac{A_t}{A} \right)^{0.9} = \left(\frac{1}{1.6} \right)^{0.9} = 0.6551$$

Therefore, the heat transfer coefficient at the inlet plane of the nozzle is

$$h_g = 8042 \left(\frac{A_t}{A} \right)^{0.9} \sigma = 8042 \times 0.6551 \times 1.066 = 5016 \text{ W m}^{-2}\text{K}^{-1}$$

At the throat plane of the nozzle, there results $M = 1$, $(A_t/A)^{0.9} = 1$, and $\sigma = 1.027$. Therefore, the heat transfer coefficient at the throat plane is

$$h_g = 8042 \times 1 \times 1.027 = 8259 \text{ W m}^{-2} \text{ K}^{-1}$$

At the exit plane of the nozzle, there results

$$\left(\frac{A_t}{A}\right)^{0.9} = \left(\frac{1}{5}\right)^{0.9} = 0.2349$$

and the Mach number has been found to be $M_e = 2.825$. After introducing this value in the following equation

$$\sigma = \frac{1}{\left[\frac{1}{2}\frac{T_{wg}}{(T_c)_{ns}}\left(1 + \frac{\gamma-1}{2}M^2\right) + \frac{1}{2}\right]^{0.68}\left[1 + \frac{\gamma-1}{2}M^2\right]^{0.12}}$$

we find $\sigma = 0.7944$.

Therefore, the heat transfer coefficient at the exit plane is

$$h_g = 8042 \times 0.2349 \times 0.7944 = 1501 \, \text{W m}^{-2} \, \text{K}^{-1}$$

Experimental data have been given graphically by NASA [5] on the thermal resistance R_d of carbon deposit on thrust chamber walls, for a rocket engine which burns liquid oxygen and RP-1, at a mixture ratio $o/f = 2.35$, and at a total absolute pressure at the nozzle inlet plane $(p_c)_{ns} = 6.895 \times 10^6 \, \text{N/m}^2$, as is the case with the present calculation. The same data can also be expressed numerically by using the following equations.

(1) For the portion of nozzle going from the throat plane to the inlet plane, where the contraction area ratio ε_c is equal to 1.6:

$$R_d = 3.397 \times 10^{-7} \times \left[\exp\left(8.079 - \frac{1.053}{\varepsilon_c}\right)\right]$$

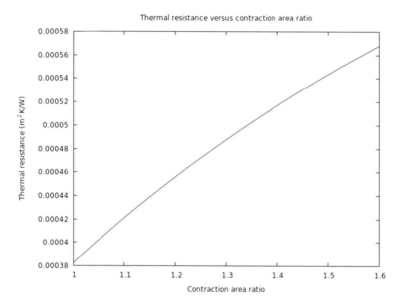

(2) For the portion of nozzle going from the throat plane to the exit plane, where
the expansion area ratio ε is equal to 5.0:

$$R_d = 3.397 \times 10^{-7} \times \left[\exp\left(7.5 - \frac{0.4749}{\varepsilon} \right) \right]$$

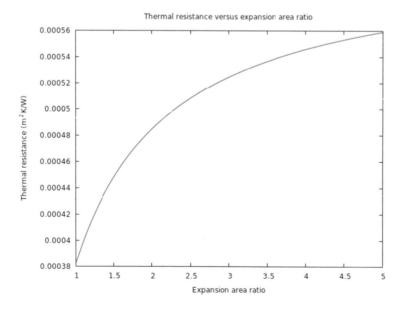

By substituting $\varepsilon_c = 1.6$, $\varepsilon = 1.0$, and $\varepsilon = 5.0$ in the two equations written above, the resulting values of thermal resistance R_d are

$$R_d = 0.0005675 \text{ m}^2\text{K/W} \quad \text{at the inlet plane}$$
$$R_d = 0.0003823 \text{ m}^2\text{K/W} \quad \text{at the throat plane}$$
$$R_d = 0.0005585 \text{ m}^2\text{K/W} \quad \text{at the exit plane}$$

The values of the overall heat transfer coefficient h_{gc} on the hot gas side result from substituting the values of h_g and of R_d determined above in the following equation

$$h_{gc} = \frac{1}{\frac{1}{h_g} + R_d}$$

By so doing, we find

$$h_{gc} = 1341 \text{ W m}^{-2}\text{K}^{-1} \quad \text{at the inlet plane}$$
$$h_{gc} = 1987 \text{ W m}^{-2}\text{K}^{-1} \quad \text{at the throat plane}$$
$$h_{gc} = 816.5 \text{ W m}^{-2}\text{K}^{-1} \quad \text{at the exit plane}$$

We describe first the regenerative cooling method, because it is the most widely used of all cooling methods, due to its advantages in terms of reliability, durability, and high performance. The following figure, due to the courtesy of NASA [43], illustrates a regeneratively cooled thrust chamber for a bell nozzle.

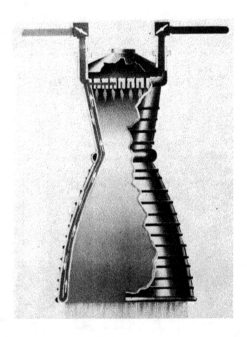

The cooling tubes or channels run longitudinally along the wall and carry cryogenic fuel, which is liquid hydrogen for the main engines of the Space Shuttle. The
cryogenic fuel coming from the fuel pump passes through the fuel valve, and then
flows downward, that is, toward the exit plane of the nozzle, as indicated by the
white arrows in the preceding figure. At this plane, the fuel reverses the direction
of its motion, and therefore flows upward in a parallel tube, is mixed with liquid
oxygen at the top, and is then ignited inside the combustion chamber. The cryogenic
fuel removes, by heat convection, the heat due to the burned gas in the combustion
chamber and in the nozzle.

The following figure, due to the courtesy of NASA [49], illustrates the regeneratively cooled plug and combustion chamber for an aerospike engine.

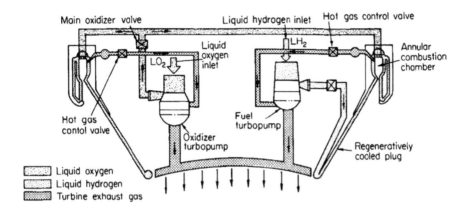

Usually, the fuel is used as the cooling fluid rather than the oxidiser, because of its
higher heat capacity, that is, because of its higher capability of removing heat from
the nozzle without vaporising. In addition, the fuel, when heated, requires a smaller
supply of energy, acquired through ignition, in order to be burned in the presence of
the oxidiser. The heat removal for the Saturn V F-1 engines was slightly different
from that described above, because these engines burned kerosene, which is not a
cryogenic fuel, but is still capable of cooling a nozzle. The kerosene came down from
the top to the bottom in one tube, turned around, and came back up in the parallel
tube.

The following figure, also due to the courtesy of NASA [43], illustrates a cross-
section of the wall of a thrust chamber, having tubes (left) or rectangular channels
(right).

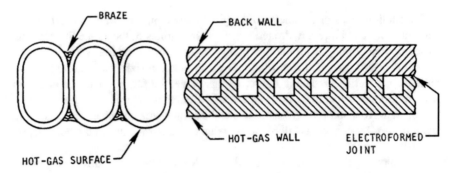

Additional information (taken from [50]) is given below on the regenerative cooling system used for each of the three liquid hydrogen-liquid oxygen main engines of the Space Shuttle. The same method has also been successfully used for the Thor, Jupiter, Atlas, H-1, J-2, F-1, RS-27, and several other US Air Force and NASA rocket engines.

The main combustion chamber of the RS-25 engine of the Space Shuttle is shown in the following figure, due to the courtesy of NASA [51].

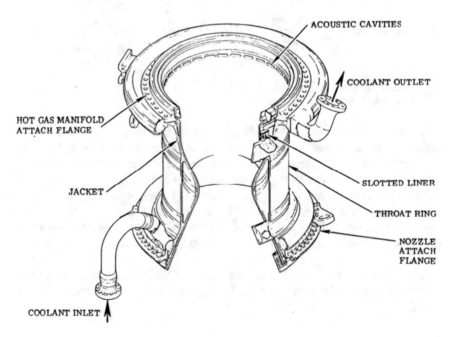

This combustion chamber includes a liner, a jacket, a throat ring, a coolant inlet manifold, and a coolant outlet manifold. The outer surface of the liner has 430 milled slots, which are closed out by electro-deposited nickel. The jacket halves are placed around the liner and welded. The coolant manifolds are welded to the jacket and the liner. The throat ring is welded to the jacket to add strength to the main combustion chamber. This creates a regeneratively-cooled combustion chamber, in which the cooling fuel makes a single up-pass through the milled slots of the liner. The liner is made of NARloy Z (North American Rockwell alloy Z), which is mostly copper, with silver and zirconium added.

The nozzle of the same engine is shown in the following figure, also due to the courtesy of NASA [51]. This nozzle consists of 1800 stainless steel tubes brazed to themselves and to a structural jacket. Nine hatbands are welded around the jacket for hoop strength. Coolant manifolds are welded to the top and to the bottom of the nozzle, along with three fuel transfer ducts and six drain lines.

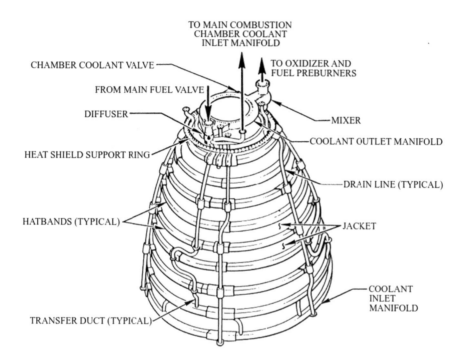

This nozzle is cooled by the fuel, which enters the diffuser and splits to flow to the main combustion chamber, to the three fuel transfer ducts, and through the chamber coolant valve to the mixer. Fuel flowing through each transfer duct splits at each steer-horn to enter the nozzle coolant inlet manifolds at six points.

The fuel then makes a single up-pass through the 1080 tubes to the outlet manifold, and then to the mixer, to join the bypass flow from the chamber coolant valve. The flow recirculation inhibitor is a porous rope-like barrier, which prevents a recirculating flow of hot exhaust gas from reaching and damaging the bellows seal located at the joint between the main combustion chamber and the nozzle. This flow recirculation inhibitor is a sleeve of braided Nextel® 321 filled with Saffil® batting. Both materials, which are composed of silica-glass ceramic fibre, can withstand working temperatures of up to 1700 K. The materials provide the required resistance to flow.

By the way, the thermal protection system used in the nozzle is an insulation system which protects the nozzle from high temperatures during launch and re-entry. During the ascent phase, the three nozzles are subject to plume radiation and convection from the three main engines and to plume radiation from the solid rocket boosters. During the re-entry phase, two of the nozzles are exposed to high heat loads due to convective aerodynamic heating.

In the following part of this paragraph, we consider in further depth the heat transfer through convection between the hot gas which moves on one side of the thrust chamber and the coolant fluid which moves on the other side. This heat passes through a series of contiguous layers, including the boundary layer along the wall on the side of the combusted gas, the thickness of the wall, and the boundary layer along the wall on the side of the coolant fluid, as shown in the following figure. Let h_{gc} (W m^{-2} K^{-1}) be the heat transfer coefficient on the hot gas side, h_c (W m^{-2} K^{-1}) be the heat coefficient on the coolant side, q (W/m^2) be the heat flux, T_{aw}, T_{wg}, T_{wc}, and T_{co} (K) be the temperatures of respectively the hot gas at the adiabatic wall, the wall on the hot gas side, the wall on the coolant side, and the coolant, k (W m^{-1} K^{-1}) be the thermal conductivity of the wall, t (m) be the thickness of the wall, and H (W m^{-2} K^{-1}) be the overall heat transfer coefficient. In the absence of deposits on the hot gas side, h_{gc} and h_g have the same value.

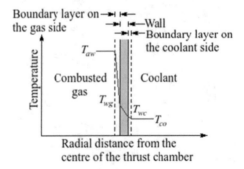

In the steady state, the heat transfer equations can be expressed as follows

$$q = h_{gc}\left(T_{aw} - T_{wg}\right) = \frac{k}{t}\left(T_{wg} - T_{wc}\right) = h_c(T_{wc} - T_{co}) = H(T_{aw} - T_{co})$$

where the overall heat transfer coefficient H results from

$$H = \left(\frac{1}{h_{gc}} + \frac{t}{k} + \frac{1}{h_c}\right)^{-1}$$

The coolant increases its temperature from its point of entry into to its point of exit from the cooling channel, depending on the heat absorbed and its rate of flow. The metals commonly used for the walls of combustion chambers are stainless steel, nickel, and Inconel®, the last of them being a high-performance austenitic nickel-chromium-based alloy. According to Huzel and Huang [5], the temperature of the wall on the hot gas side is about 1089–1255 K, and the difference of temperature between the combusted gas and the wall ranges from 1644 to 3589 K. The equations written above show that the heat flux q is the same for the three layers and depends on the temperatures and on the overall heat transfer coefficient H. The value of H, in turn, depends on the individual heat transfer coefficients relating to the boundary layers and to the wall.

Since $q = H(T_{aw} - T_{co})$, then a small value of H implies a small value of q. It is a primary objective for a designer to keep the value of h_{gc} lower than the values of h_c and t/k. By so doing, there will be a higher temperature drop $T_{aw} - T_{wg}$ in the boundary layer on the hot gas side, as shown in the preceding figure, than the temperature drops ($T_{wg} - T_{wc}$ and $T_{wc} - T_{co}$) in, respectively, the wall and the boundary layer on the coolant side.

The value of the heat transfer coefficient h_c on the coolant side depends on several factors, such as possible chemical reactions of dissociation occurring in the coolant, its pressure, and its bulk temperature. It also depends on a possible formation of bubbles due to a coolant boiling near the wall. In order to achieve a good heat-absorbing capacity of the coolant, the pressure and the velocity of the coolant in its flow are selected so that a boiling is permitted locally, but the bulk of the coolant does not reach the boiling condition [2]. In the absence of boiling, at subcritical temperatures, and in transitional or turbulent regime (in the interval $3000 \leq Re \leq 10^6$), the relation between the wall temperature and the heat flux, which depends on the heat transfer coefficient h_c, can be expressed approximately by using one of the following three equations due to Taler [52, page 4, Eqs. 18, 19, and 20]. The value of the Prandtl number indicates which of these equations is to be used.

$$Nu = 0.02155\, Re^{0.8018}\, Pr^{0.7095}\,(0.1 \leqslant Pr \leqslant 1)$$

$$Nu = 0.01253\, Re^{0.8413}\, Pr^{0.6179}\,(1 < Pr \leqslant 3)$$

$$Nu = 0.00881\, Re^{0.8991}\, Pr^{0.3911}\,(3 < Pr \leqslant 1000)$$

where $Nu = h_c d/k$ is the Nusselt number, h_c (W m^{-2} K^{-1}) is the heat transfer coefficient on the coolant side, d (m) is the hydraulic diameter (defined below) of each coolant duct at the section of interest, k (W m^{-1} K^{-1}) is the thermal conductivity of the coolant at the bulk temperature T_{co}, $Re = \rho v d/\mu$ is the Reynolds number, ρ (kg/m^3) is the density of the coolant at the bulk temperature T_{co}, v (m/s) is the velocity of the coolant, μ (Ns/m^2) is the coefficient of dynamic viscosity of the coolant at the bulk temperature T_{co}, $Pr = \mu c_p/k$ is the Prandtl number, and c_p (J kg^{-1} K^{-1}) is the specific heat of the coolant at constant pressure at the bulk temperature T_{co}.

In the conditions specified above and in turbulent or transitional regime (in the interval $3000 \leq Re \leq 5 \times 10^6$), it is also possible to use the following relation due to Gnielinski [53, page 11, Eq. 11]:

$$Nu = \frac{0.125 f_D (Re - 1000) Pr}{1 + 12.7(0.125 f_D)^{\frac{1}{2}} (Pr^{\frac{2}{3}} - 1)}$$

where Nu, Re, and Pr are defined above, and f_D is the Darcy friction factor (see below) of the cooling tubes. Some authors (see, for example, [54], Table 12, and [55], Table 8) have proposed other equations, which also express the relation between the wall temperature and the heat flux.

By hydraulic diameter of a duct, we mean the quantity $d = 4A/P$, where A (m^2) is the cross-sectional area of the duct, and P (m) is the perimeter of the wetted portion of its cross section.

When the heat is transferred through a vapour-film boundary layer, for example, when the coolant is hydrogen in supercritical conditions of pressure and temperature, then the value of h_c (W m^{-2} K^{-1}) can be estimated by means of the following correlation due to McCarthy and Wolf [56, page 95, Eq. 2]:

$$h_c = 0.025 \left(\frac{c_p \mu^{0.2}}{Pr^{0.6}} \right)_{co} \frac{G^{0.8}}{d^{0.2}} \left(\frac{T_{co}}{T_{wc}} \right)^{0.55}$$

where c_p (J kg^{-1} K^{-1}) is the isobaric specific heat of the coolant, μ (N s m^{-2}) is the coefficient of dynamic viscosity of the coolant, Pr is the Prandtl number defined above, G (kg s^{-1} m^{-2}) is the mass flow rate of the coolant per unit area to be cooled, d (m) is the diameter of the coolant passage, T_{co} (K) is the bulk temperature of the coolant, T_{wc} (K) is the temperature of the wall on the coolant side, and the subscript co indicates the bulk temperature of the coolant.

In the conditions specified above, the value of h_c (W m^{-2} K^{-1}) can also be estimated by using the following correlation due to Taylor [56, page 95, Eq. 8]:

$$h_c = 0.023 \left(\frac{c_p \mu^{0.2}}{Pr^{0.6}} \right)_{co} \frac{G^{0.8}}{d^{0.2}} \left(\frac{T_{co}}{T_{wc}} \right)^{0.57 - \frac{1.59d}{x}}$$

where d (m) is the inner diameter of each cooling tube, and x (m) is the axial distance downstream of the section of entrance of the coolant.

As mentioned above, the bulk temperature of the coolant should be kept below the critical value, because the value of the heat transfer coefficient of the vapour-film boundary layer would be too low to cool effectively the wall.

The cooling capacity Q_c (W) of a liquid coolant used in a regenerative cooling system can be estimated by means of the following equation [5]:

$$Q_c = \dot{m}_c c_p (T_{cc} - T_{ci})$$

where $\dot{m}_c$ (kg/s) is the mass flow rate of the coolant, c_p (J kg^{-1} K^{-1}) is the specific heat of the coolant at constant pressure, T_{cc} (K) is the critical temperature of the coolant, and T_{ci} (K) is the temperature of the coolant at the inlet. The allowed value of the total heat transfer rate Q (W) from the hot gas to the wall must be less than the cooling capacity Q_c (that is, $Q < Q_c$) of the coolant by a margin imposed by safety. This limitation does not affect hydrogen when used as a coolant fluid, because of its high transfer coefficient even in supercritical conditions of pressure and temperature. Hydrogen enters the coolant passage of the thrust chamber at supercritical pressure and reaches its supercritical temperature at a short distance from the inlet section. The cross-section areas of the coolant passage at various points along the wall of the thrust chamber must be designed to maintain the coolant velocity imposed by the heat transfer coefficient determined by the calculation. Possible design choices for the cooling jacket include longitudinal tubes (for engines whose thrust is equal to or greater than 13,400 N) and coaxial shells separated by helical ribs or wires (for engines of smaller thrust). In the latter case, the coolant passage is the rectangular area limited by the inner shell, the outer shell, and two adjacent ribs. The ribs are wrapped helically around the inner shell.

In case of thrust chambers of tubular shape, such as those described in Sect. 2.4, the number of tubes for the coolant fluid depends on factors such as the size of the thrust chamber, the mass flow rate of the coolant for unit area of the tubes, the maximum allowable stress for the material of which the tubes are made, and manufacturing considerations. The region of a thrust chamber which needs the maximum cooling is the throat, because there the heat flux q is also maximum. Therefore, the cooling requirements at the throat determine the number of cooling tubes for a given flow rate of the coolant used. The cross section of the cooling tubes is often circular, for easiness of manufacturing and lower stress. The mechanical and thermal stresses acting on the cooling tubes are induced by the pressure exerted by the coolant, and by the difference of temperature between the tubes and the wall. In addition, since two adjacent tubes may be subject to different pressures, then distorting stresses may also arise. As has been shown above, the region of a thrust chamber subject to the maximum stress is the throat.

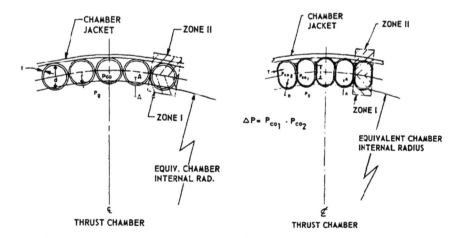

In order to evaluate the maximum combined tensile stress σ_t (N/m^2) acting on the cross section A-A of a circular cooling tube, as shown on the left-hand side of the preceding figure, due to the courtesy of NASA [5], Huzel and Huang [5] indicate the following formula

$$\sigma_t = \frac{(p_{co} - p_g)r}{t} + \frac{E\lambda q t}{2(1 - v)k} + \frac{6M_A}{t^2}$$

where p_{co} (N/m^2) is the pressure of the coolant, p_g (N/m^2) is the pressure of the combusted gas, r (m) is the radius of the cross section of the cooling tubes, t (m) is the thickness of the cooling tubes, E (N/m^2) is the modulus of elasticity of the material of which the cooling tubes are made, λ (m m^{-1} K^{-1}) is the coefficient of thermal expansion of the same material, q (W/m^2) is the quantity of heat per unit time per unit surface, v is the Poisson ratio of the same material, k (W m^{-1} K^{-1}) is the thermal conductivity of the same material, and M_A (Nm/m) is the bending moment per unit length acting at the section A-A due to the distortion induced by difference of pressure between adjacent cooling tubes or by other effects, such as discontinuities. In case of tubes of circular cross section, the bending moment per unit length M_A is caused only by discontinuity, because there is no effect of difference of pressure between adjacent tubes [5].

With reference to the preceding figure, the mean temperature in the zone I (which is on the side of the combusted gas) of each cooling tube is much higher than the mean temperature in the zone II (which is on the side of the outer shell). Therefore, the thermal expansion of the cooling tube in zone I is restrained by the low temperature in zone II. Since the mass in zone II is greater than the mass in zone I, then thermal inelastic buckling can arise in zone I in the longitudinal direction. The thermal stress σ (N/m^2) in the longitudinal direction and the critical stress σ_c (N/m^2) relating to the longitudinal inelastic buckling may be evaluated by using the following equations [5]:

$$\sigma = E\lambda\Delta T$$

$$\sigma_c = \frac{4E_t E_c t}{(E_t^{\frac{1}{2}} E_c^{\frac{1}{2}})^2 [3(1 - v^2)]^{\frac{1}{2}} r}$$

where ΔT (K) is the mean difference of temperature between zone I and zone II, E_t (N/m^2) is the tangential modulus of elasticity at the wall temperature, and E_c (N/m^2) is the tangential modulus of elasticity from the compression stress-strain curve at the wall temperature. The thermal stress σ should not be higher than $0.9\sigma_c$. The preceding figure also shows, on the right-hand side, cooling tubes of elongated cross section. The equations written above can also be applied for the purpose of computing the stresses in such tubes. The maximum combined stress acts in the section A-A. The bending moment per unit length M'_A (Nm/m) acting at this section should be computed by taking into account not only the effect of discontinuities, but also the difference of pressure between adjacent tubes, as follows

$$M'_A = M_A + K_A \frac{\ell \Delta p}{2}$$

where M_A (Nm/m) is the bending moment per unit length acting on the section A-A due only to discontinuity, K_A is a dimensionless constant whose value is in the range 0.3–0.5, ℓ (m) is the length of the flat portion on the cross-section of each tube, and Δp (N/m^2) is the difference of pressure between adjacent tubes. The value of M'_A, determined as specified above, is to be inserted (instead of M_A) in the preceding equation

$$\sigma_t = \frac{(p_{co} - p_g)r}{t} + \frac{E\lambda qt}{2(1-\nu)k} + \frac{6M_A}{t^2}$$

in order to evaluate the maximum combined tensile stress σ_t (N/m^2) acting on the cross section A-A of an elongated tube. The loads due to the pressure acting on a regeneratively cooled thrust chamber of tubular shape are borne by the chamber jacket or by tension bands wrapped around the chamber.

The following figure, due to the courtesy of NASA [57] shows a regeneratively cooled thrust chamber of the coaxial shell type, relating to an experimental rocket developing a thrust of 22,000 N.

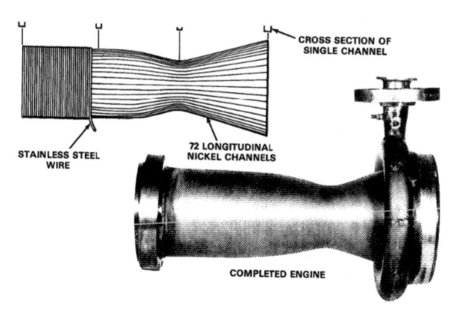

CROSS SECTION OF SINGLE CHANNEL

STAINLESS STEEL WIRE

72 LONGITUDINAL NICKEL CHANNELS

COMPLETED ENGINE

This thrust chamber, which was designed in 1957 by Edward Baehr, consists of a number of longitudinal channels of varying depth according to the velocity required for the coolant. These channels are bonded together to make up the chamber and bound by stainless steel wire wrapping which is brazed to make a fluid-tight and strong outer skin. In a thrust chamber of this type, the outer skin (the brazed wire) is subject only to the hoop stress due to the pressure exerted by the coolant. The inner shell is subject to the mechanical stress is due to the difference of pressure existing between the combusted gas and the coolant, and also to the thermal stress due to the heat transfer across the wall.

The combined maximum compressive stress σ_c (N/m^2) occurs at the inner surface of the inner shell and can be computed as follows [5]:

$$\sigma_c = \frac{(p_{co} - p_g)R}{t} + \frac{E\lambda qt}{2(1-v)k}$$

where p_{co} and p_g (N/m^2) are the pressures of respectively the coolant and the combusted gas, R (m) is the radius of the inner shell, t (m) is the thickness of the inner shell, E (N/m^2) is the modulus of elasticity of the material of which the inner shell is made, λ (m m^{-1} K^{-1}) is the coefficient of thermal expansion of the same material, q (W/m^2) is the quantity of heat per unit time per unit surface, v is the Poisson ratio of the same material, and k (W m^{-1} K^{-1}) is the thermal conductivity of the same material.

Since the heat transfer in a rocket engine occurs mainly through convection, it is desirable to reduce the pressure drop of the coolant fluid to the minimum possible value. For this purpose, it is necessary to avoid abrupt changes in the direction of the coolant fluid and also in the hydraulic diameter of the tubes. In addition, the inner surfaces of these tubes should be smooth and clean. The pressure drop Δp (N/m^2) in a tube of length L (m) and hydraulic diameter d (m) is expressed as a function of the Darcy friction factor f_D (dimensionless) of the tube by means of the Darcy-Weisbach equation, as follows

$$\Delta p = f_D \frac{L}{d}\left(\frac{1}{2}\rho v^2\right)$$

where ρ (kg/m^3) and v are respectively the density and the average velocity of the coolant flowing in the tube. The Darcy friction factor f_D depends on the Reynolds number Re defined above and also on the shape and smoothness of the tubes.

The value of f_D is determined experimentally. In practice, in laminar flow regime ($Re < 2300$), $f_D = 64/Re$. In transitional flow regime ($2300 \leq Re \leq 4000$), there are large uncertainties as to the value of f_D. In turbulent flow regime ($Re > 4000$) and in rough tubes, the value of f_D can be determined by means of the Colebrook-White relation [58], as follows

$$\frac{1}{f_D^{\frac{1}{2}}} = -2\log_{10}\left(\frac{\varepsilon}{3.7d} + \frac{2.51}{Re f_D^{\frac{1}{2}}}\right)$$

where ε (m), called absolute roughness, is the average height of the irregularities existing on the inner surface of a tube, and d (m) is the hydraulic diameter of the tube. The dimensionless ratio ε/d is called relative roughness.

The relation written above is implicit, because the unknown f_D is on both sides of the relation. It can be solved numerically, for given values of ε/d and Re, by defining an interval of search $a \le f_D \le b$ and a function $g(f_D)$ such that

$$g(f_D) \equiv \frac{1}{f_D^{\frac{1}{2}}} + 2\log_{10}\left(\frac{\varepsilon}{3.7d} + \frac{2.51}{Re\,f_D^{\frac{1}{2}}}\right)$$

with $g(a)g(b) < 0$, and then searching the value of f_D which satisfies the condition $g(f_D) = 0$ to some acceptable degree of tolerance, as has been shown in Chap. 1, Sect. 1.2. The following table, taken from [59], gives values of absolute roughness ε for some piping materials.

Material	$\varepsilon \times 10^{-3}$ (m)
Copper, lead, brass, aluminium (new)	0.001–0.002
Stainless steel	0.0015
Steel commercial pipe	0.045–0.09
Weld steel	0.045
Carbon steel (new)	0.02–0.05
Carbon steel (slightly corroded)	0.05–0.15
Carbon steel (moderately corroded)	0.15–1
Carbon steel (badly corroded)	1–3

Calculators are also available through the Internet to solve numerically the Colebrook-White relation. For example, after inserting $\varepsilon = 0.0015$ mm, $d = 18$ mm, and $Re = 10000$ in the calculator of [60], there results $f_D = 0.03101157$.

After substituting this value in the function $g(f_D)$ defined above, there results $g(0.03101157) = 3.5 \times 10^{-7}$.

As an application of the concepts given above, it is required to design the cooling tubes at the throat (which is the most stressed section) of the thrust chamber of a rocket, whose first stage burns a combination of liquid oxygen with RP-1 (kerosene), and has the data given in Sects. 2.3 and 2.4. The coolant fluid is the fuel (RP-1). The material chosen for the cooling tubes is a high-strength alloy, namely, Inconel® 718. As a result of the carbon deposits on the wall of the thrust chamber, the design temperature of the wall is assumed to be less than or equal to 811 K. In particular, in the throat region of the thrust chamber, the temperature of the wall on the hot gas side is assumed to be $T_{wg} = 660$ K.

The design value $(\overline{T}_c)_{ns}$ of the total temperature of the combustion chamber at the nozzle inlet has been found above to be

$$(\overline{T}_c)_{ns} = (T_c)_{ns}\eta_{c*}^2 = 3589 \times 0.975^2 = 3412 \text{ K}$$

where $(T_c)_{ns}$ is the theoretical value of the total temperature of the combustion chamber at the same section, and η_{c*} is the correction factor of the characteristic velocity. The design value $(\overline{T}_c)_{ns}$, multiplied by an estimated value 0.923 of the stagnation recovery factor, is used to determine the temperature T_{aw} at the adiabatic wall, as follows

$$T_{aw} = 3412 \times 0.923 = 3149 \text{ K}$$

The overall heat transfer coefficient on the hot gas side at the throat plane has also been found above to be $h_{gc} = 1987 \text{ W m}^{-2} \text{ K}^{-1}$. By substituting this value in the following equation, the heat flux q at the throat results

$$q = h_{gc}(T_{aw} - T_{wg}) = 1987 \times (3149 - 660) = 4.946 \times 10^6 \text{ W/m}^2$$

By interpolating the data of [61], we find the following data for Inconel® 718 at $T = 555$ K: coefficient of thermal expansion $\lambda = 13.8 \times 10^{-6} \text{ m m}^{-1} \text{ K}^{-1}$, modulus of elasticity $E = 1.86 \times 10^{11} \text{ N/m}^2$, thermal conductivity $k = 15.5 \text{ W m}^{-1} \text{ K}^{-1}$, and Poisson's ratio $v = 0.274$. We use circular cooling tubes of inner diameter d, whose value is to be determined. The thickness t of the cooling tubes ranges usually from 0.254 to 1.02 mm [45], depending on the combination of propellants and on the material used. We take initially $t = 0.329 \text{ mm} = 3.29 \times 10^{-4}$ m. This value will be checked against the results of the following heat transfer and mechanical stress calculations. Remembering the preceding equation

$$q = h_{gc}(T_{aw} - T_{wg}) = \frac{k}{t}(T_{wg} - T_{wc}) = h_c(T_{wc} - T_{co}) = H(T_{aw} - T_{co})$$

the temperature T_{wc} of the wall on the coolant side can be determined as follows

$$T_{wc} = T_{wg} - \frac{tq}{k} = 660 - \frac{3.29 \times 10^{-4} \times 4.946 \times 10^6}{15.5} = 555 \text{ K}$$

A cooling system based on a double pass is used, such that the coolant flows downward through alternating tubes and upward through adjacent tubes.

For each tube through which the coolant flows upward, we assume the bulk temperature of the coolant at the throat to be $T_{co} = 333$ K, which is a conservative estimate, since the coolant has previously flown through the throat region on its way downward. This temperature is much less than the critical temperature of RP-1, which is 666 K [54], Table 4, and can be expected to remain nearly constant in the remaining portion of the passage. The increase in total temperature for a typical thrust chamber is about 311 K between the inlet and the outlet of a cooling jacket [5]. The value of the heat transfer coefficient h_c on the coolant side, which is necessary to permit the heat flux $q = 4.946 \times 10^6 \text{ W/m}^2$ with the difference of temperature $T_{wc} - T_{co} =$ 555–333 K, can be computed by using the following equation

$$q = h_c(T_{wc} - T_{co})$$

This equation, solved for h_c, yields

$$h_c = \frac{q}{T_{wc} - T_{co}} = \frac{4.946 \times 10^6}{555 - 333} = 2.228 \times 10^4 \ \text{W m}^{-2} \text{K}^{-1}$$

According to Huzel and Huang [5], the number N of cooling tubes can be determined as follows

$$N = \frac{\pi[D_t + 0.8(d + 2t)]}{d + 2t}$$

where, for the engine considered here, $D_t = 0.6332$ m (see Sect. 2.4) is the diameter of the throat, d (m) is the unknown inner diameter of each tube, $t = 3.29 \times 10^{-4}$ m is the thickness of each tube, and 0.8 is a factor which takes account of the fact that the centres of the tubes are located on a circle, not on a straight line. After substitution of these values, the preceding equation becomes

$$N = \frac{\pi(0.8d + 0.6337)}{d + 0.000658}$$

For a double-pass cooling system, the velocity v (m/s) of the coolant in the tubes results from [5]:

$$v = \frac{\dot{m}}{\rho} \frac{1}{\frac{1}{2}N\left(\frac{1}{4}\pi d^2\right)}$$

where $\dot{m}$(kg/s) is the mass flow rate of the coolant, ρ (kg/m^3) is the local value of the density of the coolant, N is the number of the cooling tubes, and d (m) is the inner diameter of each cooling tube. After taking the value 375 kg/s for the mass flow rate of the coolant and substituting this value in the preceding equation, we find

$$v = \frac{375 \times 8}{\pi N d^2 \rho} = \frac{3000}{\pi N d^2 \rho}$$

We compute the density ρ, the thermal conductivity k, the dynamic viscosity μ, and the specific heat c_p at constant pressure of RP-1 at the temperature $T_{co} = 333$ K by interpolating the data tabulated by Giovanetti et al. [54, Table 4].

By so doing, we find $\rho = 776$ kg/m^3, $k = 0.0920$ W/(mK), $\mu = 0.0009652$ Ns/m^2, and $c_p = 2130$ J/(kgK) at $T_{co} = 333$ K.

We use the following correlation due to Gnielinski [53, page 11, Eq. 11]:

$$Nu = \frac{0.125 f_D (Re - 1000) Pr}{1 + 12.7(0.125 f_D)^{\frac{1}{2}} (Pr^{\frac{2}{3}} - 1)}$$

where the Darcy friction factor f_D of the cooling tubes is computed by using the following equation due to Filonenko [53, page 11, Eq. 9]:

$$f_D = \left[1.82 \log_{10}(Re) - 1.64\right]^{-2}$$

By substituting the interpolated data indicated above, $v = 3000/(\pi N d^2 \rho)$, and $N = \pi(0.8\,d + 0.6337)/(d + 0.000658)$ into $Re = \rho v d/\mu$, there results

$$Re = \frac{3000(d + 0.000658)}{0.0009652 \pi^2 d(0.8d + 0.6337)}$$

Substituting this expression of Re into $f_D = \left[1.82 \log_{10}(Re) - 1.64\right]^{-2}$ yields

$$f_D = \left[1.82 \log_{10}(Re) - 1.64\right]^{-2}$$
$$= \left\{1.82 \log_{10}\left[\frac{3000(d + 0.000658)}{0.0009652 \pi^2 d(0.8d + 0.6337)}\right] - 1.64\right\}^{-2}$$

$$Pr = \frac{\mu c_p}{k} = \frac{0.0009652 \times 2130}{0.092}$$

$$Nu = \frac{h_c d}{k} = \frac{2.228 \times 10^4 d}{0.092}$$

The quantities Re, f_D, and Nu are functions of the unknown value of d. These functions and the constant value $Pr = 0.0009652 \times 2130/0.092$ are substituted into the Gnielinski correlation

$$Nu = \frac{0.125 f_D(Re - 1000)Pr}{1 + 12.7(0.125 f_D)^{\frac{1}{2}}(Pr^{\frac{2}{3}} - 1)}$$

which is solved numerically for d. By so doing, we find $d = 0.01671$ m.

For comparison, by using the Taler correlation $Nu = 0.00881\, Re^{0.8991}\, Pr^{0.3911}$ [52, page 4, Eq. 20], we find $d = 0.01689$ m.

Substituting $d = 0.01671$ m into

$$N = \frac{\pi(0.8d + 0.6337)}{d + 0.000658}$$

yields $N = 117$. Since the number of tubes for a double-pass cooling system must be not only whole but also even, then we take $N = 118$.

By substituting $N = 118$ into the following equation

$$N = \frac{\pi(0.8d + 0.6337)}{d + 0.000658}$$

we find $d = 0.01657$ m. Therefore, the cooling system at the throat consists of 118 tubes (59 tubes for the coolant flowing downward plus 59 tubes for the coolant flowing upward), each of which is 16.57 mm in diameter and 0.329 mm in thickness. As has been found above, the density of the coolant at the temperature $T_{co} = 333$ K is $\rho = 776$ kg/m³. By substituting this value, $d = 0.01657$ m, and $N = 118$ in the following equation

$$v = \frac{3000}{\pi N d^2 \rho}$$

the velocity of the coolant results

$$v = 37.98 \, \text{m/s}$$

The Gnielinski correlation used above is valid for

$$0.5 \leq Pr \leq 2000$$
$$3000 \leq Re \leq 5 \times 10^6$$

In the present case, there results

$$Pr = \frac{\mu c_p}{k} = \frac{0.0009652 \times 2130}{0.092} = 22.35$$

$$Re = \frac{\rho v d}{\mu} = \frac{776 \times 37.98 \times 0.01657}{0.0009652} = 5.06 \times 10^5$$

The pressure of the coolant at the throat is $p_{co} = 1.034 \times 10^7$ N/m², which value results from an interpolation between the pressure at the outlet of the fuel pump and the pressure at the injector manifold. The static pressure p_t (N/m²) of the combusted gas at the throat can be computed by using the following equation of Chap. 1, Sect. 1.2:

$$(p_c)_{ns} = p_i \left(1 + \frac{\gamma - 1}{2} M_i^2 \right)^{\frac{\gamma}{\gamma - 1}}$$

where $(p_c)_{ns} = 6.895 \times 10^6$ N/m² is the total pressure of the combustion chamber at the inlet plane of the nozzle, M_i is the Mach number at the section of interest, and $\gamma \equiv c_p/c_v = 1.222$ is the specific heat ratio. In particular, at the throat ($M_t = 1$), the preceding equation, solved for p_t, yields

$$p_t = (p_c)_{ns} \left(\frac{2}{\gamma + 1} \right)^{\frac{\gamma}{\gamma - 1}} = 6.895 \times 10^6 \times \left(\frac{2}{1.222 + 1} \right)^{\frac{1.222}{1.222 - 1}} = 3.863 \times 10^6 \, \text{N/m}^2$$

The maximum combined tensile stress σ_t (N/m²) acting on the cross section A-A of a circular cooling tube can be computed by using the following equation

$$\sigma_t = \frac{(p_{co} - p_g)r}{t} + \frac{E\lambda qt}{2(1 - \nu)k} + \frac{6M_A}{t^2}$$

where $p_{co} = 1.034 \times 10^7$ N/m^2 is the pressure of the coolant at the throat, $p_g = p_t$ $= 3.863 \times 10^6$ N/m^2 is the static pressure of the combusted gas at the throat, $r = d/2 = 0.008285$ m is the radius of the cross section of the tubes, $t = 3.29 \times 10^{-4}$ m is the thickness of the tubes, $E = 1.86 \times 10^{11}$ N/m^2 is the modulus of elasticity of Inconel® 718 at $T = 555$ K, $\lambda = 13.8 \times 10^{-6}$ m m^{-1} K^{-1} is the coefficient of thermal expansion of the same alloy at the same temperature, $q = 4.946 \times 10^6$ W/m^2 is the quantity of heat per unit time per unit surface at the throat, $\nu = 0.274$ is the Poisson ratio of the same alloy at the same temperature, $k = 15.5$ W m^{-1} K^{-1} is the thermal conductivity of the same alloy at the same temperature, and M_A (Nm/m) is the bending moment per unit length acting on the section A-A due to the distortion induced by discontinuity.

After substituting these values in the preceding equation, we find

$$\sigma_t = 1.631 \times 10^8 + 1.856 \times 10^8 + 0.5543 \times 10^8 M_A$$

By keeping σ_t less than or equal to the 0.2% offset tensile yield strength, which is $\sigma_{ty} = 9.34 \times 10^8$ N/m^2 at the temperature $T = 555$ K, as recommended by the manufacturer of Inconel® 718 [61], the maximum allowable bending moment per unit length due to discontinuity results

$$(M_A)_{max} = \frac{(9.34 - 1.631 - 1.856) \times 10^8}{0.5543 \times 10^8} = 10.56 \,\text{Nm/m}$$

According to Huzel and Huang [5], the value of the bending moment per unit length due to discontinuity is, in the present case, smaller than 8.36 Nm/m, as shown by experience. Therefore, the results found above (at the throat plane, 118 cooling tubes, each of which is 16.57 mm in diameter and 0.329 mm in thickness) are confirmed.

As a further example of application of these concepts, it is required to design the cooling tubes at the throat of the thrust chamber of a rocket, whose second stage burns a combination of liquid oxygen with liquid hydrogen, and has the properties specified below. The coolant fluid is the fuel (liquid hydrogen). The oxidiser-to-fuel mixture ratio at the thrust chamber is $o/f = 5.22$, the total absolute pressure and the total temperature in the combustion chamber at the nozzle inlet are respectively $(p_c)_{ns} = 5.516 \times 10^6$ N/m^2 and $(T_c)_{ns} = 3356$ K, the molar mass of the combusted gas is $\mathcal{M} = 12$ kg/kmol, and the specific heat ratio of the combusted gas is $\gamma \equiv c_p/c_v$ $= 1.213$. The pressure at which the fuel is discharged at the turbine outlet is $p = 9.653 \times 10^6$ N/m^2, and the mass flow rate of the fuel is $\dot{m}_f = 24.72$ kg/s.

As has been found in Sects. 2.3 and 2.4 for this engine, the design value $\bar{c}^*$ of the characteristic velocity is

$$\bar{c}^* = \eta_{c^*} c^* = 0.975 \times 2342 = 2284 \,\text{m/s}$$

where $c^* = 2342$ m/s and $\eta_{c*} = 0.975$ are respectively the theoretical value and the correction factor of the characteristic velocity. In addition, the diameter D_t of the thrust chamber at the throat plane and the diameter D_e of the thrust chamber at the exit plane of the nozzle have been found to be respectively $D_t = 2R_t = 2 \times 0.1425 = 0.2850$ m and $D_e = 2R_e = 2 \times 0.9012 = 1.802$ m.

The design value $(\overline{T}_c)_{ns}$ of the total temperature in the combustion chamber at the nozzle inlet results from the corresponding theoretical value $(T_c)_{ns}$ as follows

$$(\overline{T}_c)_{ns} = \eta_{c*}^2 (T_c)_{ns} = 0.975^2 \times 3356 = 3190 \text{ K}$$

The temperature T_{aw} of the gas at the adiabatic wall is determined by multiplying $(\overline{T}_c)_{ns}$ by a recovery factor equal to 0.92, as follows

$$T_{aw} = 3190 \times 0.92 = 2935 \text{ K}$$

The radius of curvature R_{us} of the nozzle contour upstream of the throat plane is

$$R_{us} = 1.5R_t = 1.5 \times 0.1425 \text{ m} = 0.2138 \text{ m}$$

The radius of curvature R_{ds} of the nozzle contour downstream of the throat plane is

$$R_{ds} = 0.382R_t = 0.382 \times 0.1425 \text{ m} = 0.05444 \text{ m}$$

The mean radius $\overline{R}$ of curvature of the nozzle contour at the throat is

$$\overline{R} = \frac{1}{2} \times (R_{us} + R_{ds}) = 0.5 \times (0.2138 + 0.05444) = 0.1341 \text{ m}$$

The value of the constant R of the specific gas is

$$R = \frac{R^*}{M} = \frac{8314.460}{12} = 692.9 \text{ J K}^{-1} \text{ kg}^{-1}$$

Since $\gamma \equiv c_p/c_v = 1.213$ and $c_p - c_v = R = 692.9$ J K^{-1} kg^{-1}, then

$$c_p = \frac{\gamma}{\gamma - 1} R = \frac{1.213}{1.213 - 1} \times 692.9 = 3946 \text{ J K}^{-1} \text{ kg}^{-1}$$

The Prandtl number is computed as follows

$$Pr = \frac{4\gamma}{9\gamma - 5} = \frac{4 \times 1.213}{9 \times 1.213 - 5} = 0.82$$

The coefficient of dynamic viscosity is computed as follows

$$\mu = \left(1.184 \times 10^{-7}\right)\mathcal{M}^{0.5}T^{0.6} = \left(1.184 \times 10^{-7}\right) \times (12)^{0.5} \times (3190)^{0.6}$$
$$= 5.191 \times 10^{-5}\,\text{N s m}^{-2}$$

The convective heat transfer coefficient h_g is computed by using the Bartz equation written below

$$h_g = \left[\frac{0.026}{D_t^{0.2}}\left(\frac{\mu^{0.2}c_p}{Pr^{0.6}}\right)_{ns}\left(\frac{(p_c)_{ns}}{\bar{c}^*}\right)^{0.8}\left(\frac{D_t}{\bar{R}}\right)^{0.1}\right]\left(\frac{A_t}{A}\right)^{0.9}\sigma$$

After substituting $D_t = 0.2850$ m, $\mu = 5.191 \times 10^{-5}$ N s m^{-2}, $c_p = 3946$ J K^{-1} kg^{-1}, $Pr = 0.82$, $(p_c)_{ns} = 5.516 \times 10^6$ N m^{-2}, $\bar{c}^* = 2284$ m s^{-1}, and $\bar{R} = 0.1341$ m in the preceding equation, we find

$$h_g = 1.132 \times 10^4\left(\frac{A_t}{A}\right)^{0.9}\sigma \quad \text{W m}^{-2}\text{K}^{-1}$$

The value of σ is determined by means of the following equation

$$\sigma = \frac{1}{\left[\frac{1}{2}\frac{T_{wg}}{(\bar{T}_c)_{ns}}\left(1 + \frac{\gamma-1}{2}M^2\right) + \frac{1}{2}\right]^{0.68}\left[1 + \frac{\gamma-1}{2}M^2\right]^{0.12}}$$

Since there is no solid deposit on the walls of the thrust chamber, we assume an average temperature $T_{wg} = 833$ K on the hot gas side of the wall. This yields

$$\frac{T_{wg}}{(\bar{T}_c)_{ns}} = \frac{833}{3190} \approx 0.26$$

By substituting this value, $M = 1$, and $\gamma = 1.213$ into the expression of σ, we find at the throat plane

$$\sigma \approx 1.33$$

Therefore, at the throat plane, where $\sigma = 1.33$ and $A_t/A = 1$, there results

$$h_g = 1.132 \times 10^4(A_t/A)^{0.9}\sigma = 1.132 \times 10^4 \times 1^{0.9} \times 1.33 = 1.506 \times 10^4\,\text{W m}^{-2}\,\text{K}^{-1}$$

In order to avoid excessive thermal stresses in the material to be used, we want to keep the mean temperature of the tube wall below 811 K. By substituting this value of h_g in the following equation, the heat flux q at the throat results

$$q = h_g\left(T_{aw} - T_{wg}\right) = 1.506 \times 10^4 \times (2935 - 833) = 3.166 \times 10^7\,\text{W/m}^2$$

where 833 K is the value (see above) used for the temperature T_{wg} of the wall on the hot gas side. The material chosen for the cooling tubes is again Inconel® 718. By interpolating the data of [61], we find the following data for this material at $T = 833$ K: coefficient of thermal expansion $\lambda = 14.6 \times 10^{-6}$ m m^{-1} K^{-1}, modulus of elasticity $E = 1.70 \times 10^{11}$ N/m^2, thermal conductivity $k = 20.0$ W m^{-1} K^{-1}, and Poisson's ratio $\nu = 0.272$. We use cooling tubes of circular cross-section, having an internal diameter d, whose value is to be determined, and a thickness $t = 0.2$ mm $= 0.0002$ m. By solving the following equation

$$q = h_{gc}\left(T_{aw} - T_{wg}\right) = \frac{k}{t}\left(T_{wg} - T_{wc}\right) = h_c(T_{wc} - T_{co}) = H(T_{aw} - T_{co})$$

for T_{wc}, the temperature of the wall on the coolant side results

$$T_{wc} = T_{wg} - \frac{qt}{k} = 833 - \frac{3.166 \times 10^7 \times 0.0002}{20.0} = 516.4 \text{ K}$$

The mean value of the temperature of the wall on the coolant side is the mean between the temperature computed above ($T_{wc} = 516.4$ K) and the temperature ($T_{wg} = 833$ K) of the wall on the hot gas side at the throat plane, as follows

$$\frac{516.4 + 833.0}{2} = 674.7 \text{ K}$$

This value is lower than the maximum allowed value (811 K).
We assume the bulk temperature of the coolant at the throat to be $T_{co} = 75$ K. By solving the following equation

$$q = h_{gc}\left(T_{aw} - T_{wg}\right) = \frac{k}{t}\left(T_{wg} - T_{wc}\right) = h_c(T_{wc} - T_{co}) = H(T_{aw} - T_{co})$$

for h_c, there results at the throat

$$h_c = \frac{q}{T_{wc} - T_{co}} = \frac{3.166 \times 10^7}{516.4 - 75} = 7.173 \times 10^4 \text{ W m}^{-2}\text{K}^{-1}$$

We use the following correlation due to McCarthy and Wolf [56, page 95, Eq. 2]:

$$h_c = 0.025\left(\frac{c_p\mu^{0.2}}{Pr^{0.6}}\right)_{co} \frac{G^{0.8}}{d^{0.2}}\left(\frac{T_{co}}{T_{wc}}\right)^{0.55}$$

where c_p (J kg^{-1} K^{-1}) is the specific heat of the coolant (hydrogen) at constant pressure, μ (N s m^{-2}) is the coefficient of dynamic viscosity of the coolant, $Pr = c_p\mu/k$ is the Prandtl number, k (W m^{-1} K^{-1}) is the thermal conductivity of the coolant, G (kg s^{-1} m^{-2}) is the mass flow rate of the coolant per unit area to be cooled, d (m) is the diameter of the coolant passage, T_{co} (K) is the bulk temperature of the coolant, T_{wc} (K) is the temperature of the wall on the coolant side, and the subscript co indicates the bulk temperature of the coolant.

According to NIST [62], at $T_{co} = 75$ K and $p = 9.653 \times 10^6$ N/m^2, the coolant has the following properties: $c_p = 14890$ J kg^{-1} K^{-1}, $\mu = 4.9949 \times 10^{-6}$ N s m^{-2}, and $k = 0.090639$ W m^{-1} K^{-1}. Therefore, the Prandtl number in such conditions is

$$Pr = \frac{c_p \mu}{k} = \frac{14890 \times 4.9949 \times 10^{-6}}{0.090639} = 0.82055$$

which confirms the value $Pr = 0.82$ found above. A sesqui-pass (literally one-and-a-half-pass) type of design is chosen here for the cooling tubes. This term actually means a partial pass starting below the throat. This type of pass is used with coolants (such as liquid hydrogen) which must previously be heated in order to become effective, that is, in order to absorb the high heat fluxes at the throat. The extent of the partial pass results from a trade-off between thermal and gimballing requirements, since it is desirable to keep the inlet manifold forward [45].

As has been shown above, the number N of cooling tubes can be determined as follows

$$N = \frac{\pi[D_t + 0.8(d + 2t)]}{d + 2t} = \frac{\pi[0.2850 + 0.8 \times (d + 2 \times 0.0002)]}{d + 2 \times 0.0002}$$
$$= \frac{\pi(0.8d + 0.2853)}{d + 0.0004}$$

where d (whose value is to be determined) and $t = 0.0002$ m are respectively the inner diameter and the thickness of each cooling tube.

The mass flow rate of the coolant per unit area to be cooled is

$$G = \frac{\dot{m}_f}{N\pi\left(\frac{d}{2}\right)^2} = \frac{24.72 \times 4}{\pi N d^2} = \frac{24.72 \times 4 \times (d + 0.0004)}{\pi^2 d^2 (0.8d + 0.2853)}$$

After substituting the values determined above into the McCarthy-Wolf correlation, there results

$$7.173 \times 10^4 = 0.025 \times \left(\frac{14890 \times 0.0000049995^{0.2}}{0.82055^{0.6}}\right) \times \left[\frac{24.72 \times 4 \times (d + 0.0004)}{\pi^2 d^2 (0.8d + 0.2853)}\right]^{0.8}$$
$$\times \left(\frac{1}{d^{0.2}}\right) \times \left(\frac{75}{516.4}\right)^{0.55}$$

The preceding equation, solved numerically for d, yields $d = 0.0033$ m. By substituting $d = 0.0033$ into the following equation

$$N = \frac{\pi(0.8d + 0.2853)}{d + 0.0004}$$

we find $N = 244.5$. We take $N = 244$. By substituting this value of N into the preceding equation, there results $d = 0.003307$ m.

Therefore, the cooling system at the throat consists of 244 tubes, each of which is 3.307 mm in diameter and 0.2 mm in thickness.

The estimated pressure of the coolant at the throat is $p_{co} = 8.274 \times 10^6$ N/m^2. The static pressure p_t (N/m^2) of the combusted gas at the throat can be computed by using the general equation of Chap. 1, Sect. 1.2:

$$(p_c)_{ns} = p_i \left(1 + \frac{\gamma - 1}{2} M_i^2\right)^{\frac{\gamma}{\gamma - 1}}$$

where $(p_c)_{ns} = 5.516 \times 10^6$ N/m^2 is the total pressure of the combustion chamber at the inlet plane of the nozzle, M_i is the Mach number at the section of interest, and $\gamma \equiv c_p/c_v = 1.213$ is the specific heat ratio. In particular, at the throat ($M_t = 1$), the preceding equation, solved for p_t, yields

$$p_t = (p_c)_{ns} \left(\frac{2}{\gamma + 1}\right)^{\frac{\gamma}{\gamma - 1}} = 5.516 \times 10^6 \times \left(\frac{2}{1.213 + 1}\right)^{\frac{1.213}{1.213 - 1}} = 3.1 \times 10^6 \text{ N/m}^2$$

The maximum combined tensile stress σ_t (N/m^2) acting on the cross section A-A of a circular cooling tube can be computed by using the following equation

$$\sigma_t = \frac{(p_{co} - p_g)r}{t} + \frac{E\lambda qt}{2(1 - v)k} + \frac{6M_A}{t^2}$$

where $p_{co} = 8.274 \times 10^6$ N/m^2 is the pressure of the coolant at the throat, $p_g = p_t = 3.1 \times 10^6$ N/m^2 is the pressure of the combusted gas at the throat, $r = d/2 = 0.001654$ m is the radius of the cross section of the tubes, $t = 2.0 \times 10^{-4}$ m is the thickness of the tubes, $E = 1.70 \times 10^{11}$ N/m^2 is the modulus of elasticity of Inconel® 718 at $T = 833$ K, $\lambda = 14.6 \times 10^{-6}$ m m^{-1} K^{-1} is the coefficient of thermal expansion of the same alloy at the same temperature, $q = 3.166 \times 10^7$ W/m^2 is the quantity of heat per unit time per unit surface at the throat, $v = 0.272$ is the Poisson ratio of the same alloy at the same temperature, $k = 20.0$ W m^{-1} K^{-1} is the thermal conductivity of the same alloy at the same temperature, and M_A (Nm/m) is the bending moment per unit length acting on the section A-A due to the distortion induced by discontinuity.

After substituting these values in the preceding equation, we find

$$\sigma_t = 0.4279 \times 10^8 + 5.397 \times 10^8 + 1.500 \times 10^8 M_A$$

By keeping σ_t less than or equal to the 0.2% offset tensile yield strength, which is $\sigma_{ty} = 1.06 \times 10^9$ N/m^2 at the temperature $T = 675$ K, as recommended by the manufacturer of Inconel® 718 [61], the maximum allowable bending moment per unit length due to discontinuity results

$$(M_A)_{max} = \frac{(10.6 - 0.4279 - 5.397) \times 10^8}{1.500 \times 10^8} = 3.183 \, \text{Nm/m}$$

According to Huzel and Huang [5], the value of the bending moment due to discontinuity is, in the present case, smaller than 0.583 Nm/m, as shown by experience. Therefore, the results found above (at the throat plane, 244 tubes, each of which is 3.183 mm in diameter and 0.2 mm in thickness) are confirmed.

The results found in the two preceding examples concern the size of the cooling tubes at the throat, which is narrowest section of a thrust chamber. The size of the tubes changes along the longitudinal axis of a thrust chamber. The following figure, due to the courtesy of NASA [45], shows a cooling tube having a circular cross section of variable size with 6:1 maximum taper.

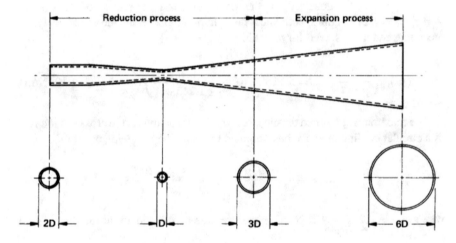

Another option consists in bifurcating the tubes by means of joints, as shown in the following figure, which is also due to the courtesy of NASA [45].

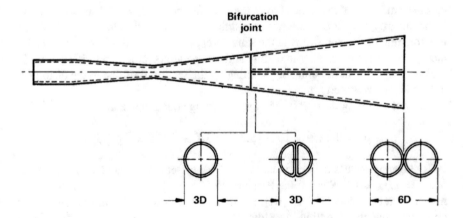

The cooling tubes are brazed to each other and to a metal shell or to hatbands for stiffening. To this end, furnace brazing is usually applied. In order to avoid the necessity of tapering, Volvo Aero Corporation has used tubes of square cross-section, which are wrapped in a spiral pattern around the inner wall of the nozzle. They are joined by gas-tungsten arc fillet welds [63]. Volvo Aero Corporation has also used a manufacturing process which consists in laser welding a close-out cover sheet on to an inner wall with milled cooling channels. When the outer sheet is laser welded to the inner sheet, the part has the form of a straight cone. Successively, the forming of a bell-shaped contour is done by expansion in a conventional expander [64].

As has been shown above, the dump cooling method is similar to the regenerative cooling, because the coolant flows, for both of these methods, through small passages on the outer side of the thrust chamber. However, in the case of dump cooling, the coolant is discharged overboard through openings at the end plane of the divergent portion of the nozzle, instead of flowing back to be discharged into the injector. This method is used in rocket engines fuelled by hydrogen at low pressures (p_c < 689500 N/m^2, where p_c is the static absolute pressure in the combustion chamber). The heated hydrogen dumped overboard gives a contribution to the total thrust. The coolant, if flowing longitudinally from the injector plane to the exit plane, can pass either in the interstice between a double wall (case A) or in tubes running along a single wall (case C). The coolant, if flowing spirally from the injector plane to the exit plane, can pass either in spiral passages existing between a double wall (case B) or in tubes wound around a single wall (case D), before being dumped overboard in the axial direction. The four cases described above are illustrated in the following figure, which is due to the courtesy of NASA [5].

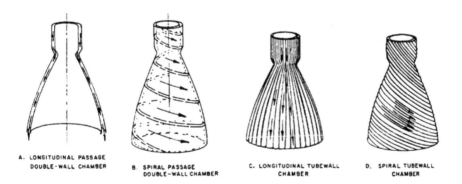

A. LONGITUDINAL PASSAGE
DOUBLE-WALL CHAMBER   B. SPIRAL PASSAGE   C. LONGITUDINAL TUBEWALL   D. SPIRAL TUBEWALL
                          DOUBLE-WALL CHAMBER   CHAMBER   CHAMBER

As has been shown above, film cooling method consists in the protection of a given surface from the harmful effects of a stream of hot gas by interposing a continuous protective film between the surface and the stream. In a liquid-propellant engine, a liquid or gaseous coolant, which is usually the fuel, is injected tangentially or at low angles into the combustion chamber along the hot gas side of the wall by means of a row of slots or orifices, as shown in the following figure, due to the courtesy of NASA [65].

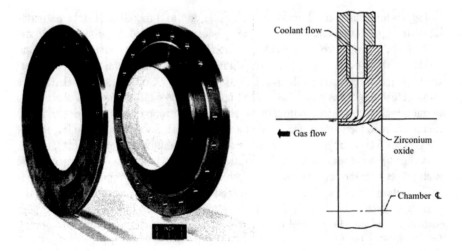

Due to phenomena of heat and mass transfer between the coolant film and the combusted gas, the thickness of the coolant film decreases in the direction of the hot stream. Therefore, in case of need, further coolant is injected through additional holes placed downstream of those placed along the first row. Film cooling is sometimes used in conjunction with regenerative cooling, for the purpose of reducing the heat transfer through the wall and, therefore, the thermal stress of the materials.

The coolant is often in the liquid state at the moment of injection, in order to absorb heat from the combusted gas by evaporating and diffusing into the main stream. This injection gives rise to a liquid film placed around the wall of the combustion chamber and containing the gaseous stream. The cooling efficiency of the injected fuel is subject to losses, due to disturbance waves on the surface of the liquid film adjacent to the combusted gas.

Zucrow and Sellers [66] indicate the following equation to be used for the design a thrust chamber cooled by a liquid film:

$$\frac{G_c}{G_g} = \frac{1}{\eta_c} \frac{H}{a(1 + b^c)}$$

where G_c (kg s^{-1} m^{-2}) is the mass flow rate of the film coolant per unit area of the wall of the thrust chamber to be cooled, G_g (kg s^{-1} m^{-2}) is the mass flow rate of the combusted gas per unit area of the cross section of the thrust chamber in the direction perpendicular to the flow, η_c is the efficiency of the film cooling, H (J/kg) is the enthalpy per unit mass of the film coolant, resulting from

$$H = \frac{c_{pvc}\left(T_{aw} - T_{wg}\right)}{c_{plc}\left(T_{wg} - T_{co}\right) + \Delta H_{vc}}$$

c_{pvc} (J kg^{-1} K^{-1}) is the average specific heat at constant pressure of the coolant in the vapour phase, c_{plc} (J kg^{-1} K^{-1}) is the average specific heat at constant pressure of the coolant in the liquid phase, T_{aw} (K) is the temperature of the gas at the adiabatic wall, T_{wg} (K) is the temperature of the wall on the hot gas side and also the temperature of the coolant film, T_{co} (K) is the bulk temperature of the coolant at the manifold, ΔH_{vc} (J/kg) is the vaporisation heat of the coolant, $a = 2V_d/(V_m f)$, V_d (m/s) is the velocity of the stream of combusted gas in the axial direction at the edge of the boundary layer, V_m (m/s) is the average velocity of the stream of combusted gas in the axial direction, f is the friction coefficient for the two-phase flow between the combusted gas and the liquid film, $b = V_g/V_d - 1$, V_g (m/s) is the velocity of the velocity of the stream of combusted gas in the axial direction at the central line of the thrust chamber, $c = c_{pvc}/c_{pg}$, and c_{pg} (J kg^{-1} K^{-1}) is the average specific heat at constant pressure of the combusted gas.

The losses mentioned above can be taken into account by means of the coefficient η_c, whose values range from 0.3 to 0.7 [5]. These values are to be determined experimentally. Hydrocarbon fuels have proven to be effective when used as coolants, because of the heat insulation properties of the carbon deposits generated by them on the walls of a thrust chamber.

When liquid hydrogen is used as a fuel, the very low critical temperature of hydrogen ($T_c = 33.18$ K, according to [62]) causes the initially liquid film to evaporate at a short distance from the point of injection.

Hatch and Papell [67] indicate the following equation to be used for the design a thrust chamber cooled by a gaseous film:

$$\frac{T_{aw} - T_{wg}}{T_{aw} - T_{co}} = \exp\left(\frac{-h_g}{G_c c_{pvc}\eta_c}\right)$$

where T_{aw} (K) is the temperature of the gas at the adiabatic wall, T_{wg} (K) is the maximum allowable temperature of the wall on the hot gas side, T_{co} (K) is the initial temperature of the coolant, h_g (W m^{-2} K^{-1}) is the heat transfer coefficient on the hot gas side, G_c (kg s^{-1} m^{-2}) is the mass flow rate of the film coolant per unit area of the wall of the thrust chamber to be cooled, c_{pvc} (J kg^{-1} K^{-1}) is the average specific heat at constant pressure of the gaseous coolant, and η_c is the efficiency of the film cooling. By means of the efficiency η_c, whose values range from about 0.25 to 0.65 [5], account is taken of the coolant which is lost into the main stream of combusted gas without producing the desired effect.

The following figure, due to the courtesy of NASA [68] shows a 37-element injector with removable film coolant ring manifold used in a small hydrogen-oxygen thrust chamber cooled by means of a hydrogen film.

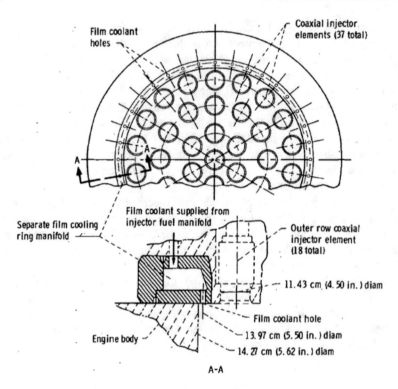

Film coolant holes

Coaxial injector elements (37 total)

Separate film cooling ring manifold

Film coolant supplied from injector fuel manifold

Outer row coaxial injector element (18 total)

11.43 cm (4.50 in.) diam

Engine body

Film coolant hole

13.97 cm (5.50 in.) diam

14.27 cm (5.62 in.) diam

A-A

The Hatch-Papell equation written above is based on the assumption of a balance between the heat coming from the wall of the combustion chamber and the heat absorbed by the coolant. The heat coming from the wall depends on the heat transfer h_g coefficient on the hot gas side and also on the difference between the temperature T_{aw} of the gas at the adiabatic wall and the initial temperature T_{co} of the coolant. The heat absorbed by the coolant depends on the heat capacity of the coolant in the interval from the initial temperature to the final temperature. When the two amounts of heat are in equilibrium, no heat is transferred through the wall. In such conditions, the inner surface of the thrust chamber reaches the temperature of the coolant corresponding to the particular location along the flow axis. Therefore, the temperature of the inner surface of the thrust chamber increases from the initial temperature of the coolant at the point of injection to the maximum allowable value of the temperature, at which point a further injection of coolant becomes necessary.

As an example of application, it is required to determine the mass flow rate G_c (kg s^{-1} m^{-2}) of film coolant per unit area of the wall of the thrust chamber to be cooled, for a rocket engine burning a combination of liquid hydrogen (fuel) with liquid fluorine (oxidiser). At the throat plane, the following quantities are known: heat transfer coefficient on the hot gas side $h_g = 3238$ W m^{-2} K^{-1}, temperature of the gas at the adiabatic wall $T_{aw} = 2911$ K, maximum allowable temperature of the wall on the hot gas side $T_{wg} = 1056$ K, initial temperature of the coolant $T_{co} = 28$ K,

average specific heat at constant pressure of the gaseous coolant $C_{pvc} = 1.507 \times 10^4$
J kg^{-1} K^{-1}, and efficiency of the film cooling $\eta_c = 0.3$.
   By substituting these data into the Hatch-Papell equation

$$\frac{T_{aw} - T_{wg}}{T_{aw} - T_{co}} = \exp\left(\frac{-h_g}{G_c C_{pvc} \eta_c}\right)$$

and solving for G_c, there results

$$G_c = \frac{-3238}{1.507 \times 10^4 \times 0.3 \times \ln\left(\frac{2911-1056}{2911-28}\right)} = 1.624 \, \text{kg s}^{-1} \, \text{m}^{-2}$$

   When the heat flux q (W/m^2), that is, the quantity of heat transferred per unit
time per unit surface, is computed for a regeneratively cooled engine with added
film cooling, then it is necessary to modify the value of the temperature T_{aw} (K)
of the adiabatic wall, before using this value in the equation $q = h_g(T_{aw}-T_{wg})$ or
in the equation $q = h_{gc}(T_{aw}-T_{wg})$. The modified value of T_{aw} must be determined
experimentally. By contrast, it is not necessary to modify the value of the heat transfer
coefficient h_g or h_{gc} (W m^{-2} K^{-1}) on the hot gas side.
   As has been shown at the beginning of this paragraph, in case of transpiration
cooling, the wall to be cooled has drilled holes or is made of a porous material, in
order for the coolant to pass through the wall into the gas flow. A protective layer
builds up on the hot gas side of the wall and insulates it from the heat carried by the
stream of combusted gas, as shown in the following figure, due to the courtesy of
NASA [69].

Hot gas

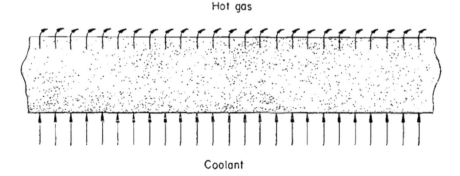

Coolant

   Since the coolant is directed away from the surface when leaving the wall, then
a counterflow is generated between the heat carried away from the surface with the
coolant stream and the heat transferred from the stream of combusted gas toward the
wall. This counterflow reduces the overall heat transfer between the hot gas and the
surface of the wall [69]. The permeable inner liner of the thrust chamber is enclosed
into an outer shell, and forms a jacket from which the coolant comes.

The cooling efficiency $(T_g - T_w)/(T_g - T_c)$ can be evaluated by using the Rannie equation [70], as follows

$$\frac{T_g - T_w}{T_g - T_{co}} = 1 - \frac{\exp\left(-36.9 Pr_g Re_d^{0.1} f\right)}{1 + \frac{c_{pc}}{c_{pg}}\left(1.18 Re_d^{0.1} - 1\right)\left[1 - \exp\left(-36.9 Pr_g Re_d^{0.1} f\right)\right]}$$

where T_g (K) is the recovery temperature of the hot gas, T_w (K) is the temperature of the wall, T_{co} (K) is the temperature of the coolant reservoir, $Pr_g = \mu_g c_{pg}/k_g$ is the Prandtl number of the hot gas, μ_g (N s m^{-2}) is the coefficient of dynamic viscosity of the hot gas, c_{pg} (J kg^{-1} K^{-1}) is the specific heat at constant pressure of the hot gas, k_g (W m^{-1} K^{-1}) is the thermal conductivity of the hot gas, $Re_d = \rho v d/\mu$ is the Reynolds number of the coolant based on the hydraulic diameter d (m), ρ (kg/m^3) is the density of the coolant at bulk temperature, v (m/s) is the velocity of the coolant, μ (N s m^{-2}) is the coefficient of dynamic viscosity of the coolant at bulk temperature, c_{pc} (J kg^{-1} K^{-1}) is the specific heat at constant pressure of the coolant at bulk temperature, $f = G_c/G_g$ is the blowing ratio, G_c (kg s^{-1} m^{-2}) is the mass flow rate of the coolant per unit area of the wall of the thrust chamber to be cooled, and G_g (kg s^{-1} m^{-2}) is the mass flow rate of the hot gas per unit area of the cross section of the thrust chamber in the direction perpendicular to the flow.

Since the Rannie equation written above indicates coolant flows slightly lower than those found necessary in experiments, then Huzel and Huang [5] recommend to use a cooling efficiency value of about 0.85. The porous material used for the walls of transpiration-cooled thrust chambers must, of course, withstand the mechanical and thermal stresses acting on such walls.

As has been shown above, ablative cooling is performed by using a protective material which covers the inner surface of a wall to be cooled. This material is made of either epoxy or, more often, phenolic resins reinforced by fibres. The protective material thermally degrades, and the products of this degradation (an endothermal pyrolysis of the resin, which leaves a pure carbon solid, called char, and releases oxygen and hydrogen) are carried away by the main stream of hot gas. The process of ablation blocks the heat flux to the outer surface of the wall. The reinforcement of the resin consists of either fibre-woven fabrics or chopped fibres of materials such as silica, graphite, or carbon.

With reference to the following figure, adapted from [71], an ablatively-cooled thrust chamber consists of a flame liner (made of a reinforced phenolic resin), a thin layer (made of silica impregnated with a phenolic resin) used for for insulation, and a high-strength structural shell. In rocket engines subject to erosion at the throat, inserts (made of either silicon carbide or JTA graphite, the latter being made of 48% graphite, 35% zirconium, 9% silicon, and 8% boron) are incorporated in the throat region.

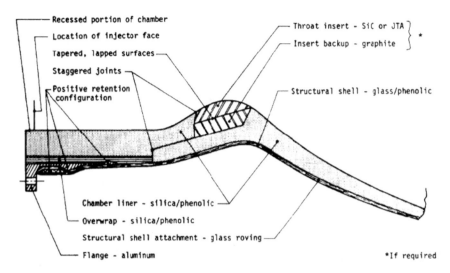

The thickness of the ablative materials varies as a function of the cross section considered. When several liner components are used, especially for the throat inserts, then joints are staggered axially to prevent a direct leak path to the outer structural shell. Such joints are located where they are not likely to open up under the action of thermal or mechanical loads. A differential thermal expansion of the components is taken into account by using materials having compatible elastic moduli or by inserting crushable or flexible materials which can expand without excessive stresses induced by strain [71].

The structural shell is made of either fibreglass or aluminium alloy or stainless steel. The thermal resistance of the inner layers keeps the outer shell at moderate temperatures.

For the design of a thrust chamber cooled by ablative materials, Huzel and Huang [5] indicate the following equation found experimentally, which expresses the char depth a (m) as a function of the heat absorbed:

$$a = c \left\{ \frac{2kt}{R_r R_v c_p \rho} \ln\left[1 + \frac{R_r R_v c_p (T_{aw} - T_d)}{L_p} \right] \right\}^{0.5} \left[\frac{(p_c)_{ns}}{6.895 \times 10^5} \right]^{0.4}$$

where c is a factor whose value is determined experimentally at the throat section and for a nozzle stagnation absolute pressure of 6.895×10^5 N/m², R_r is the weight fraction of resin content in the ablative material used, R_v is the weight fraction of the pyrolysed resin versus the total resin content R_r, c_p (J kg^{-1} K^{-1}) is the specific heat at constant pressure of the pyrolysis gases, ρ (kg/m³) is the density of the ablative material used, k (W m^{-1} K^{-1} m^{-1}) is the thermal conductivity per metre of char, t (s) is the duration of firing in the thrust chamber, L_p (J/kg) is the latent heat of pyrolysis, T_{aw} (K) is the temperature of the gas at the adiabatic wall, T_d (K) is the decomposition temperature of the resin, and $(p_c)_{ns}$ (N/m²) is the total pressure in the combustion chamber at the nozzle inlet.

The results obtained by using the preceding equation have been found to be in good agreement with the char depths measured in thrust chambers protected with Refrasil®, which is an amorphous silica woven fabric.

However, the char depths measured in sections placed downstream of the throat plane have been found to be greater than those computed by means of this equation. Therefore, Huzel and Huang [5] indicate the following equation to compute the char depth a (m) in sections downstream of the throat:

$$a = bt^{\frac{1}{2}} \exp(-0.0247\varepsilon)$$

where b (m/s$^{\frac{1}{2}}$) is a constant whose value is determined experimentally and depends on the ablative material used, t (s) is the duration of firing in the thrust chamber, $\varepsilon \equiv A_x/A_t$ is the expansion area ratio of the nozzle at the section x of interest, A_x (m^2) is the area of the section of interest, and A_t (m^2) is the area of the throat section. As an application of the concepts discussed above, it is required to determine the char depth at: (a) the combustion chamber and the throat, and (b) the cross-section of the nozzle whose area is five times the area of the throat (that is, $\varepsilon = 5$), for a rocket engine fired for a time $t = 410$ s and having the following properties: total pressure in the combustion chamber at the nozzle inlet $(p_c)_{ns} = 6.895 \times 10^5$ N/m^2, constant factor $c = 1.05$, weight fraction of resin content in the ablative material $R_r = 0.3$, weight fraction of the pyrolysed resin versus total resin content $R_v = 0.41$, specific heat at constant pressure of the pyrolysis gases $c_p = 1591$ J kg^{-1} K^{-1}, density of the ablative material $\rho = 1688$ kg/m^3, thermal conductivity per metre of char $k = 0.7327$ W m^{-1} K^{-1} m^{-1}, latent heat of pyrolysis $L_p = 1.596 \times 10^6$ J/kg, temperature of the gas at the adiabatic wall $T_{aw} = 2811$ K, decomposition temperature of the resin $T_d = 811$ K, and constant factor $b = 0.0008509$ m/s$^{\frac{1}{2}}$.

By inserting these data in the following equation

$$a = c \left\{ \frac{2kt}{R_r R_v c_p \rho} \ln\left[1 + \frac{R_r R_v c_p (T_{aw} - T_d)}{L_p} \right] \right\}^{0.5} \left[\frac{(p_c)_{ns}}{6.895 \times 10^5} \right]^{0.4}$$

we find $a = 0.021$ m $= 2.10$ cm.

Likewise, by inserting $b = 0.0008509$ m/s$^{\frac{1}{2}}$ and $\varepsilon = 5$ in the following equation

$$a = bt^{\frac{1}{2}} \exp(-0.0247\varepsilon)$$

we find $a = 0.0152$ m $= 1.52$ cm.

Radiation cooling occurs in a combustion chamber having a thin wall which is heated by the combusted gas to a temperature of thermal equilibrium. At this temperature, the heat from the wall to space equals the heat from the combusted gas to the wall. In order to prevent the inner surface of the wall from being overheated by the combusted gas, materials having high thermal conductivity are necessary.

The limits of application of radiation cooling depend on the maximum temperature which the available materials can withstand. The temperature of the combusted gas in a combustion chamber depends on such factors as the thrust level, the burning time, the pressure in the chamber, the combination of propellants, and their mixture ratios. Most portions of radiation cooled engines are subject to temperatures greater than 1500 K, with the exception of extension skirts of nozzles.

The heat-blocking properties of graphites resulting from pyrolysis of phenolic resins have been described above. Apart from these graphites, the only materials capable of meeting the requirements of a radiation cooled engine at temperatures above 1500 K are refractory materials, such as tungsten, molybdenum, and niobium. These materials are subject to oxidation caused by exhaust gases containing water vapour, carbon dioxide, and free oxygen. Therefore, refractory materials to be used in rocket engines must be protected by suitable coatings.

So far, radiation-cooled rocket engines have used a niobium alloy (C-103) with a fused silica coating (R512E). Since this coating has a limitation in temperature of 1643 K, then fuel film cooling has been necessary in rocket engines using this niobium alloy to maintain this temperature limit. Reed at alii [72] have described a material made of a rhenium substrate with an iridium oxidation-resistant coating, whose operating temperature is as high as 2473 K.

Assuming negligible difference of temperature between the alloy and the coating, a scheme of heat transfer by radiation cooling in steady state for a wall is illustrated in the following figure.

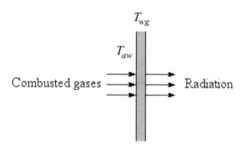

Let q (W/m^2), T_{aw} (K), and T_{wg} (K) be respectively the quantity of heat transferred per unit time per unit surface of the wall, the temperature of the combusted gases at the adiabatic wall, and the temperature of the wall on the gas side. As has been shown at the beginning of this section, the heat flux q from the combusted gases is proportional to the difference of temperature $T_{aw}-T_{wg}$, as follows

$$q = h_{gc}\left(T_{aw} - T_{wg}\right)$$

where h_{gc} (W m^{-2} K^{-1}) is the overall heat transfer coefficient on the hot gas side.

The heat flux q, due to the radiant energy emitted from the surface of the wall, is expressed by the Stefan-Boltzmann law, as follows

$$q = \varepsilon\sigma T_{wg}^4$$

where ε $(0 < \varepsilon < 1)$ is the emissivity of the outer surface of the material of which the wall is made, and $\sigma = 5.670367 \times 10^{-8}$ W m^{-2} K^{-4} [73] is the Stefan-Boltzmann constant.

In conditions of thermal equilibrium, the following equation holds

$$h_{gc}\left(T_{aw} - T_{wg}\right) = \varepsilon\sigma T_{wg}^4$$

The design of a radiation cooling system consists in determining a value of the temperature T_{wg} of the combusted gases at the adiabatic wall which satisfies the equation written above and meets the requirements posed by the material in the operating conditions.

As an application of the concepts discussed above, it is required to determine the temperature T_{wg} of the wall on the hot gas side for a nozzle extension of a rocket engine such that the overall heat transfer coefficient on the hot gas side is $h_{gc} = 206.6$ W/(m^2 K), the temperature of the combusted gases at the adiabatic wall is $T_{aw} = 2722$ K, and the emissivity of the outer surface of the wall is $\varepsilon = 0.95$.

By substituting these data in the following equation

$$h_{gc}\left(T_{aw} - T_{wg}\right) = \varepsilon\sigma T_{wg}^4$$

there results

$$206.6 \times \left(2722 - T_{wg}\right) = 0.95 \times 5.67 \times 10^{-8} \times T_{wg}^4$$

This equation can be solved numerically for T_{wg}. The result, with four significant figures, is $T_{wg} = 1478$ K. Consequently, the heat flux on the hot gas side of the wall is

$$q = h_{gc}\left(T_{aw} - T_{wg}\right) = 206.6 \times (2722 - 1478) = 2.57 \times 10^5 \text{ W/m}^2$$

2.6   Injectors

The position of a typical injector in the combustion chamber of a rocket engine is shown in the following figure, due to the courtesy of NASA [49].

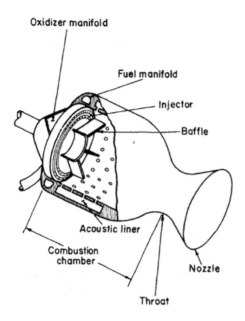

As has been shown in Sect. 2.1, the injector is the part of a rocket engine in which the liquid fuel and the liquid oxidiser are admitted in the combustion chamber, broken up into particles or droplets in order to increase the areas of their contact surfaces, mixed one with the other, and left to vaporise before reacting in the combustion process. An injector terminates with a perforated plate which marks the beginning of the combustion chamber.

Huzel and Huang [5] have identified some requirements to be taken into account in the design of an injector. They are indicated below.

- Stability in the combustion process, meaning by that a smooth combustion not only during the steady state operation but also in the start and stop transients. For this purpose, it is necessary to prevent unburned propellants from accumulating in the combustion chamber before ignition, in order to avoid an excess of pressure. It is also necessary to maintain a mixture of propellants rich in fuel at the moment of shutting the engine off, in order to avoid overheating. The propagation of local detonations due to combustion instabilities can be prevented by using damping devices placed either on the injector face (baffles) or along the wall of the combustion chamber (acoustic liners). Full details on baffles, acoustic absorbers, and other stabilisation devices are given in Sect. 2.9 and also in [74]. The correct amount and mixture ratio of propellants are maintained by means of valves.
- Performance of an injector in the combustion process, which depends on factors such as distribution of the mass of propellants, local value of the mixture ratio, mixing of the propellants, size and vaporisation of the droplets, heat flow, and velocity of the chemical reactions. Experience has shown that droplets of small size vaporise at high rates, and therefore injectors having a large number of

injecting elements have also a better performance, for the same volume of the combustion chamber, than those having a small number.

- Thermal and structural integrity of the injector, meaning by that its capability of withstanding the thermal and mechanical loads to which it is subject during the various phases of operation of the engine. In particular, an adequate system of cooling it is necessary to protect the injector from overheating.
- Proper sizing of the orifices, in order to obtain desired values of droplet size and also of pressure drops at specific flow rates.
- Protection of the whole combustion chamber from heat, meaning by that a complete mixing of the propellants, in order to avoid the formation of hot spots near the wall. To this end, the value of the mixture ratio *o/f* is kept low near the wall by placing a set of fuel holes around the periphery of the injector.
- Capability of operating in special conditions, for example, at low thrust levels during throttling, or at values of the mixture ratio which differ from the nominal values.

The working principle of an injector is illustrated in the following figure, due to the courtesy of Boeing-Rocketdyne [50], which shows the main injector of the Space Shuttle main engine.

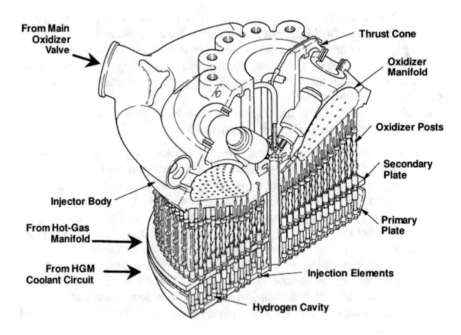

This injector admits into the main combustion chamber a combination of hot, fuel-rich gas from the two pre-burners, cold gaseous hydrogen from the cooling circuits, and and cold liquid oxygen from the high-pressure oxidiser turbo-pump. Passageways are formed for these fluids to enter the proper cavities in the injector, by welding the injector into the centre of the hot-gas manifold (HGM). The injector consists of 600 coaxial elements, which inject liquid oxygen from the oxidiser manifold through their centre posts. Each element also injects, through its annulus, the hot, fuel-rich gas entering the cavity between the heat shield and the secondary plate. Cold gaseous hydrogen, which had previously migrated through the double walls of the hot-gas manifold, enters the slot between the secondary plate and the lip of the primary plate. Both of these plates are porous and are transpiration-cooled by the cold gaseous hydrogen which flows through them. The flow shields are bolted to the outer row of elements and protect them from damage and erosion caused by the high-velocity gas. An augmented spark ignition (ASI) system chamber is located in the centre of the injector. Small quantities of hydrogen and oxygen are continuously injected into this chamber and initially ignited by two spark igniters located therein. This flame then ignites the propellants flowing through the injector elements into the combustion chamber. The thrust cone is a mounting pad for the gimbal bearing, which in turn attaches the engine to the vehicle [50].

The elements of the injector described above are of the coaxial type, which has been chosen because the combination of propellants is liquid oxygen and hydrogen. In order to meet the requirements posed by other combinations of propellants, elements of different types can be chosen, as will be shown below.

In many cases, the injected streams of propellants are made to impinge at a prede-termined distance from the injector face, in order to obtain a good mixture. The impingement distance depends on the heat transfer conditions. The type of elements in which all impingement points are at the same distance from the injector face is called uni-planar impingement. The types of elements in which the impingement points are at two and at more than two distances from the injector face are called respectively bi-planar and multi-planar impingement. The angles between impinging streams are usually chosen in the range from 0.3491 rad to 0.7854 rad.

The following figures, due to the courtesy of NASA [75], show some types of injector elements used for liquid/liquid injection.

(a) Unlike doublet, in which each stream of oxidiser is made to impinge on each stream of fuel in a pair. In this arrangement, the angle Φ between the two streams is variable within the range indicated above. It provides a good mixing, but its performance is sensitive to continuous throttling.

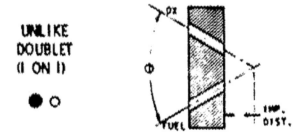

(b) Unlike triplet, in which two streams of one propellant impinge symmetrically on one stream of the other propellant. It provides a good mixing, but its performance is sensitive to continuous throttling.

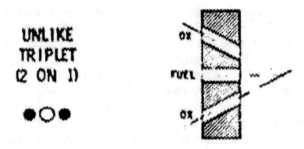

(c) Unlike quadruplet, in which two streams of one propellant impinge symmetrically on two streams of the other propellant. It can be used near the wall of the combustion chamber, but is difficult to manifold.

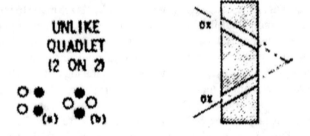

(d) Unlike quintuplet, in which four streams of one propellant impinge symmetrically on one stream of the other propellant. It provides a good mixing, but is difficult to manifold.

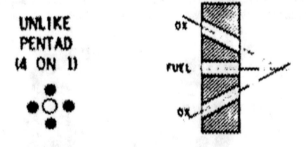

(e) Concentric tube with swirling, in which two concentric tubes are used to inject coaxially the two propellants, in the presence of a swirling device. It provides a good mixing, but is unstable when throttled.

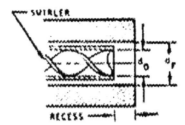

CONCENTRIC
TUBE
(WITH
SWIRLER)

(f) Concentric tube without swirling, in which two concentric tubes are used to inject coaxially the two propellants, without swirling devices. It has a very good compatibility with the wall of the combustion chamber, but provides a poor mixing.

CONCENTRIC
TUBE
(WITHOUT
SWIRLER)

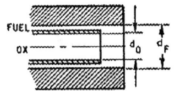

(g) Like doublet, in which two streams of one propellant are made to impinge on two streams of the other propellant. It provides a good mixing, but requires an increased axial distance to mix the fuel with the oxidiser.

LIKE
DOUBLET
(1 ON 1)

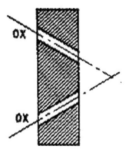

(h) Shower head, in which non-impinging streams of oxidiser and fuel enter the combustion chamber perpendicularly to the injector face. In this arrangement, the mixing of one propellant with the other is due exclusively to the turbulence in the combustion chamber. It is excellent for boundary layer cooling, but provides a poor mixing.

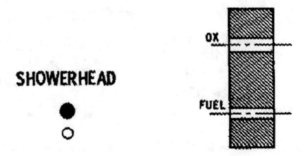

SHOWERHEAD

(i)   Variable area (movable pintle). It can be throttled and is simple to manufacture, but poses problems of compatibility with the wall of the combustion chamber.

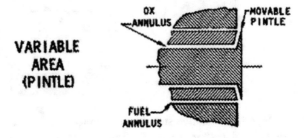

VARIABLE
AREA
(PINTLE)

(j)   Splash plate, in which the streams of the injected propellants are deflected by splash plates, These plates are kept cool by the impinging liquid propellants, which ignite only after leaving the plates. It can be throttled, but poses problems of compatibility with the wall of the combustion chamber.

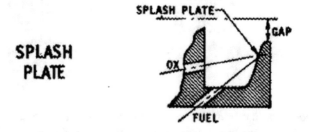

SPLASH
PLATE

The following figures, also due to the courtesy of NASA [75], show some types of injector elements used for gas/liquid injection.

(a)   Unlike triplet, in which two streams of the liquid propellant impinge symmetrically on one stream of the gaseous propellant. It provides a good mixing, but its performance is sensitive to continuous throttling.

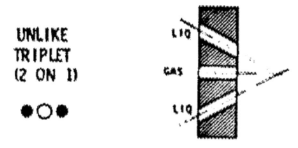

(b) Unlike quintuplet, in which four streams of the liquid propellant impinge symmetrically on one stream of the gaseous propellant. It provides a good mixing, but its performance is sensitive to continuous throttling.

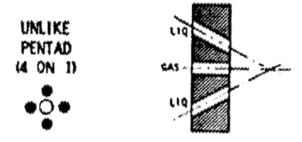

(c) Concentric tube with swirling, in which two concentric tubes are used to inject coaxially the two propellants, in the presence of a swirling device. It provides an excellent mixing, but tends to become unstable when throttled.

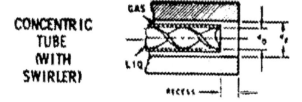

(d) Concentric tube without swirling, in which two concentric tubes are used to inject coaxially the two propellants, without swirling devices. It has a very good compatibility with the wall of the combustion chamber, but tends to become unstable when throttled.

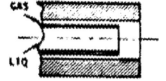

(e) Like doublet, in which two streams of the gaseous (or liquid) propellant are made
to impinge on two streams of the same propellant. It provides a good mixing,
but requires an increased axial distance to mix the fuel with the oxidiser.

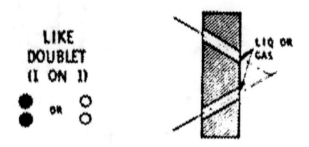

By throttling of a liquid-propellant rocket engine we mean a the variation of the
thrust of that engine with respect to the 100% rated power level.

For example, the main engine of the Space Shuttle has a rated power level (100%)
of thrust amounting to 2094223 N in vacuo and 1675200 N at the sea level. This
thrust can be varied from 1408307 N (67%) to 2281493 N (109%) in increments
of approximately 20907 N (1%). These three levels of thrust are called respectively
rated power level, minimum power level, and full power level. Throttling is obtained
by varying the output of the pre-burners, thus varying the speed of the high-pressure
turbo-pumps, and therefore the mass flow rates of the propellants [50]. Throttling
can also by obtained by using a variable-area injector. In particular, a pintle injector
has a single central pintle, which can be moved to vary the area of the orifices of the
injector, as shown in the following figure, due to the courtesy of NASA [49].

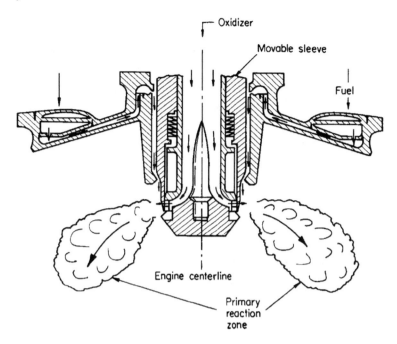

The maximum thrust of the engine is obtained by fully opening the orifices. When the area of the orifices is reduced with respect to its maximum value, the pressure in the combustion chamber and the thrust of the engine are also reduced. A pintle injector has been used in the Apollo Lunar Module Descent Engine [76], a scheme of which is shown in the following figure, due to the courtesy of NASA [77].

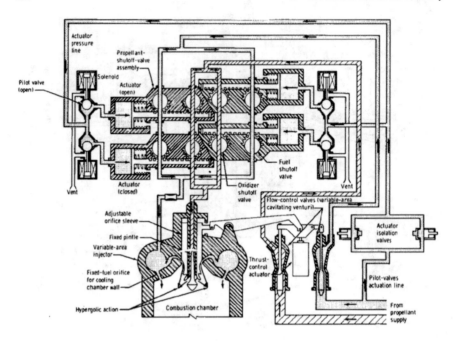

A pintle injector has also been used in the Merlin rocket engine, which burns liquid oxygen and RP-1 as propellants in a gas-generator power cycle. The Merlin engine has been developed by Space Exploration Technologies Corp. (SpaceX) for its Falcon rocket family [78].

In order to design an injector, it is necessary to take account of some factors. One of them is the injection velocity v (m/s), which is the velocity at which the oxidiser, or the fuel, is injected into the combustion chamber. The injection velocity can be determined as follows

$$v = \frac{\dot{m}}{A\rho}$$

where $\dot{m}$ (kg/s) is the mass flow rate of the propellant, A (m^2) is the area of the orifices of the injector, and ρ (kg/m^3) is the density of the propellant.

Another factor is the drop of injection pressure Δp_i (N/m^2), which can be calculated as follows

$$\Delta p_i = \frac{1}{2}\rho\left(\frac{v}{C_d}\right)^2 = \frac{1}{2\rho}\left(\frac{\dot{m}}{AC_d}\right)^2$$

where C_d is a dimensionless coefficient of discharge, whose value (usually in the range $0.5 \le C_d \le 0.92$) is determined experimentally by means of water flow tests. An orifice having a well-round entrance and a smooth bore has also a high value of C_d, and gives rise to a low value of drop of injection pressure, for the same value of injection velocity v.

Another factor is the resultant angle β of two impinging streams of propellant. With reference to the following figure, let $\dot{m}_1$ and $\dot{m}_2$(kg/s) be the mass flow rates of two impinging streams. Let α_1 and α_2 be the angles which these streams form with the axis of symmetry of the thrust chamber. Let v_1 and v_2 (m/s) be the injection velocities of the two streams, determined as has been shown above.

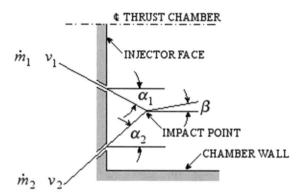

The resultant angle β of these impinging streams is defined by Huzel and Huang [5] by means of the following equation

$$\beta = \arctan\left(\frac{\dot{m}_1 v_1 \sin\alpha_1 - \dot{m}_2 v_2 \sin\alpha_2}{\dot{m}_1 v_1 \cos\alpha_1 + \dot{m}_2 v_2 \cos\alpha_2}\right)$$

The resultant angle β is defined in such a way as to be positive when the resulting stream after the impact is directed toward the chamber wall, and negative when the stream is directed toward the axis of symmetry of the thrust chamber.

In case of an engine burning a combination of hypergolic propellants, a small positive value (ranging from 0.03491 rad to 0.08727 rad) of the β angle is advantageous for the purpose of mixing the liquid propellants along the wall of the combustion chamber.

In case of an engine burning a combination of cryogenic propellants, their mixing occurs principally in the gaseous state. In these conditions, a positive value of the β angle can generate hot streaks on the wall of the combustion chamber. A negative value of the β angle should be chosen to avoid this undesirable effect.

Another factor is the injection momentum ratio, which is defined as follows

$$R_m = \frac{\dot{m}_o v_o}{\dot{m}_f v_f}$$

where $\dot{m}_o$(kg/s) and $\dot{m}_f$(kg/s) are the mass flow rates of respectively the oxidiser and the fuel, and v_o (m/s) and v_f (m/s) are their respective injection velocities. The injection momentum ratio is an index of performance. For the design of oxygen-hydrogen injectors, Huzel and Huang [5] indicate the following values of injection momentum ratio: $1.5 \leq R_m \leq 3.5$ when liquid hydrogen is injected, and $0.5 \leq R_m \leq 0.9$ when gaseous hydrogen id injected.

Still another factor is the structural load, which is due to the pressures (measured in N/m^2) exerted by the propellants behind the face of the injector (p_f) and in the injector manifolds (p_m).

In the steady state, the pressure p_f behind the face of the injector is equal to the drop of injection pressure Δp_i, defined above, as follows

$$p_f = \Delta p_i$$

The pressure p_m in the injector manifolds results from summing the pressure $(p_c)_i$ of the combustion chamber at the injector end to the drop Δp_i of injection pressure, as follows

$$p_m = (p_c)_i + \Delta p_i$$

At the start of the engine, the pressure behind the face of the injector may be much greater than the corresponding pressure in the steady state. When the valves of the propellants are opened rapidly, the pressure of the propellant can cause a hydraulic ram. Let p_p (N/m^2) be the pressure of the propellant at the moment of opening the valves. Huzel and Huang [5] indicate the following empirical formula to estimate the pressure load:

$$p_f = p_m = 4p_p$$

As an example of application, it is required to determine the size of the orifices of the injector for the two propellants, the injection velocity, and the injection momentum ratio for a rocket engine of given properties.

Let us consider again the rocket engine described in Sects. 2.3, 2.4, and 2.5. This engine burns a combination of liquid oxygen (oxidiser) with RP-1 (fuel). The density of the oxidiser is $\rho_o = 1141$ kg/m^3 and its mass flow rate in the thrust chamber is $\dot{m}_o = 880.4$ kg/s. The density of the fuel is $\rho_f = 808.1$ kg/m^3 and its mass flow rate in the thrust chamber is $\dot{m}_f = 375.1$ kg/s. The drop of injection pressure is $\Delta p_i = 1.379 \times 10^6$ N/m^2 for both of the propellants. A coefficient of discharge $C_d = 0.75$ is taken for both of them.

The total area A_o of the orifices for the oxidiser in the injector results from the following equation

$$\Delta p_i = \frac{1}{2\rho_o} \left(\frac{\dot{m}_o}{A_o C_d} \right)^2$$

which, solved for A_o, yields

$$A_o = \frac{\dot{m}_o}{C_d (2\rho_o \Delta p_i)^{\frac{1}{2}}} = \frac{880.4}{0.75 \times (2 \times 1141 \times 1.379 \times 10^6)^{\frac{1}{2}}} = 0.02093 \, \text{m}^2$$

Likewise, the total area A_f of the orifices for the fuel in the injector results from

$$A_f = \frac{\dot{m}_f}{C_d (2\rho_f \Delta p_i)^{\frac{1}{2}}} = \frac{375.1}{0.75 \times (2 \times 808.1 \times 1.379 \times 10^6)^{\frac{1}{2}}} = 0.01059 \, \text{m}^2$$

In the present case (liquid/liquid injection), an injector of the like-doublet type can be used, in which oxidiser and fuel jets are made to impinge in pairs, as has been shown above.

In case of using 700 doublets for the oxidiser and 700 doublets for the fuel, the area of each orifice for the oxidiser is

$$a_o = \frac{A_o}{2 \times 700} = \frac{0.02093}{1400} = 0.1495 \times 10^{-4} \, \text{m}^2 = 0.1495 \, \text{cm}^2$$

Hence, the diameter of each orifice for the oxidiser is

$$d_o = \left(\frac{4a_o}{\pi}\right)^{\frac{1}{2}} = \left(\frac{4 \times 0.1495 \times 10^{-4}}{3.1416}\right)^{\frac{1}{2}} = 0.004362 \, \text{m} = 4.362 \, \text{mm}$$

Likewise, the area of each orifice for the fuel is

$$a_f = \frac{A_f}{2 \times 700} = \frac{0.01059}{1400} = 7.564 \times 10^{-6} \, \text{m}^2$$

Hence, the diameter of each orifice for the fuel is

$$d_f = \left(\frac{4a_f}{\pi}\right)^{\frac{1}{2}} = \left(\frac{4 \times 7.564 \times 10^{-6}}{3.1416}\right)^{\frac{1}{2}} = 0.003103 \, \text{m} = 3.103 \, \text{mm}$$

The mean velocities of injection for respectively the oxidiser (v_o) and the fuel (v_f) result from the following equation

$$v = \frac{\dot{m}}{A\rho}$$

For the oxidiser, we find

$$v_o = \frac{\dot{m}_o}{A_o \rho_o} = \frac{880.4}{0.02093 \times 1141} = 36.87 \, \text{m/s}$$

For the fuel, we find

$$v_f = \frac{\dot{m}_f}{A_f \rho_f} = \frac{375.1}{0.01059 \times 808.1} = 43.83 \text{ m/s}$$

The injection momentum ratio results from

$$R_m = \frac{\dot{m}_o v_o}{\dot{m}_f v_f} = \frac{880.4 \times 36.87}{375.1 \times 43.83} = 1.974$$

2.7   Gas-generator and Other Engine Cycles

A gas generator is the part of a liquid-propellant rocket engine which supplies energy to drive the turbo-pumps. A turbo-pump is a rotating machine which takes one of the liquid propellants at low pressure from a tank and supplies it to the combustion chamber at the required mass flow rate and injection pressure. The energy absorbed by the turbine is provided by the expansion of compressed gases, which are usually mixtures of the propellants burned in the engine [79]. There are several ways to use the propellants for this purpose. Therefore, there are several arrangements of components, which are called rocket engine cycles. Some of these cycles are briefly described below.

The gas-generator cycle is shown in the following figure, due to the courtesy of NASA [80].

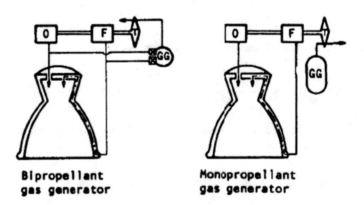

Bipropellant gas generator   **Monopropellant gas generator**

In the bi-propellant gas-generator cycle (left), the working fluid for the turbine (T) is derived by combustion of the oxidiser (O) with the fuel (F) in the gas generator (GG) at a temperature below the turbine temperature limits. A gas generator consists of a propellant valve, an injector, and a combustion chamber. The propellants for the gas generator are tapped from the turbo-pump discharge lines, injected into the combustor through the gas-generator injector, burned, and converted to gas. This gas is expanded through the turbine which drives the pumps. Since the operating temperature limit of the current turbine materials is about 1090 K, the gas generator is operated with excess fuel, in order for the temperature to remain within this limit. A small amount (about 3% in the J-2X engine) of the propellants is used to keep the engine running, whereas the remaining part of the propellants is used to generate thrust. A gas generator (GG) is a separate, small combustion chamber which produces gases. These gases are used to drive the turbines connected to the pumps. The following figure, due to the courtesy of NASA [81], shows a scheme of the J-2X engine, which is based on the gas-generator cycle. This engine uses liquid hydrogen and liquid oxygen as, respectively, fuel and oxidiser (OXID).

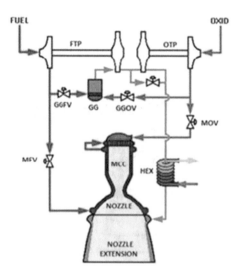

These propellants go immediately from their tanks into their turbo-pumps, which are the fuel turbo-pump (FTP) and the oxidiser turbo-pump (OTP). There, the mechanical energy of the rotating pumps is used to put the liquid propellants under pressure. At the exit from the pumps, a small amount of each propellant is tapped off to supply the gas generator (GG), which is substantially a small engine included into the principal engine. This small engine generates hot combustion products (which are, in the present case, steam and gaseous hydrogen) at high pressure. These gases are used to drive first the turbine connected to the fuel pump and then the turbine connected to the oxidiser pump. After driving the two turbines, the hot gases are used to warm the helium flowing through the heat exchanger (HEX), which is used to pressurise the oxygen tank of the stage, and are then dumped along the walls of the nozzle extension, to keep them cool. The remaining liquid oxygen, which does not go to the gas generator, is directed to the main injector. The hydrogen coming from the fuel pump is used to regeneratively cool the walls of the nozzle and the walls of the main combustion chamber (MCC), and is then directed to the injector of the main combustion chamber. A very small amount of the warm gaseous hydrogen is tapped off before entering the main injector, and is routed back to pressurise the hydrogen tank, as is the case with the helium flowing through the heat exchanger on the oxygen side. The exhaust gases coming from the turbines are dumped along the nozzle extension. These gases generate some thrust, but not as efficiently as the gases which are accelerated by flowing through the throat and the nozzle. The loss of effectiveness is the price paid for this simple engine cycle. On the other hand, the turbines of an engine using a gas-generator cycle are not subject to counter-pressures which would arise if the exhaust gases where injected into the combustion chamber. This fact makes it easy to design the turbines and the pipes. The other items shown in the preceding scheme are the main fuel valve (MFV), the main oxidiser valve (MOV), the gas-generator fuel valve (GGFV), and the gas-generator oxidiser valve (GGOV). These and other minor valves are used to control the engine during the start and shutdown transients [81]. The gas-generator cycle is used in the Vulcain, HM7B, Merlin, RS-68, RS-27A, J-2X, J-2, F-1, RD-107, and CE-20 engines.

The thrust chamber tap-off cycle is shown in the following figure, due to the courtesy of NASA [80].

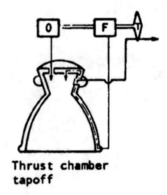

Thrust chamber tapoff

In the tap-off cycle, the working fluid for the turbine is tapped off near the face of the injector at a location in which a sufficiently cool gas is available [82]. The thrust chamber tap-off cycle is used in the J-2S engine.

The following figure, due to the courtesy of NASA [80], shows the expander cycle, which is used in the RL10, Vinci, RD-0146, YF-75D, LE-5A/5B (expander bleed cycle), LE-9, and MB-60 engines.

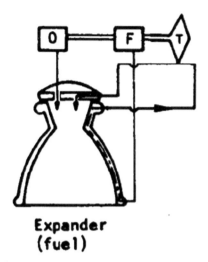

**Expander
(fuel)**

In the expander cycle, also called hot-fuel tap-off cycle, hydrogen (F) is evaporated and heated in the jacket of the thrust chamber, and is then used to drive the turbines. The exhaust gas coming from the turbines is fed to the combustion chamber [82].

The following figure, due to the courtesy of NASA [80], shows the staged-combustion cycle.

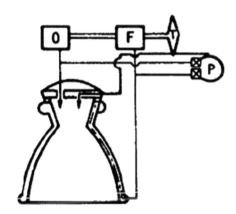

Staged combustion

In the staged-combustion cycle, most of the fuel (F), except a small quantity used as coolant, and a small amount of the oxidiser (O) are pre-burned in a pre-burner (P) at an extremely fuel-rich mixture. The resulting fuel-rich hot gas is used to drive the turbo-pump turbine, and is then injected into the main combustion chamber together with the remaining oxidiser and the coolant fuel, where all of them are finally burned.

The following figure, due to the courtesy of Boeing-Rocketdyne [50], shows the staged-combustion cycle, used in the RS-25 engine (which is the main engine of the Space Shuttle) and in the RS-170/180 engine.

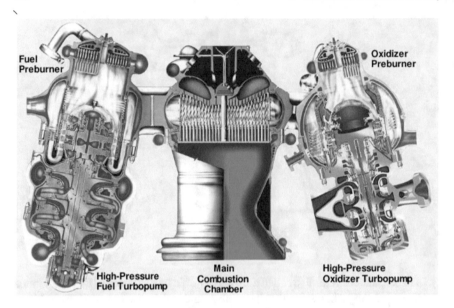

Liquid-propellant rocket engines may also use solid-propellant gas generators for turbine spinners at the engine start or for other applications of brief duration. This is because the temperature of the gases generated by solid-propellant gas generators is usually above 1366 K, which fact makes them unfit for non-cooled components over long times. These gas generators are only used to start, but the pressure and the gas flow should continue after the burnout of the solid propellant and the takeover of the liquid propellants.

The following figure, due to the courtesy of NASA [5] shows one of such devices, which is a disposable solid-propellant gas generator used to drive the turbines at the engine start. This device is a cartridge bolted to a flange at a liquid-propellant gas generator, as will be shown below. This particular cartridge can be used only once.

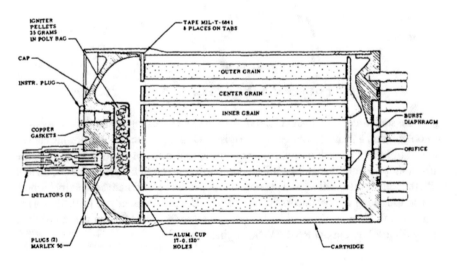

A rocket engine may also be started by using a start tank containing a gas (for example, hydrogen or helium) stored under high pressure. This device, also known as start bottle, is shown in the following figure, due to the courtesy of NASA [17], where the J-2 engine is illustrated.

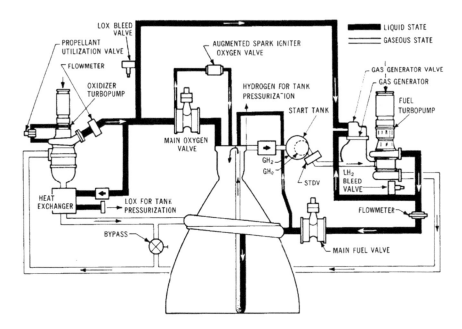

The J-2 engine was used for the Saturn IB and Saturn V launch vehicles for the NASA Apollo programme. When the engine is started, the gas is allowed to leave the start tank and impinge on the blades of the turbines which drive the pumps.

Solid propellants are mixtures of fuels with oxidisers. Their exposed surfaces burn uniformly at rates depending on the temperature and on the pressure of the combusted gases. According to Huzel and Huang [5], the combustion rate R (m/s) of a given solid propellant can be expressed as follows

$$R = k_1 \left(\frac{p_c}{6.895 \times 10^6} \right)^n$$

where k_1 (m/s) is a constant quantity, which expresses the constant burning rate of a given propellant, at a given initial temperature, and at a pressure of 6.895×10^6 N/m^2 in the combustion chamber, p_c (N/m^2) is the pressure in the combustion chamber, and n is a constant quantity, whose value depends on the sensitivity of the burning rate of the propellant to changes of pressure, at a given temperature.

The mass flow rate $\dot{m}_g$ (kg/s) through a solid-propellant gas generator can be calculated by means of the following equation of [5]:

$$\dot{m}_g = A_b R \rho_p$$

where A_b (m^2) is the burning area of the propellant, R (m/s) is the combustion rate, and ρ_p (kg/m^3) is the density of the propellant.

In order for the mass flow rate of a solid-propellant gas generator to be constant, the burning area of the propellant must also be constant.

The total area A_o (m^2) of the orifices of a solid-propellant gas generator must be such as to satisfy the following equation of [5]:

$$p_c = k_2 \left(\frac{A_b}{A_o} \right)^{\frac{1}{1-n}}$$

where p_c (N/m^2) is the pressure in the combustion chamber, k_2 (N/m^2) is a constant quantity for a given propellant at a given temperature, A_b (m^2) is the burning area of the propellant, and n is the constant quantity mentioned above.

Liquid mono-propellants may also be used to generate gases. The following figure, due to the courtesy of NASA [83], shows two of these gas generators using (left) hydrazine and (right) hydrogen peroxide.

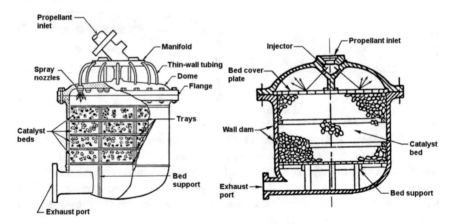

Gas generators using mono-propellants are simple and easy to control. They require a separate tank, unless the gas-generating mono-propellant is also used to feed the main engine.

Liquid-propellant gas generators are frequently chosen for liquid-propellant rocket engines, because the same propellants are used in the gas generators and in such engines. The liquid-propellant gas generator used in the H-1 rocket engine is shown in the following figure, due to the courtesy of NASA [5]. The H-1 engine was developed by Rocketdyne for use in the S-I and S-IB first stages of respectively the Saturn I and the Saturn I-B rockets, which were used for the NASA Apollo programme. The propellants used in this engine are liquid oxygen and RP-1 (kerosene). The liquid-propellant gas generator of the H-1 engine produces combusted gases during steady-state operation to drive the two-stage turbine.

The turbine supplies power through a gear reduction train to drive the propellant pumps. The liquid-propellant gas generator of the H-1 engine consists of a gas generator control valve, an injector assembly, a combustor, and two squib-less igniters. These components are briefly described below. The propellants entering the liquid-propellant gas generator are ignited by a solid-propellant gas generator (SPGG in the following figure) and by the two squib-less igniters during engine start. The hot gases from the solid-propellant gas generator ignite the squib-less igniters before the liquid propellants enter the liquid-propellant gas generator. The igniters burn for 2.5–3 s to ensure the ignition of the liquid propellants. The control valve of the liquid-propellant gas generator is a normally closed valve containing two poppets, which admit the bootstrap propellants into the combustor of the gas generator during engine operation.

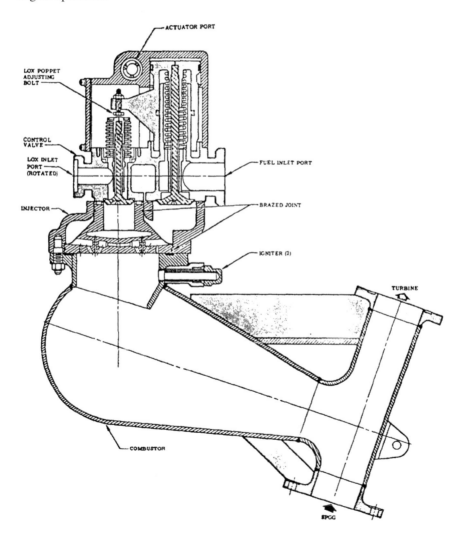

The combustion pressure in the thrust chamber actuates the control valve by means of a piston, which opens the fuel poppet first. A yoke integral with the piston opens the oxidiser poppet. The cracking pressures relative to the atmosphere are $(7.239 \pm 1.379) \times 10^5$ N/m^2 for the fuel poppet and $(13.79 \pm 1.379) \times 10^5$ N/m^2 for the oxidiser poppet. The fully-open operating pressure relative to the atmosphere is $(18.96 \pm 1.724) \times 10^5$ N/m^2 for the control valve. The bootstrap propellant mass flow rates are 6.133 kg/s for the fuel and 2.091 kg/s for the oxidiser. A bellows assembly enclosing the liquid oxygen poppet stem and the closure spring, and the seals on the actuator piston prevent leakage of fuel and oxidiser into the control valve actuator. A drain line directs any fuel leakage into the valve actuator to the fuel drain manifold, where it is dumped overboard into the engine exhaust stream. The control valve is designed to ensure a fuel-rich cut-off to prevent an excessive increase in temperature in the combustor with consequent turbine burning. The spring pressure closes the control valve at engine cut-off, when the pressure in the thrust chamber decreases.

The fuel and the oxidiser from the gas generator control valve enter the injector and flow through passages which provide a uniform mixture ratio $o/f = 0.3409$. The injector cavity is designed to permit an oxidiser lead into the combustor during start to prevent detonation. From the injector, two fuel streams impinge on a single oxidiser stream. The injector has 44 impingement points. The fuel which enters the combustor through 36 holes around the periphery of the impingements provides film coolant for the injector. During countdown, the oxidiser injector of the gas generator receives a purge with ambient gaseous nitrogen to prevent entrance of contaminants from the solid-propellant gas generator into the injector.

The bootstrap propellants burn in the combustor and exit to the gas turbine. Two squib-less igniters installed in the combustor just below the injector of the gas generator assure propellant ignition during start. The combustor is a welded assembly with flanges for installation of the solid-propellant gas generator and for attachment to the gas turbine. The operating temperature is 920.9 ± 264.3 K, and the operating absolute pressure is $(4.456 \pm 0.03378) \times 10^6$ N/m^2.

Two squib-less igniters installed in the injector mounting flange on the combustor burn for 2.5–3 s after their ignition by the solid-propellant gas generator. They ensure ignition of fuel and oxidiser if the solid-propellant gas generator should have expired before bootstrap propellant entry into the combustor.

A gas generator burns the propellants just as the main engine does, the only difference being the capability possessed by the former of varying the mixture ratio o/f, whose value is adjusted in order for the combusted gas to have the desired temperature and chemical properties.

The guidelines for designing a gas generator for a liquid bi-propellant rocket engine are the same as those relating to the thrust chamber of the main engine. When the characteristic length L^* is calculated, the volume to be taken into account goes from the injector to the throat of the nozzle of the turbine. Account is also taken of the maximum temperature which the materials of which the turbine is made can withstand. Consequently, the design temperature of the turbine is usually kept under 1255 K [5].

2.8 Igniters

An igniter is a device which releases energy (usually in form of heat) in a rocket engine, for the purpose of initiating the combustion of the main propellants. This combustion, after initiation, is capable of sustaining itself without the necessity of receiving heat. The source of energy which stimulates an igniter to release heat is either inside (as is the case with a solid propellant) or outside (as is the case with a spark arrangement) the igniter itself. The principal types of igniters used in rocket engines are described below.

Pyrotechnic igniters are devices containing mixtures of fuels, oxidants, and often other materials. These devices are used in rocket engines (thrust chambers and gas generators) to release energy in form of heat. Pyrotechnic mixtures are in a state of meta-stability, meaning by this term, stability and non-reactivity under some conditions, and release of thermal energy originally stored in chemical form after receiving an external stimulus, which is usually given to such mixtures by addition of heat. The amount of energy required to stimulate this release of heat is called activation energy. The net amount of thermal energy released in a pyrotechnic reaction is called heat of reaction. The heats of reaction for some binary pyrotechnic compositions are given in [84]. The following figure, due to the courtesy of NASA [85], shows an electrically-triggered initiator, or squib, used to ignite a rocket engine.

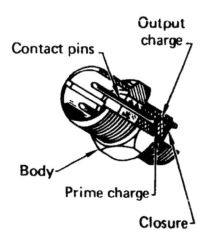

The initiator shown in the preceding figure is a body of metal containing a prime charge, an output charge, and two contact pins used as electrodes. The prime charge is the head of an electric match, which contains a short wire (a bridge wire), 0.0127–0.127 mm in diameter, made of a material (usually nichrome, which is an alloy 80% nickel and 20% chromium, by mass). This wire glows red-hot when a voltage is applied to its ends. The bridge wire is placed between the two electrodes, and is surrounded by a pyrotechnic composition (materials used for this purpose include lead styphnate, lead azide, diazodinitrophenol, and zirconium-ammonium perchlorate) which is sensitive to heat. A scheme of an electric match is shown in the following figure (re-drawn from [86].

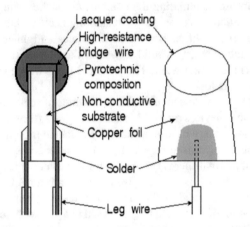

The pyrotechnic composition specified above may be surrounded by a second, less sensitive composition, which protects the first. The second composition, in turn, is covered by a nitrocellulose lacquer. A scheme of a squib, including an electric match, is shown in the following figure (re-drawn from [86].

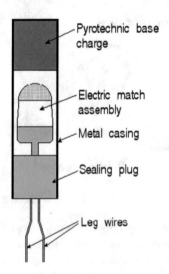

When a small voltage (28 V or less) is applied to the electrodes, the heating of the bridge wire causes the deflagration of the pyrotechnic material in contact with it. The output charge is also made of pyrotechnic material, as will be shown below. The addition of the output charge to the prime charge greatly magnifies the ignition effect.

The initiator supplies the energy, in form of heat, which is necessary to ignite the propellants. The solid-propellant charge burns with a hot flame within the combustion chamber. The igniter can be designed to fit directly onto the injector. This method of ignition can only be used once. Therefore, a rocket engine having a pyrotechnic igniter cannot be restarted, as has been shown in Sect. 2.1.

The most effective materials used for output charges in initiators are metal-oxidant pyrotechnic formulations. Some examples of these formulations are shown in the following table, due to the courtesy of NASA [85].

Application /Designation	Fuel	Oxidants	Binders
Mk 247, Mk 265 (igniters)	Boron, 23.7%	KNO_3, 70,7%	Laminac, 5,6%
XM-6 & XM-8 (EBW)			
MB-1 (500-V initiator)	Zirconium, 66.3%	$NH4ClO_4$, 32.7%	Nitrocellulose, 1.0%
FA-878	Zirconium, 40%	$BaNO_3$, 20%	
(ign. elements Mk 10,		PbO_2, 20%	
Mk 11, Mk 13, Mk 17)		PETN, 20%	
M2 squib	$Pb(SCN)_2$, 32% Charcoal, 18%	$KClO_3$, 40%	Egyptian lacquer, 10%
NOTS Model 39	Magnesium, 60%	Polytetrafluoro-ethylene, 40%	

Pyrotechnic igniters can be mounted in recesses as plugs of the screw-in type, in case of gas generators and small thrust chambers.

Hypergolic igniters are based on the property of some bi-propellant combinations which ignite spontaneously at room temperature when the two components (the fuel and the oxidiser) of a combination get in touch one with the other. Therefore, hypergolic combinations do not require an external source of ignition. They need only a valve to mix the fluids for the purpose of initiating the combustion. This method of ignition reduces the components of an ignition system, and therefore its chances of failure.

As has been shown in Chap. 1, Sect. 1.2, the most common fuels used in a combination of hypergolic propellants are hydrazine (H_2NNH_2), monomethyl hydrazine ($CH_3(NH)NH_2$), unsymmetrical dimethyl hydrazine ($H_2NN(CH_3)_2$), and Aerozine 50, the last being an equal mixture of hydrazine and unsymmetrical dimethyl hydrazine. The oxidisers used in combination with these fuels are usually nitrogen tetroxide (N_2O_4) and various blends of nitrogen tetroxide with nitric oxide (NO).

Hypergolic igniters have been used in liquid bi-propellant engines. An example has been cited in Sect. 2.1, which describes a mixture of gaseous fluorine with liquid oxygen used to start an engine which burns liquid hydrogen and liquid oxygen. Two further examples are cited below. The RD-180 engine of the Atlas V rocket burns a mixture of kerosene with liquid oxygen. This engine has a device, placed in the fuel line, for hypergolic ignition. This device consists of two cartridges which contain a mixture of 15% triethylaluminium ($Al_2(C_2H_5)_6$, also known as TEA, which is a pyrophoric substance) with 85% triethylborane (($C_2H_5)_3B$, also known as TEB). A scheme of the RD-180 engine is shown in the following figure, which is due to the courtesy of the United Launch Alliance [87].

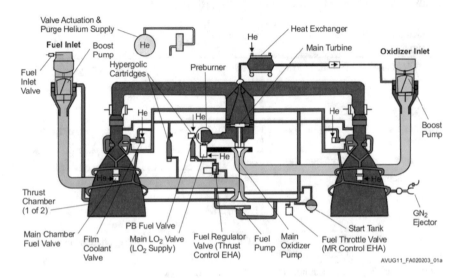

As shown in the preceding figure, the TEA-TEB mixture is stored in two closed cylindrical cartridges, one of which is in the fuel line directly ahead of the pre-burner, and the other is in either of the main fuel inlet to the two main combustion chambers. The cartridges have burst diaphragms to prevent this mixture from coming in contact with atmospheric oxygen. The pre-burner illustrated above is the gas generator which drives the main turbine.

In order to start the engine, a start tank of spherical shape, shown in the lower part of the preceding figure, is connected to both of the fuel lines by means of tubes and associated valves. The start tank is filled with fuel and put under pressure by using gas. When the valves to the fuel lines and the cartridges are opened, the high pressure of the fuel drives pistons which are in the cartridges. This causes the TEA-TEB mixture to be released into the gas generator and into the main combustion chambers, which in turn have been filled with oxygen after the liquid oxygen valves are opened, and the tank pressure causes oxygen to enter the engine. The TEA-TEB mixture comes in touch with oxygen, and therefore self-ignites and starts the combustion inside the gas generator and the main combustion chambers. The combustion is sustained by the fuel entering the gas generator and the main combustion chambers just after the TEA-TEB mixture. Now, the gas generator is running, and therefore the main turbine drives the boost pumps (shown on the two sides of the preceding figure), which in turn discharge the propellants under pressure to the gas generator and to the main combustion chambers.

The F-1 engine of the Saturn V rocket burns a mixture of RP-1 (kerosene) with liquid oxygen. This engine also has an igniter containing a cartridge of hypergolic fluid (which is, again, a mixture of 85% triethylborane with 15% triethylaluminium) with burst diaphragms at either end. This igniter is in the fuel line, and the hypergolic fluid has its own orifices in the injector of the main combustion chamber. The hypergolic fluid, followed by the fuel, enters the main combustion chamber. There, this fluid ignites spontaneously in contact with the liquid oxygen already injected into the main combustion chamber. This method of ignition is also known as hypergolic starting slug [88].

The hypergolic igniters described in the examples cited above can only be used once, and therefore the rocket engines using them cannot be restarted. However, Hulka et al [89] have described a modified version of the Russian NK-33 rocket engine, called AJ26-59, which has the capability of being restarted. This engine burns liquid oxygen and kerosene, which of course are not hypergolic propellants, and also uses the TEA-TEB mixture for hypergolic ignition. Its capability of being restarted is obtained by carrying on board further cartridges of the TEA-TEB mixture than those necessary to the first start.

A rocket engine designed to burn exclusively hypergolic propellants does not require any special apparatus to be restarted as many times as necessary, because its propellants ignite spontaneously when coming in mutual contact. Such is the case, for example, with the AJ10-190 engine (whose fuel and oxidiser are respectively monomethyl hydrazine and nitrogen tetroxide) used in the orbital manoeuvring system of the Space Shuttle. An engine using other propellants than those of hypergolic type requires a re-usable igniter (a spark plug, for example) and a system of valves in order to be started and stopped when necessary.

Electric spark igniters based on spark plugs are often used in liquid-propellant rockets when multiple starts are necessary. They are efficient and reliable, and can be used in either direct spark igniters or augmented spark igniters.

In case of a direct spark igniter, a high-voltage electric circuit is used to generate a spark across a gap, in order to expose the vaporised propellants in the combustion chamber to a ionising electric discharge, as shown in the following scheme, due to the courtesy of NASA [90].

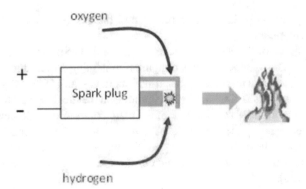

Direct spark igniters are used is small combustion chambers, because the electric discharge is confined to a very small zone. One of the six spark plugs used in the main engine of the Space Shuttle is shown in the following figure, due to the courtesy of Boeing-Rocketdyne [50].

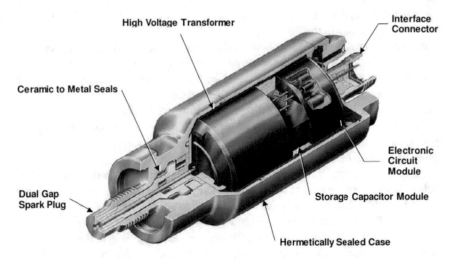

In the main engine of the Space Shuttle, an augmented spark igniter (also known as torch igniter), is used. In this type of igniter, a small quantity of propellants (in the present case, hydrogen and oxygen) is swirled into a very small zone, where a spark plug generates an electric discharge. This causes the ionisation of the gasified propellants, which become very hot and generate a flame front which propagates toward the combustion zone, just when the rest of the propellants reaches the injector. The following figure, due to the courtesy of Boeing-Rocketdyne [50], illustrates an augmented spark igniter used in the main engine of the Space Shuttle.

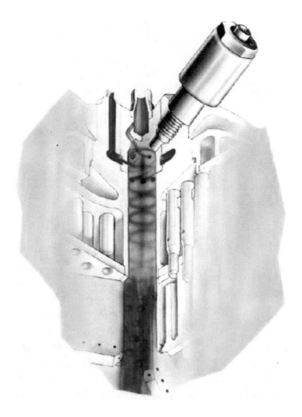

The J-2 rocket engine, which was used for the Saturn IB and Saturn V launch vehicles for the NASA Apollo programme, had an augmented spark igniter. Each of the main engines of the Space Shuttle had three augmented spark igniters, one for the main combustion chamber and two for the pre-burners. The chamber of each augmented spark igniter was located in the centre of the injector, as shown in the preceding figure. Two spark plugs were used, for redundancy, in each augmented spark igniter.

A further scheme of this type of igniter is shown in the following figure, due to the courtesy of NASA [91], which illustrates an augmented spark igniter used in a 445 N reaction control engine burning a mixture of liquid oxygen (LO_2) with liquid methane (LCH_4).

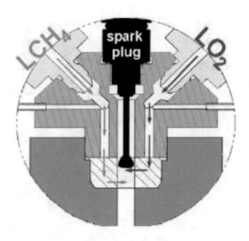

In case of rocket engines using liquid mono-propellants (for example, hydrogen peroxide), catalysts are used to initiate and sustain the reaction of decomposition of the mono-propellant. Liquid catalysts, if used for this purpose, require complex systems of valves and interlocking devices. This can be avoided by using solid catalysts. These catalysts have also been used in liquid bi-propellant rocket engines. One of such engines is the AR2-3, which was developed by Rocketdyne in the 1950s, and is briefly described below. The AR2-3 is one of the family of AR engines, where AR stands for aircraft rocket. The first engine of this family was the AR-1, which operated at a fixed thrust of 25577 N. It was proven in flight on the FJ-4 aircraft. The AR2 series of engines includes the AR-2, the AR2-1, the AR2-2, and the AR2-3. All the engines of the AR2 series have a main-stage thrust of 29358 N and are variable down to 14679 N of thrust. These engines use 90% hydrogen peroxide and kerosene. They have been used oh the FJ-4, F-86, and NF104A aircraft. They are liquid-propellant pump-fed engines designed to provide aircraft thrust augmentation. The AR2-3 rocket engine supplies hydrogen peroxide and kerosene propellants to the thrust chamber by means of two centrifugal pumps (one for the oxidiser and one for the fuel). These pumps are directly driven by a single turbine. The pumps and the turbine are mounted on the same shaft. The oxidiser flows from the pump outlet through the pressure-actuated oxidiser valve, through the thrust chamber cooling jacket, and into the main thrust chamber, through the silver-plated catalytic screen pack, where it is decomposed into super-heated steam and oxygen. The fuel flows from the pump outlet through the chamber-pressure-actuated fuel valve, into the concentric annular-ring type fuel injector, and is injected into the hot, oxygen-rich gases, where it burns and is exhausted through the 12:1 area ratio nozzle. The auto-ignition of the fuel eliminates the necessity of an ignition system. A small oxidiser flow, of about 3% from the oxidiser pump discharge, is delivered and metered through the thrust control valve into a catalytic gas generator, where it is decomposed into super-heated steam and oxygen to drive the turbine.

A scheme of the engine flow described above is shown in the following figure, due to the courtesy of NASA [92].

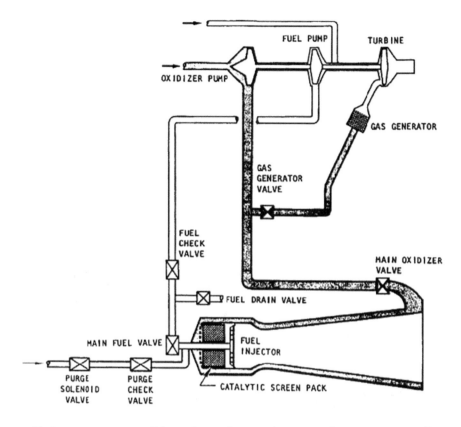

Under emergency conditions, the engine may be operated as a mono-propellant engine using the oxidiser. The engine operates at a moderate chamber pressure and provides 29358 N of thrust in vacuo and 246 s specific impulse.

A timely ignition of the propellants is essential to the safe accomplishment of a mission. This holds in particular in case of manned missions. For this purpose, a safety device is required to control the supply of fuel or oxidiser or both to a liquid-propellant engine. A delayed ignition in the combustion chamber causes an accumulation of fuel within the chamber. This accumulation can cause an explosion and consequently the destruction of the engine or even of the whole vehicle on which the engine is mounted. These destructive effects can be avoided by interrupting or reducing the flow of either fuel, or oxidiser, or both to the combustion chamber when the combustion does not take place within a proper period of time. When a malfunction is detected, the control system switches the engine to a safe lock-up mode or shuts the engine down.

In order to assure a timely and even ignition of propellants and a smooth and quick increase in their flow to the rated value, the initial values of flow and mixture ratio are different from the corresponding operational values. A low initial value of flow prevents an excessive quantity of unburned fuel from accumulating in the combustion chamber. An initial value of mixture ratio close to the stoichiometric value results in a high release of heat per unit mass of propellant. This fact, in turn, makes it possible to the hot gases in the combustion chamber to reach an equilibrium more rapidly than would be the case with other values of mixture ratio. By contrast, as has been shown in Chap. 1, Sect. 1.4, the mixture ratio in operating conditions is usually fuel-rich, in order to obtain a high value of specific impulse.

According to Sutton and Biblarz [2], the total time required to start a liquid-propellant rocket engine comprises ideally the following times:

(1)   time (from 0.002 s to more than 1 s) taken by the valves, which control the admittance of the propellants in the combustion chamber, to move from the closed position to the fully open position;
(2)   time taken by the propellants to go from the valves to the injector face;
(3)   time taken by the propellants to form and mix discrete streams of droplets into the combustion chamber;
(4)   time (from 0.02 to 0.05 s) taken by the droplets of propellant to vaporise and ignite;
(5)   time taken by the flame front to propagate downstream of the ignition section along the axis of the combustion chamber; and
(6)   time taken by the ignited propellant to raise the temperature and the pressure within the combustion chamber to values so high as to make the combustion self-sustaining.

There are overlaps in these intervals of time, because some of them may occur simultaneously. Apart from interval (4), the length of these intervals increases with the diameter of the combustion chamber.

The propellant valves are designed in such a way as to operate in a desired sequence, in order to admit one of the propellant before the other in the combustion chamber, and to also control the flow and the mixture ratio. Such valves are often partially opened at the ignition time, for the purpose of avoiding the accumulation of unburned propellants in the combustion chamber. After a signal indicating a successful ignition has been received by the controller, the valves are moved to the fully open position and the full flow of propellants arrives at the combustion chamber.

The signal of successful ignition can be generated by using several devices. Some of such devices are indicated below. They are:

(a)   Photocells for visual detection.
(b)   Optical detectors, such as cells, mounted inside the combustion chamber, which are sensitive to the visible or infrared radiation which is emitted by the combusted gases.
(c)   Pyrometers, which determine the temperature of the combusted gases by measuring the wavelength of the emitted radiation.
(d)   Fusible wire links, which are based on a fusible wire strung between two parallel conductors placed near the exit section of the thrust chamber, downstream of the throat. This wire is also a conductor, which is fused by the ignition flame, when the combustion reaction has been initiated correctly. The control system is substantially an electric motor, which is connected to a source of electric

current (for example, to an accumulator). When the wire is melted, due to the heat generated by a correct ignition, then the motor circuit is interrupted and the propellant valves reach the fully open position. By contrast, when the wire is not melted, because the combustion reaction has not been initiated correctly, then the electric motor continues to operate, and moves the propellant valves to the closed position. Before the rocket engine is started, the propellant valves are in the closed position and the electric circuit carries no current. At the starting moment, the closure of a switch causes the electric motor to be fed by current, and therefore the propellant valves are moved toward the open position. A device like this was invented in 1952 by Prentiss [93].

(e) Pressure-sensing devices, apt to detect the pressure rise in the combustion chamber due to a correct ignition of the propellants. Such devices, if capable of withstanding very high pressures, can also be used for multi-start engines.

(f) Resistance wires, capable of measuring different values of electric resistance depending on whether a correct ignition has, or has not, taken place in the combustion chamber. Such devices, if capable of withstanding high temperatures without melting, can also be used for multi-start engines.

(g) Electric devices used in spark plugs. Such devices detect the correct operation of spark plugs by sensing the ionisation near the electrodes.

2.9 Combustion Instability

As has been shown in Chap. 1, Sect. 1.6, by combustion instability we mean the presence of self-sustaining pressure oscillations in the combustion chamber of a rocket engine. This phenomenon can appear in thermal devices when an unsteady release of heat is coupled with pressure oscillations. In a rocket engine, the heat released by the substances which react chemically is coupled with acoustic fluctuations. When the unsteady heat release and the acoustic fluctuations are in phase, then even a small perturbation will amplify according to the criterion enunciated by Lord Rayleigh. This criterion states that "If heat be given to the air at the moment of greatest condensation, or be taken from it at the moment of greatest rarefaction, the vibration is encouraged. On the other hand, if heat be given at the moment of greatest rarefaction, or abstracted at the moment of greatest condensation, the vibration is discouraged". In addition, "When the transfer of heat takes place at the moment of greatest condensation or of greatest rarefaction, the pitch is not affected … the pitch is raised if heat be communicated to the air a quarter period before the phase of greatest condensation; and the pitch is lowered if heat be communicated a quarter period after the phase of greatest condensation" [94]. In other terms, the oscillations in a combustor are self-sustained when the rate of heat release and the pressure fluctuations are in phase. When this happens, pressure oscillations of large amplitude occur in a combustion chamber at frequencies of hundreds of hertz. Acoustic waves cause unsteadiness in a combustion process. An unsteady combustion process, in turn, generates new acoustic waves. This fact can be represented graphically by means of the following feedback scheme.

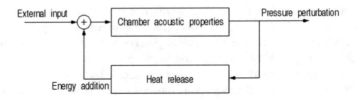

An unsteady release of heat adds energy to the acoustic waves in the combustion chamber, when the unsteady heat release rate is in phase with the pressure oscillations at the location of the heat source, in accordance with the criterion of Lord Rayleigh. Consequently, the flame in the chamber may be extinguished or may come back to the injector face and burn part of the injection system. The amplitude of these oscillations can become so large as to damage the structure of the combustion chamber or destroy the engine.

These pressure waves can travel at sonic velocities up and down the length of the combustion chamber, around it, across it, or radially in and out. The following figure, due to the courtesy of NASA [95], shows the fundamental modes of acoustic combustion instability.

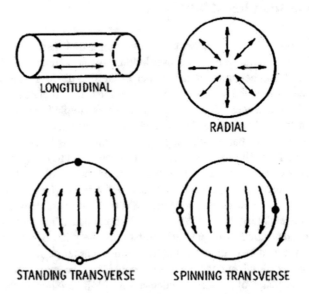

The F-1 engine, which was designed in the 1950s for the Saturn V rocket, was subject to combustion instability. This instability was observed for more than seven years during the development of the engine. Twenty of the forty-four tests of the first F-1 engine showed pressure oscillations due to combustion instability with peak amplitudes greater than or comparable to the average pressure in the combustor. The pressure oscillations caused erosion and burning of the injector face, as a result of large radial and tangential motions of the gas.

In order to solve the instability problems, a programme called Project First was created to develop a stable F-1 engine. This programme lasted for four years: from 1962 to 1965. During this time, the engine was subject to more than 2000 full-scale tests [96]. Copper baffles were added to the injector plate to restrain the propagation of the pressure waves. The recovered injector plate of an F-1 engine (on display at the Museum of Flight in Seattle, Washington, U.S.A.) is shown in the following figure, due to the courtesy of Wikipedia [97].

Another example of passive control is the Pratt & Whitney FT8 gas turbine, which uses 13 Helmholtz resonators. These devices act as vibration absorbers, whose shape is chosen so that the absorption is tuned to a particular frequency, as the sequel will show.

As has been shown above, the symmetry of the combustion chamber lends itself to certain natural acoustic modes which are driven into resonance because of their coupling with the rate of heat release. Several methods of breaking the symmetry of the chamber have been proven effective in quenching the combustion instabilities.

Combustion instabilities have been classified according to their respective frequency ranges, even though there are no sharp limiting lines to separate the so-called low, intermediate, and high frequency ranges.

The low-frequency (also called chugging) class includes pressure waves whose frequencies are less than 180 hertz [5]. In other terms, the low-frequency class includes waves whose lengths are much larger than the characteristic length (defined in Sect. 2.4) of the combustion chamber. The instabilities belonging to this class begin with a sinusoidal wave, whose amplitude grows linearly with time. Such instabilities result from a coupling between the combustion process and the feed system [47]. They are usually eliminated by increasing the pressure drop in the injector, or by increasing the length-to-diameter ratio in the injector, or by decreasing the volume of the combustion chamber [49].

The high-frequency (also called screaming) class includes pressure waves whose frequencies are equal to or greater than 1000 hertz [5]. The following figure, due to the courtesy of NASA [98], shows pressure oscillation in a combustion chamber of a rocket engine in the presence of a high-frequency instability.

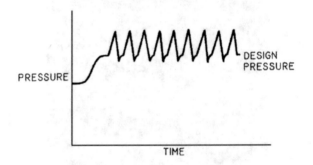

This is type of combustion instability which causes the most destructive effects. An oscillatory source of energy is required to sustain this particular instability. This energy comes from the combustion of the propellants, and is supplied according to the manner in which the combustion takes place. Each pressure wave interacts with the burning gases so strongly as to receive the sustaining energy directly from them, which means within a time interval no longer than the semi-period of the wave itself. Some of the means by which the combustion process supplies energy to the pressure waves are loss of ignition, delays due to chemical preparation or other causes, detonations, anomalous behaviour of droplets heated above their critical temperature and pressure, et c.

The two principal methods used to eliminate the combustion instability of the high-frequency class are the following: (1) changing the environment in which the combustion process takes place, by changing either the geometric properties of the combustion chamber or the manner (pattern of the injection holes, or size of the injector holes, or pressure drop) in which the propellants are injected into the combustion chamber; and (2) increasing the damping characteristics of the combustion chamber. Some means of passive control, which can be classified in the second of these methods, are baffles, acoustic liners, and resonators installed in the combustion chamber. They are briefly described below.

Baffles are mechanical devices meant to regulate or restrain the gas flow in the vicinity of the injector face, as shown in the following figure, due to the courtesy of NASA [49]. Baffles have been used successfully to prevent transverse acoustic modes of combustion instability. The transverse (that is, tangential and radial) modes are characterised by oscillatory pressure waves and gas particle motion parallel to the injector face [74].

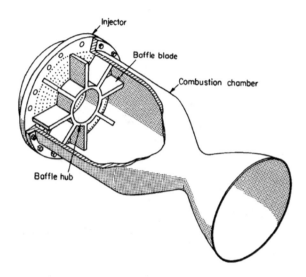

Another device for passive control is an acoustic liner installed within the combustion chamber. An acoustic liner consists of a number of cavities, each of which has a narrow opening or neck, as shown in the following figure, due to the courtesy of NASA [49].

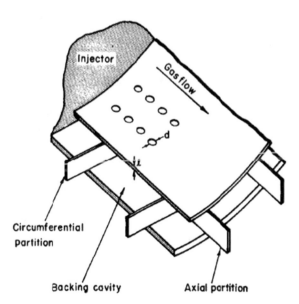

In acoustic liners, energy is dissipated because of jet formation in the flow of gas through their orifices. The following figure, due to the courtesy of NASA [49],

shows a water-cooled acoustic liner having an array of resonator cavities and used
as a combustion stabiliser for a 66723 N thrust throttleable engine burning nitrogen
tetroxide (oxidiser) and Aerozine 50 (fuel).

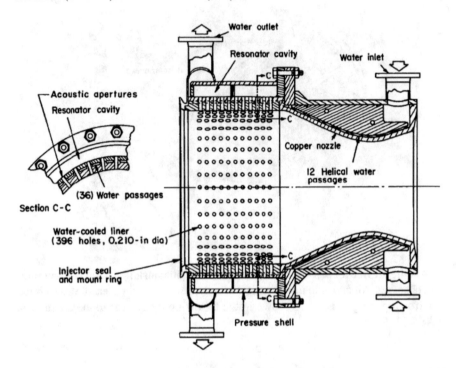

This combustion chamber consists of a composite cylindrical brazed shell with
integral coolant passages. The liner apertures are drilled in spaces between the rectangular passages. The resonator cavities have a damping effect, because they allow a
normal velocity at the wall which has a component in phase with the pressure oscillation. This means that work is done over each cycle in moving the fluid back and
forth at the boundary. This work is equal to the energy dissipated due to jet formation
and friction [49].

Two common types of acoustic resonators are Helmholtz resonators and quarter-
wavelength tubes.

A Helmholtz resonator is a rigid-walled cavity of volume V (m^3), which is
connected through a circular orifice of diameter d (m) and length ℓ (m) to the
combustion chamber, where undesirable pressure oscillations are expected to occur.
A Helmholtz resonator is shown in the following figure.

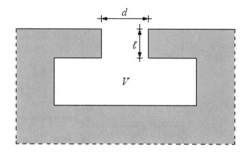

For the frequencies of interest, the wavelength λ of the pressure oscillations is assumed to be much greater than the dimensions of the resonator, that is,

$$\lambda \gg \ell$$

$$\lambda \gg d$$

$$\lambda \gg V^{\frac{1}{3}}$$

By choosing properly the values of these dimensions, this device can be made resonant at any desired frequency. The resonance frequency f_0 (Hz) of a Helmholtz resonator is [99]:

$$f_0 = \frac{a}{2\pi} \left[\frac{S}{V(\ell + \Delta\ell)} \right]^{\frac{1}{2}} = \frac{a}{2\pi} \left[\frac{\pi d^2}{4V(\ell + \Delta\ell)} \right]^{\frac{1}{2}}$$

where $a = (\gamma RT)^{\frac{1}{2}}$ (m/s) is the speed of sound, $S = \pi d^2/4$ (m²) is the area of the orifice, and $\Delta\ell = 0.85d$ (m) is the length correction, which takes account of the flow effects in the vicinity of the orifice ends.

As has been shown by several authors (see, for example, [100], a Helmholtz resonator is analogous to a mechanical system comprising a body of mass m, a spring of elastic constant k, and a dashpot of mechanical resistance R_m, which is subject to a force $F(t) = A \cos \omega t$. This system is shown in the following figure, which is due to the courtesy of NASA (adapted from [49]).

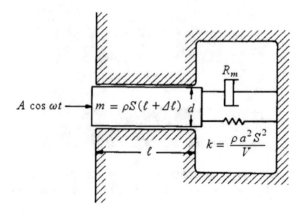

In this mechanical system, the elastic constant k of the spring corresponds to $\rho a^2 S^2/V$, and the mass m of the body corresponds to $\rho S(\ell + \Delta\ell)$, where ρ is the density of the unperturbed gas. The differential equation governing the motion of the body of mass m in the direction x of the driving force $F(t) = A \cos \omega t$ is

$$m\frac{d^2x}{dt^2} + R_m\frac{dx}{dt} + kx = A \cos \omega t$$

As is well known, the resonance frequency of this mechanical system is

$$f_0 = \frac{1}{2\pi}\left(\frac{k}{m}\right)^{\frac{1}{2}}$$

This equation accounts for the expression $f_0 = [a/(2\pi)]\{S/[V(\ell + \Delta\ell)]\}^{1/2}$ written above for the resonance frequency of a Helmholtz resonator.

A Helmholtz resonator can be tuned to a resonant frequency which corresponds to one of the natural frequencies of vibration of the combustion chamber. These natural frequencies can be determined as will be shown below.

Another type of acoustic resonator commonly used to damp pressure oscillations in a combustion chamber is a quarter-wavelength tube. It is similar to a Helmholtz resonator, but has no cavity, as shown in the following figure.

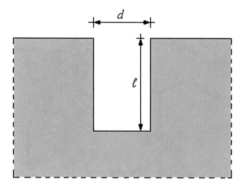

The resonance frequency f_0 (Hz) of a quarter-wavelength tube is [99]:

$$f_0 = \frac{a}{4(\ell + \Delta\ell)}$$

where $a = (\gamma RT)^{1/2}$ (m/s) is the speed of sound, ℓ (m) is the length of the quarter-wavelength tube, and $\Delta\ell = 0.85d$ (m) is the correction of the tube length ℓ.

In order to determine the natural modes of vibration for a given combustion chamber, Frendi et al [99] placed a loudspeaker at the centre of the inlet boundary of their combustion chamber. The loudspeaker emitted plane harmonic or random acoustic waves. A random acoustic disturbance was generated by the loudspeaker, and the sound pressure level (dB) was determined as a function of frequency at several points in the combustion chamber. By so doing, several peaks corresponding to the various vibration modes of the combustion chamber were determined. In order to determine the mode shapes, the individual modes were excited by introducing a plane, harmonic acoustic wave at the given frequency. After determining the natural frequencies (512 and 1880 Hz) of their combustion chamber, Frendi et al [99] selected the frequency of 1880 Hz, corresponding to one of the natural frequencies found by them, for extensive tests. Consequently, the resonant frequency f_0 of the acoustic devices indicated above was tuned to 1880 Hz.

Other types of acoustic resonators than those described above have slots instead of circular orifices, as shown in the following figure, due to the courtesy of NASA [74].

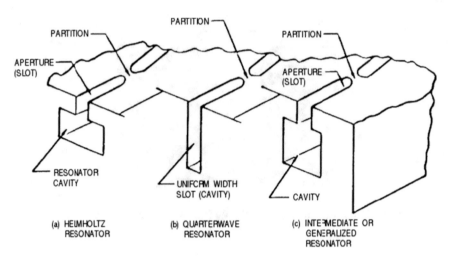

(a) HELMHOLTZ
    RESONATOR

(b) QUARTERWAVE
    RESONATOR

(c) INTERMEDIATE OR
    GENERALIZED
    RESONATOR

They are said to be of the Helmholtz, quarter-wavelength, or intermediate type, according to whether the width of their cavities is much larger, or equal, or slightly larger in comparison with the width of their slots.

The medium-frequency (also called buzzing) class includes pressure waves whose frequencies are in the range 180–1000 hertz [5]. The beginning of a medium-frequency instability usually shows a growing coherence of the combustion noise at a particular frequency with slowly increasing amplitude. There is usually wave motion in the propellant feed system. There may also be wave motion in the combustion chamber, but this motion does not usually correspond in phase and frequency to a natural mode of the chamber itself. The shape of the pressure wave is very nearly sinusoidal, and one or both of the propellant feed system may be highly coupled. Buzzing-type instabilities are not particularly damaging if they remain at low amplitudes, but may degrade performance, total impulse, or thrust vector. In some case, the amplitude increases to such an extent as to triggering a high-frequency mode [49]. This type of instability occurs more frequently in medium-size engines (whose thrust is in the range 2000–250,000 N) than in large engines [2].

Finally, many rocket vehicles propelled by engines burning liquid propellants are subject to longitudinal vibrations, because of an instability arising from interaction of the vehicle structure with the propulsion system. This phenomenon has been nicknamed pogo instability, because of its analogy with the motion of a pogo jumping stick. Pogo instabilities can occur in the propellant feed lines of large vehicles, such as space launch vehicles or ballistic missiles [2].

The oscillations due to the pogo instability occur principally in the first longitudinal mode of the structure during operation of the first stage (burning liquid propellants) of a launch vehicle. The effects of this instability are shown in the following figure, due to the courtesy of NASA [101].

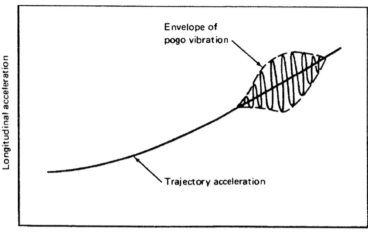

The pogo vibrations begin spontaneously, then intensify, and finally die away in a period of time ranging from 10 to 40 s. The frequency of vibration follows the frequency of the first structural mode, which increases with the consumption of propellants. Pogo vibrations have also occurred, but less often, in higher modes of longitudinal oscillation. More periods than one of instability have also been observed, each of which with its own mode of vibration, during operation of a single stage of a rocket vehicle. Vibrations have occurred in the range from 5 to 60 Hz, and vibration amplitudes (from zero to the peak values) have reached 17 g_0 at the input to the payload and 34 g_0 at the engine [101], where $g_0 = 9.80665$ m/s^2 is the acceleration of gravity at the surface of the Earth.

The pogo vibrations can damage the crew and the equipment, and can also overload the structure of a rocket vehicle. In addition, they impair the propulsion performance of the vehicle.

In order to suppress these vibrations, hydraulic accumulators have been installed in the propulsion systems of launch vehicles such as the Saturn V and the Space Shuttle.

As an example, each of the three main engines of the Space Shuttle has a pogo suppression system accumulator, which is attached to the low-pressure oxidiser duct. The accumulator, when pressurised with gaseous oxygen coming from the heat exchanger coil, dampens the pressure oscillations due to the oxidiser feed system. This accumulator is shown in the following figure, due to the courtesy of Boeing-Rocketdyne [50].

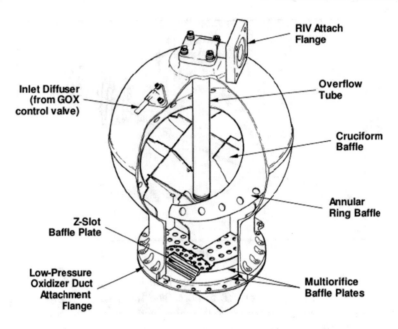

It is a hollow metallic sphere, 32 cm in diameter, which is mounted by a flange to the low-pressure oxidiser duct. During engine operation, the accumulator is pressurised with gaseous oxygen. The gaseous oxygen is a compliant medium in direct contact with the liquid oxygen, and is therefore capable of smoothing the oxidiser flow by absorbing the pressure oscillations. The pressure in the accumulator is maintained by a constant flow of gaseous oxygen into, through, and out of the accumulator. The gaseous oxygen comes from the heat exchanger coil, flows through the gaseous oxygen control valve into the accumulator, exits through the bottom of the inverted standpipe (overflow tube), and returns to the oxidiser feed system to be re-condensed in the liquid state [50].

The following figure, due to the courtesy of NASA [102], shows schematically how the accumulator described above is used to suppress the pogo vibrations in the Space Shuttle.

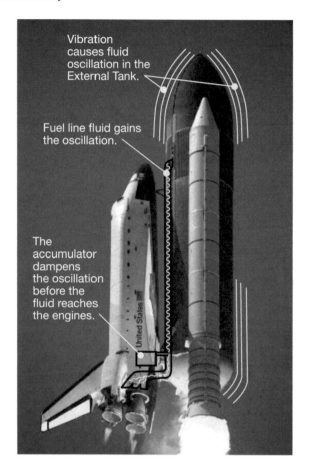

Vibration causes fluid oscillation in the External Tank.

Fuel line fluid gains the oscillation.

The accumulator dampens the oscillation before the fluid reaches the engines.

References

1. West J, Westra D, Richardson BR, Tucker PK (2014) Designing liquid rocket engine injectors for performance, stability, and cost. In: Conference paper. NASA Marshall Space Flight Centre, 16 Nov 2014, 13 p. https://ntrs.nasa.gov/archive/nasa/casi.ntrs.nasa.gov/20150002585.pdf
2. Sutton GP, Biblarz O (2001) Rocket propulsion elements. 7th edn, Wiley, New York. ISBN 0-471-32642-9
3. NASA (2015) Space Travel, NASA Tests Methane-Powered Engine Components for Next Generation Landers. Marshall Space Flight Centre, 28 Oct 2015. https://www.nasa.gov/sites/default/files/thumbnails/image/img_3190.jpg
4. Tomazic WA, Bartoo ER, Rollbuhler RJ (1960) Experiments with hydrogen and oxygen in regenerative engines at chamber pressures from 100 to 300 pounds per square inch absolute. NASA TM X-253, 1st Apr 1960, 37 p. https://ntrs.nasa.gov/archive/nasa/casi.ntrs.nasa.gov/19660024054.pdf

5. Huzel DK, Huang DH (1967) Design of liquid propellant rocket engines. NASA SP-125, NASA. 2nd edn Washington, D.C., 472 p. https://ntrs.nasa.gov/archive/nasa/casi.ntrs.nasa.gov/19710019929.pdf

6. Bilstein RE (1996) Stages to Saturn. NASA SP-4206. https://ia802605.us.archive.org/19/items/stagestosaturnte00bilsrich/stagestosaturnte00bilsrich.pdf

7. NASA, Math and science at work, space shuttle roll maneuver student edition. http://www.nasa.gov/pdf/519348main_AP_ST_Phys_RollManeuver.pdf

8. NASA. https://spaceflightsystems.grc.nasa.gov/education/rocket/rktth1.html

9. NASA. http://exploration.grc.nasa.gov/education/rocket/rktcontrl.html

10. NASA. http://www.hq.nasa.gov/office/pao/History/conghand/orient.htm

11. Edwards SS, Parker GH (1958) An investigation of the jetevator as a means of thrust vector control. LMSD-2630, Lockheed, Sunnyvale, California, USA, Feb 1958, 44 p. http://www.dtic.mil/dtic/tr/fulltext/u2/a950112.pdf

12. Spinardi G (1994) From Polaris to trident: the development of US fleet ballistic missile technology. Cambridge University Press. ISBN 0-521-41357-5

13. NASA dryden fact sheet—highly manoeuvrable aircraft technology. http://www.nasa.gov/sites/default/files/images/110536main_HiMAT_3-view.jpg

14. Benson T (ed) Practical rocketry, NASA, Glenn research centre. http://exploration.grc.nasa.gov/education/rocket/TRCRocket/practical_rocketry.html

15. McGlinchey LF (1974) Viking orbiter 1975 thrust vector control system accuracy. NASA-CR-140705, JPL Technical Memorandum 33–703, 15 Oct 1974. https://ntrs.nasa.gov/archive/nasa/casi.ntrs.nasa.gov/19750002097.pdf

16. NASA Space Science Data Coordinated Archive, Viking 1 Orbiter. https://nssdc.gsfc.nasa.gov/nmc/spacecraftDisplay.do?id=1975-075A

17. NASA (1968) Saturn V news reference, J-2 engine fact sheet, Dec 1968. https://www.nasa.gov/centers/marshall/pdf/499245main_J2_Engine_fs.pdf

18. NASA. Beginner's guide to rockets, guidance system. https://spaceflightsystems.grc.nasa.gov/education/rocket/guidance.html

19. Noll RB, Zvara J, Deyst JJ (1971) Spacecraft attitude control during thrusting manoeuvres, NASA SP-8059, Feb 1971, 51 p. https://ntrs.nasa.gov/archive/nasa/casi.ntrs.nasa.gov/19710016722.pdf

20. Shamnas MS, Balakrishnan SR, Balaji S (2013) Effects of secondary injection in rocket nozzle at various conditions. Int J Eng Res Technol (IJERT) 2(6):373–379

21. Anonymous (1946) Augmented flow—an interesting method of fluid-flow augmentation with attractive possibilities. Flight, 15 Aug 1946, pp 174–175

22. Blake BA (2009) Numerical investigation of fluidic injection as a means of thrust modulation. Final thesis report, 2009, School of Engineering and Information Technology, University of New South Wales at the Australian Defence Force Academy, Canberra, Australia, 29 p. Article available at the http://seit.unsw.adfa.edu.au/ojs/index.php/juer/article/viewFile/256/155

23. Strykowski PJ, Krothapalli A, Forliti DJ (1996) Counterflow thrust vectoring of supersonic jets. AIAA J 34(11):2306–2314

24. Hunter CA, Deere KA (1999) Computational investigation of fluidic counterflow thrust vectoring, AIAA 99-2669. In: 35th AIAA/ASME/SAE/ASEE joint propulsion conference & exhibit, Los Angeles, CA, 20–23 June 1999, 13 p

25. Deere KA, Berrier BL, Flamm JD, Johnson SK (2003) Computational study of fluidic thrust vectoring using separation control in a nozzle, AIAA-2003-3803. In: 21st AIAA applied aerodynamics conference, 23–26 June 2003, Orlando, Florida, USA, 11 p

26. Wing DJ, Giuliano VJ (1997) Fluidic thrust vectoring of an axisymmetric exhaust nozzle at static conditions, FEDSM97-3228. In: 1997 ASME fluids engineering division summer meeting, 22–26 June 1997, 6 p

27. Heppenheimer TA (1999) The space shuttle decision: NASA search for a reusable space vehicle. NASA SP-4221. 1999, 554 p. https://ntrs.nasa.gov/archive/nasa/casi.ntrs.nasa.gov/19990056590.pdf

28. Hyde JC, Gill GS (1976) Liquid rocket engine nozzles, NASA SP-8120. NASA, Washington, D.C., July 1976, 123 p. https://ntrs.nasa.gov/archive/nasa/casi.ntrs.nasa.gov/19770009165. pdf
29. Crown JC (1948) Supersonic nozzle design. NACA Technical Note No. 1651, NACA, Washington, D.C., June 1948, 35 p https://ntrs.nasa.gov/archive/nasa/casi.ntrs.nasa.gov/199300 82268.pdf
30. Rao GVR (1958) Exhaust nozzle contour for optimum thrust. J Jet Propul 28(6):377–382
31. Kuo KK, Acharya R (2012) Applications of turbulent and multiphase combustion. Wiley, Hoboken, New Jersey. ISBN 978-1-118-12756-8
32. Ahlberg JH, Hamilton S, Migdal D, Nilson EN (1961) Truncated perfect nozzles in optimum nozzle design. ARS J 31(5):614–620
33. Rao GVR (1960) Approximation of optimum thrust nozzle contour. ARS J 30(6):581
34. Braeunig RA, Rocket and space technology—rocket propulsion. http://braeunig.us/space/pro puls.htm
35. Newlands R (2017) The thrust optimised parabolic nozzle. Aspirespace, 18 Apr 2017. http:// www.aspirespace.org.uk/downloads/Thrust%20optimised%20parabolic%20nozzle.pdf
36. Aukerman CA (1991) Plug nozzles—the ultimate customer driven propulsion system. NASA-CR-187169, Aug 1991, 27 p. https://ntrs.nasa.gov/archive/nasa/casi.ntrs.nasa.gov/199200 13861.pdf
37. Corda S, Neal BA, Moes TR, Cox TH, Monaghan RC, Voelker LS, Corpening GP, Larson RR, Powers BG (1998) Flight testing the linear aerospike SR-71 experiment (LASRE), NASA/TM-1998-206567. NASA Dryden Space Research Centre, Sept 1998, 26 p. https://www.nasa.gov/ centers/dryden/pdf/88598main_H-2280.pdf
38. NASA. Linear aerospike SR-71 experiment (LASRE) Project. https://www.nasa.gov/centers/ armstrong/news/FactSheets/FS-043-DFRC.html
39. NASA. https://www.hq.nasa.gov/pao/History/x-33/xrs2200.htm
40. Anonymous (1981) Liquid rocket propulsion technology: an evaluation of NASA's program, NASA-CR-164550, Jan 1981, 67 p. https://ntrs.nasa.gov/archive/nasa/casi.ntrs.nasa.gov/198 10018653.pdf
41. Shyne RJ, Keith Th G Jr (1990) Analysis and design of optimised truncated scarfed nozzles subject to external flow field effects. NASA Technical Memorandum 105175, Jan 1990, 12 p. https://ntrs.nasa.gov/archive/nasa/casi.ntrs.nasa.gov/19900015790.pdf
42. O'Leary RA, Beck JE (1992) Nozzle design, Threshold, Pratt & Whitney Rocketdyne's engineering journal of power technology. Spring 1992. http://www.rocket-propulsion.info/resour ces/articles/NozzleDesign.pdf
43. Wieseneck HC (1970) Regenerative cooling of the space shuttle engine. NASA, July 1970, 32 p. https://ntrs.nasa.gov/archive/nasa/casi.ntrs.nasa.gov/19700030313.pdf
44. Turner MJL (2006) Rocket and spacecraft propulsion, 2nd edn. Springer, Berlin. ISBN 3-540-22190-5
45. Van Huff NE, Fairchild DA (1972) Liquid rocket engine fluid-cooled combustion chambers. NASA SP-8087, Apr 1972, 120 p. https://ntrs.nasa.gov/archive/nasa/casi.ntrs.nasa.gov/197 30022965.pdf
46. Bartz DR (963) Turbulent boundary-layer heat transfer from rapidly accelerating flow of rocket combustion gases and of heated air, NASA CR-62615. Jet Propulsion Laboratory, California Institute of Technology, Pasadena, California. USA, Dec 1963, 128 p. https://ntrs. nasa.gov/archive/nasa/casi.ntrs.nasa.gov/19650013685.pdf
47. Hill PhG, Peterson CR (1965) Mechanics and thermodynamics of propulsion, addison-wesley, reading. USA, Massachusetts
48. Wang Q, Wu F, Zeng M, Luo L, Sun, J (2006) Numerical simulation and optimization on heat transfer and fluid flow in cooling channel of liquid rocket engine thrust chamber. Eng Comput 23(8):907–921
49. Harrje DT, Reardon FH (ed) (1972) Liquid propellant rocket combustion instability. NASA SP-194, Jan 1972, 657 p. https://ntrs.nasa.gov/archive/nasa/casi.ntrs.nasa.gov/19720026079. pdf

50. Anonymous (1998) Space Shuttle main engine orientation. Boeing-Rocketdyne Propulsion & Power, Presentation BC98-04, June 1998, 105 p. http://large.stanford.edu/courses/2011/ph240/nguyen1/docs/SSME_PRESENTATION.pdf

51. Anonymous (1993) Heat transfer in rocket engine combustion chambers and regeneratively cooled nozzles. NASA-CR-193949, 16 Nov 1993, 169 p. https://ntrs.nasa.gov/archive/nasa/casi.ntrs.nasa.gov/19940019998.pdf

52. Taler D, Taler J (2017) Simple heat transfer correlations for turbulent tube flow. In: E3S Web of conferences 13, 02008 (2017), 7 pages. https://www.e3s-conferences.org/articles/e3sconf/pdf/2017/01/e3sconf_wtiue2017_02008.pdf

53. Gnielinski V (1975) Neue Gleichungen für den Wärme- und den Stoffübergang in turbulent durchströmten Rohren und Kanälen. Forsch Ingenieurwes 41:8–16

54. Giovanetti AJ, Spadaccini LJ, Szetela EJ (1983) Deposit formation and heat transfer in hydrocarbon rocket fuels. NASA-CR-168277, 1983, 168 p. https://ntrs.nasa.gov/archive/nasa/casi.ntrs.nasa.gov/19840004157.pdf

55. Haidn OJ (2008) Advanced rocket engines. In: Advances on propulsion technology for high-speed aircraft, NATO RTO-EN-AVT-150, 2008, Neuilly-sur-Seine (France), 40 p. Available at the http://www.macro-inc.com/NASADocs/AdvanceRocketEnginesEN-AVT-150-06.pdf

56. Locke JM, Landrum DB (2008) Study of heat transfer correlations for supercritical hydrogen in regenerative cooling channels. J Propul Power 24(1):94–103

57. Sloop JL (1978) Liquid hydrogen as a propulsion fuel. 1945-1959, NASA SP-4404, Jan 1978, 341 p. Available at the https://ntrs.nasa.gov/archive/nasa/casi.ntrs.nasa.gov/19790008823.pdf

58. Colebrook CF (1939) Turbulent flow in pipes, with particular reference to the transition region between smooth and rough pipe laws. J Inst Civil Eng 11(4):133–156

59. Nuclear Power, Darcy Friction Factor, Relative Roughness of Pipe. http://www.nuclear-power.net/nuclear-engineering/fluid-dynamics/major-head-loss-friction-loss/relative-roughness-of-pipe/

60. AJ Design Software, Colebrook Equations Formulas Calculator. https://www.ajdesigner.com/php_colebrook/colebrook_equation.php

61. Special Metals Corporation, Inconel® alloy 718, Pub. No. SMC-045. http://www.specialmetals.com/assets/smc/documents/inconel_alloy_718.pdf

62. National Institute of Standards and Technology (NIST), NIST Chemistry WebBook, SRD 69, Hydrogen. http://webbook.nist.gov/cgi/cbook.cgi?ID=C1333740&Mask=4, see also Thermophysical Properties of Fluid Systems. https://webbook.nist.gov/chemistry/fluid/

63. Stenholm T, Pekkari LO, Andersson T (1988) The development of the Vulcain nozzle extension. In: 34th AIAA/ASME/SAE/ASEE joint propulsion conference & exhibit, 13–15 July 1988, Cleveland, Ohio, USA

64. Rydén R, Johansson L, Palmnäs U (2004) The Volvo Aero laser welded sandwich nozzle for booster engines. In: 55th International Astronautical Congress 2004—Vancouver, Canada, Paper number IAC-04-S.3.07. https://iafastro.directory/iac/archive/browse/IAC-04/S/3/1259/

65. Lucas JG, Golladay RL (1963) An experimental investigation of gaseous-film cooling of a rocket motor. NASA TN D-1988, Oct 1963, 24 p. https://ntrs.nasa.gov/archive/nasa/casi.ntrs.nasa.gov/19630012015.pdf

66. Zucrow MJ, Sellers JP (1961) Experimental investigation of rocket motor film cooling. ARS J 668

67. Hatch JE, Papell SS (1959) Use of a theoretical flow model to correlate data for film cooling or heating an adiabatic wall by tangential injection of gases of different fluid properties. NASA TN D-130, Nov 1959, 44 p. https://ntrs.nasa.gov/archive/nasa/casi.ntrs.nasa.gov/19890068390.pdf

68. Hannum N, Roberts WE, Russell LM (1977) Hydrogen film cooling of a small hydrogen-oxygen thrust chamber and its effect on erosion rates of various ablative materials. NASA-TP-1098, Dec 1977, 29 p. https://ntrs.nasa.gov/archive/nasa/casi.ntrs.nasa.gov/19780005181.pdf

69. Eckert ERG, Livingood JNB (1954) Comparison and effectiveness of convection-, transpiration-, and film-cooling methods with air as a coolant. NACA-TR-1182, Jan 1954, 17 p. https://ntrs.nasa.gov/archive/nasa/casi.ntrs.nasa.gov/19930092205.pdf

70. Rannie W (1947) A simplified theory for porous wall cooling. Progress Report No. 4–50, Jet Propulsion Laboratory, Pasadena, California, USA., 25 Nov 1947, 24 p. https://trs.jpl.nasa. gov/bitstream/handle/2014/45706/17-0944.pdf?sequence=1&isAllowed=y

71. Ewen RL, Evensen HM (1977) Liquid rocket engine self-cooled combustion chambers. NASA SP-8124, Sept 1977, 130 p. https://ntrs.nasa.gov/archive/nasa/casi.ntrs.nasa.gov/197 80013268.pdf

72. Reed BD, Biaglow JA, Schneider SJ (1997) Iridium-coated rhenium radiation-cooled rockets, NASA TM-107453, July 1997, 14 p. https://ntrs.nasa.gov/archive/nasa/casi.ntrs.nasa.gov/199 70036365.pdf

73. National Institute of Standards and Technology (NIST). https://physics.nist.gov/cgi-bin/cuu/ Value?sigma

74. Combs LP et al (1974) Liquid rocket engine combustion stabilisation devices. NASA SP-8113, Nov 1974, 127 p. https://ntrs.nasa.gov/archive/nasa/casi.ntrs.nasa.gov/19750020175. pdf

75. Gill GS, Nurick WH, Keller RB Jr (ed) Liquid rocket engine injectors. NASA SP-8089, Mar 1976, 130 p. https://ntrs.nasa.gov/archive/nasa/casi.ntrs.nasa.gov/19760023196.pdf

76. Casiano MJ, Hulka JR, Yang V (2010) Liquid-propellant rocket engine throttling: a comprehensive review. J Propu Power 26(5):897–923. http://www.yang.gatech.edu/publications/Jou rnal/JPP%20(2010,%20Casiano).pdf

77. Hammock WR Jr, Currie EC, Fisher AE (1973) Apollo experience report—descent propulsion system. NASA TN D-7143, Mar 1973, 36 p. https://ntrs.nasa.gov/archive/nasa/casi.ntrs.nasa. gov/19730011150.pdf

78. Wikipedia, SpaceX rocket engines. https://en.wikipedia.org/wiki/SpaceX_rocket_engines

79. Stangeland ML (1988) Turbopumps for liquid rocket engines, Threshold, Pratt & Whitney Rocketdyne's engineering journal of power technology, Summer 1988. http://www.pwreng ineering.com/articles/turbopump.htm

80. Macaluso SB, Keller RB Jr (ed) (1974) Liquid rocket engine turbines. NASA SP-8110, Jan 1974, 160 p. https://ntrs.nasa.gov/archive/nasa/casi.ntrs.nasa.gov/19740026132.pdf

81. Greene WD, Inside the J-2X doghouse: the gas-generator cycle engine. NASA. https://blogs. nasa.gov/J2X/tag/gas-generator/

82. Sobin AJ, Bissel WR, Keller RB Jr (ed) (1974) Turbopump systems for liquid rocket engines. NASA SP-8107, Aug 1974, 168 p. https://ntrs.nasa.gov/archive/nasa/casi.ntrs.nasa.gov/197 50012398.pdf

83. Zehetner HW, Keller RB Jr. (ed) Liquid propellant gas generators. NASA SP-8081, Mar 1972, 110 p. https://ntrs.nasa.gov/archive/nasa/casi.ntrs.nasa.gov/19730018978.pdf

84. Kosanke KL, Kosanke BJ, Pyrotechnic ignition and propagation: a review. In: Kosanke KL, Kosanke BJ (eds) Selected pyrotechnic publications of, pp 420–431. http://www.jpyro.com/ wp-content/uploads/2012/08/Kos-420-431.pdf

85. Barrett DH, Keller RB Jr (ed) (1971) Solid rocket motor igniters. NASA SP-8051, Mar 1971, 112 p. https://ntrs.nasa.gov/archive/nasa/casi.ntrs.nasa.gov/19710020870.pdf

86. Kosanke KL, Kosanke BJ (2012) Electric matches and squibs. In: KL Kosanke, BJ Kosanke (eds) Selected pyrotechnic publications of, pp. 257–259. http://www.jpyro.com/wp-content/ uploads/2012/08/Kos-257-259.pdf

87. United Launch Alliance. http://www.ulalaunch.com/uploads/docs/AtlasVUsersGuide2010. pdf, Public Domain. https://commons.wikimedia.org/w/index.php?curid=48196878

88. Clark JD (1972) Ignition!: an informal history of liquid rocket propellants. Rutgers University Press, New Brunswick, New Jersey. ISBN 0-8135-0725-1

89. Hulka J, Forde JS, Werling RE, Anisimov VS, Kozlow VA, Kositsin IP (1998) Modification and verification testing of a Russian NK-33 rocket engine for reusable and restartable applications. AIAA 98-3361, 1998, 26 p. http://lpre.de/resources/articles/AIAA-1998-3361. pdf

90. Greene WD (2014) Inside the LEO doghouse: light my fire!. NASA, 24 Feb 2014. https:// blogs.nasa.gov/J2X/tag/ignition/

91. Kleinhenz J, Sarmiento Ch, Marshall W (2012) Experimental investigation of augmented spark ignition of a LO$_2$/LCH$_4$ reaction control engine at altitude conditions. NASA/TM-2012-217611, June 2012, 43 p. https://ntrs.nasa.gov/archive/nasa/casi.ntrs.nasa.gov/201200 11145.pdf

92. Anderson WE, Butler K, Crocket D, Lewis T, McNeal C (2000) Peroxide propulsion at the turn of the century, 13 Mar 2000, 59 p. https://ntrs.nasa.gov/archive/nasa/casi.ntrs.nasa.gov/20000033615.pdf

93. Prentiss SS (1956) Safety device, including fusible member for rocket engine starting control, U.S. patent No. 2,741,085 granted on the 10th of Apr 1956. https://patents.google.com/patent/US2741085

94. Rayleigh L (1878) The theory of sound, vol 2. Macmillan & Co., London. http://gallica.bnf.fr/ark:/12148/bpt6k95131k.image

95. Burrows MC, Exploring in aerospace rocketry, 4—Thermodynamics, NASA TM X-52391, Jan 1968, 21 p. https://ntrs.nasa.gov/archive/nasa/casi.ntrs.nasa.gov/19680010371.pdf

96. Ellison LR, Moser MD, Combustion instability analysis and the effect of drop size on acoustic driving rocket flow, 13 May 2004, 6 p. https://ntrs.nasa.gov/archive/nasa/casi.ntrs.nasa.gov/20040075603.pdf

97. Wikipedia, by Loungeflyz (CC BY-SA 4.0. https://creativecommons.org/licenses/by-sa/4.0) from Wikimedia Commons

98. Jonash ER, Tomazic WA (1962) Current research and development on thrust chambers. In: Proceedings of the NASA-university conference on the science and technology of space exploration, vol 2. NASA SP-11, Chicago, Illinois, 1–3 Nov 1962, pp. 43–52. http://www.dtic.mil/dtic/tr/fulltext/u2/a396443.pdf

99. Frendi A, Nesman T, Canabal F (2005) Control of combustion instabilities through various passive devices, NASA, 25 p. https://ntrs.nasa.gov/archive/nasa/casi.ntrs.nasa.gov/200501 82054.pdf

100. Kinsler LE, Frey AR, Coppens AB, Sanders JV (2000) Fundamentals of acoustics. 4th edn, Wiley, New York. ISBN 0-471-84789-5

101. Rubin S, Prevention of coupled structure-propulsion instability (pogo). NASA SP-8055, Oct 1970, 52 p. https://ntrs.nasa.gov/archive/nasa/casi.ntrs.nasa.gov/19710016604.pdf

102. Modlin T, Space Shuttle pogo—NASA eliminates "bad vibrations". Eng Innov Struct Des 270–285 https://www.nasa.gov/centers/johnson/pdf/584733main_Wings-ch4g-pgs270-285.pdf

Chapter 3
Feed Systems Using Gases Under Pressure

3.1 Fundamental Concepts

As has been shown in Chap. 1, Sect. 1.4, a feed system is necessary to transfer either a mono-propellant or an oxidiser and a fuel from their respective tanks to the thrust chamber of a rocket engine. This chapter describes feed systems using gases stored at high pressures for this purpose. A scheme of such systems is shown in the following figure, due to the courtesy of NASA [1].

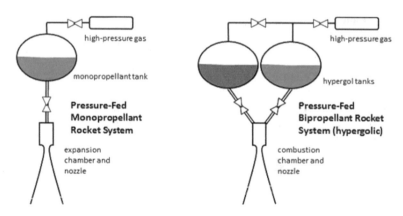

A gas stored under pressure makes it possible to control the pressure in the ullage space in the tanks of propellant. By ullage space we mean the volume by which a container or a tank falls short of being full of liquid.

In a rocket engine fed by a gas stored under pressure, the pressure in the ullage space is used to force the propellants through the lines of the feed system and into the combustion chamber at the required pressures and flow rates.

A. de Iaco Veris, *Fundamental Concepts of Liquid-Propellant Rocket Engines*, Springer Aerospace Technology, https://doi.org/10.1007/978-3-030-54704-2_3

In a rocket engine fed by pumps, the pressure in the ullage space is used to supply the propellants to the inlet section of the pumps at the required pressures, and the pumps in turn deliver the propellants to the combustion chamber at the required pressures and flow rates.

As a general rule, the choice of a feed system, independently of whether it uses gases under pressure or pumps, depends on several factors, which are principally the mission which the rocket vehicle is meant to accomplish, the size and the mass of this vehicle, the intensity and the duration of the thrust, the space available to the engine, the degree of reliability of the propulsion system, and so on. In practice, according to Lee et al. [2], a pump-fed propulsion system is chosen when:

- the engine must produce more than 4500 N of thrust, or
- the total mass exceeds 9100 kg.

In contrast, a pressure-fed propulsion system, which may be either a mono-propellant or a bi-propellant system, is chosen when:

- the mission duty cycle is a pulse mode, or
- the propellant mass is less than 3600 kg.

For those cases in which the choice of the feed system is not obvious, a weight-and-cost trade-off study is made to determine the better system.

The types of gas used in the feed systems described here are the following:

- gases stored for bi-propellants;
- gases obtained by evaporation of two propellants;
- gases stored for mono-propellants; and
- gases obtained as combustion products.

Some criteria for choosing one or another of these types are briefly presented below. A gas under pressure must be chemically compatible with the propellants and also with the materials of which the tanks and the lines are made, at the temperatures of operation. The feed system using this gas must be simple and reliable. The gas used in the feed system must have a low molar mass, in order for the value of the mass ratio m_0/m_u (where m_0 is the mass of the rocket vehicle at launch, and m_u is its mass at burnout) to be low. The materials used for the feed system must have a low specific mass, for the same reason. These requirements will be discussed in the following paragraphs.

3.2 Requirements for Gases Stored Under Pressure

When the time of operation of a rocket engine is short, or when the temperature of the gas stored under pressure is less than or equal to the temperature of the propellants, then the required mass m_g (kg) of that gas can be computed by means of the law of perfect gases, as follows

$$m_g = \frac{p_T V_T}{R_g T_g}$$

where p_T (N/m^2) is the pressure of the gas in the tank, V_T (m^3) is the total volume of the empty tank, R_g (N m kg^{-1} K^{-1}) is the constant of the specific gas used, and T_g (K) is the mean temperature of that gas. As has been shown in Chap. 1, Sect. 1.1, the constant R_g of a specific gas is equal to the universal gas constant $R* = 8314.460$ N m kmol^{-1} K^{-1} divided by the molar mass $\mathcal{M}$ (kg/kmol) of that gas, as follows

$$R_g = \frac{R*}{\mathcal{M}}$$

When the conditions indicated above are not satisfied, the required mass m_g of the gas under pressure can be computed as follows.

We consider first a rocket engine which is started only once, and assume the heat transfer from the tank walls to be negligible. In such conditions, the total heat Q (J) transferred from the gas under pressure to the vaporised propellant is [3]:

$$Q = H A t (T_u - T_e)$$

where H (W m^{-2} K^{-1}) is the total heat transfer coefficient, whose value is determined experimentally, at the interface between liquid and gas, A (m^2) is the area of the interface between liquid and gas, t (s) the duration of operation, T_u (K) is the temperature of the gases at burnout, and T_e (K) is the temperature of the propellant. The temperatures T_e and T_u are assumed to have constant values at the interface between liquid and gas.

The total heat Q defined above is assumed to have heated and vaporised the propellant, as indicated by the following equation

$$Q = m_v \left[c_{pl}(T_v - T_e) + h_v + c_{pv}(T_u - T_v) \right]$$

where m_v (kg) is the total mass of vaporised propellant, c_{pl} and c_{pv} (J kg^{-1} K^{-1}) are the specific heats at constant pressure of respectively the liquid propellant and the vaporised propellant, h_v (J/kg) is the heat of vaporisation per unit mass of the propellant, and T_v (K) is the temperature of vaporisation of the liquid propellant.

The equations $Q = H A t (T_u - T_e)$ and $Q = m_v \left[c_{pl}(T_v - T_e) + h_v + c_{pv}(T_u - T_v) \right]$ make it possible to compute the value of m_v for an assumed value of T_u.

The partial volume V_v (m^3) occupied by vaporised propellant is [3]:

$$V_v = \frac{m_v Z R_p T_u}{p_T}$$

where m_v (kg) is the total mass of vaporised propellant, Z is the compressibility factor evaluated at the total pressure p_T (N/m^2) and at the temperature T_u (K) of the gaseous mixture at burnout, and R_p (N m kg^{-1} K^{-1}) is the constant of the specific propellant vapour.

The remaining volume V_g (m^3) of the tank at burnout, in the absence of residual propellants, can be assumed to be occupied by the gas under pressure, as follows

$$V_g = V_T - V_v$$

where V_T (m^3) is the total volume of the empty tank.

The mass m_g (kg) of the gas under pressure can be computed by means of the law of perfect gases, as follows

$$m_g = \frac{p_T V_g}{R_g T_u}$$

In order for the heat balance to be maintained, the total heat Q (J) must be such that

$$Q = m_g c_{pg}(T_g - T_u)$$

where c_{pg} (J kg^{-1} K^{-1}) is the specific heat at constant pressure of the gas under pressure. The preceding equation makes it possible to determine the mean value T_g (K) of the gas under pressure for the assumed value T_u (K) of the gases at burnout.

In case of the value of T_g being predetermined, m_g, m_v, and T_u must have values such that

$$m_g c_{pg}(T_g - T_u) = m_v\left[c_{pl}(T_v - T_e) + h_v + c_{pv}(T_u - T_v)\right]$$

The preceding discussion has neglected the heat transfer through the tank walls. On the contrary, when the gas under pressure, the propellants, and the tank walls have a considerable difference of temperature, then the total heat transferred from one to another of them must be taken into account for the purpose of determining the propellants vaporised at burnout. In these conditions, the preceding equation

$$Q = m_v\left[c_{pl}(T_v - T_e) + h_v + c_{pv}(T_u - T_v)\right]$$

can be re-written as follows

$$Q \pm Q_{w1} = m_v\left[c_{pl}(T_v - T_e) + h_v + c_{py}(T_u - T_v)\right]$$

where Q_{w1} (J) is the total heat transferred between the tank walls and the liquid or gaseous propellants during a given mission. The sign (plus or minus) in front of Q_{w1} indicates whether the tank walls add or subtract heat.

Likewise, the preceding equation

$$Q = m_g c_{pg}(T_g - T_u)$$

can be re-written as follows

$$Q = m_g c_{pg} (T_g - T_u) \pm Q_{w2}$$

where Q_{w2} (J) is the total heat transferred between the gas under pressure and the tank walls during a given mission. The sign (plus or minus) in front of Q_{w2} indicates whether the tank walls add or subtract heat.

Consequently, the equation expressing the heat balance when the gas under pressure, the propellants, and the tank walls have a considerable difference of temperature can be written as follows

$$m_g c_{pg} (T_g - T_u) \pm Q_{w2} = m_v \left[c_{pl} (T_v - T_e) + h_v + c_{pv} (T_u - T_v) \right] - (\pm Q_{w1})$$

When the mission to be accomplished by a space vehicle includes periods of propelled flight and periods of coasting, then the heat transfer across the gas-liquid interface must be determined by considering the total time of the mission. Therefore, the preceding equation

$$Q = H A t (T_u - T_e)$$

must be re-written as follows

$$Q = H A t_m (T_m - T_e)$$

where t_m (s) is the total time of the mission including periods of propelled flight and periods of coasting, and T_m (K) is the mean temperature of the gas under pressure during the total time of the mission. The temperature T_m depends on several factors, which are the duration of the coasting periods, the heat transfer between the gas and the walls, etc.

Other effects (vapour condensation, stability of the gas under pressure in the propellants, and chemical reactions between this gas and the propellants) can be taken into account, in case of availability of experimental data.

As an application of the concepts discussed above, we consider a rocket engine whose oxidiser and fuel are respectively nitrogen tetroxide (N_2O_4) and hydrazine (N_2H_4). These propellants are fed to the thrust chamber by using helium (He) as the gas under pressure. Let $V_T = 3.37$ m^3 be the total volume of the empty tank of nitrogen tetroxide, $A = 1.86$ m^2 be the area of the average cross section of the tank, $p_T = 1.14 \times 10^6$ N/m^2 be the absolute pressure in the tank, and $T_e = 289$ K be the temperature of the oxidiser. As a first case, we want to calculate the total mass m_g (kg) of the gas under pressure and its temperature T_g (K) at the inlet section of the tank, knowing that the duration of operation is $t = 500$ s, and that the heat transfer coefficient, determined experimentally, at the liquid-gas interface is $H = 40.9$ W/(m^{2}K). We assume the temperature of the gas at burnout to be $T_u = 389$ K, in the absence of heat transfer at the walls of the tanks.

The following data are available for nitrogen tetroxide (N_2O_4) at the pressure indicated above ($p_T = 1.14 \times 10^6$ N/m²): vaporisation temperature $T_v = 357$ K, heat of vaporisation $h_v = 414000$ J/kg, mean value of specific heat at constant pressure in liquid state $c_{pl} = 1760$ J/(kg K), mean value of specific heat at constant pressure in vapour state $c_{pv} = 754$ J/(kg K), compressibility factor $Z = 0.95$, and molar mass $\mathcal{M} = 2 \times 14 + 4 \times 16 = 92$ kg/kmol. The specific heat at constant pressure of helium (He) at 293 K and atmospheric pressure is $c_{pg} = 5190$ J/(kg K) [4] and its molar mass is $\mathcal{M} = 4$ kg/kmol.

The total heat transferred at the gas-liquid interface results from

$$Q = H A t (T_u - T_e) = 40.9 \times 1.86 \times 500 \times (389 - 289) = 3.804 \times 10^6 \text{J}$$

This value of Q and the values of the quantities c_{pl}, c_{pv}, h_v, T_v, T_e, and T_u are substituted into the following equation

$$Q = m_v \left[c_{pl} (T_v - T_e) + h_v + c_{pv} (T_u - T_v) \right]$$

This yields

$$3.804 \times 10^6 = m_v [1760 \times (357 - 289) + 414000 + 754 \times (389 - 357)]$$

The preceding equation, solved for the total mass m_v of vaporised propellant, yields

$$m_v = \frac{3.804 \times 10^6}{1760 \times (357 - 289) + 414000 + 754 \times (389 - 357)} = 6.820 \text{ kg}$$

The partial volume occupied by vaporised propellant (N_2O_4, whose molar mass is $\mathcal{M} = 92$ kg/kmol) results from

$$V_v = \frac{m_v Z R_p T_u}{p_T} = \frac{6.820 \times 0.95 \times 8314.46 \times 389}{92 \times 1.14 \times 10^6} = 0.1998 \approx 0.2 \text{ m}^3$$

The partial volume V_g occupied by the gas under pressure (helium) results from the difference between the total volume V_T of the empty tank and the partial volume V_v computed above, as follows

$$V_g = V_T - V_v = 3.37 - 0.2 = 3.17 \text{ m}^3$$

The required mass m_g of the gas under pressure (helium, whose molar mass is $\mathcal{M} = 4$ kg/kmol) results from the equation of perfect gases, as follows

$$m_g = \frac{p_T V_g}{R_g T_u} = \frac{1.14 \times 10^6 \times 3.17 \times 4}{8314.46 \times 389} = 4.469 \text{ kg}$$

In order for the heat balance to be maintained, the total heat Q must be such that

$$Q = m_g c_{pg} (T_g - T_u)$$

The preceding equation, solved for T_g, yields the required temperature of the gas under pressure at the inlet section of the tank, as follows

$$T_g = T_u + \frac{Q}{m_g c_{pg}} = 389 + \frac{3.804 \times 10^6}{4.469 \times 5190} = 553 \text{ K}$$

Now, as a second case, we want to calculate the total mass m_g (kg) of the gas under pressure and its temperature T_g (K) at the inlet section of the tank, for a mission which includes periods of propelled flight and periods of coasting, and has a total duration $t_m = 18000$ s. The mean temperature of the gas under pressure during this mission is $T_m = 292$ K. The total heat transferred between the tank walls and the propellant is $Q_{w1} = -2.11 \times 10^6$ J, where the minus sign indicates that the tank walls subtract heat from the propellant. The total heat transferred between the gas under pressure and the tank walls is $Q_{w2} = -0.633 \times 10^6$ J, where the minus sign indicate that the tank walls subtract heat from the gas under pressure. The temperature of the gases at burnout is $T_u = 367$ K.

The total heat Q transferred from the gas under pressure to the vaporised propellant at the gas-liquid interface is

$$Q = H A t_m (T_m - T_e) = 40.9 \times 1.86 \times 18000 \times (292 - 289) = 4.108 \times 10^6 \text{J}$$

This value of Q and the values of the quantities Q_{w1}, c_{pl}, c_{pv}, h_v, T_v, T_e, and T_u are substituted into the following equation

$$Q \pm Q_{w1} = m_v [c_{pl}(T_v - T_e) + h_v + c_{pv}(T_u - T_v)]$$

Consequently, the total mass m_v of vaporised propellant is

$$m_v = \frac{4.108 \times 10^6 - 2.11 \times 10^6}{1760 \times (357 - 289) + 414000 + 754 \times (367 - 357)} = 3.692 \text{ kg}$$

The partial volume occupied by vaporised propellant (N_2O_4, whose molar mass is $M = 92$ kg/kmol) results from

$$V_v = \frac{m_v Z R_p T_u}{p_T} = \frac{3.692 \times 0.95 \times 8314.46 \times 367}{92 \times 1.14 \times 10^6} = 0.102 \text{ m}^3$$

The partial volume V_g occupied by the gas under pressure (helium) results from the difference between the total volume V_T of the empty tank and the partial volume V_v computed above, as follows

$$V_g = V_T - V_v = 3.37 - 0.102 = 3.268\,\text{m}^3$$

The mass m_g of the required gas under pressure (helium, whose molar mass is $\mathcal{M}$ $= 4$ kg/kmol) results from the equation of perfect gases, as follows

$$m_g = \frac{p_T V_g}{R_g T_u} = \frac{1.14 \times 10^6 \times 3.268 \times 4}{8314.46 \times 367} = 4.884\,\text{kg}$$

By substituting this value of m_g into the following equation

$$Q = m_g c_{pg}\left(T_g - T_u\right) \pm Q_{w2}$$

and solving for T_g, the required temperature of the gas under pressure at the inlet section of the tank results

$$T_g = T_u + \frac{Q - Q_{w2}}{m_g c_{pg}} = 367 + \frac{(4.108 + 0.633) \times 10^6}{4.884 \times 5190} = 554\,\text{K}$$

3.3 Feed Systems Using Gases Stored for Bi-Propellants

The gas used in these systems, which is usually nitrogen or helium, is stored in a vessel at a pressure going from 2.1×10^7 to 3.4×10^7 N/m^2 [3]. The following figure, due to the courtesy of NASA [5], shows the position of the spherical vessel of helium in the descent stage of the Apollo lunar module.

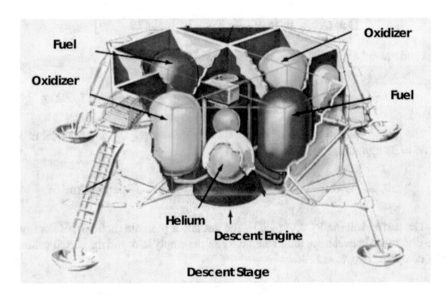

Helium is preferred to other gases for the following reasons:

- low molar mass ($\mathcal{M} = 4$ kg/kmol) and therefore low total mass;
- low boiling point (4.22 K at atmospheric pressure); and
- absence of chemical reactivity.

In case of a bi-propellant engine using liquid hydrogen as its fuel, gaseous hydrogen has also been used as the pressurising agent.

Helium or other pressurised gases can be stored in a vessel with or without a system for thermal conditioning. A thermal conditioning (that is, heating or cooling or both) of the gas can lead to low-mass systems. On the other hand, a thermal conditioning of the gas implies high complexity and cost. A successful system which did not condition thermally the stored gas (nitrogen) was the Lunar Orbiter velocity and reaction control system. By contrast, in the descent stage illustrated above of the Apollo lunar module, helium under pressure was stored under supercritical conditions and then heated. This led to a reduction in mass of 60% over that of an ambient-temperature high-pressure storage without thermal conditioning. These two systems of storage are briefly described below.

The velocity and reaction control system of the Lunar Orbiter IV is shown schematically in the following figure, due to the courtesy of NASA [6].

This system was designed so that its 445-N-thrust engine, using a hypergolic combination of oxidiser (N_2O_4) and fuel (Aerozine-50), could impart a velocity change of 1017 m/s to the 387 kg spacecraft for mid-course corrections, injection into an initial lunar orbit, and successive injection into a photographic orbit. The source of ullage gas was a spherical tank containing gaseous nitrogen (N_2) without provision for conditioning the temperature of this gas. The tank was made of Ti–6Al–4V alloy, with a mass of 9.9 kg and an internal volume of 0.0259 m^3. A mass of 6.56 kg of nitrogen was stored into this sphere at a pressure of 2.4×10^7 N/m^2 and at a temperature of 294 K. Two parallel-redundant normally-closed squib valves isolated the pressurising gas in the storage vessel from the rest of the system until the gas was required for its first use. The gas in the storage vessel was also the source for the cold-gas reaction control system, which is shown in the lower portion of the preceding figure. Only part (from 0.91 to 1.36 kg) of the nitrogen mass in the storage vessel was used to generate and maintain the pressure in the propellant tanks for the velocity control system; the remaining part was used for the reaction control system and for reserve of pressurising gas. When the squib valves were opened, the pressurising gas flowed through a normally-open shut-off squib valve to the pressure regulator of the velocity control system, which reduced the pressure of the gas to 1.3×10^6 N/m^2. The gas was then routed through the check valves and into the oxidiser and fuel tanks. In the velocity control system, a single regulator was used to pressurise both of the propellant tanks, in order to preclude undesired in-flight shifts of propellant mixture ratio; such shifts could occur if each tank had its own pressure regulator and the regulator set-point of one changed during flight. With a single regulator, the set-point could still change, but each tank could have the same ullage pressure. Though each tank had a bladder, check valves were used to isolate the ullage gases of the fuel and oxidiser tanks. If any propellant vapour permeated one of such bladders, the check valves prevented the vapour from entering the other system. After the velocity control system completed its function, the normally-open shut-off squib valve was set to the closed position, in order to isolate the velocity control system regulator and tankage from the gas in the storage vessel. Each ullage had a pressure relief system to protect the system from over-pressures. The pressurisation system of the Lunar Orbiter velocity and reaction control system was subject to leakages only two times during twenty-eight engine firings in five different missions [2].

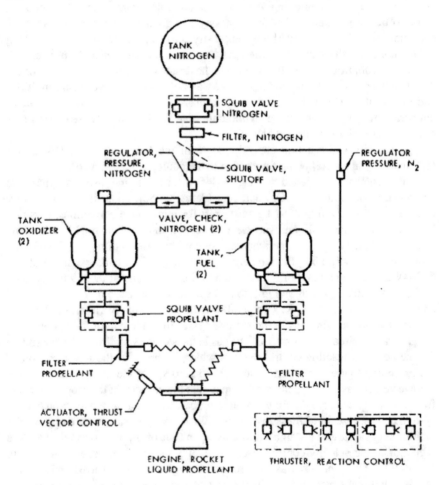

The descent stage of the Apollo lunar module used just the same combination of hypergolic propellants (N_2O_4 as the oxidiser and Aerozine-50 as the fuel) as that used by the Lunar Orbiter velocity and reaction control system. In the descent stage of the Apollo lunar module, a double-walled, Mylar®-insulated, high-pressure, cryogenic vessel was used to store 22 kg of supercritical helium for the pressurisation system. The helium tank was loaded with liquid helium at 4.4 K and topped with high-pressure gaseous helium, which increased the system temperature to approximately 6–7 K. During the 131-h (maximum) standby period, the helium pressure and temperature were increased by incoming heat leak. The maximum temperature reached by helium prior to outflow was 28 K, and the rate of pressure rise ranged from 9.6 to 19.2 N/(m²s). The feed system for the descent stage of the Apollo lunar module is illustrated schematically in the following figure, due to the courtesy of NASA [5].

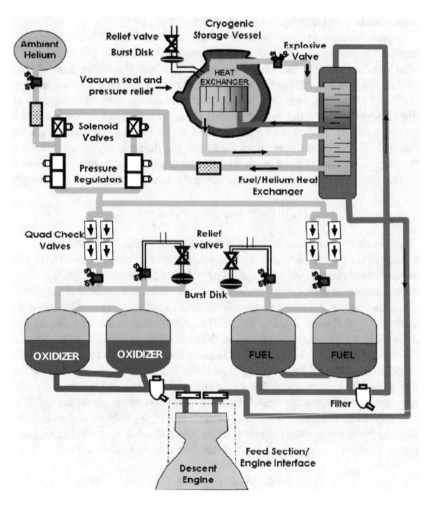

With reference to the colours shown in the preceding figure, the helium fluid (cyan) passed at a maximum flow rate of 0.000668 kg/s through the first loop of the external two-pass fuel-to-helium heat exchanger, where it absorbed heat from the fuel (red). Then, the helium (dark green) was warmed and routed back through the internal helium-to-helium heat exchanger inside the cryogenic storage vessel. The warm helium transferred heat to the remaining supercritical helium in the cryogenic storage vessel, and caused an increase in pressure, so that continuous expulsion of helium was ensured throughout the period of operation. After the helium passed through the internal helium-to-helium heat exchanger, where it was cooled, it was routed back (light green) through the second loop of the fuel-to-helium heat exchanger, and was heated to approximately 278 K before being delivered (medium green) as the pressurising agent for the fuel (red) and oxidiser (blue) tanks of the propulsion system. The use of a supercritical-helium storage tank and passive control configuration for the descent stage of the Apollo lunar module reduced the number of components required and resulted in a high degree of reliability [2].

The pressurisation systems described above are only two particular examples of component combinations. Further combinations can be found in [7].

The requirements for the gas under pressure, which have been considered in Sect. 3.2, apply only to the net or effective mass m_g of gas necessary to pressurise the tanks of propellant. However, the gross mass m_s of the stored gas depends also on the system design, on the gas expansion during operation, and on the range of the environmental temperature. The gross mass m_s of the stored gas results from a sum, whose addends are the net mass m_g of the stored gas, the mass m_{sv} of the residual gas in the storage vessel, the mass m_d of the residual gas in the lines downstream of the regulator, the mass m_{ex} of the residual mass in the heat exchanger, etc., as follows

$$m_s = m_g + m_{sv} + m_d + m_{ex} + \ldots$$

A parameter which takes account of the additional terms indicated above is the use factor of the pressurising gas. This parameter, whose value is greater than unity, is defined as the ratio m_s/m_g of the gross mass to the net mass of the pressuring gas. The lowest value of pressure in a storage vessel depends on the values of the pressure drops in the various components (regulator, heat exchanger, etc.) which make up a feed system and also on a safety value. The pressurising action of a stored gas is considered to have come to an end when the pressure in its storage vessel decays to 2.758×10^6 N/m^2 [3]. When a source of heat is present inside the storage vessel, then the expansion of the gas in the vessel is assumed to be polytropic ($pV^n =$ constant), that is, such as to involve a transfer of heat and work. By contrast, when no source of heat is present in the storage vessel, then the expansion of the gas in the vessel can be assumed to be isentropic ($pV^\gamma =$ constant), that is, frictionless and without heat transfer between the gas and the vessel walls. The following equation can be used to calculate the final temperature T_2 (K) of the gas in the storage vessel

$$\frac{T_2}{T_1} = \left(\frac{p_2}{p_1}\right)^{\frac{n-1}{n}}$$

where T_1 (K) is the initial temperature of the gas in the storage vessel, n is the exponent of the polytropic expansion, and p_2 (N/m^2) and p_1 (N/m^2) are respectively the final pressure and the initial pressure in the storage vessel.

The value of the exponent n is first estimated, and then verified experimentally. In case of an isentropic expansion of helium, $n = \gamma = c_p/c_v = 1.667$ [4].

In most cases, the expansion process through the regulator and the lines can be considered adiabatic, so that the total temperature remains constant. This expansion comes to an end in the propellant tanks. The temperature of the propellants in the tanks is assumed to be equal to the temperature at the outlet section of the heat exchanger. For a specified range of temperatures in a pressure-feed system, the lower value can be used to determine the mass of the pressurising gas, and the upper value can be used to determine the volume of the storage vessel, for a given storage pressure [3].

In the following example of application, an isentropic expansion process is assumed to occur in the oxidiser (N_2O_4) tank of a pressure-fed system having the following properties: temperature range 278–311 K in the storage vessel at the system start, pressure $p_s = 3.1 \times 10^7$ N/m² in the storage vessel at the system start, pressure $p_u = 2.76 \times 10^6$ N/m² in the storage vessel at burnout, volume $V_d = 0.0113$ m³ in the gas lines downstream of the regulator, volume $V_{ex} = 0.0283$ m³ in the heat exchanger, negligible residual volume in the gas lines, and a 2% reserve of pressurising gas. We want to compute the gross mass of the pressurising gas (helium), the volume of the gas storage vessel, and the use factor of the pressurising gas, by using the values relating to the first case (single start) of the example of Sect. 3.2. These values are also given below for convenience: absolute pressure in the oxidiser tank $p_T = 1.14 \times 10^6$ N/m², vaporisation temperature of the oxidiser $T_v = 357$ K, heat of vaporisation of the oxidiser $h_v = 414000$ J/kg, mean value of the specific heat of the oxidiser at constant pressure in liquid state $c_{pl} = 1760$ J/(kg K), mean value of the specific heat of the oxidiser at constant pressure in vapour state $c_{pv} = 754$ J/(kg K), compressibility factor of the oxidiser $Z = 0.95$, and molar mass of the oxidiser $M = 2 \times 14 + 4 \times 16 = 92$ kg/kmol. The temperature of the gas in the ullage space at burnout is $T_u = 389$ K. The specific heat at constant pressure of the pressurising gas (helium) at 293 K and atmospheric pressure is $c_{pg} = 5190$ J/(kg K) and its molar mass is $M = 4$ kg/kmol.

In Sect. 3.2, the net mass of the pressurising gas has been found to be $m_g = 4.469$ kg. The temperature and the pressure of the residual pressurising gas in the lines downstream of the regulator after shutdown are assumed to have the same values as those of the ullage gases in the oxidiser tank at burnout, which are respectively $T_u = 389$ K and $p_T = 1.14 \times 10^6$ N/m². Since the volume in the gas lines downstream of the regulator is $V_d = 0.0113$ m³ (see above), then the mass m_d of the residual pressurising gas in the lines downstream of the regulator results from the law of perfect gases, as follows

$$m_d = \frac{p_T V_d}{R_g T_u} = \frac{1.14 \times 10^6 \times 0.0113 \times 4}{8314.46 \times 389} = 0.0159 \, \text{kg}$$

The temperature of the residual pressurising gas in the heat exchanger is assumed to be the same as the temperature of the pressurising gas at the inlet section of the oxidiser tank. In Sect. 3.2, this temperature has been found to be $T_g = 554$ K. Like-wise, the pressure of the residual pressurising gas in the heat exchanger is assumed to be the same as the pressure ($p_u = 2.76 \times 10^6$ N/m^2) in the storage vessel at burnout. Therefore, the mass m_{ex} of the residual pressurising gas in the heat exchanger results from the law of perfect gases, as follows

$$m_{ex} = \frac{p_u V_{ex}}{R_g T_g} = \frac{2.76 \times 10^6 \times 0.0283 \times 4}{8314.46 \times 554} = 0.0678 \text{ kg}$$

By using the following equation

$$\frac{T_2}{T_1} = \left(\frac{p_2}{p_1}\right)^{\frac{n-1}{n}}$$

and solving for T_2, where T_1 is the lower limit (311 K) of the system operating temperatures, and $p_1 = p_s = 3.1 \times 10^7$ N/m^2 and $p_2 = p_u = 2.76 \times 10^6$ are respectively the initial pressure and the final pressure of the helium ($n = \gamma \equiv c_p/c_v = 1.667$), the temperature of the residual helium in the storage vessel at burnout results

$$T_2 = T_1 \left(\frac{p_2}{p_1}\right)^{\frac{n-1}{n}} = 311 \times \left(\frac{2.76 \times 10^6}{3.1 \times 10^7}\right)^{\frac{1.667-1}{1.667}} = 118 \text{ K}$$

Since the pressure in the storage vessel at burnout is $p_u = 2.76 \times 10^6$ N/m^2 and the temperature of the residual helium in the storage vessel at burnout is $T_2 = 118$ K, then the mass m_{sv} (kg) in the storage vessel at burnout can be determined by using the equation of perfect gases, as follows

$$m_{sv} = \frac{p_u V_L}{R_g T_2} = \frac{2.76 \times 10^6 \times 4}{8314.46 \times 118} V_L$$

where V_L is the volume (m^3) of the pressurising gas necessary to put the oxidiser tank under pressure.

Likewise, since the pressure in the storage vessel at start is $p_s = 3.1 \times 10^7$ N/m^2 and the higher temperature of the operating range is $T_s = 311$ K, then the mass m_s (kg) in the storage vessel at start can be determined by using the equation of perfect gases, as follows

$$m_s = \frac{p_s V_L}{R_g T_s} = \frac{3.1 \times 10^7 \times 4}{8314.46 \times 311} V_L$$

where, again, V_L is the volume (m^3) of the pressurising gas necessary to put the oxidiser tank under pressure. Remembering the preceding equation

$$m_s = m_g + m_{sv} + m_d + m_{ex} + \ldots$$

truncated after m_{ex}, and substituting the values found above, there results

$$\frac{3.1 \times 10^7 \times 4}{8314.46 \times 311} V_L = 4.469 + \frac{2.76 \times 10^6 \times 4}{8314.46 \times 118} V_L + 0.0159 + 0.0679$$

This equation, solved for V_L, yields

$$V_L = 0.124 \, \text{m}^3$$

By substituting this value into

$$m_s = \frac{p_s V_L}{R_g T_s} = \frac{3.1 \times 10^7 \times 4}{8314.46 \times 311} V_L$$

the gross mass of the pressurising gas results

$$m_s = 5.949 \, \text{kg}$$

By adding the 0.2% reserve of pressurising gas to the value computed above, the gross mass becomes

$$m_s = 5.949 \times 1.02 = 6.068 \, \text{kg}$$

By introducing $m_s = 6.068$ kg, $R^* = 8314.460$ N m kmol^{-1} K$^-$, $\mathcal{M} = 4$ kg/kmol, $p_s = 3.1 \times 10^7$ N/m^2 and $T_s = 311$ K in the equation of perfect gases, the volume V_u of the storage vessel for the pressurising gas results

$$V_u = \frac{m_s R^* T_s}{\mathcal{M} p_s} = \frac{6.068 \times 8314.46 \times 311}{4 \times 3.1 \times 10^7} = 0.1265 \, \text{m}^3$$

According to the definition given above, the use factor of the pressurising gas has the following value

$$\frac{m_s}{m_g} = \frac{6.068}{4.469} = 1.358$$

Now we want to compute the same quantities as those determined above, relating to the second case (multi-start) of the example of Sect. 3.2. There, the net mass of the pressurising gas has been found to be $m_g = 4.883$ kg. The volume in the gas lines downstream of the regulator is $V_d = 0.0113$ m^3 (see above). The temperature and the pressure of the pressurising gas in the lines downstream of the regulator are assumed to be the same as those in the oxidiser tank at burnout ($T_u = 367$ K and $p_u = 1.14 \times 10^6$ N/m^2). Therefore, the mass m_d of the residual pressurising gas in the lines downstream of the regulator results from the law of perfect gases, as follows

$$m_d = \frac{p_u V_d}{R_g T_u} = \frac{1.14 \times 10^6 \times 0.0113 \times 4}{8314.46 \times 367} = 0.0169\,\text{kg}$$

The temperature of the residual gas under pressure in the heat exchanger is assumed to be the same as the temperature of the gas under pressure at the inlet section of the tank. In Sect. 3.2, this temperature has been found to be $T_g = 554$ K. The pressure of the same residual gas in the heat exchanger is assumed to be the same as the pressure of the gas in the storage vessel at burnout, which is (see above) $p_u = 2.76 \times 10^6$ N/m^2. The volume in the heat exchanger is (see above) $V_{ex} = 0.0283$ m^3. By introducing these values in the equation of perfect gases, the mass of the residual gas in the heat exchanger results

$$m_{ex} = \frac{p_u V_{ex}}{R_g T_g} = \frac{2.76 \times 10^6 \times 0.0283 \times 4}{8314.46 \times 554} = 0.0678\,\text{kg}$$

In the first case, the temperature of the residual helium in the storage vessel at burnout has been found to be $T_2 = 118$ K, and the mass of this gas in the storage vessel at burnout has been found to be

$$m_{sv} = \frac{p_u V_L}{R_g T_2} = \frac{2.76 \times 10^6 \times 4}{8314.46 \times 118} V_L$$

where V_L (m^3) is the volume of the pressurising gas necessary to put the oxidiser tank under pressure.

Again, in the first case, the mass m_s (kg) of gas in the storage vessel at start has been found to be

$$m_s = \frac{p_s V_L}{R_g T_s} = \frac{3.1 \times 10^7 \times 4}{8314.46 \times 311} V_L$$

Remembering the equation

$$m_s = m_g + m_{sv} + m_d + m_{ex} + \ldots$$

truncated after m_{ex}, and substituting the values found above, there results

$$\frac{3.1 \times 10^7 \times 4}{8314.46 \times 311} V_L = 4.883 + \frac{2.76 \times 10^6 \times 4}{8314.46 \times 118} V_L + 0.0169 + 0.0678$$

This equation, solved for V_L, yields

$$V_L = 0.135\,\text{m}^3$$

By substituting this value into

$$m_s = \frac{p_s V_L}{R_g T_s} = \frac{3.1 \times 10^7 \times 4}{8314.46 \times 311} V_L$$

the gross mass of the pressurising gas results

$$m_s = 6.474 \, \text{kg}$$

By adding the 0.2% reserve of pressurising gas to the value computed above, the gross mass becomes

$$m_s = 6.474 \times 1.02 = 6.603 \, \text{kg}$$

According to the definition given above, the use factor of the pressurising gas has the following value

$$\frac{m_s}{m_g} = \frac{6.603}{4.883} = 1.352$$

We want to compute the gross mass m_s (kg) of the gas under pressure in the oxidiser tank, the volume V_u (m^3) of the storage vessel, and the use factor of the pressurising gas for the first case (single start) of the example of Sect. 3.2, in the absence of a system for thermal conditioning. Since the pressurising gas is not heated, then its bulk temperature T_g at burnout can be assumed to be the average of the initial temperature (278 K, a datum) and the final temperature (which has been found to be $T_2 = 118$ K) in the storage vessel, as follows

$$T_g = \frac{278 + 118}{2} = 198 \, \text{K}$$

Since this temperature is lower than the temperature ($T_e = 289$ K) of the propellant, then the pressurising gas does not warm the propellant. The net mass m_g of the required gas under pressure results from the law of perfect gases, as follows

$$m_g = \frac{p_T V_T}{R_g T_g} = \frac{1.14 \times 10^6 \times 3.37 \times 4}{8314.46 \times 198} = 9.335 \, \text{kg}$$

The mass m_d of the residual pressurising gas in the lines downstream of the regulator results from the law of perfect gases, as follows

$$m_d = \frac{p_T V_d}{R_g T_g} = \frac{1.14 \times 10^6 \times 0.0113 \times 4}{8314.46 \times 198} = 0.0313 \, \text{kg}$$

In the first case, the temperature of the residual helium in the storage vessel at burnout has been found to be $T_2 = 118$ K, and the mass of this gas in the storage vessel at burnout has been found to be

$$m_{sv} = \frac{p_u V_L}{R_g T_2} = \frac{2.76 \times 10^6 \times 4}{8314.46 \times 118} V_L$$

where V_L (m^3) is the volume of the pressurising gas necessary to put the oxidiser tank under pressure.

The gross mass m_s (kg) of gas in the storage vessel at start is

$$m_s = \frac{p_s V_L}{R_g T_s} = \frac{3.1 \times 10^7 \times 4}{8314.46 \times 278} V_L$$

Remembering the equation

$$m_s = m_g + m_{sv} + m_d + m_{ex} + \ldots$$

where, in the present case, $m_{ex} = 0$, and substituting the values found above, there results

$$\frac{3.1 \times 10^7 \times 4}{8314.46 \times 278} V_L = 9.334 + \frac{2.76 \times 10^6 \times 4}{8314.46 \times 118} V_L + 0.0313$$

This equation, solved for V_L, yields

$$V_L = 0.2209 \, \text{m}^3$$

Therefore, the gross mass m_s (kg) of gas in the storage vessel at start is

$$m_s = \frac{p_s V_L}{R_g T_s} = \frac{3.1 \times 10^7 \times 4 \times 0.2209}{8314.46 \times 278} = 11.85 \, \text{m}^3$$

By adding the 0.2% reserve of pressurising gas to the value computed above, the gross mass becomes

$$m_s = 11.85 \times 1.02 = 12.09 \, \text{kg}$$

According to the definition given above, the use factor of the pressurising gas has the following value

$$\frac{m_s}{m_g} = \frac{12.09}{9.335} = 1.295$$

Finally, it is required to compute the gross mass m_s (kg) of the gas under pressure in the oxidiser tank, the volume V_u (m^3) of the storage vessel, and the use factor of the pressurising gas for the first case (single start) of the example of Sect. 3.2, in the presence of a system for thermal conditioning, assuming a polytropic ($pV^n = $ constant, with $n = 1.2$) expansion of helium in the storage vessel.

By using the following equation

$$\frac{T_2}{T_1} = \left(\frac{p_2}{p_1}\right)^{\frac{n-1}{n}}$$

and solving for T_2, where T_1 is the initial temperature (278 K, a datum) of the helium, and $p_1 = p_s = 3.1 \times 10^7$ N/m^2 and $p_2 = p_u = 2.76 \times 10^6$ are respectively the initial pressure and the final pressure of the same gas ($n = 1.2$), the temperature of the residual helium in the storage vessel at burnout results

$$T_2 = T_1 \left(\frac{p_2}{p_1}\right)^{\frac{n-1}{n}} = 278 \times \left(\frac{2.76 \times 10^6}{3.1 \times 10^7}\right)^{\frac{1.2-1}{1.2}} = 186 \text{ K}$$

The bulk temperature T_g of the helium at burnout can be assumed to be the average of the initial temperature $T_1 = 278$ K and the final temperature $T_2 = 186$ K in the storage vessel, as follows

$$T_g = \frac{278 + 186}{2} = 232 \text{ K}$$

The net mass m_g of the required gas under pressure results from the law of perfect gases, as follows

$$m_g = \frac{p_T V_T}{R_g T_g} = \frac{1.14 \times 10^6 \times 3.37 \times 4}{8314.46 \times 232} = 7.967 \text{ kg}$$

The mass m_d of the residual gas under pressure in the lines downstream of the regulator results from the law of perfect gases, as follows

$$m_d = \frac{p_T V_d}{R_g T_g} = \frac{1.14 \times 10^6 \times 0.0113 \times 4}{8314.46 \times 232} = 0.0267 \text{ kg}$$

The mass of this gas in the storage vessel at burnout is

$$m_{sv} = \frac{p_u V_L}{R_g T_2} = \frac{2.76 \times 10^6 \times 4}{8314.46 \times 186} V_L$$

where V_L (m^3) is the volume of the pressurising gas necessary to put the oxidiser tank under pressure.

The gross mass m_s (kg) of gas in the storage vessel at start is

$$m_s = \frac{p_s V_L}{R_g T_s} = \frac{3.1 \times 10^7 \times 4}{8314.46 \times 278} V_L$$

Remembering the equation

$$m_s = m_g + m_{sv} + m_d + m_{ex} + \ldots$$

where, in the present case, $m_{ex} = 0$, and substituting the values found above, there results

$$\frac{3.1 \times 10^7 \times 4}{8314.46 \times 278} V_L = 7.966 + \frac{2.76 \times 10^6 \times 4}{8314.46 \times 186} V_L + 0.0267$$

This equation, solved for V_L, yields

$$V_L = 0.1719 \, \text{m}^3$$

Therefore, the gross mass m_s (kg) of gas in the storage vessel at start is

$$m_s = \frac{p_s V_L}{R_g T_s} = \frac{3.1 \times 10^7 \times 4 \times 0.1719}{8314.46 \times 278} = 9.220 \, \text{kg}$$

By adding the 0.2% reserve of pressurising gas to the value computed above, the gross mass becomes

$$m_s = 9.220 \times 1.02 = 9.404 \, \text{kg}$$

According to the definition given above, the use factor of the pressurising gas has the following value

$$\frac{m_s}{m_g} = \frac{9.412}{7.967} = 1.180$$

Gases under pressure are usually stored in spherical vessels, because of the structural efficiency of the spherical shape, which implies a lower mass than is possible with other shapes. Such vessels are usually of monocoque design, operate at high stress levels, are mounted within the rocket vehicle, and are insulated from deflection of the vehicle structure by appropriately designed mountings [2]. By the way, monocoque (a French word meaning single shell) is a structural technique in which a body supports loads through its external skin, with no internal frame to hold the body rigid. The alloy most often used to construct storage vessels is Ti–6Al–4V [8].

An estimate for the mass m (kg) of a spherical vessel of thickness t (m), inside diameter d (m), and made of a material having uniform density ρ (kg/m^3) can be made as follows

$$m = \rho V = \frac{4}{3} \pi \rho \left[\left(\frac{d}{2} + t \right)^3 - \left(\frac{d}{2} \right)^3 \right]$$

Another estimate, suggested by Huzel and Huang [3], for the same mass m (kg) can be made by assuming a vessel made of two hemispherical shells, such that the thickness of the weld lands is taken into account by assuming a band of width w (m) and thickness equal to one-half the wall thickness placed over the weld seam. This estimate is

$$m = \pi d^3 \rho \left(\frac{p}{4\sigma} \right) + 39.37\pi w d^2 \rho \left(\frac{p}{8\sigma} \right)$$

where p (N/m^2) is the maximum pressure at which the gas is stored, and σ (N/m^2) is the allowable working stress of the material of which the vessel is made.

The storage vessel must be capable of containing the gas at high pressure for long periods of time without losses due to leakage. Gases having low molar masses, such as hydrogen and helium, are less subject to leak through homogeneous metals of good quality than is the case with those having high molar masses. Leakages can occur through porous metals, for example, through castings and welded joints.

The pressure-fed systems cited above (used for the descent stage of the Apollo lunar module and for the velocity and reaction control system of the Lunar Orbiter) have, each of them, a pressure regulator for the gases (respectively, helium and nitrogen) stored under pressure. A regulator maintains the desired values of flow and pressure to the propellant tanks as the pressure of the stored gas decreases.

The example described below refers to a pressure regulator used for the Space Storable Propulsion Module, which was used for a Jupiter Orbiter mission.

With reference to the following figure, due to the courtesy of NASA [9], this pressure regulator has an unbalanced poppet, which achieves sealing to the seat by means of the upstream pressure force. In addition, the poppet is held against the seat by an axial guidance flexure.

The actuation mechanism consists of a bellows, which is exposed to space internally and to the regulator downstream pressure externally, and which is held in the null position by one or more coil springs. If the regulator downstream pressure decreases from the preset pressure, the reference spring force overcomes the bellows pressure force and the actuator exerts a net opening force on the poppet through the lever arrangement. When this net force is greater than the poppet seating force, the poppet opens and allows the gas (helium) under pressure to pressurise the downstream side of the regulator. When the downstream regulator pressure rises back up to the set pressure, the actuator returns to the null position and the poppet is caused to return to the seat by the axial guidance flexure spring force and the difference of pressure across the poppet [9].

The heat exchanger, which warms the gas under pressure, may be designed to form an integral part of the diverging portion of the nozzle, as shown in the following figure. In this case, the heat exchanger is a tube wound around the contour of the nozzle. Other types of design for heat exchangers will be shown in Sect. 3.4.

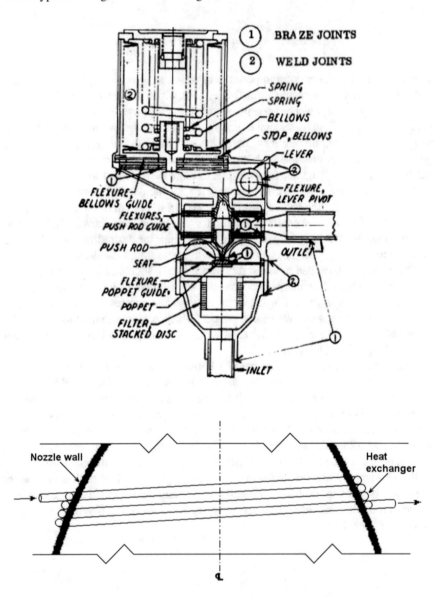

The heat transfer coefficient h_g (W m^{-2} K^{-1}) on the side of the combusted gas can be determined as has been shown in Chap. 2, Sect. 2.5. The heat transferred by conduction from the nozzle wall to the heat exchanger can be assumed to be entirely absorbed by the pressurising gas which circulates in the heat exchanger, and therefore the temperature of the pressurising gas increases. Likewise, the determination of the heat transfer coefficient h_h (W m^{-2} K^{-1}) on the side of the pressurising gas and the design of the heat exchanger can be done as has been shown in Chap. 2, Sect. 2.5. The number of turns in the heat exchanger around the nozzle depends on the required increase in temperature of the pressurising gas and also on the position of the heat exchanger along the nozzle. The various temperatures in the elements of the heat exchanger can be expressed by means of the following equation [3]:

$$\dot{m}_h c_p (T_o - T_i) = A \left(\frac{1}{h_g} + \frac{t}{k} + \frac{1}{h_h} \right)^{-1} \left(T_{aw} - \frac{T_i + T_o}{2} \right)$$

where $\dot{m}_h$ (kg/s) is the mass flow rate of the pressurising gas, c_p (J kg^{-1} K^{-1}) is the specific heat at constant pressure of the pressurising gas, T_i (K) is the mean temperature of the pressurising gas at the inlet section of the heat exchanger, T_o (K) is the mean temperature of the pressurising gas at the outlet section of the heat exchanger, A (m^2) is the effective area of the heat exchanger, h_g (W m^{-2} K^{-1}) is the heat transfer coefficient on the side of the combusted gas, t (m) is the thickness of the tube of the heat exchanger, k (W m^{-1} K^{-1}) is the thermal conductivity of the material of which this tube is made, and T_{aw} (K) is the adiabatic wall temperature on the side of the combusted gas. The temperature of the pressurising gas which leaves the heat exchanger at any time also depends on the temperature of the same gas at the exit section of the storage vessel. The material, of which the tube of the heat exchanger is made, is to be chosen by the designer bearing in mind the necessity of attaching firmly by brazing the tube to the wall of the nozzle. The width of this tube depends on the thermal and mechanical loads to which the tube is subject.

The following example of application concerns the design of a heat exchanger of the type illustrated above, used in parallel with other heat exchangers and placed at the cross-section of area ratio $A/A_t = 10$ of a nozzle extension for a rocket engine of known characteristics.

Let $A_t = 0.02606$ m^2 be the area at the throat, $\dot{m}_h = 0.0109$ kg/s be the mass flow rate of the pressurising gas (helium), and $p_{co} \equiv p_h = 3.1026 \times 10^7$ N/m^2 its pressure in the storage vessel. The values of the specific heat ratio and of the specific heat at constant pressure of helium can be either taken from [4] ($\gamma \equiv c_p/c_v = 1.667$ and $c_p = 5190$ J kg^{-1} K^{-1} at 293 K and atmospheric pressure), or determined at a given temperature and pressure, as will be shown below.

Let $T_i = 192$ K be the mean temperature of the helium at the inlet section of the heat exchanger, $T_o = 554$ K be the mean temperature of the helium at the outlet section of the heat exchanger, $T_{aw} = 2722$ K be the adiabatic wall temperature on the side of the combusted gas, and $h_g = 167.8$ W m^{-2} K^{-1} be the heat transfer coefficient on the side of the combusted gas. We want to determine the dimensions and the number of turns for the tube of the heat exchanger, assuming this tube to be made of an alloy Ti–13V–11Cr–3Al, aged at 763 K, having the following thermal and mechanical characteristics [10]: tensile yield strength $\sigma_y = 1.3 \times 10^9$ N/m^2, modulus of elasticity $E = 1.1 \times 10^{11}$ N/m^2, thermal conductivity (at 698 K) $k = 17.1$ W m^{-1} K^{-1}, coefficient of thermal expansion (in the range 293–923 K) $\lambda = 10.44 \times 10^{-6}$ m m^{-1} K^{-1}, and Poisson's ratio $\nu = 0.30$. For each chosen section of the heat exchanger, the temperature of the tube at the wall depends on the bulk temperature of the pressurising gas (helium) at that section. Since the pressurising gas absorbs heat from the nozzle, the maximum temperature occurs at the outlet section of the heat exchanger. We assume the maximum allowable temperature at the outlet section of the heat exchanger on the combusted gas side to be $T_{wg} = 778$ K. Remembering the following equation of Chap. 2, Sect. 2.5

$$q = h_g\left(T_{aw} - T_{wg}\right)$$

where q (W/m^2) is the quantity of heat transferred per unit time per unit surface through convection at the outlet section of the heat exchanger, there results

$$q = 167.8 \times (2722 - 778) = 3.262 \times 10^5 \text{ W/m}^2$$

We take the preliminary value $t = 0.00127$ m for the thickness of the tube. This value will be checked against the results of the following calculations.

Remembering the following equation of Chap. 2, Sect. 2.5

$$T_{wc} = T_{wg} - \frac{tq}{k}$$

and substituting $T_{wg} = 778$ K, $t = 0.00127$ m, $q = 3.262 \times 10^5$ W/m^2, and $k = 17.1$ W m^{-1} K^{-1}, the mean temperature of the wall on the helium side is

$$T_{wc} = 778 - \frac{0.00127 \times 3.262 \times 10^5}{17.1} = 754 \text{ K}$$

Remembering again the following equation of Chap. 2, Sect. 2.5

$$q = h_c(T_{wc} - T_{co})$$

where $T_{co} = T_o = 554$ K, and solving for h_c, the heat transfer coefficient on the helium side can be computed as follows

$$h_h \equiv h_c = \frac{q}{T_{wc} - T_{co}} = \frac{3.262 \times 10^5}{754 - 554} = 1631 \text{ W m}^{-2} \text{ K}^{-1}$$

According to Wang et al. [11, page 911, Eq. 4], the Prandtl number Pr can be computed approximately as follows

$$Pr = \frac{4\gamma}{9\gamma - 5}$$

Since in the present case $\gamma = c_p/c_v = 1.667$, then

$$Pr = \frac{4 \times 1.667}{9 \times 1.667 - 5} = 0.6667$$

The result found above for the Prandtl number takes account only of the specific heat ratio ($\gamma = 1.667$ at 293 K and atmospheric pressure, according to [4]) of helium. The pressure and the temperature of helium are not taken into account. This result can be checked by using the thermal data given by NIST [12] or by Petersen [13].

The data given by NIST at the pressure $p = 1.241 \times 10^6$ N/m^2 (regulated pressure of helium) and at the temperature $T_{co} = 554$ K are: $c_p = 5192.1$ J/(kg K), $c_v = 3116.7$ J/(kg K), $\mu = 3.0488 \times 10^{-5}$ N s m^{-2}, and $k = 0.23952$ W m^{-1} K^{-1}. Therefore, the Prandtl number in these conditions is

$$Pr = \frac{\mu c_p}{k} = \frac{3.0488 \times 10^{-5} \times 5192.1}{0.23952} = 0.6609$$

The mass flow rate G of the coolant (helium) per unit area to be cooled can be computed as follows

$$G = \frac{\dot{m}_h}{\pi \left(\frac{d}{2}\right)^2} = \frac{0.0109 \times 4}{\pi d^2} \text{ kg s}^{-1} \text{ m}^{-2}$$

where d (m) is the inside diameter of the helium passage. In the present case, the heat is transferred through a vapour-film boundary layer, because the coolant is helium in supercritical conditions of pressure and temperature.

Therefore, the heat transfer coefficient on the helium side computed above ($h_h = 1631$ W m^{-2} K^{-1}) satisfies the following correlation due to McCarthy and Wolf [14, page 95, Eq. 2]:

$$h_h = 0.025 \left(\frac{c_p \mu^{0.2}}{Pr^{0.6}}\right)_{co} \frac{G^{0.8}}{d^{0.2}} \left(\frac{T_{co}}{T_{wc}}\right)^{0.55}$$

where μ (N s m^{-2}) is the coefficient of dynamic viscosity of helium, c_p (J kg^{-1} K^{-1}) is the isobaric specific heat of helium, Pr is the Prandtl number of helium, T_{co} (K) is the bulk temperature of helium, T_{wc} (K) is the temperature of the wall on the helium side, and the subscript co indicates the bulk temperature ($T_{co} = 554$ K) of the coolant (helium). By substituting the quantities computed above into the McCarthy-Wolf correlation, there results

$$1631 = 0.025 \left(\frac{5192 \times 0.0000304 9^{0.2}}{0.6609^{0.6}}\right) \left(\frac{0.0109 \times 4}{\pi d^2}\right)^{0.8} \frac{1}{d^{0.2}} \left(\frac{554}{754}\right)^{0.55}$$

The preceding equation, solved numerically for d, yields

$$d = 0.01205 \, \text{m}$$

which is the inside diameter of the helium passage in the heat exchanger.

The maximum tensile combined pressure and thermal stress σ_t (N/m^2) at the outlet section of the heat exchanger can be determined by using the following equation of Chap. 2, Sect. 2.5

$$\sigma_t = \frac{(p_{co} - p_g)r}{t} + \frac{E\lambda qt}{2(1-v)k} + \frac{6M_A}{t^2}$$

where p_{co} (N/m^2) is the pressure of the coolant, p_g (N/m^2) is the pressure of the combusted gas, r (m) is the radius of the cross section of the cooling tubes, t (m) is the thickness of the cooling tubes, E (N/m^2) is the modulus of elasticity of the material of which the cooling tubes are made, λ (K^{-1}) is the coefficient of thermal expansion of the same material, q (W/m^2) is the quantity of heat per unit time per unit surface, v is the Poisson ratio of the same material, k (W m^{-1} K^{-1}) is the thermal conductivity of the same material, and M_A (Nm/m) is the bending moment per unit length acting on a cross section A-A of the circular cooling tube due to the distortion induced by discontinuities. In the present case, we assume $p_{co} - p_g \approx p_{co} = 3.103 \times 10^7$ N/m^2 and neglect the third term (containing M_A) in comparison with the other two terms on the right-hand side of the preceding equation. Thus, the maximum stress at the outlet section of the heat exchanger is

$$\begin{aligned}
\sigma_t &= \frac{3.103 \times 10^7 \times 0.01205}{2 \times 0.00127} \\
&+ \frac{1.1 \times 10^{11} \times 10.44 \times 10^{-6} \times 3.262 \times 10^5 \times 0.00127}{2 \times (1 - 0.30) \times 17.1} \\
&= 1.671 \times 10^8 \text{N/m}^2
\end{aligned}$$

This value is lower than the tensile yield strength ($\sigma_y = 1.3 \times 10^9$ N/m^2) of the alloy Ti–13V–11Cr–3Al.

Now, in order to determine the maximum tensile combined pressure and thermal stress σ_t (N/m^2) at the inlet section of the heat exchanger, the difference between the maximum temperature of the wall on the combusted gas side (778 K) and the bulk temperature of the helium in the heat exchanger is assumed to remain approximately constant throughout the heat exchanger. The bulk temperature of the helium results from the difference $T_o - T_i = 554 - 192 = 362$ K. Therefore, the mean temperature of the wall on the combusted gas side at the inlet section of the heat exchanger is

$$T_{wg} = 778 - 362 = 416 \, \text{K}$$

Remembering again the following equation of Chap. 2, Sect. 2.5

$$q = h_g \left(T_{aw} - T_{wg} \right)$$

where q (W/m^2) is the quantity of heat transferred per unit time per unit surface through convection at the inlet section of the heat exchanger, there results

$$q = 167.8 \times (2722 - 416) = 3.869 \times 10^5 \, \text{W/m}^2$$

By using again the equation of Chap. 2, Sect. 2.5

$$\sigma_t = \frac{(p_{co} - p_g)r}{t} + \frac{E\lambda qt}{2(1-v)k} + \frac{6M_A}{t^2}$$

with $p_{co} - p_g \approx p_{co} = 3.103 \times 10^7$ N/m^2 and neglecting the third term (containing M_A) in comparison with the other two terms on the right-hand side of the preceding equation, the maximum tensile combined pressure and thermal stress σ_t (N/m^2) at the inlet section of the heat exchanger results

$$\sigma_t = \frac{3.103 \times 10^7 \times 0.01205}{2 \times 0.00127}$$
$$+ \frac{1.1 \times 10^{11} \times 10.44 \times 10^{-6} \times 3.869 \times 10^5 \times 0.00127}{2 \times (1 - 0.30) \times 17.1}$$
$$= 1.708 \times 10^8 \, \text{N/m}^2$$

This value is lower than the tensile yield strength ($\sigma_y = 1.3 \times 10^9$ N/m^2) of the alloy Ti–13V–11Cr–3Al.

Therefore, it is safe to choose a tube made of the alloy Ti–13V–11Cr–3Al, 0.00127 m in thickness and 0.01205 m in inner diameter, for the heat exchanger. The margin of safety is sufficiently large, should the temperature at the inlet section of the heat exchanger be higher than the maximum temperature considered above.

Now, in order to determine the number of turns for the tube of the heat exchanger, the effective area A (m^2) of the heat exchanger is determined by means of the following equation

$$\dot{m}_h c_p (T_o - T_i) = A\left(\frac{1}{h_g} + \frac{t}{k} + \frac{1}{h_h}\right)^{-1}\left(T_{aw} - \frac{T_i + T_o}{2}\right)$$

We first substitute $\dot{m}_h = 0.0109$ kg/s, $c_p = 5192$ J/(kg K), $T_o = 554$ K, $T_i = 192$ K, $h_g = 167.8$ W m^{-2} K^{-1}, $t = 0.00127$ m, $k = 17.1$ W/(m K), $h_h = 1631$ W m^{-2} K^{-1}, and $T_{aw} = 2722$ K, and then solve the preceding equation for A. This yields

$$A = \frac{0.0109 \times 5192 \times (554 - 192)}{\left(2722 - \frac{192+554}{2}\right)}$$
$$\times \left(\frac{1}{167.8} + \frac{0.00127}{17.1} + \frac{1}{1631}\right) = 0.05797 \, \text{m}^2$$

Since the area of the thrust chamber at the throat is $A_t = 0.02606 \text{ m}^2$ (see above), then the corresponding diameter at the throat is

$$D_t = \left(\frac{4A_t}{\pi}\right)^{\frac{1}{2}} = \left(\frac{4 \times 0.02606}{3.1416}\right)^{\frac{1}{2}} = 0.1822 \text{ m}$$

and the diameter of the thrust chamber at the section of area ratio $\varepsilon \equiv A/A_t = 10$ is

$$D = D_t(\varepsilon)^{\frac{1}{2}} = 0.1822 \times (10)^{\frac{1}{2}} = 0.5761 \text{ m}$$

Assuming that a portion of 40% of the internal surface of N turns of tube is the effective area $A = 0.05797 \text{ m}^2$ of the heat exchanger, A is the area of a rectangle of base πD and altitude $0.4 \, N\pi d$. Therefore, the number N of turns of the tube of the heat exchanger results from

$$N = \frac{A}{(\pi D)(0.4\pi d)} = \frac{0.05797}{3.1416^2 \times 0.5761 \times 0.01205 \times 0.4} = 2.115$$

where $d = 0.01205$ m is the inside diameter of the tube, and $D = 0.5761$ m is the diameter of the thrust chamber at the section of area ratio $\varepsilon = 10$. The diameter D is assumed to remain constant along the altitude of the rectangle indicated above.

3.4   Feed Systems Using Evaporation of Two Propellants

The feed systems described here have been used prevalently, if not exclusively, in rocket engines using cryogenic or near-cryogenic propellants. A description and a list of these propellants are given in Chap. 1, Sect. 1.4.

In particular, these feed systems have been used in large rocket vehicles whose propulsion systems are fed by pumps. They have been used especially for oxidisers, because most fuels tend to boil violently when heated. However, liquid hydrogen, which is a fuel, is an exception to this rule. Since the oxidiser vapour has usually a high molar mass, feed systems using evaporation of two propellants require a mass of pressurising gas for the oxidiser tank higher than the mass required by systems using inert gases, such as those described in Sect. 3.3. This undesirable effect is counterbalanced by the elimination of storage vessels, because in feed systems using evaporation of two propellants the pressurising agent is stored as a liquid in the tank containing the main propellant.

Feed systems based on evaporation of two propellants used for pump-fed rocket engines take propellants tapped off downstream of the pump and vaporised in a heat exchanger. These propellants are then used to pressurise the main propellant tank from which they have been taken. There are many types of heat exchangers. One of them is the helical-coil type, which is described below. A functional scheme of this heat exchanger is shown in the following figure, due to the courtesy of Wikimedia

[15], which illustrates a heat exchanger designed by US Army Colonel Scott S. Haraburda.

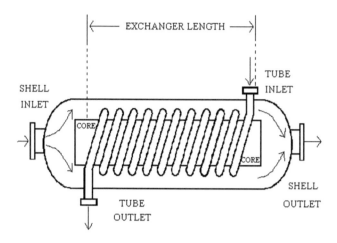

A helical-coil heat exchanger consists essentially of a shell (which is a container, usually cylindrical, designed to hold a fluid at a pressure much higher than the ambient pressure) and a tube wound around a core inside the shell. One of two fluids runs through the tube, from the inlet to the outlet section, and the other fluid flows over the tube, from the inlet to the outlet section of the shell, in order to transfer heat from one fluid to the other.

The following figure, due to the courtesy of NASA [2], illustrates another type (shell-and-tube) of heat exchanger used to vaporise liquid cryogenic propellants for evaporated-propellant pressurisation, as will be shown below.

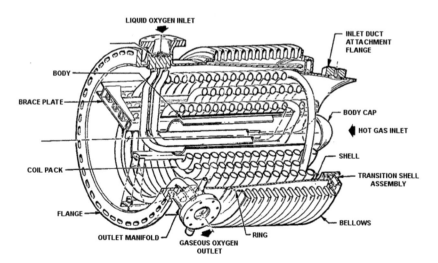

This particular heat exchanger, which has been used for the J-2 rocket engine, is a shell assembly, consisting of a duct, bellows, flanges, and coils. It is mounted in the turbine exhaust duct between the oxidiser turbine discharge manifold and the thrust chamber. It heats and expands gaseous helium for use in the third stage of the Saturn V rocket or converts liquid oxygen to gaseous oxygen for the second stage for maintaining pressurisation in the oxidiser tank of the vehicle. During engine operation, either liquid oxygen is tapped off the oxidiser high-pressure duct or helium is provided from the vehicle stage and routed to the heat exchanger coils [16].

Heat exchangers have also been employed on the Titan rockets to vaporise liquid nitrogen tetroxide to be used as the pressurising agent [2].

As has been shown above, evaporated-propellant pressurisation is successful with cryogenic propellants and also with the storable oxidiser nitrogen tetroxide, whose boiling point is 294.2 K. The simplest method of evaporated-propellant pressurisation is self-pressurisation (flash boiling) in a propellant tank during feed-out. By the way, flash boiling is the phenomenon which occurs in a heated liquid whose pressure is lower than the saturation vapour pressure of the liquid. When the pressure falls sufficiently below the saturation vapour pressure, then rapid boiling of the liquid can result.

The self-pressurisation method is reliable, but the mass requirements of the pressurising agent are high, because of its low temperature, and hence of its high density. In addition, pre-pressurisation from a separate system may be necessary to meet engine start requirements, as was the case with the Centaur rocket vehicle [2]. In this case, the vapour pressure of bulk propellants (hydrogen and oxygen) boiling in the tanks was used to provide the modest pressure required at the inlet of tank-mounted boost pumps. In systems requiring higher pressures, superheated vapour is obtained by passing the propellant through an engine heat exchanger or some other heat source, as will be shown below.

More favourable conditions exist in the ullage space when the pressurising agent is heated to the maximum temperature compatible with structural and propellant requirements, and when the temperature of the gas in the ullage space is stratified, that is, controlled by heating or cooling at the tank wall and upper bulkhead [2].

In the S-II (meaning by this name the second stage of the Saturn V rocket), which used evaporated-propellant pressurisation in both oxidiser and fuel systems, propellant vapour was superheated to reduce the mass of gas required to the minimum. For main stage pressurisation, the oxidiser tank of the S-II was pressurised with warm gaseous oxygen (275 $\pm$ 8 K at the maximum mixture ratio, and 250 $\pm$ 8 K at low mixture ratio). As shown in the following figure, due to the courtesy of NASA [2], the pressurising agent for the oxygen was obtained by extracting a portion of the liquid oxygen (at a temperature of approximately 94 K) leaving the discharge area of the pump and routing this fluid through the shell-and-tube heat exchanger, which has been shown in the preceding figure. The turbine outlet gas provided the heat source (at 598 $\pm$ 30 K at the maximum mixture ratio, and 548 $\pm$ 35 K at low mixture ratio) for the heat exchanger. Within the heat exchanger, the liquid oxygen was vaporised and then routed into a collector. From there, the gaseous oxygen was routed into the ullage space of the oxygen tank through a flow restrictor and gas distributor.

The fuel tank of the S-II was pressurised with gaseous hydrogen extracted from the cooling jacket of the thrust chamber, where this fluid was used as a coolant. The temperature of the pressurising hydrogen ranged from 111 ± 11 K at the maximum mixture ratio to 72 ± 17 K at low mixture ratio. The pressurising hydrogen was collected from the four outboard engines and was routed to the ullage space via the hydrogen-tank flow-control orifice and gas distributor.

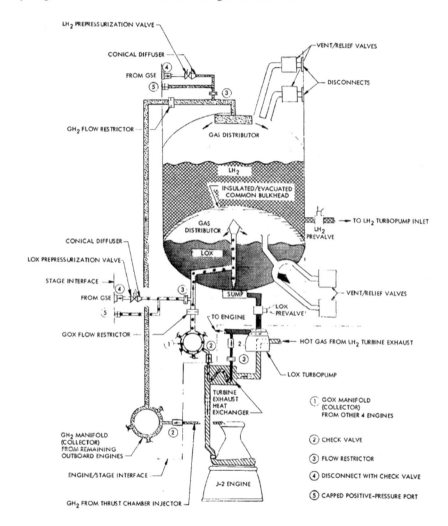

The pressure in both of the propellant tanks can be adjusted by means of valves or regulators, as has been shown above. The mass flow rate of each propellant (oxidiser or fuel) required for vaporisation and pressurisation depends on the mass flow rate of that propellant at the pump inlet (or at the tank outlet) and also on the heat and mass transfer processes which occur in the propellant tanks. These processes, in turn, depend on the temperature of the environment and on the temperature of the pressurising agent.

Let $\dot{m}$ (kg/s) be the propellant mass flow rate at the pump inlet for each of the N engines of a rocket stage, p_T (N/m^2) be pressure of one of the propellants (oxidiser or fuel) in the tank, ρ (kg/m^3) be the density of the liquid propellant, R (N m kg^{-1} K^{-1}) be the specific gas constant of the vaporised propellant, T (K) be the temperature of the gas in the ullage space of the tank, $\dot{m}_e$ (kg/s) be the mass flow rate of evaporation of the propellant, and $\dot{m}_v$ (kg/s) be the average flow rate of the propellant through the tank vent. When the values of these quantities are known, then the required flow rate of propellant $\dot{m}_p$ (kg/s), for each engine, tapped off for tank pressurisation can be determined by means of the following equation [3]:

$$\dot{m}_p = \frac{\dot{m}\, p_T}{\rho R T} - \frac{\dot{m}_e - \dot{m}_v}{N}$$

For example, we consider a rocket stage having the following properties: oxidiser (liquid oxygen) mass flow rate at the pump inlet for each engine $\dot{m} = 131.8$ kg/s, pressure of the oxidiser in the tank $p_T = 3.103 \times 10^5$ N/m^2, density of the liquid oxidiser $\rho = 1141$ kg/m^3, specific gas constant of the vaporised oxidiser $R = R^*/\mathcal{M}$ $= 8314.46/32 = 259.8$ N m kg^{-1} K^{-1}, temperature of the gas in the ullage space of the oxidiser tank $T = 122$ K, mass flow rate of evaporation of the oxidiser $\dot{m}_e = 0.7527$ kg/s, average flow rate of the oxidiser through the tank vent $\dot{m}_v = 0.7439$ kg/s, and $N = 4$ engines in the rocket stage.

By substituting these data in the preceding equation, the required mass flow rate of oxidiser $\dot{m}_o$, for each engine, tapped off for tank pressurisation results

$$\dot{m}_o = \frac{131.8 \times 3.103 \times 10^5}{1141 \times 259.8 \times 122} - \frac{0.7527 - 0.7439}{4} = 1.129 \,\text{kg/s}$$

The same calculation can be repeated for the other propellant (liquid hydrogen) of the same rocket stage, knowing the following data: fuel mass flow rate at the pump inlet for each engine $\dot{m} = 27.12$ kg/s, pressure of the fuel in the tank $p_T = 2.62 \times 10^5$ N/m^2, density of the liquid fuel $\rho = 70.8$ kg/m^3, specific gas constant of the vaporised fuel $R = R^*/\mathcal{M} = 8314.46/4 = 2078.6$ N m kg^{-1} K^{-1}, temperature of the gas in the ullage space of the fuel tank $T = 67$ K, mass flow rate of evaporation of the fuel $\dot{m}_e = 1.905$ kg/s, and average flow rate of the fuel through the tank vent $\dot{m}_v = 2.994$ kg/s. The required mass flow rate of fuel $\dot{m}_f$, for each engine, tapped off for tank pressurisation results

$$\dot{m}_f = \frac{27.12 \times 2.62 \times 10^5}{70.8 \times 2078.6 \times 67} - \frac{1.905 - 2.994}{4} = 0.9929 \,\text{kg/s}$$

3.5 Feed Systems Using Gases Stored for Mono-Propellants

These feed systems have not been employed frequently so far, because of the low specific impulses (230–240 s) which can be obtained by using traditional mono-propellants. One of these system was employed for the orbit adjust subsystem (OAS) of the Landsat satellites, which used a hydrazine (N_2H_4) propellant tank with a bladder and nitrogen (N_2) as the pressurising gas [2].

They have recently attracted considerable interest during continuous efforts aimed at replacing hydrazine with one of the so-called green propellants, which are less toxic and easier to handle and store than hydrazine.

The chemical substances considered for this purpose are ionic compounds used in concentrated aqueous solutions. Each of these compounds contains an ionic liquid (oxidiser) and a fuel (reducer). By ionic liquids we mean ionic compounds (salts) which are liquid below 373 K. More commonly, ionic liquids have melting points below room temperature [17].

According to Fahrat et al. [18], some of these oxidisers are:

- HAN (hydroxyl ammonium nitrate) $[NH_3OH]^+[NO_3]^-$
- ADN (ammonium dinitramide) $[NH_4]^+[N(NO_2)_2]^-$
- HFN (hydrazinium nitroformate) $[N_2H_5]^+[C(NO_2)_3]^-$
- AN (ammonium nitrate) $[NH_4]^+[NO_3]^-$
- HN (hydrazinium nitrate) $[N_2H_5]^+[NO_3]^-$

According to the same authors [18], ionic or molecular fuels associated with these oxidisers are:

- TEAN (trisethanol ammonium nitrate) $[NH(C_2H_4OH)_3]^+[NO_3]^-$
- AA (ammonium azide) $[NH_4]^+[N_3]^-$
- HA (hydrazinium azide) $[N_2H_5]^+[N_3]^-$
- HEHN (2-hydroxyethilhydrazinium nitrate) $[HO-C_2H_4-N_2H_4]^+[NO_3]^-$
- Methanol, ethanol, glycerol, glycine, urea

As has been shown in Chap. 1, Sect. 1.4, of all the oxidisers indicated above, those which have been attracted particular interest are hydroxyl ammonium nitrate and ammonium dinitramide.

Hydroxyl ammonium nitrate, also known as HAN, is a salt derived from hydroxyl amine (NH_2OH) and nitric acid (HNO_3). The Air Force Research Laboratory at Edwards Air Force Base in California, USA, has developed a hydroxyl ammonium nitrate-based propellant known as AF-M315E. This propellant is less toxic and easier to handle than hydrazine, and has a specific impulse $I_s = 257$ s, which is about 12% greater than the specific impulse of hydrazine, the latter being $I_s = 230$ s. It requires a catalyst bed preheating at a temperature exceeding 558 K to be ready for general operation [19].

Ammonium dinitramide, also known as ADN, whose chemical formula is $NH_4N(NO_2)_2$, is the ammonium (NH_4^+) salt of the dinitraminic acid ($HN(NO_2)_2$), and was invented in the 1970s in the former Soviet Union and independently invented again in 1989 in the United States by SRI International. Gaseous ammonium dinitramide decomposes under heat into ammonia (NH_3), nitrous oxide (N_2O), and nitric acid (HNO_3). The Swedish company EURENCO Bofors produces a liquid mono-propellant, called LMP-103S, as a substitute for hydrazine by dissolving 65% ammonium dinitramide in 35% water solution of methanol (CH_3OH) and ammonia (NH_3). LMP-103S has 6% higher specific impulse and 30% higher impulse density (see below) than hydrazine mono-propellant [20]. LMP-103S has been tested in the PRISMA (Prototype Research Instruments and Space Mission technology Advancement, COSPAR designation 2010-028B and 2010-028F) mission in 2010.

The following table, adapted from [21], shows some performance data of the LMP-103S and AF-M315E mono-propellants in comparison with hydrazine.

	LMP-103S	AF-M315E	Hydrazine
Flame Temperature	1873 K	2173 K	873 K
I_s	252 s (theor.) 235 s (delivered)*	266 s (theor.) ~250 s (delivered)	220-224 s (1N thrusters)
Density (kg/m³)	1240*	1465	1010
Density I_s Increase over N_2H_4	30%*	50%	-
Preheat Temperature	573 K nominal	643 K nominal	588 K Capable of cold starts (278 K)
Minimum Operational Temperature	283 - 323 K	< 273 K System dependent – Propellant becomes viscous, but no precipitation or phase change occurs.	278 - 323 K
Minimum Storage Temperature	266 K	Very low (< 295 K) Forms a glass – no crystallization occurs	274 K (Freezing Point) Reheated for re-use

Anflo, K. and Mollerberg. R., "Flight Demonstration of New Thruster and Green Propellant Technology on the PRISMA Satellite" ACTA Astronautica, Vol. 65, Nov.-Dec. 2009
Distribution A: Approved for public release; distribution unlimited

The following table, from [21], shows some compatibility and handling properties of the LMP-103S and AF-M315E mono-propellants in comparison with hydrazine.

Propellant	LMP-103S	AF-M315E
Thruster Materials Compatibility	High combustion temperature and oxidative environment - high temperature, corrosion resistant refractory metal (Ir and Re) chamber materials needed.	High combustion temperature and oxidative environment - high temperature, corrosion resistant refractory metal (Ir and Re) chamber materials needed.
System Materials Compatibility	Compatible with most COTS materials currently used for N_2H_4 systems. Propellant is basic – compatible with many metals.	Limited material compatibility driven primarily by HAN content and acidity
Handling & Safety	Significantly reduced toxicity removes the need for SCAPE suits	Low toxicity and vapor pressure allow handling with only basic PPE Propellant will not crystallize if concentrated

The following figure (re-drawn from [22]) illustrates the feed system used in a spacecraft during the PRISMA mission.

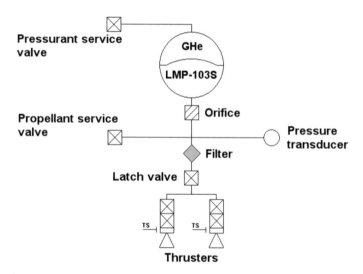

This system consists of one propellant tank with an internal flexible diaphragm, two service valves, one pressure transducer, one filter system, one isolation latch valve, and two 1-N mono-propellant thrusters. The tank contains 5.6 kg of LMP-103S propellant. All of these components are of the commercial off-the-shelf (COTS) type used for hydrazine thrusters. The tubes are 6 mm in diameter and are made of corrosion-resistant steel. All the components are welded to the tubes. The feed pressure decreases in proportion to the consumed propellant. The initial pressure is 2.2×10^6 N/m^2 at 293 K. This value decreases to 0.5×10^6 N/m^2 when all propellant is consumed. The thrust decreases from an initial value of 1 N to a final value of 0.25 N. The propellant and the gas (helium) under pressure are stored in the same tank and are separated by means of a flexible diaphragm. The gas acts on this diaphragm by pushing the propellant through the system filter to the control valve, which controls the flow of propellant to the thruster. The thruster requires pre-heating before firing. When the firing command is given, the series redundant flow control valve opens and admits the propellant in the thrust chamber. There, the propellant decomposes and ignites in the pre-heated reactor bed. This generates hot gases and hence thrust. The pressure transducer and the tank temperature sensor are used to determine the correct amount of propellant. All components are powered by a remote terminal unit with 28 ± 1 V of direct current. The temperature of this unit is controlled by a thermal control remote terminal unit, and is kept in the range of 283–323 K during the entire mission [22].

The following figure (re-drawn from [19]) illustrates the feed system to be used in the GPIM (Green Propellant Infusion Mission).

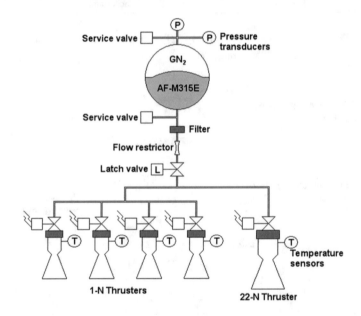

In this mission [23], scheduled for 2018, NASA wants to demonstrate the practical capabilities of the AF-M315E propellant. This feed system consists principally of one propellant tank with an internal flexible diaphragm, two service valves, two pressure transducers, one filter, one flow restrictor, one isolation latch valve, four 1-N thrusters for attitude control, and one 22-N primary divert thruster. Each of these thrusters has a single-seat valve and a temperature sensor. The feed system operates in blow-down mode, which means that the pressure decreases in proportion to the amount of consumed propellant. The gas stored under pressure over the diaphragm is nitrogen. The system components used for AF-M315E are in many cases those used for hydrazine. The 1-N thrusters and the 22-N thruster require a pre-heating of the catalyst bed before firing. The power requirements relating to this pre-heating are high, due to the elevated minimum start temperature of the thrusters. These thrusters use the LCH-240 high-temperature long-life catalyst developed by Aerojet

Rocketdyne. The use of refractory alloys, necessary to withstand the high flame temperature of the AF-M315E propellant, is confined to the thrust chamber, the nozzle, and an upper thermal isolation structure. The other parts of the thrusters can be made of conventional alloys used for hydrazine thrusters [19].

The BCP100 spacecraft, manufactured by Ball Aerospace, and its propulsion system described above are shown in the following figure, due to the courtesy of Aerojet Rocketdyne [24].

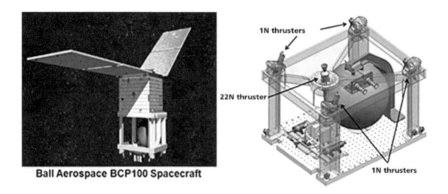

Ball Aerospace BCP100 Spacecraft

The criteria of design and the methods of calculation for the feed systems described here are similar to those described in Sect. 3.3.

3.6 Feed Systems Using Combustion Products

In a feed system using combustion products, the pressurising gas is obtained by combustion of propellants in a turbine gas generator, or by combustion in a solid-propellant gas generator, or by injection of a hypergolic fluid into the main propellant tank. This system has been used to supply pressurising gas in several small military vehicles, and also to pressurise the fuel tanks of the Titan stages. It has rarely been used, because the pressurising gas is chemically incompatible with the propellants, or has a temperature too high, or has condensible elements. A scheme of the feed system used in the Titan stages is shown in the following figure, due to the courtesy of NASA [2].

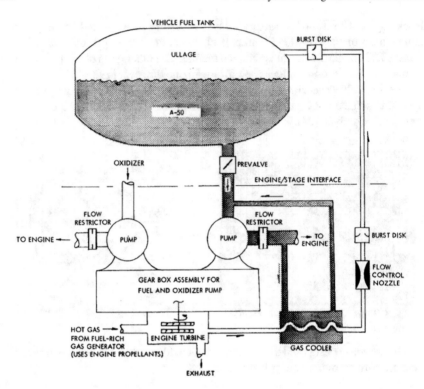

The rocket engine uses nitrogen tetroxide (N_2O_4) as the oxidiser and Aerozine 50 as the fuel. The gas generator uses the engine propellants to produce fuel-rich exhaust gases which drive the engine turbo-pumps. The combustion temperature in this gas generator is about 1256 K at a fuel-rich mixture ratio $o/f = 0.085$. The temperature at the turbine outlet is about 1183 K. The pressurising gas is tapped off at the turbine outlet, passed through a gas cooler, and then routed to the fuel tank. It is injected into the fuel tank within a temperature range of 361 K to 417 K. The flow of the pressurising gas is controlled by a flow-control nozzle located downstream of the gas cooler [2].

The following figure, due to the courtesy of NASA [3], shows a solid-propellant gas generator, which is a small device burning a solid propellant to supply hot gas under pressure to the main liquid propellants.

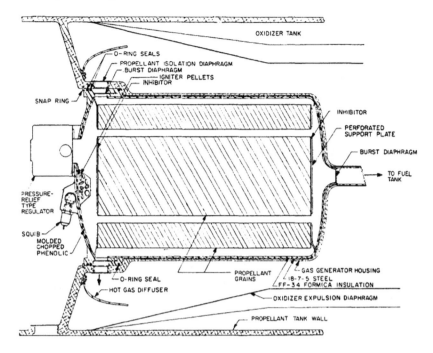

A gas generator like this can be employed to pressurise tanks of rocket engines using storable liquid propellants. This gas generator consists of two electrically-fired initiators (squibs), one charge of igniter pellets, some safety and arming devices, one regulator of the pressure-relief type, and two solid-propellant grains. There may also be, in some cases, a device to cool the hot gas produced. The gas generator illustrated above is contained in an insulated housing made of steel. This housing, in turn, is contained in the main propellant tank, which is made of an aluminium alloy. The gas outlets are hermetically sealed by means of burst diaphragms which assure a safe storage of the solid propellant over long periods of time. At the proper moment, the propellant grains are ignited by the igniter pellets. The gases generated by the combustion process pressurise the main propellants for the duration of the propellant grains. Such grains are designed to produce the desired pressures and flow rates within a given range of temperature. In case of the upper limit of temperature being exceeded, a regulator discharges overboard the excess gases.

Some types of solid-propellant gas generators used for feed systems are briefly described below. The following figure illustrates two solid-propellant gas generators, one of which without cooling and the other with a solid coolant.

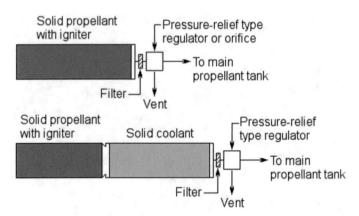

The gas generator shown in the upper part of the preceding figure consists of a solid-propellant charge with igniter, a filter, and a hot-gas regulator or an orifice. This type of gas generator can be used only for short periods of time. When the solid propellant is ignited, the hot gases are passed through the filter, regulated, and directed to the main propellant tanks. The regulator or the orifice mentioned above dumps overboard the excess gas, and therefore a vent line must be present for this purpose.

The gas generator shown in the lower part of the preceding figure consists of a solid-propellant charge with igniter, a sublimating solid coolant, a filter, and a regulator. When the solid propellant is ignited, the hot gases are cooled by passing through a solid material which decomposes or sublimates. This cooling process generates additional gases. The mixture of gases generated in this manner is passed through the filter, regulated, and directed to the main propellant tanks.

Huzel and Huang [3] describe a gas generator using ammonium nitrate (NH_4NO_3) as the solid propellant, which has a flame temperature of about 1544 K. The ammonium nitrate is used with a solid coolant made of oxalic acid (HOOCCOOH or $(COOH)_2$) compressed into pellets. Oxalic acid decomposes endothermically, at a temperature above 394 K, and produces a mixture of gases consisting of carbon monoxide (CO), carbon dioxide (CO_2), and water (H_2O). The desired temperature is obtained by adjusting the ratio of the propellant to the coolant. By so doing, a temperature of 478 K has been reached [3].

The following figure illustrates a solid-propellant gas generator having an azide cooling pack.

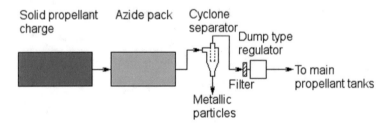

In this gas generator, the hot gas resulting from the combustion of a solid propellant is cooled by passing through an azide material. An example of this material is sodium azide (NaN_3), which decomposes after being heated at or above approximately 573 K into gaseous nitrogen (N_2) and sodium (Na) particles, as follows

$$2NaN_3 \rightarrow 2Na + 3N_2$$

These particles are removed when the gas containing them passes through a cyclone separator. The gas coming from this separator is filtered to remove further particles, regulated, and directed to the main propellant tanks. A gas generator like this has made it possible to obtain gaseous nitrogen at a temperature of about 590 K [3].

Generally speaking, a cyclone separator is a device which removes solid particles from a gaseous or liquid stream without using filters. The solid particles are removed because they are denser than the other molecules of the stream. A fluid mixed with solid particles enters tangentially in a cyclone, which consists of a cylindrical body, a conical outlet for the particles, and a top axial pipe outlet for the clean fluid. An outer vortex is created due to the centrifugal force applied to the molecules, and the fluid circles down to the bottom end of the cone. Particles of high density are pushed against the wall and separated from the fluid. In the conical part, the fluid reverses the direction of its motion, goes up via the central part of the cyclone (inner vortex), and exits on the top by the fluid outlet pipe. The solid particles travel down the wall, and are collected into a receptacle at the bottom of the conical part. This device is shown in the following figure, due to the courtesy of the US Government [25].

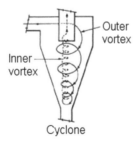

Cyclone

Still another type of solid-propellant gas generator is the helium system with solid-propellant gas generator heating. This system consists of a spherical vessel storing helium at high pressure, a solid-propellant gas generator mounted inside this vessel, a filter, and a pressure regulator, as shown in the following figure.

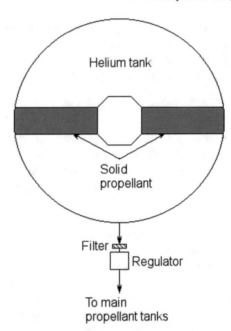

This solid-propellant gas generator provides heat, which causes the helium to expand, and also additional pressurising gas. This system requires a large vessel to store the solid propellant and the helium at high pressure.

3.7   Control Systems for Liquid-Propellant Gas Generators

As has been shown in Sect. 3.6, liquid-propellant gas generators require a careful design, in order to avoid problems concerning the chemical compatibility of the pressurising gas with the propellants, or the temperature of this gas, or the presence of condensible elements into it. These problems can be solved by using control systems, which are described in the present paragraph.

As far as the chemical compatibility of the pressurising gas is concerned, hydrazine mono-propellant is considered satisfactory in view of the chemical characteristics and the molar masses of the gases (nitrogen and hydrogen) resulting from its decomposition in the presence of a catalyst. These gases contain no carbon, and therefore generate no deposits which could lower the performance of a heat exchanger. In case of gas generators using a combination of two liquid propellants, the gases generated by them can be made chemically compatible with the propellants by varying the value of the oxidiser-to-fuel mixture ratio o/f with respect to the stoichiometric value. By so doing, the same combination of propellants can be used to generate a pressurising gas compatible with both the oxidiser and the fuel. An example is given by the hypergolic combination of nitrogen tetroxide and Aerozine 50, which was used for the first and second stages of the Titan II Intercontinental Ballistic Missile and Titan space launch vehicles [26]. As has been shown in Sect. 3.6, the gas generator of the Titan engine uses the engine propellants to produce fuel-rich exhaust gases which drive the engine turbo-pumps.

As far as the temperature of the pressurising gas is concerned, more than one option can be chosen to assure compatibility. One of such options consists in varying of the value of the oxidiser-to-fuel mixture ratio to lower the combustion temperature, as has been shown above. Another option consists in injecting a non-reacting liquid into the gas generator, in order to subtract heat from the combusted gas in the evaporation of this liquid. A third option consists in cooling the gas by means of a heat exchanger, through which the gas flows with one of the liquid propellants. Of course, this option can be chosen only when the propellant can safely absorb the heat coming from the combusted gas.

The presence of condensible elements into the combusted gas can be avoided by regulating the combustion temperature, as will be shown below. In order for the molar mass of the combusted gas to be low, the combustion temperature of a fuel-rich mixture should reach a value (approximately 811 K) sufficient to break the bonds of the complex molecules of the combustion products and give rise to substances having low-mass molecules. Of course, this result can be obtained only when the fuel injected in excess into the gas generator does not reduce the combustion temperature below about 811 K. This is because, when the temperature is too low, some molecules of the combusted gas have a high molar mass, and these molecules are subject to condense.

Some typical control systems used for liquid-propellant gas generators are briefly described below.

One of these control systems uses a single gas generator, which provides gas under pressure to both the fuel tank and the oxidiser tank, with injection cooling. A scheme of this system is shown in the following figure.

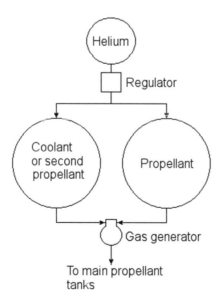

The liquid substances employed may be either a mono-propellant and an inert coolant or a combination of two propellants. In the second case, the cooling is obtained by injecting the second propellant in excess with respect to the quantity necessary for the stoichiometric mixture ratio. The mono-propellant and the coolant or the two propellants are kept under pressure by means of gaseous helium stored at a high pressure in a vessel. Downstream of this vessel, the pressure is kept to the desired value by using valves and a pressure regulator. This system makes it possible to cool the gas coming from the gas generator to a temperature compatible with the propellants. A combination of liquid substances must be studied carefully, in order to satisfy not only the temperature requirement, but also the other requirements, and in particular the one concerning the chemical compatibility of the pressurising gas with both of the propellants. To this regard, fuel-rich gases have been found apt to pressurise storable oxidisers. Examples of such oxidisers are given in [3].

Another of these control systems uses a single gas generator and a heat exchanger, as shown in the following scheme, which refers to the case of a gas generator using a combination of two propellants.

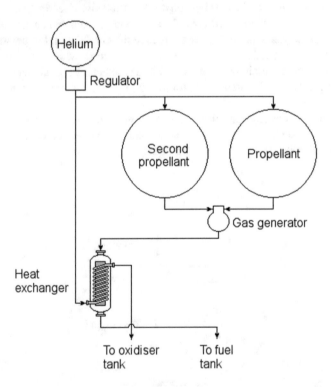

The hot gas coming from the gas generator is directed to a heat exchanger, in order to transfer heat to the cold helium. The helium expands after being heated and is used to pressurise the tank of the main oxidiser. The gas cooled through the gas generator is used to pressurise the tank of the main fuel.

Still another of these control systems uses a double gas generator, which provides gas under pressure to both the fuel tank and the oxidiser tank, with injection cooling. A scheme of this system is shown in the following figure.

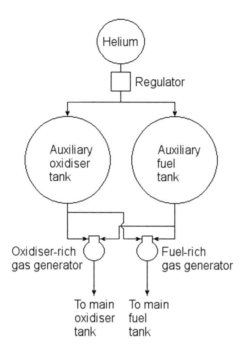

In this system, the fuel and the oxidiser are fed to two gas generators by using the pressure exerted by the helium. One of the gas generators operates with injection of excess oxidiser, and produces a cool, oxidiser-rich gas, which is used to pressurise the main oxidiser tank. The other gas generator operates with injection of excess fuel, and produces a cool, fuel-rich gas, which is used to pressurise the main fuel tank. This system requires a balance in the output of the two gas generators and also a pressure control in both of the main propellant tanks. By so doing, temperatures as low as 589 K have been reached in the gas generated [3].

3.8   Feed Systems Using Direct Injection into the Main Propellant Tanks

In these systems, a small quantity of oxidiser is injected directly into the main tank containing the fuel, and a small quantity of fuel is injected directly into the main tank containing the oxidiser. An hypergolic reaction, which takes place between the oxidiser and the fuel, produces gases used to pressurise the oxidiser and the fuel into their respective main tanks. Two versions (parallel and serial) of the direct injection system are shown in the following figure.

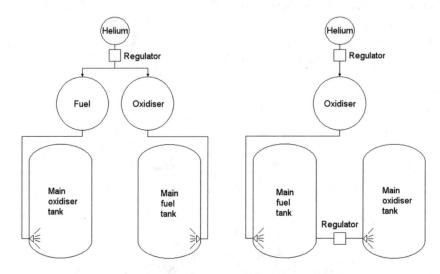

The parallel version (left) of the system consists of a small vessel containing helium stored at high pressure, a helium pressure regulator, two small auxiliary propellant tanks (one for the fuel and the other for the oxidiser), and two main propellant tanks. The fuel is directly injected from its auxiliary tank into the main tank containing the oxidiser, and the oxidiser is directly injected from its auxiliary tank into the main tank containing the fuel.

The serial version (right) of the system consists of a small vessel containing helium stored at high pressure, a helium pressure regulator, only one small auxiliary propellant tank containing one of the two propellants, two main propellant tanks, and a regulator of the difference of pressure between the two main propellant tanks. The serial version takes advantage of the possibility of exerting a lower pressure in one of the two main propellant tanks than in the other. In the scheme shown on the right-hand side of the preceding figure, a small quantity of the fuel contained in its main tank passes through a regulator and is injected into the main tank containing the oxidiser.

3.9   Choice of a Feed System Using Gases Under Pressure

The choice of one of another system which uses gases stored under pressure is usually the result of a preliminary study. Huzel and Huang [3] have suggested four criteria to be considered in this study. They are:

- mission requirements for the rocket vehicle;
- chemical compatibility of the propellants with the materials;

- system reliability; and
- system performance.

The mission requirements concern the possibility of using storable propellants, the need to start the engine one or more times, and the capability of controlling the values of pressure in the ullage space.

The chemical compatibility of the propellants with the materials concern the absence of reactivity between the propellants and the materials of which their containers are made, the absence of condensible or soluble particles in the combusted gas, and the maintenance of a desired value of temperature in the gas under pressure.

The system reliability depends on the degree of complexity of the system, on the number of its failure modes, and on the number of its components. The reliability of each component, in turn, is to be evaluated considering its time of development and the funds available. To this regard, components expected to require greater efforts of development to reach desirable degrees of reliability are gas generators, heat exchangers forming an integral part of the thrust chamber, storage vessels subject to high pressures, and pipes and regulators for gases at high temperatures.

The system performance depends on the gross mass of the gas stored for pressurisation, which in turn depends on its molar mass and also on the mass of the system of pressurisation. The latter mass is part of the mass of the rocket vehicle at burnout.

References

1. Greene WD (2013) Inside the LEO doghouse, start me up! NASA, 19 Dec 2013. https://blogs.nasa.gov/J2X/2013/12/19/inside-the-leo-doghouse-start-me-up/
2. Lee JC, Ramirez P, Keller RB Jr (1975) Pressurization systems for liquid rockets. NASA SP-8112, 175pp. https://ntrs.nasa.gov/archive/nasa/casi.ntrs.nasa.gov/19760015212.pdf
3. Huzel DK, Huang DH (1967) Design of liquid propellant rocket engines, 2nd edn. NASA SP-125, NASA, Washington, D.C., 472pp. https://ntrs.nasa.gov/archive/nasa/casi.ntrs.nasa.gov/19710019929.pdf
4. The Engineering Toolbox, Specific heat and individual gas constant of gases. https://www.engineeringtoolbox.com/specific-heat-capacity-gases-d_159.html
5. Interbartolo M (2009) Apollo lunar module propulsion systems overview. Slide presentation, NASA, 26pp. https://ntrs.nasa.gov/archive/nasa/casi.ntrs.nasa.gov/20090016298.pdf
6. Boeing (1968) Lunar Orbiter IV—photographic mission summary, vol 1. NASA CR-1054, 128pp. https://ntrs.nasa.gov/archive/nasa/casi.ntrs.nasa.gov/19680017342.pdf
7. Childs FW, Horowitz TR, Jenisch W Jr, Sugarman B (1962) Design guide for pressurization system evaluation liquid propulsion rocket engines, vol I. Aerojet-General Corporation, 30 Sept 1962, 137pp. https://ia801401.us.archive.org/18/items/nasa_techdoc_19630008160/19630008160.pdf
8. Wagner WA, Keller RB Jr (eds) (1974, May) Liquid rocket metal tanks and tank components. NASA SP-8088, 165pp. https://ntrs.nasa.gov/archive/nasa/casi.ntrs.nasa.gov/19750004950.pdf
9. Wichmann H (1972, March) Design for pressure regulating components—final report. NASA-CR-139300, 271pp. https://ntrs.nasa.gov/archive/nasa/casi.ntrs.nasa.gov/19740022135.pdf
10. MatWeb Material property data, Titanium Ti–13V–11Cr–3Al (Ti–13-11-3) Aged 490 °C. http://www.matweb.com/search/datasheet.aspx?MatGUID=ba6943e32b354789aa61b7517ca16b00

11. Wang Q, Wu F, Zeng M, Luo L, Sun J (2006) Numerical simulation and optimization on heat transfer and fluid flow in cooling channel of liquid rocket engine thrust chamber. Eng Comput 23(8):907–921
12. National Institute of Standards and Technology (NIST), NIST Chemistry WebBook, Thermophysical Properties of Fluid Systems. https://webbook.nist.gov/chemistry/fluid/
13. Petersen H (1970, September) The properties of helium: density, specific heats, viscosity, and thermal conductivity at pressures from 1 to 100 bar and from room temperature to about 1800 K. Danish Atomic Energy Commission, Research Establishment Risø, Risø Report No. 224, 42pp. http://orbit.dtu.dk/files/52768509/ris_224.pdf
14. Locke JM, Landrum DB (2008) Study of heat transfer correlations for supercritical hydrogen in regenerative cooling channels. J Propul Power 24(1):94–103
15. Wikimedia. https://commons.wikimedia.org/wiki/FILE:HCHE.jpg
16. NASA (1968, December) Saturn V news reference, J-2 engine fact sheet. https://www.nasa.gov/centers/marshall/pdf/499245main_J2_Engine_fs.pdf
17. Yamamoto T (2011) Ionic liquids, Seminar, 5 Nov 2011, 40pp. http://www.f.u-tokyo.ac.jp/~kanai/seminar/pdf/Lit_T_Yamamoto_M2.pdf
18. Fahrat K, Batonneau Y, Brahmi R, Kappenstein C (2011, September) Application of ionic liquids to space propulsion. In: Handy ST (ed) Applications of ionic liquids in science and technology, InTech, Rijeka, Croatia. ISBN 978-953-307-605-8. http://cdn.intechweb.org/pdfs/20222.pdf
19. Spores RA, Masse R, Kimbrel S, McLean C (2013) GPIM AF-M315E propulsion system. In: 50th AIAA/ASME/SAE/ASEE joint propulsion conference & exhibit, 28–30 July 2013, Cleveland, Ohio, USA, 12pp. https://ntrs.nasa.gov/archive/nasa/casi.ntrs.nasa.gov/20140012587.pdf
20. Sjöberg P, Skifs H, Thormählen P, Anflo K (2009) A stable liquid mono-propellant based on ADN. In: Insensitive munitions and energetic materials technology symposium, Tucson, USA, 11–14 May 2009. http://www.dtic.mil/ndia/2009/insensitive/8Asjoberg.pdf
21. Brand A (2011) Reduced toxicity, high performance monopropellant. Briefing slides, Air Force Research Laboratory (AFMC), 31pp. http://www.dtic.mil/dtic/tr/fulltext/u2/a554667.pdf
22. Anflo K, Bergman G, Hasanof T, Kuzavas L, Thormählen P, Åstrand B (2007) Flight demonstration of new thruster and green propellant technology on the PRISMA satellite. In: 21st annual AIAA/USU conference on small satellites, 13–16 Aug 2007, Logan, Utah, USA, 21pp. https://digitalcommons.usu.edu/cgi/viewcontent.cgi?article=1495&context=smallsat
23. Green Propellant Infusion Mission (GPIM) overview, NASA. https://www.nasa.gov/mission_pages/tdm/green/overview.html
24. Aerojet Rocketdyne. http://www.rocket.com/green-monopropellant-propulsion
25. US government, Public domain, via Wikimedia Commons. https://commons.wikimedia.org/wiki/File:SCyclone.jpg
26. Nufer B (2010) Hypergolic propellants: the handling hazards and lessons learned from use, Technical report, Joint JANNAF Interagency Propulsion Committee 25th Safety and Environmental Protection Joint Subcommittee Meeting, 6–10 Dec 2010, Orlando, Florida, USA, 100pp. https://ntrs.nasa.gov/archive/nasa/casi.ntrs.nasa.gov/20100042352.pdf

Chapter 4
Feed Systems Using Turbo-Pumps

4.1 Fundamental Concepts

In a liquid-propellant rocket engine fed by turbo-pumps, one or more turbines are used to drive pumps. The pumps, in turn, are used to increase the pressures of the liquid propellants above the values of storage in the tanks, for the purpose of injecting the propellants at high pressures into the main combustion chamber. In other words, the pumps take the liquid propellants from their tanks at low pressures, and supply them to the main combustion chamber at the required mass flow rates and injection pressures.

As has been shown in Chap. 2, Sect. 2.7, the energy supplied to a turbine is provided by the expansion of a compressed gas, which is usually a mixture of the propellants burned in the engine [1]. The advantages offered by feed systems using turbo-pumps over feed systems using gases stored under pressure are lower mass and higher performance.

A feed system using turbo-pumps consists of the following components:

- pumps for propellants;
- one or more turbines to drive the pumps;
- a source of energy for the turbines;
- mounts for the turbo-pumps; and
- auxiliary components.

The components indicated above are described in the following sections.

© The Editor(s) (if applicable) and The Author(s), under exclusive license
to Springer Nature Switzerland AG 2021
A. de Iaco Veris, *Fundamental Concepts of Liquid-Propellant Rocket Engines*,
Springer Aerospace Technology,
https://doi.org/10.1007/978-3-030-54704-2_4

4.2   Pumps for Propellants

In general terms, a pump is a machine which imparts energy to a fluid. In specific terms, a turbo-pump for a liquid-propellant rocket engine is a turbine-driven roto-dynamic pump, that is, a particular type of pump in which the motion of a rotating element converts mechanical energy into hydraulic energy supplied to each of the propellants (fuel and oxidiser) to be injected under high pressure into the combustion chamber of the engine. There are three types of turbo-pumps: axial-flow pumps, centrifugal-flow pumps, and mixed-flow pumps. They are described below.

According to the definition given by Scheer et al. [2], an axial-flow pump consists of a set of discs or a cylinder carrying aerofoil-shaped blades on the periphery, which rotates at high speed within a casing or housing which contains fixed blades (stator vanes) placed between the rotor blades. Small clearances at the blade tips and between rotor blades and stator vanes are maintained under all operating conditions. The flow is nearly parallel to the pump shaft, and head rise (defined below) is produced by summation of increases in pressure produced as the fluid traverses each set (stage) of rotor blades and stator vanes. Axial-flow pumps have small diameters but give modest pressure increases. Although multiple compression stages are needed, axial-flow pumps work well with low-density fluids. The following figure, due to the courtesy of NASA [2], is a cross-sectional view of the axial-flow Mark 15-fuel (liquid hydrogen) turbo-pump used for the J-2 engine.

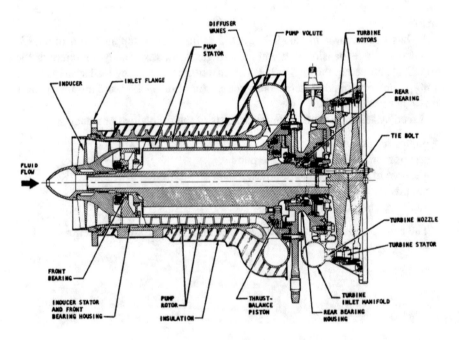

A centrifugal-flow pump, also known as radial-flow pump, consists of a bladed rotor (impeller) shaped so as to sweep the fluid roughly at right angles outward from the pump shaft. This action imparts a high spiral velocity to the fluid. Much of this velocity is converted to pressure increase by the diffuser which surrounds the impeller and collects the fluid. The rotating blades are typically enclosed within a casing, and are used to impart energy to a fluid through centrifugal force. A centrifugal-flow pump has two principal parts: (1) a rotating element, which includes the impeller and the shaft; and (2) a fixed element, which is made up of a casing (volute), a stuffing box to prevent the pumped fluid from coming out of the volute casing from the location of the shaft, and bearings, which absorb the radial and axial loads transmitted through the shaft, keep the rotating element in position and in correct alignment with the fixed element, and permit the shaft to rotate with the least possible amount of friction. Centrifugal-flow pumps are far more powerful than axial-flow pumps for high-density fluids, but require large diameters for low-density fluids. The following figure, due to the courtesy of NASA [3] is a cross-sectional view of a typical centrifugal-flow pump.

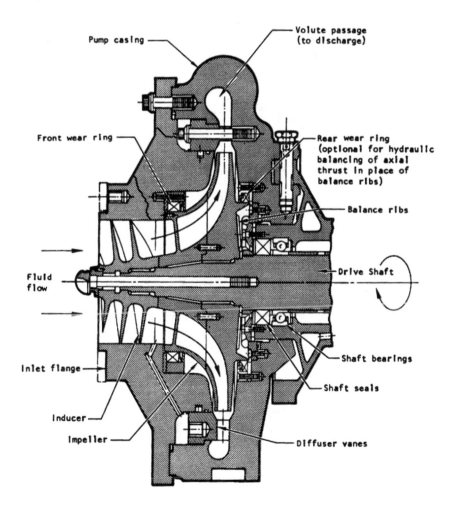

The inducer, shown in the preceding figures, is the axial inlet portion of the pump rotor, and has the function of raising the inlet head (see below) by an amount sufficient to preclude cavitation (see below) in the following stage [4].

The following figure, also due to the courtesy of NASA [5], shows a typical unshrouded (that is, non-covered) impeller of a liquid-hydrogen centrifugal-flow pump. This particular impeller has vanes directed radially at the outlet.

Mixed-flow pumps are a compromise between axial-flow and centrifugal-flow pumps. A mixed-flow pump is similar to a centrifugal pump, but has a mixed-flow impeller, that is, an impeller which transports the fluid diagonally, halfway between the axial direction and the radial direction. A mixed-flow pump pushes the fluid out away from the pump shaft. Pressure is developed in the pump partly by centrifugal force and partly by the lifting action of the mixed-flow impeller. Of the three types (axial-flow, centrifugal-flow, and mixed-flow) of turbo-pumps described above, those of the centrifugal-flow type are the most frequently used. However, when the pumped fluid is liquid hydrogen, multistage axial-flow turbo-pumps are frequently used, due to the lowest temperature and density of this fluid. Three impellers used for respectively centrifugal-flow, axial-flow, and mixed-flow pumps are illustrated sequentially, from left to right, in the following figure, which is re-drawn from [6].

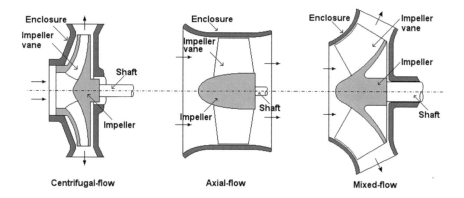

| Centrifugal-flow | Axial-flow | Mixed-flow |

By head we mean the energy per unit weight of a fluid. The total head H (m) in a given point of a fluid is the sum of the following parts

$$H = z + \frac{p}{\rho g_0} + \frac{v^2}{2g_0}$$

By head rise $H_2 - H_1$ (m) of a pump operating between a point 1 (suction) and a point 2 (discharge) we mean the following difference of energy per unit weight

$$H_2 - H_1 = \left(z_2 + \frac{p_2}{\rho g_0} + \frac{v_2^2}{2g_0} \right) - \left(z_1 + \frac{p_1}{\rho g_0} + \frac{v_1^2}{2g_0} \right)$$

In the preceding equation, H (m) is the total head in the chosen point, z (m) is the elevation of the chosen point with respect to a reference plane, $p/(\rho g_0)$ (m) is the pressure head in the chosen point, $v^2/(2g_0)$ (m) is the velocity head in the chosen point, p (N/m^2) is the absolute pressure in the fluid in the chosen point, ρ (kg/m^3) is the density of the fluid in all points at the given temperature, $g_0 = 9.80665$ m/s^2 [7] is the acceleration of gravity near the surface of the Earth, and v (m/s) is the velocity of the fluid in the chosen point along a stream line.

Care must be taken to avoid cavitation in rotodynamic pumps. Cavitation is the spontaneous formation and accumulation of vapour bubbles in a flowing liquid as its static pressure drops below its vapour pressure. The burst or collapse of the bubbles gives rise to shock waves of high energy inside the liquid. The shock waves cause erosion and serious damage to the impeller and to the housing of a pump. The wreckage, due to cavitation, of the impeller of a centrifugal-flow pump is shown in the following figure, due to the courtesy of Milan, through Wikimedia [8].

For the purpose of avoiding cavitation in a pump, it is necessary to be careful about the value taken by a quantity called available Net Positive Suction Head ($NPSH_A$), as will be shown below. Cavitation is likely to occur on the inlet side of a pump, particularly when the pump is situated at a level well above the surface of the liquid in the supply reservoir. By writing the equation of energy, in steady conditions, between the surface of the liquid in the supply reservoir (point 2) and the inlet flange of the pump (point 1), there results

$$\frac{p_{surf}}{\rho g_0} + \Delta z - \Delta H_f = \frac{p_i}{\rho g_0} + \frac{v_i^2}{2g_0}$$

where p_{surf} (N/m^2) is the absolute pressure at the surface of the liquid in the reservoir, ρ (kg/m^3) is the density in all points of the liquid at the operating temperature, g_0 = 9.80665 m/s^2 is the acceleration of gravity near the surface of the Earth, $\Delta z = z_2 - z_1$ (m) is the difference of elevation between the surface of the liquid in the reservoir and the inlet flange of the pump, ΔH_f (m) is the head loss due to friction in the suction line, p_i (N/m^2) is the absolute pressure in the liquid at the inlet flange of the pump, and v_i (m/s) is the velocity of the liquid at the inlet flange of the pump. The preceding equation may be re-written as follows

$$\frac{p_i}{\rho g_0} = \frac{p_{surf}}{\rho g_0} + \Delta z - \Delta H_f - \frac{v_i^2}{2g_0}$$

In order for cavitation not to occur at the impeller, the following condition must be satisfied

$$p_i > p_v$$

where p_v (N/m^2) is the vapour pressure of the liquid at the given temperature.

The head loss ΔH_f (m), due to the friction in the suction line, can be computed by means of the Darcy-Weisbach equation (see Chap. 2, Sect. 2.5), as follows

$$\Delta H_f = f_D \frac{L}{d} \frac{v^2}{2g_0}$$

where f_D, L (m), and d (m) are respectively the dimensionless Darcy friction factor, the length, and the hydraulic diameter of the suction line, and v (m/s) is the mean velocity of the liquid in this line.

The available Net Positive Suction Head $NPSH_A$ (m) is defined as follows

$$NPSH_A = \frac{p_{surf}}{\rho g_0} \pm \Delta z - \Delta H_f + \frac{v_i^2}{2g_0} - \frac{p_v}{\rho g_0}$$

where Δz (m) is positive and is called suction head when the centre of the impeller is below the level of the fluid being pumped, and is negative and is called suction lift when the centre of the impeller is above the level of the fluid being pumped. The equation written above, whose term $v_i^2/(2g_0)$ is often neglected, makes it possible to calculate the NPSH available ($NPSH_A$) at the inlet flange of a pump. The curves determined experimentally by the manufacturer of a given pump (see, for example, [9, 10]) specify the NPSH required ($NPSH_R$) in the operating conditions. For good operation of a pump, the following condition must be satisfied, with a margin of at least 1 m:

$$NPSH_A > NPSH_R$$

In the particular case of a pump for a rocket engine, the NPSH available ($NPSH_A$) is the difference, at the pump inlet, between the head due to the total fluid pressure and the head due to the propellant vapour pressure [11], as follows

$$NPSH_A = \frac{p_i + \frac{1}{2}\rho v_i^2 - p_v}{\rho g_0}$$

The critical value of the $NPSH_R$, under which cavitation occurs at the pump inlet, is usually taken as the value for which the head rise is 2% less than the non-cavitation value [12]. In case of a value insufficient to meet the design requirements for a rocket engine, the following methods have been used:

(1) increasing the pressure in the tank;
(2) decreasing the design speed of the pump; and

(3) re-designing the pump inlet by increasing the diameter and lowering the flow coefficient $\varphi_1 = c_{m1}/u_1$, where c_{m1} (m/s) is the meridian velocity of the fluid at the impeller inlet, and u_1 (m/s) is the tangential velocity of the blade at the inlet tip.

The first method increases the NPSH available, but also increases the wall thickness and the mass of the tank. The second method decreases the NPSH required, but also decreases the pump efficiency and increases the turbo-pump mass. The third method decreases the NPSH required, but can decrease the pump efficiency [12]. Further parameters, whose values have to be taken into account in the design of a pump for a rocket engine, are pump head coefficient, pump specific speed, suction specific speed, and Thoma's cavitation factor. They are defined below.

The non-dimensional pump head coefficient ψ is defined as follows

$$\psi = \frac{g_0 \Delta H}{u_2^2}$$

where ΔH (m) is the head rise at the best efficiency point (BEP), u_2 (m/s) is the tangential velocity of the blade at the tip of the impeller, and g_0 (m/s^2) is the gravitational acceleration of the Earth. The values of ψ for centrifugal-flow pumps vary from approximately 0.35 to more than 0.70 [3].

Pump specific speed N is another non-dimensional parameter, defined as follows [11]:

$$N = \frac{\omega q^{\frac{1}{2}}}{(g_0 \Delta H)^{\frac{3}{4}}}$$

where ω (rad/s) is the angular velocity of the shaft, and q (m^3/s) and ΔH (m) are respectively the volume flow rate and the head rise at the BEP. Axial-flow pump stages for rocket engines have specific speeds N ranging from approximately 1.17 to 4.02. It should be noted that these are stage characteristics, and that considerable difference occurs when the entire pump is examined [2]. Current flight-proven centrifugal-flow pumps range from 0.16 to 0.77 in specific speed. The specific speed of the Mark-14 pump for the Atlas vehicle has reached the value of 1.1 [3].

Suction specific speed S is also a non-dimensional parameter, defined as follows [11]:

$$S = \frac{\omega q^{\frac{1}{2}}}{(g_0 NPSH_R)^{\frac{3}{4}}}$$

where q (m^3/s) and $NPSH_R$ (m) are respectively the volume flow rate and the head rise at the BEP. The suction specific speed is used as an index for cavitation performance of pumps. Rocket pumps operate up to a suction specific speed of about 18.3 [13]. The design suction specific speed of a low-pressure oxidiser pump inducer used in the Space Shuttle main engine is 25.6 [13].

For example, the liquid-hydrogen turbo-pump for the M-1 rocket engine (burning liquid oxygen and liquid hydrogen) is a ten-stage, axial-flow pump. For this pump, the angular velocity of the shaft is $\omega = 1385$ rad/s, the volume flow rate of the liquid hydrogen is $q = 3.931$ m³/s, and the design value of the Net Positive Suction Head is $NPSH_R = 101.5$ m [14]. Introducing these values and $g_0 = 9.807$ m/s² in the preceding equation yields

$$S = \frac{1385 \times 3.931^{\frac{1}{2}}}{(9.807 \times 101.5)^{\frac{3}{4}}} = 15.5$$

Thoma's cavitation factor σ_{TH} is another non-dimensional parameter, defined as follows [11]:

$$\sigma_{TH} = \frac{p_1^T - p_v}{p_2^T - p_v} = \left(\frac{N}{S}\right)^{\frac{4}{3}}$$

where $p_2^T - p_1^T \, (\text{N/m}^2)$ is the total pressure rise across the pump, and $p_v \, (\text{N/m}^2)$ is the vapour pressure of the fluid at the operating temperature. Thoma's cavitation factor is used to ascertain whether cavitation is, or is not, occurring in a pump. For this purpose, the actual value of σ_{TH} is compared with a critical value $(\sigma_{TH})_{cr}$ for a given pump. When σ_{TH} is lower than $(\sigma_{TH})_{cr}$, cavitation may occur. The value of $(\sigma_{TH})_{cr}$ may be either computed through empirical relations or determined experimentally.

4.3 Turbines Driving the Pumps

A turbine driving a pump for propellants is a mechanical device which extracts energy from a moving gas. This energy is used to drive the pump. Turbines used for this purpose are, with a few exceptions, of the axial type, meaning by this, that the gas flow strikes the rotor in a direction parallel to the shaft axis.

The following figure, due to the courtesy of NASA [15], illustrates (left) the Mark 15-fuel (liquid hydrogen) turbine, and (right) the Mark 15-oxidiser (liquid oxygen) turbine, both of which are installed in the J-2 engine.

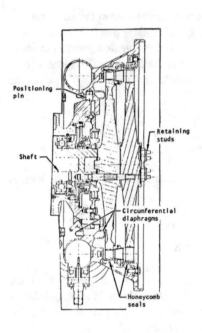
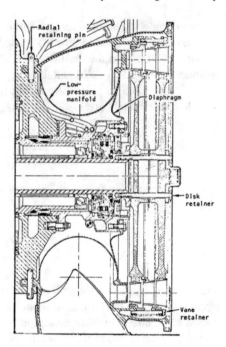

The Mark 15-fuel is the high-pressure turbine in the series-installed turbines for the J-2 engines used in the Saturn S-II and S-IVB stages. The turbine exhaust gas from this unit is used subsequently as working fluid in the low-pressure Mark 15-oxidiser turbine. The Mark 15-fuel is mounted to the turbo-pump with a short-coupled diaphragm containing radial pins for axial positioning. A circumferential "hat" section and U-diaphragm perform the sealing function at the radial joint. Honeycomb rotor tip seals are carried by the stator ring and retainer. Turbine torque transmission and disc retention are accomplished with a stud drive. The Mark 15-oxidiser is the low-pressure turbine in the series-installed turbines for the J-2 engines. This turbine uses exhaust gas from the Mark 15-fuel turbine as working fluid, and incorporates a low-pressure torus-and-shroud configuration which does not require the numerous welds of a separate torus and nozzle. Radially installed roll pins retain the manifold at the outer diameter, and a conventional diaphragm at the inner diameter provides a seal at that juncture [15].

Generally speaking, the main parts of a gas turbine are:

- a row of static vanes, called nozzles or stators, which guide the gas to flow in the desired direction at the desired velocity;
- the rotor, which is the moving part of a turbine and has blades attached to it by means of dove-tail or fir-tree joints;
- the shaft, to which the rotor is keyed;
- the blades, on which the flowing gas impinges and causes the rotor to turn; and
- the casing, which is the outer part which covers the turbine and contains the rotor, the shaft, and the blades.

The following figure, due to the courtesy of NASA [16], shows how a turbine blade is attached to a rotor by means of a fir-tree joint.

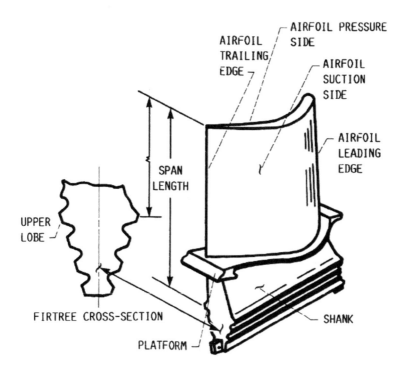

There are two principal types of turbines: impulse turbines and reaction turbines.

In an impulse turbine, the gas flows through the nozzle and impinges on the moving blades of the rotor. These blades convert the kinetic energy of the gas flow into mechanical energy. They also direct the gas flow either to the next stage, in case of a multi-stage turbine, or to the exhaust, in case of a single-stage turbine. There is no expansion or pressure drop of the gas over the blades. The area of passage through the blades remains constant. An impulse turbine can have one or more stages.

In a reaction turbine, the gas flows firstly through the fixed blades of a stator and then through the moving blades of a rotor. A pressure drop takes place in the stator blades and also in the rotor blades. The area of passage through the blades varies continuously, in order to allow the gas to expand continuously over the blades. A reaction turbine is usually multi-stage.

The types of turbine used to drive pumps of rocket engines are briefly described below. A single-stage, single-rotor impulse turbine has a single rotor, which is a revolving disc having a row of blades attached to its circumference. The following figure, re-drawn from [17], shows the nozzles and the rotor blades of a single-stage, single-rotor impulse turbine.

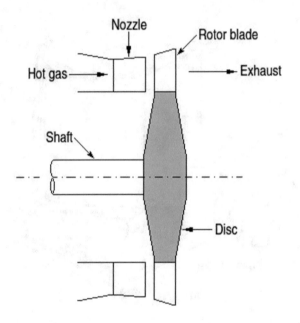

The hot gas flows to the rotor through the nozzles. The cross-sectional area of each nozzle increases gradually towards the mouth, in order for the gas to expand and gain velocity. The pressure drop occurs in the nozzles, and the velocity drop occurs on the rotor blades. In the nozzles, the pressure energy (pressure head) possessed by the gas is converted into kinetic energy (velocity head), and therefore the static pressure of the gas decreases. When the gas flow impinges on the blades of the rotor, the kinetic energy possessed by the gas is imparted to the rotor, and is converted into energy of rotation. The following figure, re-drawn from [18], shows how the gas pressure and velocity vary in a single-stage, single-rotor impulse turbine. In this figure, N and R indicate respectively a nozzle and a rotor blade.

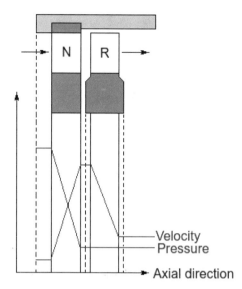

In case of a large pressure drop, it is practically impossible to use a turbine having only one stage. This is because a single-stage turbine having a large pressure drop has also a high peripheral velocity $U = \omega r$ (m/s), where ω (rad/s) is the angular velocity, and r (m) is the radius of the rotor. Therefore, turbines operating with a large pressure drop have more than one stage, as will be shown below.

The compounding of impulse turbines is obtained by using more than one set of nozzles, blades, and rotors in a series. The moving blades are keyed to a common shaft. By so doing, either the gas velocity or the gas pressure is absorbed by the turbine in stages.

The two main types of compounded impulse turbines are:

- velocity-compounded impulse turbines; and
- pressure-compounded impulse turbines.

A velocity-compounded impulse turbine was first proposed by Curtis to solve the problem of the high rotational speed of a single-stage impulse turbine used in a gas flow of high pressure and temperature. With reference to the following figure, re-drawn from [18], in a velocity-compounded impulse turbine, the two rings R_1 and R_2 of moving blades are separated by a ring F of fixed vanes.

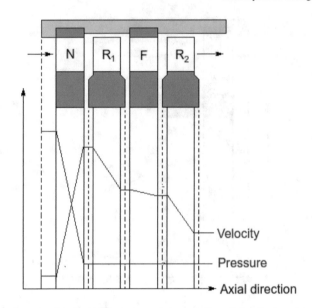

The moving blades are keyed to the turbine shaft, and the fixed vanes are fixed to the casing. The high-pressure gas expands in the nozzles N first. The nozzles convert the pressure energy of the gas flow into kinetic energy. The pressure drop occurs in the nozzles. Downstream of the nozzles, the pressure remains constant. This high-velocity flow is directed from the nozzles N on to the first ring R_1 of moving blades. The gas, when flowing over the moving blades of R_1, imparts some of its momentum to them and loses some velocity. Only a part of the kinetic energy lost by the gas is absorbed by the moving blades of R_1, and therefore the gas possesses residual kinetic energy at the moment of being exhausted on to the next ring F of fixed vanes. The function of the fixed vanes F is to redirect the flow coming from the first ring R_1 of moving blades to the second ring R_2 of moving blades. There is very little change in the velocity of the gas as it passes through the fixed vanes. The gas then enters the second ring R_2 of moving blades. This process may be repeated until practically all the kinetic energy possessed by the gas has been absorbed by the moving blades of the rotors.

Velocity-compounded turbines have the following disadvantages:

- since there is only one drop of pressure in all the stages, then the nozzles require a careful design and an expensive manufacturing, because they have to be of the converging-diverging type in order for the velocity of the gas to be supersonic; and
- the high velocity of the gas at the exit section of the nozzles implies high cascade losses, and shock waves are generated in the supersonic flow.

In order to avoid these disadvantages, the total pressure drop is divided into a series of stages. Since the pressure drop is lower in each stage, then the gas flows at subsonic velocity in the nozzles. With reference to the following figure, re-drawn from [18], a pressure-compounded (or Rateau) impulse turbine has alternate rows of nozzles N and moving blades R.

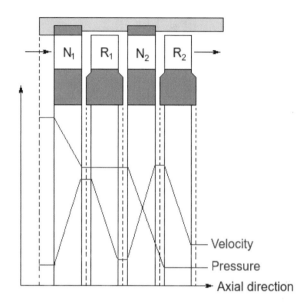

The nozzles are fitted to the casing, and the moving blades are keyed to the shaft. In this type of turbine, the gas expands by degrees when passing through each row of nozzles. The gas coming from the gas generator is fed to the first row N_1 of nozzles, where it loses a part of its pressure and increases its velocity. Then, the gas passes through the first row R_1 of moving blades, where it loses velocity at a constant pressure. Then, the gas passes through the second row N_2 of nozzles, where it loses a further part of its pressure and increases its velocity. Then, the gas is fed to the second row R_2 of moving blades, where it loses velocity at a constant pressure. This process may be repeated until practically all the kinetic energy possessed by the gas has been absorbed by the moving blades. In particular, the preceding figure shows how the gas pressure and velocity vary in a two-stage, pressure-compounded impulse turbine.

In a multi-stage reaction (or Parsons) turbine, the gas pressure decreases continually over the rings of the fixed vanes (F_1, F_2, ...) and also over the rings of the moving blades (R_1, R_2, ...), as shown in the following figure, re-drawn from [18], which illustrates a two-stage reaction turbine and the variation of pressure and velocity of the gas through the stages of a reaction turbine.

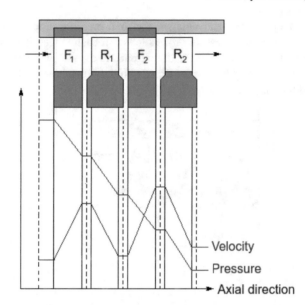

In each stage of a reaction turbine, the pressure drop is smaller than in each stage of an impulse turbine. Therefore, the gas velocity is smaller in a reaction turbine than in an impulse turbine. Reaction turbines are aerodynamically more efficient than impulse turbines. However, reaction turbines have higher losses due to tip leakage, because of the higher difference of pressure across the rotor blades.

The specific heat at constant pressure c_p (J kg^{-1} K^{-1}) and the specific heat ratio γ of the working fluid of a turbine have a large influence on the isentropic spouting velocity C_0 (m/s) and, therefore, on the design of a turbine, as will be shown below. The isentropic spouting velocity is defined as that velocity which is obtained during an isentropic expansion of the working fluid between the entry and the exit pressures of the stages of a turbine, as follows

$$C_0 = (2\Delta h_T)^{\frac{1}{2}}$$

where Δh_T (J/kg) is the enthalpy drop per unit mass of the working fluid of a turbine, that is, the available energy content per unit mass of the working fluid. The enthalpy drop per unit mass of the working fluid is the difference

$$\Delta h_T = h_0 - h_e$$

where h_0 (J/kg) is the enthalpy per unit mass of the working fluid at the turbine inlet, and h_e (J/kg) is the enthalpy per unit mass of the working fluid at the exhaust pressure, assuming an isentropic expansion, as follows

$$\Delta h_T = c_p (T_0 - T_e) = c_p T_0 \left[1 - \left(\frac{p_e}{p_0} \right)^{\frac{\gamma-1}{\gamma}} \right]$$

where T_0 (K) is the total temperature of the working fluid at the turbine inlet, T_e (K) is the static temperature of the working fluid at the turbine exhaust, p_0 (N/m^2) is the total pressure of the working fluid at the turbine inlet, and p_e (N/m^2) is the static pressure of the working fluid at the turbine exhaust.

The ratio p_e/p_0 can be expressed as the reciprocal of the turbine pressure ratio R_T, which is a parameter frequently used in turbine design, as follows

$$\Delta h_T = c_p T_0 \left[1 - \left(\frac{1}{R_T} \right)^{\frac{\gamma-1}{\gamma}} \right]$$

Two-stage pressure-compounded turbines are used for low-energy-fuel combinations (for example, liquid oxygen with RP-1, and nitrogen tetroxide with Aerozine 50), and two-row velocity-compounded turbines are used for high-energy hydrogen-fuel combinations (for example, liquid oxygen with liquid hydrogen). Low-energy-fuel turbines which are exceptions to this rule are the Redstone A-7 and the F-1 turbines, which were two-row velocity-compounded turbines, because they had low peripheral velocities U (m/s) of their rotors. Two-stage (two-row for velocity-compounded) turbines are generally chosen because they offer higher efficiency than single-stage turbines without the added complexity and mass of turbines with three or more stages (three rows for velocity-compounded). Two-row velocity-compounded turbines are either as efficient or nearly as efficient as two-stage pressure-compounded turbines when the velocity ratio U/C_0 is less than 0.2. For velocity ratios between 0.2 and 0.34, two-stage pressure-compounded impulse turbines are generally used. This is because they are more efficient up to 0.3 and, even though slightly less efficient between 0.3 and 0.34, they have less axial thrust than reaction turbines, since there is little or no pressure drop across the rotor. For velocity ratios above 0.34, reaction turbines are generally used, because they are more efficient than impulse turbines. The following figure, due to the courtesy of NASA [12], shows approximate regions of application for various types of rocket engine turbines.

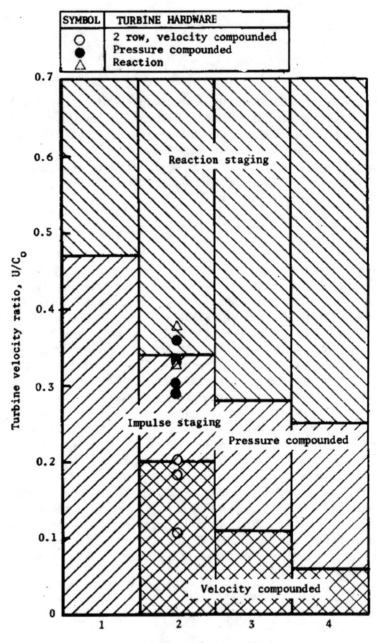

Number of rotor blade rows

4.4   Sources of Energy for Turbines

A turbine used in a rocket engine is driven by the expansion of a compressed gas, which is produced by either the decomposition of a mono-propellant or the combustion of a fuel with an oxidiser. The gas necessary for this purpose is generated in a rocket engine according to one of the cycles which have been described at length in Chap. 2, Sect. 2.7. For convenience of the reader, these cycles are mentioned below. They are the gas-generator cycle (used for either a mono-propellant or a couple of propellants), the tap-off cycle, the expander cycle, and the staged-combustion cycle.

   The present section is meant to show how the energy resulting from gases which expand in one or two turbines, as indicated above, is used to drive the pumps. In other words, we describe here the manners of coupling the turbines with the pumps. The drive arrangements depend on the propellants burned in the rocket and also on the general design of its engine. The term turbo-pump arrangement indicates the physical relation between the pumps and the turbines for both the fuel and the oxidiser. Examples of these arrangements are shown in the following figure, due to the courtesy of NASA [12].

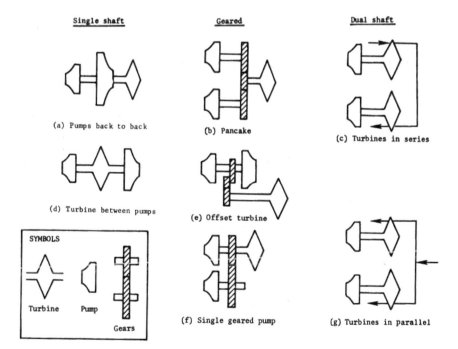

When the densities of the two propellants differ only a little (such is the case, for example, with liquid oxygen and RP-1), then the single-stage arrangement (a and d) is advantageous in terms of low complexity and mass, because it requires only one turbine. In this arrangement, the turbine may be mounted either at one end (a, as in the F-1 engine) or between the pumps (d, as in the A-7 engine).

When the turbine has very short blades, or a very large diameter, or both, then it is possible to place the two pumps on the same shaft and the turbine on a separate shaft, which is geared to the pump shaft (e). This arrangement permits the turbine to operate at a higher rotational speed than that of the pumps. The MA-5 booster and sustainer engines and the YLR-91-AJ-7 engines had turbo-pumps of this type.

The "pancake" arrangement (b) consists in gearing separately the two pumps to a single turbine. This arrangement, which permits each of the three components of a turbo-pump to operate at optimum rotational speed, was chosen for the YLR-87-AJ-7 and YLR-81-BA-11 engines.

The single-geared pump arrangement (f) may be chosen to avoid a low efficiency resulting from a small area of admission in the oxidiser turbine. This may happen in small hydrogen-fuelled upper-stage engines. This arrangement was chosen for the RL-10 engine.

The dual-shaft arrangement (c and g) permits each pump to operate at its optimum speed, which leads to a high overall efficiency of the pumps. This arrangement is chosen for hydrogen-oxygen engines, because these propellants have very different values of density. In this arrangement, the turbines may operate either in series (c) or in parallel (g). When the turbines operate in series, the initial turbo-pump can rotate faster than the other, which reduces the turbine flow rate requirements. The J-2 and J-2S engines used this arrangement. When the turbines operate in parallel, great flexibility is obtained for off-design modes of operation, such as starting, throttling, and using variable values of mixture ratio. The parallel operation was chosen for the RS-25, which is one of the three main engines of the Space Shuttle.

By the way, the efficiency η of a machine is the ratio of the useful power P_d (W) delivered by the machine to the power P_s (W) supplied to it:

$$\eta = \frac{P_d}{P_s}$$

A mass m (kg) of fluid, when raised through a height $h = z_2 - z_1$, gains an energy (measured in joules) equal to mg_0h, where $g_0 = 9.80665$ m/s^2 is the acceleration of gravity near the surface of the Earth.

When this gain of energy occurs in a time t (s), the power P (W) needed to that effect is

$$P = \frac{mg_0h}{t}$$

For a fluid in motion, as is the case with a propellant for a rocket engine, the mass flow rate (kg/s) is

$$\dot{m} = \frac{\rho V}{t} = \rho q$$

where ρ (kg/m^3) is the density, V (m^3) is the volume, and q (m^3/s) is the volume flow rate of the fluid. Therefore, the power P (W) needed to obtain a head rise ΔH (m) in the fluid is

$$P = \frac{mg_0 \Delta H}{t} = \rho g_0 q \Delta H$$

In case of a turbine, the efficiency is

$$\eta_T = \frac{P_d}{\rho g_0 q \Delta H}$$

Likewise, in case of a pump, the efficiency is

$$\eta_P = \frac{\rho g_0 q \Delta H}{P_s}$$

4.5 Description of a Typical Turbo-Pump

The present section describes a typical turbo-pump used in a liquid-propellant rocket engine. The turbo-pump described here is the Mark 3 pump, which was used in the Atlas, Thor, and Saturn IB boosters. The turbo-pump arrangement chosen here, marked with (e) in the preceding figure, consists in placing the two pumps on the same shaft and the turbine on a separate shaft, which is geared to the pump shaft, as will be shown below. The following figure, due to the courtesy of NASA [19], shows the Mark 3 turbo-pump mounted on the H-1 engine of the Saturn IB booster.

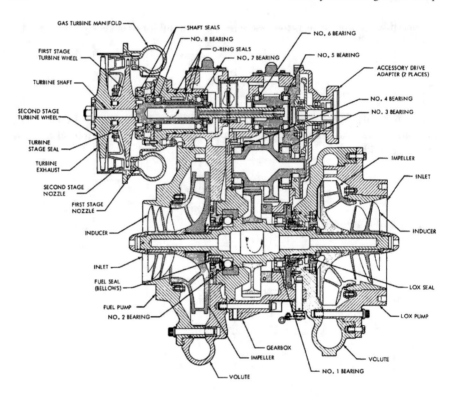

This turbo-pump consists of an oxidiser (liquid oxygen) pump, a fuel (RP-1)
pump, a reduction gearbox, an accessory drive adapter, and a gas turbine. The turbo-
pump is mounted on the side of the thrust chamber, and requires only two short,
high-pressure ducts connecting the volute outlets to the main propellant valves. The
turbine is started by a solid-propellant cartridge, and is powered by a liquid-propellant
gas generator during main operation. The high-speed turbine drives the turbo-pump
through a series of reduction gears which drive the main shaft. On the H-1 engine of
the Saturn IB booster, the turbo-pump gears and bearings are cooled and lubricated
by means of a fuel additive blender unit, which mixes the fuel with an oxidation
and corrosion inhibitor additive (Oronite® 262, that is, zinc dialkyl dithiophosphate)
and injects the mixture (98% RP-1 plus 2% additive) at the flow rate 0.000347 m^3/s
into the turbo-pump gearbox [12]. On other engines, a separate tank for lubricating
oil is provided. During main operation, the turbo-pump supplies oxidiser and fuel,
at the required pressures and flow rates, to the main thrust chamber and to the gas
generator.

  As shown in the preceding figure, both of the pumps are of the centrifugal-flow
type. They are mounted back-to-back on the same shaft. Each of them consists
of an axial-flow inducer, a centrifugal-flow impeller with backward-curved blades,
a diffuser with static vanes, and a volute, according to the scheme shown in the
following figure, due to the courtesy of NASA [3].

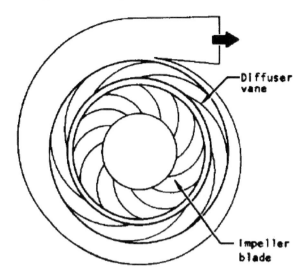

The static diffuser vanes provide uniform pressure distribution, reduction of fluid velocity around the impeller, and reduction of fluid turbulence in the pump volute. Each of the propellants enters the pump in the axial direction through the inlet plane of the casing, passes from the inducer to the rotating blades of the impeller, flows outwards whilst picking up the tangential velocity of the impeller blades, flows from the impeller to the static vanes of the diffuser, and finally flows from the diffuser to the expanding volute. Each propellant gains both velocity and pressure when passing through the impeller. The diffuser decelerates the flow and further increases the pressure of the fluid. Balance ribs are placed on the back side of the impeller to absorb the axial thrust acting on the pump shaft. Balance ribs are blades located on the back of the impeller shroud, as shown in the following figure, due to the courtesy of NASA [3].

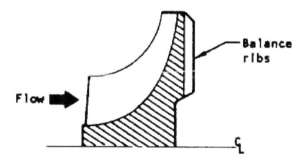

The ribs on the back-face of the impeller cause the pumped fluid to spin with the impeller. This rotational motion decreases the pressure exerted by the fluid with the radial distance from the impeller tips [12]. Different methods for absorbing the axial thrust acting on a pump shaft have also been used, either alone or in combination, in other turbo-pumps. These methods will be described and illustrated in Sect. 4.9.

The gearbox of the Mark 3 turbo-pump is shown in the following figure, due to the courtesy of NASA [19].

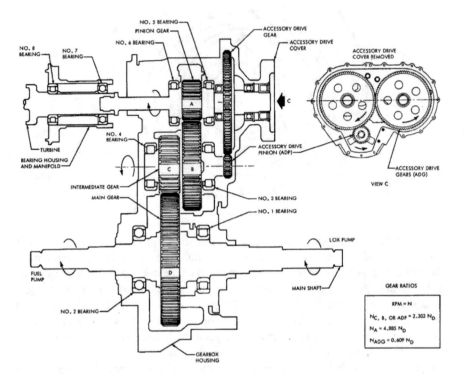

In the preceding figure, the letters A, B, C, and D identify the gears. In particular, A indicates the high-speed pinion gear, B and C indicate the intermediate gear, and D indicates the main shaft gear. An accessory drive pinion, integral with the intermediate gear, drives the counter-rotating accessory drive gears. The lower accessory drive gear operates the outboard engine hydraulic pump. The pitch diameters of the gears of the Mark 3 turbo-pump are $d_A = 76.2$ mm, $d_B = 161.544$ mm, $d_C = 104.775$ mm, and $d_D = 241.3$ mm. By pitch diameter of a gear we mean the diameter of the circle described by the midpoint of the length of the teeth around the gear. The gear ratios are indicated in the preceding figure. The gear material is alloy steel AMS 6225 [20].

The turbine which drives the pumps is of the impulse, two-stage, pressure-compounded type. It is bolted to the fuel pump casing, and consists of an inlet manifold, wheels and nozzles for the two stages, a turbine shaft, and a splined quill shaft which connects the turbine shaft to the high-speed (A) pinion gear. In the preceding figure, the turbine shaft inboard bearing (No. 7) is a split race ball bearing, and the outboard bearing (No. 8) is a roller bearing.

Carbon ring shaft seals prevent hot gas leakage into the bearings, as shown in the following figure, due to the courtesy of NASA [19].

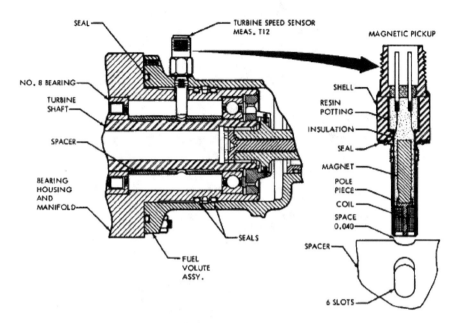

The gas coming from the liquid-propellant gas generator enters the turbine inlet manifold, which directs it to the inlet nozzles of the first stage. After passing through the blades of the first stage, the gas increases its velocity by passing through the nozzles of the second stage, and is directed to the blades of the second stage. The gas leaves the turbine into the exhaust duct. The turbine stage seal prevents the gas from by-passing the nozzles of the second stage.

According to [15], the input rotational speed of this turbo-pump is 3434.8 rad/s, the speed reduction ratio input/pump-shaft is 4.885 for both of the pumps, and the power transmitted is 2988 kW. The Mark 3 turbo-pump described above and the H-1 rocket engine, on which this turbo-pump is mounted, are shown in the following figure, adapted from Wikimedia [21].

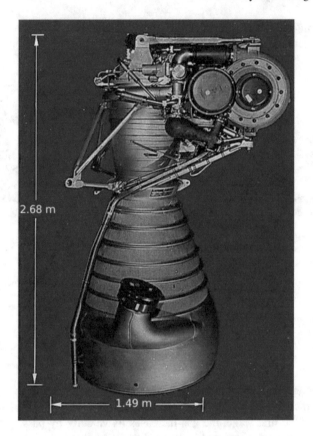

2.68 m

1.49 m

4.6   Turbo-Pump Performance

The performance of a turbo-pump may be evaluated on the basis of the payload which can be carried by a rocket vehicle whose engine has a given thrust level. The performance of a turbo-pump affects the payload for the following reasons. Firstly, because the mass of a turbo-pump contributes to the mass of a rocket stage at burnout. Therefore, a high mass of a turbo-pump implies a low mass of the payload, for a given mass of a rocket stage at burnout.

Secondly, because the pressure required at the inlet flange of a pump determines the pressure of the pumped propellant in its tank. Therefore, when the pressure required at the inlet flange of a pump is high, the mass of the system of pressurisation in a propellant tank is also high, and consequently the payload mass is low for a given mass of a rocket stage at burnout.

Thirdly, because gases coming from the exhaust duct of a turbine are expelled at a lower specific impulse than gases coming from the thrust chamber. Such is the case, for example, with a rocket engine using the gas-generator cycle or the tap-off cycle, which are, both of them, open cycles. When this happens, the overall specific impulse of a rocket engine decreases as a result of the lower specific impulse of the gases coming from the exhaust duct of the turbine. Therefore, a decrease in performance, in terms of lower specific impulse, implies a lower mass of the payload, for a given mass of a rocket stage at burnout.

A quantity, called Equivalent Mass Factor of a turbo-pump (EMF_{TP}), measured in seconds, is used to take account of the low value of the specific impulse possessed by the gas coming from the exhaust duct of a turbine. Sobin et al. [12] define this quantity as follows

$$EMF_{TP} = \frac{\partial m_{PL}}{\partial (I_s)_E} \left[1 - \frac{(I_s)_T}{(I_s)_E} \right] \frac{g_0 (I_s)_E^2}{F}$$

where m_{PL} (kg) is the mass of the payload, $(I_s)_E$ (s) is the specific impulse of the gas exhausted by the thrust chamber of the engine, $(I_s)_T$ (s) is the specific impulse of the gas exhausted by the turbine, g_0 (m/s^2) is the acceleration of gravity, and F (N) is the thrust of the engine. The value of the partial derivative on the right-hand side of the preceding equation depends on the velocity increment required by the mission, the system gross mass, the system specific impulse, and the stage propellant mass fraction $R_p = m_p/m_0$ (see Chap. 1, Sect. 1.3). For a rocket engine using a gas-generator cycle, the value of EMF_{TP} ranges from 5 s for booster engines to 200 s for upper-stage engines [12]. For a rocket engine using a staged-combustion cycle, $(I_s)_T = (I_s)_E$, and therefore the corresponding EMF_{TP} is essentially zero [12].

The effect of a turbo-pump on a stage payload is taken into account by adding the mass m_{TP} (kg) of the turbo-pump to the product resulting from multiplying the EMF_{TP} (s) by the mass flow rate $\dot{m}_T$ (kg/s) of the gas coming from the exhaust duct of the turbine, as follows

$$EM_T = m_{TP} + EMF_{TP} \cdot \dot{m}_T$$

The quantity EM_T, measured in kg, is called the equivalent mass of a turbo-pump. This quantity is an index of performance of a turbo-pump, because the best turbo-pump has the lowest value of equivalent mass.

The value of equivalent mass can be reduced by decreasing either the mass m_{TP} of the turbo-pump or the mass flow rate $\dot{m}_T$ of the turbine exhaust gas. The mass flow rate $\dot{m}_T$ of the turbine exhaust gas, in turn, can be reduced by increasing the turbo-pump efficiency, the turbine inlet temperature, and the turbine pressure ratio p_e/p_0 (see Sect. 4.3). For a staged-combustion cycle, the turbine inlet temperature and the turbo-pump efficiency have a direct effect upon the engine mass and the delivered payload. High turbine and pump efficiencies reduce the required pump discharge pressure and also reduce the engine system mass. A high turbine inlet temperature, within the limits of the materials of which the turbine is made, also decreases the required pump discharge pressure and the engine mass [12].

As an example of application, it is required to compute the equivalent mass of a turbo-pump for a rocket engine having the following properties: mass of the turbo-pump $m_{TP} = 861.8$ kg, mass flow rate of the gas exhausted by the turbine $\dot{m}_T = 41.7$ kg/s, and equivalent mass factor of the turbo-pump $EMF_{TP} = 55$ s. After substituting these values in the preceding equation, we find

$$EM_T = m_{TP} + EMF_{TP} \cdot \dot{m}_T = 861.8 + 55 \times 41.7 = 3155.3 \text{ kg}$$

4.7 Turbo-Pump Design Parameters

Several parameters are to be considered carefully by the designer of a turbo-pump. Some of these parameters have been identified by Huzel and Huang [22]. They are:

- properties of the propellants;
- heads and mass flow rates of the pumps;
- specific speeds of the pumps;
- net positive suction heads of the pumps;
- efficiencies of the pumps;
- overall performance and operating efficiency of the turbine;
- system cycle efficiency of the turbo-pump; and
- calibration and off-design characteristics of the turbo-pump.

These parameters are discussed below.

The first of them concerns the properties of the propellants. The following table, adapted from [22], shows physical properties of some liquid substances.

The first ten of these substances are storable and cryogenic propellants (in the following order, nitrogen tetroxide, hydrogen peroxide, anhydrous hydrazine, RP-1, ethanol, unsymmetrical dimethyl hydrazine, anhydrous hydrazine mixed in equal parts with unsymmetrical dimethyl hydrazine, liquid oxygen, liquid fluorine, and liquid hydrogen). The remaining two are inert substances (liquid nitrogen and water), which are cited here because they are used to calibrate pumps. The physical properties indicated in the table are temperature T (K), vapour pressure p_v (N/m^2), density ρ (kg/m^3), and coefficient of dynamic viscosity μ (Ns/m^2). The minimum and the maximum values of density are given for RP-1, which is a mixture of hydrocarbons.

Liquid substance	T (K)	p_v (N/m^2)	ρ (kg/m^3)	μ (Ns/m^2)
N_2O_4	288.706	76531.8	1452.87	43.9196×10^{-5}
H_2O_2 (90%)	288.706	179.264	1406.42	128.794×10^{-5}
N_2H_4	288.706	1089.37	1013.97	102.732×10^{-5}
RP-1	288.706	213.737	797.719–813.738	222.011×10^{-5}
C_2H_5OH (95%)	288.706	4274.75	807.330	153.064×10^{-5}
$N_2H_3(CH_3)_2$	288.706	12617.4	795.477	58.0539×10^{-5}
$N_2H_3(CH_3)_2/N_2H_4$ (50%)	288.706	12203.7	907.606	95.0098×10^{-5}
Liquid O_2	90.0389	101325	1140.03	19.0640×10^{-5}
Liquid F_2	84.8167	101325	1509.10	24.3385×10^{-5}
Liquid H_2	20.4278	101325	70.9618	1.43411×10^{-5}
Liquid N_2	77.3722	101325	807.971	15.5822×10^{-5}
H_2O	288.706	1765.06	999.071	113.074×10^{-5}

The cryogenic propellants pose problems concerning the construction materials, seals, bearings, and lubricants used in a turbo-pump. They also pose the problem of avoiding the formation of ice.

The range of total temperatures in which the components of a turbo-pump for cryogenic propellants operate is about 16–89 K at the pumps and 920–1200 K at the turbine [22]. This fact causes gradients of temperature, which in turn pose problems of thermal expansion or contraction of the materials of which the pumps and the turbines are made.

A structural method or a mechanical method can be adopted to ensure unrestrained thermal growth of the turbine casing, as will be shown below.

The structural method uses a long cylinder for turbo-pump housing. This results in a cone smaller in diameter on the chilled side and larger in diameter on the heated side. Theoretically, the temperature should be equal to ambient temperature somewhere along this cone, and therefore the diameter at that location should remain constant. At this point, the cylindrical housing is flanged and bolted onto the pump casing. The ambient temperature interface solves the problem of differential thermal growth. This method has been chosen for the Mark-29-fuel (liquid hydrogen) turbo-pump, which is illustrated in the following figure, due to the courtesy of NASA [12]. It has also been used for the uprated F-1 turbo-pump.

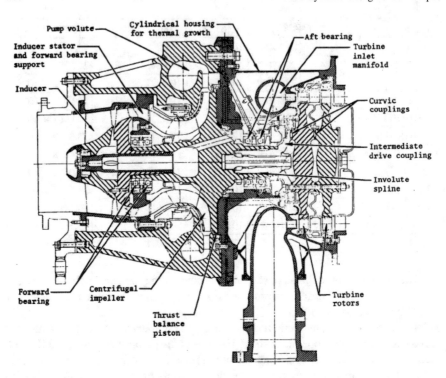

The mechanical method uses an arrangement of radial pins or keys, on which the hot turbine casing can slide unrestrained without losing its concentric position with respect to the pump. A drawback of this method is the necessity of withstanding the full reactive torque of the turbine by means of the radial pins or keys. This may introduce strains and deformation in the mounting brackets in which the radial pins are anchored. Many mechanical problems have arisen when the mounting brackets containing the pins were attached directly to the hot-gas manifold or torus [12]. This method has been chosen for the J-2 axial-flow fuel turbo-pump, which is illustrated in Sect. 4.2, for the J-2 centrifugal-flow oxidiser turbo-pump, which is illustrated in the following figure, due to the courtesy of NASA [12], and also for the F-1 turbo-pump.

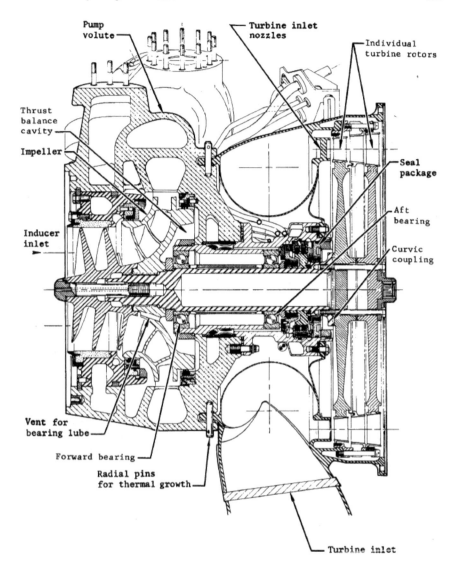

The saturation temperatures of the propellants (that is, their boiling temperatures at a given pressure) affect the choice of the materials and the requirements of thermal conditioning for a turbo-pump. In a turbine, the specific heats, the specific heat ratio, and the molar mass of the combusted gas affect the heights and the tip velocities of the blades, the type of staging, and the number of stages.

The values of the vapour pressure p_v (N/m^2) of the propellants in ordinary operating conditions are used by the designer to determine the total pressure of suction at the pump inlet, because this value contributes to form the available net positive suction head, as has been shown in Sect. 4.2.

Density is the physical property of a propellant which affects to the greatest extent the design of a turbo-pump. The densities of the propellants indicated in the preceding table range from 1509 kg/m^3 for liquid fluorine to 70.48 kg/m^3 for liquid hydrogen at the given temperatures. The effects of this wide range are shown in the following table, adapted from [12], in which the properties of the J-2 liquid oxygen (LO$_2$, density $\rho = 1134$ kg/m^3) pump are compared with the properties of the J-2 liquid hydrogen (LH$_2$, density $\rho = 70.48$ kg/m^3) pump.

Pump property	LO$_2$ ($\rho = 1134$ kg/m^3)	LH$_2$ ($\rho = 70.48$ kg/m^3)
Pump type	Centrifugal	Axial
Number of stages	1	7 + inducer
Mass flow rate (kg/s)	208.8	37.92
Volume flow rate (m^3/s)	0.1842	0.5382
Rotational speed (rad/s)	916.6	2841
Pressure rise (N/m^2)	7.412 × 10^6	8.329 × 10^6
Inducer head rise (m)	No inducer	1539
Pump head rise (m)	666.0	11580
NPSH (m)	5.486	22.86
Impeller discharge tip speed (m/s)	118.9	263.7
Turbo-pump mass (kg)	138.3	167.4
Power (W)	1.758 × 10^6	5.948 × 10^6

As shown in the preceding table, the head rise in the impeller alone of the liquid hydrogen pump is 1539 m, which is more than twice the head rise (666 m) of the whole liquid oxygen pump. The whole head rise of the liquid hydrogen pump is 11580 m, which is almost twenty times the value (666 m) of the liquid oxygen pump, even though the pressure rises of the two pumps are of the same order of magnitude. As a result of this difference in head rise, the impeller discharge tip speed of the liquid hydrogen pump is 263.7 m/s, which is more than twice as great as the impeller discharge tip speed (118.9 m/s) of the liquid oxygen pump. If the liquid hydrogen pump were of the single-stage centrifugal type (as is the case with the liquid oxygen pump), the tip speed of the former would have to be four times the tip speed of the latter [12]. The mass (167.4 kg) of the liquid hydrogen pump is greater than the mass (138.3 kg) of the liquid oxygen pump. Finally, the liquid hydrogen pump requires a power of 5.948 × 10^6 W, which is more than three times the power (1.758 × 10^6 W) required by the liquid oxygen pump, even though the mass flow rate (208.8 kg/s) of the latter is more than five times the mass flow rate (37.92 kg/s) of the former.

Viscosity is a physical property which measures the degree of internal resistance opposed by a fluid to motion. The fluids for which the rate of deformation is proportional to the shear stress applied to them are called Newtonian fluids, after Isaac Newton. Such is the case with water, oil, and the liquid propellants cited in the table given at the beginning of this section. In case of a one-dimensional motion of a Newtonian fluid along a direction x (planar Couette flow), this linear relationship can be expressed as follows

$$\tau_{xy} = \mu \frac{dv}{dy}$$

where τ_{xy} (N/m^2) is the shear stress, dv/dy (s^{-1}) is the rate of shear strain, that is, the velocity gradient in the direction y perpendicular to the direction x of motion, $v \equiv v(y)$ (m/s) is the velocity of motion in the direction x, and μ (kg m^{-1} s^{-1}) is the coefficient of dynamic viscosity of the fluid.

The viscosity of the pumped fluid plays an important role in the efficiency of a pump. Since a rotodynamic pump operates at motor speed, then its efficiency decreases as viscosity increases, due to the increased frictional losses within the pump.

As has been shown by several authors (see, for example, [23–25]), the losses which affect the performance of a rotodynamic pump can be classified as follows

- hydraulic losses, due to friction and vortex dissipation in all components (inlet casing, impeller, diffuser, volute, and outlet casing) located between the suction flange and the discharge nozzle;
- mechanical losses, due to friction in the radial bearings, in the axial bearing, and in the shaft seals;
- leakage losses, which are losses of capacity (m^3/s) through the running clearances between the rotating element and the static parts of the pump, due to differences of static pressure in the working fluid upstream and downstream of the impeller; and
- disc friction losses, due to shear stresses arising on the two (front and rear) circular surfaces of the disc-shaped impeller which rotates in the working fluid.

According to Stepanoff [23], disc friction loss is by far the most important of all mechanical losses. Stepanoff also indicates the following equation to compute the power P_d (W) absorbed by disc friction

$$P_d = K \frac{2\pi \omega^3 r^5}{5}$$

where K is a factor, whose value is determined experimentally, based on the Reynolds number, and ω (rad/s) and r (m) are respectively the angular velocity and the radius of the impeller.

Some liquid propellants, for example, liquid fluorine and hydrazine, are highly reactive chemically and thermally unstable above some temperatures. Therefore, when using such propellants, special care must be taken in choosing the materials of which a pump is made. This also holds with the seals, the bearings, and the insulation materials which protect the pump from the heat coming from the turbine.

4.8   Heads and Mass Flow Rates of Pumps

The concept of head rise $\Delta H = H_2 - H_1$ supplied by a pump has been discussed in
Sect. 4.2. The pump of a rocket engine operates to increase the pressure of a liquid
propellant, from p_1 at the point of suction to p_2 at the point of discharge, by an amount
Δp (N/m^2). The relation between head rise ΔH and pressure rise Δp is

$$\Delta H = \frac{\Delta p}{\rho g_0}$$

where ρ (kg/m^3) is the density of the propellant, and g_0 (m/s^2) is the gravitational
acceleration of the Earth.

The head level H_2 (m) required by a rocket engine to work at the desired thrust level
depends on the useful head and on the sum of the head losses affecting its hydraulic
circuit. Such losses include not only those relating to the pumps (see Sect. 4.7), but
also those relating to the injectors, the thrust chamber manifold, the cooling jacket,
the valves, and the duct. An additional head margin is allowed for system calibration
[22].

The following figure is a pump performance diagram, that is, a plot showing the
performance of a pump in terms of head rise ΔH (m) and efficiency η_P as a function
of discharge q (l/s).

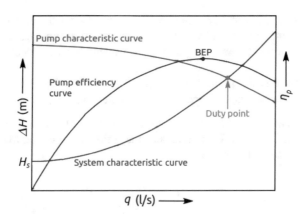

The system characteristic curve defines the amount of head ΔH necessary to:

- raise the pumped fluid through a certain height H_s (m), called the static head; and
- overcome the head losses ΔH_L (m), which depend on the discharge q (l/s).

Therefore, the head rise results from

$$\Delta H = H_s + \Delta H_L$$

In other words, the system characteristic curve defines the head rise ΔH required to push a fluid at any given volume flow rate q (l/s) through a system. This curve is a graphical presentation of the energy equation. The static head H_s (m) is the vertical distance between the level of suction and the level of discharge when the pump is shut off, as shown in the following figure, re-drawn from [26].

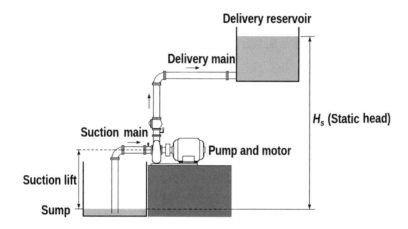

The head losses ΔH_L, which are due to friction in the pipe and in the fittings, are approximately proportional, through a coefficient a, to the square of the volume flow rate q (l/s), and therefore the system characteristic curve can be represented by a parabola, as follows

$$\Delta H = H_s + aq^2$$

The pump characteristic curve defines the head rise ΔH (m) which a particular pump can deliver at any given volume flow rate q (l/s). When the system curve and the characteristic curve are plotted together, a pump can only operate at the point, called the duty point, where these curves intersect. The duty point defines the volume flow rate q (l/s) and the head rise ΔH (m) at which a particular pump can operate in a particular system. In practice, it is desirable for a pump to operate so that its duty point should be close to a point of maximum efficiency.

As has been shown in Sect. 4.4, the efficiency η_P of a pump is the ratio of the useful power delivered by the pump to the power supplied to it, as follows

$$\eta_P = \frac{P_d}{P_s} = \frac{WP}{BP}$$

where WP, called water power and measured in kW, is the useful power imparted to water by a pump, and is computed as follows

$$WP = \rho g q \Delta H = \frac{(1000 \text{ kg/m}^3)\,(9.80665 \text{ m/s}^2)\,(\text{l/s})\,(\text{m})}{(1000 \text{ l/m}^3)\,(1000 \text{ W/kW})} \approx \frac{q \Delta H}{102}$$

and BP, called brake power and measured in kW, is the power supplied to the pump shaft, and results from

$$BP = \frac{WP}{\eta_P}$$

The value of the pump efficiency η_P at various values of volume flow rate is given by the manufacturer. This value varies with the diameter of the impeller. Values of efficiency ranging from 0.60 to 0.80 (or from 60 to 80%) are considered normal for a pump [27].

The performance of a pump is measured by the variation of the volume flow rate q (l/s) with respect to the head rise ΔH (m), that is, by the pump characteristic curve. This performance depends on the following properties:

- pump type;
- pump size;
- impeller size; and
- rotational speed.

The performance of a pump can be shown by either a single curve for an impeller of fixed diameter, or multiple curves for impellers of various diameters in one casing. The performance curves given by most manufacturers show the total head rise ΔH (m) generated by the pump, the brake power BP (kW) required to drive it, the derived efficiency $\eta_P = WP/BP$, and the $NPSH_R$ (m), which is the net positive suction head required to drive the pump over a range of volume flow rates at a constant rotational speed of the impeller. The following figure, adapted from [28], is a typical plot of performance characteristics given by a manufacturer.

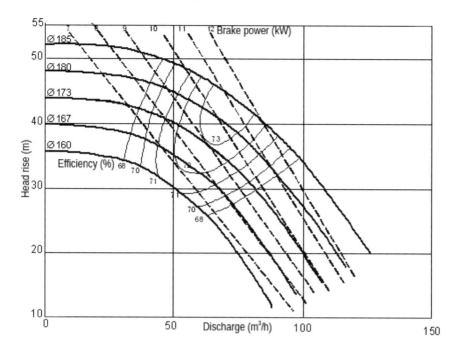

This plot shows curves for various impeller diameters. Each diameter (185 mm, 180 mm, and so on) is indicated above the corresponding curve. The plot also shows pump efficiency curves. Each efficiency value (68%, 70%, and so on) is placed near the corresponding curve. The point at which the maximum efficiency is achieved for each impeller diameter is called the Best Efficiency Point (BEP). Consequently, a pump has several BEP points, each of which is associated with an individual performance curve. It is a good practice to select or design a pump whose impeller can be increased in diameter, to permit a future increase in head rise and volume flow rate [27].

Volume flow rates, head rises, and brake powers can be calculated in a pump for variable rotational speed ω (rad/s) or variable impeller diameter d (m), by using the affinity laws, as will be shown below. The affinity laws are equations which can be used to estimate changes in the performance of a pump as a result of a change in either rotational speed or impeller diameter.

There are two sets of affinity laws. The first set assumes a pump having an impeller of constant diameter (d = constant). The second set assumes a pump having a constant rotational speed (ω = constant).

The first set states that, for a given pump having an impeller of constant diameter, the volume flow rate is directly proportional to the rotational speed, the head rise is directly proportional to the square of the rotational speed, and the required brake power is directly proportional to the cube of the rotational speed. This statement is expressed by the following three equations

$$\frac{q_1}{q_2} = \frac{\omega_1}{\omega_2}$$

$$\frac{\Delta H_1}{\Delta H_2} = \left(\frac{\omega_1}{\omega_2}\right)^2$$

$$\frac{BP_1}{BP_2} = \left(\frac{\omega_1}{\omega_2}\right)^3$$

There is usually no change in efficiency (η_P = constant) when the rotational speeds are within the operational range. Therefore, the affinity laws belonging to the first set can be considered to be sufficiently reliable and accurate.

For example, let us consider a pump, equipped with an impeller of constant diameter, which delivers water at a volume flow rate $q_1 = 6.309$ l/s against a head rise $\Delta H_1 = 30.48$ m, operating at a rotational speed $\omega_1 = 371.8$ rad/s and having a brake power requirement $BP_1 = 2.634$ kW.

When the rotational speed decreases from $\omega_1 = 371.8$ rad/s to $\omega_2 = 334.6$ rad/s, this pump delivers water at a volume flow rate

$$q_2 = \frac{\omega_2}{\omega_1}q_1 = \frac{334.6}{371.8} \times 6.309 = 5.678 \text{ l/s}$$

against a head rise

$$\Delta H_2 = \left(\frac{\omega_2}{\omega_1}\right)^2 \Delta H_1 = \left(\frac{334.6}{371.8}\right)^2 \times 30.48 = 24.69 \text{ m}$$

and requires a brake power

$$BP_2 = \left(\frac{\omega_2}{\omega_1}\right)^3 BP_1 = \left(\frac{334.6}{371.8}\right)^3 \times 2.634 = 1.920 \text{ kW}$$

The second set states that, for a given pump operating at a constant rotational speed, the volume flow rate is directly proportional to the impeller diameter, the head rise is directly proportional to the square of the impeller diameter, and the required brake power is directly proportional to the cube of the impeller diameter. This statement is expressed by the following three equations

$$\frac{q_1}{q_2} = \frac{d_1}{d_2}$$

$$\frac{\Delta H_1}{\Delta H_2} = \left(\frac{d_1}{d_2}\right)^2$$

$$\frac{BP_1}{BP_2} = \left(\frac{d_1}{d_2}\right)^3$$

There are usually changes in efficiency associated with changes in impeller diameter, because the impeller diameter is variable, but the other parts of the pump remain the same. Therefore, the results can be affected by inaccuracies.

Huzel and Huang point out that the order of magnitude of such inaccuracies is usually 2% or 3% within a reasonable range from the rated design point, and that they affect the power requirements rather than the relationship between head rise and volume flow rate. Therefore, the pump affinity laws hold quite well in most cases [22].

For example, let us consider a centrifugal pump operating at a constant rotational speed $\omega_1 = 183.3$ rad/s, equipped with an impeller 0.280 m in diameter, and requiring a brake power $BP_1 = 10.67$ kW to deliver water at a volume flow rate $q_1 = 18.93$ l/s against a head rise $\Delta H_1 = 33.83$ m. We want to calculate the volume flow rate, the head rise, and the required brake power, if the impeller diameter were reduced from $d_1 = 0.280$ m to $d_2 = 0.230$ m.

When the impeller is trimmed as indicated above, this pump delivers water at a volume flow rate

$$q_2 = \frac{d_2}{d_1} q_1 = \frac{0.23}{0.28} \times 18.93 = 15.55 \text{ l/s}$$

against a head rise

$$\Delta H_2 = \left(\frac{d_2}{d_1}\right)^2 \Delta H_1 = \left(\frac{0.23}{0.28}\right)^2 \times 33.83 = 22.83 \text{ m}$$

and requires a brake power

$$BP_2 = \left(\frac{d_2}{d_1}\right)^3 BP_1 = \left(\frac{0.23}{0.28}\right)^3 \times 10.67 = 5.914 \text{ kW}$$

The performance of a pump is also expressed by its specific speed N, which has been defined in Sect. 4.2. This parameter is used to classify inducers and impellers according to their performance and their geometric proportions, independently of their size or of the speed at which they operate, as shown in the following figure, due to the courtesy of NASA [29].

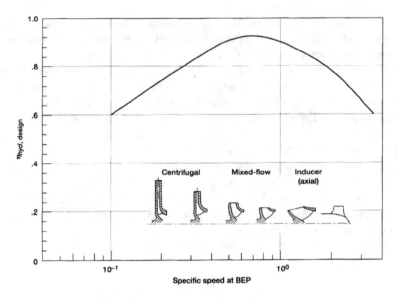

As has been shown in Sect. 4.2, the specific speed of a pump is computed with reference to the best efficiency point (BEP) of that pump. For a given rotational speed, a low value of specific speed correspond to a high head rise ΔH and to a low volume flow rate q. The reverse is true for a high value of specific speed. For an impeller of the centrifugal-flow type (the first and the second from left), the head rise is due principally to centrifugal forces. This type of impeller is used for a head rise greater than 61 m. The corresponding specific speed ranges from 0.18 to 0.44. The ratio r_2/r_1 of the impeller radius at outlet to the impeller radius at inlet ranges from 2 to 3. For an impeller of the Francis type (the third and the fourth from left), the specific speed ranges from 0.44 to 0.88, and the ratio r_2/r_1 ranges from 1.3 to 1.8. For an impeller of the mixed-flow type (the fifth from left), the head rise is due partly to centrifugal forces and partly to axial forces. The specific speed ranges from 0.80 to 1.3. For an impeller of the axial-flow type (the sixth from left), the head rise is due only to axial forces. The specific speed ranges from 1.1 to 2.2 for multi-stage impellers, and from 2.2 to 4.4 for inducers [22].

As shown in Sects. 4.3 and 4.4, the efficiency η_T of a turbine is the ratio

$$\eta_T = \frac{P_d}{P_s} = \frac{P_d}{\dot{m}_T \Delta h_T}$$

where P_d (W) is the useful power delivered to the shaft of the turbine, P_s (W) is the power supplied to it by the working fluid, $\dot{m}_T$ (kg/s) is the mass flow rate through the turbine, and Δh_T (J/kg) is the enthalpy drop per unit mass of the working fluid, that is, the available energy content per unit mass of the working fluid. The enthalpy drop per unit mass of the working fluid is the difference

$$\Delta h_T = h_0 - h_e$$

where h_0 (J/kg) is the enthalpy per unit mass of the working fluid at the turbine inlet, and h_e (J/kg) is the enthalpy per unit mass of the working fluid at the exhaust pressure, assuming an isentropic expansion, as follows

$$\Delta h_T = c_p(T_0 - T_e) = c_p T_0 \left[1 - \left(\frac{p_e}{p_0} \right)^{\frac{\gamma-1}{\gamma}} \right]$$

where T_0 (K) is the total temperature of the working fluid at the turbine inlet, T_e (K) is the static temperature of the working fluid at the turbine exhaust, p_0 (N/m²) is the total pressure of the working fluid at the turbine inlet, p_e (N/m²) is the static pressure of the working fluid at the turbine exhaust, and c_p (J kg^{-1} K^{-1}) and γ are respectively the specific heat at constant pressure and the specific heat ratio of the working fluid. The ratio p_e/p_0 can be expressed as the reciprocal of the turbine pressure ratio R_T, as follows

$$\Delta h_T = c_p T_0 \left[1 - \left(\frac{1}{R_T} \right)^{\frac{\gamma-1}{\gamma}} \right]$$

Therefore, the efficiency of a turbine can be expressed as follows

$$\eta_T = \frac{P_d}{\dot{m}_T c_p T_0 \left[1 - \left(\frac{1}{R_T} \right)^{\frac{\gamma-1}{\gamma}} \right]}$$

Most of the working fluid used in gas turbines are fuel-rich gases generated by the combustion of liquid bi-propellants. The following table, from [22], shows total temperature T (K) at turbine inlet, specific isobaric heat c_p (J kg^{-1} K^{-1}), specific gas constant R (J kg^{-1} K^{-1}), and mixture ratio o/f for fuel-rich gases resulting from the combustion of three combinations of liquid bi-propellants.

Propellants	T (K)	c_p (J kg^{-1} K^{-1})	γ	R (J kg^{-1} K^{-1})	o/f
LO$_2$/RP-1	866.5	2659	1.097	233.0	0.303
LO$_2$/RP-1	894.3	2675	1.100	242.7	0.320
LO$_2$/RP-1	922.0	2692	1.106	253.4	0.337
LO$_2$/RP-1	949.8	2705	1.111	315.3	0.354
LO$_2$/RP-1	977.6	2713	1.115	271.2	0.372
LO$_2$/RP-1	1005	2726	1.119	278.7	0.390
LO$_2$/RP-1	1033	2734	1.124	288.4	0.408
LO$_2$/RP-1	1061	2742	1.128	298.1	0.425
LO$_2$/RP-1	1089	2751	1.132	312.1	0.443
LO$_2$/RP-1	1116	2759	1.137	317.4	0.460
LO$_2$/RP-1	1144	2763	1.140	326.6	0.478

(continued)

(continued)

Propellants	T (K)	c_p (J kg^{-1} K^{-1})	γ	R (J kg^{-1} K^{-1})	o/f
LO$_2$/RP-1	1172	2767	1.144	335.7	0.497
LO$_2$/RP-1	1200	2772	1.148	344.3	0.516
N$_2$O$_4$/H$_2$NN(CH$_3$)$_2$	1033	1591	1.420	470.8	0.110
N$_2$O$_4$/H$_2$NN(CH$_3$)$_2$	1089	1666	1.420	492.8	0.165
N$_2$O$_4$/H$_2$NN(CH$_3$)$_2$	1144	1742	1.420	514.9	0.220
N$_2$O$_4$/H$_2$NN(CH$_3$)$_2$	1200	1817	1.420	537.5	0.274
N$_2$O$_4$/H$_2$NN(CH$_3$)$_2$	1255	1892	1.420	559.6	0.328
N$_2$O$_4$/H$_2$NN(CH$_3$)$_2$	1311	1968	1.420	582.1	0.382
LO$_2$/LH$_2$	810.9	8583	1.374	2335	0.785
LO$_2$/LH$_2$	922.0	8122	1.364	2168	0.903
LO$_2$/LH$_2$	1033	7787	1.354	2034	1.025
LO$_2$/LH$_2$	1144	7536	1.343	1926	1.143
LO$_2$/LH$_2$	1255	7243	1.333	1808	1.273
LO$_2$/LH$_2$	1366	7076	1.322	1722	1.410

The value of the temperature at the turbine inlet is limited by the thermal resistance of the materials of which a turbine is made. In practice, a limit of 1200 K is used in the design of a turbine [22].

The efficiency of a turbine is limited by the following losses:

(1) Nozzle losses, which are due to phenomena of flow turbulence, friction in the fluid, and loss of heat, which occur when the combusted gas passes through the nozzles of a turbine. These losses decrease the velocity at which the combusted gas leaves the nozzles.

(2) Blade losses, which are due to the residual velocity of the gas as it leaves the rotor blades, to the angle between the direction of the gas flow through the nozzles and the direction of motion of the rotor blades, to turbulence in the flow, and to friction in the fluid. As has been shown in Sect. 4.3, the losses due to the residual velocity of the gas can be reduced by selecting carefully the value of the blade-to-gas velocity ratio U/C_0, where U (m/s) is the peripheral velocity of the rotor blades, and C_0 (m/s) is the isentropic spouting velocity, which in turn depends on the enthalpy drop per unit mass of the gas Δh_T (J/kg) as follows: $C_0 = (2\Delta h_T)^{\frac{1}{2}}$. The losses due to turbulence in the flow can be reduced by designing carefully the blades and the nozzles.

(3) Leakage losses, which are due to the clearances between the tips of the rotor blades and the turbine casing. These clearances permit part of the gas to leak past the blades without impinging on them, which fact causes an energy loss. Losses of the same type are due to gas leakage from one stage to the next in a multi-stage pressure-compounded turbine, because of the clearances existing between the shaft and the sealing diaphragm.

(4) Disk friction losses, which due to the friction between the rotor disc surface and the gas which is in contact with this surface. They are also due to the centrifugal action impressed by the rotor disc on the gas. As a result of this action, part of the gas flows radially toward the casing, and is dragged along the internal surface of the casing by the rotor blades.

(5) Mechanical losses, which are due to the friction in the bearings and in the seals.

The ratio $P_d/\dot{m}_T$ (Ws/kg) of the useful power delivered to the shaft to the mass flow through a turbine is also known as the operating efficiency of the turbine.

For liquid-propellant rocket engines, the most frequently used turbo-pumps are those of the impulse type, whose working principle has been described in Sect. 4.3. In case of low values (0.2 or less) of the blade-to-gas velocity ratio U/C_0, single-shaft, two-stage, velocity-compounded turbines are often used. Turbo-pumps like these have the advantages of simplicity and low mass. By contrast, the presence of a gearbox between the pumps and the turbine makes it possible to reach a higher value of the blade-to-gas velocity ratio U/C_0 than would be possible with a direct drive and also to use a pressure-compounded turbine in place of a velocity-compounded turbine. Turbo-pumps like these have the advantage of a higher efficiency over those mentioned above.

The power level of a turbine can be regulated by controlling the total pressure p_0 (N/m^2) of the gas at the turbine inlet and its mass flow rate $\dot{m}_T$ (kg/s) through the turbine [22].

As an example of application, the following data are known for the turbo-pump mounted on a rocket engine. The propellants burned in the engine are liquid oxygen and RP-1, at a mixture ratio $o/f = 0.408$. The oxidiser pump requires a brake power $BP_o = 1.107 \times 10^7$ W. The fuel pump requires a brake power $BP_f = 8.792 \times 10^6$ W. The total temperature of the gas at the turbine inlet is $T_0 = 1033$ K. The total pressure of the gas at the turbine inlet is $p_0 = 4.413 \times 10^6$ N/m^2. The static pressure of the gas at the turbine exhaust is $p_e = 1.862 \times 10^5$ N/m^2. The mass flow rate of the gas through the turbine is $\dot{m}_T = 41.73$ kg/s. The rotational speed of the turbine shaft is $\omega_T = 733.0$ rad/s. The torque at the shaft of the turbine is $M_T = 2.763 \times 10^4$ Nm. It is required to compute the efficiency η_T of the turbine in percent and the useful power P_d delivered to the shaft of the turbine per unit of mass flow $\dot{m}_T$.

In case of the propellant combination LO$_2$/RP-1 being used at a total temperature $T_0 = 1033$ K at the turbine inlet and at a mixture ratio $o/f = 0.408$, the preceding table indicates a specific heat at constant pressure $c_p = 2734$ J kg^{-1} K^{-1} and a specific heat ratio $\gamma = 1.124$ for the combusted gas.

The enthalpy drop per unit mass of the gas is computed by using the following equation

$$\Delta h_T = c_p T_0 \left[1 - \left(\frac{p_e}{p_0} \right)^{\frac{\gamma-1}{\gamma}} \right] = 2734 \times 1033 \times \left[1 - \left(\frac{1.862 \times 10^5}{4.413 \times 10^6} \right)^{\frac{1.124-1}{1.124}} \right]$$

$$= 8.325 \times 10^5 \text{ J/kg}$$

The useful power delivered at the shaft of the turbine is computed as follows

$$P_d = M_T \, \omega_T = 2.763 \times 10^4 \times 733.0 = 2.025 \times 10^7 \text{ W}$$

The difference between the useful power P_d delivered to the shaft of the turbine and the total brake power $BP_o + BP_f$ required by the two pumps is

$$\Delta P = P_d - \left(BP_o + BP_f\right) = (2.025 - 1.107 - 0.879) \times 10^7 = 3.9 \times 10^5 \text{ W}$$

This difference between the useful power generated by the turbine and the power required by the pumps can be used for auxiliary drives.

The turbine efficiency is computed by substituting the results found above into the following equation

$$\eta_T = \frac{P_d}{\dot{m}_T \, \Delta h_T}$$

This yields

$$\eta_T = \frac{2.027 \times 10^7}{41.73 \times 8.325 \times 10^5} = 0.5835 = 58.35\%$$

The useful power delivered to the shaft of the turbine per unit of mass flow rate results from

$$\frac{P_d}{\dot{m}_T} = \frac{2.027 \times 10^7}{41.73} = 4.857 \times 10^5 \text{ Ws/kg}$$

The system cycle efficiency η_c of a turbo-pump is the ratio of the specific impulse $(I_s)_{eng}$ of the complete engine to the specific impulse $(I_s)_{tc}$ of the thrust chamber alone, as follows

$$\eta_c = \frac{(I_s)_{eng}}{(I_s)_{tc}}$$

The value of η_c measures the energy losses due to the turbo-pump and their effect on the overall engine performance. An engine fed by turbo-pumps may use either a cycle of the open type (which includes the gas-generator cycle and the tap-off cycle) or a cycle of the closed type (which includes the expander cycle and the staged-combustion cycle). In a cycle of the open type, the gas exhausted by the turbine is either discharged separately from the gas exhausted by the thrust chamber or ducted into the low-pressure portion of the thrust chamber. In a cycle of the closed type, the gas exhausted by the turbine is ducted into the high-pressure portion of the thrust chamber.

The cycles of the open type have a lower discharge pressure of the propellant pumps and a higher pressure ratio of the turbine. Therefore, they have the advantages of simplicity and low mass, but also the disadvantage of an inefficient use of the gas exhausted by the turbine for the purpose of generating thrust. In other words, a cycle of the open type is less efficient than a cycle of the closed type. The system cycle efficiency η_{co} of a cycle of the open type (subscript o) can be expressed as follows

$$\eta_{co} = \frac{(I_s)_{eng}}{(I_s)_{tc}} = \frac{\dot{m}_{tc}(I_s)_{tc} + \dot{m}_T(I_s)_{Te}}{\dot{m}_{eng}(I_s)_{tc}} = \frac{F_{tc} + F_{Te}}{g_0\dot{m}_{eng}(I_s)_{tc}}$$

where $(I_s)_{eng}$ (s) is the total specific impulse of the engine, $(I_s)_{tc}$ (s) is the specific impulse of the gas exhausted by the main thrust chamber, $(I_s)_{Te}$ (s) is the specific impulse of the gas exhausted by the turbine, $\dot{m}_{eng}$ (kg/s) is the total mass flow rate of the propellants, $\dot{m}_{tc}$ (kg/s) is the mass flow rate of the gas exhausted by the thrust chamber, $\dot{m}_T$ (kg/s) is the mass flow rate of the gas exhausted by the turbine, F_{tc} (N) is the thrust generated by the gas exhausted by the thrust chamber, F_{Te} (N) is the thrust generated by the gas exhausted by the turbine, and g_0 (m/s²) is the gravitational acceleration of the Earth.

The system cycle efficiency of a cycle of the open type can be increased by attaching a nozzle of high expansion ratio to the turbine exhaust duct, in case of an engine for an upper stage, or by post-combusting the gas exhausted by the turbine with additional propellant into an after-burner, as is the case with the engine of a supersonic jet aircraft. The system cycle efficiency of a cycle of the open type ranges from 0.96 to 0.99 [22].

In a cycle of the closed type, the gas exhausted by the turbine is firstly ducted into the main combustion chamber in order to be burned again with one of the propellants, and then ejected through the nozzle attached to the main combustion chamber. A cycle of the closed type requires a pressure of the gas at the turbine exhaust higher than the pressure of the gas in the combustion chamber. Therefore, in a cycle of the closed type, the pressures of the propellants at the discharge of the pumps are higher and the turbine pressure ratio is much lower than is the case with a cycle of the open type. In most cases, a gas composed of 100% of combustion products of one propellant and part of the other propellant is used in a cycle of the closed type as the working fluid which drives the turbine.

The system cycle efficiency η_{cc} of a cycle of the closed type (subscript c) can be expressed approximately as follows

$$\eta_{cc} = \frac{(I_s)_{eng}}{(I_s)_{tc}} = \left[1 - \frac{E_p}{c_p(T_c)_{ns}}\right]^{\frac{1}{2}} = \left[1 - \frac{g_0\left(\frac{o}{f}\Delta H_o + \Delta H_f\right)}{\left(1 + \frac{o}{f}\right)c_p(T_c)_{ns}}\right]^{\frac{1}{2}}$$

where $(I_s)_{eng}$ (s) is the total specific impulse of the engine, $(I_s)_{tc}$ (s) is the specific impulse of the gas exhausted by the main thrust chamber, E_p (J/kg) is the ideal energy per unit mass required to increase the total pressure of 1 kg of propellant from the value at the pump inlet to the value at the main combustion chamber, $(T_c)_{ns}$ (K) is the total temperature of the gas in the combustion chamber at the nozzle inlet, c_p (J kg^{-1} K^{-1}) is the isobaric specific heat of the gas in the main combustion chamber, g_0 (m/s^2) is the gravitational acceleration of the Earth, o/f is the oxidiser-to-fuel mixture ratio, ΔH_o (m) is the pressure head rise based on the difference Δp_o between total pressure of the gas in the combustion chamber at the nozzle inlet and the static pressure of the oxidiser at the pump inlet, and ΔH_f (m) is the pressure head rise based on the difference Δp_f between the total pressure of the gas in the combustion chamber at the nozzle inlet and the static pressure of the fuel at the pump inlet.

The system cycle efficiency of a cycle of the closed type ranges from 0.996 to 0.9996 [22].

As an example of application, the following data are known for the turbo-pump mounted on a rocket engine. The mass flow rate of oxidiser (liquid oxygen) in the thrust chamber is 880.4 kg/s. The mass flow rate of fuel (RP-1) in the thrust chamber is 375.1 kg/s. The total pressure of the gas in the combustion chamber at the nozzle inlet is $(p_c)_{ns} = 6.895 \times 10^6$ N/m^2. The thrust generated by the gas exhausted by the thrust chamber is $F_{tc} = 3.324 \times 10^6$ N. The engine cycle is the gas-generator cycle, which is a cycle of the open type. The mass flow rate of oxidiser in the gas generator is 12.11 kg/s. The mass flow rate of fuel in the gas generator is 29.62 kg/s. The thrust generated by the gas exhausted by the turbine is $F_{Te} = 1.201 \times 10^4$ N. It is required to determine the cycle efficiency of the turbo-pump system, the specific impulse of the thrust chamber and the thrust of the engine system at sea level, and the oxidiser-to-fuel mixture ratio for the overall engine system at rated conditions.

The required oxidiser-to-fuel mixture ratio is, by definition, the ratio of oxidiser mass flow rate to fuel mass flow rate. By using the values given above, we find

$$\frac{o}{f} = \frac{880.4}{375.1} = 2.347$$

The total mass flow rate of propellant in the thrust chamber is the sum of those due to the two propellants, as follows

$$\dot{m}_{t_c} = 880.4 + 375.1 = 1255.5 \text{ kg/s}$$

The specific impulse of the gas exhausted by the thrust chamber at sea level is

$$(I_s)_{tc} = \frac{F_{tc}}{g_0 \dot{m}_{tc}} = \frac{3.324 \times 10^6}{9.807 \times 1.256 \times 10^3} = 269.9 \text{ s}$$

The total mass flow rate of propellant exhausted by the turbine is the sum of those due to the two propellants, as follows

$$\dot{m}_{Te} = 12.11 + 29.62 = 41.73 \text{ kg/s}$$

The specific impulse of the gas exhausted by the turbine at sea level is

$$(I_s)_{Te} = \frac{F_{Te}}{g_0 \dot{m}_{Te}} = \frac{1.201 \times 10^4}{9.807 \times 41.73} = 29.35 \text{ s}$$

The total mass flow rate of the engine system results from

$$\dot{m}_{eng} = \dot{m}_{tc} + \dot{m}_{Te} = 1.256 \times 10^3 + 41.73 = 1.297 \times 10^3 \text{ kg/s}$$

By substituting the results found above into the following equation

$$\eta_{co} = \frac{F_{tc} + F_{Te}}{g_0 \dot{m}_{eng} (I_s)_{tc}}$$

we find the following value of the cycle efficiency of the turbo-pump system in case of a gas-generator cycle

$$\eta_{co} = \frac{3.324 \times 10^6 + 1.201 \times 10^4}{9.807 \times 1.297 \times 10^3 \times 269.9} = 0.9717 = 97.17\%$$

By introducing this value of η_{co} and the value of $(I_s)_{tc}$ computed above into the following equation

$$\eta_{co} = \frac{(I_s)_{eng}}{(I_s)_{tc}}$$

we find the following value of the specific impulse of the engine system at sea level

$$(I_s)_{eng} = \eta_{co}(I_s)_{tc} = 0.9717 \times 269.9 = 262.3 \text{ s}$$

The thrust of the engine system at sea level results from

$$F_{eng} = (I_s)_{eng} g_0 \dot{m}_{eng} = 262.3 \times 9.807 \times 1.297 \times 10^3 = 3.336 \times 10^6 \text{ N}$$

The oxidiser-to-fuel mixture ratio for the engine system is

$$\frac{o}{f} = \frac{880.4 + 12.11}{375.1 + 29.62} = 2.205$$

Now we suppose to use, for the same engine indicated above, the staged-combustion cycle, which is a cycle of the closed type. We want to compute the value of the cycle efficiency of the turbo-pump system, assuming no change in: (a) the total pressure of the gas in the combustion chamber at the nozzle inlet, (b) the properties of the combusted gas, and (c) the performance of the thrust chamber due to the changes in mixture ratio and to the two-stage combustion. We also know the static pressure of the oxidiser at the inlet of its pump to be 3.792×10^5 N/m^2, and the static pressure of the fuel at the inlet of its pump to be 3.103×10^5 N/m^2.

In Chap. 2, Sect. 2.5, we have found the following properties of the combusted gas in this type of engine: total temperature of the combustion chamber at the nozzle inlet $(T_c)_{ns}$ = 3589 K, molar mass $\mathcal{M}$ = 22.5 kg/kmol, specific heat at constant pressure c_p = 2034 J kg^{-1} K^{-1}, and specific heat ratio γ = 1.222. According to the first table of Sect. 4.7, at the conditions indicated there, the density of liquid oxygen is ρ_o = 1140 kg/m^3, and the mean density of RP-1 is ρ_f = 0.5 × (797.7 + 813.7) = 805.7 kg/m^3. In addition, as indicated above, the total pressure of the gas in the combustion chamber at the nozzle inlet is $(p_c)_{ns}$ = 6.895 × 10^6 N/m^2.

Therefore, we can compute the pressure head rise ΔH_o, based on the difference Δp_o between the total pressure of the gas in the combustion chamber at the nozzle inlet and the static pressure of the oxidiser at the pump inlet, as follows

$$\Delta H_o = \frac{\Delta p_o}{\rho_o g_0} = \frac{6.895 \times 10^6 - 3.792 \times 10^5}{1140 \times 9.807} = 582.8 \text{ m}$$

Likewise, we can compute the pressure head rise ΔH_f, based on the difference Δp_f between the total pressure of the gas in the combustion chamber at the nozzle inlet and the static pressure of the fuel at the pump inlet, as follows

$$\Delta H_f = \frac{\Delta p_f}{\rho_f g_0} = \frac{6.895 \times 10^6 - 3.103 \times 10^5}{805.7 \times 9.807} = 833.3 \text{ m}$$

By substituting the values found above into the following equation

$$\eta_{cc} = \left[1 - \frac{g_0 \left(\frac{o}{f} \Delta H_o + \Delta H_f \right)}{\left(1 + \frac{o}{f} \right) c_p (T_c)_{ns}} \right]^{\frac{1}{2}}$$

we find the following value of the cycle efficiency of the turbo-pump system in case of a staged-combustion cycle

$$\eta_{cc} = \left[1 - \frac{9.807 \times (2.205 \times 582.8 + 833.3)}{(1 + 2.205) \times 2034 \times 3589} \right]^{\frac{1}{2}} = 0.9996 = 99.96\%$$

By introducing this value of η_{cc} and the value of $(I_s)_{tc}$ computed above into the following equation

$$\eta_{cc} = \frac{(I_s)_{eng}}{(I_s)_{tc}}$$

we find the following value of the specific impulse of the engine system at sea level

$$(I_s)_{eng} = \eta_{cc} (I_s)_{tc} = 0.9996 \times 269.9 = 269.8 \text{ s}$$

The performance of an individual rocket engine depends on its thrust, specific impulse, mass flow rate, and oxidiser-to-fuel mixture ratio. Each engine is designed to operate at a performance level defined by the values of the quantities indicated above. Another engine, even of the same type, may operate at performance parameters which are slightly different from the design values. When the deviations of the actual from the nominal performance values are greater than a few percent, then a rocket vehicle may be incapable of accomplishing its mission.

These deviations are due to several reasons. Sutton and Biblarz [30] have identified some of them, which are specified below. Firstly, the unavoidable tolerances on the construction materials may alter the impingement of the propellants in the injector, and this change may affect the efficiency of the combustion process. Secondly, a change in mixture ratio, even small, may cause an increase in unused propellant at the end of the burn. The quantity of residual propellant ranges usually from 0.5 to 2% of the total quantity [31]. Thirdly, variations in either the composition or the storage temperature of the propellants may induce changes in their density and viscosity. Fourthly, regulator setting tolerances or changes in flight acceleration can also cause deviations of the performance parameters from the design values.

The engine calibration system is a process which corrects an engine system, by adjusting the values of its parameters, in order to obtain the desired performance within the allowed limits of tolerance.

The engine components which need calibration are

- hydraulic and pneumatic components, such as pumps, pipes, valves, expansion joints, etc.;
- high-temperature components, such as thrust chamber, turbines, gas generators or pre-burners, etc.; and
- components for cryogenic propellants, such as pumps, valves etc.

Hydraulic and pneumatic components are tested by means of water flowing through them in flow benches. This makes it possible to determine their pressure drops at rated conditions. Components operating at high temperatures are tested through hot firing. Likewise, components operating at low temperature are tested by using cryogenic propellants or liquid nitrogen.

The engine properties are estimated by summing the corrected values of the pressure drops due to its individual components at the desired mass flow rates. In addition, the actual measured mixture ratio $\dot{m}_o / \dot{m}_f$ must be equal to the desired mixture ratio o/f.

Two methods can be employed to control the performance parameter of an engine. The first of them uses an automatic system with feedback and a digital computer which controls the deviations of the parameters in real time. The other method uses an initial static calibration of the engine system. In particular, the pressures supplied to the propellants by means of the respective pumps (or by means of gases under pressure, in case of gas-pressurised propellants) must be balanced against the pressure drops and the pressure in the combustion chamber. This balancing is necessary in order for an engine to operate at the desired values of the mass flow rates and of the oxidiser-to-fuel mixture ratio o/f.

The following figure shows a diagram of pressure balancing, expressed in terms of head rise ΔH (m) against volume flow rate q (m³/s), for one of the two propellant lines of a bi-propellant rocket engine fed by a turbo-pump system.

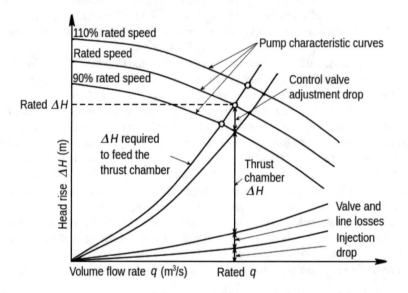

The rotation speed of a pump is a variable quantity. The pump calibration curves can only be drawn after obtaining data resulting from tests. Such tests are executed by varying the impeller diameter (impeller trimming) and the hydraulic resistances at the discharge, in order to obtain the properties required by engine system.

The volume flow rate of the propellants to either a gas generator or a pre-burner also needs calibration. The power balance requires the power delivered at the turbine shaft to be equal to the sum of the power needed to drive the pumps and the power losses in the bearings, seals, etc.

In a system balance of a turbo-pump at the initial point of design, known quantities (such as engine type, pressure in the thrust chamber, expansion area ratio in the nozzle, oxidiser-to-fuel mixture ratio, and thrust) are combined with assumed quantities (such as pump efficiencies, turbine efficiencies, turbine pressure ratio, turbine inlet temperature, and various indicators of engine performance), in order to predict the requirements relating to the pump head rise, the pump flow rate, and the turbine flow rate. In making this balance, the power delivered to the turbine shaft is equated to all the sources of power absorption, as has been shown above.

A system balance of a turbo-pump is also necessary at the final point of design. This is because refinements and changes, which may be made in the detailed design of a rocket engine, can affect the turbo-pump operation. For example, during the development of the engine of the MA-5 sustainer, the efficiencies of the pumps and of the turbine were found, when tested, below the initial values assumed for them in the design balance of the engine system. This fact required changes to increase the value of the turbine drive pressure available from the gas generator, in order for the engine to meet its rated thrust values.

An off-design operating condition of a turbo-pump may be more demanding for a particular component than the nominal design condition. Consequently, the whole operating range of a rocket engine is to be considered carefully. Otherwise, a pump which operates satisfactorily in design conditions may perform unsatisfactorily or even fail in off-design conditions.

In order to define the actual operating range of a turbo-pump, it is necessary to determine the planned operating range. Then, a system balance including the predicted operating characteristics of the turbo-pump is used to determine the whole range of operation over the planned engine throttling and the variation of the oxidiser-to-fuel mixture ratio. For this purpose, it is necessary to take account not only of the ordinary variations in the operating conditions, but also of the variations due to manufacturing tolerances. This requires the availability of data on known tolerances of components from existing liquid-propellant engines. These data are used to estimate tolerances on hydraulic resistance of lines, valves, and passage channels of the coolant in the nozzle, tolerances on the head rise/flow rate and efficiency/flow rate curves of the pump, etc.

The tolerances on all the components are assumed to have a Gaussian distribution. The effects of these tolerances are treated statistically, by taking the square root of the sum of the squares of the effects of all the components. The effect of each component is determined either by performing an engine balance assuming that the given component is at one of the extreme points of its range of tolerance, or by linearising the equations which govern the engine system in order to determine the influence coefficients for each of the components considered.

The control points used for regulating the thrust and the oxidiser-to-fuel mixture ratio can affect the discharge pressure of the turbo-pump and therefore its system design. Either calibration orifices or a pressure regulator have been used to control either the gas flow rate from the gas generator to the turbine or the oxidiser flow rate to the gas generator. For the J-2, F-1, and H-1 rocket engines, calibration orifices have been used in the turbine-gas generator lines to set the power level. For the RL10 engine, heated fuel (hydrogen) has been used to control the turbine power. For this purpose, the fuel was heated by passing through the cooling jacket of the thrust chamber before being ducted to the turbine through a control system which sensed the thrust chamber pressure and adjusted a turbine bypass valve. By contrast, the MA-5 and MB-5 rocket engines used a pressure regulator to control the flow of the oxidiser to the gas generator.

The control of the oxidiser-to-fuel mixture ratio has been obtained by using either the main propellant valves or a valve controlling the bypass of a propellant around its pump. The engine of the MA-5 sustainer used the control of the main propellant valves, whereas the J-2 engine used a control valve around the oxidiser pump to regulate the oxidiser-to-fuel mixture ratio.

Engines which use orifices in the gas generator feed lines for thrust calibration or a turbine bypass for mixture ratio regulation are subject to large changes in thrust level, and therefore in pump discharge pressure, when their mixture ratio is varied.

This type of control system is unsuitable for high-pressure staged-combustion cycles, because of the high variations in the pump discharge pressure as a result of the series arrangement of the thrust chamber, the turbine, and the turbine hot-gas generator [12].

4.9 Design of Centrifugal-Flow Pumps

Centrifugal-flow pumps have been illustrated and described qualitatively in the preceding sections. Here, their design is described quantitatively.

The impeller of a centrifugal pump is a bladed rotor shaped so as to sweep the pumped fluid roughly $\pi/2$ rad (or 90°) outward from the pump shaft. The fluid enters the pump in the axial direction and is swept radially by the rotating blades of the impeller. The following figure, due to the courtesy of Wikimedia [32], illustrates a rubber impeller shaped as a paddle-wheel, which has been used for the water pump of an outboard engine.

Let r_1 (m) and r_2 (m), such that $r_1 < r_2$, be the radii defining the endpoints of an impeller blade at the points of respectively entrance and exit of the fluid. The tangential component of the velocity vector acting on a fluid particle increases, as the particle moves radially from the point of radius r_1 to the point of radius r_2, due to the centrifugal force acting on the particle. Assuming a constant angular velocity ω (rad/s) of the shaft, and in the absence of energy losses, the ideal head rise ΔH_{ic} (m) of the fluid between the entrance and the exit of the impeller is

$$\Delta H_{ic} = \frac{\omega^2 \left(r_2^2 - r_1^2\right)}{2g_0}$$

where g_0 (m/s^2) is the gravitational acceleration of the Earth.

In order to reach a high performance, the vanes of most impellers used in centrifugal-flow pumps for rocket engines are shrouded and curved backward (such that $\beta_2 < \pi/2$ rad), as shown in the following figure, re-drawn from [22].

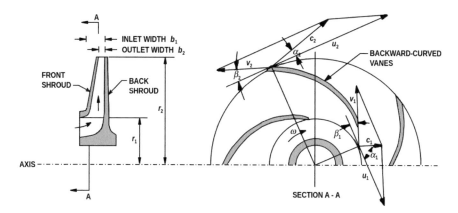

As shown in the preceding figure, the impeller width decreases towards the periphery, in order to keep the cross-sectional area of the radial flow path nearly constant. A velocity diagram can be drawn to show the direction of the fluid flow vector in various points of the impeller. In drawing these diagrams, the following simplifying assumptions are made:

- there are no losses;
- the impeller passages are completely filled, at all times, with flowing fluid;
- the flow is two-dimensional; and
- the pumped fluid leaves the impeller passages tangentially to the vane surfaces.

The subscripts 1 and 2 are used in the following figure to indicate respectively the inlet point and the outlet point of the impeller.

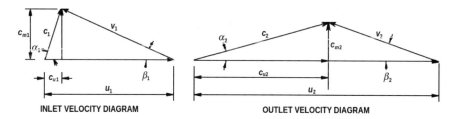

The c vectors indicate the flow velocities (m/s) with respect to the duct and the casing. The v vectors indicate the flow velocities (m/s) with respect to the inducer or the impeller. The u vectors indicate the flow velocities (m/s) of points on the inducer or the impeller. The angle between the vectors c and u is indicated with α. The angle between a tangent to the impeller vane and a line in the direction of the motion of the vane is indicated with β. The latter is also the angle between v and the extension of u. The following equations can be deduced from the velocity diagrams shown in the preceding figure:

$$\Delta H_{ip} = \frac{u_2^2 - u_1^2 + v_1^2 - v_2^2}{2g_0}$$

$$\Delta H_i = \frac{u_2^2 - u_1^2 + v_1^2 - v_2^2 + c_2^2 - c_1^2}{2g_0} = \frac{u_2 c_{u2} - u_1 c_{u1}}{g_0}$$

$$q_{imp} = c_{m1} A_1 = c_{m2} A_2$$

$$c_{u1} = u_1 - \frac{c_{m1}}{\tan \beta_1}$$

$$c_{u2} = u_2 - \frac{c_{m2}}{\tan \beta_2}$$

where ΔH_{ip} (m) is the ideal static pressure head rise of the fluid flowing through the impeller due to centrifugal forces and to a decrease in flow velocity relative to the impeller, ΔH_i (m) is the ideal total pressure head rise of the fluid flowing through the impeller (that is, the ideal head rise due to the pump impeller), q_{imp} (m³/s) is the impeller volume flow rate at the design point, A_1 (m²) is the area of the section perpendicular to the radial flow at the impeller inlet, A_2 (m²) is the area of the section perpendicular to the radial flow at the impeller outlet, r_1 (m) is the radius of the vanes at the impeller inlet, r_2 (m) is the radius of the vanes at the impeller outlet (that is, the outside radius of the impeller), ω (rad/s) is the angular velocity of the impeller, $u_1 = \omega r_1$ (m/s) is the impeller peripheral velocity at inlet, $u_2 = \omega r_2$ (m/s) is the impeller peripheral velocity at outlet (that is, the velocity of the impeller tip), v_1 (m/s) is the flow velocity at inlet relative to the impeller, v_2 (m/s) is the flow velocity at outlet relative to the impeller, c_1 (m/s) is the absolute flow velocity at inlet, c_{u1} (m/s) is the tangential component of the absolute flow velocity at inlet, c_{m1} (m/s) is the meridian component of the absolute flow velocity at inlet (the meridian component is, in the present case, also the radial component, because the impeller of a centrifugal-flow pump sweeps the pumped fluid perpendicularly to the pump shaft), c_2 (m/s) is the absolute flow velocity at outlet, c_{u2} (m/s) is the tangential component of the absolute flow velocity at outlet, c_{m2} (m/s) is the meridian (or radial) component of the absolute flow velocity at outlet, β_1 is the angle of the impeller vane at inlet, and β_2 is the angle of the impeller vane at outlet.

In case of low-density propellants (for example, liquid hydrogen, whose density is 70.85 kg/m³ at a temperature of 20.369 K and atmospheric pressure, according to [33] straight radial vanes (such that $\beta_2 = \pi/2$ rad) are used in impellers of centrifugal-flow pumps, as shown in the following figure, re-drawn from [22].

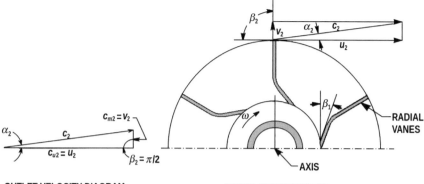

OUTLET VELOCITY DIAGRAM   **RADIAL VANE IMPELLER**

Radial vane impellers used in centrifugal-flow pumps make it possible to reach high values of head coefficient $\psi = g_0 \Delta H / u_2^2$ (see Sect. 4.2). The preceding figure also shows the outlet velocity diagram for a centrifugal-flow pump having an impeller with radial vanes. At the outlet, there results $\beta_2 = \pi/2$ rad and $c_{u2} = u_2$. After substituting $c_{u2} = u_2$ into the preceding equation $\Delta H_i = (u_2 c_{u2} - u_1 c_{u1})/g_0$, there results

$$\Delta H_i = \frac{u_2^2 - u_1 c_{u1}}{g_0}$$

In case of a centrifugal-flow pump having no inducer, a proper choice of the angle β_1 of the impeller vane at inlet or the provision of guide vanes at inlet results in the minimum value of the tangential component c_{u1} of the absolute flow velocity c_1 at inlet. For best efficiency, the component c_{u1} should be equal to zero.

This is called no pre-rotation. In this case, the angle α_1 between the vectors c_1 and u_1 is $\pi/2$ rad (or 90°), and the equation $\Delta H_i = (u_2 c_{u2} - u_1 c_{u1})/g_0$ becomes

$$\Delta H_i = \frac{u_2 c_{u2}}{g_0}.$$

In practice, the centrifugal pumps used for rocket engines have an inducer placed upstream of the impeller and connected in series with it, as shown in the following figure, due to the courtesy of NASA [4].

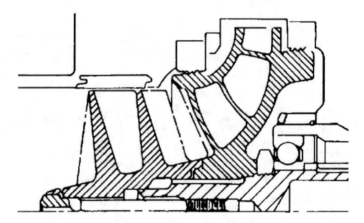

This figure illustrates an inducer having a low head coefficient ($\psi = 0.11$), a cylindrical tip, and a tapered hub. This particular inducer has been used in the Mark-15 liquid-oxygen pump for the J-2 rocket engine. The complete Mark-15 liquid-oxygen pump is illustrated in Sect. 4.7.

The presence of an inducer modifies the flow entering the impeller of a pump. In particular, the flow enters the impeller at an angle β_1' greater than the impeller inlet vane angle β_1, and leaves the impeller at an angle β_2' smaller than the impeller discharge vane angle β_2.

This fact and the presence of hydraulic losses change the relative flow velocities v_1 and v_2 to respectively v_1' and v_2'. For the same reasons, the absolute flow velocities c_1 and c_2 change to respectively c_1' and c_2', and their absolute tangential components c_{u1} and c_{u2} change to respectively c_{u1}' and c_{u2}'. By contrast, since the areas A_1 and A_2 of the sections perpendicular to the radial flow and the volume flow rates q_1 and q_2 remain constant, then the meridian or radial components c_{m1} and c_{m2} remain constant. The dashed lines in the following figure, redrawn from [22], show the changed vectors and their components.

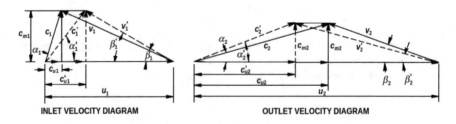

INLET VELOCITY DIAGRAM   OUTLET VELOCITY DIAGRAM

Consequently, the preceding equation $\Delta H_i = (u_2 c_{u2} - u_1 c_{u1})/g_0$ becomes

$$\Delta H_{imp} = \frac{u_2 c_{u2}' - u_1 c_{u1}'}{g_0}$$

where ΔH_{imp} (m) is the actual total pressure head rise of the fluid flowing through the impeller, c'_{u1} (m/s) is the tangential component of the design absolute flow velocity at inlet, and c'_{u2} (m/s) is the tangential component of the design absolute flow velocity at outlet.

The impeller vane coefficient e_v is the ratio of the tangential component c'_{u2} of the design absolute flow velocity to the tangential component c_{u2} of the ideal absolute flow velocity at the outlet of the impeller, as follows

$$e_v = \frac{c'_{u2}}{c_{u2}}$$

Typical values for e_v range from 0.65 to 0.75 [22].
The preceding equation $c_{u2} = u_2 - c_{m2}/\tan \beta_2$ becomes

$$c'_{u2} = u_2 - \frac{c_{m2}}{\tan \beta'_2}$$

The required total pressure head rise ΔH_{imp} (m) of the fluid flowing through the impeller can be determined as follows

$$\Delta H_{imp} = \Delta H + \Delta H_e - \Delta H_{ind}$$

where ΔH (m) is the rated design head rise of the pump, ΔH_{ind} (m) is the required head rise of the inducer at the rated design point, and ΔH_e (m) is the hydraulic head loss in the volute. Typical design values of ΔH_e range from 0.10 ΔH to 0.30 ΔH [22].

The required volume flow rate q_{imp} (m³/s) through the impeller can be determined as follows

$$q_{imp} = q + q_e$$

where q (m³/s) is the rated flow rate of the pump, and q_e (m³/s) is the impeller loss due to leakage (see Sect. 4.7). Typical design values of the leakage loss q_e range from 0.01 q_{imp} to 0.05 q_{imp} [22].

In the design of a centrifugal-flow pump, the values of some quantities, such as head rise ΔH (m), volume flow rate q (m³/s), rotating speed ω (rad/s), specific speed N, and suction specific speed S, are chosen by the designer. This done, the impeller for this pump is designed by choosing the velocities and the angles of its vanes in such a way as to obtain the desired properties with the maximum efficiency. For this purpose, it is possible to use design or experimental data on pump head coefficient ψ, impeller vane coefficient e_v, and impeller loss due to leakage q_e (m³/s). Finally, the impeller is designed according to the chosen values of the quantities mentioned above.

The values of some principal quantities are chosen as will be shown below.

The radial component c_{m1} (m/s) of the absolute flow velocity at the impeller entrance depends on the inlet conditions, and in particular on the velocity of discharge at the outlet of the inducer, and on the diameter of the duct. Typical design values of c_{m1} range from 3.05 to 18.3 m/s [22].

The radial component c_{m2} (m/s) of the absolute flow velocity at the impeller exit depends on the outlet conditions, and in particular on the tip velocity u_2 (m/s) of the impeller blades and on the flow coefficient $\varphi = c_{m2}/u_2$ at the impeller exit. Typical design values of c_{m2} range from 0.01 u_2 to 0.15 u_2 [22].

The diameter d_1 (m) of the impeller at the point of entrance of the vanes depends on the design of the inducer, on the diameter of the impeller shaft, and on the diameter of the impeller hub.

The tip velocity u_2 (m/s) of the impeller blades at the impeller exit can be computed by means of the equation $\psi = g_0 \Delta H / u_2^2$, for given values of the head rise ΔH (m) and of the head coefficient ψ. The maximum design value of u_2 depends on the strength of the material used. The latter, in turn, determines the maximum head rise which can be obtained with a single-stage impeller. Typical design values of u_2 range from 61 to 457 m/s [22]. When the rotating speed ω (rad/s) and the tip velocity u_2 (m/s) have known values, then the impeller diameter d_2 (m) at the point of exit can be computed as follows $d_2 = 2u_2/\omega$.

The inlet vane angle β_1 depends on the flow conditions at the vane inlet. The angle β_1 should be made equal or close to the inlet flow angle β_1'. Remembering the preceding equation

$$c_{u1} = u_1 - \frac{c_{m1}}{\tan \beta_1}$$

it is possible to calculate approximately the angle β_1' as follows

$$\tan \beta_1' = \frac{c_{m1}}{u_1 - c_{u1}'}$$

Typical design values of β_1 range from 0.1396 to 0.5236 rad [22].

The outlet vane angle β_2 is equal to $\pi/2$ rad in the particular case of an impeller having radial vanes, as has been shown above. In case of an impeller having backward-curved vanes, β_2 is to be determined by considering the curves expressing efficiency and head rise against volume flow rate. For a given value of the impeller peripheral velocity u_2 at outlet, the head rise and the volume flow rate increase with increasing values of β_2. Typical design values of β_2 range from 0.2967 to 0.4887 rad, with an average value of 0.3927 rad for most specific speeds [22]. The following figure (re-drawn from [22]), shows a typical radial-flow impeller with backward-curved vanes.

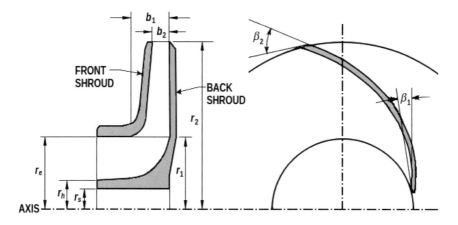

In the preceding figure, $r_s = d_s/2$ (m), $r_1 = d_1/2$ (m), $r_2 = d_2/2$ (m), are the radii of respectively the impeller shaft, the impeller blade at entrance, and the impeller blade at exit, and b_1 (m) and b_2 (m) are the widths of the impeller blade at the points of respectively entrance and exit. In addition, $r_h = d_h/2$ (m) is the impeller hub radius, and $r_e = d_e/2$ is the eye radius.

The shaft diameter d_s can be determined by means of the following equations given in [22]:

$$\sigma_s = \frac{16T}{\pi d_s^3}$$

$$\sigma_t = \frac{32M}{\pi d_s^3}$$

$$\sigma_{sw} = \frac{1}{2}\left(4\sigma_s^2 + \sigma_t^2\right)^{\frac{1}{2}}$$

$$\sigma_{tw} = \frac{1}{2}\sigma_t + \frac{1}{2}\left(4\sigma_s^2 + \sigma_t^2\right)^{\frac{1}{2}}$$

where σ_s (N/m^2) is the shear stress due to the shaft torque, T (Nm) is the shaft torque, σ_t (N/m^2) is the shaft shear stress due to the tensile stress due to the bending moment, M (Nm) is the shaft bending moment, σ_{sw} (N/m^2) is the allowable working shear stress of the shaft material, and σ_{tw} (N/m^2) is the allowable working tensile stress of the shaft material.

The shaft torque and the shaft bending moment correspond to either the yield load or the ultimate load. The yield load is determined by multiplying the design limit load by the factor 1.1, and the ultimate load is determined by multiplying the design limit load by the factor 1.5.

The maximum tensile stress σ_{tmax} (N/m^2) in an impeller due to the centrifugal forces is the tangential stress at the edge of the shaft hole. It can be determined by means of the following equation [22]:

$$\sigma_{tmax} = \frac{\rho\, u_{2max}^2 (3+\nu)}{4}\left[1 + \frac{1-\nu}{3+\nu}\left(\frac{d_s}{d_2}\right)^2\right]K_s$$

where ρ (kg/m^3) and ν are respectively density and Poisson's ratio of the material of which the impeller is made, u_{2max} (m/s) is the maximum allowable peripheral speed of the impeller, and K_s is a design factor whose value is determined experimentally. Huzel and Huang [22] indicate 0.4 and 1.0 as the range on which the value of K_s varies, according to the shape of the impeller. In case of pumps for rocket engines, the maximum allowable peripheral speed u_{2max} of the impeller can be computed by multiplying the design value u_2 (m/s) by the factor 1.25. Of course, σ_{tmax} must be smaller than the maximum allowable working stress of the impeller material.

The widths b_1 (m) and b_2 (m) of the impeller at respectively entrance and exit can be computed as follows [22]:

$$b_1 = \frac{q_{imp}}{\pi\, d_1\, c_{m1}\, \varepsilon_1}$$

$$b_2 = \frac{q_{imp}}{\pi\, d_2\, c_{m2}\, \varepsilon_2}$$

where q_{imp} (m^3/s) is the impeller volume flow rate at the rated design point, d_1 (m) and d_2 (m) are the impeller diameters at respectively entrance and exit, c_{m1} (m/s) and c_{m2} (m/s) are the meridian components of the absolute flow velocity at respectively entrance and exit, and ε_1 ($0.75 \le \varepsilon_1 \le 0.90$) and ε_2 ($0.85 \le \varepsilon_2 \le 0.95$) are the contraction factors at respectively entrance and exit. These contraction factors take account of the reduction of the effective flow area due to the thickness of the vanes and to local circulatory flows.

According to Huzel and Huang [22], the number Z_2 of backward-curved vanes at the impeller outlet is to be chosen between 5 and 12, by applying the following empirical rule:

$$Z_2 = \frac{60}{\pi}\beta_2$$

where β_2 (rad) is the vane angle at the impeller outlet.

According to Furst and Keller [3], impellers having from 6 to 48 blades have been tested, as shown in the following table (adapted from [3]), which indicates the discharge blade angle β_2 (radians), the number Z_2 of blades at the impeller outlet, the ratio Z_2/β_2, the tip diameter d_2 (m) of the blades, the tip width b_2 (m) of the blades, the specific speed N of the pump at the best efficiency point, and the best pump efficiency η_P for oxidiser or fuel pumps mounted on several liquid-propellant rocket engines. The table given below has been modified with respect to the original, in order for d_2, b_2, and β_2 to be expressed in metric units, and N to be truly dimensionless, in accordance with the definition given in Sect. 4.2.

Pump identification[a]	Discharge blade angle β_2, rad	Number of blades Z_2	Z_2/β_2	Tip diameter, m	Tip width, m	Best efficiency specific speed	Best pump efficiency
Titan							
87-5 fuel	0.611	12	19.6	0.273	0.0188	0.413	0.72
87-5 oxidizer	0.489	9	18.4	0.239	0.0254	0.239	0.75
91-5 fuel	0.489	8	16.4	0.125	0.0112	0.640	0.74
91-5 fuel (exptl)	0.489	9	18.4	0.121	0.0122	0.581	0.68
91-5 oxidizer	0.611	12	19.6	0.223	0.0135	0.346	0.62
87-3 fuel	0.393	8	20.4	0.279	0.0170	0.358	0.55
87-3 oxidizer	0.393	8	20.4	0.251	0.0239	0.603	0.65
91-3 fuel	0.393	6	15.3	0.110	0.0102	0.682	0.60
91-3 oxidizer	0.393	8	20.4	0.112	0.0163	0.427	0.65
Titan IIA fuel	0.489	9	18.4	0.177	0.0122	0.527	0.68
NERVA							
Mark III Mod III	1.57	18	11.5	0.312	0.0124	0.334	0.65
Mark III Mod IV	1.57	48	30.6	0.312	0.0124	0.351	0.70
Mark III Mod IV	1.57	24	15.3	0.312	0.0124	0.366	0.70
M-1							
M-I oxygen	0.611	12	19.6	0.272	0.0206	0.411	0.66
Atlas and H-1							
Mark 3 fuel	0.436	10	22.9	0.362	0.0216	0.278	0.72
F-I							
Mark 10 oxidizer	0.436	6	13.8	0.495	0.0686	0.783	0.74
Mark 10 fuel	0.436	6	13.8	0.594	0.0432	0.439	0.76
J-2							
Mark 15-0 oxygen	0.436	6	13.8	0.259	0.0188	0.585	0.81
X-8							
Mark 19 hydrogen	1.57	24	15.3	0.279	0.0109	0.245	0.67
J-2S							

(continued)

(continued)

Pump identification[a]	Discharge blade angle β_2, rad	Number of blades Z_2	Z_2/β_2	Tip diameter, m	Tip width, m	Best efficiency specific speed	Best pump efficiency
Mark 29-F hydrogen	1.05	24	22.9	0.292	0.0135	0.366	0.76

[a]87-5 = LR-87-AJ-5 engine system
91-5 = LR-91-AJ-J engine system
87-3 = LR-87-AJ-3 engine system
91-3 = LR-91-AJ-3 engine system

From the point of view of manufacturing, impellers with a low number of blades are desirable. Impellers with blades curved backward ($\beta_2 < \pi/2$ rad) and low head coefficient $\psi = g_0 \Delta H / u_2^2$ tend to have a wider and more stable operating range than impellers with radial blades ($\beta_2 = \pi/2$ rad). However, the range of high efficiency for pumps with impellers of low head coefficient is smaller than the corresponding range for pumps with impellers of high head coefficient [3].

Impellers are fabricated by casting, machining, and diffusion bonding. Impellers having highest tip speeds (much above 300 m/s) are used for pumping liquid hydrogen. These impellers are of the open-face type, and are machined from high-strength forged materials, usually from the Ti-5Al-2.5Sn titanium alloy, extra-low-interstitial (ELI) grade. A figure illustrating one of such impellers for liquid-hydrogen pumps has been given in Sect. 4.2. Shrouded impellers for dense fuels or oxidisers are cast, since the properties of cast materials are adequate for the required tip speeds of less than 300 m/s. Open-faced impellers for oxidiser pumps may be cast or machined. The following figure, due to the courtesy of NASA [5], illustrates unshrouded machined impellers for low-density propellants. These impellers have been used in the Titan II rocket.

Shrouded impellers for liquid-hydrogen pumps have been machined from forgings. They have also been fabricated by generating all the components separately from a titanium alloy, diffusion bonding them, and finishing internal passages by chemical milling [3].

Impeller materials which have been used successfully with liquid propellants for rockets are shown in the following table, due to the courtesy of NASA [3]. FLOX is a mixture of liquid fluorine (LF_2) with liquid hydrogen (LH_2). IRFNA stands for Inhibited Red Fuming Nitric Acid (HNO_3). UDMH stands for Unsymmetrical Di-Methyl Hydrazine ($H_2NN(CH_3)_2$). Inconel 718 is a trade name of Special Metals for a precipitation-hardening nickel-chromium-iron alloy. "K" Monel indicates the MONEL® alloy K-500, which is a trade name of Special Metals for a precipitation-hardenable alloy containing nickel, copper, and aluminium.

The materials indicated in the following table are chemically compatible with the pumped fluid, have satisfactory strength and ductility at the operating temperatures, and can be used to fabricate impellers with existing technology.

Impeller material	Pumped fluid						
	LH_2,CH_4	IRFNA, N_2O_4	LOX	LF_2	FLOX	RP-1	N_2H_4, UDMH, or 50/50 mixture
Aluminium							
A356 (cast)		X	X				X
A357 (cast)			X		X	X	
20I4-T6	X						
6061-T6			X			X	X
7075-T73	X	X	X			X	X
7079			X				
Steel							
AM 350		X					X
304L (cast)		X	X				
304L		X					X
310			X			X	
347 (cast)			X				
Inconel 718 (cast)			X	X	X		
"K" Monel			X				
Ti-5A1-2.5Sn	X					X	

Note X indicates that the material was used successfully with the fluid shown: absence of X means either that no data on the use are available or that the material was incompatible with the fluid. Materials not shown as cast were forged

The following section of the present section is meant to give criteria for designing an inducer, which is the axial inlet portion of a pump rotor and has the function of raising the inlet head of the pumped fluid by an amount sufficient to preclude cavitation in the following stages. An inducer may be an integral part of a pump rotor, but may also be matched separately to the pump shaft upstream of the pump impeller. The principal objective in the design of an inducer is the attainment of high suction performance while maintaining structural integrity under all operating conditions [4]. An inducer is designed to operate at low inlet pressure under conditions of incipient cavitation, which fact permits a low pressure in the propellant tank. The inducer provides a sufficient head rise to avoid cavitation in the following stages of the pump [34]. The head rise due to the inducer ranges from 5 to 20% of the head rise of the pump. The required head rise ΔH_{ind} (m) due to the inducer at the design point for a given pump is expressed by the following equation

$$\Delta H_{ind} = (NPSH_R)_{imp} - (NPSH_R)_{ind}$$

where $(NPSH_R)_{imp}$ (m) is the required net positive suction head due to the impeller, and $(NPSH_R)_{ind}$ (m) is the required net positive suction head due to the inducer. The dimensionless parameters specific speed N and suction specific speed S have been defined in Sect. 4.2 as follows

$$N = \frac{\omega \, q^{\frac{1}{2}}}{(g_0 \, \Delta H)^{\frac{3}{4}}}$$

$$S = \frac{\omega \, q^{\frac{1}{2}}}{(g_0 \, NPSH_R)^{\frac{3}{4}}}$$

After solving these equations for respectively ΔH and $NPSH_R$, and then substituting into the preceding equation $\Delta H_{ind} = (NPSH_R)_{imp} - (NPSH_R)_{ind}$, there results

$$\frac{1}{g_0} \left(\frac{\omega \, q^{\frac{1}{2}}}{N_{ind}} \right)^{\frac{4}{3}} = \frac{1}{g_0} \left(\frac{\omega \, q^{\frac{1}{2}}}{S_{imp}} \right)^{\frac{4}{3}} - \frac{1}{g_0} \left(\frac{\omega \, q^{\frac{1}{2}}}{S_{ind}} \right)^{\frac{4}{3}}$$

where g_0 (m/s^2) is the gravitational acceleration of the Earth at sea level, ω (rad/s) is the angular velocity of the shaft (the same for the inducer and for the impeller), q (m^3/s) is the rated volume flow rate of the pump, N_{ind} is the specific speed of the inducer, S_{imp} is the suction specific speed of the impeller, and S_{ind} is the suction specific speed of the inducer.

The principal increase in static pressure in an inducer occurs at the leading edge of its vanes, because the velocity of the flow decreases there with respect to the vanes, and also because the angle of attack between the inlet flow and the vanes is small. The capability of an inducer to avoid cavitation depends on its flow coefficient $\varphi_0 = c_{m0}/u_0$, where c_{m0} (m/s) is the meridian velocity of the fluid at the inducer inlet, and u_0 (m/s) is the tangential velocity of the blade at the inlet tip. The value of the flow coefficient φ_0 must be low for a pump of high specific speed N, in order for the pump to have a high value of suction specific speed S. To this end, the values of the angles θ_t and θ_h between the vanes and a plane perpendicular to the axis of rotation must

be small, as shown in the following figure, re-drawn from [22], which illustrates an
inducer of low head rise, having three vanes and a non-tapered hub.

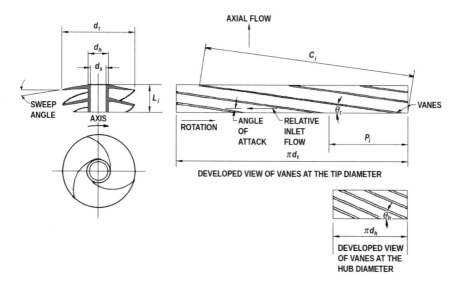

In general terms, the angle θ of an inducer vane varies in the radial direction of
the vane according to the following equation

$$c = d \tan \theta = d_t \tan \theta_t = d_h \tan \theta_h$$

where c is a constant. The angle θ also varies axially, as a result of the axial or
meridian component c_m of the absolute velocity of the flow.

The flow coefficient φ_{ind} at inducer inlet, the inducer diameter ratio $r_d = d_h/d_t$ of
the hub diameter d_h to the tip diameter d_t, and the suction specific speed S_{ind} of the
inducer are related by the following equation given in [22]:

$$\frac{S_{ind}}{\left(1 - r_d^2\right)^{\frac{1}{2}}} = \frac{2.9803\left(1 - 2\varphi_{ind}^2\right)^{\frac{3}{4}}}{\varphi_{ind}}$$

The three vanes of the inducer shown in the preceding figure are placed at an equal tip distance P_i (m). This common tip distance, also known as pitch, can be expressed as follows

$$P_i = \frac{\pi d_t}{Z_0}$$

where d_t (m) is the tip diameter of the inducer, and Z_0 is the number of vanes of the inducer. With reference to the preceding figure, let C_i (m) be the length of the chord between the tips of the vanes. The ratio S_v of the chord length C_i to the pitch distance P_i (m) defined above

$$S_v = \frac{C_i}{P_i}$$

is called solidity of a vane at the tip. The ratio L_i/d_t of the inducer length L_i (m) to the inducer tip diameter d_t (m) is another parameter which describes the proportions of an inducer.

Sometimes, it is necessary, for better performance, to vary the length of either the hub diameter d_h, or the blade tip diameter d_t, or both. The following figure, due to the courtesy of NASA [4], shows (left) an inducer having a constant hub diameter and a constant blade tip diameter, (centre) an inducer having a varying hub diameter and a constant blade tip diameter, and (right) an inducer having a varying hub diameter and a varying blade tip diameter.

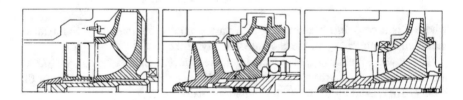

Due to structural reasons, the inducer vanes are sometimes designed to lean forward instead of being perpendicular to the axis of rotation. The angle between a canted vane and a plane perpendicular to the rotation axis is called the sweep angle. Forward-leaning vanes are shown in two impellers of the preceding figure. These vanes are also shown in the following figure, re-drawn from [22], where

$$d_h = \frac{d_{0h} + d_{1h}}{2}$$

$$d_t = \frac{d_{0t} + d_{1t}}{2}$$

are the mean diameters of respectively the hub and the tip of the blades, and the subscripts 0 and 1 refer to respectively the inlet and the outlet of the inducer.

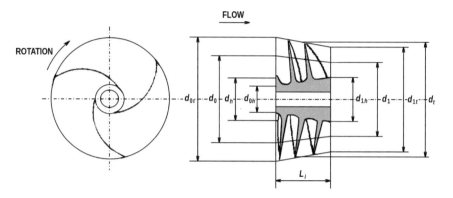

The following table, adapted from [22], gives typical design values of parameters or variables for inducers.

Parameter or variable	Typical design values
Specific speed, N_{ind}	2.2–4.4
Suction specific speed, S_{ind}	7.3–18.3
Head coefficient, ψ_{ind}	0.06–0.15
Inlet flow coefficient, φ_{ind}	0.06–0.20
Inlet vane angle, θ (radians)	0.1396–0.2792, in a plane normal to the axis
Angle of attack, i (radians)	0.05224–0.1396
Diameter ratio, $r_d = d_h/d_t$	0.2–0.5
Vane solidity, $S_v = C_i/P_i$	1.5–3 at the tip
Number of vanes, Z_0	3–5
Hub contour	Cylindrical—$\pi/12$ rad tapered
Tip contour	Cylindrical—$\pi/12$ rad tapered
Vane loading	Leading edge loading, channel lead
Leading edge	Swept forward, radial, swept backward
Sweep angle	Normal to shaft—$\pi/12$ rad forward
Vane thickness	7–30% of the chord length C_i
Tip clearance (tip—casing)	0.5–1% of the inducer outside diameter
Length to tip diameter ratio, L_i/d_t	0.3–0.6

The following figure shows the velocity diagrams at the inlet and at the outlet of the inducer.

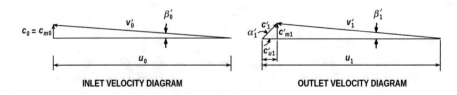

INLET VELOCITY DIAGRAM   **OUTLET VELOCITY DIAGRAM**

With reference to the preceding figures, the following equations hold for cylindrical or conical hubs and for non-tapered or tapered inducer vanes, as specified below.

(a) For cylindrical hubs and non-tapered inducer vanes (assuming $c'_{u0} = 0$):

$$d_0^2 = d_1^2 = \frac{d_h^2 + d_t^2}{2}$$

$$c'_0 = c_{m0} = c_{m1} = \frac{4q_{ind}}{\pi \left(d_t^2 - d_h^2\right)}$$

$$u_0 = u_1 = \frac{1}{2}\omega d_0 = \frac{1}{2}\omega d_1$$

(b) For conical hubs and tapered inducer vanes:

$$d_1^2 = \frac{d_{1h}^2 + d_{1t}^2}{2}$$

$$d_1^2 = \frac{d_{1h}^2 + d_{1t}^2}{2}$$

$$d_t = \frac{d_{0t} + d_{1t}}{2}$$

$$d_h = \frac{d_{0h} + d_{1h}}{2}$$

$$c'_0 = c_{m0} = \frac{4q_{ind}}{\pi \left(d_{0t}^2 - d_{0h}^2\right)} \quad \text{(no pre-rotation)}$$

$$c_{m1} = \frac{4q_{ind}}{\pi \left(d_{1t}^2 - d_{1h}^2\right)}$$

$$u_0 = \frac{1}{2}\omega d_0$$

$$u_1 = \frac{1}{2}\omega d_1$$

(c) For all inducers:

$$q_{ind} = q + q_{ee} + \frac{1}{2}q_e$$

$$r_d = \frac{d_h}{d_t}$$

$$u_t = \frac{1}{2}\omega d_t$$

$$u_{0t} = \frac{1}{2}\omega d_{0t}$$

$$\Delta H_{ind} = \frac{u_t^2 \psi_{ind}}{g_0} = \frac{u_1 c'_{u1}}{g_0}$$

$$\Delta H_{indt} = \frac{\Delta H_{ind}}{\eta_{ind}} = \frac{u_1 c_{u1}}{g_0}$$

$$\varphi_{ind} = \frac{c_{m0}}{u_{0t}}$$

In the preceding equations, q (m^3/s) is the rated volume flow rate of the pump, q_{ind} (m^3/s) is the required volume flow rate of the inducer at the rated design point, q_e (m^3/s) is the volume flow rate due to the leakage loss of the impeller at the rated design point, q_{ee} (m^3/s) is the volume flow rate due to the leakage loss of the inducer through the tip clearance (usually q_{ee} ranges from 0.02 q to 0.06 q), ΔH_{ind} (m) is the required head rise of the inducer at rated conditions, η_{ind} is the efficiency of the inducer, d_t (m) is the mean tip diameter of the inducer, d_{0t} (m) is the tip diameter of the inducer at the inlet, d_{1t} (m) is the tip diameter of the inducer at the outlet, d_h (m) is the mean diameter of the hub, d_{0h} (m) is the diameter of the hub at the inlet, d_{1h} (m) is the diameter of the hub at the outlet, d_0 (m) is the mean effective diameter of the inducer at the inlet, d_1 (m) is the mean effective diameter of the inducer at the outlet, u_0 (m/s) is the peripheral velocity of the inducer at the mean effective inlet diameter (d_0), u_1 (m/s) is the peripheral velocity of the inducer at the mean effective outlet diameter (d_1), u_t (m/s) is the mean tip velocity of the inducer, u_{0t} (m/s) is the tip velocity of the inducer at the inlet, u_{1t} (m/s) is the tip velocity of the inducer at the outlet, v'_0 (m/s) is the velocity of the flow at the inlet relative to the inducer, v'_1 (m/s) is the velocity of the flow at the outlet relative to the inducer, c'_0 (m/s) is the absolute velocity of the flow at the inlet, c'_{u0} (m/s) is the tangential component of the absolute velocity of the flow at the inlet, c_{m0} (m/s) is the meridian or axial component of the absolute velocity of the flow at the inlet, c_1 (m/s) is the ideal absolute velocity of the flow at the outlet, c'_1 (m/s) is the absolute velocity of the flow at the outlet, c'_{u1} (m/s) is the tangential component of the absolute velocity of the flow at the outlet, c_{m1} (m/s) is the meridian or axial component of the absolute velocity of the flow at the outlet, $r_d = d_h/d_t$ is the ratio of the diameter of the hub to the diameter of the tip, ψ_{ind} is the head coefficient of the inducer ($0.06 \leq \psi_{ind} \leq 0.15$), and φ_{ind} is the flow coefficient of the inducer at the inlet ($0.06 \leq \varphi_{ind} \leq 0.20$).

The materials of which inducers are made are usually chosen from alloys of stainless steel (for example, 304 or 347), titanium (for example, Ti-5Al-2.5Sn ELI or Ti-6Al-4 V), and aluminium (for example, 7075-T73). The respective densities of these materials are approximately 8000, 4480, and 2810 kg/m^3, and their strengths vary about in the same order as their densities. The choice of a specific material from those indicated above depends on their strength-to-weight ratio, chemical reactivity, cavitation erosion, and the need for special properties, such as ductility, notch toughness, etc. [4]. A cast aluminium alloy is the proper choice for an inducer subject to low stress when low cost is important. This is because aluminium is easy to forge and machine, but not apt to be welded or brazed. Titanium alloy parts are more expensive to machine than comparable parts of aluminium alloys, but less expensive than alloys such as Inconel® 718 or René® 41. Titanium alloys can be welded by gas-tungsten arc or electron-beam processes. Proper welding procedures are to be observed for titanium alloys [4].

An increase in the suction performance of a pump may be obtained by pre-whirl of the inlet flow. For this purpose, the inlet flow is mixed with high-velocity fluid taken from the pump discharge. This flow is injected through a ring of orifices in an axial direction upstream of the inducer. Pre-whirl introduced in this manner has improved flow distribution and reduced flow instabilities which may occur when a pump is throttled [4].

Backflow occurs in an inducer at low flow (about 90% or less of the design value) as a result of local head breakdown in the tip region of the vanes. In order to reduce backflow or control its effects, a backflow deflector can be incorporated in the inlet line, as shown in the following figure, due to the courtesy of NASA [4].

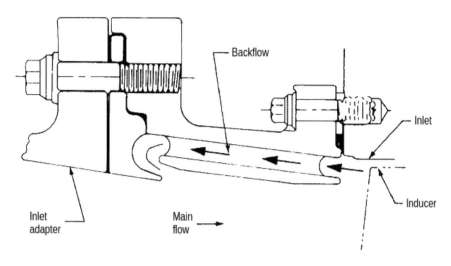

Below nominal flow, (between 20 and 90% of the design flow) where back-flow becomes important, the deflector results in increased head, reduced critical NPSH, and lower amplitudes of low-frequency oscillations. Above nominal flow, the deflector has a detrimental effect [4]. Further information on backflow control can be found in [35].

As a practical application of the concepts discussed above, it is required to design an inducer and an impeller for a centrifugal-flow pump to be used for the oxidiser (liquid oxygen) in the engine of the first stage of a rocket. The design data are indicated below. The required head rise of the pump is $\Delta H = 893.1$ m. The required volume flow rate of the pump is $q = 0.7836$ m^3/s. The angular velocity of the pump shaft is $\omega = 733.0$ rad/s. The specific speed of the pump is $N = 0.7241$. The required net positive suction head of the inducer is $(NPSH_R)_{ind} = 17.68$ m. The suction specific speed of the pump is $S = 13.61$. The head coefficient of the pump is $\psi = 0.46$. The inducer has a varying hub diameter and a varying blade tip diameter, as has been shown for one of the three schemes illustrated above. The head coefficient of the inducer is $\psi_{ind} = 0.06$. In the inducer, the ratio of the diameter of the hub to the diameter of the tip is $r_d = d_h/d_t = 0.3$, and the ratio of the length to the diameter of the tip is $L_i/d_t = 0.4$. The maximum angle of attack at the tip of a vane of the inducer at the inlet is $i = 0.06981$ rad. The semi-aperture angle of the inducer at the tip of its blades is 0.1222 rad. The semi-aperture angle of the hub of the inducer is 0.2443 rad. The solidity of the vanes of the impeller at the mean tip diameter tip is $S_v = 2.2$. The volume flow rate due to the leakage loss of the inducer through the clearance at the tip is $q_{ee} = 0.032\, q = 0.032 \times 0.7836 = 0.02508$ m^3/s. The impeller chosen for this pump is of the radial-flow type with mixed-flow vanes at the impeller entrance, as shown in the following figure, adapted from [12], which illustrates the impeller of the oxidiser pump used for the Atlas MA-5 booster.

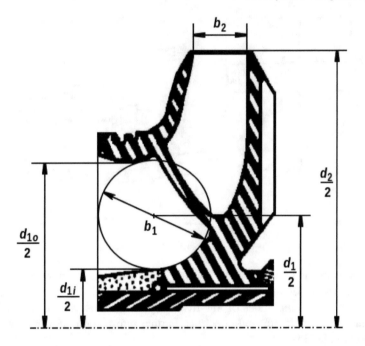

The suction specific speed of the impeller is $S_{imp} = 4.023$. The vanes of the impeller have an angle $\beta_2 = 0.4189$ rad at the outlet. The contraction factors of the impeller are $\varepsilon_1 = 0.82$ at the inlet and $\varepsilon_2 = 0.88$ at the outlet. The impeller vane coefficient is $e_y = c'_{u2}/c_{u2} = 0.74$. The volume flow rate due to the leakage loss of the impeller is $q_e = 0.035\,q = 0.035 \times 0.7836 = 0.02743$ m³/s. The head loss in the volute of the pump is $\Delta H_e = 0.19\,\Delta H = 0.19 \times 893.1 = 169.7$ m.

We calculate first the quantities concerning the inducer and then those concerning the impeller. The required net positive suction head of the impeller can be computed by using the following equation which defines the suction specific speed

$$S = \frac{\omega\, q^{\frac{1}{2}}}{(g_0\, NPSH_R)^{\frac{3}{4}}}$$

Since $g_0 = 9.807$ m/s², $\omega = 733.0$ rad/s, $q = 0.7836$ m³/s, and $S_{imp} = 4.023$, then the required net positive suction head of the impeller is

$$(NPSH_R)_{imp} = \frac{1}{g_0}\left(\frac{\omega\, q^{\frac{1}{2}}}{S_{imp}}\right)^{\frac{4}{3}} = \frac{1}{9.807} \times \left(\frac{733.0 \times 0.7836^{\frac{1}{2}}}{4.023}\right)^{\frac{4}{3}} = 89.52 \text{ m}$$

By substituting this value and $(NPSH_R)_{ind} = 17.68$ m into the following equation

$$\Delta H_{ind} = (NPSH_R)_{imp} - (NPSH_R)_{ind}$$

we compute the required head rise of the inducer as follows

$$\Delta H_{ind} = 89.52 - 17.68 = 71.84 \text{ m}$$

By substituting this value, $\psi_{ind} = 0.06$, and $g_0 = 9.807$ m/s^2 into the following equation

$$\Delta H_{ind} = \frac{u_t^2 \psi_{ind}}{g_0}$$

and solving for u_t, we compute the mean velocity of the vanes of the inducer at the tip as follows

$$u_t = \left(\frac{g_0 \Delta H_{ind}}{\psi_{ind}}\right)^{\frac{1}{2}} = \left(\frac{9.807 \times 71.84}{0.06}\right)^{\frac{1}{2}} = 108.4 \text{ m/s}$$

By substituting this value and $\omega = 733.0$ rad/s into the following equation, the mean diameter of the vanes of the inducer is

$$d_t = \frac{2u_t}{\omega} = \frac{2 \times 108.4}{733.0} = 0.2958 \text{ m}$$

Since the ratio of the inducer length to the diameter of its vanes is $L_i/d_t = 0.4$, then the inducer length is

$$L_i = 0.4 d_t = 0.4 \times 0.2958 = 0.1183 \text{ m}$$

Since the semi-aperture angle of the inducer at the tip of its blades is 0.1222 rad, then the tip diameter of the blades of the inducer at the inlet is

$$d_{0t} = d_t + 2\frac{L_t}{2}\tan(0.1222) = 0.2958 + 0.1183 \times 0.1228 = 0.3103 \text{ m}$$

Likewise, the tip diameter of the blades of the inducer at the outlet is

$$d_{1t} = d_t - 2\frac{L_t}{2}\tan(0.1222) = 0.2958 - 0.1183 \times 0.1228 = 0.2813 \text{ m}$$

The mean diameter of the hub of the inducer results from

$$d_h = d_t r_d = 0.2958 \times 0.3 = 0.08874 \text{ m}$$

Since the semi-aperture angle of the hub of the inducer is 0.2443 rad, then the diameter of this hub at the inlet is

$$d_{0h} = d_h - 2\frac{L_t}{2}\tan(0.2443) = 0.08874 - 0.1183 \times 0.2493 = 0.05925 \text{ m}$$

Likewise, the diameter of the same hub at the outlet is

$$d_{1h} = d_h + 2\frac{L_t}{2}\tan(0.2443) = 0.08874 + 0.1183 \times 0.2493 = 0.1182 \text{ m}$$

By substituting $q = 0.7836 \text{ m}^3/\text{s}$, $q_{ee} = 0.02508 \text{ m}^3/\text{s}$, and $q_e = 0.02743 \text{ m}^3/\text{s}$ into the following equation

$$q_{ind} = q + q_{ee} + \frac{1}{2}q_e$$

the required volume flow rate of the inducer at the rated design point is

$$q_{ind} = 0.7836 + 0.02508 + 0.5 \times 0.02743 = 0.8224 \text{ m}^3/\text{s}$$

Assuming $c_{u0} = 0$ and $\alpha'_0 = 0$, the absolute velocity c'_0 of the flow at the inlet of the inducer can be computed by means of the following equation

$$c'_0 = c_{m0} = \frac{4q_{ind}}{\pi\left(d_{0t}^2 - d_{0h}^2\right)} = \frac{4 \times 0.8224}{3.14 \times \left(0.3103^2 - 0.05925^2\right)} = 11.29 \text{ m/s}$$

The meridian component c_{m1} of the absolute velocity of the flow at the outlet of the inducer can be computed as follows

$$c_{m1} = \frac{4q_{ind}}{\pi\left(d_{1t}^2 - d_{1h}^2\right)} = \frac{4 \times 0.8224}{3.14 \times \left(0.2813^2 - 0.1182^2\right)} = 16.07 \text{ m/s}$$

The mean effective diameter d_0 at the inlet of the inducer can be computed as follows

$$d_0^2 = \frac{d_{0h}^2 + d_{0t}^2}{2}$$

By substituting $d_{0h} = 0.05925 \text{ m}$ and $d_{0t} = 0.3103 \text{ m}$ into the preceding equation, there results

$$d_0 = \left(\frac{0.05925^2 + 0.3103^2}{2}\right)^{\frac{1}{2}} = 0.2234 \text{ m}$$

The peripheral velocity u_0 of the inducer vanes at $d_0 = 0.2234 \text{ m}$ can be computed as follows

$$u_0 = \frac{1}{2}\omega d_0 = 0.5 \times 733.0 \times 0.2234 = 81.88 \text{ m/s}$$

The mean effective diameter d_1 at the outlet of the inducer can be computed as follows

$$d_1^2 = \frac{d_{1h}^2 + d_{1t}^2}{2}$$

By substituting $d_{1h} = 0.1182$ m and $d_{1t} = 0.2813$ m into the preceding equation, there results

$$d_1 = \left(\frac{0.1182^2 + 0.2813^2}{2}\right)^{\frac{1}{2}} = 0.2158 \text{ m}$$

The peripheral velocity u_1 of the inducer vanes at $d_1 = 0.2158$ m can be computed as follows

$$u_1 = \frac{1}{2}\omega d_1 = 0.5 \times 733.0 \times 0.2158 = 79.09 \text{ m/s}$$

By solving the following equation

$$\Delta H_{ind} = \frac{u_t^2 \psi_{ind}}{g_0} = \frac{u_1 c'_{u1}}{g_0}$$

for c'_{u1}, the tangential component of the absolute velocity of the flow at the outlet of the inducer is

$$c'_{u1} = \frac{g_0 \Delta H_{ind}}{u_1} = \frac{9.807 \times 71.84}{79.09} = 8.908 \text{ m/s}$$

By inspection of the velocity diagrams at the inlet and at the outlet of the inducer, it is possible to compute geometrically the following quantities.

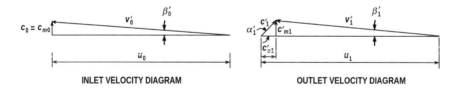

INLET VELOCITY DIAGRAM   OUTLET VELOCITY DIAGRAM

The inlet velocity of the flow relative to the inducer results from

$$v'_0 = \left(c_{m0}^2 + u_0^2\right)^{\frac{1}{2}} = \left(11.29^2 + 81.88^2\right)^{\frac{1}{2}} = 82.65 \text{ m/s}$$

The inlet flow angle relative to the inducer results from

$$\sin \beta'_0 = \frac{c_{m0}}{v'_0}$$

Hence

$$\beta_0' = \arcsin\left(\frac{c_{m0}}{v_0'}\right) = \arcsin\left(\frac{11.29}{82.65}\right) = 0.1370 \text{ rad}$$

The absolute velocity of the flow at the outlet of the inducer results from

$$c_0' = \left(c_{m1}^2 + c_{u1}'^2\right)^{\frac{1}{2}} = \left(16.07^2 + 8.908^2\right)^{\frac{1}{2}} = 18.37 \text{ m/s}$$

The outlet flow angle of the inducer results from

$$\tan \alpha_1' = \frac{c_{m1}}{c_{u1}'}$$

Hence

$$\alpha_1' = \arctan\left(\frac{c_{m1}}{c_{u1}'}\right) = \arctan\left(\frac{16.07}{8.908}\right) = 1.081 \text{ rad}$$

The outlet velocity of the flow relative to the inducer results from

$$v_1' = \left[c_{m1}^2 + \left(u_1 - c_{u1}'\right)^2\right]^{\frac{1}{2}} = \left[16.07^2 + (79.09 - 8.908)^2\right]^{\frac{1}{2}} = 72.00 \text{ m/s}$$

The inducer relative outlet flow angle results from

$$\tan \beta_1' = \frac{c_{m1}}{u_1 - c_{u1}'}$$

Hence

$$\beta_1' = \arctan\left(\frac{c_{m1}}{u_1 - c_{u1}'}\right) = \arctan\left(\frac{16.07}{79.09 - 8.908}\right) = 0.2336 \text{ rad}$$

The tip velocity of the inducer at the inlet results from

$$u_{0t} = \frac{1}{2}\omega d_{0t} = 0.5 \times 733.0 \times 0.3103 = 113.7 \text{ m/s}$$

The relative flow angle at the inducer inlet tip results from

$$\tan \beta_{0t}' = \frac{c_{m0}}{u_{0t}}$$

Hence

$$\beta'_{0t} = \arctan\left(\frac{c_{m0}}{u_{0t}}\right) = \arctan\left(\frac{11.29}{113.7}\right) = 0.09897 \text{ rad}$$

By choosing a vane angle $\theta_{0t} = 0.1571$ rad at the inducer inlet tip, the angle of attack at the inlet tip is

$$\theta_{0t} - \beta'_{0t} = 0.1571 \text{ rad} - 0.09897 \text{ rad} = 0.05813 \text{ rad}$$

which is less than the maximum value of 0.06981 rad, as desired.

As has been shown above, the vane angle θ of an inducer varies in the radial direction of the vane according to the following equation

$$c = d \tan \theta = d_t \tan \theta_t = d_h \tan \theta_h$$

where c is a constant. Therefore, the vane angle θ_0 at the inducer inlet mean effective diameter d_0 is such that

$$\tan \theta_0 = \frac{d_{0t}}{d_0} \tan \theta_{0t} = \frac{0.3103}{0.2234} \tan 0.1571 = 0.2200$$

Hence

$$\theta_0 = \arctan(0.2200) = 0.2166 \text{ rad}$$

For the same reason, the vane angle θ_{0h} at the inducer inlet hub diameter d_{0h} is such that

$$\tan \theta_{0h} = \frac{d_{0t}}{d_{0h}} \tan \theta_{0t} = \frac{0.3103}{0.05925} \tan 0.1571 = 0.8295$$

Hence

$$\theta_{0h} = \arctan(0.8295) = 0.6925 \text{ rad}$$

Remembering the equation

$$\varphi_{ind} = \frac{c_{m0}}{u_{0t}}$$

and substituting $c_{m0} = 11.29$ m/s and $u_{0t} = 113.7$ m/s, the flow coefficient of the inducer at the inlet is

$$\varphi_{ind} = \frac{11.29}{113.7} = 0.09930$$

The suction specific speed of the inducer S_{ind} is computed by means of the following equation

$$\frac{S_{ind}}{\left(1 - r_d^2\right)^{\frac{1}{2}}} = \frac{2.9803\left(1 - 2\varphi_{ind}^2\right)^{\frac{3}{4}}}{\varphi_{ind}}$$

After substituting $\varphi_{ind} = 0.09930$, $r_d = d_h/d_t = 0.3$, and solving for S_{ind}, we find

$$S_{ind} = \left[\frac{2.9803 \times \left(1 - 2 \times 0.09930^2\right)^{\frac{3}{4}}}{0.09930}\right] \times \left(1 - 0.3^2\right)^{\frac{1}{2}} = 28.21$$

which is greater than the suction specific speed of the pump ($S = 13.61$).

We choose a value of 0.2531 rad for the angle θ_1 of the inducer vane at the mean effective diameter $d_1 = 0.2158$ m. By using this value, the difference between θ_1 and the inducer relative outlet flow angle $\beta_1' = 0.2336$ rad is

$$\theta_1 - \beta_1' = 0.2531 - 0.2336 = 0.0195 \text{ rad}$$

which allows for the effect due to local circulatory flow at the boundary.

The vane angle θ_{1t} at the diameter $d_{1t} = 0.2813$ m of the inducer outlet tip can be computed by using again the following equation

$$c = d \tan\theta = d_t \tan\theta_t = d_h \tan\theta_h$$

where c is a constant. Therefore, we have

$$\tan\theta_{1t} = \frac{d_1}{d_{1t}} \tan\theta_1 = \frac{0.2158}{0.2813} \tan 0.2531 = 0.1984$$

Hence

$$\theta_{1t} = \arctan(0.1984) = 0.1959 \text{ rad}$$

Likewise, the vane angle θ_{1h} at the diameter $d_{1h} = 0.1182$ m of the inducer outlet hub can be computed as follows

$$\tan\theta_{1h} = \frac{d_1}{d_{1h}} \tan\theta_1 = \frac{0.2158}{0.1182} \tan 0.2531 = 0.4722$$

Hence

$$\theta_{1h} = \arctan(0.4722) = 0.4412 \text{ rad}$$

We choose an inducer with three vanes ($Z_0 = 3$). Therefore, the pitch (that is, the equal tip distance) of the vanes results from the following equation

$$P_i = \frac{\pi d_t}{Z_0} = \frac{3.1416 \times 0.2958}{3} = 0.3076 \text{ m}$$

The length C_i of the chord between the tips of the vanes can be computed as follows

$$C_i = \frac{L_i}{\sin\left(\frac{\theta_{0t} + \theta_{1t}}{2}\right)} = \frac{0.1183}{\sin\left(\frac{0.1571 + 0.1959}{2}\right)} = 0.6737 \text{ m}$$

The solidity S_v (that is, the ratio of the chord length C_i to the pitch distance P_i) results from

$$S_v = \frac{C_i}{P_i} = \frac{0.6737}{0.3076} = 2.190$$

A summary of the results obtained for the inducer is given below.
The required head rise is $\Delta H_{ind} = 71.84$ m.
The required volume flow rate is $q_{ind} = 0.8224$ m^3/s.
The inlet velocity diagram, at the mean effective diameter $d_0 = 0.2234$ m, can be drawn by using the following quantities:
$\alpha'_0 = \pi/2 = 1.571$ rad, $\beta'_0 = 0.1370$ rad, $u_0 = 81.88$ m/s, $v'_0 = 82.65$ m/s, $c'_0 = c_{m0} = 11.29$ m/s, and $c_{u0} = 0$ m/s.
The outlet velocity diagram, at the mean effective diameter $d_1 = 0.2158$ m, can be drawn by using the following quantities:
$\alpha'_1 = 1.081$ rad, $\beta'_1 = 0.2336$ rad, $u_1 = 79.09$ m/s, $v'_1 = 72.00$ m/s, $c'_1 = 18.37$ m/s, $c'_{u1} = 8.908$ m/s, and $c_{m1} = 16.07$ m/s.
The axial length of the inducer is $L_i = 0.1183$ m.
The semi-aperture angle of the inducer at the tip of its blades is 0.1222 rad.
The semi-aperture angle of the hub is 0.2443 rad.
The diameters of the inducer at the inlet are $d_{0t} = 0.3103$ m, $d_{0h} = 0.05925$ m, and $d_0 = 0.2234$ m.
The vane angles of the inducer are $\theta_{0t} = 0.1571$ rad at d_{0t}, $\theta_{0h} = 0.6925$ rad at d_{0h}, and $\theta_0 = 0.2166$ rad at d_0.
The diameters of the inducer at the outlet are $d_{1t} = 0.2813$ m, $d_{1h} = 0.1182$ m, and $d_1 = 0.2158$ m.
The vane angles of the inducer are $\theta_{1t} = 0.1959$ rad at d_{1t}, $\theta_{1h} = 0.4412$ rad at d_{1h}, and $\theta_1 = 0.2531$ rad at d_1.
The number of vanes of the inducer is $Z_0 = 3$.
The solidity of the vanes is $S_v = 2.190$.

The flow coefficient of the inducer at the inlet is $\varphi_{ind} = 0.09930$.

So much for the inducer. Now we consider the impeller of the pump. Since the head rise and the head coefficient of the pump are known to be respectively $\Delta H = 893.1$ m and $\psi = 0.46$, then from the following equation

$$\psi = \frac{g_0 \, \Delta H}{u_2^2}$$

we compute the peripheral velocity u_2 of the impeller at the outlet as follows

$$u_2 = \left(\frac{g_0 \, \Delta H}{\psi}\right)^{\frac{1}{2}} = \left(\frac{9.807 \times 893.1}{0.46}\right)^{\frac{1}{2}} = 138.0 \text{ m/s}$$

Since the angular velocity of the shaft is know to be $\omega = 733.0$ rad/s, then the diameter of the impeller at the outlet results from

$$d_2 = \frac{2u_2}{\omega} = \frac{2 \times 138.0}{733.0} = 0.3765 \text{ m}$$

As has been shown above, the required total pressure head rise ΔH_{imp} of the propellant flowing through the impeller can be determined as follows

$$\Delta H_{imp} = \Delta H + \Delta H_e - \Delta H_{ind}$$

where $\Delta H = 893.1$ m is the required head rise of the pump, $\Delta H_{ind} = 71.84$ m is the required head rise of the inducer, and $\Delta H_e = 169.7$ m is the hydraulic head loss in the volute. After substituting these values in the preceding equation, we find

$$\Delta H_{imp} = 893.1 + 169.7 - 71.84 = 991.0 \text{ m}$$

The required volume flow rate q_{imp} through the impeller results from

$$q_{imp} = q + q_e$$

where $q = 0.7836$ m^3/s is the required flow rate of the pump, and $q_e = 0.02743$ m^3/s is the impeller loss due to leakage. After substituting these values in the preceding equation, we find

$$q_{imp} = 0.7836 + 0.02743 = 0.8110 \text{ m}^3/\text{s}$$

The tangential component c'_{u2} of the design absolute flow velocity at the outlet of the impeller can be computed by means of the following equation

$$\Delta H_{imp} = \frac{u_2 c'_{u2} - u_1 c'_{u1}}{g_0}$$

which is solved for c'_{u2}, as follows

$$c'_{u2} = \frac{g_0 \Delta H_{imp} + u_1 c'_{u1}}{u_2} = \frac{9.807 \times 991.0 + 79.09 \times 8.908}{138.0} = 75.53 \text{ m/s}$$

Since the impeller vane coefficient is known to be $e_v = c'_{u2}/c_{u2} = 0.74$, then the tangential component c_{u2} of the ideal absolute flow velocity at the outlet of the impeller results from

$$c_{u2} = \frac{c'_{u2}}{e_v} = \frac{75.53}{0.74} = 102.1 \text{ m/s}$$

The vanes of the impeller are known to form an angle $\beta_2 = 0.4189$ rad at the outlet with the tangential direction. By inspection of the velocity diagrams at the inlet and at the outlet of the impeller, it is possible to compute geometrically the following quantities.

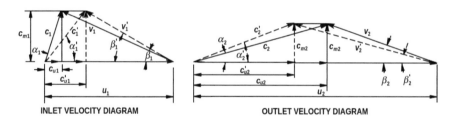

INLET VELOCITY DIAGRAM   OUTLET VELOCITY DIAGRAM

The meridian component c_{m2} of the design absolute flow velocity at the outlet of the impeller results from

$$c_{m2} = (u_2 - c_{u2}) \tan \beta_2 = (138.0 - 102.1) \tan 0.4189 = 15.98 \text{ m/s}$$

The design absolute flow velocity c'_2 at the outlet of the impeller results from

$$c'_2 = \left(c'^2_{u2} + c^2_{m2}\right)^{\frac{1}{2}} = \left(75.53^2 + 15.98^2\right)^{\frac{1}{2}} = 77.20 \text{ m/s}$$

The angle α'_2 between the design absolute flow velocity c'_2 at the outlet of the impeller and the peripheral velocity u_2 at the same point results from

$$\tan \alpha'_2 = \frac{c_{m2}}{c'_{u2}}$$

Hence

$$\alpha'_2 = \arctan\left(\frac{c_{m2}}{c'_{u2}}\right) = \arctan\left(\frac{15.98}{77.20}\right) = 0.2041 \text{ rad}$$

The design flow velocity v'_2 relative at the outlet of the impeller results from

$$v'_2 = \left[(u_2 - c'_{u2})^2 + c^2_{m2}\right]^{\frac{1}{2}} = \left[(138.0 - 75.53)^2 + 15.98^2\right]^{\frac{1}{2}} = 64.48 \text{ m/s}$$

The angle β'_2 at which the fluid leaves the impeller results from

$$\tan \beta'_2 = \frac{c_{m2}}{u_2 - c'_{u2}} = \frac{15.98}{138.0 - 75.53} = 0.2558$$

Hence

$$\beta'_2 = \arctan(0.2558) = 0.2504 \text{ rad}$$

The widths b_1 and b_2 of the impeller at respectively entrance and exit result from the following equations

$$b_1 = \frac{q_{imp}}{\pi \, d_1 \, c_{m1} \, \varepsilon_1}$$
$$b_2 = \frac{q_{imp}}{\pi \, d_2 \, c_{m2} \, \varepsilon_2}$$

Since the contraction factors of the impeller are known to be $\varepsilon_1 = 0.82$ at the inlet and $\varepsilon_2 = 0.88$ at the outlet, and the other quantities have been found to be $q_{imp} = 0.8110 \text{ m}^3/\text{s}$, $d_1 = 0.2158 \text{ m}$, $d_2 = 0.3765 \text{ m}$, $c_{m1} = 16.07 \text{ m/s}$, and $c_{m2} = 15.98 \text{ m/s}$, then there results

$$b_1 = \frac{0.8110}{3.1416 \times 0.2158 \times 16.07 \times 0.82} = 0.09078 \text{ m}$$
$$b_2 = \frac{0.8110}{3.1416 \times 0.3765 \times 15.98 \times 0.88} = 0.04876 \text{ m}$$

The flow coefficient at the impeller exit results from

$$\varphi = \frac{c_{m2}}{u_2}$$

Since $c_{m2} = 15.98$ m/s and $u_2 = 138.0$ m/s, then

$$\varphi = \frac{15.98}{138.0} = 0.1158$$

The number Z_2 of vanes of the impeller can be determined by means of the following empirical rule suggested by Huzel and Huang [22]:

$$Z_2 = \frac{60}{\pi} \beta_2$$

where β_2 (rad) is the vane angle at the impeller outlet. In the present case, this angle is known to be $\beta_2 = 0.4189$ rad, and therefore

$$Z_2 = \frac{60 \times 0.4189}{3.1416} = 8$$

A summary of the results obtained for the impeller is given below.

The required head rise is $\Delta H_{imp} = 991.0$ m.

The required volume flow rate is $q_{imp} = 0.8110$ m³/s.

The inlet velocity diagram, at the mean effective diameter $d_1 = 0.2158$ m, can be drawn by using the following quantities: $\alpha'_1 = 1.081$ rad, $\beta'_1 = 0.2336$ rad, $u_1 = 79.09$ m/s, $v'_1 = 72.00$ m/s, $c'_1 = 18.37$ m/s, $c'_{u1} = 8.908$ m/s, and $c_{m1} = 16.07$ m/s.

The outlet velocity diagram, at the outside diameter $d_2 = 0.3765$ m, can be drawn by using the following quantities: $\alpha'_2 = 0.2041$ rad, $\beta'_2 = 0.2504$ rad, $u_2 = 138.0$ m/s, $v'_2 = 64.48$ m/s, $c'_2 = 77.20$ m/s, $c'_{u2} = 75.53$ m/s, and $c_{m2} = 15.98$ m/s.

The dimensions at the inlet are: diameter of the impeller eye $d_{1t} = 0.2813$ m, diameter of the impeller hub $d_{1h} = 0.1182$ m, mean effective diameter $d_1 = 0.2158$ m, vane angle at the mean effective diameter $\beta'_1 = 0.2336$ rad, and vane width $b_1 = 0.09078$ m.

The dimensions at the outlet are: outside diameter $d_2 = 0.3765$ m, vane angle $\beta_2 = 0.4189$ rad, and vane width $b_2 = 0.04876$ m.

The number of vanes of the impeller is $Z_2 = 8$.

The following part of the present section deals with the design of the housing, which is the envelope containing the pump. According to Furst and Keller [3], the housing consists of:

- the casing, which is the part of the pump containing the impeller;
- the diffusing system and the volute for single-stage pumps; and
- the crossover system for multi-stage pumps.

An example of a multi-stage centrifugal-flow pump is the Centaur (RL-10) liquid-hydrogen pump, which has two stages mounted with the impellers back-to-back with volute housing and an external diffusing passage between the two stages, as shown in the following figure, due to the courtesy of NASA [3].

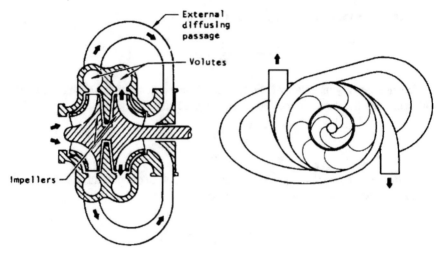

External diffusing passage

The diffusing system may include vaned or vaneless diffusers upstream of the volute and one or more conical diffusers downstream of the volute. The following figure, also due to the courtesy of NASA [3], shows three types of diffusing systems.

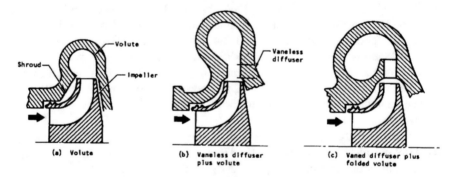

The housing also contains the bearings, which support the rotating element, and the seals, which prevent leakage of the pumped fluid.

The diffusing system and the volute form the part of the pump in which the kinetic energy impressed to the fluid by the rotating element is converted into pressure energy. The head loss due to friction at the inlet of a pump is very small, due to the short path and to the low velocity of the flow. The conical or cylindrical shape of the suction duct depends on whether the tip contour is or is not tapered. The best results have been obtained by tapering the contour of both the tip and the hub [22]. This is because the gradual reduction of the suction area upstream of the impeller eye makes the flow steady.

The rotating parts of oxidiser pumps must not be allowed to rub metal to metal. This rubbing may cause dangerous explosions in liquid-oxygen pumps. Practically attained minimum values of the inducer clearance-to-blade length ratio are 5% for fuel and 20% for oxidiser pumps. When closer clearances are required for oxidiser pumps, a liner made of PCTFE (Poly-Chloro-Tri-Fluoro-Ethylene, whose trade name is Kel-F® or Neoflon®) should be used for the inducer casing. The chemical formula of this polymer is $(CF_2CClF)_n$, where n is the number of monomer units in its molecule. A layer of this material, 2.54 mm in thickness, is inserted between the tips of the inducer blades and the wall of the suction duct, as shown in the following figure, due to the courtesy of NASA [4].

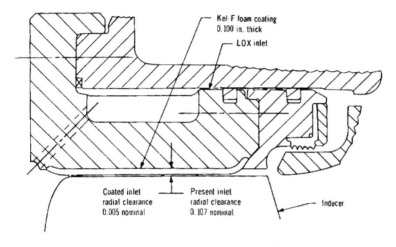

As has been shown here and also in Sect. 4.5, the diffusing system may, or may not, have static vanes. In case of a vaneless diffuser, the pumped fluid at the outlet of the impeller is discharged into a single volute channel, whose cross section increases gradually, and the conversion from velocity head to pressure head takes place principally in the conical nozzle at the outlet of the pump.

In case of a vaned diffuser, the pumped fluid at the outlet of the impeller is discharged into a diffuser provided with vanes, and the same conversion takes place prevalently in the channels between the diffusing vanes, before the fluid has reached the volute channel. A vaneless diffuser has the advantage of simplicity, but a vaned diffuser is more efficient than the other.

The principal design parameters of a vaneless-diffuser, single outlet, centrifugal-flow pump are shown in the following figure, re-drawn from [22].

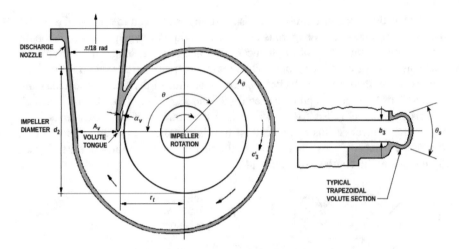

These parameters are the area A_v (m²) of the volute at the throat, the flow area A_θ (m²), the angle θ_s (rad) between the side walls of the volute, the volute tongue angle α_v (rad), the radius r_t (m) at which the volute tongue starts, and the width b_3 (m) of the volute. By volute tongue angle α_v we mean the camber angle formed by the tongue with the circumferential direction [25]. The direction of the flow at the exit of the impeller must match the angle α_v, in order to avoid turbulence and wear downstream of the tongue. For this purpose, the value of the volute tongue angle α_v is chosen close to the value of the angle α'_2 between the absolute velocity vector c'_2 and the peripheral velocity vector u_2 at the outlet of the impeller, that is, $\alpha_v \approx \alpha'_2$. The values of the other parameters are discussed below, in accordance with the indications given in [22].

The methods used to determine the cross-sectional area A_θ of the volute at an angular distance θ (rad) from the volute tongue are: (a) constant moment of momentum per unit mass; and (b) constant mean velocity.

In the method (a), the circumferential component c'_{u3} of the absolute velocity of the pumped fluid in the volute is assumed to be inversely proportional to the radius r ($c'_{u3}\, r$ = constant), which would be true only in the absence of friction losses. After the shape of the volute has been determined, the volute cross-sectional area satisfying the given flow requirement is determined at each circumferential station. This method was used for the design of the pump housing of Titan I and Titan II. It was also used, corrected for friction losses, for the design of the J-2S Mark 29 fuel pump (illustrated in Sect. 4.7), which experienced very light radial bearing loads [3].

In the method (b), the mean absolute velocity c'_3 (m/s) of the pumped fluid in the volute is assumed to be constant at all cross sections of the volute over the circumference, and therefore the area A_θ of each cross section is increased proportionally to the increase in the central angle θ of the volute. In other terms, the constant mean absolute velocity c'_3 is made equal to

$$c_3' = \frac{q}{A_v} = \frac{\theta}{2\pi} \frac{q}{A_\theta}$$

where q (m^3/s) is the volume flow rate of the pump, A_v (m^2) is the area of the volute at the throat (where $\theta = 0$), and A_θ (m^2) is the area of the volute at an angular distance θ (rad) from the volute tongue. This method was conceived to simplify the volute design.

There are only small differences in pump efficiency between the two methods. The unsymmetrical pressure (with its associated radial hydraulic forces upon the impeller) around the volute passage has been found to be higher in pumps designed according to the constant-mean-velocity method [3].

The design value of the mean absolute velocity c'_3 (m/s) can be determined from the following correlation

$$c_3' = K_v (2g_0 \Delta H)^{\frac{1}{2}}$$

where K_v is a design factor, whose value (determined experimentally and ranging from 0.15 to 0.55) is lower for pumps of higher specific speed, ΔH (m) is the design head rise of the pump, and $g_0 = 9.80665$ m/s^2 is the acceleration of gravity near the surface of the Earth.

The radius r_t at which the volute tongue starts should be from 1.05 to 1.1 times the outer radius r_2 of the impeller, in order to avoid turbulence.

The width b_3 of the volute at the bottom of its trapezoidal cross section is chosen to be 2.0 times b_2 (where b_2 is the impeller width at the outlet) in case of pumps of low specific speeds, and from 1.6 to 1.75 times b_2 in case of pumps of high specific speeds.

The maximum value of the angle θ_s between the side walls of the volute should be about $\pi/3$ radians. A value smaller than this should be chosen for pumps of high specific speeds or high flow angles α'_2 at the impeller outlet.

The pressure exerted by the pumped fluid on the walls of the volute cannot be kept uniform under all operating conditions. Consequently, the impeller shaft is subject to a radial force. In order to eliminate or reduce this force, double-volute housings are used. A typical double-volute, single-outlet housing of a centrifugal-flow pump is shown in the following figure, re-drawn from [22].

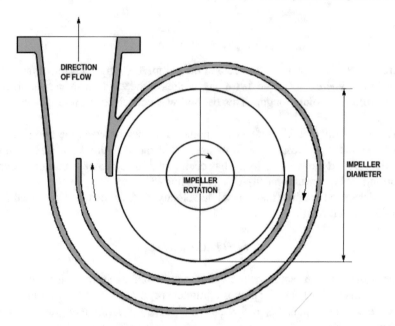

In the double-volute housing illustrated above, the flow forms two streams as a result of a second tongue placed at an angular distance of π radians from the first. This symmetry in the flow may reduce the total resultant radial force.

A diffuser provided with static vanes has been illustrated in Sect. 4.5. The shape of the volute of a vaned diffuser is the same as that of a vaneless diffuser, but a vaned diffuser has a plurality of passages through which the pumped fluid flows from the impeller to the discharge nozzle. This fact makes it possible to convert the velocity head possessed by the fluid into pressure head in a narrower space than is the case with a vaneless diffuser. High performance is obtained by a small radial distance (from 0.8 to 3.0 mm, depending on the impeller diameter) between the impeller and the inlet tips of the static vanes. The width of a vaned diffuser at the inlet may be chosen to be from 1.6 to 2.0 times b_2 (where b_2 is the impeller width at the outlet). With reference to the following figure, re-drawn from [22], the vane angle α_3 at the inlet should be chosen equal or close to the angle α'_2 between the vectors c'_2 and u_2 at the outlet of the impeller.

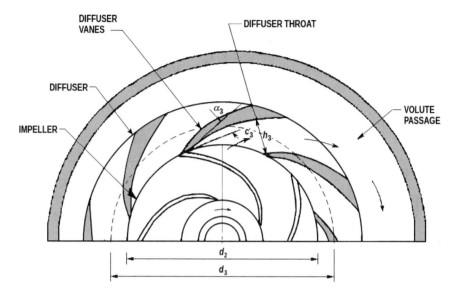

The mean value of the absolute velocity c'_3 (m/s) of the flow at the throats of the diffuser can be determined approximately as follows

$$c'_3 = \frac{d_2}{d_3} c'_2$$

where d_2 (m) is the diameter of the impeller at the outlet, d_3 (m) is the pitch diameter of the diffuser at the throats, and c'_2 (m/s) is the absolute velocity of the flow at the outlet of the impeller.

Let Z_3 be the number of vanes in the diffuser. Since each passage between two contiguous vanes can be assumed to carry an equal fraction of the total volume flow rate q (m³/s), then the following equation holds

$$b_3 h_3 Z_3 = \frac{q}{c'_3}$$

where b_3 (m) and h_3 (m) are respectively the width and the height at the throat of each passage in the diffuser.

After the number Z_2 of backward-curved rotating vanes at the impeller outlet has been determined by means of the empirical rule indicated above, that is, by

$$Z_2 = \frac{60}{\pi} \beta_2$$

where β_2 (rad) is the vane angle at the impeller outlet, then the number Z_3 of static vanes in the diffuser should be chosen so as to avoid phenomena of resonance.

The height h_3 (m) and the width b_3 (m) at the throat of each passage between two contiguous vanes in a diffuser are chosen so that $h_3 = b_3$. The angle of divergence of each passage downstream of the throat is chosen in the range from $\pi/18$ to $\pi/15$ rad. The velocity of the flow downstream of the diffuser is kept a little higher than its velocity in the discharge line of the pump [22].

Materials successfully used in housings of centrifugal-flow pumps are indicated in the following table, due to the courtesy of NASA [3].

Material	Pumped fluid				
	LH_2	LOX	RP-1	N_2O_4	50/50 UDMH/N_2H_4
Aluminium					
A356 (cast)	x	x	x	x	x
A357 (cast)	x	x	x		
6061	x	x			
7075	x	x	x	x	X
7079	x	x			
Steel					
AM350				x	x
304L (cast)	x	x			
310 (cast)	x				
310	x	x	x		
347 (cast)		x			
Iriconel 718	x	x			
"K" Monel	x	x	x		
"KR" Monel		x	x		
T1-5AI-2 5Sn (ELI)	x				

Note X means That the material has been used successfully with the fluid shown; absence of an X means either that no data on the specific use are available or that the material cannot be used with the fluid Materials not shown as cast were wrought

In the preceding table, "K" Monel and "KR" Monel are trade names indicating nickel-based alloys. The pump housings have been successfully fabricated by using one or the other of the two following processes:

(a) casting in one or more pieces;
(b) welding together forged, formed, cast, or machined elements.

Materials used for this purpose are cast aluminium alloys, cast stainless steels, and high-strength wrought aluminium alloys and steels. Diffuser vanes can be integral or separate. Reinforcing bolts through diffuser or guide vanes have been used to provide structural aid. Housing structures must bear mounting loads and loads due to internal pressure. External loads on the volute can be reduced by using flexible ducts [3].

Huzel and Huang [22] suggest to use the following formula for a rough estimate of the hoop tensile stress σ_t (N/m^2) at a given section of a housing

$$\sigma_t = p \frac{A}{A'}$$

where p (N/m^2) is the local absolute pressure inside the housing or the difference of pressure across the wall of the housing, A (m^2) is the projected area of the surface on which the pressure acts, and A' (m^2) is the area of the surface of the material subject to the force pA.

The stress acting really on a housing structure is higher than the one expressed above, due to discontinuities and deformation of the wall, and also to thermal stresses resulting from temperature gradients across the wall [22].

As an application of the concepts given above, it is required to design the housing of a double-volute, single-discharge, centrifugal-flow pump, whose inducer and impeller have been determined above. The second tongue of this double-volute pump is separated from the first by an angular distance of π radians, according to the figure given above. The value of the design factor K_v, which appears in the equation $c'_3 = K_v(2g_0 \Delta H)^{\frac{1}{2}}$, is assumed to be 0.337.

By substituting $K_v = 0.337$, $g_0 = 9.807$ m/s^2, and $\Delta H = 893.1$ m in the following equation, the average value of the flow velocity in the volute results

$$c'_3 = K_v(2g_0\Delta H)^{\frac{1}{2}} = 0.3370 \times (2 \times 9.807 \times 893.1)^{\frac{1}{2}} = 44.60 \text{ m/s}$$

We assume the area A_θ of each cross section of the volute to vary with the central angle θ so that this value of c'_3 is constant. Then, we use the equation

$$c'_3 = \frac{\theta}{2\pi} \frac{q}{A_\theta}$$

where $q = 0.7836$ m^3/s is the volume flow rate of the pump. This yields for each of the two volutes

$$A_\theta = \frac{\theta}{2\pi} \frac{q}{c'_3} = \frac{\theta}{2\pi} \frac{0.7836}{44.60}$$

At $\theta = \frac{1}{4}\pi$, $A_{\pi/4} = 0.002196$ m^2; at $\theta = \frac{1}{2}\pi$, $A_{\pi/2} = 2A_{\pi/4} = 0.004392$ m^2; at $\theta = \frac{3}{4}\pi$, $A_{3\pi/4} = 3A_{\pi/4} = 0.006588$ m^2; and at $\theta = \pi$, $A_\pi = 4A_{\pi/4} = 0.008784$ m^2.

The total area A_v of the cross section of the volute at the inlet of the discharge nozzle is

$$A_v = 2A_\pi = 2 \times 0.008784 = 0.01757 \text{ m}^2$$

In order to avoid turbulence and wear downstream of the tongue, the value of the volute tongue angle α_v is taken approximately equal to α'_2. Since α'_2 has been found above to be equal to 0.2041 rad, then we take $\alpha_v = 0.2041$ rad $\approx \pi/15$ rad.

The radius r_t at which the volute tongue starts is taken slightly greater than the outer radius r_2 of the impeller, in order to avoid turbulence. For this purpose, we choose $r_t = 1.05\ r_2$. Since the outer diameter of the impeller has been found to be $d_2 = 0.3765$ m, then the value of r_t can be determined as follows

$$r_t = 1.05 \times \frac{0.3765}{2} = 0.1977 \text{ m}$$

The cross section of the volute is assumed to be of trapezoidal shape. As has been shown above, the width b_3 of the volute at the bottom of its trapezoidal cross section is chosen to be 1.75 times the width b_2 of the impeller at the outlet. Since b_2 has been found to be equal to 0.04876 m, then

$$b_3 = 1.75 \times 0.04876 = 0.08533 \text{ m}$$

An area slightly greater than $A_v = 0.01757$ m^2 is necessary for the circle at the inlet of the discharge nozzle, in order for the shape of the cross section of the volute to change gradually from trapezoidal to circular.

Therefore, we compute the area A_i and the diameter d_i of the circle at the inlet of the discharge nozzle as follows

$$A_i = 1.13 \times 0.01757 = 0.01985 \text{ m}^2$$

$$d_i = \left(\frac{4 \times 0.01985}{3.1416} \right)^{\frac{1}{2}} = 0.1590 \text{ m}$$

Assuming the discharge nozzle to be a cone frustum whose length is 0.254 m and whose semi-aperture angle is $\pi/36$ rad, the diameter d_e and the area A_e of the circle at the outlet of the discharge nozzle are

$$d_e = 0.16 + 2 \times 0.254 \times \tan\left(\frac{\pi}{36}\right) = 0.2044 \text{ m}$$

$$A_e = \pi \left(\frac{d}{2} \right)^2 = 3.1416 \times \left(\frac{0.2044}{2} \right)^2 = 0.03281 \text{ m}^2$$

The velocities c_i and c_e of the flow at respectively the inlet and the outlet of the discharge nozzle are

$$c_i = \frac{q}{A_i} = \frac{0.7836}{0.01985} = 39.48 \text{ m/s}$$

$$c_e = \frac{q}{A_e} = \frac{0.7836}{0.03281} = 23.88 \text{ m/s}$$

As has been shown in Sect. 4.5, a thrust balance system is necessary to counteract the axial loads generated by changes of axial momentum and of pressure in the pumped fluid. The residual value of the axial force, after the application of a thrust balance system, must be so low as to be safely absorbed by the pump bearings [36].

These loads can be reduced by mounting the two propellant pumps back to back. This, together with the mounting of radial ribs, has been done in the Mark 3 turbo-pump, which has been described in detail in Sect. 4.5. Other methods, used either alone or in conjunction one with another, are described below.

A method consists in installing a wearing ring which provides a balance chamber at the back shroud of the impeller. This balance chamber is a space located between the back wearing ring (of diameter d_{br}) and the shaft seal (of diameter d_s), as shown in the following figure, adapted from [3].

The axial loads are balanced by choosing properly the diameter d_{br}, the diameter d_s, and the value of the pressure p_c of the fluid admitted into the balance chamber. The value of the pressure p_c can be controlled by adjusting the clearances and the leakages of the back wearing ring and of the shaft seals.

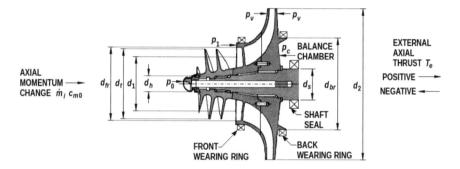

For this purpose, Huzel and Huang [22] indicate the following equation

$$p_c\pi\left(d_{br}^2 - d_s^2\right) = p_v\pi\left(d_{br}^2 - d_{fr}^2\right) + p_1\pi\left(d_{fr}^2 - d_t^2\right) + p_0\pi d_h^2 + 4\dot{m}_i c_{m0} \pm T_e$$

where p_c (N/m^2) is the pressure in the balance chamber, p_v (N/m^2) is the average net pressure in the space between the shrouds of the impeller and the walls of the casing, p_1 (N/m^2) is the static pressure at the outlet of the inducer, p_0 (N/m^2) is the static pressure at the inlet of the inducer, d_s (m) is the effective diameter of the shaft seal, d_h (m) is the diameter of the hub at the inlet of the inducer, d_t (m) is the diameter of the inducer at the tip of its blades, d_{fr} (m) is the diameter of the front wearing ring, d_{br} (m) is the diameter of the back wearing ring, $\dot{m}_i$ (kg/s) is the mass flow rate of the pumped propellant at the inducer, c_{m0} (m/s) is the axial velocity of the flow at the inlet of the inducer, and T_e (N) is the external axial force resulting from the unbalanced axial loads due to the other propellant and/or the turbine. The external axial force T_e is positive when the impeller is pulled away from the suction side, and negative otherwise.

The value of the static pressure p_1 (N/m^2) at the outlet of the inducer can be measured in tests or approximated by multiplying the value of the static pressure p_0 (N/m^2) at the inlet of the inducer by a design factor k_i, as follows

$$p_1 = k_i \, p_0$$

where the value of k_i, determined experimentally, ranges from 1.1 to 1.8 [22]. After p_1 has been determined, the average net pressure p_v (N/m^2) in the space between the shrouds of the impeller and the walls of the casing can be computed approximately by using the following equation given in [22]:

$$p_v = p_1 + \frac{3}{4}\left[\frac{1}{2}\rho\left(u_2^2 - u_1^2\right)\right]$$

where ρ (kg/m^3) is the density of the pumped propellant, u_2 (m/s) is the peripheral velocity of the impeller at the outlet diameter d_2, and u_1 (m/s) is the peripheral velocity of the impeller at the mean effective inlet diameter d_1. The value of the pressure p_c (N/m^2) in the balance chamber can be adjusted by conducting tests on single components of a pump. In these tests, it is possible to vary the clearances at the back wearing ring and at the shaft seals. A disadvantage of this method is the increase in the leakage losses.

Another method, also described in Sect. 4.5, consists in installing straight radial ribs at the back shroud of the impeller, as shown in the following figure, adapted from [3].

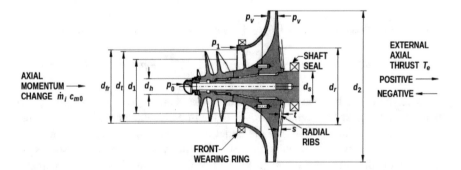

The radial ribs reduce the static pressure in the space between the back shroud of the impeller and the wall of the casing, because the pumped fluid spins in this space with the impeller. This rotational motion decreases the pressure exerted by the fluid with the radial distance from the impeller tips [12]. Holes may be provided through the impeller into the inside diameter region of the radial ribs to vent that region statically and to provide a positive coolant flow into the radial ribs to prevent cavitation caused by fluid heating which results from the pumping work of the radial ribs [3].

According to Huzel and Huang [22], the reduction F_a (N) of the axial forces acting on the back shroud of the impeller, due to the installation of radial ribs, can be computed approximately as follows

$$F_a = \frac{3\pi}{32}\left(d_r^2 - d_s^2\right)\left[\frac{1}{2}\rho\left(u_r^2 - u_s^2\right)\right]\left(\frac{s+t}{2s}\right)$$

where d_r (m) is the outside diameter of the radial ribs, d_s (m) is the inside diameter of the radial ribs, u_r (m/s) is the peripheral velocity of the radial ribs at the diameter d_r, u_s (m/s) is the peripheral velocity of the radial ribs at the diameter d_s, ρ (kg/m^3) is the density of the pumped fluid, t (m) is the height or the thickness of the radial ribs, and s (m) is the average distance between the impeller back shroud and the wall of the casing.

The required reduction F_a (N) of the axial forces acting of the back shroud of the impeller can be computed by using the following equation [22]:

$$p_v\pi\left(d_{fr}^2 - d_s^2\right) - 4F_a = p_1\pi\left(d_{fr}^2 - d_t^2\right) + p_0\pi d_h^2 + 4\dot{m}_i c_{m0} \pm T_e$$

where p_v (N/m^2) is the average net pressure in the space between the shrouds of the impeller and the walls of the casing, d_{fr} (m) is the diameter of the front wearing ring, d_s (m) is the inside diameter of the radial ribs, p_1 (N/m^2) is the static pressure at the outlet of the inducer, d_t (m) is the diameter of the inducer at the tip of its blades, p_0 (N/m^2) is the static pressure at the inlet of the inducer, d_h (m) is the diameter of the hub at the inlet of the inducer, $\dot{m}_i$ (kg/s) is the mass flow rate of the pumped propellant at the inducer, c_{m0} (m/s) is the axial velocity of the flow at the inlet of the inducer, and T_e (N) is the external axial force resulting from the unbalanced axial loads due to the other propellant and/or the turbine.

The pressures p_1 (N/m^2) and p_v (N/m^2) can be computed approximately by means of the preceding equations which are re-written below for convenience

$$p_1 = k_i\, p_0$$

$$p_v = p_1 + \frac{3}{4}\left[\frac{1}{2}\rho\left(u_2^2 - u_1^2\right)\right]$$

The gear-driven pumps of the Titan launch vehicles had open-face impellers with radial ribs in the impeller hub, in order to reduce the axial force to a value which could be sustained by a split-inner-race ball bearing. The turbo-pump of the F-1 engine had shrouded impellers. In order to control the axial force, the impeller of the oxidiser pump of the F-1 engine had an inlet wear ring and radial ribs, whereas the impeller of the fuel pump of the same engine had inlet and hub wear rings. The pump forces balanced the direct-drive turbine forces, so that tandem split-inner-race ball bearings could sustain the unbalanced axial force [3].

As an example of application, it is required to compute the reduction F_a (N) of the axial forces acting of the back shroud of the impeller of a given centrifugal-flow pump, due to the installation of radial ribs having the following dimensions: outside diameter of the radial ribs $d_r = 0.3759$ m, inside diameter of the radial ribs $d_s = 0.1219$ m, height of the radial ribs $t = 0.005334$ m, and average distance $s = 0.006350$ m between the impeller back shroud and the wall of the casing. The angular velocity of the shaft of the pump is $\omega = 733.0$ rad/s. The density of the pumped fluid (liquid oxygen) is $\rho = 1141$ kg/m^3 in the operating conditions.

The peripheral velocity u_r at the outside diameter $d_r = 0.3759$ m of the radial ribs results from

$$u_r = \frac{1}{2}\omega \, d_r = 0.5 \times 733.0 \times 0.3759 = 137.8 \text{ m/s}$$

The peripheral velocity u_s at the inside diameter $d_s = 0.1219$ m of the radial ribs results from

$$u_s = \frac{1}{2}\omega \, d_s = 0.5 \times 733.0 \times 0.1219 = 44.68 \text{ m/s}$$

The reduction F_a of the axial forces acting of the back shroud of the impeller, due to the presence of the radial ribs, results from

$$
F_a = \frac{3\pi}{32}\left(d_r^2 - d_s^2\right)\left[\frac{1}{2}\rho\left(u_r^2 - u_s^2\right)\right]\left(\frac{s+t}{2s}\right) = \frac{3 \times 3.1416}{32} \times \left(0.3759^2 - 0.1219^2\right)
$$
$$
\times \left[0.5 \times 1141 \times \left(137.8^2 - 44.68^2\right)\right] \times \left(\frac{0.006350 + 0.005334}{2 \times 0.006350}\right) = 3.321 \times 10^5 \text{ N}
$$

Still another method to control axial thrust in a centrifugal-flow pump uses a self-compensating balance piston. This method was used in the Mark 48-O (liquid oxygen) turbo-pump, which was designed to achieve a balance of axial thrust between the pump and the turbine. For this purpose, the back side of the pump impeller contains a balance piston, which is shown in the following figure, due to the courtesy of NASA [37]. The rotating member of the balance piston is the rear shroud of the impeller. To operate the piston, high-pressure liquid oxygen from the impeller discharge passes through a high-pressure orifice located at the outer diameter of the impeller into the balance piston cavity. From this cavity, the liquid oxygen passes through a low-pressure orifice near the impeller hub into the balance piston sump. From there, the liquid oxygen returns to the eye of the impeller through axial passages in the diffuser vanes and radial holes in the diffuser and inlet [38].

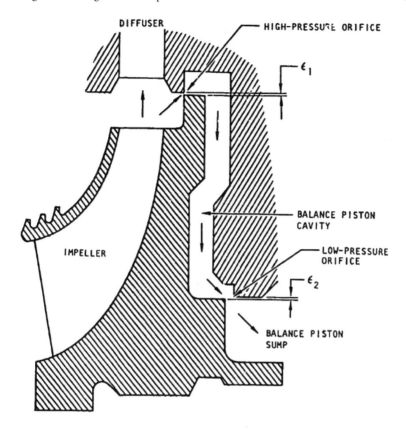

The thrust-compensating effect is obtained because the openings of the high-pressure orifice and of the low-pressure orifice vary with the axial position of the rotor. Therefore, the axial force due to the pressure on the rear shroud of the impeller varies correspondingly. For example, an unbalanced load toward the pump inlet causes a reduction in the high-pressure orifice gap and an increase in the low-pressure orifice gap. This, in turn, causes a reduction in the pressure force of the impeller rear shroud, introducing a compensating load change [38].

Materials used for thrust balance systems are chosen taking account of their compatibility with the propellant, strength at the required rotating speed, and safety in the event of an inadvertent rub. Plastic materials such as PCTFE (Poly-Chloro-Tri-Fluoro-Ethylene, whose trade name is Kel-F® or Neoflon®), PTFE (Poly-Tetra-Fluoro-Ethylene, whose trade name is Teflon®), and fluorinated PVC (Poly-Vinyl-Chloride) resist burning if rubbed on the impeller in liquid-oxygen pumps, and therefore are safe stationary sealing materials when the difference of the seal pressure is not excessive. These materials are also satisfactory for liquid-hydrogen pumps.

Materials for balance-piston orifices are chosen to resist galling if rubbed against balance-piston rotor or impeller materials. In case of balance pistons used in liquid-oxygen pumps, materials used for orifices must resist explosion or ignition upon impact, and must also be physically and chemically stable in liquid oxygen. The following table, due to the courtesy of NASA [3], indicates materials used for thrust balance systems.

Component	Material				
Balance ribs	Same material as impeller				
Anti-vortex vanes	Same material as housing				
Impeller seals	KEL-F[a], stainless steels[a], fiberglas-reinforced Teflon, silver[a], leaded bronze, impeller materials, housing materials				
Balance piston	Al 2024 anodized	Al 7075-T73	Inconel 718[a]	"K" Monel[a]	Ti-5AI-2.5Sn
Balance piston orifice	Flame-plated tungsten carbide on 310 stainless	304 stainless	Silver-plated 310 stain-less, silver	Leaded bronze	Leaded bronze

[a]Material Suitable for use with LOX

4.10   Design of Axial-Flow Pumps

Apart from the case of inducers, axial-flow pumps have been used in rocket engines only when the pumped propellant is liquid hydrogen, because this fluid requires high values of volume flow rate and head rise. For this purpose, such pumps have been used in the multi-stage configuration only. The head rise ΔH made available by a typical single-stage centrifugal-flow pump, when the pumped fluid is liquid hydrogen, is about 19,800 m, which corresponds to a pressure rise Δp of about $70.85 \times 9.807 \times 19,800 = 1.376 \times 10^7$ N/m². This is due to the very low density $\rho = 70.85$ kg/m³ of liquid hydrogen. When higher values then these are required for ΔH and Δp, then a multi-stage axial-flow pump is chosen. A pump of the latter type makes it possible to obtain a head rise ranging from about 1520 to 2740 m for each stage [22].

The volume flow rate q of axial-flow pumps used for liquid hydrogen cannot be lower than about 0.347 m³/s, because the height of the rotor or stator vanes of such pumps cannot be lower than about 12.7 mm, for manufacturing reasons. For values greater than 0.347 m³/s of volume flow rate and in the range from 9140 to 19,800 m of head rise, the design requirements of a liquid-hydrogen pump can be satisfied by either a single-stage centrifugal flow pump or a multi-stage axial flow pump. In practice, a centrifugal-flow pump for liquid hydrogen is chosen when the volume flow rate required is lower than 0.0158 m³/s [22].

The following table, adapted from [2], indicates (in metric units) delivered volume flow rate (m³/s), total head rise (m), rotational speed (rad/s), number of stages, and application for five multi-stage liquid-hydrogen pumps (Mark 9, Mark 15-F, Mark 25, Mark 26, and M-1) used in rocket engines.

Pump	Delivered flow, m³/s	Headrise,[a] m	Speed, rad/s	Number of stages	Application
Mark 9	0.64541	15,697	3434.8	Inducer plus six main stages	Phoebus (Development)
Mark 15-F	0.57172	12,283	2960.0	Inducer plus seven main stages	J-2 engine (Operational)
Mark 25	1.1672	18,898	3560.5	Tandem inducer plus four main stages	Phoebus (Development)
Mark 26	0.56781	12,192	2513.3	Inducer plus seven main stages	J-2 engine (Experimental)
M-1	3.9305	17,221	1384.9	Inducer plus transition plus eight main stages	M-1 engine (Development)

[a]Overall headrise—inducer inlet to volute discharge

The meridian or axial component c_m of the absolute velocity vector c of the flow is assumed to have a constant value in all the stages (rotor and stator) of an axial-flow pump. This assumption requires a constant area of the cross section of the pump, in order for the equation of continuity of flow to be satisfied. By so doing, the effects of frictional drag at the walls of the housing and at the vanes are neglected.

The rotor of an axial-flow pump imparts kinetic energy to the pumped fluid by increasing the axial component c_m of the absolute velocity vector c of the flow. This transfer of kinetic energy occurs by using aerofoil-shaped vanes. These vanes are considered below in three particular sections of the cylindrical pump: at the impeller tip diameter d_t (m), at the impeller hub diameter d_h (m), and at the mean effective diameter d_m (m) of the pump, defined as follows

$$d_m = \left(\frac{d_t^2 + d_h^2}{2}\right)^{\frac{1}{2}} = \left[\frac{d_t^2\left(1 + r_d^2\right)}{2}\right]^{\frac{1}{2}}$$

where $r_d = d_h/d_t$ is the impeller hub ratio. Let P_r (m) be the common distance (also known as pitch) separating two contiguous vanes of the rotor, measured along the circumference, as shown in the following figure, re-drawn from [22].

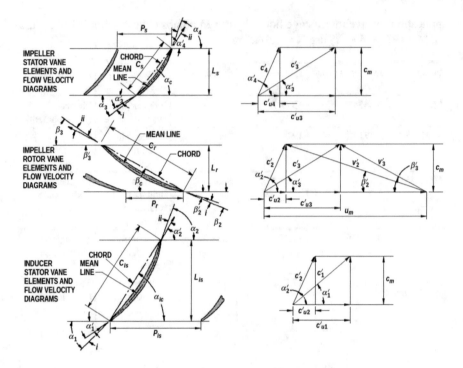

Given the mean effective diameter d_m (m) and the number Z_r of the rotor vanes, the distance P_r results from

$$P_r = \frac{\pi d_m}{Z_r}$$

The solidity S_r of the rotor vanes, at the mean effective diameter d_m, is

$$S_r = \frac{C_r}{P_r}$$

where C_r (m) is the length of the chord of each rotor vane. The solidity S_r of a rotor vane usually increases from the tip diameter d_t (m) to the hub diameter d_h (m), for structural reasons. The profile of each vane is represented in the preceding figure by the mean line of the vane itself. The thickness of each vane varies along its mean line. The angle formed in each point by the tangent to the mean line of a rotor vane with the tangential (perpendicular to the axial) direction varies from β_2 to β_3. We consider here an axial-flow pump, where the pumped fluid flows upward, and the rotor turns clockwise (from left to right), as shown in the following figure, adapted from [39], which illustrates the Mark 15-fuel pump. This pump is also illustrated in Sect. 4.2.

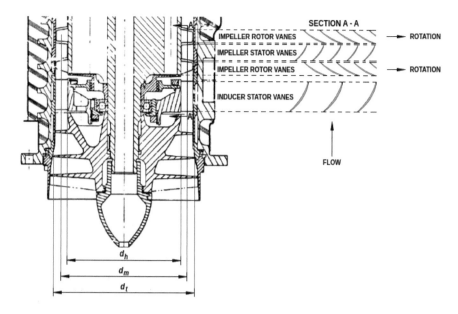

According to Huzel and Huang [22], the mean line of each vane of the rotor can be approximated by a circular arc, and the following equations hold at the mean effective diameter d_m:

$$\beta_c = \frac{\beta_2 + \beta_3}{2}$$

$$C_r = 2R_r \sin\left(\frac{\beta_3 - \beta_2}{2}\right) = \frac{L_r}{\sin \beta_c}$$

where β_c (rad) is the angle formed by the chord of each rotor vane with the tangential direction, β_2 (rad) is the angle formed by the tangent of each rotor vane at the inlet with the tangential direction, β_3 (rad) is the angle formed by the tangent of each rotor vane at the outlet with the tangential (perpendicular to the axial) direction, C_r (m) is the length of the chord of each rotor vane, R_r (m) is the radius of curvature of each rotor vane, and L_r (m) is the length of each rotor vane in the axial direction.

The relative velocity vector v'_2 of the flow at the inlet of each rotor vane forms an angle β'_2 (such that $\beta'_2 < \beta_2$) with the tangential direction. The angular difference $i = \beta_2 - \beta'_2$ is the incidence (or attack) angle at the inlet of each rotor vane. Likewise, the relative velocity vector v'_3 of the flow at the outlet of each rotor vane forms an angle β'_3 (such that $\beta'_3 < \beta_3$) with the tangential direction. The angular difference $ii = \beta_3 - \beta'_3$ is the angle allowed for circulatory flow at the outlet of each rotor vane. The velocity diagrams at the inlet and at the outlet of the impeller rotor vanes can be drawn according to the following equations

$$\beta_2 = \beta'_2 + i$$

$$\beta_3 = \beta'_3 + ii$$

$$c_m = \frac{q_{imp}}{\frac{\pi}{4}\left(d_t^2 - d_h^2\right)\varepsilon}$$

$$u_m = \omega\frac{d_m}{2}$$

$$\Delta H_{imp} = (\Delta H)_1 + \Delta H_e = \frac{u_m c'_{u3} - u_m c'_{u2}}{g_0}$$

$$(\psi)_1 = \frac{g_0(\Delta H)_1}{u_m^2}$$

$$c_m = c'_{u2}\tan\alpha'_2 = c'_{u3}\tan\alpha'_3 = c'_2\sin\alpha'_2 = c'_3\sin\alpha'_3 = v'_2\sin\beta'_2$$
$$= v'_3\sin\beta'_3$$

where i (rad) is the attack angle at the inlet of each vane, ii (rad) is the angle allowed for circulatory flow at the outlet of each vane, β'_2 (rad) and β'_3 (rad) are the angles formed by the relative velocity vector v with the tangential direction at respectively the inlet and the outlet of each rotor blade, α'_2 (rad) and α'_3 (rad) are the angles formed by the absolute velocity vector c with the tangential direction at respectively the inlet and the outlet of each rotor blade, c_m (m/s) is the meridian or axial component of the absolute velocity vector c, u_m (m/s) is the peripheral velocity of the rotor blades at the mean effective diameter d_m (m), c'_2 (m/s) and c'_3 (m/s) are the design magnitudes of the absolute velocity vector c at respectively the inlet and the outlet of each rotor blade, v'_2 (m/s) and v'_3 (m/s) are the design magnitudes of the relative velocity vector v at respectively the inlet and the outlet of each rotor blade, q_{imp} (m³/s) is the required volume flow rate of the impeller at the rated design point, q (m³/s) is the rated volume flow rate of the pump, q_e (m³/s) is the loss of volume flow rate of the impeller due to leakage (q_e varies from $0.02q$ to $0.1q$), ε is the contraction factor of the vane passage (ε varies from 0.85 to 0.95), ΔH_{imp} (m) is the required head rise for each stage of the impeller, $(\Delta H)_1$ (m) is the rated design head rise for each stage of the axial-flow pump, ΔH_e (m) is the hydraulic head loss for each stage of the impeller stator, and $(\psi)_1$ is the head coefficient for each stage of the axial-flow pump. All the quantities indicated above refer to the mean effective diameter d_m.

At various cylindrical sections of the pump between the hub diameter d_h and the tip diameter d_t, the following relations hold between the vane angles and the flow velocities

$$d_m\tan\beta_2 = d_t\tan\beta_{2t} = d_h\tan\beta_{2h} = d_x\tan\beta_{2x}$$

$$d_m \tan \beta_3 = d_t \tan \beta_{3t} = d_h \tan \beta_{3h} = d_x \tan \beta_{3x}$$

$$\frac{u_m}{d_m} = \frac{u_t}{d_t} = \frac{u_h}{d_h} = \frac{u_x}{d_x}$$

$$\frac{c'_{u2}}{d_m} = \frac{c'_{u2t}}{d_t} = \frac{c'_{u2h}}{d_h}$$

$$\frac{c'_{u3}}{d_m} = \frac{c'_{u3t}}{d_t} = \frac{c'_{u3h}}{d_h}$$

where β_{2t} (rad) and β_{2h} (rad) are the vane angles at the inlet of the rotor relating to respectively the tip diameter and the hub diameter, β_{3t} (rad) and β_{3h} (rad) are the vane angles at the outlet of the rotor relating to respectively the tip diameter and the hub diameter, u_t (m/s) and u_h (m/s) are the peripheral velocities of the rotor vanes at respectively the tip diameter and the hub diameter, c'_{u2t} (m/s) and c'_{u2h} (m/s) are the tangential components of the design absolute velocity vector c'_2 at the inlet of the rotor relating to respectively the tip diameter and the hub diameter, and c'_{u3t} (m/s) and c'_{u3h} (m/s) are the tangential components of the design absolute velocity vector c'_3 at the outlet of the rotor relating to respectively the tip diameter and the hub diameter.

The stator of an axial-flow pump converts much of the head of the pumped fluid due to the tangential component c'_{u3} of its absolute velocity vector c'_3 at the outlet of the rotor vanes into pressure head. For this purpose, the curvature of the stator vanes is such as to reduce the tangential component of the absolute velocity vector of the flow. By contrast, the axial component c'_{m3} of the vector c'_3 does not change, because the passages between stator and rotor perpendicular to the axial direction have the same cross-sectional area. The diameters d_t (m) and d_h (m) of the stator at respectively the tip and the hub can be considered equal to the diameters of the rotor at respectively the tip and the hub. The value of the chord-to-pitch ratio C_s/P_s of the stator vanes increases as their diameter increases from d_h to d_t. The length L_s (m) of the stator vanes in the axial direction is usually chosen equal to the length L_r (m) of the rotor vanes in the same direction.

The velocity diagrams at the inlet and at the outlet of the stator vanes are drawn assuming the absolute flow velocities and the angles to be equal to the absolute velocities and to the angles at the inlet and at the outlet of the rotor vanes. The vane angles α_3 at the inlet of the stator vanes are chosen slightly greater than the angles α'_3 formed there by the absolute velocity vector c'_3 with the tangential direction ($\alpha_3 = \alpha'_3 + i$), in order for the fluid to be deflected effectively. Likewise, the vane angles α_4 at the outlet of the stator vanes are chosen slightly greater than the angles α'_4 formed there by the absolute velocity vector c'_4 with the tangential direction ($\alpha_4 = \alpha'_4 + ii$), in order for circulatory flow to be allowed. The velocity diagrams at the inlet and at the outlet of the stator vanes of the impeller are drawn according to the following equations

$$P_s = \frac{\pi d_m}{Z_s}$$

$$S_s = \frac{C_s}{P_s}$$

$$\alpha_c = \frac{\alpha_3 + \alpha_4}{2}$$

$$C_s = 2R_s \sin\left(\frac{\alpha_4 - \alpha_3}{2}\right) = \frac{L_s}{\sin \alpha_c}$$

$$\alpha_3 = \alpha_3' + i$$

$$\alpha_4 = \alpha_4' + ii$$

$$c_m = c_{u3}' \tan \alpha_3' = c_{u4}' \tan \alpha_4' = c_3' \sin \alpha_3' = c_4' \sin \alpha_4'$$

$$d_m \tan \alpha_3 = d_t \tan \alpha_{3t} = d_h \tan \alpha_{3h} = d_x \tan \alpha_{3x}$$

$$d_m \tan \alpha_4 = d_t \tan \alpha_{4t} = d_h \tan \alpha_{4h} = d_x \tan \alpha_{4x}$$

where P_s (m) is the pitch or the space between two contiguous stator vanes, Z_s is the number of the stator vanes, C_s (m) is the length of the chord of the stator vanes, $S_s = C_s/P_s$ is the solidity of the stator vanes, α_c (rad) is the angle formed by the chord of each stator vane with the tangential direction, α_3 (rad) and α_4 (rad) are the angles formed by the tangent to each stator vane at respectively the inlet and the outlet with the tangential direction, R_s (m) is the radius of curvature of each stator vane, L_s (m) is the length of each stator vane in the axial direction, i (rad) is the attack angle at the inlet of each stator vane, ii (rad) is the angle allowed for circulatory flow at the outlet of each stator vane, α'_3 (rad) and α'_4 (rad) are the angles formed by the absolute velocity vectors c'_3 and c'_4 with the tangential direction at respectively the inlet and the outlet of each stator vane, c_m (m/s) is the meridian or axial component of the absolute velocity vector c_m, c'_3 (m/s) and c'_4 (m/s) are the magnitudes of the absolute velocity vectors c'_3 and c'_4 at respectively the inlet and the outlet of each stator vane, c'_{u3} (m/s) and c'_{u4} (m/s) are the tangential components of the absolute velocity vectors c'_3 and c'_4 at respectively the inlet and the outlet of each stator vane, α_{3t} (rad), α_{3h} (rad), and α_{3x} (rad) are the inlet angles formed by the tangent to each stator vane at respectively the tip diameter, the hub diameter, and any intermediate diameter with the tangential direction, and α_{4t} (rad), α_{4h} (rad), and α_{4x} (rad) are the outlet angles formed by the tangent to each stator vane at respectively the tip diameter, the hub diameter, and any intermediate diameter with the tangential direction. Unless specified otherwise, the quantities indicated above refer to the mean effective diameter d_m.

Typical design values for these quantities are given below for liquid-hydrogen axial-flow pumps, according to [22]. The ratio $r_d = d_h/d_t$ of the impeller hub diameter d_h to the impeller tip diameter d_t ranges from 0.76 to 0.86. The specific speed for each stage $(N)_1$ ranges from 1.1 to 1.8. The head coefficient for each stage $(\psi)_1 = g_0(\Delta H)_1/u_m^2$ ranges from 0.25 to 0.35. The solidity of the rotor vanes $S_r = C_r/P_r$ at the mean effective diameter ranges from 1.0 to 1.3. The solidity of the stator vanes $S_s = C_s/P_s$ at the mean effective diameter ranges from 1.5 to 1.8. The number of vanes of the rotor Z_r ranges from 14 to 20. The number of vanes of the stator Z_s ranges from 35 to 45. The value of Z_r should have no common factor with the value of Z_s, in order to avoid resonance phenomena. The head rise produced by the moving vanes of an impeller rotor depends on the value of the angular difference $\beta_3 - \beta_2$, called vane curvature.

The design method used for a multi-stage axial-flow pump is similar to the method described in Sect. 4.9, which relates to a single-stage centrifugal pump. The difference between the two methods consists in the determination of the number of stages of the axial-flow pump. The latter method is described below. The system requirements relating to the rocket engines give the values of some basic quantities, such as the total head rise ΔH (m) of the pump, the volume flow rate q (m³/s), and the required net positive suction head $(NPSH)_R$ (m). The rotating speed ω (rad/s) of the pump is determined by choosing an inducer having a given suction specific speed $(S)_{ind}$. This done, the specific speed $(N)_1$ for each of the n stages and the total head rise ΔH (m) of the pump are determined by using the following equations

$$(N)_1 = \frac{\omega \, q^{\frac{1}{2}}}{\left[g_0(\Delta H)_1\right]^{\frac{3}{4}}}$$

$$\Delta H = \Delta H_{ind} + \Delta H_{ee} + n(\Delta H)_1$$

where $(\Delta H)_1$ (m) is the head rise for each stage, $g_0 = 9.80665$ m/s² is the acceleration of gravity of the Earth at the sea level, ΔH (m) is the total head rise of the pump, ΔH_{ind} (m) is the head rise of the inducer, and ΔH_{ee} (m) is the hydraulic head loss at the stator of the inducer. After the specific speed $(N)_i$ for each stage has been determined, the values of other quantities (such as the impeller hub ratio $r_d = d_h/d_t$, the solidities $S_r = C_r/P_r$ and $S_s = C_s/P_s$ of respectively the rotor vanes and the stator vanes, the numbers of vanes Z_r and Z_s of respectively the rotor and the stator, the head coefficient $(\psi)_1 = g_0(\Delta H)_1/u_m^2$ for each stage, etc.) are chosen by using data taken from pumps previously designed having about the same values of specific speed for each stage. Finally, the diameters of the rotor and stator vanes of the impeller, the velocity diagrams, and the vane profiles are determined by using the equations given above, as the sequel will show.

The blades of an axial-flow pump are usually machined integrally with discs or a rotor drum. However, the M-1 main-stage blades had dove-tail attachments. The blade attachment must be designed to carry the centrifugal load of the blade and to transmit the aerofoil steady-state and vibratory bending loads to the rotor structure [2]. Three rotor-structure assemblies have been used in axial-flow pumps. They are shown in the following figure, due to the courtesy of NASA [2].

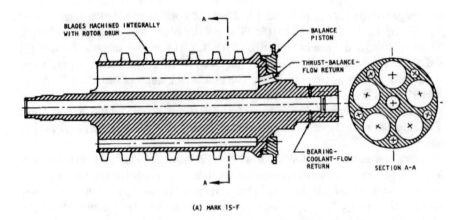

(A) MARK 15-F

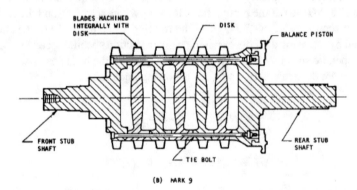

(B) MARK 9

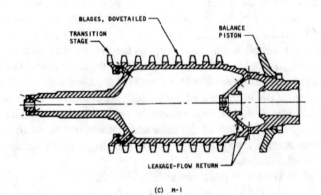

(C) M-1

A one-piece rotor structure, machined from a single forging, has been used for the Mark 15-F (see part (A) of the preceding figure) and Mark 26 pump rotors. Axial holes have been machined in the forging to provide a return flow path for thrust balance system and bearing coolant flows. The Mark 9 (see part (B) of the preceding figure) and Mark 25 pump rotors have been fabricated according to the build-up concept with the disks and stub shafts clamped together through bolts. Rabbets have been used to attain relative radial positioning, the torque loads being transmitted by shear in the tie bolts. The M-1 rotor (see part (C) of the preceding figure) is a hollow structure fabricated from four forged and machined ring components TIG-welded together [2].

The materials used for the principal components of the axial-flow pumps named above, all of which had liquid hydrogen as their pumped fluid, are indicated in the following table, due to the courtesy of NASA [2].

Component	Pump Configuration				
	Mark 9	Mark 15-F	Mark 25	Mark 26	M-1
	Material				
Rotor	310	K-Monel	K-Monel	Same as Mark 15-F	Inconel 718
Blades	310	K-Monel	K-Monel		Mainstage: Inconel 718; transition: Ti A110-AT-ELI
Volute	310	310	310		304 ELC
Stator housing	310	(Integral with volute)	310		304
Vanes	310	310	310		Inconel 718
Front bearing housing	310	310	310		347
Rear bearing housing	310	310	310		304
Balance piston	Al 2024	K-Monel	Inconel 718		Al 7075-T73
Balance piston orifice	Flame-plated tungsten carbide on 310	Leaded bronze	Silver-plated 310		304

The choice of the materials indicated in the preceding table has taken account of the properties of these materials at the temperature (20.28 K) of liquid hydrogen.

The distance d_a (m) between a rotor vane and the contiguous stator vane ranges from 0.02 d_t to 0.05 d_t, where d_i (m) is the impeller tip diameter, as shown in the following figure, adapted from [39].

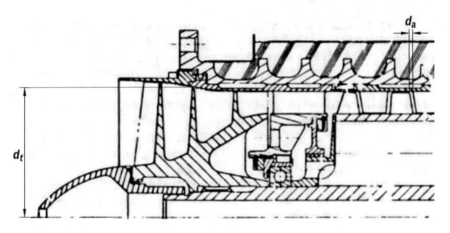

The tip clearance c (m), shown in the following figure, ranges from 0.000127 to 0.000254 m [22].

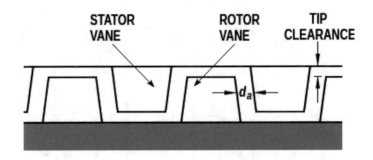

The method used to design an inducer for an axial-flow pump does not differ from the method described in Sect. 4.9, which relates to an inducer for a centrifugal-flow pump. The inducer used for an axial-flow pump has often a cylindrical casing, vanes of a constant tip diameter, and a highly tapered hub. In other words, the hub diameter increases from a small value at the inlet of the inducer to a greater value at the outlet of the inducer, at which point the hub diameter is close to the impeller diameter. This type of inducer, which has been used in the Mark 15-F pump for the J2 engine, is shown in the following figure, due to the courtesy of NASA [4].

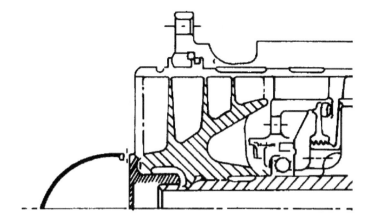

A stationary part of the inducer (inducer stator) is placed downstream of the rotating part. The inducer stator is aimed at converting much of the velocity head, imparted to the pumped fluid by the rotating part of the inducer, into pressure head. The inducer stator is also aimed at discharging the fluid so that its absolute velocity vector c'_2 should have a magnitude c'_2 and a direction α'_2 as close as possible to respectively the magnitude c'_4 and the direction α'_4 of the absolute velocity vector c'_4 of the fluid at the outlet of the impeller stator.

The inducer stator has the same hub diameter d_h and tip diameter d_t as those of the impeller, and has therefore the same effective cross-sectional area in the passage perpendicular to the meridian or axial component c_m of the absolute velocity vector of the flow.

The velocity diagrams at the inlet and at the outlet of the inducer stator vanes can be drawn by using the following equations:

$$P_{is} = \frac{\pi d_m}{Z_{is}}$$

$$S_{is} = \frac{C_{is}}{P_{is}}$$

$$\alpha_{ic} = \frac{\alpha_1 + \alpha_2}{2}$$

$$C_{is} = 2R_{is}\sin\left(\frac{\alpha_2 - \alpha_1}{2}\right) = \frac{L_{is}}{\sin\alpha_{ic}}$$

$$\alpha_1 = \alpha'_1 + i$$

$$\alpha_2 = \alpha'_2 + ii$$

$$c_m = c'_{u1}\tan\alpha'_1 = c'_{u2}\tan\alpha'_2 = c'_1\sin\alpha'_1 = c'_2\sin\alpha'_2$$

$$d_m \tan \alpha_1 = d_t \tan \alpha_{1t} = d_h \tan \alpha_{1h} = d_x \tan \alpha_{1x}$$

$$d_m \tan \alpha_2 = d_t \tan \alpha_{2t} = d_h \tan \alpha_{2h} = d_x \tan \alpha_{2x}$$

where P_{is} (m) is the pitch or the space between two contiguous inducer stator vanes, Z_{is} is the number of the inducer stator vanes, C_{is} (m) is the length of the chord of the inducer stator vanes, $S_{is} = C_{is}/P_{is}$ is the solidity of the inducer stator vanes, α_{ic} (rad) is the angle formed by the chord of each inducer stator vane with the tangential direction, α_1 (rad) and α_2 (rad) are the angles formed by the tangent to each inducer stator vane at respectively the inlet and the outlet with the tangential direction, R_{is} (m) is the radius of curvature of each inducer stator vane, L_{is} (m) is the length of each inducer stator vane in the axial direction, i (rad) is the attack angle at the inlet of each inducer stator vane, ii (rad) is the angle allowed for circulatory flow at the outlet of each inducer stator vane, α'_1 (rad) and α'_2 (rad) are the angles formed by the absolute velocity vectors $\boldsymbol{c}'_1$ and $\boldsymbol{c}'_2$ with the tangential direction at respectively the inlet and the outlet of each inducer stator vane, c_m (m/s) is the meridian or axial component of the absolute velocity vector $\boldsymbol{c}_m$, c'_1 (m/s) and c'_2 (m/s) are the magnitudes of the absolute velocity vectors $\boldsymbol{c}'_1$ and $\boldsymbol{c}'_2$ at respectively the inlet and the outlet of each inducer stator vane, c'_{u1} (m/s) and c'_{u2} (m/s) are the tangential components of the absolute velocity vectors $\boldsymbol{c}'_1$ and $\boldsymbol{c}'_2$ at respectively the inlet and the outlet of each inducer stator vane, α_{1t} (rad), α_{1h} (rad), and α_{1x} (rad) are the inlet angles formed by the tangent to each inducer stator vane at respectively the tip diameter, the hub diameter, and any intermediate diameter with the tangential direction, and α_{2t} (rad), α_{2h} (rad), and α_{2x} (rad) are the outlet angles formed by the tangent to each inducer stator vane at respectively the tip diameter, the hub diameter, and any intermediate diameter with the tangential direction. Unless specified otherwise, the quantities indicated above refer to the mean effective diameter d_m.

According to Huzel and Huang [22], the solidity of the inducer stator vanes $S_{is} = C_{is}/P_{is}$ ranges from 1.5 to 1.8. The number of the inducer stator vanes Z_{is} ranges from 15 to 20. The value of Z_{is} should have no common factor with the value of the impeller rotor vanes Z_r, in order to avoid resonance phenomena.

As has been shown in Sect. 4.2, an axial-flow pump consists essentially of two portions:

- a cylindrical portion, which contains the inducer stage and the impeller stages; and
- a volute portion, placed downstream of the last rotor stage of the impeller, which contains radial guide vanes.

The volute portion of an axial-flow pump is designed so that the flow from the last axial stage is ducted to an exit passage, and is gradually turned toward a plane normal to the rotation axis of the pump, before being collected in the volute proper. Vanes are used in the exit passage to guide and diffuse the flow, and also to tie the volute walls together structurally. The volute portion of the Mark 15-F pump and its diffuser vanes are shown in the following figure, adapted from [2].

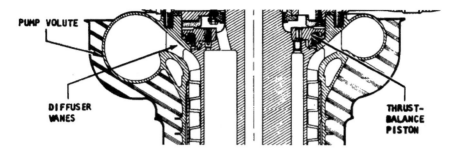

With the exception of the Mark 9 pump, the volute portions of axial-flow pumps have had some degree of "fold-over", as shown in the following figure, due to the courtesy of NASA [2].

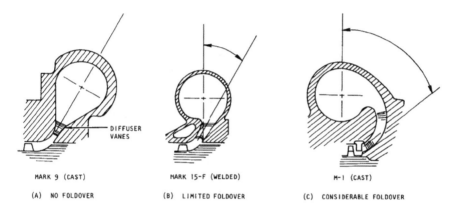

A folded volute permits a smaller overall housing envelope, which results in a lower weight. In addition, a folded volute can be used to obtain a single-vortex rather than double-vortex motion in the volute. This kind of motion is particularly effective to increase efficiency in a volute-exit conical diffuser and to stabilise performance [2].

The number of diffuser vanes ranges usually from 17 to 23 [22]. This number should have no common factor with the number of impeller rotor vanes, in order to avoid resonance phenomena. The vane angle α_v, shown in the following figure, adapted from [2], can be determined by drawing the velocity diagrams for the diffuser vanes. The line, which forms the angle α_v with the chord line, is normal to the plane of the row of diffuser vanes.

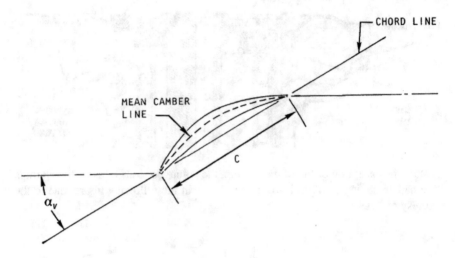

The areas of the various sections of the volute for an axial-flow pump can be computed by using one of the two methods (either $c'_{u3}\, r = $ constant or $c'_3 = $ constant) described in Sect. 4.9 for a centrifugal-flow pump. For example, when the average velocity c'_3 (m/s) of the flow is assumed to be constant in all sections of the volute, then the area A_θ (m^2) of each cross section of the volute is computed by using the following equations

$$c'_3 = \frac{\theta}{2\pi} \frac{q}{A_\theta}$$

$$c'_3 = K_v (2g_0 \Delta H)^{\frac{1}{2}}$$

where θ (rad) is the central angle corresponding to the area A_θ of interest, q (m^3/s) is the rated design volume flow rate of the pump, K_v is a design factor, whose value is determined experimentally and ranges from 0.15 to 0.55, ΔH (m) is the design head rise of the pump, and $g_0 = 9.80665$ m/s^2 is the acceleration of gravity near the surface of the Earth.

In case of axial-flow pumps for liquid hydrogen, the design velocity of flow in the volute ranges from 30.5 to 45.7 m/s [22]. The cross section of the volute has usually a circular shape, due to the high pressure. The boil-off of liquid hydrogen is limited as far as possible by using high-quality thermal insulating materials applied to the external surface of the pump.

In an axial-flow turbo-pump, thrust bearings and compensating balance pistons are used to counteract the high axial thrust loads due to pressure forces and fluid momentum changes in the pump and in the turbine [2].

The following figure, due to the courtesy of NASA [36], shows the balance piston and bearing scheme chosen for the Mark 15-F turbo-pump.

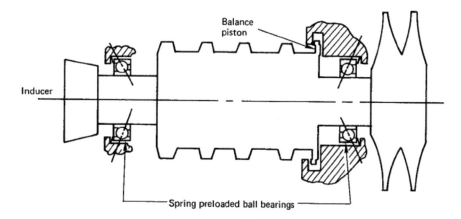

The balance piston, which is part of the rotor assembly, is a disk having variable orifices with a pair of seal rubs located one on the front side (left in the figure) and the other on the rear side (right in the figure) of the disk. High-pressure fluid coming from the pump discharge is introduced into the thrust balance system and flows through the two variable orifices in series with a low-pressure area.

When an axial thrust moves the rotor assembly toward the inducer of the pump, then the balance piston closes the high-pressure orifice at the front seal rub and opens the low-pressure orifice at the rear seal rub, thereby reducing the pressure in the control chamber placed between the two seal rubs. This reduction of pressure counteracts the hydraulic axial thrust acting on the rotor assembly toward the inducer, and restrains the motion of the rotor assembly in that direction. Likewise, when a reverse axial thrust moves the rotor assembly towards the discharge of the pump, then the balance piston opens the high-pressure orifice at the front seal rub and closes the low-pressure orifice at the rear seal rub, thereby increasing the pressure in the control chamber placed between the two seal rubs. This increase in pressure counteracts the reverse axial thrust acting on the rotor assembly toward the discharge, and causes the rotor assembly to return to its neutral position. The working principle described above is shown in the following figure, due to the courtesy of NASA [2], which illustrates the series-flow thrust-balance system used for the Mark 15-F pump.

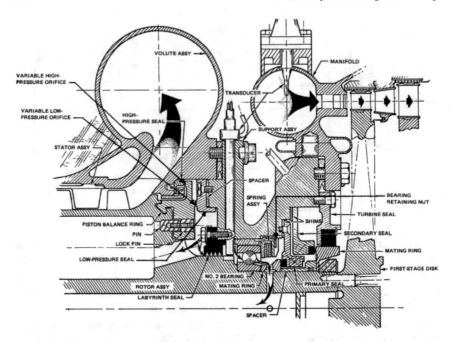

Let ΔT_a (N) be the reduction of axial thrust due to the thrust-balance system described above. According to Huzel and Huang [22], the value of ΔT_a can be computed as follows

$$\Delta T_a = \Delta p_c \frac{\pi}{4}\left(d_f^2 - d_r^2\right)$$

where Δp_c (N/m^2) is the variation of pressure of the fluid in the control chamber, d_f (m) is the diameter of the balance piston at the front seal rub, and d_r (m) is the diameter of the balance piston at the rear seal rub.

As an application of the concepts discussed above, the following requirements are given for a multi-stage axial-flow pump, similar to the Mark 15-F pump for the J2 engine, having an inducer of cylindrical casing, with vanes of a constant tip diameter, and a highly tapered hub. The pumped fluid is liquid hydrogen.

Rated design head rise of the pump $\Delta H = 13660$ m.

Rated design volume flow rate of the pump $q = 0.3836$ m^3/s.

Required Net Positive Suction Head $NPSH_R = 41.15$ m.

Suction specific speed of the inducer $(S)_{ind} = 19.53$.

Maximum allowed value of the flow coefficient at the inlet of the inducer $\varphi_{ind} = 0.09$.

Head coefficient of the inducer $\psi_{ind} = 0.307$.

Leakage loss rate of the inducer $q_{ee} = 0.03\, q = 0.03 \times 0.3836 = 0.01151$ m^3/s.

Head loss due to the inducer stator $\Delta H_{ee} = 0.08\, \Delta H_{ind}$.

Solidity of the vanes of the inducer stator $S_{is} = C_{is}/P_{is} = 1.53$.

Number of vanes in the inducer stator $Z_{is} = 17$.
Specific speed for each stage of the pump $(N)_1 = 1.188$.
Head coefficient for each stage of the pump $(\psi)_1 = 0.304$.
Impeller hub ratio $r_d = d_h/d_t = 0.857$.
Leakage loss rate of the impeller $q_e = 0.06\, q = 0.06 \times 0.3836 = 0.02302$ m³/s.
Head loss for each stage of the impeller stator $\Delta H_e = 0.08\,(\Delta H)_1$.
Solidity of the impeller rotor vanes $S_r = C_r/P_r = 1.05$ at the mean effective diameter.
Number of impeller rotor vanes $Z_r = 16$.
Solidity of the impeller stator vanes $S_s = C_s/P_s = 1.61$ at the mean effective diameter.
Contraction factor for the stator and rotor vane passage $\varepsilon = 0.88$.
Angle of attack at the vane inlet $i = \pi/45 = 0.06981$ rad.
Angle allowed for circulatory flow at the vane outlet $ii = \pi/36 = 0.08727$ rad.

We want to design the axial-flow pump corresponding to the data given above, including the stator of the inducer, and the rotor and the stator of the impeller.

As to the stator of the inducer, we remember the following equation (see Sect. 4.2), which defines the suction specific speed

$$S = \frac{\omega\, q^{\frac{1}{2}}}{(g_0\, NPSH_R)^{\frac{3}{4}}}$$

After substituting $S \equiv (S)_{ind} = 19.53$, $q = 0.3836$ m³/s, $NPSH_R = 41.15$ m, and $g_0 = 9.807$ m/s² and solving for ω, the rotating speed of the pump results

$$\omega = \frac{19.53 \times (9.807 \times 41.15)^{\frac{3}{4}}}{0.3836^{\frac{1}{2}}} = 2839 \text{ rad/s}$$

Then, the head rise $(\Delta H)_1$ for each stage of the pump results from solving the following equation

$$(N)_1 = \frac{\omega\, q^{\frac{1}{2}}}{\left[g_0\,(\Delta H)_1\right]^{\frac{3}{4}}}$$

for $(\Delta H)_1$. After substituting $\omega = 2839$ rad/s, $q = 0.3836$ m³/s, and $(N)_1 = 1.188$, we find

$$(\Delta H)_1 = \frac{q^{\frac{2}{3}}}{g_0}\left[\frac{\omega}{(N)_1}\right]^{\frac{4}{3}} = \frac{0.3836^{\frac{2}{3}}}{9.807}\left(\frac{2839}{1.188}\right)^{\frac{4}{3}} = 1720 \text{ m}$$

Since the head coefficient for each stage of the axial-flow pump is $(\psi)_1 = 0.304$, we substitute this value in the following equation

$$(\psi)_1 = \frac{g_0 \, (\Delta H)_1}{u_m^2}$$

and compute the peripheral velocity u_m of the rotor blades at the mean effective diameter d_m of the impeller as follows

$$u_m = \left[\frac{g_0 (\Delta H)_1}{(\psi)_1} \right]^{\frac{1}{2}} = \left(\frac{9.807 \times 1720}{0.304} \right)^{\frac{1}{2}} = 235.6 \text{ m/s}$$

The mean effective diameter d_m of the impeller, in turn, results from substituting $u_m = 235.6$ m/s and $\omega = 2839$ rad/s in the following equation

$$u_m = \omega \frac{d_m}{2}$$

This yields

$$d_m = \frac{2u_m}{\omega} = \frac{2 \times 235.6}{2839} = 0.1660 \text{ m}$$

Substituting this value of d_m and $r_d = d_h/d_t = 0.857$ (where r_d is the impeller hub ratio) in the following equation

$$d_m = \left(\frac{d_t^2 + d_h^2}{2} \right)^{\frac{1}{2}} = \left[\frac{d_t^2 \left(1 + r_d^2 \right)}{2} \right]^{\frac{1}{2}}$$

we find the following value of the tip diameter of the impeller rotor

$$d_t = \left(\frac{2}{1 + r_d^2} \right)^{\frac{1}{2}} d_m = \left(\frac{2}{1 + 0.857^2} \right)^{\frac{1}{2}} \times 0.1660 = 0.1783 \text{ m}$$

Since $r_d = d_h/d_t = 0.857$ and $d_t = 0.1783$ m, then the hub diameter of the impeller rotor (which has a cylindrical casing) is

$$d_h = r_d d_t = 0.857 \times 0.1783 = 0.1528 \text{ m}$$

The height h_v of each rotor vane results from

$$h_v = \frac{d_t - d_h}{2} = \frac{0.1783 - 0.1528}{2} = 0.01275 \text{ m}$$

Remembering the following equation

$$\Delta H = \Delta H_{ind} + \Delta H_{ee} + n(\Delta H)_1$$

where $(\Delta H)_1 = 1720$ m is the head rise for each stage, $\Delta H = 13{,}660$ m is the total head rise of the pump, ΔH_{ind} (m) is the minimum head rise of the inducer, n is the number of stages, and $\Delta H_{ee} = 0.08\,\Delta H_{ind}$ is the hydraulic head loss at the stator of the inducer, and choosing $n = 7$, there results

$$\Delta H - n(\Delta H)_1 = \Delta H_{ind}(1 - 0.08)$$

Hence, the minimum head rise of the inducer is

$$\Delta H_{ind} = \frac{\Delta H - n(\Delta H)_1}{1 - 0.08} = \frac{13{,}660 - 7 \times 1720}{1 - 0.08} = 1761 \text{ m}$$

The tip diameter of the vanes of the inducer is taken equal to $d_t = 0.1783$ m, which is the tip diameter of the vanes of the impeller. Therefore, the peripheral speed of the vanes of the inducer is

$$u_t = \omega\frac{d_t}{2} = 2839 \times \frac{0.1783}{2} = 253.1 \text{ m/s}$$

Since the head coefficient of the inducer is known to be $\psi_{ind} = 0.307$, and the peripheral speed of the inducer vanes results $u_t = 253.1$ m/s, then we use the following equation

$$\psi_{ind} = \frac{g_0\,\Delta H_{ind}}{u_t^2}$$

to compute the head rise due to the inducer as follows

$$\Delta H_{ind} = \frac{\psi_{ind}u_t^2}{g_0} = \frac{0.307 \times 253.1^2}{9.807} = 2005 \text{ m}$$

This value is greater than the minimum head rise of the inducer $(\Delta H_{ind} = 1761$ m) computed above.

As has been shown in Sect. 4.9, the required volume flow rate at the impeller q_{imp} and the required volume flow rate at the inducer q_{ind} are expressed by the following equations

$$q_{imp} = q + q_e$$

$$q_{ind} = q + q_{ee} + \frac{1}{2}q_e$$

By substituting $q = 0.3836$ m^3/s, $q_e = 0.02302$ m^3/s, and $q_{ee} = 0.01151$ m^3/s in the preceding equations, we find

$$q_{imp} = 0.3836 + 0.02302 = 0.4066 \text{ m}^3/\text{s}$$
$$q_{ind} = 0.3836 + 0.01151 + 0.5 \times 0.02302 = 0.4066 \text{ m}^3/\text{s}$$

We take a hub whose diameter at the inlet of the inducer is $d_{0h} = 0.07366$ m. Since the casing of the inducer is cylindrical, then the magnitude c'_0 of the absolute velocity vector c'_0 of flow at the inlet of the inducer can be computed by means of the following equation of Sect. 4.9

$$c'_0 = c_{m0} = \frac{4q_{ind}}{\pi \left(d_{0t}^2 - d_{0h}^2\right)}$$

In the present case, $d_{0t} = d_t = \text{constant} = 0.1783$ m, because the inducer casing is cylindrical, and therefore the preceding equation can be re-written as follows

$$c'_0 = c_{m0} = \frac{4q_{ind}}{\pi \left(d_t^2 - d_{0h}^2\right)}$$

Since $q_{ind} = 0.4066$ m^3/s, $d_t = 0.1783$ m, and $d_{0h} = 0.07366$ m, then

$$c'_0 = c_{m0} = \frac{4q_{ind}}{\pi \left(d_t^2 - d_{0h}^2\right)} = \frac{4 \times 0.4066}{3.1416 \times \left(0.1783^2 - 0.07366^2\right)} = 19.64 \text{ m/s}$$

As a check, we compute the flow coefficient φ_{ind} at the inlet of the inducer, as follows

$$\varphi_{ind} = \frac{c_{m0}}{u_t} = \frac{19.64}{253.1} = 0.07760$$

which is less than the maximum allowed value $\varphi_{ind} = 0.09$.

We assume the meridian component c_m of the absolute flow velocity to have equal values at the outlet of the inducer, through the stators, and through the rotors. As has been shown above in the present section, the value of c_m can be computed by means of the following equation

$$c_m = \frac{q_{imp}}{\frac{\pi}{4}\left(d_t^2 - d_h^2\right)\varepsilon}$$

Since $q_{imp} = 0.4066$ m^3/s, $d_t = 0.1783$ m, $d_h = 0.1528$ m, and $\varepsilon = 0.88$, then

$$c_m = \frac{q_{imp}}{\frac{\pi}{4}\left(d_t^2 - d_h^2\right)\varepsilon} = \frac{4 \times 0.4066}{3.1416 \times \left(0.1783^2 - 0.1528^2\right) \times 0.88} = 69.68 \text{ m/s}$$

Remembering the following equation of Sect. 4.9

$$c_{m1} = \frac{4q_{ind}}{\pi \left(d_{1t}^2 - d_{1h}^2\right)}$$

where $c_{m1} = c_m = 69.68$ m/s, $d_{1t} = d_t = 0.1783$ m, and $q_{ind} = 0.4066$ m^3/s, and solving for d_{1h}, there results

$$d_{1h} = \left(d_t^2 - \frac{4q_{ind}}{\pi c_m}\right)^{\frac{1}{2}} = \left(0.1783^2 - \frac{4 \times 0.4066}{3.1416 \times 69.68}\right)^{\frac{1}{2}} = 0.1561 \text{ m}$$

Since the mean effective diameter d_1 at the outlet of the inducer is defined as follows

$$d_1^2 = \frac{d_{1h}^2 + d_{1t}^2}{2}$$

and $d_{1t} = d_t = 0.1783$ m, $d_{1h} = 0.1561$ m, then

$$d_1 = \left(\frac{0.1561^2 + 0.1783^2}{2}\right)^{\frac{1}{2}} = 0.1676 \text{ m}$$

The peripheral speed u_1 of the blades of the inducer at the mean effective diameter d_1 is

$$u_1 = \omega \frac{d_1}{2} = 2839 \times \frac{0.1676}{2} = 237.9 \text{ m/s}$$

Remembering the following equation of Sect. 4.9

$$\Delta H_{ind} = \frac{u_t^2 \psi_{ind}}{g_0} = \frac{u_1 c_{u1}'}{g_0}$$

and solving for c'_{u1}, the tangential component c'_{u1} of the absolute velocity vector c'_1 at the inlet of the inducer stator is

$$c_{u1}' = \frac{g_0 \Delta H_{ind}}{u_1} = \frac{9.807 \times 2005}{237.9} = 82.65 \text{ m/s}$$

The magnitude c'_1 of of the absolute velocity vector c'_1 at the inlet of the inducer stator results from

$$c_1' = \left(c_{u1}'^2 + c_m^2\right)^{\frac{1}{2}} = \left(82.65^2 + 69.68^2\right)^{\frac{1}{2}} = 108.1 \text{ m/s}$$

The angle α'_1 which the vector c'_1 forms with the tangential direction at the inlet of the inducer stator is

$$\alpha_1' = \arctan\left(\frac{c_m}{c_{u1}'}\right) = \arctan\left(\frac{69.68}{82.65}\right) = 0.7005 \text{ rad}$$

In the inducer stator, the flow area in the axial direction is chosen to be equal to the flow area in the stators and in the rotors of the impeller. In other words, the diameters d_t, d_h, and d_m, and the contraction factor ε of the stator vanes in the inducer are chosen to be $d_t = 0.1783$ m, $d_h = 0.1528$ m, and $d_m = 0.1660$ m, and $\varepsilon = 0.88$. We also assume equal conditions for the flow at the inlet and at the outlet of the inducer stator.

Since $\alpha'_1 = 0.7005$ rad and the angle of attack at the inlet of the inducer stator vane is $i = 0.06981$ rad, then the vane angle α_1 at the inlet of the inducer stator vane, at the mean effective diameter d_m, is

$$\alpha_1 = \alpha_1' + i = 0.7005 + 0.06981 = 0.7703 \text{ rad}$$

We want the angle α'_2 between the vectors c'_2 and u_2 at the outlet of the inducer stator to be $\alpha'_2 = 13\pi/36 = 1.134$ rad, as a design choice. For this purpose, the vane angle α_2 at the outlet of the inducer stator must be

$$\alpha_2 = \alpha_2' + ii = 1.134 + 0.08727 = 1.221 \text{ rad}$$

where $ii = 0.08727$ rad is the angle allowed for circulatory flow at the vane outlet of the inducer stator.

The tangential component c'_{u2} of the absolute velocity vector c'_2 at the outlet of the inducer stator results from

$$c_{u2}' = \frac{c_m}{\tan \alpha_2'} = \frac{69.68}{\tan 1.134} = 32.53 \text{ m/s}$$

The magnitude c'_2 of the absolute velocity vector c'_2 at the outlet of the inducer stator results from

$$c_2' = \left(c_{u2}'^2 + c_m^2\right)^{\frac{1}{2}} = \left(32.53^2 + 69.68^2\right)^{\frac{1}{2}} = 76.90 \text{ m/s}$$

The pitch P_{is} of the stator vanes (that is, the distance, in the circumferential direction, between two contiguous stator vanes) results from the following equation

$$P_{is} = \frac{\pi d_m}{Z_{is}}$$

where $d_m = 0.1660$ m is the mean effective diameter, and $Z_{is} = 17$ is the number of vanes in the stator of the inducer. Hence

$$P_{is} = \frac{3.1416 \times 0.1660}{17} = 0.03068 \text{ m}$$

By using this value and $S_{is} = C_{is}/P_{is} = 1.53$, the length C_{is} of the chord of the vanes in the stator of the inducer at the mean effective diameter results

$$C_{is} = 0.03068 \times 1.53 = 0.04694 \text{ m}$$

The chord angle α_{ic} of the inducer stator vanes results from

$$\alpha_{ic} = \frac{\alpha_1 + \alpha_2}{2} = \frac{0.7703 + 1.221}{2} = 0.9957 \text{ rad}$$

Remembering the equation

$$C_{is} = 2R_{is} \sin\left(\frac{\alpha_2 - \alpha_1}{2}\right) = \frac{L_{is}}{\sin \alpha_{ic}}$$

the axial length L_{is} of the inducer stator vanes, at the mean effective diameter, is

$$L_{is} = C_{is} \sin \alpha_{ic} = 0.04694 \times \sin 0.9957 = 0.03939 \text{ m}$$

The radius of curvature R_{is} of the inducer stator vanes results from

$$C_{is} = 2R_{is} \sin\left(\frac{\alpha_2 - \alpha_1}{2}\right) = \frac{L_{is}}{\sin \alpha_{ic}}$$

This equation, solved for R_{is}, yields

$$R_{is} = \frac{C_{is}}{2 \sin\left(\frac{\alpha_2 - \alpha_1}{2}\right)} = \frac{0.04694}{2 \times \sin\left(\frac{1.221 - 0.7703}{2}\right)} = 0.1050 \text{ m}$$

A summary of the results found above for the inducer stator is given below. The velocity diagram is shown in the following figure.

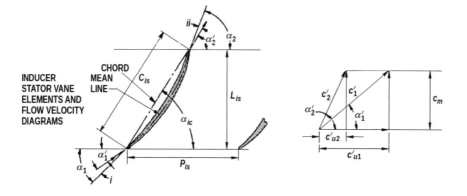

The velocity diagram shown above can be drawn by using the following values, which refer to the mean effective diameter $d_m = 0.1660$ m, unless specified otherwise. The angle between the vectors c'_1 and u_1 at the inlet of the inducer stator is $\alpha'_1 = 0.7005$ rad. The magnitude of the absolute velocity vector c'_1 is $c'_1 = 108.1$ m/s. The meridian (axial) component of c'_1 is $c_m = 69.68$ m/s. The tangential component of c'_1 is $c'_{u1} = 82.65$ m/s. The angle between the vectors c'_2 and u_2 at the outlet of the inducer stator is $\alpha'_2 = 13\pi/36 = 1.134$ rad. The magnitude of the absolute velocity vector c'_2 is $c'_2 = 76.90$ m/s. The meridian (axial) component of c'_2 is $c_m = 69.68$ m/s. The tangential component of c'_2 is $c'_{u2} = 32.53$ m/s. The elements of the inducer stator vanes are: vane angle at the inlet $\alpha_1 = 0.7703$ rad; vane angle at the outlet $\alpha_2 = 1.221$ rad; chord angle $\alpha_{ic} = 0.9957$ rad; vane solidity $S_{is} = C_{is}/P_{is} = 1.53$; number of vanes $Z_{is} = 17$; vane pitch $P_{is} = 0.03068$ m; vane chord length $C_{is} = 0.04694$ m; vane axial length $L_{is} = 0.03939$ m; and vane radius of curvature $R_{is} = 0.1050$ m.

Now, as to the stator and rotor of the impeller, we assume the absolute velocity vector at the impeller rotor inlets and at the impeller stator outlets to be the same as the absolute velocity vector at the inducer stator outlet.

This assumption can be understood by considering the following figure.

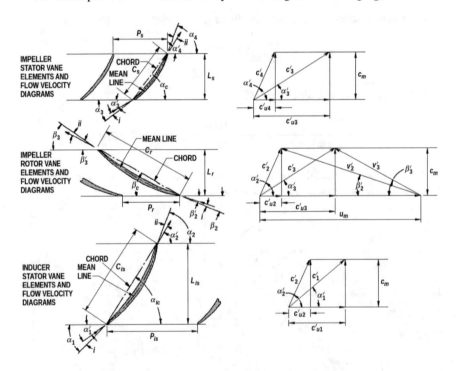

In specific terms, with reference to the preceding figure, we assume:

$$\alpha_2' = \alpha_4' = 13\pi/36 = 1.134 \text{ rad}$$
$$c_2' = c_4' = 76.90 \text{ m/s}$$
$$c_m = 69.68 \text{ m/s}$$
$$c_{u2}' = c_{u4}' = 32.53 \text{ m/s}$$

A simple inspection of the preceding figure makes it possible to compute the angle β'_2, which the relative velocity vector v'_2 forms with the tangential direction at the inlets of the impeller rotor vanes, as follows

$$\tan \beta_2' = \frac{c_m}{u_m - c_{u2}'} = \frac{69.68}{235.6 - 32.53} = 0.3431$$

Hence

$$\beta_2' = \arctan(0.3431) = 0.3305 \text{ rad}$$

Likewise, the magnitude v'_2 of the relative velocity vector v'_2 can be computed as follows

$$v_2' = \left[\left(u_m - c_{u2}' \right)^2 + c_m^2 \right]^{\frac{1}{2}} = \left[(235.6 - 32.53)^2 + 69.68^2 \right]^{\frac{1}{2}} = 214.7 \text{ m/s}$$

The angle β_2 formed by the tangent of each rotor vane at the inlet with the tangential direction results from

$$\beta_2 = \beta_2' + i = 0.3305 + 0.06981 = 0.4003 \text{ rad}$$

where $i = 0.06981$ rad is the angle of attack at the vane inlet.

Remembering the following equation

$$\Delta H_{imp} = (\Delta H)_1 + \Delta H_e = \frac{u_m c_{u3}' - u_m c_{u2}'}{g_0}$$

where ΔH_{imp} is the required head rise for each stage of the impeller, $(\Delta H)_1$ is the rated design head rise for each stage of the axial-flow pump, ΔH_e is the hydraulic head loss for each stage of the impeller stator, u_m is the peripheral velocity of the rotor blades at the mean effective diameter d_m, c'_2 and c'_3 are the design magnitudes of the absolute velocity vector c at respectively the inlet and the outlet of each rotor blade, $g_0 = 9.807$ m/s^2 is the gravitational acceleration on the surface of the Earth, and also remembering that, in the present case, $(\Delta H)_1 = 1720$ m, and $\Delta H_e = 0.08$ $(\Delta H)_1 = 0.08 \times 1720 = 137.6$ m, the required head rise for each stage of the impeller results

$$\Delta H_{imp} = 1720 + 137.6 = 1858 \text{ m}$$

In addition, from the same equation

$$\Delta H_{imp} = (\Delta H)_1 + \Delta H_e = \frac{u_m c'_{u3} - u_m c'_{u2}}{g_0}$$

solved for c'_3, there results

$$c'_{u3} = \frac{g_0 \Delta H_{imp}}{u_m} + c'_{u2} = \frac{9.807 \times 1858}{235.6} + 32.53 = 109.9 \text{ m/s}$$

The magnitude c'_3 of the absolute velocity vector c'_3 at the outlet of the impeller rotor vanes results from

$$c'_3 = \left(c'^2_{u3} + c^2_m\right)^{\frac{1}{2}} = \left(109.9^2 + 69.68^2\right)^{\frac{1}{2}} = 130.1 \text{ m/s}$$

The angle α'_3, which the absolute velocity vector c'_3 forms with the tangential direction at the outlets of the impeller rotor vanes, results from

$$\tan \alpha'_3 = \frac{c_m}{c'_{u3}} = \frac{69.68}{109.9} = 0.6340$$

Hence

$$\alpha'_3 = \arctan(0.6340) = 0.5651 \text{ rad}$$

The magnitude v'_3 of the relative velocity vector v'_3 at the outlets of the impeller rotor vanes results from

$$v'_3 = \left[\left(u_m - c'_{u3}\right)^2 + c^2_m\right]^{\frac{1}{2}} = \left[(235.6 - 109.9)^2 + 69.68^2\right]^{\frac{1}{2}} = 143.7 \text{ m/s}$$

The angle β'_3, which the relative velocity vector v'_3 forms with the tangential direction at the outlets of the impeller rotor vanes, results from

$$\tan \beta'_3 = \frac{c_m}{u_m - c'_{u3}} = \frac{69.68}{235.6 - 109.9} = 0.5543$$

Hence

$$\beta'_3 = \arctan(0.5543) = 0.5062 \text{ rad}$$

Remembering the equation

$$\beta_3 = \beta'_3 + ii$$

where β_3 is the angle formed by the tangent of each rotor vane at the outlet with the tangential direction, and $ii = 0.08727$ rad is the angle allowed for circulatory flow at the vane outlet, there results

$$\beta_3 = 0.5062 + 0.08727 = 0.5934 \text{ rad}$$

The pitch distance P_r separating two contiguous vanes of the rotor, measured along the circumference, results from

$$P_r = \frac{\pi d_m}{Z_r}$$

where, in the present case, the mean effective diameter and the number of the rotor vanes are respectively $d_m = 0.1660$ m and $Z_r = 16$. Hence

$$P_r = \frac{3.1416 \times 0.1660}{16} = 0.03259 \text{ m}$$

The length C_r of the chord of each rotor vane at the mean effective diameter results from

$$S_r = \frac{C_r}{P_r}$$

where, in the present case, the solidity of the rotor vanes at the mean effective diameter is $S_r = 1.05$. Hence

$$C_r = 1.05 \times 0.03259 = 0.03422 \text{ m}$$

The angle β_c formed by the chord of each rotor vane with the tangential direction results from

$$\beta_c = \frac{\beta_2 + \beta_3}{2}$$

Since, in the present case, $\beta_2 = 0.4003$ rad and $\beta_3 = 0.5934$ rad, then

$$\beta_c = \frac{0.4003 + 0.5934}{2} = 0.4969 \text{ rad}$$

The length L_r of the rotor vanes in the axial direction results from

$$L_r = C_r \sin \beta_c = 0.03422 \times \sin 0.4969 = 0.01631 \text{ m}$$

The radius of curvature R_r of the rotor vanes at the mean effective diameter results from

$$C_r = 2R_r \sin\left(\frac{\beta_3 - \beta_2}{2}\right)$$

Since, in the present case, $C_r = 0.03422$ m, $\beta_3 = 0.5934$ rad, and $\beta_2 = 0.4003$ rad, then

$$R_r = \frac{C_r}{2\sin\left(\frac{\beta_3 - \beta_2}{2}\right)} = \frac{0.03422}{2 \times \sin\left(\frac{0.5934 - 0.4003}{2}\right)} = 0.1775 \text{ m}$$

Now, we assume the absolute velocity vector at the impeller stator inlets to be the same as the absolute velocity vector at the impeller rotor outlets.

Remembering the following equation

$$\alpha_3 = \alpha'_3 + i$$

where α_3 is the vane angle at the inlets of the impeller stator, $\alpha'_3 = 0.5651$ rad is the angle formed by the absolute velocity vector c'_3 with the tangential direction at the inlets of the impeller stator, and $i = 0.06981$ rad is the angle of attack at the inlets of the impeller stator, there results at the mean effective diameter

$$\alpha_3 = 0.5651 + 0.06981 = 0.6349 \text{ rad}$$

Remembering the following equation

$$\alpha_4 = \alpha'_4 + ii$$

where α_4 is the vane angle at the outlets of the impeller stator, and $\alpha'_4 = 13\pi/36$ rad is the angle formed by the absolute velocity vector c'_4 with the tangential direction at the outlets of the impeller stator, and $ii = \pi/36$ rad is the angle allowed for circulatory flow at the outlets of the impeller stator, there results at the mean effective diameter

$$\alpha_4 = \frac{13}{36}\pi + \frac{1}{36}\pi = \frac{7}{18}\pi = 1.222 \text{ rad}$$

At the mean effective diameter, the axial length L_s of the stator vanes is equal to the axial length $L_r = 0.01631$ m of the rotor vanes, and therefore

$$L_s = 0.01631 \text{ m}$$

Remembering the following equation

$$\alpha_c = \frac{\alpha_3 + \alpha_4}{2}$$

where α_c is the angle formed by the chord of each stator vane with the tangential direction, $\alpha_3 = 0.6349$ rad and $\alpha_4 = 1.222$ rad are the angles formed by the tangent to each stator vane at respectively the inlet and the outlet with the tangential direction, there results

$$\alpha_c = \frac{0.6349 + 1.222}{2} = 0.9285 \text{ rad}$$

Remembering the following equation

$$C_s = 2R_s \sin\left(\frac{\alpha_4 - \alpha_3}{2}\right) = \frac{L_s}{\sin \alpha_c}$$

the length C_s of the chord and the radius of curvature R_s of each stator vane, at the mean effective diameter, are

$$C_s = \frac{L_s}{\sin \alpha_c} = \frac{0.01631}{\sin 0.9285} = 0.02037 \text{ m}$$

$$R_s = \frac{C_s}{2 \sin\left(\frac{\alpha_4 - \alpha_3}{2}\right)} = \frac{0.02037}{2 \times \sin\left(\frac{1.222 - 0.6349}{2}\right)} = 0.03520 \text{ m}$$

Remembering the following equation

$$S_s = \frac{C_s}{P_s}$$

where P_s is the pitch or the space between two contiguous stator vanes, $C_s = 0.02037$ m is the length of the chord of the stator vanes, and $S_s = 1.61$ is the solidity of each stator vane, there results

$$P_s = \frac{C_s}{S_s} = \frac{0.02037}{1.61} = 0.01265 \text{ m}$$

Remembering the following equation

$$P_s = \frac{\pi d_m}{Z_s}$$

where $P_s = 0.01265$ m is the pitch, $d_m = 0.1660$ m is the mean effective diameter, and Z_s is the number of the impeller stator vanes, there results

$$Z_s = \frac{\pi \, d_m}{P_s} = \frac{3.1416 \times 0.1660}{0.01265} = 41$$

A summary of the results found above for the impeller stator and rotor is given below, with reference to the following velocity diagrams.

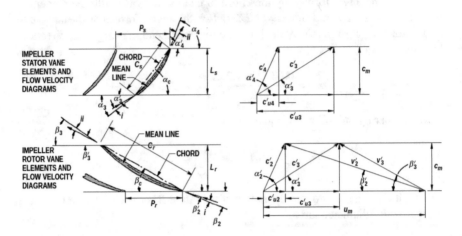

At the inlet of the impeller rotor vanes, at the mean effective diameter $d_m = 0.1660$ m, we find: $\alpha'_2 = 13\pi/36 = 1.134$ rad, $\beta'_2 = 0.3305$ rad, $u_m = 235.6$ m/s, $v'_2 = 214.7$ m/s, $c'_2 = 76.90$ m/s, $c'_{u2} = 32.53$ m/s, and $c_m = 69.68$ m/s.

At the outlet of the impeller rotor vanes, at the mean effective diameter $d_m = 0.1660$ m, we find: $\alpha'_3 = 0.5651$ rad, $\beta'_3 = 0.5062$ rad, $u_m = 235.6$ m/s, $v'_3 = 143.7$ m/s, $c'_3 = 130.1$ m/s, $c'_{u3} = 109.9$ m/s, and $c_m = 69.68$ m/s.

At the inlet of the impeller stator vanes, at the mean effective diameter $d_m = 0.1660$ m, we find: $\alpha'_3 = 0.5651$ rad, and $c'_3 = 130.1$ m/s.

At the outlet of the impeller stator vanes, at the mean effective diameter $d_m = 0.1660$ m, we find: $\alpha'_4 = 13\pi/36 = 1.134$ rad, $c'_4 = 76.90$ m/s, $c'_{u4} = 32.53$ m/s, and $c_m = 69.68$ m/s.

Tip diameter for the impeller rotor and stator vanes $d_t = 0.1783$ m.

Hub diameter for the impeller rotor and stator vanes $d_h = 0.1528$ m.

Height for the impeller rotor and stator vanes $h_v = 0.01275$ m.

Mean diameter for the impeller rotor and stator vanes $d_m = 0.1660$ m.

Elements of the impeller rotor vanes: $\beta_2 = 0.4003$ rad, $\beta_3 = 0.5934$ rad, $\beta_c = 0.4969$ rad, $S_r = 1.05$, $Z_r = 16$, $P_r = 0.03259$ m, $C_r = 0.03422$ m, $L_r = 0.01631$ m, and $R_r = 0.1775$ m.

Elements of the impeller stator vanes: $\alpha_3 = 0.6349$ rad, $\alpha_4 = 7\pi/18 = 1.222$ rad, $\alpha_c = 0.9285$ rad, $S_s = 1.61$, $Z_s = 41$, $P_s = 0.01265$ m, $C_s = 0.02037$ m, $L_s = 0.01631$ m, and $R_s = 0.03520$ m.

4.11 Design of Turbines

Pumps for rocket engines are usually driven by impulse turbines, whose working principle has been described in Sect. 4.3. This is because impulse turbines are simpler and lighter than reaction turbines. The following figure, due to the courtesy of NASA [15], illustrates the Mark 10, single-stage, two-row, velocity-compounded impulse turbine, which has been used for the F-1 engine.

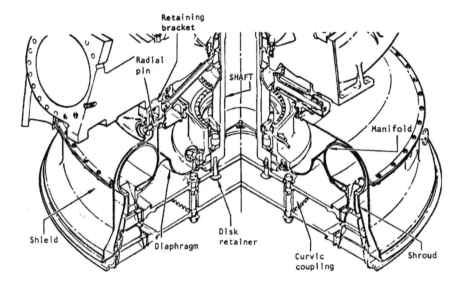

The design of a turbine for a rocket engine is based on the following considerations. The number of stages depends on the power required.

As has been shown in Sect. 4.3, the design isentropic velocity ratio determines the range of efficiency of a turbine. The isentropic velocity ratio for a single-stage turbine is U/C_0, where U (m/s) is the peripheral velocity of the blades, $C_0 = (2 \Delta h_T)^{1/2}$ (m/s) is the isentropic spouting velocity, and Δh_T (J/kg) is the enthalpy drop per unit mass of the working fluid through a turbine. For a N-stage turbine, the isentropic velocity ratio is $(U_1^2 + U_2^2 + \cdots + U_N^2)^{1/2}/C_0$. A plot of isentropic velocity ratio versus turbine efficiency η_T is given in the following figure, due to the courtesy of NASA [15], which illustrates the variation in peak efficiency for 1-, 2-, and 3-row impulse staging turbines and for reaction staging turbines.

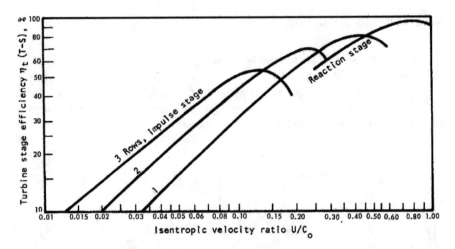

These curves make it possible to establish and compare initial estimates of efficiency for turbines of new design. In practice, design isentropic velocity ratios for two-row turbines used in gas-generator and tap-off engine cycles (see Sect. 4.4) range from 0.1 to 0.4, and corresponding efficiencies vary from 35 to 65% [15]. Staged-combustion and expander-cycle turbines reach efficiencies above 80%, because the lower values of the turbine system pressure ratio p_0/p_e permit operation in a range of velocity ratios corresponding to higher efficiency, as has been shown quantitatively in Sect. 4.8.

Peripheral velocities U of current turbine blades vary from 305 to 457 m/s, and associated inlet temperatures for liquid oxygen/RP-1 and for liquid oxygen/liquid hydrogen turbines range from 811 to 1089 K. Rocket engine turbines have been operated successfully without cooling at inlet temperatures of 1200 K [15].

The number of stages in a turbine of new design depends on the quantity of energy possessed by the unit mass of the working fluid and made available to the turbine. It also depends on a compromise between a gain in turbine efficiency and the added mass due to the further turbine stages required to increase performance. In a study of staging, pressure ratios and enthalpies per unit mass are varied in several possible configurations, to determine the design point corresponding to the highest efficiency with the smallest pitch diameter and the least number of stages. Current turbines are usually of the single-row or two-row axial type. This is due principally to size and mass restrictions affecting turbines used for rocket engines. Impulse-type staging is usually specified in turbines of new design with isentropic velocity ratios below 0.35. These turbines operate with high-energy working fluids and at pressure ratios which raise the stage spouting velocity into the supersonic domain. Reaction turbines are used with low pressure ratios (<2) and isentropic velocity ratios above 0.45 [15].

In a direct-drive arrangement, when the pumps and the turbine are mounted on a single shaft, the rotating speed ω (rad/s) of the turbine and therefore the isentropic velocity ratio U/C_0 have values lower than the ideal ones. In this arrangement, a single-stage, two-row, velocity-compounded impulse turbine is generally chosen, as is the case with the Mark 10 turbo-pump illustrated in the following figure, adapted from [5].

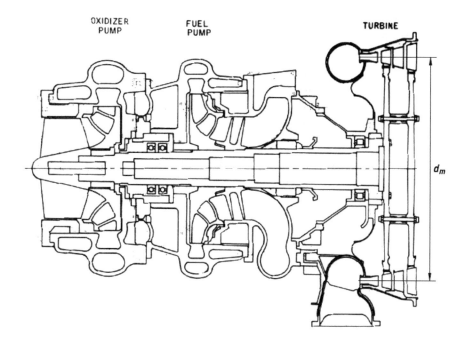

By contrast, in a geared, offset-turbine arrangement, the presence of a gearbox makes it possible to operate the turbine at rotational speeds much higher than those which are possible with a direct-drive arrangement. For example, for the Mark 3 turbo-pump described at length in Sect. 4.5, the input rotational speed is 3434.8 rad/s, and the speed reduction ratio input/pump-shaft is 4.885 for both of the pumps [15]. Consequently, the Mark 3 turbo-pump has a two-stage, pressure-compounded impulse turbine.

After the type of impulse turbine and the number of its stages have been determined, it is necessary to determine the size of its rotor. To this end, it is necessary to know the properties (inlet temperature T_0, specific heat ratio γ, etc.) of the working fluid, the pressure ratio p_0/p_e of the turbine, and the rotation speed ω of the pumps.

The rotor tip diameter d_t (m) is chosen by taking account that a large diameter results in a high isentropic velocity ratio U/C_0 and a high efficiency η_T, but also in high size, high mass, and high stresses in operating conditions. Therefore, the rotor tip diameter results from a compromise.

The useful power P_d (W) delivered by the turbine shaft must be equal to the sum of the net power required by the pumps, the power loss in the gearbox (if any), and the power required by auxiliary drives. After the power resulting from this sum has been determined, the required mass flow rate $\dot{m}_T$ (kg/s) of the working fluid can be determined by using the following equation of Sect. 4.8:

$$\eta_T = \frac{P_d}{\dot{m}_T c_p T_0 \left[1 - \left(\frac{1}{R_T}\right)^{\frac{\gamma-1}{\gamma}}\right]}$$

where η_T is the efficiency of the turbine, c_p (J kg^{-1} K^{-1}) is the specific heat at constant pressure of the working fluid, T_0 (K) is the temperature of the working fluid at the inlet of the turbine, $R_T = p_0/p_e$ is the pressure ratio of the turbine, p_0 (N/m^2) is the total pressure of the working fluid at the turbine inlet, p_e (N/m^2) is is the static pressure of the working fluid at the turbine exhaust, and γ is the specific heat ratio of the working fluid. For this purpose, the value of the turbine efficiency η_T can be estimated by using the preceding plot which gives η_T as a function of the isentropic velocity ratio U/C_0.

This done, the dimensions of the nozzles, rotor blades, and stator blades can be determined according to the properties and the mass flow rate $\dot{m}_T$ (kg/s) of the working fluid through the turbine. These properties for some combinations of propellants are given in the following table, adapted from [15].

Working fluid[a]	Chamber pressure of working-fluid source, N/m^2	Turbine inlet temperature (T_{t1}), K	Ratio of specific heats (γ)	Specific heat at constant pressure (C_p), J/(kg K)	Specific gas constant (R), J/(kg K)
LOX/RP-1	6.895×10^6	866.5	1.097	2659	236.7
		977.6	1.115	2713	279.8
		1089	1.132	2747	317.4
		1144	1.140	2763	338.9
		1200	1.148	2772	360.5
LOX/LH$_2$	1.034×10^7	810.9	1.374	8365	2276
		922.0	1.364	7997	2130
		1089	1.348	7545	1948
		1200	1.337	7277	1835
		1311	1.326	7063	1738
LF$_2$/LH$_2$	5.171×10^6	755.4	1.395	11,580	3368
		866.5	1.388	10,920	3088

(continued)

(continued)

Working fluid[a]	Chamber pressure of working-fluid source, N/m²	Turbine inlet temperature (T_{t1}), K	Ratio of specific heats (γ)	Specific heat at constant pressure (C_p), J/(kg K)	Specific gas constant (R), J/(kg K)
		977.6	1.380	10,500	2846
		1089	1.373	9705	2642
		1200	1.366	9177	2459
FLOX/CH₄	5.171×10^6	977.6	1.176	3588	538.0
		1089	1.189	3580	570.3
		1200	1.215	3492	607.9
		1311	1.233	3391	624.1
		1422	1.242	3312	629.5
Hydrogen		644.3	1.396	14,570	4121
		699.8	1.394	14,610	4121
		755.4	1.392	14,660	4121
		810.9	1.390	14,720	4121
		866.5	1.388	14,780	4121
$\frac{N_2O_4}{50-50}$[b]	1.034×10^7	1200	1.248	3040	602.6
		1228	1.249	3010	597.2
		1255	1.254	2964	597.2
		1283	1.260	2906	597.2

[a]Products of combustion of propellant combinations listed
[b]Hydrazine-UDMH

The shape of most nozzles for a turbine of a rocket engine is like that of a thrust chamber, that is, of the converging-diverging type, as shown in the preceding figure. The nozzles convert the potential energy of the working fluid into kinetic energy by reducing the temperature and the pressure of the fluid as it passes through the vane channels. The kinetic energy is then converted to work as the working fluid passes through the rotor blades [15]. The path followed by the working fluid through a two-row impulse turbine as that of the Mark 10 turbo-pump is shown in the following figure, re-drawn from [22].

FLOW

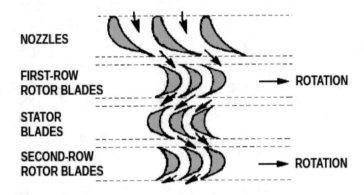

NOZZLES

FIRST-ROW
ROTOR BLADES   → ROTATION

STATOR
BLADES

SECOND-ROW
ROTOR BLADES   → ROTATION

The expansion of a gas flowing through an actual nozzle is not isentropic, due to the viscosity of the working fluid, friction, boundary-layer effects, etc. The loss of kinetic energy due to friction and turbulence in the flow causes an increase in temperature (reheat) in the gas flowing through a nozzle. Therefore, the actual velocity C_1 of the gas at the exit of a nozzle is lower than its spouting velocity C_0, which is calculated assuming an isentropic expansion. In addition, the effective area of a nozzle is less than its geometric area, due to boundary-layer effects and circulatory flow. These phenomena can be taken into account by means of the coefficients defined below.

The nozzle velocity coefficient k_n is defined as the ratio of the actual velocity C_1 (m/s) of the gas at the exit of a nozzle to its spouting velocity C_0 (m/s), as follows

$$k_n = \frac{C_1}{C_0}$$

where C_0 is calculated assuming an isentropic expansion of the gas from the total pressure at the inlet of the nozzle to the static pressure at the inlet of the rotor blade, as will be shown below.

The nozzle efficiency η_n is defined as the ratio of the output power (kinetic energy, based on C_1, per unit time) of the gas at the outlet of the nozzle to the input power (kinetic energy, based on C_0, per unit time) of the same gas at the inlet of the nozzle, as follows

$$\eta_n = \frac{\frac{1}{2}\dot{m}_T C_1^2}{\frac{1}{2}\dot{m}_T C_0^2} = k_n^2$$

The nozzle throat area coefficient ε_{nt} is defined as the ratio of the effective area A_{1t} (m^2) of the nozzle at the throat to the geometric area A_{0t} (m^2) of the nozzle at the throat, as follows

$$\varepsilon_{nt} = \frac{A_{1t}}{A_{0t}}$$

Due to the drop of enthalpy per unit mass Δh_{0-1} (J/kg) which occurs in the nozzle

$$\Delta h_{0-1} = c_p T_0 \left[1 - \left(\frac{p_1}{p_0} \right)^{\frac{\gamma-1}{\gamma}} \right]$$

the actual velocity C_1 (m/s) of the gas at the exit of a nozzle results

$$C_1 = k_n C_0 = k_n (2\Delta h_{1-2})^{\frac{1}{2}} = k_n \left\{ 2c_p T_0 \left[1 - \left(\frac{p_1}{p_0} \right)^{\frac{\gamma-1}{\gamma}} \right] \right\}^{\frac{1}{2}}$$

where Δh_{0-1} (J/kg) is the drop of enthalpy per unit mass of the gas in the nozzle, c_p (J kg^{-1} K^{-1}) is the specific heat at constant pressure of the gas, γ is the specific heat ratio of the gas, T_0 (K) is the total temperature of the gas at the inlet of the nozzle, p_1 (N/m^2) is the static pressure of the gas at the outlet of the nozzle, and p_0 (N/m^2) is the total pressure of the gas at the inlet of the nozzle.

The reheat q_{nr} (m^2/s^2 or J/kg) of the nozzle is the difference of kinetic energy per unit mass between the inlet and the outlet of the nozzle, as follows

$$q_{nr} = \frac{1}{2}C_0^2 - \frac{1}{2}C_1^2 = \frac{(1 - k_n^2)C_1^2}{2k_n^2} = \frac{(1 - \eta_n)C_1^2}{2\eta_n}$$

This is because $C_0^2 = C_1^2/k_n^2 = C_1^2/\eta_n$.
Remembering the following equation of Chap. 1, Sect. 1.2:

$$\frac{\dot{m}}{A_t} = \frac{p_c \gamma^{\frac{1}{2}}}{(RT_c)^{\frac{1}{2}}} \left[\frac{2}{\gamma + 1} \right]^{\frac{\gamma+1}{2(\gamma-1)}}$$

where A_t (m^2) is the area at the throat of a converging-diverging nozzle, the total area A_{nt} (m^2) required at the throat of a turbine nozzle is

$$A_{nt} = \frac{\dot{m}_T}{\varepsilon_{nt} p_0 \left[\frac{\gamma}{RT_0} \left(\frac{2}{\gamma+1} \right)^{\frac{\gamma+1}{\gamma-1}} \right]^{\frac{1}{2}}}$$

where $R = R^*/M$ (N m K^{-1} kg^{-1}) is the constant of the specific gas, $R^* = 8314.46$ N m kmol^{-1} K^{-1} is the universal gas constant, M (kg/kmol) is the molar mass of the gas, and $\dot{m}_T$ (kg/s) is the mass flow rate of the gas through the turbine.

The values of the coefficients k_n, η_n, and ε_{nt} defined above depend on the geometric and thermodynamic properties of a turbine. Design values of these coefficients, determined experimentally or by experience acquired from existing rocket turbines, are given below. Design values of the nozzle velocity coefficient k_n range from 0.89 to 0.98. Consequently, design values of the nozzle efficiency $\eta_n = k_n^2$ vary from 0.79 to 0.96. Design values of the nozzle throat area coefficient ε_{nt} vary from 0.95 to 0.99 [22].

The cross section of the nozzles in a rocket turbine may be either a square or a rectangle, according to whether the width b_{nt} (m) normal to the flow direction at the throat of a nozzle is equal to or different from the radial height h_{nt} (m) at the same point, as shown in the following figure, re-drawn from [22].

The nozzles are closely spaced on either a circular arc which extends over a part of the circumference or the whole circumference. With reference to the following figure, the direction of the gas expanding through a nozzle is approximately axial at the inlet of the nozzle, and changes so as to form an angle α_1 (rad) with the plane of rotation at the outlet of the nozzle. Therefore, the gas flow turns in the nozzle through an angle of $\pi/2 - \alpha_1$. Let θ_n (rad) be the angle which the central line of the nozzle forms with the plane of rotation at the outlet of the nozzle. The choice of the value of θ_n is left to the designer. Usually, this value ranges from $\pi/12$ to $\pi/6$ radians [22]. The angle α_1 is generally greater than θ_n, due to the unsymmetrical shape of the nozzle at the outlet.

The nozzle area A_{nt} (m^2) at the throat results from

$$A_{nt} = Z_n \, b_{nt} \, h_{nt}$$

where Z_n is the number of nozzles, b_{nt} (m) is the width normal to the flow direction at the throat of a nozzle, and h_{nt} (m) is the radial height at the same point.

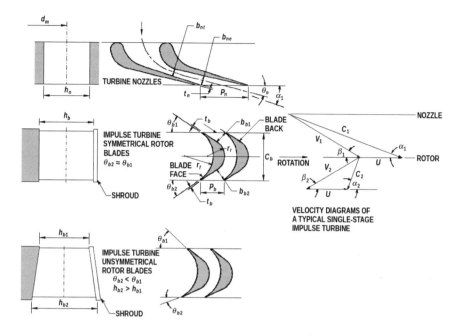

The total nozzle area A_{ne} (m^2) at the outlet results from

$$A_{ne} = Z_n b_{ne} h_{ne} = Z_n h_{ne} (P_n \sin \theta_n - t_n) = \frac{\dot{m}_T}{\rho_1 C_1 \varepsilon_{ne}}$$

where b_{ne} (m) is the width normal to the flow direction at the exit of a nozzle, h_{ne} (m) is the radial height at the same point, P_n (m) is the nozzle pitch (the distance, measured in the direction of rotation at the mean diameter d_m, between corresponding points of two adjacent blades of the nozzle). θ_n (rad) is the angle which the central line of the nozzle forms with the plane of rotation at the outlet of the nozzle, t_n (m) is the thickness of the nozzle at the exit, $\dot{m}_T$ (kg/s) is the mass flow rate of the gas through the turbine, ρ_1 (kg/m^3) is the density of the gas at the outlet of the nozzle, C_1 (m/s) is the spouting velocity of the gas at the outlet of the nozzle, and ε_{ne} is the area coefficient at the exit of the nozzle.

The nozzle pitch P_n (m) results from

$$P_n = \frac{\pi d_m}{Z_n}$$

where d_m (m) is the mean diameter of the nozzle or rotor blades, and Z_n is the number of nozzles.

The materials of which the nozzles, manifolds, rotor blades, rotor discs, and stator vanes for gas turbines for rocket engines are usually made are indicated in the following table, adapted from [15].

Turbine designation	Manifold and nozzle[a]	Rotor blades[b]	Rotor disks	Stator vanes[b]
Mark 3	Hastelloy B, cast	Stellite 21	16-25-6	Hastelloy B, cast
Mark 4	M: 19-9DL N: Hastelloy B, cast	Stellite 21	16-25-6	Hastelloy B, casl
Mark 9	Inconel X-750	Inconel X-750	Inconel X-750	Inconel X-750
Mark 10 ($d_m = 0.889$ m)	Rene 41	Alloy 713C	Rene 41	Stellite 21
Mark 10 ($d_m = 0.762$ m)	M: Hastelloy C N; Stellite 21	Alloy 713C	Inconel 718	Stellite 21
Mark 15-O	310	Stellite 21	16-25-6	Stellite 21
Mark 15-F	Hastelloy C	Alloy 713C	Inconel 718	Stellite 21
Mark 25	Inconel X-750	Inconel 718	Inconel 718	Inconei X-750
Mark 29-O	321	Stellite 21	16-25-6	Stellite 21
Mark 29-F	Hastelloy C	Alloy 713C	Inconel 718	Hastelloy C
Summary (10 turbines)	Hastelloy B, cast 19-9DL Inconel X-750 Rene 41 Hastelloy C Stellite 21 310 321	Stellite 21 Inconel X-750 Alloy 713C Inconel 718	16-25-6 Inconel X-750 Rene 41 Inconel 718	Hastelloy B, cast Inconel X-750 Stellite 21 Hastelloy C

[a]The manifolds in most designs are welded from more than one alloy; the principal alloy used for the manifold is listed above. The nozzle is part of the welded manifold assembly for all the listed turbines. M designates manifold material; N, nozzle material

[b]For rotor blades and stator vanes
• Stellite 21 and alloy 713C arc available only in the cast form (casting alloys)
• In the Mark 3 and 4 turbines, the blades were attached to the disk by welding; in the Mark 9 and 25, integrally machined with the disk. In the other listed turbines, fir-tree blade attachments were used

Stellite® 21 is a cobalt-chromium alloy, which is designed to be resistant to wear, corrosion, and heat. The alloys of the Stellite group may also have some portions of tungsten or molybdenum and some small but critical amounts of carbon. Stellite is a trademarked name of Deloro Stellite Company supplying Stellite alloys like Stellite 3, Stellite 6, Stellite 12 and Stellite 21.

Inconel® (comprising 713C, 718, X-750, …) is a family of austenitic nickel-chromium based super-alloys having excellent resistance to oxidation and thermal fatigue, as has been shown in Chap. 2, Sect. 2.5.

Hastelloy® is a registered trademark name of Haynes International, Inc., which is applied as the prefix name of a range of corrosion-resistant metal alloys under the term super-alloys or high-performance alloys. The principal ingredient of these alloys is nickel. Other ingredients are molybdenum, chromium, cobalt, iron, copper,

manganese, titanium, zirconium, aluminium, carbon, and tungsten in various percentages. The primary quality of the Hastelloy alloys is resistance to high thermal and
mechanical stresses.

The rotor blades of an impulse turbine transform as much as possible of the kinetic
energy possessed by the gas coming from the nozzles into mechanical energy. This
process is subject to losses due to friction, eddy current, boundary layer effects,
and reheat, as has been shown above. The velocity vector diagram illustrated in the
following figure, re-drawn from [22], shows the gas flow through the rotor blades of
a single-stage, single-rotor impulse turbine, based on the mean diameter d_m of the
stator or rotor blades.

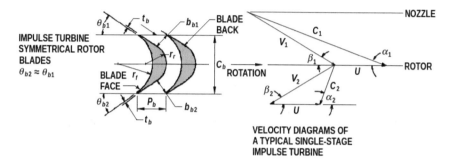

VELOCITY DIAGRAMS OF
A TYPICAL SINGLE-STAGE
IMPULSE TURBINE

The gas flow at the nozzle outlet is represented by the absolute velocity vector
C_1. This vector forms an angle α_1 with the plane of rotation of the rotor. The vector
U represents the peripheral speed of the rotor blades at their mean diameter d_m. The
vectors V_1 and V_2 represent the relative velocities of the gas flow at respectively
the inlet and the outlet of the rotor. The magnitudes V_1 and V_2 of these vectors are
such that $V_1 > V_2$, due to the friction losses. The gas flow at the outlet of the rotor
blades is represented by the absolute velocity vector C_2. It is desirable for C_2 to be
in magnitude be as small as possible and in direction as close as possible to the axial
direction. The force F acting on the rotor blades is equal to the rate of change of
momentum of the gas flowing through the turbine.

With reference to the preceding figure, the tangential component F_t (N) of the
force F acting on the rotor blades at their mean diameter d_m (m) is

$$F_t = \dot{m}_T(C_1 \cos \alpha_1 + C_2 \cos \alpha_2) = \dot{m}_T(V_1 \cos \beta_1 + V_2 \cos \beta_2)$$

where $\dot{m}_T$ (kg/s) is the mass flow rate of gas through the turbine.

The power P_b (W) which the gas flow transfers to the rotor blades at their mean
diameter d_m is

$$P_b = F_t U = \dot{m}_T(C_1 \cos \alpha_1 + C_2 \cos \alpha_2)U = \dot{m}_T(V_1 \cos \beta_1 + V_2 \cos \beta_2)U$$

where U (m/s) is the peripheral velocity of the rotor blades at their mean diameter
d_m, as follows

$$U = \frac{1}{2}\omega\, d_m$$

and ω (rad/s) is the angular velocity of the rotor. Again, with reference to the preceding figure, the angle β_1 (rad), which the relative velocity vector V_1 forms with the plane of rotation at the inlet of the rotor, results from

$$\beta_1 = \arctan\left(\frac{C_1 \sin \alpha_1}{C_1 \cos \alpha_1 - U}\right)$$

The axial component F_a (N) of the force F acting on the rotor blades at d_m is

$$F_a = \dot{m}_T(C_1 \sin \alpha_1 - V_2 \sin \beta_2)$$

The blade velocity coefficient k_b is defined as follows

$$k_b = \frac{V_2}{V_1}$$

The blade efficiency η_b is defined as the ratio of the power P_b which the gas flow transfers to the rotor blades to the input power $1/2\dot{m}_T C_1^2$ due to the gas flow through the turbine, as follows

$$\eta_b = \frac{P_b}{\frac{1}{2}\dot{m}_T C_1^2}$$

In the ideal case, the blade efficiency η_b is maximum for a single-stage, single-rotor impulse turbine when

$$U = \frac{1}{2}C_1 \cos \alpha_1 = \frac{1}{2}C_{1t}$$

where C_{1t} is the tangential component of the absolute velocity vector C_1 of the gas at the outlet of the nozzle. In this ideal case, the power P_b which the gas flow transfers to the rotor blades

$$P_b = F_t U = \dot{m}_T(C_1 \cos \alpha_1 + C_2 \cos \alpha_2)U = \dot{m}_T(V_1 \cos \beta_1 + V_2 \cos \beta_2)U$$

can be expressed as follows

$$P_b = \frac{1}{4}\dot{m}_T C_1^2\left(1 + k_b \frac{\cos \beta_2}{\cos \beta_1}\right)\cos^2 \alpha_1$$

and therefore the blade efficiency η_b defined above becomes

$$\eta_b = \frac{1}{2}\left(1 + k_b \frac{\cos \beta_2}{\cos \beta_1}\right)\cos^2 \alpha_1$$

This is because $k_b = V_2/V_1$, $V_1 \cos \beta_1 + U = C_1 \cos \alpha_1$, and $U = \frac{1}{2} C_1 \cos \alpha_1$, as shown in the velocity diagram of the preceding figure.

When there is some reaction or expansion of the gas flowing through the blades of the rotor, then the relative velocity V_2 (m/s) of the gas at the outlet of the rotor can be determined as follows

$$V_2 = \left(k_b^2 V_1^2 + 2\eta_n \Delta h_{1-2}\right)^{\frac{1}{2}}$$

where Δh_{1-2} (J/kg) is the isentropic drop of enthalpy per unit mass of the gas flowing through the rotor blades due to the reaction or expansion. In the absence of reaction or expansion, $\Delta h_{1-2} = 0$.

The reheat q_{br} (m^2/s^2 or J/kg) of the rotor blades is the difference of kinetic energy per unit mass between the inlet and the outlet of the rotor blades, as follows

$$q_{br} = \frac{\left(1 - k_b^2\right)V_1^2}{2} + (1 - \eta_n)\Delta h_{1-2}$$

The rotor blade efficiency η_b of a single-stage, single-rotor impulse turbine can be expressed as follows

$$\eta_b = \frac{P_b}{\dot{m}_T c_p T_0 \left[1 - \left(\frac{p_e}{p_0}\right)^{\frac{\gamma-1}{\gamma}}\right]}$$

where P_b (W) is the power which the gas transfers to the rotor blades, $\dot{m}_T$ (kg/s) is the mass flow rate through the turbine, c_p (J kg^{-1} K^{-1}) and γ are respectively the specific heat at constant pressure and the specific heat ratio of the gas, T_0 (K) is the total temperature of the gas at the turbine inlet, p_0 (N/m^2) is the total pressure of the gas at the turbine inlet, and p_e (N/m^2) is is the static pressure of the gas at the turbine exhaust.

The total efficiency η_T of a single-stage, single-rotor impulse turbine can be expressed by the following product

$$\eta_T = \eta_n \eta_b \eta_m$$

where η_n and η_b are respectively the nozzle efficiency and the rotor blade efficiency defined above, and η_m is the machine efficiency, which takes account of the mechanical, leakage, and disc-friction losses. In this case, the rotor blade efficiency

$$\eta_b = \frac{1}{2}\left(1 + k_b \frac{\cos \beta_2}{\cos \beta_1}\right)\cos^2 \alpha_1$$

is higher when β_2 becomes much smaller than β_1. The reduction of β_2 without decreasing the flow area at the outlet of the rotor blades can be obtained by using unsymmetrical blades for the rotor, as shown in the following figure, re-drawn from [22].

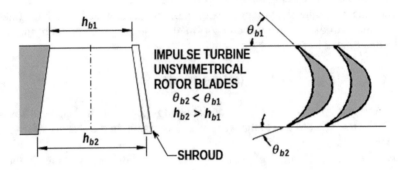

In case of unsymmetrical blades, the radial height h_b (m) of each blade increases from h_{b1} at the inlet to h_{b2} at the outlet. In practice, for the purpose of avoiding flow separation and reducing centrifugal forces, the value of the angle β_2 of an unsymmetrical blade is determined as follows

$$\beta_2 = \beta_1 - \Delta\beta$$

where $\Delta\beta$ ranges from $\pi/36$ to $\pi/12$ rad [22]. The angle θ_{b1} is formed by the plane tangent to the pressure surface at the inlet of a blade and the plane of rotation. The angle θ_{b2} is formed by the plane tangent to the pressure surface at the outlet of a blade and the plane of rotation. In case of unsymmetrical blades, the angles θ_{b1} and θ_{b2} are such that $\theta_{b2} < \theta_{b1}$, as shown in the preceding figure. The equation

$$\eta_b = \frac{1}{2}\left(1 + k_b \frac{\cos\beta_2}{\cos\beta_1}\right)\cos^2\alpha_1$$

also shows that the blade efficiency η_b increases when the angle α_1 decreases. The angle α_1 is formed by the absolute velocity vector C_1 with the plane of rotation.

Design values of the blade velocity coefficient k_b range from 0.80 to 0.90. Design values of the blade efficiency η_b range from 0.70 to 0.92 [22].

The radial height h_b at the inlet of a rotor blade is usually greater, by a factor ranging from 1.05 to 1.10, than the radial height h_n of the nozzle. The centrifugal force acting on a rotor blade depends on its radial height h_b and on its peripheral velocity U.

The mean diameter d_m (m) of the rotor blades is defined [22] as follows

$$d_m = d_t - h_b$$

where d_t (m) is the rotor tip diameter. Further elements of a rotor blade than those described above are shown in the following figure, due to the courtesy of NASA [40].

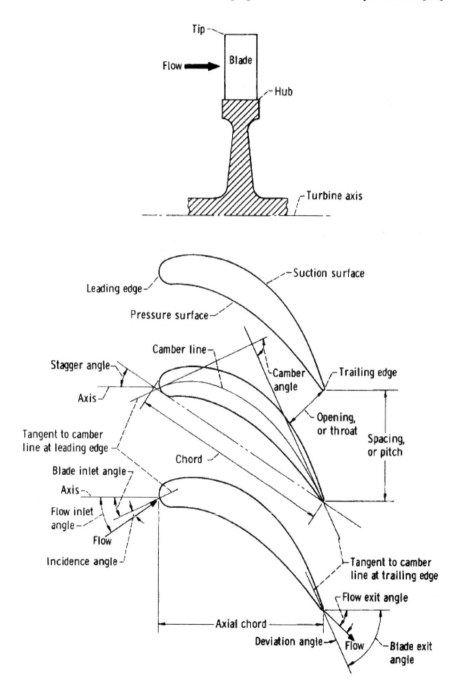

The tip and the hub are respectively the outermost section and the innermost section of a blade.

The radial height h_b (m) of a blade is the difference between its radius at the tip and its radius at the hub.

The blade pitch P_b (m) or blade spacing, measured at the mean diameter d_m, is the distance in the direction of rotation between corresponding points of two adjacent blades.

The pressure surface is the concave surface of a blade, where the pressure reaches its maximum value.

The suction surface is the convex surface of a blade, where the pressure reaches its minimum value.

The leading edge and the trailing edge are respectively the front (or nose) and the rear (or tail) of a blade.

The chord line is the line between the two points in which the front and the rear of a blade section touch a flat surface, on which the blade section is placed with its convex surface upside.

The chord C_b (m) is the length of the perpendicular projection of the blade profile onto the chord line. This length is approximately equal to the distance between the leading edge and the trailing edge.

The axial chord is the length of the projection of a blade, as set in the turbine, onto a line parallel to the turbine axis.

The camber line is the mean line of a blade profile, and extends from the leading edge to the trailing edge, halfway between the pressure surface and the suction surface.

The camber angle is the external angle formed by the intersection of the tangents to the camber line at the leading edge and at the trailing edge.

The stagger angle is the angle between the chord line and the axial direction of the turbine.

The blade aspect ratio is the ratio of the height h_b to the axial length C_b of a blade. Design values of the aspect ratio range from 1.3 to 2.5 [22].

The solidity of a blade is the ratio of its chord C_b to the pitch P_b of the blade. Design values of a blade solidity range from 1.4 to 2.0 [22].

The nozzle pitch P_n and the blade pitch P_b can be chosen independently from each other, as long as the number Z_b of the rotor blades is sufficient to direct the flow. This number depends on the aspect ratio and on the solidity of the blades, and should not have any common factor with the number of the nozzles or with the number of the stator blades.

A blade profile has segments of straight lines combined with circular arcs or parabolic arcs or both. The following figure, adapted from [15], shows a blade profile based on straight-line segments combined with circular arcs.

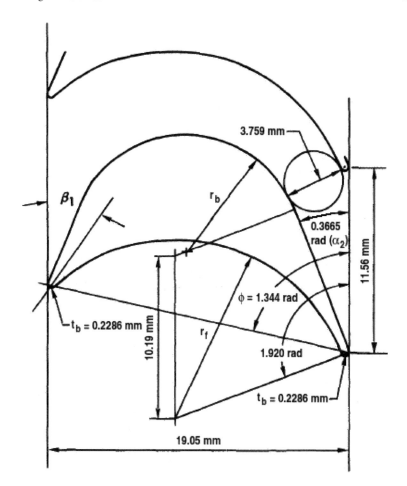

The pressure surface of the blade is concave, and the radius of its circular arc is designated with r_f. The suction side of the blade is convex, and the radius of its circular arc is designated with r_b. The two tangents at the two endpoints of the circular arc on the pressure side form two angles (θ_{b1} and θ_{b2} at respectively the inlet and the outlet of the blade) with the direction of rotation. The thickness of the blade at its leading edge and at its trailing edge is designated with t_b.

As shown in the preceding figure, the angle θ_{b1} should be sightly greater than the angle β_1 formed by the relative velocity vector V_1 with the direction of rotation at the inlet of the rotor blades. If it were otherwise ($\theta_{b1} < \beta_1$), the flowing gas would impinge on the blades just at the inlet, with consequent losses. The angle θ_{b2} is usually chosen equal to the angle β_2 formed by the relative velocity vector V_2 with the direction of rotation at the outlet of the rotor blades.

The mass flow rate $\dot{m}_T$ (kg/s) of gas through a turbine is assumed constant. The areas A_{b1} (m^2) at the blade inlet and A_{b2} (m^2) at the blade outlet, which are needed for the gas flow, can be determined by means of the following equations

$$\dot{m}_T = \rho_1 V_1 A_{b1} \varepsilon_{b1} = \rho_2 V_2 A_{b2} \varepsilon_{b2}$$

$$A_{b1} = Z_b b_{b1} h_{b1} = Z_b h_{b1} (P_b \sin \theta_{b1} - t_b)$$

$$A_{b2} = Z_b b_{b2} h_{b2} = Z_b h_{b2} (P_b \sin \theta_{b2} - t_b)$$

$$P_b = \frac{\pi d_m}{Z_b}$$

where ρ_1 (kg/m^3) and ρ_2 (kg/m^3) are the values of density of the gas at respectively the inlet and the outlet of the rotor blades, V_1 (m/s) and V_2 (m/s) are the magnitudes of the relative velocity vectors at respectively the inlet and the outlet of the rotor blades, ε_{b1} and ε_{b2} are the dimensionless area coefficients at respectively the inlet and the outlet of the rotor blades, Z_b is the number of the rotor blades, b_{b1} (m) and b_{b2} (m) are the widths (perpendicular to the direction of the flow) of the passages at respectively the inlet and the outlet of the rotor blades, h_{b1} (m) and h_{b2} (m) are the radial heights of the passages at respectively the inlet and the outlet of the rotor blades, P_b (m) is the pitch of the rotor blades, d_m (m) is the mean diameter of the rotor blades, θ_{b1} (rad) and θ_{b2} (rad) are the angles formed by the tangents to the circular arc on the pressure side with the direction of rotation at respectively the inlet and the outlet of the rotor blades, and t_b (m) is the thickness of the edges of the rotor blades.

As shown in the following figure, due to the courtesy of NASA [40], the tips of turbine rotor or stator blades are usually covered with shrouds.

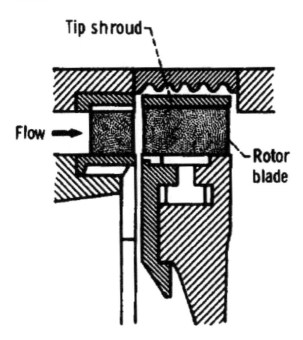

For example, in the model A3-3 pressure-compounded, two-stage turbine used in the RL 10 engine, the blades are fully shrouded, and labyrinth seals are incorporated to reduce blade and interstage leakage. In this turbine, shown in Sect. 4.13, the aluminium shrouds are brazed to the blade tips and to the vane roots.

Experimental test data have shown that turbine rotors with shrouded blades, when compared to rotors with unshrouded blades, are capable of providing efficiency advantages in impulse turbines. This increased performance is attributed to the reduced amount of blade-tip unloading and leakage of the shrouded blades. Turbines developing powers equal to or greater than 373 kW, including the turbines for the J-2, F-1, and M-1 engines, have shrouded blades [15]. The shroud forms frequently an integral part of the blade, and the shroud of each blade fits closely to the shroud of the contiguous blade, as shown in the following figure, adapted from [15], which illustrates an interlocking rubbing shroud, providing damping and resistance to blade torsional motion.

The shroud may also form a continuous ring, in which case it is either attached or welded to the blade tips. The roots of the blades may be either welded to the disc or attached to it by using dove-tail or fir-tree joints, as has been shown in Sect. 4.3.

The loads acting on the rotor blades of a turbine are due to the following causes. The first cause is the centrifugal force acting on each blade. This force produces a tensile stress which reaches its maximum value at the root section of a blade. In order to reduce this stress, a blade can be tapered by making its section at the tip narrower than its section at the root. In addition, the centres of mass of the sections of a blade at various radii do not belong to a radial line. Therefore, the centrifugal forces acting in these offset centres of mass of a blade generate bending moments, whose value is also maximum at the root section of the blade.

The second cause is the bending moment due to the gas flow through the blades. The momentum change of the gas flow generates a force, which has a radial component, which drives the turbine, and an axial component. This force, in turn, generates a bending moment acting on the blade. For the purpose of determining this bending moment, the force due to the momentum change of the flowing gas can be considered as acting at the mid-height of the blade.

The third cause is the bending moment due to a vibration force. This is because the gas flow through the blades is not uniform, but varies periodically between a maximum and a minimum value. The frequency of this force should not be equal to any of the natural frequencies of the blade, in order to avoid bending stresses of considerable magnitude. A careful design of the blades can offset part of the bending moment due to the gas flow by using the bending moment due to the centrifugal force.

The tensile stress σ_c (N/m^2) at the root section of a blade of constant cross-sectional area, without shroud, due to centrifugal forces, can be expressed as follows [41, p. 927, Eq. J-2]:

$$\sigma_c = \frac{1}{2}\rho_b\, h_b\, d_m\, \omega^2$$

The tensile stress σ_{ct} (N/m^2) at the root section of a tapered blade, without shroud, due to centrifugal forces, can be expressed as follows [41, p. 927, Eq. J-5]:

$$\sigma_{ct} = \frac{1}{2}\rho_b\, h_b\, d_m\, \omega^2\left[1 - \frac{1}{2}\left(1 - \frac{A_t}{A_r}\right)\left(1 + \frac{h_b}{3d_m}\right)\right]$$

The bending moment M_g (Nm) at the root section of a blade, due to the gas flow, can be expressed as follows [22]:

$$M_g = \frac{h_b}{2Z_b}\left(F_t^2 + F_a^2\right)^{\frac{1}{2}}$$

where ρ_b (kg/m^3) is the density of the material of which the blade is made, h_b (m) is the average height of the blade, d_m (m) is the mean diameter of the rotor blades, ω (rad/s) is the angular velocity of the turbine, A_r (m^2) is the area of the cross section of the blade at the root, A_t (m^2) is the area of the cross section of the blade at the tip, Z_b is the number of the rotor blades, and F_t (N) and F_a (N) are respectively the tangential component and the axial component of the force $\mathbf{F}$ acting on the rotor blade at its mean diameter. These components, as has been shown above, can be expressed as follows

$$F_t = \dot{m}_T(C_1\cos\alpha_1 + C_2\cos\alpha_2) = \dot{m}_T(V_1\cos\beta_1 + V_2\cos\beta_2)$$

$$F_a = \dot{m}_T(C_1\sin\alpha_1 - V_2\sin\beta_2)$$

where $\dot{m}_T$ (kg/s) is the mass flow rate through the turbine.

The bending stresses at the root can be computed from the resultant bending moment.

The vibration stresses can be estimated from data on blades designed previously. When a rotor blade is fitted with a separation shroud, the centrifugal force due to the shroud produces additional stress at the root. In this case, the total stress due to centrifugal forces results from summing the stress due to the blade to the stress due to the shroud.

The stresses acting on the disc of a turbine are due to the following causes. The first cause is the stress due to the rotor blades. The second cause is the stress due to the centrifugal forces acting on the material of which the disc is made. The third cause is the stress due to the torque acting on the disc.

The discs of turbines are usually thick near the axis and thin near the rim where the blades are attached. The following equation [22] can be used to estimate the tensile stress σ_d (N/m^2) acting on a uniform disc of turbine, without the contribution due to the blades

$$\sigma_d = \frac{1}{8}\rho_d \frac{d_d^2\,\omega^2}{\ln\left(\frac{t_0}{t_r}\right)}$$

where ρ_b (kg/m^3) is the density of the material of which the disc is made, d_d (m) is the diameter of the disc, ω (rad/s) is the angular velocity of the turbine, t_0 (m) is the thickness of the disc near the axis, and t_r (m) is the thickness of the disc near the rim.

The following equation [22] can be used to estimate the tensile stress σ_d (N/m^2) acting on any disc of turbine, without the contribution due to the blades

$$\sigma_d = \frac{1}{2}m_d r_i \frac{\omega^2}{A_d}$$

where m_d (kg) is the mass of the disc, r_i (m) is the distance of the centre of gravity of the half disc from the axis, ω (rad/s) is the angular velocity of the turbine, and A_d (m^2) is the area of the cross section of the disc. The tensile stress σ_d computed at the maximum allowable angular velocity by using the preceding equation should be equal to a fraction ranging from 0.75 to 0.80 of the yield stress of the material.

The following discussion deals specifically with the design of a single-stage, two-row, velocity-compounded impulse turbine. We assume the expansion of gas to take place only in the nozzle, without further expansion in the rotor blades. Since the gas possesses a considerable amount of kinetic energy at the outlet of the first row of rotating blades, then a row of stationary blades (stator) is necessary to direct the flow to the inlet of a second row of rotating blades, which extract a further amount of kinetic energy from the gas. At the outlet of the second row of rotating blades, the gas flows at low velocity in a direction close to the axial direction of the turbine. The following figure, re-drawn from [22], shows a velocity diagram of a typical single-stage, two-row, velocity-compounded impulse turbine at the mean diameter.

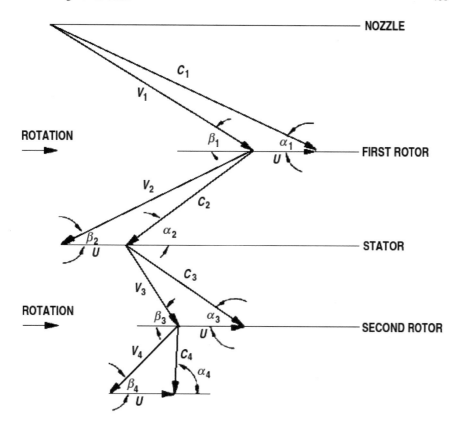

The vector U in the preceding figure is the peripheral velocity of the rotor blades at the mean diameter d_m. The vector C_1 is the absolute velocity of the gas flow at the inlet of the first row of rotating blades. The vector C_1 forms an angle α_1 with the plane of rotation. The vectors V_1 and V_2 are the relative velocities of the gas flow at respectively the inlet and the outlet of the first row of rotating blades. The vector C_2 is the absolute velocity of the gas flow at the inlet of the stator. The vector C_2 forms an angle α_2 with the plane of rotation. The vector C_3 is the absolute velocity of the gas flow at the inlet of the second row of rotating blades. The vector C_3 forms an angle α_3 with the plane of rotation. The vectors V_3 and V_4 are the relative velocities of the gas flow at respectively the inlet and the outlet of the second row of rotating blades. The velocity of the flow at the outlet of any row of blades, be they rotating or stationary, is lower in magnitude than the velocity of the flow at the inlet, because of losses due to friction. The blade velocity coefficient k_b is assumed to have the same value for any row of blades, so that

$$k_b = \frac{V_2}{V_1} = \frac{C_3}{C_2} = \frac{V_4}{V_3}$$

The total power, which the flowing gas transfers to the rotating blades of a turbine, results from the sum of the powers transferred to each row of rotating blades. Therefore, the power P_{2b} (W), which the flowing gas transfers to the blades of a two-row turbine, is

$$
\begin{aligned}
P_{2b} &= \dot{m}_T U \left(C_1 \cos\alpha_1 + C_2 \cos\alpha_2 + C_3 \cos\alpha_3 + C_4 \cos\alpha_4 \right) \\
&= \dot{m}_T U \left(V_1 \cos\beta_1 + V_2 \cos\beta_2 + V_3 \cos\beta_3 + V_4 \cos\beta_4 \right)
\end{aligned}
$$

The nozzle-blade efficiency η_{nb} of a two-row turbine can be expressed as follows

$$
\eta_{nb} = \frac{P_{2b}}{\dot{m}_T \, \Delta h_T} = \frac{P_{2b}}{\dot{m}_T c_p T_0 \left[1 - \left(\frac{p_e}{p_0} \right)^{\frac{\gamma-1}{\gamma}} \right]}
$$

where P_{2b} (W) is the power indicated above, $\dot{m}_T$ (kg/s) is the mass flow rate through the turbine, Δh_T (J/kg) is the total isentropic drop of enthalpy per unit mass in the turbine, c_p (J kg^{-1} K^{-1}) and γ are respectively the specific heat at constant pressure and the specific heat ratio of the gas, T_0 (K) is the total temperature of the gas at the turbine inlet, p_0 (N/m^2) is the total pressure of the gas at the turbine inlet, and p_e (N/m^2) is is the static pressure of the gas at the turbine exhaust.

As has been shown above, in case of a single-stage, single-rotor impulse turbine, the total efficiency η_T of a turbine can be expressed by the following product

$$
\eta_T = \eta_n \, \eta_b \, \eta_m
$$

where η_n and η_b are respectively the nozzle efficiency and the rotor blade efficiency, and η_m is the machine efficiency, which takes account of the mechanical, leakage, and disc-friction losses.

In case of a single-stage, two-rotor, velocity-compounded impulse turbine, the total efficiency η_T of a turbine can be expressed by the following product

$$
\eta_T = \eta_{nb} \, \eta_m
$$

where η_{nb} is the nozzle-blade efficiency, whose expression has been given above. In the ideal case, the nozzle-blade efficiency η_{nb} is maximum for a single-stage, two-rotor, velocity-compounded impulse turbine when

$$
U = \frac{1}{4} C_1 \cos\alpha_1 = \frac{1}{4} C_{1t}
$$

where C_{1t} is the tangential component of the absolute velocity vector C_1 of the gas at the outlet of the nozzle.

The design of a single stage, two-row velocity-compounded impulse turbine is similar to that of a single-stage, single-rotor impulse turbine, which has been described above. The difference is in the fact that the velocity vectors change in magnitude and in direction for each row of blades. Therefore, the height h_b of symmetrical blades in the radial direction is constant in each row, but increases from one row to the next in the direction of the exhaust.

The reheat must be taken into account to determine the gas density at each section. For this purpose, the reheat q_{br} (m^2/s^2 or J/kg) at each row of blades can be determined by using the following equation

$$q_{br} = \frac{\left(1 - k_b^2\right)V_1^2}{2} + (1 - \eta_n)\Delta h_{1-2}$$

In case of unsymmetrical blades used in single-stage, two-row, velocity-compounded impulse turbines, the radial height h_{b2} (m) at the outlet of each row is determined first by means of the following equation

$$A_{b2} = Z_b\, b_{b2}\, h_{b2} = Z_b\, h_{b2}(P_b \sin\theta_{b2} - t_b)$$

where A_{b2} (m^2) is the area needed for the gas flow at the blade outlet, Z_b is the number of the blades, b_{b2} (m) is the width (perpendicular to the direction of the flow) of the passage at the outlet of the blades, P_b (m) is the pitch of the blades, θ_{b2} (rad) is the angle formed by the tangent to the circular arc on the pressure side with the direction of rotation at the outlet of the blades, and t_b (m) is the thickness of the edges of the blades.

Then, the radial height h_{b1} (m) at the inlet of each row is chosen greater, by a factor if about 1.08, than the radial height at the outlet of the preceding row.

The following discussion deals specifically with the design of a two-stage, two-rotor, pressure-compounded impulse turbine. The following figure, re-drawn from [22], shows a velocity diagram of a typical two-stage, two-rotor, pressure-compounded impulse turbine at the mean diameter.

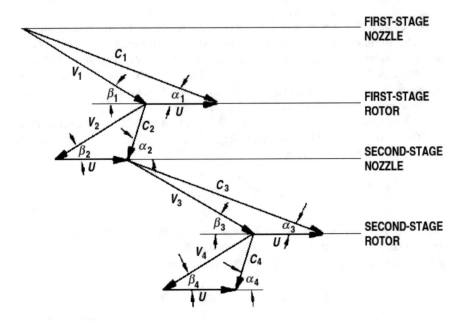

Each stage of a pressure-compounded impulse turbine can be considered separately from the other stages. The vectors C_1 and C_3 are the absolute velocity vectors, whose magnitudes C_1 and C_3 are the spouting velocities of the gas at the outlet of respectively the first-stage nozzle and the second-stage nozzle. The vectors C_1 and C_3 form the angles respectively α_1 and α_3 with the plane of rotation. The magnitudes C_1 and C_3 of C_1 and C_3 are chosen to be about the same.

The vectors V_1 and V_2 are the relative velocity vectors of the gas at respectively the inlet and the outlet of the first-stage rotor. The vectors V_1 and V_2 form the angles respectively β_1 and β_2 with the plane of rotation. Likewise, the vectors V_3 and V_4 are the relative velocity vectors of the gas at respectively the inlet and the outlet of the second-stage rotor. The vectors V_3 and V_4 form the angles respectively β_3 and β_4 with the plane of rotation.

The nozzle of the second stage is designed in such a manner as to receive the gas flow coming from the rotor of the first stage in the direction of the absolute velocity vector C_2, and to change the direction of this flow from α_2 to α_3. At the same time, the gas flow not only changes its direction, but also expands at a lower pressure and increases the magnitude of its absolute velocity from C_2 to C_3. Now, the gas flow enters the rotor of the second stage in the direction of the absolute velocity vector C_3, and exits from this rotor in the direction of the absolute velocity vector C_4. The vector U is the peripheral velocity of the rotors at the mean diameter.

The total power extracted from the gas expanding through the turbine results from the sum of the power extracted in the first stage and the power extracted in the second stage. A turbine can be designed in such a way as to extract power in equal parts in the two stages. In this case, the velocity diagram of the first stage is equal to the velocity diagram of the second stage, so that the vectors C_1 and C_3 are equal in direction and in magnitude, and the vectors C_2 and C_4 are also equal in direction and in magnitude.

Part of the heat generated in the first stage of a pressure-compounded impulse turbine due to friction con be converted into useful work in the second stage. In addition, part of the kinetic energy possessed by the gas flow at the outlet of the first stage of this turbine can be used in the second stage for the same purpose. This part of kinetic energy is measured by the carry-over ratio r_c, whose value ranges from 0.4 to a number close to unity. The carry-over ratio is the part of kinetic energy possessed by the gas flow leaving the first stage which is converted into useful work in the second stage to the total kinetic energy possessed by the gas flow leaving the first stage.

In order to determine the enthalpy drop in each stage which results in an equal amount of useful work extracted in the two stages, the reheat must be evaluated. For this purpose, it is possible to use the following equations

$$q_{nr} = \frac{1}{2}C_0^2 - \frac{1}{2}C_1^2 = \frac{(1 - k_n^2)C_1^2}{2k_n^2} = \frac{(1 - \eta_n)C_1^2}{2\eta_n}$$

$$q_{br} = \frac{(1 - k_b^2)V_1^2}{2} + (1 - \eta_n)\Delta h_{1-2}$$

where the values of the velocity coefficients k_n and k_b (relating to respectively the nozzles and the rotor blades) must be known in order to compute the amounts of reheat in the nozzles (q_{nr}) and in the rotor blades (q_{br}). The values of these coefficients can be made available experimentally or by experience gained in turbines previously designed.

Some of the equations considered above, which apply to the case of single-stage turbines, also apply to the case of two-stage turbines. Further equations to be used in the latter case are given below.

The total temperature T_{2t} (K) of the gas at the inlet of the nozzle of the second stage results from

$$T_{2t} = T_2 + \frac{C_2^2}{2c_p}$$

where T_2 (K) is the static temperature of the gas at the inlet of the nozzle of the second stage, C_2 (m/s) is the magnitude of the absolute velocity vector of the gas at the outlet of the rotor of the first stage, and c_p (J kg^{-1} K^{-1}) is the specific heat of the gas at constant pressure.

The total pressure p_{2t} (N/m^2) of the gas at the inlet of the nozzle of the second stage results from

$$p_{2t} = p_2 \left(\frac{T_{2t}}{T_2} \right)^{\frac{\gamma}{\gamma - 1}}$$

where p_2 (N/m^2) is the static pressure of the gas at the inlet of the nozzle of the second stage, and $\gamma = c_p/c_v$ is the specific heat ratio of the gas.

The spouting velocity C_3 (m/s) at the outlet of the nozzle of the second stage results from

$$C_3 = k_n \left\{ 2c_p T_{2t} \left[1 - \left(\frac{p_3}{p_{2t}} \right)^{\frac{\gamma-1}{\gamma}} \right] \right\}^{\frac{1}{2}} = k_n \left(r_c C_2^2 + 2\Delta h_{2-3} \right)^{\frac{1}{2}}$$

where k_n is the velocity coefficient of the nozzle, (N/m^2) is the static pressure of the gas at the outlet of the nozzle of the second stage, r_c is the carry-over ratio of kinetic energy in the second stage, and Δh_{2-3} (m^2/s^2 or J/kg) is the isentropic drop of enthalpy per unit mass of the gas expanding through the nozzle of the second stage.

The total area $(A_{nt})_2$ (m^2) required for the gas flow through the nozzle of the second stage results from

$$(A_{nt})_2 = \frac{\dot{m}_T}{\varepsilon_{nt} p_{2t} \left[\frac{\gamma}{RT_{2t}} \left(\frac{2}{\gamma+1} \right)^{\frac{\gamma+1}{\gamma-1}} \right]^{\frac{1}{2}}}$$

where $\dot{m}_T$ (kg/s) is the mass flow rate of gas through the turbine, ε_{nt} is the nozzle throat area coefficient, and R (J kg^{-1} K^{-1}) is the specific gas constant.

The following example of application concerns the design of a single-stage, two-row, velocity-compounded impulse turbine for a rocket engine. The following data are known. The propellants are liquid oxygen (oxidiser) and RP-1 (fuel) mixed in the ratio $o/f = 0.408$. The specific heat of the combusted gas at constant pressure is $c_p = 2734$ J kg^{-1} K^{-1}. The specific heat ratio of the combusted gas is $\gamma = c_p/c_v = 1.124$. The specific gas constant is $R = c_p(\gamma - 1)/\gamma = 301.6$ J kg^{-1} K^{-1}. The total temperature of the combusted gas at the inlet of the turbine is $T_0 = 1033$ K. The total pressure of the combusted gas at the inlet of the turbine is $p_0 = 4.413 \times 10^6$ N/m^2. The static pressure of the combusted gas exhausted by the turbine is $p_e = 0.1862 \times 10^6$ N/m^2. The total isentropic enthalpy drop per unit mass of the combusted gas which occurs in the turbine results from the following equation of Sect. 4.3

$$\Delta h_T = c_p T_0 \left[1 - \left(\frac{p_e}{p_0} \right)^{\frac{\gamma-1}{\gamma}} \right] = 2734 \times 1033 \times \left[1 - \left(\frac{0.1862 \times 10^6}{4.413 \times 10^6} \right)^{\frac{1.124-1}{1.124}} \right]$$

$$= 0.8325 \times 10^6 \text{ J/kg}$$

The mass flow rate of the combusted gas through the turbine is $\dot{m}_T = 41.76$ kg/s. The angular velocity of the turbine shaft is $\omega = 733.0$ rad/s. The total efficiency of the velocity-compounded turbine is $\eta_T = 0.582$ (or 58.2%). The aspect ratio of the nozzle is 9.7. The velocity coefficient of the nozzle is $k_n = 0.96$. The area coefficient at the throat of the nozzle is $\varepsilon_{nt} = 0.97$. The area coefficient at the outlet of the nozzle is $\varepsilon_{ne} = 0.95$. The velocity coefficient of the rotor and stator blades is $k_b = 0.89$. The exit area coefficient of the rotor and stator blades is $\varepsilon_{b2} = 0.95$. The chord length of the rotor and stator blades is $C_b = 0.03556$ m. The thickness at the exit of the nozzles and of the blades is $t_n = t_b = 0.00127$ m. The solidity of the blades of the first rotor is $S_{br1} = 1.82$. The solidity of the blades of the stator is $S_{bs} = 1.94$. The solidity of the blades of the second rotor is $S_{br2} = 1.67$.

We want to determine the velocity diagrams and the principal dimensions of this turbine assuming a reaction (that is, a drop of pressure) of 6% in the rotor blades and in the stator blades downstream of the nozzles.

At the inlet of the nozzle (subscript 0), the total temperature of the gas is $T_0 = 1033$ K, the total pressure of the gas is $p_0 = 4.413 \times 10^6$ N/m^2, and no drop of enthalpy per unit mass of the gas has occurred. The nozzle efficiency is $\eta_n = k_n^2 = 0.96^2 \approx 0.92$.

At the outlet of the nozzle (subscript 1), a drop Δh_{0-1} of 6% of the total drop of enthalpy per unit mass $\Delta h_T = 0.8325 \times 10^6$ J/kg has occurred. Therefore, the partial drop of enthalpy per unit mass, which has occurred in the nozzle, is

$$\Delta h_{0-1} = 0.8325 \times 10^6 \times (1 - 0.06) = 782550 \text{ J/kg}$$

By substituting this value of Δh_{0-1} in the following equation

$$\Delta h_{0-1} = c_p T_0 \left[1 - \left(\frac{p_1}{p_0} \right)^{\frac{\gamma-1}{\gamma}} \right]$$

and solving for p_1, the static pressure of the gas at the outlet of the nozzle results

$$p_1 = p_0 \left(1 - \frac{\Delta h_{0-1}}{c_p T_0} \right)^{\frac{\gamma}{\gamma-1}} = 4.413 \times 10^6 \times \left(1 - \frac{782550}{2734 \times 1033} \right)^{\frac{1.124}{1.124-1}}$$
$$= 0.2330 \times 10^6 \text{ N/m}^2$$

Also, by substituting $k_n = 0.96$ and $\Delta h_{0-1} = 782{,}550$ J/kg in the following equation

$$C_1 = k_n (2\Delta h_{0-1})^{\frac{1}{2}}$$

we find the following value of the spouting velocity at the outlet of the nozzle

$$C_1 = 0.96 \times (2 \times 782550)^{\frac{1}{2}} = 1201 \text{ m/s}$$

The amount of reheat occurring in the nozzle results from

$$q_{nr} = \frac{(1 - k_n^2) C_1^2}{2 k_n^2}$$

After substituting $C_1 = 1201$ m/s and $k_n^2 = 0.96^2 \approx 0.92$ in the preceding equation, we find

$$q_{nr} = \frac{(1 - 0.92) \times 1201^2}{2 \times 0.92} = 6.271 \times 10^4 \text{ J/kg}$$

If the expansion of the gas in the nozzle were isentropic (subscript is), then the theoretical value $(T_1)_{is}$ of the gas temperature at the exit of the nozzle would be

$$(T_1)_{is} = T_0 - \frac{\Delta h_{0-1}}{c_p} = 1033 - \frac{782550}{2734} = 746.8 \text{ K}$$

By contrast, due to the reheat q_{nr} computed above, the actual value T_1 of the gas temperature at the outlet of the nozzle is higher than $(T_1)_{is}$, as follows

$$T_1 = (T_1)_{is} + \frac{q_{nr}}{c_p} = 746.8 + \frac{6.271 \times 10^4}{2734} = 769.7 \text{ K}$$

At the pressure $p_1 = 0.2330 \times 10^6$ N/m^2 and at the temperature $T_1 = 769.7$ K, the density of the gas at the outlet of the nozzle results from the law of perfect gases ($p = \rho RT$), as follows

$$\rho_1 = \frac{p_1}{RT_1} = \frac{0.2330 \times 10^6}{301.6 \times 769.7} = 1.004 \text{ kg/m}^3$$

We choose the value 0.4363 rad for the angle α_1, which the absolute velocity vector C_1 of the gas flow forms with the plane of rotation at the outlet of the nozzle.

As has been shown above, in the ideal case, the nozzle-blade efficiency η_{nb} for a single-stage, two-rotor, velocity-compounded impulse turbine is maximum when

$$U = \frac{1}{4} C_1 \cos \alpha_1$$

The preceding equation makes it possible to compute the peripheral velocity of the blades at the mean diameter of the rotor, as follows

$$U = \frac{1}{4} C_1 \cos \alpha_1 = \frac{1}{4} \times 1201 \times \cos 0.4363 = 272.1 \text{ m/s}$$

The mean diameter d_m of the rotor results from the following equation

$$U = \frac{1}{2} \omega \, d_m$$

After substituting $U = 272.1$ m/s and $\omega = 733.0$ rad/s and solving for d_m, we find

$$d_m = \frac{2U}{\omega} = \frac{2 \times 272.1}{733.0} = 0.7424 \text{ m}$$

The angle β_1, which the relative velocity vector V_1 forms with the plane of rotation at the inlet of the rotor, results from the following equation

$$\beta_1 = \arctan\left(\frac{C_1 \sin \alpha_1}{C_1 \cos \alpha_1 - U}\right)$$

After substituting $C_1 = 1201$ m/s, $\alpha_1 = 0.4363$ rad, and $U = 272.1$ m/s in the preceding equation, we find

$$\beta_1 = \arctan\left(\frac{1201 \times \sin 0.4363}{1201 \times \cos 0.4363 - 272.1}\right) = 0.5562 \text{ rad}$$

The magnitude V_1 of the relative velocity vector V_1 can be determined through a simple inspection of the velocity diagram shown in the following figure.

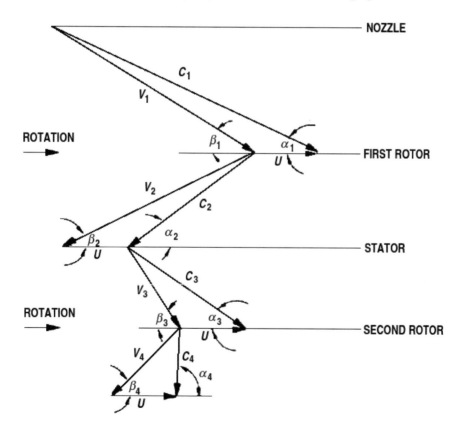

This yields

$$C_1 \sin \alpha_1 = V_1 \sin \beta_1$$

Hence

$$V_1 = \frac{C_1 \sin \alpha_1}{\sin \beta_1} = \frac{1201 \times \sin 0.4363}{\sin 0.5562} = 961.3 \text{ m/s}$$

At the inlet of the stator (subscript 2), the gas flow has had a further drop of enthalpy per unit mass Δh_{1-2} due to the blades of the first rotor. We suppose the total drop of enthalpy per unit mass ($0.06 \, \Delta h_T = 0.06 \times 0.8325 \times 10^6$ J/kg), to be equally divided in the first rotor, in the stator, and in the second rotor. Therefore, the partial drop of enthalpy per unit mass, which occurs in the first rotor, is

$$\Delta h_{1-2} = \frac{0.06 \times 0.8325 \times 10^6}{3} = 1.665 \times 10^4 \text{ J/kg}$$

The relative velocity V_2 of the gas at the outlet of the first rotor can be determined as follows

$$V_2 = \left(k_b^2 V_1^2 + 2\eta_n \Delta h_{1-2}\right)^{\frac{1}{2}}$$

After substituting $k_b = 0.89$, $V_1 = 961.3$ m/s, $\eta_n = k_n^2 = 0.96^2 \approx 0.92$, and $\Delta h_{1-2} = 1.665 \times 10^4$ J/kg in the preceding equation, we find

$$V_2 = \left(0.89^2 \times 961.3^2 + 2 \times 0.92 \times 1.665 \times 10^4\right)^{\frac{1}{2}} = 873.3 \text{ m/s}$$

The amount of reheat which occurs in the blades of the first rotor results from

$$q_{br1} = \frac{\left(1 - k_b^2\right)V_1^2}{2} + (1 - \eta_n)\Delta h_{1-2} = \frac{\left(1 - 0.89^2\right) \times 961.3^2}{2} + (1 - 0.92)$$
$$\times 16650 = 9.739 \times 10^4 \text{ J/kg}$$

The static pressure of the gas at the outlet of the first rotor results from the following equation

$$\Delta h_{1-2} = c_p T_1 \left[1 - \left(\frac{p_2}{p_1}\right)^{\frac{\gamma-1}{\gamma}}\right]$$

This equation, solved for p_2, yields

$$p_2 = p_1 \left(1 - \frac{\Delta h_{1-2}}{c_p T_1}\right)^{\frac{\gamma}{\gamma-1}} = 0.2331 \times 10^6 \times \left(1 - \frac{1.665 \times 10^4}{2734 \times 769.7}\right)^{\frac{1.124}{1.124-1}}$$

$$= 0.2169 \times 10^6 \text{ N/m}^2$$

If the expansion of the gas in the first rotor were isentropic (subscript is), then the theoretical value $(T_2)_{is}$ of the gas temperature at the outlet of the first rotor would be

$$(T_2)_{is} = T_1 - \frac{\Delta h_{1-2}}{c_p} = 769.7 - \frac{1.665 \times 10^4}{2734} = 763.6 \text{ K}$$

By contrast, due to the reheat q_{br1} computed above, the actual value T_2 of the gas temperature at the outlet of the first rotor is higher than $(T_2)_{is}$, as follows

$$T_2 = (T_2)_{is} + \frac{q_{br1}}{c_p} = 763.6 + \frac{9.739 \times 10^4}{2734} = 799.2 \text{ K}$$

At the pressure $p_2 = 0.2169 \times 10^6$ N/m^2 and at the temperature $T_2 = 799.2$ K, the density of the gas at the outlet of the first rotor results from the law of perfect gases $(p = \rho RT)$, as follows

$$\rho_2 = \frac{p_2}{RT_2} = \frac{0.2169 \times 10^6}{301.6 \times 799.2} = 0.8999 \text{ kg/m}^3$$

We choose the value 0.4363 rad for the angle β_2, which the relative velocity vector V_2 of the gas flow forms with the plane of rotation at the outlet of the first rotor (unsymmetrical blades). The angle α_2, which the absolute velocity vector C_2 of the gas flow forms with the plane of rotation at the outlet of the first rotor, is determined through a simple inspection of the velocity diagram shown in the preceding figure. This yields

$$\tan \alpha_2 = \frac{V_2 \sin \beta_2}{V_2 \cos \beta_2 - U}$$

Hence

$$\alpha_2 = \arctan\left(\frac{V_2 \sin \beta_2}{V_2 \cos \beta_2 - U}\right) = \arctan\left(\frac{873.3 \times \sin 0.4363}{873.3 \times \cos 0.4363 - 272.1}\right)$$
$$= 0.6178 \text{ rad}$$

The velocity diagram of the preceding figure also shows that the magnitude C_2 of the absolute velocity vector C_2 results from

$$C_2 \sin \alpha_2 = V_2 \sin \beta_2$$

Hence

$$C_2 = \frac{V_2 \sin \beta_2}{\sin \alpha_2} = \frac{873.3 \times \sin 0.4363}{\sin 0.6178} = 637.1 \text{ m/s}$$

At the outlet of the stator (subscript 3), the gas flow has had a further drop of enthalpy per unit mass $\Delta h_{2\text{-}3}$ due to the blades of the stator. As has been shown above, there results

$$\Delta h_{2\text{-}3} = \Delta h_{1\text{-}2} = 1.665 \times 10^4 \text{ J/kg}$$

The magnitude C_3 of the absolute velocity vector $\boldsymbol{C_3}$ at the outlet of the stator results from

$$C_3 = \left(k_b^2 C_2^2 + 2\eta_n \Delta h_{2\text{-}3}\right)^{\frac{1}{2}} = \left(0.89^2 \times 637.1^2 + 2 \times 0.93 \times 1.665 \times 10^4\right)^{\frac{1}{2}}$$
$$= 593.4 \text{ m/s}$$

The amount of reheat which occurs in the blades of the stator results from

$$q_{bs} = \frac{\left(1 - k_b^2\right) C_2^2}{2} + (1 - \eta_n)\Delta h_{2\text{-}3} = \frac{\left(1 - 0.89^2\right) \times 637.1^2}{2} + (1 - 0.92)$$
$$\times 16650 = 4.352 \times 10^4 \text{ J/kg}$$

The static pressure of the gas at the outlet of the stator results from the following equation

$$\Delta h_{2\text{-}3} = c_p T_2 \left[1 - \left(\frac{p_3}{p_2}\right)^{\frac{\gamma-1}{\gamma}} \right]$$

This equation, solved for p_2, yields

$$p_3 = p_2 \left(1 - \frac{\Delta h_{2\text{-}3}}{c_p T_2} \right)^{\frac{\gamma}{\gamma-1}} = 0.2169 \times 10^6 \times \left(1 - \frac{1.665 \times 10^4}{2734 \times 799.2} \right)^{\frac{1.124}{1.124-1}}$$
$$= 0.2024 \times 10^6 \text{ N/m}^2$$

If the expansion of the gas in the stator were isentropic (subscript *is*), then the theoretical value $(T_3)_{is}$ of the gas temperature at the outlet of the stator would be

$$(T_3)_{is} = T_2 - \frac{\Delta h_{2\text{-}3}}{c_p} = 799.2 - \frac{1.665 \times 10^4}{2734} = 793.1 \text{ K}$$

By contrast, due to the reheat q_{bs} computed above, the actual value T_3 of the gas temperature at the outlet of the stator is higher than $(T_3)_{is}$, as follows

$$T_3 = (T_3)_{is} + \frac{q_{bs}}{c_p} = 793.1 + \frac{4.352 \times 10^4}{2734} = 809.0 \text{ K}$$

At the pressure $p_3 = 0.2024 \times 10^6$ N/m^2 and at the temperature $T_3 = 809.0$ K, the density of the gas at the outlet of the stator results from the law of perfect gases ($p = \rho R T$), as follows

$$\rho_3 = \frac{p_3}{R T_3} = \frac{0.2024 \times 10^6}{301.6 \times 809.0} = 0.8295 \text{ kg/m}^3$$

We choose the value 0.6109 rad for the angle α_3, which the absolute velocity vector C_3 of the gas flow forms with the plane of rotation at the outlet of the stator. By so doing, α_3 does not differ very much from α_2, which has been found to be equal to 0.6178 rad.

Since the value of α_3 has been chosen, then a simple inspection of the velocity diagram shown in the preceding figure makes it possible to determine β_3 (the angle formed by the relative velocity vector V_3 of the gas flow with the plane of rotation at the outlet of the stator), as follows

$$\tan \beta_3 = \frac{C_3 \sin \alpha_3}{C_3 \cos \alpha_3 - U}$$

After substituting $\alpha_3 = 0.6109$ rad, $C_3 = 593.4$ m/s, and $U = 272.1$ m/s in the preceding equation, we find

$$\beta_3 = \arctan\left(\frac{593.4 \times \sin 0.6109}{593.4 \times \cos 0.6109 - 272.1} \right) = 1.038 \text{ rad}$$

By inspecting again the velocity diagram shown in the preceding figure, there results

$$C_3 \sin \alpha_3 = V_3 \sin \beta_3$$

Hence, the magnitude V_3 of the relative velocity vector V_3 results

$$V_3 = \frac{C_3 \sin \alpha_3}{\sin \beta_3} = \frac{593.4 \times \sin 0.6109}{\sin 1.038} = 395.1 \text{ m/s}$$

At the outlet of the second rotor (subscript 4), the gas flow has had a further drop of enthalpy per unit mass Δh_{3-4} due to the blades of the second rotor. As has been shown above, there results

$$\Delta h_{3-4} = \Delta h_{1-2} = 1.665 \times 10^4 \text{ J/kg}$$

The relative velocity V_4 of the gas at the outlet of the second rotor can be determined as follows

$$V_4 = \left(k_b^2 V_3^2 + 2\eta_n \Delta h_{3-4}\right)^{\frac{1}{2}}$$

After substituting $k_b = 0.89$, $V_3 = 395.1$ m/s, $\eta_n = k_n^2 = 0.96^2 \approx 0.92$, and $\Delta h_{3-4} = 1.665 \times 10^4$ J/kg in the preceding equation, we find

$$V_4 = \left(0.89^2 \times 395.1^2 + 2 \times 0.92 \times 1.665 \times 10^4\right)^{\frac{1}{2}} = 392.8 \text{ m/s}$$

The amount of reheat which occurs in the blades of the second rotor results from

$$q_{br2} = \frac{\left(1 - k_b^2\right)V_3^2}{2} + (1 - \eta_n)\Delta h_{3-4} = \frac{\left(1 - 0.89^2\right) \times 395.1^2}{2} + (1 - 0.92) \times 16650$$
$$= 1.756 \times 10^4 \text{ J/kg}$$

The static pressure of the gas at the outlet of the second rotor results from the following equation

$$\Delta h_{3-4} = c_p T_3 \left[1 - \left(\frac{p_4}{p_3}\right)^{\frac{\gamma-1}{\gamma}}\right]$$

This equation, solved for p_4, yields

$$p_4 = p_3 \left(1 - \frac{\Delta h_{3-4}}{c_p T_3}\right)^{\frac{\gamma}{\gamma-1}} = 0.2024 \times 10^6 \times \left(1 - \frac{1.665 \times 10^4}{2734 \times 809.0}\right)^{\frac{1.124}{1.124-1}}$$
$$= 0.1890 \times 10^6 \text{ N/m}^2$$

This static pressure is slightly higher than the static pressure of the gas exhausted by the turbine ($p_e = 0.1862 \times 10^6$ N/m²) as a result of the reheat.

If the expansion of the gas in the second rotor were isentropic (subscript *is*), then the theoretical value $(T_4)_{is}$ of the gas temperature at the outlet of the second rotor would be

$$(T_4)_{is} = T_3 - \frac{\Delta h_{3-4}}{c_p} = 809.0 - \frac{1.665 \times 10^4}{2734} = 802.9 \text{ K}$$

By contrast, due to the reheat q_{br2} computed above, the actual value T_4 of the gas temperature at the outlet of the second rotor is higher than $(T_4)_{is}$, as follows

$$T_4 = (T_4)_{is} + \frac{q_{br2}}{c_p} = 802.9 + \frac{1.756 \times 10^4}{2734} = 809.3 \text{ K}$$

At the pressure $p_4 = 0.1890 \times 10^6$ N/m^2 and at the temperature $T_4 = 809.3$ K, the density of the gas at the outlet of the second rotor results from the law of perfect gases $(p = \rho R T)$, as follows

$$\rho_4 = \frac{p_4}{RT_4} = \frac{0.1890 \times 10^6}{301.6 \times 809.3} = 0.7743 \text{ kg/m}^3$$

We choose the value 0.7679 rad for the angle β_4, which the relative velocity vector V_4 of the gas flow forms with the plane of rotation at the outlet of the second rotor. The angle α_4, which the absolute velocity vector C_4 of the gas flow forms with the plane of rotation at the outlet of the second rotor, is determined through a simple inspection of the velocity diagram shown in the preceding figure. This yields

$$\tan \alpha_4 = \frac{V_4 \sin \beta_4}{V_4 \cos \beta_4 - U}$$

Hence

$$\alpha_4 = \arctan\left(\frac{V_4 \sin \beta_4}{V_4 \cos \beta_4 - U}\right) = \arctan\left(\frac{392.8 \times \sin 0.7679}{392.8 \times \cos 0.7679 - 272.1}\right) = 1.518 \text{ rad}$$

The velocity diagram of the preceding figure also shows that the magnitude C_4 of the absolute velocity vector C_4 results from

$$C_4 \sin \alpha_4 = V_4 \sin \beta_4$$

Hence

$$C_4 = \frac{V_4 \sin \beta_4}{\sin \alpha_4} = \frac{392.8 \times \sin 0.7679}{\sin 1.518} = 273.2 \text{ m/s}$$

The total area A_{nt} required at the throat of the turbine nozzle is

$$A_{nt} = \frac{\dot{m}_T}{\varepsilon_{nt} p_0 \left[\frac{\gamma}{RT_0}\left(\frac{2}{\gamma+1}\right)^{\frac{\gamma+1}{\gamma-1}}\right]^{\frac{1}{2}}}$$

After substituting $\dot{m}_T = 41.76$ kg/s, $\varepsilon_{nt} = 0.97$, $p_0 = 4.413 \times 10^6$ N/m^2, $\gamma = 1.124$, $R = 301.6$ J kg^{-1} K^{-1}, and $T_0 = 1033$ K in the preceding equation, we find

$$A_{nt} = \frac{41.76}{0.97 \times 4.413 \times 10^6 \times \left[\frac{1.124}{301.6 \times 1033} \times \left(\frac{2}{1.124+1}\right)^{\frac{1.124+1}{1.124-1}}\right]^{\frac{1}{2}}} = 0.008598 \text{ m}^2$$

Since the aspect ratio of the nozzle is 9.7 and we choose the value 0.03810 m for the radial height h_{nt} of the nozzle at the throat, then the width of the nozzle at the throat is

$$b_{nt} = \frac{0.03810}{9.7} = 0.003928 \text{ m}$$

The number of nozzles results from

$$Z_n = \frac{A_{nt}}{b_{nt} h_{nt}} = \frac{0.008598}{0.003928 \times 0.0381} = 57$$

The pitch of the nozzles results from

$$P_n = \frac{\pi d_m}{Z_n} = \frac{3.1416 \times 0.7424}{57} = 0.04092 \text{ m}$$

We choose the value 0.03490 rad for the difference $\alpha_1 - \theta_n$, where $\alpha_1 = 0.4363$ rad is the angle which the absolute velocity vector C_1 of the gas flow forms with the plane of rotation at the outlet of the nozzle, and θ_n is the angle which the central line of the nozzle forms with the plane of rotation at the outlet of the nozzle, as shown in the following figure.

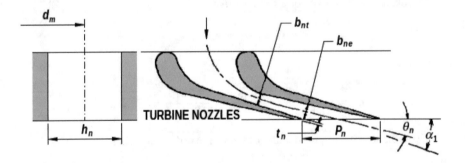

Therefore

$$\theta_n = 0.4363 - 0.03490 = 0.4014 \text{ rad}$$

The total area A_{ne}, the height h_{ne}, and the width b_{ne} of the exit section of the nozzle result, all of them, from the following equation

$$A_{ne} = Z_n b_{ne} h_{ne} = Z_n h_{ne} (P_n \sin \theta_n - t_n) = \frac{\dot{m}_T}{\rho_1 \, C_1 \, \epsilon_{ne}}$$

Firstly, after substituting $\dot{m}_T = 41.76$ kg/s, $\rho_1 = 1.004$ kg/m^3, $C_1 = 1201$ m/s, and $\varepsilon_{ne} = 0.95$, we find

$$A_{ne} = \frac{41.76}{1.004 \times 1201 \times 0.95} = 0.03646 \text{ m}^2$$

Then, since the value of the total area A_{ne} has been determined, the height h_{ne} and the width b_{ne} of the exit section of the nozzle can be computed as follows

$$h_{ne} = \frac{A_{ne}}{Z_n(P_n \sin\theta_n - t_n)} = \frac{0.03646}{57 \times (0.04092 \times \sin 0.4014 - 0.00127)} = 0.04346 \text{ m}$$

$$b_{ne} = \frac{A_{ne}}{Z_n h_{ne}} = \frac{0.03646}{57 \times 0.04363} = 0.01466 \text{ m}$$

The dimensions of the blades of the first rotor, at the mean diameter $d_m = 0.7424$ m, are determined below. Since the solidity $S_{br1} = C_b/P_{br1}$ of these blades is equal to 1.82, where $C_b = 0.03556$ m is the chord length of the rotor and stator blades, then the pitch of the blades of the first rotor is

$$P_{br1} = \frac{C_b}{S_{br1}} = \frac{0.03556}{1.82} = 0.01954 \text{ m}$$

The number of the blades of the first rotor results from

$$Z_{br1} = \frac{\pi d_m}{P_{br1}} = \frac{3.1416 \times 0.7424}{0.01954} = 119$$

We choose the value 0.03720 rad for the difference $\theta_{b1r1} - \beta_1$, where θ_{b1r1} is the angle which the plane tangent to the pressure surface forms with the plane of rotation at the inlet of the blades of the first rotor, and $\beta_1 = 0.5562$ rad is the angle which the relative velocity vector V_1 of the gas flow forms with the plane of rotation at the inlet of the blade of the first rotor, as shown in the following figure.

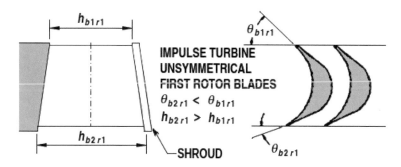

Since

$$\theta_{b1r1} - 0.5562 \text{ rad} = 0.03720 \text{ rad}$$

Then

$$\theta_{b1r1} = 0.5934 \text{ rad}$$

We also choose the value of the angle θ_{b2r1}, which the plane tangent to the pressure surface forms with the plane of rotation at the outlet of the blades of the first rotor, so that $\theta_{b2r1} = \beta_2 = 0.4363$ rad, where β_2 is the angle which the relative velocity vector V_2 of the gas flow forms with the plane of rotation at the outlet of the first rotor.

We choose the blade height h_{b1r1} at the inlet of the first rotor so that $h_{b1r1} = 1.08$ h_{ne}, where $h_{ne} = 0.04346$ m is the height at the exit section of the nozzle. Therefore

$$h_{b1r1} = 1.08 \times 0.04346 = 0.04694 \text{ m}$$

The width b_{b1r1} of the blade passage at the inlet of the first rotor results from

$$b_{1r1} = P_{br1} \sin \theta_{b1r1} - t_b = 0.01954 \times \sin 0.5934 - 0.00127 = 0.009656 \text{ m}$$

where $P_{br1} = 0.01954$ m is the pitch of the blades of the first rotor, $\theta_{b1r1} = 0.5934$ rad is the angle which the plane tangent to the pressure surface forms with the plane of rotation at the inlet of the blades of the first rotor, and $t_b = 0.00127$ m is the thickness of the blades at the inlet of the first rotor.

The total area A_{b2r1} at the outlet of the blades of the first rotor results from the following equation

$$\dot{m}_T = \rho_2 V_2 A_{b2r1} \, \varepsilon_{b2}$$

Substituting $\dot{m}_T = 41.76$ kg/s, $\rho_2 = 0.8999$ kg/m³, $V_2 = 873.3$ m/s, and $\varepsilon_{b2} = 0.95$, and solving for A_{b2r1}, we find

$$A_{b2r1} = \frac{\dot{m}_T}{\rho_2 \, V_2 \, \varepsilon_{b2}} = \frac{41.76}{0.8999 \times 873.3 \times 0.95} = 0.05593 \text{ m}^2$$

The radial height h_{b2r1} of a blade of the first rotor at the outlet can be computed by using the following equation

$$A_{b2r1} = Z_{br1} b_{b2r1} \, h_{b2r1} = Z_{br1} \, h_{b2r1} (P_{br1} \sin \theta_{b2r1} - t_b)$$

Substituting $A_{b2r1} = 0.05593$ m², $Z_{br1} = 119$, $P_{br1} = 0.01954$ m, $\theta_{b2r1} = 0.4363$ rad, and $t_b = 0.00127$ m, and solving for h_{b2r1}, we find

$$h_{b2r1} = \frac{A_{b2r1}}{Z_{br1}(P_{br1} \sin \theta_{b2r1} - t_b)} = \frac{0.05593}{119 \times (0.01954 \times \sin 0.4363 - 0.00127)}$$
$$= 0.06726 \text{ m}$$

The width b_{b2r1} of the blade passage at the outlet of the first rotor results from

$$b_{b2r1} = P_{br1} \sin \theta_{b2r1} - t_b = 0.01954 \times \sin 0.4363 - 0.00127 = 0.006987 \text{ m}$$

The mean radial height of a blade of the first rotor results from

$$h_{br1} = \frac{h_{b1r1} + h_{b2r1}}{2} = \frac{0.04694 + 0.06726}{2} = 0.05710 \text{ m}$$

For the blades of the first rotor, we use the TimkenSteel 4140HW alloy steel, whose density is $\rho_b = 7850$ kg/m^3 [42]. The blades are tapered and have shrouds at their tips. The centrifugal tensile stress at the root of each blade is computed by considering a blade of constant radial height $h_{br1} = 0.05710$ m without shroud, as follows

$$\sigma_{cr1} = \frac{1}{2} \rho_b \, h_{br1} \, d_m \, \omega^2 = \frac{7850 \times 0.0571 \times 0.7424 \times 733^2}{2} = 0.8934 \times 10^8 \text{ N/m}^2$$

The dimensions of the blades of the stator, at the mean diameter $d_m = 0.7424$ m, are determined below. Since the solidity $S_{bs} = C_b/P_{bs}$ of these blades is equal to 1.94, where $C_b = 0.03556$ m is the chord length of the rotor and stator blades, then the pitch of the blades of the stator is

$$P_{bs} = \frac{C_b}{S_{bs}} = \frac{0.03556}{1.94} = 0.01833 \text{ m}$$

The number of the blades of the stator results from

$$Z_{bs} = \frac{\pi d_m}{P_{bs}} = \frac{3.1416 \times 0.7424}{0.01833} = 127$$

We choose the value 0.028 rad for the difference $\theta_{b1s} - \alpha_2$, where θ_{b1s} is the angle which the plane tangent to the pressure surface forms with the plane of rotation at the inlet of the blades of the stator, and $\alpha_2 = 0.6179$ rad is the angle which the absolute velocity vector C_2 of the gas flow forms with the plane of rotation at the inlet of the blades of the stator, as shown in the following figure.

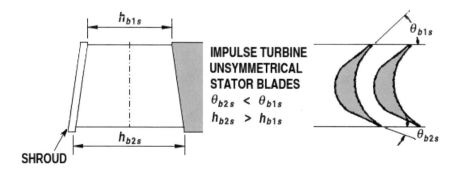

Therefore

$$\theta_{b1s} = 0.6178 + 0.028 = 0.6458 \text{ rad}$$

We choose the angle θ_{b2s} (which the plane tangent to the pressure surface forms with the plane of rotation at the outlet of the blades of the stator) equal to the angle $\alpha_3 = 0.6109$ rad (which the absolute velocity vector C_3 forms with the plane of rotation at the outlet of the blades of the stator). Therefore

$$\theta_{b2s} = \alpha_3 = 0.6109 \text{ rad}$$

As has been shown above, the radial height of a blade at the inlet of each row is chosen greater, by a factor if about 1.08, than the radial height of a blade at the outlet of the preceding row. Since the radial height of a blade of the first rotor at the outlet has been found to be $h_{b2r1} = 0.06726$ m, then the radial height of a blade at the inlet of the stator is chosen to be

$$h_{b1s} = 1.08 \times 0.06726 = 0.07264 \text{ m}$$

The width b_{b1s} of the blade passage at the inlet of the stator results from the following equation

$$b_{b1s} = P_{bs} \sin \theta_{b1s} - t_b = 0.01833 \times \sin 0.6458 - 0.00127 = 0.009762 \text{ m}$$

The total area A_{b2s} at the outlet of the blades of the stator results from the following equation

$$\dot{m}_T = \rho_3 C_3 A_{b2s} \epsilon_{b2}$$

After substituting $\dot{m}_T = 41.76$ kg/s, $\rho_3 = 0.8295$ kg/m^3, $C_3 = 593.4$ m/s, and $\varepsilon_{b2} = 0.94$, and solving for A_{b2s}, we find

$$A_{b2s} = \frac{\dot{m}_T}{\rho_3 \, C_3 \, \epsilon_{b2}} = \frac{41.76}{0.8295 \times 593.4 \times 0.95} = 0.08930 \text{ m}^2$$

The radial height h_{b2s} of a blade of the stator at the outlet can be computed by using the following equation

$$A_{b2s} = Z_{bs} \, b_{b2s} \, h_{b2s} = Z_{bs} \, h_{b2s}(P_{bs} \sin \theta_{b2s} - t_b)$$

Substituting $A_{b2s} = 0.08930$ m^2, $Z_{bs} = 127$, $P_{bs} = 0.01833$ m, $\theta_{b2s} = 0.6109$ rad, and $t_b = 0.00127$ m, and solving for h_{b2s}, we find

$$h_{b2s} = \frac{A_{b2s}}{Z_{bs}(P_{bs} \sin \theta_{b2s} - t_b)} = \frac{0.08930}{127 \times (0.01833 \times \sin 0.6109 - 0.00127)}$$

$$= 0.07606 \text{ m}$$

The width b_{b2s} of the blade passage at the outlet of the stator results from

$$b_{b2s} = P_{bs} \sin \theta_{b2s} - t_b = 0.01833 \times \sin 0.6109 - 0.00127 = 0.009244 \text{ m}$$

The dimensions of the blades of the second rotor, at the mean diameter $d_m = 0.7424$ m, are determined below. Since the solidity $S_{br2} = C_b/P_{br2}$ of these blades is equal to 1.67, where $C_b = 0.03556$ m is the chord length of the rotor and stator blades, then the pitch of the blades of the second rotor is

$$P_{br2} = \frac{C_b}{S_{br2}} = \frac{0.03556}{1.67} = 0.02129 \text{ m}$$

The number of the blades of the second rotor results from

$$Z_{br2} = \frac{\pi d_m}{P_{br2}} = \frac{3.1416 \times 0.7424}{0.02129} = 109$$

We choose the value 0.009 rad for the difference $\theta_{b1r2} - \beta_3$, where θ_{b1r2} is the angle which the plane tangent to the pressure surface forms with the plane of rotation at the inlet of the blades of the second rotor, and $\beta_3 = 1.038$ rad is the angle which the relative velocity vector V_3 of the gas flow forms with the plane of rotation at the inlet of the blade of the second rotor, as shown in the following figure.

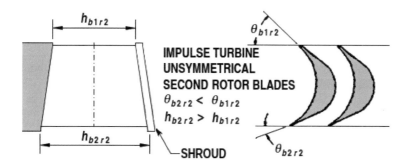

Since

$$\theta_{b1r2} - 1.038 \text{ rad} = 0.009 \text{ rad}$$

Then

$$\theta_{b1r2} = 1.047 \text{ rad}$$

We also choose the value of the angle θ_{b2r2}, which the plane tangent to the pressure surface forms with the plane of rotation at the outlet of the blades of the second rotor, so that $\theta_{b2r2} = \beta_4 = 0.7679$ rad, where β_4 is the angle which the relative velocity vector V_4 of the gas flow forms with the plane of rotation at the outlet of the second rotor.

As has been shown above, we choose the blade height h_{b1r2} at the inlet of the second rotor so that $h_{b1r2} = 1.08 \, h_{b2s}$, where $h_{b2s} = 0.07606$ m is the height of the blade at the exit section of the stator. Therefore

$$h_{b1r2} = 1.08 \times 0.07606 = 0.08214 \text{ m}$$

The width b_{b1r2} of the blade passage at the inlet of the second rotor results from

$$b_{b1r2} = P_{br2} \sin \theta_{b1r2} - t_b = 0.02129 \times \sin 1.047 - 0.00127 = 0.01717 \text{ m}$$

where $P_{br2} = 0.02129$ m is the pitch of the blades of the second rotor, $\theta_{b1r2} = 1.047$ rad is the angle which the plane tangent to the pressure surface forms with the plane of rotation at the inlet of the blades of the second rotor, and $t_b = 0.00127$ m is the thickness of the blades at the inlet of the second rotor.

The total area A_{b2r2} at the outlet of the blades of the second rotor results from the following equation

$$\dot{m}_T = \rho_4 V_4 A_{b2r2} \epsilon_{b2}$$

Substituting $\dot{m}_T = 41.76$ kg/s, $\rho_4 = 0.7743$ kg/m³, $V_4 = 392.8$ m/s, and $\varepsilon_{b2} = 0.95$, and solving for A_{b2r2}, we find

$$A_{b2r2} = \frac{\dot{m}_T}{\rho_4 \, V_4 \, \epsilon_{b2}} = \frac{41.76}{0.7743 \times 392.8 \times 0.95} = 0.1445 \text{ m}^2$$

The radial height h_{b2r2} of a blade of the second rotor at the outlet can be computed by using the following equation

$$A_{b2r2} = Z_{br2} \, b_{b2r2} \, h_{b2r2} = Z_{br2} \, h_{b2r2} (P_{br2} \sin \theta_{b2r2} - t_b)$$

Substituting $A_{b2r2} = 0.1445$ m², $Z_{br2} = 109$, $P_{br2} = 0.02129$ m, $\theta_{b2r2} = 0.7679$ rad, and $t_b = 0.00127$ m, and solving for h_{b2r2}, we find

$$h_{b2r2} = \frac{A_{b2r2}}{Z_{br2}(P_{br2} \sin \theta_{b2r2} - t_b)} = \frac{0.1445}{109 \times (0.02129 \times \sin 0.7679 - 0.00127)}$$
$$= 0.09806 \text{ m}$$

The width b_{b2r2} of the blade passage at the outlet of the second rotor results from

$$b_{b2r2} = P_{br2} \sin \theta_{b2r2} - t_b = 0.02129 \times \sin 0.7679 - 0.00127 = 0.01352 \text{ m}$$

The mean radial height of a blade of the second rotor results from

$$h_{br2} = \frac{h_{b1r2} + h_{b2r2}}{2} = \frac{0.08214 + 0.09806}{2} = 0.09010 \text{ m}$$

For the blades of the second rotor, we use again the TimkenSteel 4140HW alloy steel, whose density is $\rho_b = 7850 \text{ kg/m}^3$ [43]. The blades are tapered and have shrouds at their tips. The centrifugal tensile stress at the root of each blade is computed by considering a blade of constant radial height $h_{br2} = 0.09010$ m without shroud, as follows

$$\sigma_{cr2} = \frac{1}{2}\rho_b\, h_{br2}\, d_m\omega^2 = \frac{7850 \times 0.09010 \times 0.7424 \times 733^2}{2} = 1.411 \times 10^8 \text{ N/m}^2$$

As has been shown above, the combined nozzle-blade efficiency η_{nb} of a two-rotor turbine can be expressed as follows

$$\eta_{nb} = \frac{P_{2b}}{\dot{m}_T\, \Delta h_T}$$

where $\dot{m}_T = 41.76$ kg/s is the mass flow rate of combusted gas through the turbine, $\Delta h_T = 0.8325 \times 10^6$ J/kg is the total enthalpy drop per unit mass of the combusted gas which occurs in the turbine, and P_{2b} is the power which the flowing gas transfers to the blades of the turbine. The power P_{2b} results from the following equation

$$P_{2b} = \dot{m}_T U (C_1 \cos\alpha_1 + C_2 \cos\alpha_2 + C_3 \cos\alpha_3 + C_4 \cos\alpha_4)$$

Therefore, after dropping the common factor $\dot{m}_T$, there results

$$\eta_{nb} = \frac{U (C_1 \cos\alpha_1 + C_2 \cos\alpha_2 + C_3 \cos\alpha_3 + C_4 \cos\alpha_4)}{\Delta h_T}$$

After substituting $U = 272.1$ m/s, $C_1 = 1201$ m/s, $\alpha_1 = 0.4363$ rad, $C_2 = 637.1$ m/s, $\alpha_2 = 0.6178$ rad, $C_3 = 593.4$ m/s, $\alpha_3 = 0.6109$ rad, $C_4 = 273.2$ m/s, and $\alpha_4 = 1.518$ rad, the numerator of the preceding fraction results

$$272.1 \times (1201 \times \cos 04363 + 637.1 \times \cos 0.6178 + 593.4$$
$$\times \cos 0.6109 + 273.2 \times \cos 1.518) = 0.5737 \times 10^6 \text{ J/kg}$$

Hence, the combined nozzle-blade efficiency is

$$\eta_{nb} = \frac{0.5737 \times 10^6}{0.8325 \times 10^6} = 0.689$$

Remembering that, in case of a single-stage, two-rotor, velocity-compounded impulse turbine, the total efficiency η_T of a turbine is expressed by the following product

$$\eta_T = \eta_{nb}\,\eta_m$$

where, in the present case, $\eta_{nb} = 0.689$ is the combined nozzle-blade efficiency, and $\eta_T = 0.582$ is the total efficiency of the turbine, then the machine efficiency η_m results

$$\eta_m = \frac{\eta_T}{\eta_{nb}} = \frac{0.582}{0.689} = 0.845$$

The velocity diagram, at the mean diameter $d_m = 0.7424$ m, can be drawn by using the following results.

The peripheral velocity of the rotor blades is $U = 272.1$ m/s.

The absolute velocity vector C_1 of the combusted gas at the outlet of the nozzle is $C_1 = 1201$ m/s in magnitude and forms an angle $\alpha_1 = 0.4363$ rad with the plane of rotation. The relative velocity vector V_1 of the combusted gas at the outlet of the nozzle is $V_1 = 961.3$ m/s in magnitude and forms an angle $\beta_1 = 0.5562$ rad with the plane of rotation.

The absolute velocity vector C_2 of the combusted gas at the outlet of the first rotor is $C_2 = 637.1$ m/s in magnitude and forms an angle $\alpha_2 = 0.6178$ rad with the plane of rotation. The relative velocity vector V_2 of the combusted gas at the outlet of the first rotor is $V_2 = 873.3$ m/s in magnitude and forms an angle $\beta_2 = 0.4363$ rad with the plane of rotation.

The absolute velocity vector C_3 of the combusted gas at the outlet of the stator is $C_3 = 593.4$ m/s in magnitude and forms an angle $\alpha_3 = 0.6109$ rad with the plane of rotation. The relative velocity vector V_3 of the combusted gas at the outlet of the stator is $V_3 = 395.1$ m/s in magnitude and forms an angle $\beta_3 = 1.038$ rad with the plane of rotation.

The absolute velocity vector C_4 of the combusted gas at the outlet of the second rotor is $C_4 = 273.2$ m/s in magnitude and forms an angle $\alpha_4 = 1.518$ rad with the plane of rotation. The relative velocity vector V_4 of the combusted gas at the outlet of the second rotor is $V_4 = 392.8$ m/s in magnitude and forms an angle $\beta_4 = 0.7679$ rad with the plane of rotation.

The partial isentropic drops of enthalpy per unit mass of the combusted gas are $\Delta h_{0-1} = 782{,}550$ J/kg in the nozzle, $\Delta h_{1-2} = 16{,}550$ J/kg in the first rotor, $\Delta h_{2-3} = 16{,}550$ J/kg in the stator, and $\Delta h_{3-4} = 16{,}550$ J/kg in the second rotor. The total isentropic drop of enthalpy per unit mass of the combusted gas in the turbine is $\Delta h_T = 832{,}500$ J/kg.

The efficiencies are $\eta_n = 0.92$ for the nozzle, $\eta_{nb} = 0.689$ for the nozzle-blade combination, $\eta_m = 0.845$ for the machine, and $\eta_T = 0.582$ for the whole turbine.

The nozzle, at the mean diameter, has an aspect ratio of 9.7, and $Z_n = 57$ blades. These blades have a pitch $P_n = 0.4092$ m, and form an angle $\theta_n = 0.4014$ rad between their central lines and the plane of rotation at the outlet of the nozzle. The radial height and the width of the nozzle at the throat are respectively $h_{nt} = 0.03810$ m and $b_{nt} = 0.003928$ m. The radial height and the width of the nozzle at the exit section are respectively $h_{ne} = 0.04346$ m and $b_{ne} = 0.01466$ m.

The first rotor, at the mean diameter, has $Z_{br1} = 119$ blades. These blades have a solidity $S_{br1} = 1.82$, a chord length $C_b = 0.03556$ m, and a pitch $P_{br1} = 0.01954$ m. The pressure surface of each blade forms an angle $\theta_{b1r1} = 0.5934$ rad at the inlet and an angle $\theta_{b2r1} = 0.4363$ rad at the outlet with the plane of rotation. The radial height of each blade is $h_{b1r1} = 0.04694$ m at the inlet and $h_{b2r1} = 0.06726$ m at the outlet. The width of each blade passage is $b_{b1r1} = 0.009656$ m at the inlet and $b_{b2r1} = 0.006987$ m at the outlet.

The stator, at the mean diameter, has $Z_{bs} = 127$ blades. These blades have a solidity $S_{bs} = 1.94$, a chord length $C_b = 0.03556$ m, and a pitch $P_{bs} = 0.01833$ m. The pressure surface of each blade forms an angle $\theta_{b1s} = 0.6458$ rad at the inlet and an angle $\theta_{b2s} = 0.6109$ rad at the outlet with the plane of rotation. The radial height of each blade is $h_{b1s} = 0.07264$ m at the inlet and $h_{b2s} = 0.07606$ m at the outlet. The width of each blade passage is $b_{b1s} = 0.009762$ m at the inlet and $b_{b2s} = 0.009244$ m at the outlet.

The second rotor, at the mean diameter, has $Z_{br2} = 109$ blades. These blades have a solidity $S_{br2} = 1.67$, a chord length $C_b = 0.03556$ m, and a pitch $P_{br2} = 0.02129$ m. The pressure surface of each blade forms an angle $\theta_{b1r2} = 1.047$ rad at the inlet and an angle $\theta_{b2r2} = 0.7679$ rad at the outlet with the plane of rotation. The radial height of each blade is $h_{b1r2} = 0.08214$ m at the inlet and $h_{b2r2} = 0.09806$ m at the outlet. The width of each blade passage is $b_{b1r2} = 0.01717$ m at the inlet and $b_{b2r2} = 0.01352$ m at the outlet.

The following example of application concerns the design of a two-stage, two-rotor, pressure-compounded impulse turbine for a rocket engine. The data are for this turbine are the same as those for the velocity-compounded turbine described in the preceding example. As has been shown above, a pressure-compounded turbine can be designed in such a way as to extract power from the expanding gas in equal parts in the two stages. To this end, it is necessary to execute a procedure of trial and error to determine the partial isentropic drops of enthalpy per unit mass in the nozzle and in the rotor of each stage which result in about the same amount of power extracted in each stage. In the present example, the stage carry-over ratio is assumed to be $r_c = 0.91$.

As a result of the procedure mentioned above, the partial isentropic drops of enthalpy per unit mass are supposed to be those indicated below.

At the nozzle of the first stage, the partial drop Δh_{0-1} is 50% of the total drop $\Delta h_T = 832500$ J/kg, such that $\Delta h_{0-1} = 0.5 \times 832{,}500 = 416{,}250$ J/kg.

At the rotor of the first stage, the partial drop Δh_{1-2} is 3% of the total drop $\Delta h_T = 832{,}500$ J/kg, such that $\Delta h_{1-2} = 0.03 \times 832{,}500 = 24{,}975$ J/kg.

At the nozzle of the second stage, the partial drop Δh_{2-3} is 44% of the total drop $\Delta h_T = 832{,}500$ J/kg, such that $\Delta h_{2-3} = 0.44 \times 832{,}500 = 366{,}300$ J/kg.

At the rotor of the second stage, the partial drop Δh_{3-4} is 3% of the total drop $\Delta h_T = 832{,}500$ J/kg, such that $\Delta h_{3-4} = 0.03 \times 832{,}500 = 24{,}975$ J/kg.

Let us consider the following figure.

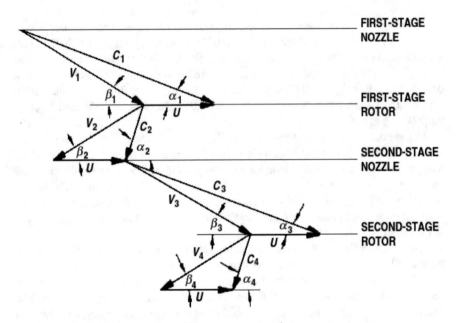

At the inlet of the nozzle of the first stage (subscript 0), the total temperature of the gas is $T_0 = 1033$ K, and the total pressure of the gas is $p_0 = 4.413 \times 10^6$ N/m². The nozzle efficiency is $\eta_n = k_n^2 = 0.96^2 \approx 0.92$.

At the outlet of the nozzle of the first stage (subscript 1), the spouting velocity C_1 of the combusted gas results from

$$C_1 = k_n(2\Delta h_{0-1})^{\frac{1}{2}} = 0.96 \times (2 \times 416250)^{\frac{1}{2}} = 875.9 \text{ m/s}$$

We choose the value 0.4363 rad for the angle α_1, which the absolute velocity vector C_1 of the combusted gas forms with the plane of rotation.

For the maximum efficiency, the peripheral velocity U of the rotor blades at the mean diameter should be such that

$$U = \frac{1}{2}C_1 \cos \alpha_1$$

Hence, in the present case, there results

$$U = \frac{1}{2} \times 875.9 \times \cos 0.4363 = 396.9 \text{ m/s}$$

Since C_1, α_1, and U have known values, the angle β_1, which the relative velocity vector V_1 forms with the plane of rotation at the outlet of the nozzle of the first stage, results from the following equation

$$\beta_1 = \arctan\left(\frac{C_1 \sin\alpha_1}{C_1 \cos\alpha_1 - U}\right)$$

After substituting $C_1 = 875.9$ m/s, $\alpha_1 = 0.4363$ rad, and $U = 396.9$ m/s in the preceding equation, we find

$$\beta_1 = \arctan\left(\frac{875.9 \times \sin 0.4363}{875.9 \times \cos 0.4363 - 396.9}\right) = 0.7505 \text{ rad}$$

The magnitude V_1 of the relative velocity vector V_1 can be determined through a simple inspection of the velocity diagram shown in the preceding figure. This yields

$$C_1 \sin\alpha_1 = V_1 \sin\beta_1$$

Hence

$$V_1 = \frac{C_1 \sin\alpha_1}{\sin\beta_1} = \frac{875.9 \times \sin 0.4363}{\sin 0.7505} = 542.7 \text{ m/s}$$

At the inlet of the nozzle of the second stage (subscript 2), the relative velocity V_2 of the gas at the outlet of the rotor of the first stage can be determined as follows

$$V_2 = \left(k_b^2 V_1^2 + 2\eta_n \Delta h_{1-2}\right)^{\frac{1}{2}}$$

After substituting $k_b = 0.89$, $V_1 = 542.7$ m/s, $\eta_n = k_n^2 = 0.96^2 \approx 0.92$, and $\Delta h_{1-2} = 24975$ J/kg in the preceding equation, we find

$$V_2 = \left(0.89^2 \times 542.7^2 + 2 \times 0.92 \times 24975\right)^{\frac{1}{2}} = 528.4 \text{ m/s}$$

We choose the value 0.6632 rad for the angle β_2, which the relative velocity vector V_2 of the gas flow forms with the plane of rotation at the outlet of the rotor of the first stage. The angle α_2, which the absolute velocity vector C_2 of the gas flow forms with the plane of rotation at the outlet of the rotor of the first stage, is determined through a simple inspection of the velocity diagram shown in the preceding figure. This yields

$$\tan\alpha_2 = \frac{V_2 \sin\beta_2}{V_2 \cos\beta_2 - U}$$

Hence

$$\alpha_2 = \arctan\left(\frac{V_2 \sin\beta_2}{V_2 \cos\beta_2 - U}\right) = \arctan\left(\frac{528.4 \times \sin 0.6632}{528.4 \times \cos 0.6632 - 396.9}\right) = 1.511 \text{ rad}$$

The velocity diagram of the preceding figure also shows that the magnitude C_2 of the absolute velocity vector C_2 results from

$$C_2 \sin\alpha_2 = V_2 \sin\beta_2$$

Hence

$$C_2 = \frac{V_2 \sin\beta_2}{\sin\alpha_2} = \frac{528.4 \times \sin 0.6632}{\sin 1.511} = 325.9 \text{ m/s}$$

At the outlet of the nozzle of the second stage (subscript 3), the spouting velocity C_3 of the combusted gas results from the following equation

$$C_3 = k_n\left(r_c C_2^2 + 2\Delta h_{2-3}\right)^{\frac{1}{2}} = 0.96 \times \left(0.91 \times 325.9^2 + 2 \times 366300\right)^{\frac{1}{2}} = 874.2 \text{ m/s}$$

This value is very close to the value of C_1 (875.9 m/s). Therefore, we choose the value 0.4363 rad for the angle α_3, which the absolute velocity vector C_3 of the combusted gas forms with the plane of rotation at the exit of the nozzle of the second stage. By so doing, $\alpha_3 = \alpha_1 = 0.4363$ rad.

The angle β_3, which the relative velocity vector V_3 forms with the plane of rotation at the outlet of the nozzle of the second stage, results from

$$\beta_3 = \arctan\left(\frac{C_3 \sin\alpha_3}{C_3 \cos\alpha_3 - U}\right) = \arctan\left(\frac{874.2 \times \sin 0.4363}{874.2 \times \cos 0.4363 - 396.9}\right) = 0.7514 \text{ rad}$$

Therefore, $\beta_3 = 0.7514$ rad. The magnitude V_3 of the relative velocity vector V_3 results from

$$V_3 = \frac{C_3 \sin\alpha_3}{\sin\beta_3} = \frac{874.2 \times \sin 0.4363}{\sin 0.7514} = 541.2 \text{ m/s}$$

At the outlet of the rotor of the second stage (subscript 4), the relative velocity V_4 of the gas can be determined as follows

$$V_4 = \left(k_b^2 V_3^2 + 2\eta_n \Delta h_{3-4}\right)^{\frac{1}{2}} = \left(0.89^2 \times 541.2^2 + 2 \times 0.92 \times 24975\right)^{\frac{1}{2}} = 527.2 \text{ m/s}$$

We choose the value 0.6632 rad for the angle β_4, which the relative velocity vector V_4 of the gas flow forms with the plane of rotation at the outlet of the rotor of the second stage. Therefore, $\beta_4 = \beta_2 = 0.6632$ rad. The angle α_4, which the absolute velocity vector C_4 of the gas flow forms with the plane of rotation at the outlet of the rotor of the second stage, results from

$$\alpha_4 = \arctan\left(\frac{V_4 \sin \beta_4}{V_4 \cos \beta_4 - U}\right) = \arctan\left(\frac{527.2 \times \sin 0.6632}{527.2 \times \cos 0.6632 - 396.9}\right) = 1.514 \text{ rad}$$

The magnitude C_4 of the absolute velocity vector C_4 of the combusted gas at the outlet of the rotor of the second stage results from

$$C_4 = \frac{V_4 \sin \beta_4}{\sin \alpha_4} = \frac{527.2 \times \sin 0.6632}{\sin 1.514} = 325.1 \text{ m/s}$$

The mean diameter d_m of the rotor results from the following equation

$$U = \frac{1}{2}\omega\, d_m$$

After substituting $U = 396.9$ m/s and $\omega = 733.0$ rad/s and solving for d_m, we find

$$d_m = \frac{2U}{\omega} = \frac{2 \times 396.9}{733.0} = 1.083 \text{ m}$$

As has been shown above, the combined nozzle-blade efficiency η_{nb} of a two-rotor turbine can be expressed as follows

$$\eta_{nb} = \frac{U(C_1 \cos \alpha_1 + C_2 \cos \alpha_2 + C_3 \cos \alpha_3 + C_4 \cos \alpha_4)}{\Delta h_T}$$

After substituting $U = 396.9$ m/s, $C_1 = 875.9$ m/s, $\alpha_1 = 0.4363$ rad, $C_2 = 325.9$ m/s, $\alpha_2 = 1.511$ rad, $C_3 = 874.2$ m/s, $\alpha_3 = 0.4363$ rad, $C_4 = 325.1$ m/s, and $\alpha_4 = 1.514$ rad, the numerator of the preceding fraction results

$$396.9 \times (875.9 \times \cos 0.4363 + 325.9 \times \cos 1.511 + 874.2 \times \cos 0.4363 + 325.1 \times \cos 1.514) = 0.6446 \times 10^6 \text{ J/kg}$$

Hence, the combined nozzle-blade efficiency is

$$\eta_{nb} = \frac{0.6446 \times 10^6}{0.8325 \times 10^6} = 0.7743$$

Assuming the machine efficiency of this pressure-compounded turbine to be the same as that ($\eta_m = 0.845$) of the velocity-compounded turbine considered above, the total efficiency η_T of this pressure-compounded turbine is

$$\eta_l = \eta_{nb}\, \eta_m = 0.7743 \times 0.845 = 0.6543$$

The velocity diagram for this turbine, at the mean diameter $d_m = 1.083$ m, can be drawn by using the following results.

The peripheral velocity of the rotor blades is $U = 396.9$ m/s.

The absolute velocity vector C_1 of the combusted gas at the outlet of the nozzle of the first stage is $C_1 = 875.9$ m/s m/s in magnitude and forms an angle $\alpha_1 = 0.4363$ rad with the plane of rotation. The relative velocity vector V_1 of the combusted gas at the outlet of the nozzle of the first stage is $V_1 = 542.7$ m/s in magnitude and forms an angle $\beta_1 = 0.7505$ rad with the plane of rotation.

The absolute velocity vector C_2 of the combusted gas at the outlet of the rotor of the first stage is $C_2 = 325.9$ m/s in magnitude and forms an angle $\alpha_2 = 1.511$ rad with the plane of rotation. The relative velocity vector V_2 of the combusted gas at the outlet of the rotor of the first stage is $V_2 = 528.4$ m/s in magnitude and forms an angle $\beta_2 = 0.6632$ rad with the plane of rotation.

The absolute velocity vector C_3 of the combusted gas at the outlet of the nozzle of the second stage is $C_3 = 874.2$ m/s in magnitude and forms an angle $\alpha_3 = 0.4363$ rad with the plane of rotation. The relative velocity vector V_3 of the combusted gas at the outlet of the nozzle of the second stage is $V_3 = 541.2$ m/s in magnitude and forms an angle $\beta_3 = 0.7514$ rad with the plane of rotation.

The absolute velocity vector C_4 of the combusted gas at the outlet of the rotor of the second stage is $C_4 = 325.1$ m/s in magnitude and forms an angle $\alpha_4 = 1.514$ rad with the plane of rotation. The relative velocity vector V_4 of the combusted gas at the outlet of the rotor of the second stage is $V_4 = 527.2$ m/s in magnitude and forms an angle $\beta_4 = 0.6632$ rad with the plane of rotation.

The partial isentropic drops of enthalpy per unit mass of the combusted gas are $\Delta h_{0-1} = 416250$ J/kg in the nozzle of the first stage, $\Delta h_{1-2} = 24975$ J/kg in the rotor of the first stage, $\Delta h_{2-3} = 366300$ J/kg in the nozzle of the second stage, and $\Delta h_{3-4} = 24975$ J/kg in the rotor of the second stage. The total isentropic drop of enthalpy per unit mass of the combusted gas in the turbine is $\Delta h_T = 832500$ J/kg.

The efficiencies are $\eta_n = 0.92$ for the nozzle, $\eta_{nb} = 0.7743$ for the nozzle-blade combination, $\eta_m = 0.845$ for the machine, and $\eta_T = 0.6543$ for the whole turbine.

The two examples considered above show that a pressure-compounded turbine has a higher total efficiency than that of a velocity-compounded turbine of the same properties. However, the greater mean diameter of the former (1.083 m against 0.7424 m) implies greater size and mass.

4.12   Bearings for Turbo-Pumps

The shaft of a turbo-pump is supported by two or more bearings. They are placed as closely as possible to major rotating components, in order to reduce radial movements and to control critical speeds. Ball bearings and cylindrical roller bearings are used because of their capabilities in terms of load, speed, stiffness, and misalignment tolerances. Tapered roller bearings, needle bearings, and pure-thrust ball bearings have not been used in rocket engine turbo-pumps, due to their limited speed capabilities [36]. As has been shown in Sect. 4.2, the bearings are used to absorb the radial and axial loads transmitted through the shaft, to keep the rotating parts of the turbo-pump in position and in correct alignment with the fixed parts, and to permit the shaft to rotate with the least possible amount of friction. The following figure, due to the courtesy of NASA [12] shows the bearings (two forward bearings and one aft bearing) which support the shaft of the Mark 10 turbo-pump.

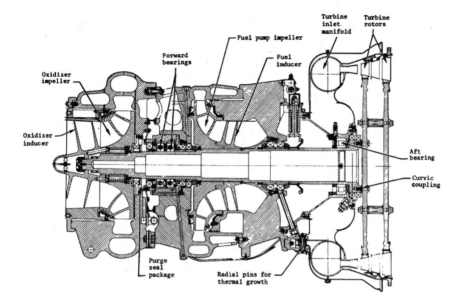

Some of the forces acting on the shaft of a turbo-pump are directed radially, that is, perpendicularly to the shaft axis; others are directed along the shaft axis.

The radial forces are due to the weights of components (such as shafts, impellers of pumps, rotors of turbines, and gears), to centrifugal forces acting on rotating parts, to forces of inertia caused by accelerations, to non-uniform distribution of pressure in the discharge volute of pumps, and to tangential forces caused by the presence of rotating gears.

The axial forces are due to weights of rotating parts mounted on a vertical shaft, to unbalanced axial thrust caused by the pumps, and to axial thrust acting on the blades of turbine rotors.

As has been shown above, ball bearings and cylindrical roller bearings are generally used for turbo-pumps of liquid-propellant rocket engines. This is because they have the following advantages over fluid-film bearings:

- ability to operate independently of external pressurising systems;
- ability to operate satisfactorily after ingesting foreign material;
- tolerance for short periods of coolant or lubricant absence;
- high rate of radial spring; and

- low generation of heat and low consumption of coolant or lubricant.

A hybrid ball bearing, with races made of 60NiTi alloy and balls made of Si_3N_4, is shown in the following figure, due to the courtesy of NASA [43]. By race we mean the track or channel in which the rolling elements ride.

National Aeronautics and Space Administration
John H. Glenn Research Center at Lewis Field

A ball bearing comprises an inner ring, an outer ring, and a plurality of balls placed between the two rings and held in their proper positions by a cage or separator. By contrast, in a cylindrical roller bearing, the rolling elements are cylinders of slightly greater length than diameter.

Conrad-type ball bearings (so named after their inventor, Robert Conrad) are often used, because they can support a combined radial and axial load, a thrust load in both directions along the shaft axis, and moment loading. Their capacity depends on the number of balls used for them. This number is limited by the assembly method. They require a two-piece cage, which limits their speed capability [36].

Split-inner-ring ball bearings and angular-contact bearings, which have 30% more capability to support thrust than Conrad-type bearings, are used singly or in tandem to support heavy axial loads. Tandem duplexed ball bearings are used when the thrust load is greater than the capability of a single bearing. Two or three bearings are used to share the loading [36].

Cylindrical roller bearings have several times the capacity to support radial loads and the stiffness of ball bearings. They are used to support only radial loads. Axial loads are supported by ball bearings axially connected to the roller bearing shaft. The following figure, due to the courtesy of NASA [36], shows how a thrust-carrying ball bearing (right) is used to eliminate axial loads from a roller bearing (left).

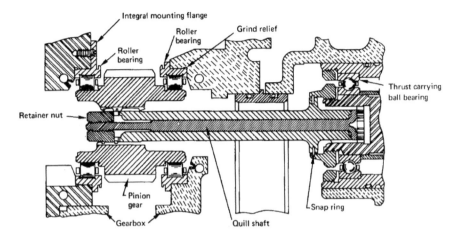

A parameter often used to measure a bearing speed is DN, which is the product of the bearing bore (the inner diameter of the inner ring) expressed in mm and the shaft speed expressed in rpm. Table III in [36] indicates the approximate maximum values of DN which have been achieved for bearings in operational turbo-pumps with specified coolants. These values range from 1.6×10^6 (for Conrad-type ball bearings and cylindrical roller bearings) to 2.06×10^6 (for angular-contact ball bearings). The value of DN is proportional to the tangential velocity of the bearing at the inside diameter of the inner race. When the maximum allowable limit of DN is exceeded, then lubrication and cooling are insufficient, and the bearing may fail due to overheating, contact wear, and fatigue. Since the value of DN is proportional to the shaft diameter, it depends on the bearing location. The following figure, due to the courtesy of NASA [12], shows the most common locations for bearings.

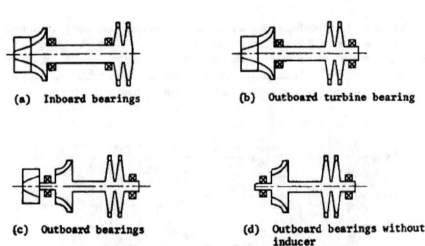

(a) Inboard bearings   (b) Outboard turbine bearing

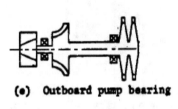

(c) Outboard bearings   (d) Outboard bearings without
                                              inducer

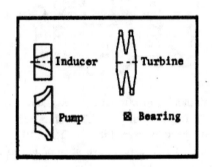

(e) Outboard pump bearing

The arrangement (a) of the preceding figure has frequently been used for rocket engine turbo-pumps in which the pump and the turbine are mounted on the same shaft. The J-2, RL10, and F-1 engines are examples of this arrangement, which has the advantage of avoiding the additional supporting structure and the separate lubrication and sealing systems which are required for the arrangements (b), (c), and (d).

For the arrangement (a), the shaft is often sized so that the turbo-pump speed is below the lowest value of the critical speed, which is that due to the simple shaft bending. This sizing is particularly important if the turbo-pump must operate over a wide range of operating speeds (throttling). Sobin et al. [12] indicate the following equation to estimate this limiting speed of the shaft:

$$\omega = \frac{1.053 \, (DN)^{\frac{4}{3}}}{K_{DN} P^{\frac{1}{3}}}$$

where ω (rad/s) is the rotational speed of the shaft, P (W) is the power on the shaft, and K_{DN} is a coefficient whose value, determined empirically, is 325 for one pump on the shaft and low-density propellant (liquid hydrogen), 374 for one pump on the shaft and dense propellant, and 478 for two pumps on the shaft and dense propellant.

Some increase in the speed limit can be obtained by placing the pump bearing between the inducer and the pump impeller, according to the arrangement (e) of the preceding figure, and using the inducer stator to support it.

By placing the turbine bearing overboard, according to the arrangement (b) of the preceding figure, the shaft diameter can be sized by torsional stress. This results generally in a smaller shaft diameter, and therefore the turbo-pump can be designed for a higher speed. Sobin et al. [12] indicate the following equation to express the speed ω (rad/s) as a function of the torsional stress which would occur in a solid shaft equal in diameter to the shaft diameter at the bearing:

$$\omega = \frac{4.388 \times 10^{-6} \tau_{eq}^{\frac{1}{2}} (DN)^{\frac{3}{2}}}{P^{\frac{1}{2}}}$$

where τ_{eq} (N/m^2) is the torsional stress of a solid shaft of the same outside diameter at the bearing. Sobin et al. [12] have found a good correlation with the final design by using a value of 1.724×10^8 N/m^2 for the equivalent-shaft stress.

Even higher speeds may be obtained by moving the bearing on the pump end to a location between the inducer and the impeller, according to the arrangement (c) of the preceding figure. This placement reduces the required bearing diameter by reducing the torque transmitted through the bearing bore.

With a boost pump, a main pump inducer may not be needed, because the boost pump would provide sufficient net positive suction head for the main pump impeller. In this case, the pump bearing can be placed outboard, according to the arrangement (d) of the preceding figure.

The coolant and the lubricant used for the bearings of turbo-pump are often the propellants themselves. This practice has the advantage of eliminating a separate system of cooling and lubrication. A small quantity of the pumped propellant is bled from a region of high pressure of the pump, flows through the bearings, and re-enter into a region of low pressure of the pump. For this purpose, the propellants must be thermally stable at the operating temperature, chemically inert with the materials of which the bearings are made, and sufficiently viscous.

The design data concerning the capacity to carry loads, the operating speed, and the life in service possessed by bearings are usually specified by their manufacturers. Table II in [36] contains a summary of capabilities possessed by bearings used for turbo-pumps of liquid-propellant rocket engines. The required life in service for a turbo-pump ranges usually from 1 to 2 h [36]. Bearings are designed for this life service by specifying a calculated B_{10} fatigue life, including speed effects, which exceeds the turbo-pump life by an arbitrary arbitrary factor, whose value is usually set to 10. This value is chosen not only to prevent fatigue failures, but also to compensate for other unquantifiable life-reducing factors, such as the low lubricity of most propellants and coolants. The B_{10} fatigue life is the operating time which 90% of identical bearings can exceed without fatigue failure. It is calculated by relating the stress cycles of a new design to the statistically analysed results of repeated fatigue tests on controlled test bearings [36].

4.13   Seals for Turbo-Pumps

The seals considered in this section are devices used in turbo-pumps of rocket engines to prevent or reduce the leakage of propellants or other fluids between the rotating parts (shaft, pump impeller, and turbine blades) and the stationary parts (stators and casing) of a turbo-pump. Therefore, only rotating seals are considered here. Following the criterion proposed by Zuk [44], these seals can be classified into the following categories.

The first category include the seals which depend on the selection and control of the sealing region. This category includes the following subcategories:

- positive rubbing contact seals, which in turn include mechanical face seals, circumferential shaft riding seals, lip seals, and soft packing seals;
- seals which operate at close clearances (ranging from 0.025 to 0.25 cm), which in turn include hydrodynamic seals, hydrostatic seals, and floating bushing seals; and
- fixed clearance seals (where the seal is in the gap between the rotating shaft and the stationary sleeve), which in turn include fixed bushing seals and labyrinth seals.

The second category includes the seals which depend on the control of properties of the fluid. This category includes the following subcategories:

- controlled heating and cooling seals; and
- ferromagnetic seals.

The third category includes the seals which depend on the control of forces acting on the fluid. This category includes the following subcategories:

- centrifugal seals;
- screw pump seals; and
- magnetic seals.

This section describes briefly only those of the seals indicated above which have most frequently been used in turbo-pumps of rocket engines, according to the data given in Table 1 of [45]. The seals used in valves will be described in Chap. 5, Sect. 5.8.

Two types of mechanical face seal are shown in the following figure, due to the courtesy of NASA [44].

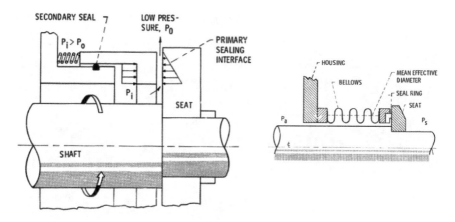

On the left-hand side, a rotating seal seat is mounted to the shaft and is held in close vicinity of a non-rotating sealing ring. The sealing ring is held in close vicinity of the seat by a mechanical spring. The sealing ring is allowed to move axially to follow the axial motion of the seal seat, such as run-out. Anti-rotation lugs, not shown in the preceding figure, prevent sealing ring rotation. Relative motion of the sealing ring and the stationary housing occurs across a secondary seal, which is shown as an O-ring in the preceding figure. The secondary seal may also be a piston ring, in case of high temperatures. The secondary seal fulfils the function of allowing the sealing ring to track axial motions of the seal seat. Since it is practically impossible to locate the seal seat to be perfectly perpendicular to the axis of rotation, then axial run-out occurs. In order to reduce the leakage due to this run-out, small axial movement of the sealing ring is allowed. In other words, the spring forces the sealing ring against the rotating seal seat to follow the run-out or wobble. The secondary seal and ring function may be combined into one integral unit, as shown on the right-hand side of the preceding figure, where a bellows is used. The seal shown on the left is internally pressurised. For this purpose, the pressurising fluid is located on the inner diameter of the seal and is separated from the low-pressure environment located at the outside diameter. By so doing, the pressurising fluid cannot leak to the external environment. The same seal is also pressure-balanced. At the primary sealing interface, leakage occurs and the pressure of the fluid decreases. In case of a laminar flow of a viscous fluid, the pressure decrease is linear, as shown in the preceding figure. The force due to this pressure decrease is called the seal separating force. The spring force and a hydrostatic force, which act against the sealing ring, counterbalance the seal separating force and hold the sealing ring against the seal seat.

Two types of circumferential shaft riding seal are shown in the following figure, due to the courtesy of NASA [44].

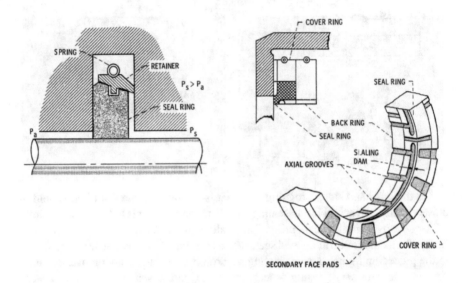

The seal ring shown on the left-hand side of the preceding figure consists usually of three segments, in order to permit radial misalignment of the shaft and also the motion of the shaft, and still maintain a very small gap. A retainer cover prevents leakage of the segment joints, and a garter spring keeps the segmented seal rings in close vicinity of the seal shaft. The type of seal shown on the left-hand side of the preceding figure is completely unbalanced. By contrast, the other type shown on the right-hand side of the same figure is partially pressure-balanced, because it has pressure-relief slots which result in only a small net unbalanced force in both the axial and the radial direction. Circumferential seals cannot be completely balanced. In addition, they are very sensitive to the installation. They are used when the shaft is subject to large axial movement, because they are more tolerant of pressure reversals than face seals.

A lip seal is shown in the following figure, due to the courtesy of NASA [44].

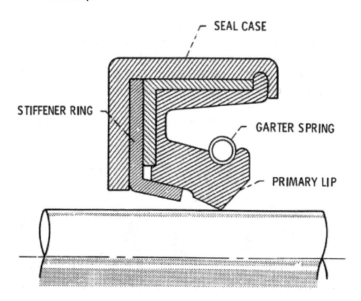

A lip seal consists of the seal case, the stiffener ring, and the primary lip, which is held in close vicinity of the shaft by a garter spring. This type of seals is generally used for low or zero differences of pressure. For example, it is used to prevent an oil mist, which lubricates a bearing, from leaking to the external environment. Tip seals are cheap, compact, and easy to install. However, since they are made of elastomeric materials, the elastomer may be incompatible with the sealed fluid, and may also be subject to stress relaxation.

Six types of labyrinth seal are shown in the following figure, due to the courtesy of NASA [45]. They are fixed clearance devices which restrict fluid leakage by dissipating the kinetic energy of the flow through a series of constrictions and cavities, which accelerate or decelerate the fluid or change abruptly the direction of flow, to create the maximum friction and turbulence.

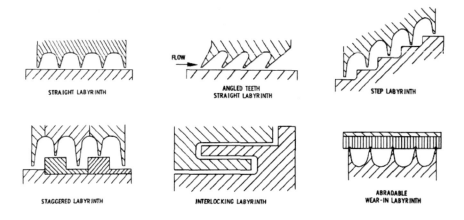

The ideal labyrinth transforms all the kinetic energy at each throttling into internal energy in each cavity. In practice, a labyrinth transfers a considerable amount of kinetic energy from one throttling to the next [45].

The following figure, due to the courtesy of NASA [15], shows the location of some of the seals described above in the A3-3 turbine of the RL-10 engine.

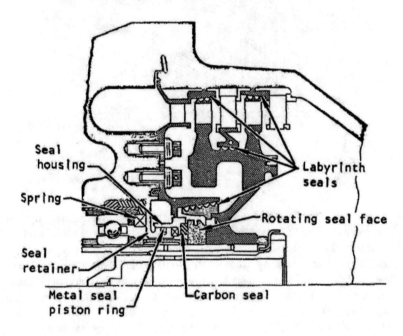

4.14 Gears for Turbo-Pumps

According to Hartman et al. [20], gear drives are used in turbo-pumps of rocket engines, with power transmitted ranging from 75 to 3356 kW, when:

- pumps handling propellants whose densities are equal to or greater than 700 kg/m^3 are to be driven by small, high-speed turbines; and
- a single turbine is to drive pumps handling propellants with greatly differing densities that require different pump speeds.

The geared turbo-pumps or engines cited as examples in [20] are the Mark 3 (used for the Atlas, Thor, and Saturn IB boosters), the Mark 4 (used for the Atlas sustainer), the LR-87-AJ-5 (used for the first stage of the Titan II), the LR-91-AJ-5 (used for the second stage of the Titan II), the RL10 (used for the Centaur), and the LR81-BA-11 (used for the Agena). The gears used in these engines are shown in the following figure, adapted from [20].

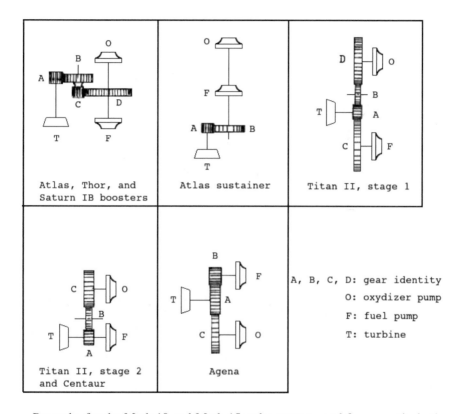

Recently, for the Mark 10 and Mark 15 turbo-pumps, used for respectively the F-1 and J-2 engines, the high-speed technology has made it no longer necessary to use gearing to couple a high-volume pump with an efficient, high-speed turbine. Therefore, turbo-pumps of latest design have the pump connected directly to the turbine [12]. This is because, for rocket engines using propellants which do not differ much in density (for example, liquid oxygen and RP-1), the rotational speeds of the two pumps are about the same, and therefore the fuel pump and the oxidiser pump can be mounted on the same shaft, as in the F-1 engine. For rocket engines using propellants which differ widely in density (for example, liquid oxygen and liquid hydrogen), the two pumps are mounted on separate shafts, each of them being driven by its own turbine, so that each pump can operate at its optimum rotational speed, as in the J-2 engine.

A gear drive may also be used to remove the inducer speed limit due to cavitation. This can be done by placing a low-speed pre-inducer to provide the required inlet pressure to the main inducer, as shown in the following figure, due to the courtesy of NASA [12].

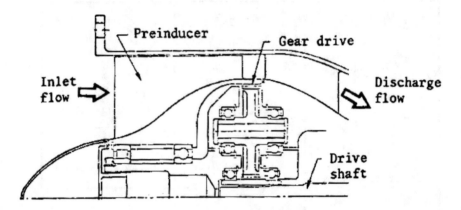

A pre-inducer may be driven not only by a gear drive, but also by other means, such as a through-flow hydraulic turbine drive, a re-circulated flow hydraulic turbine drive, an electric drive, and a gas turbine drive [12].

In general, geared turbo-pumps are restricted to those of small size, in which the turbine speed must be high enough to obtain reasonable blade heights, and in which multiple turbines would require excessive control elements in terms of size and mass [12].

The following figure, due to the courtesy of NASA [20] illustrates three methods for mounting gears. In this figure, D is the pitch diameter of the gear, d is the overhang distance, and S is the bearing span.

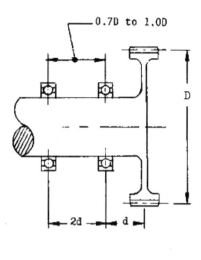

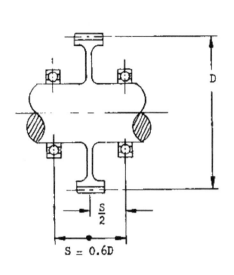

(a)   Straddle mount   (b)   Overhung mount

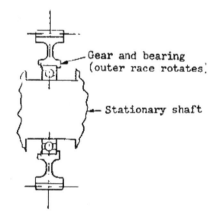

(c) Mount for idler gear
    with single bearing

The first method (straddle mount) is used whenever possible to reduce deflections under load, in order to avoid changes in centre distances and misalignment.

The second method (overhung mount) is sometimes used to save space. An example is the turbine shaft of the Mark 4 turbo-pump used for the Atlas sustainer.

The third method (mount for idler gear with single bearing) has been used on the RL10 gears. This method tolerates misalignment by rocking the ball bearing, so that the teeth of the idler gear line up with the teeth of the driving gear.

The materials used in gears for turbo-pumps are similar to those used in gears for aircraft. Deep-carburised case-hardened steels similar to vacuum-melted AISI

9310 (also known as AMS 6265, which is a nickel-chromium-molybdenum case-hardening steel, with good strength and toughness properties) have been found to have the best combination of properties for power gears. This is because the hardened outer surface of this steel resists compressive stress and wear, and its tough ductile core resists shock loads and bending-stress cycle fatigue. The fatigue life of gears has been extended by using vacuum-melted steel.

A summary of the materials used in gears for turbo-pumps is given in the following table, due to the courtesy of NASA [20].

Application	Gear material	Lubricant/coolant	Comments
Critical, highly loaded power gears	AMS 6265	M1L-L-60S6 oil MIL-L-7S0B oil MIIA-25336 oil Fuel-additive[a]	Used where high capacity and reliability are required; corrosion protection requited
Moderately loaded gears	AMS 6260 AISI 9310 AISI 4620 AISI 8620 AMS 6470 (nitrided)	Same as above	Used in applications less critical than those above Used for wear resistance; must have smooth edge radii to avoid edge chipping; must have protection from moisture corrosion
Lightly loaded gears	Any of the above, plus AISI 4340 AISI 4140	Same as above	Used for accessory gears
Propellant-cooled gears, moderately or lightly loaded	AMS 6260 AMS 6265	LH_2, LO_2, RP-1	Gear material must be protected from corrosion by moisture.
	AISI 440C	LH_2, LO_2, IRFNA, N_2O_4	Gear material very brittle; has some corrosion resistance; in experimental status only
	Beryllium copper (Berylco 25)	Ethylene diamine, UDMH, N_2H_4, LH_4, GH_2	Gear material low in hardness; has inherent corrosion resistance; in experimental status

[a]RP-1 plus Oronite 262 (2–3% concentration in service, up to 10% during run-ins)

References

1. Stangeland ML (1988) Turbopumps for liquid rocket engines. Threshold Pratt & Whitney Rocketdyne's Eng J Power Technol. http://www.pwrengineering.com/articles/turbopump.htm

2. Scheer DD, Huppert MC, Viteri F, Farquhar J, Keller RB Jr (eds) (1978) Liquid rocket engine axial-flow turbopumps. NASA SP-8125, NASA Lewis Research Centre, Cleveland, OH, 127pp. https://ntrs.nasa.gov/archive/nasa/casi.ntrs.nasa.gov/19780023221.pdf

3. Furst RB, Keller RB Jr (eds) (1973) Liquid rocket engine centrifugal flow turbopumps. NASA SP-8109, NASA Lewis Research Centre, Cleveland, OH, 116pp. https://ntrs.nasa.gov/archive/nasa/casi.ntrs.nasa.gov/19740020848.pdf

4. Jakobson JK, Keller RB Jr (eds) (1971) Rocket engine turbopump inducers. NASA SP-8052, 107pp. https://ntrs.nasa.gov/archive/nasa/casi.ntrs.nasa.gov/19710025474.pdf

5. Campbell WE, Farquhar J (1972) Centrifugal pumps for rocket engines. NASA conference paper, 34pp. https://ntrs.nasa.gov/archive/nasa/casi.ntrs.nasa.gov/19750003130.pdf

6. Burgoyne D et al (2003) Radial, mixed, and axial flow pumps. Introduction, Engineering Science Data Unit, Publication No. 80030, 61pp. http://www.idmeb.org/contents/resource/80030b_15_23.pdf

7. National Institute of Standards and Technology (NIST), U.S. Department of Commerce. https://physics.nist.gov/cgi-bin/cuu/Value?gn

8. Milan J-J (2004) Wikimedia, 19:25, 24 Nov 2004 (UTC). https://commons.wikimedia.org/wiki/File:Usure_par_cavitation_d%27un_impulseur_de_pompe_centrifuge_01.jpg

9. Paugh JJ How to compute Net Positive Suction Head for centrifugal pumps, 3pp. http://www.warrenpumps.com/resources/npsh.pdf

10. Evans J A brief introduction to centrifugal pumps, 12pp. http://www.pacificliquid.com/pumpintro.pdf

11. Brennen ChE (1994) Hydrodynamics of pumps. Concepts ETI and Oxford University Press. ISBN 0-19-856442-2. https://authors.library.caltech.edu/25019/3/pumbook.pdf

12. Sobin AJ, Bissel WR, Keller RB Jr (eds) (1974) Turbopump systems for liquid rocket engines. NASA SP-8107, 168pp. https://ntrs.nasa.gov/archive/nasa/casi.ntrs.nasa.gov/1975001 2398.pdf

13. Lakshminarayana B (1996) Fluid dynamics and heat transfer of turbomachinery. Wiley, New York. ISBN 0-471-85546-4

14. Farquahr J, Lindley BK (1966) Hydraulic design of the M-1 hydrogen turbopump. NASA CR-54822, 101pp. https://ntrs.nasa.gov/archive/nasa/casi.ntrs.nasa.gov/19660023044.pdf

15. Macaluso SB, Keller RB Jr (eds) (1974) Liquid rocket engine turbines. NASA SP-8110, NASA, 160pp. https://ntrs.nasa.gov/archive/nasa/casi.ntrs.nasa.gov/19740026132.pdf

16. Moss LA, Smith TE (1987) SSME single crystal turbine blade dynamics. NASA CR-179644, 27pp. https://ntrs.nasa.gov/archive/nasa/casi.ntrs.nasa.gov/19870016951.pdf

17. Hill PhG, Peterson CR (1965) Mechanics and thermodynamics of propulsion. Addison-Wesley, Reading, MA

18. Yahya, SM Wikipedia, Axial turbine, Turbomachinery, Public domain. https://en.wikipedia.org/wiki/Axial_turbine

19. Anonymous (1972) Skylab Saturn IB flight manual. NASA TM-X 70137, NASA, George C. Marshall Space Flight Centre, 273pp. https://ntrs.nasa.gov/archive/nasa/casi.ntrs.nasa.gov/197 40021163.pdf

20. Hartman MA, Butner MF, Keller RB Jr (eds) (1974) Liquid rocket engine turbopump gears. NASA SP-8100, 117pp. https://ntrs.nasa.gov/archive/nasa/casi.ntrs.nasa.gov/19750002094.pdf

21. Wikimedia. https://commons.wikimedia.org/wiki/File:H-1_rocket_engine_diagram_image.jpg

22. Huzel DK, Huang DH (1967) Design of liquid propellant rocket engines, 2nd ed. NASA SP-125, NASA, Washington, DC, 472pp. https://ntrs.nasa.gov/archive/nasa/casi.ntrs.nasa.gov/197 10019929.pdf

23. Stepanoff AJ (1957) Centrifugal and axial flow pumps, 2nd edn. Wiley, New York. ISBN 0-471-82137-3

24. Hole G (1994) Fluid viscosity effects on centrifugal pumps. Pumps and Systems Magazine, 4pp. http://www.warrenpumps.com/brochures/Fluid%20Viscosity%20Effects.PDF

25. Gülich JF (2008) Centrifugal pumps. Springer, Berlin. ISBN 978-540-73694-3

26. Arasmith S (2009) Introduction to small water systems. Chapter 6, Introduction to pumping systems. ACR Publications. https://dec.alaska.gov/Water/OPCert/Docs/Chapter6.pdf
27. Satterfield Z (2013) Reading centrifugal pump curves. Tech Brief 12(1):5pp. http://www.nesc.wvu.edu/pdf/dw/publications/ontap/tech_brief/tb55_pumpcurves.pdf
28. Wikimedia, by LEMEN, Wikimedia Commons. https://commons.wikimedia.org/wiki/File:Courbe_pompe.TIF
29. Veres JP (1995) Centrifugal and axial pump design and off-design performance prediction. NASA TM 106745, 24pp. https://ntrs.nasa.gov/archive/nasa/casi.ntrs.nasa.gov/19950013379.pdf
30. Sutton GP, Biblarz O (2001) Rocket propulsion elements, 7th edn. Wiley, New York. ISBN 0-471-32642-9
31. Anonymous. Engine systems, control and integration. Washington State University, 32pp. https://wsuwp-uploads.s3.amazonaws.com/uploads/sites/44/2014/10/20.-Hybrid-engine-system-control-integration.pdf
32. Wikimedia, by LittleGun—Own work, CC BY-SA 3.0. Wikimedia Commons. https://commons.wikimedia.org/w/index.php?curid=15696087
33. National Institute of Standards and Technology (NIST), U.S. Department of Commerce, Thermophysical Properties of Fluid Systems. https://webbook.nist.gov/chemistry/fluid/
34. Urasek DC (1696) Investigation of flow range and stability of three inducer-impeller pump combinations operating in liquid hydrogen. NASA TM X-1727, 22pp. https://ntrs.nasa.gov/archive/nasa/casi.ntrs.nasa.gov/19690006881.pdf
35. Sloteman DP, Cooper P, Dussord JL (1984) Control of backflow at the inlets of centrifugal pumps and inducers. In: Proceedings of the first international pump symposium. Turbomachinery Laboratories, Department of Mechanical Engineering, Texas A&M University, pp 9–22. http://hdl.handle.net/1969.1/164378
36. Butner MF, Keller RB Jr (eds) (1971) Liquid rocket engine turbopump bearings. NASA SP-8048, 85pp. https://ntrs.nasa.gov/archive/nasa/casi.ntrs.nasa.gov/19710018535.pdf
37. Csomor A, Sutton R (1977) Small, high-pressure, liquid oxygen turbopump, interim report. NASA CR-135211, 289pp. https://ntrs.nasa.gov/archive/nasa/casi.ntrs.nasa.gov/19770020459.pdf
38. Csomor A (1979) Small, high-pressure, liquid oxygen turbopump. NASA CR-159509, 156pp. https://ntrs.nasa.gov/archive/nasa/casi.ntrs.nasa.gov/19790009050.pdf
39. Huppert MC, Rothe K (1974) Axial pumps for propulsion systems. Conference paper, Pennsylvania State University Fluid Mechanics, Acoustics, and Design of Turbomachinery, Pt. 2, pp 629–654. https://ntrs.nasa.gov/archive/nasa/casi.ntrs.nasa.gov/19750003129.pdf
40. Glassman AJ (ed) (1994) Turbine design and application, vols 1–3. NASA SP-290, NASA Lewis Research Centre, 390pp. https://ntrs.nasa.gov/archive/nasa/casi.ntrs.nasa.gov/19950015924.pdf
41. Mattingly JD (2005) Elements of gas turbine propulsion. Tata McGraw-Hill, New Delhi. ISBN 978-0-07-060628-9
42. TimkenSteel Corporation. 4140 Alloy Steel Technical Data. http://www.timkensteel.com/~/media/TTCEW14201106474140HWAlloySteelTechDatap9082514.ashx
43. DellaCorte Ch (2014) Novel super-elastic materials for advanced bearing applications. In: 2014 CIMTEC international ceramics conference, Florence, Italy, 8–13 Jun 2014. NASA, 9pp. https://ntrs.nasa.gov/archive/nasa/casi.ntrs.nasa.gov/20140010477.pdf
44. Zuk J (1976) Dynamic sealing principles. NASA TM X-71851. In: Conference on theory and practice of lubrication, Dayton, OH, United States, 26–29 Apr 1976, 57pp. https://ntrs.nasa.gov/archive/nasa/casi.ntrs.nasa.gov/19760010311.pdf
45. Burcham RE, Keller RB Jr (eds) (1978) Liquid rocket engine turbopump rotating-shaft seals. NASA SP-8121, 168pp. https://ntrs.nasa.gov/archive/nasa/casi.ntrs.nasa.gov/19780022641.pdf

Chapter 5
Control Systems and Valves

5.1 Fundamental Concepts on Control Systems

The engine of a rocket vehicle is a particular case of a dynamical system. By system we mean a collection of parts or elements or components which work together to attain an object. By dynamical system we mean a system whose behaviour changes with time in response to an external stimulus or force. In case of an engine of a rocket vehicle, the object of the system is the generation of a thrust vector having the desired magnitude and direction.

A dynamical system has inputs (or signals coming in), outputs (or signals going out), and an internal processor which transforms inputs into outputs. The part of a dynamical system which transforms inputs into outputs, that is, the processor of signals, is also known as the plant. The plant is the part of a dynamical system which fulfils the function of receiving, handling, and emitting signals.

A dynamical system is usually represented graphically by means of a block diagram, in which lines indicate input or output signals, and boxes indicate the plant or other components of the system, as shown in the following figure.

In case of a rocket vehicle, the success or the failure of the mission of the vehicle depends on the outputs of the various components or subsystems of which the system is made. Therefore, a rocket vehicle belongs to a particular class of dynamical systems, which are called controlled systems.

Controlled systems exist in several fields of engineering. Controlling a dynamical system means regulating, or commanding, or governing that system in order to reach

A. de Iaco Veris, *Fundamental Concepts of Liquid-Propellant Rocket Engines*,
Springer Aerospace Technology,
https://doi.org/10.1007/978-3-030-54704-2_5

a desired goal. A control system is an arrangement of components connected one to another for the purpose of regulating the system itself or another system to be controlled by the first.

A control system which cannot adjust itself or another system to input signals received is called an open-loop system. An open-loop system is shown in the following figure, re-drawn from [1], which illustrates two systems which are connected between them, because the output signals u of the first (controller) are also the input signals of the second, which is the controlled system. Therefore, the first system controls the second system, but is not controlled by it. There is no mutual control of one system over the other. In other words, a system whose output signal y has no effect upon the input to the control process is called an open-loop control system.

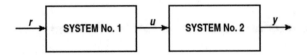

A control system which can adjust itself or another system to input signals is called a closed-loop or feedback system. A closed-loop system is shown in the following figure, re-drawn from [1], which illustrates two systems which have mutual control one over the other.

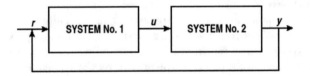

In other words, a closed-loop control system is a control system in which some function of the output y of some part of the system is fed back as a secondary input which adds to the primary input r to the system, so as to affect the response of the system itself. As an example of a control system for a rocket vehicle, Lorenzo and Musgrave [2] describe a pressure-fed bi-propellant engine, in which the propellants are kept under pressure in their respective tanks and then supplied to the main combustion chamber through appropriate feed lines, control valves, and injector elements, as has been shown in Chap. 3. The propellants are to be delivered to the main combustion chamber in a determined mixture ratio

$$\frac{o}{f} \equiv \frac{\dot{m}_o}{\dot{m}_f}$$

where $\dot{m}_o$ and $\dot{m}_f$ are the mass flow rates of respectively the oxidiser and the fuel, and at a total mass flow rate $\dot{m}_t = \dot{m}_o + \dot{m}_f$ related to the desired thrust. The control system of this rocket engine regulates the thrust by controlling the pressure p_c of the gas in the combustion chamber and the mixture ratio o/f of the two propellants, as shown in the following figure, adapted from [2].

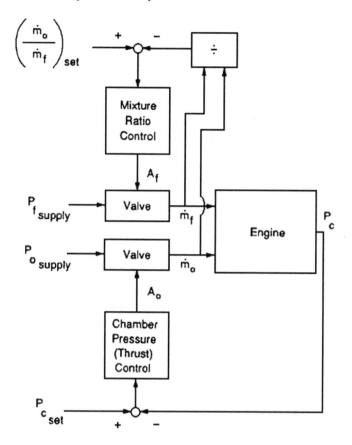

The preceding scheme has two control loops. One of them controls the mixture ratio o/f of the propellants, and the other controls the pressure p_c of the gas in the combustion chamber. The first loop operates usually, but not necessarily, on the propellant supplied at higher mass flow rate than the other. This loop is tuned to be the fast loop. This reduces excursions in the mixture ratio away from the set point, which in turn keeps the temperatures of the gas and of the metal at the design conditions. The second loop, which controls the pressure of the gas in the combustion chamber, is the slow loop, and its bandwidth is set by thrust response requirements. The type of control shown above requires three measurements (p_c, $\dot{m}_o$ and $\dot{m}_f$) and two control inputs for the valve areas A_o and A_f [2].

An open-loop control system for a rocket engine is calibrated to a fixed set of conditions, and uses orifices and on-off command devices to correct deviations of some parameters from their design values. For example, orifices of proper size are inserted into the flow lines to command pressure drops, and the mass flow rates of the propellants are controlled by opening or closing valves. An open-loop control system is simple, but is also limited to a specific set of parameters, and cannot compensate for variable conditions. The sequence times in an open-loop control system are often established by means of interlocks.

A closed-loop control system uses sensors, computers to detect errors (which are differences between a given reference signal r and the output y of the system), and actuation commands u generated by the computers to correct the errors. Therefore, a closed-loop control system does not require calibration for a specific set of conditions. It requires sensors and computers to detect errors and take appropriate steps to correct them. A closed-loop control system is often used in a rocket engine to control the mixture ratio o/f of the propellants, and the thrust vector F in magnitude and direction.

A closed-loop control system in a rocket engine may operate in one of the modes indicated below.

(1)   On-off control, which can be described as follows:

$$u = \begin{cases} u_{\max} & \text{if } e > 0 \\ u_{\min} & \text{if } e < 0 \end{cases}$$

where the control error $e = r - y$ is the difference between the reference signal r and the output y of the system, and u is the actuation command. The preceding equation does not define a value of the actuation command u when the control error e is equal to zero. In practice, u is taken equal to zero $(u = 0)$ when the control error e is in a narrow band centred around $e = 0$. The on-off control is used for a rocket engine, for example, when a pressure switch opens or closes a valve which regulates the pressure in a tank.

(2)   Proportional-integral-derivative (PID) control, which can be described as follows:

$$u = \begin{cases} u_{\max} & \text{if } e \geq e_{\max} \\ k_p e & \text{if } e_{\min} < e < e_{\max} \\ u_{\min} & \text{if } e \leq e_{\min} \end{cases}$$

where k_p is the controller gain, $e_{\min} = u_{\min}/k_p$, and $e_{\max} = u_{\max}/k_p$.

The interval $(e_{\min}, e_{\max})$ is called the proportional band, because the behaviour of the controller is linear when the error $e = r - y$ is in this interval, as follows

$$u = k_p(r - y) = k_p e \quad \text{if } e_{\min} < e < e_{\max}$$

In a proportional control, the process variable often deviates from a reference value. When some level of the control signal u is required for the system to maintain a desired value, then the control error e must be other than zero to generate the required value. For this purpose, the control signal u is made proportional to the integral of the error over a given time interval, as follows

$$u(t) = k_i \int_0^t e(\tau)d\tau$$

where k_i is the integral gain. This type of control is called integral control. A controller with integral control has zero steady-state error [1].

Unfortunately, there may not always be a steady state, because the system may be subject to oscillations. In this case, a controller can be made able to predict the error e some time T_d ahead of the present time t, by using the following linear extrapolation

$$e(t + T_d) \approx e(t) + T_d \frac{de}{dt}$$

By combining proportional, integral, and derivative control, it is possible to obtain a PID controller based on the following equation

$$u(t) = k_p e(t) + k_i \int_0^t e(\tau) d\tau + k_d \frac{de}{dt}$$

By so doing, the control signal $u(t)$ results from the sum of three terms: the term relating to the present time is proportional, through the coefficient k_p, to the error at the present time; the term relating to the past time is proportional, through the coefficient k_i, to the error cumulated over an interval before the present time; and the term relating to the future time is proportional, through the coefficient k_d, to the error resulting from a linear extrapolation of the error at some time ahead of the present time [1]. The proportional-integral-derivative control is used for a rocket engine in several cases, for example, to control the pressure in the combustion chamber (proportional term), or the mixture ratio of the propellants (integral term), or the direction of the thrust vector with phase lead (derivative term). An example of a closed-loop control system for a rocket engine has been given by Huzel and Huang [3] by means of the following scheme, re-drawn from [3].

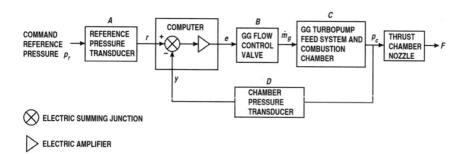

This control system illustrated above has the purpose of maintaining the value p_c of the pressure in the combustion chamber equal to a desired value p_r of reference, by using a valve which controls the value $\dot{m}_g$ of the mass flow rate in the gas generator (GG). The magnitude F of the thrust vector is controlled indirectly, by regulating the pressure p_c in the combustion chamber.

This control system consists of a sensor (chamber pressure transducer), a computer containing an electric summing junction and an electric amplifier, and a controller (gas generator flow control valve) which regulates the value of the mass flow rate $\dot{m}_g$ in the gas generator. The computer compares the reference input signal r with the input signal y coming from the sensor (chamber pressure transducer). The signal r is related to the reference pressure p_r, and the signal y is related to the actual pressure p_c in the combustion chamber through differential equations, which describe the behaviour of the components of the system.

The letters A, B, C, and D in the preceding figure indicate the relations existing between the input and the output of the respective components. The equations which govern this control systems can be written symbolically as follows

$$r = Ap_r$$

$$e = r - y$$

$$\dot{m}_g = Be$$

$$p_c = C\dot{m}_g$$

$$y = Dp_c$$

The behaviour of this control system results from the solution of the preceding equations. These equations are usually solved by using the method of the Laplace transformation (see, for example, [4]). The values to be given to the gains and to the response lags of a control system are to be chosen carefully, sometimes by trial and error, in order not to introduce overshoots or other causes of instability, which could give rise to large oscillations.

5.2 Control Systems for Rocket Engines

The control systems described in general terms in the preceding section are used in liquid-propellant rocket engines to perform some or all of the following tasks:

- engine start;
- engine shutdown;
- engine restart, in case of engines having restart capability;
- execution of a given plan of operation;
- change of the given plan of operation, or even engine shutdown, in the presence of a malfunction;
- propellant tank filling up;

- draining excess of propellants after filling;
- in case of cryogenic propellants, chilling pipes, pumps, cooling jackets, injectors, and valves by bleeding the cold propellants through them;
- check-out the proper operation of critical components before flight; and
- in case of recoverable and reusable engines, recycling and refurbishing the engines to put them in conditions of readiness for a new use.

In the specific case of liquid-propellant rocket engines, most of the actuators used for control functions are valves, regulators, pressure switches, and flow controls. Special computers for automatic control in large engines are also commonly used [5].

In addition, safety controls are used to protect personnel and equipment in case of malfunctions, and check-out controls make it possible to test the operation of critical components of a rocket engine without actually firing the engine.

The start sequence of a rocket engine has the purpose of controlling the engine from the moment in which the start signal is given to the full operation of the main stage. This sequence includes the steps of preparation (thrust chamber purging and chill-down of propellant transfer lines), application of start energy (start tanks and turbine spinner), and introduction followed by ignition of the propellants in the main combustion chamber. Secondary start sequences may be needed for some subsystems, as is the case with a gas generator or a pre-burner. A start sequence is regulated by means of interlocks, and by monitoring each step of the sequence. The opening sequence of the propellant valves may perform either an oxidiser-lead or a fuel-lead start, depending on the combination of propellants chosen, on the method used to ignite the propellants, and on the method used to cool the thrust chamber. The start delay time is the time necessary to purge the engine, open the valves, initiate the combustion, and increase the pressure in the combustion chamber to the rated value. This time is usually small (from 0.003 to 0.015 s in small thrusters) for an engine fed by gas under pressure, in which the pressurisation system has to be activated and the ullage volume has to be put under pressure before the start. An engine fed by turbo-pumps requires more time to start (from 1 to 5 s), because it is necessary not only to execute the operations indicated above for a pressure-fed engine, but also to start a gas generator or a pre-burner, and to increase the speed of the turbo-pumps to a level in which the combustion can be firstly self-sustained and then brought to its full extent. When the combination of propellants used in a rocket engine is not hypergolic, additional time is necessary for the igniter to work and for the control function to confirm the proper operation of the igniter. The ignition methods used for liquid-propellant rocket engines have been discussed in Chap. 2, Sects. 2.7 and 2.8.

The shutdown sequence is executed either in normal operating conditions or in cases of emergency. This is done by shutting off the flow to the gas generator (or to the pre-burner) and to the main combustion chamber. In case of test firings, the shutdown sequence also includes purges and flushes for post-firing safety. The control system regulates the valve closing sequence in such a way as to provide a fuel-rich cut-off in the main combustion chamber. This prevents high picks of temperature and results in a smooth and rapid termination of the thrust. The valves close in a fixed sequence. The valve controlling the gas generator or the pre-burner closes first. The pressurisation in the propellant tanks is stopped. The pumps slow down, as a result of the decrease in the gas flow through the turbine. The pressures and the mass flow rates of the propellants decrease quickly and reduce to zero.

In the three main engines (RS-25) of the Space Shuttle, the characteristics of start, run, and shutdown are established by the combined actions of the main fuel valve, the main oxidiser valve, the oxidiser pre-burner oxidiser valve, the fuel pre-burner oxidiser valve, and the chamber coolant valve. These valves are powered by hydraulic actuators which receive positioning signals from the engine controller, which in turn uses performance data gathered by sensors located throughout the engine. A functional scheme showing the propellant flow through the RS-25 engine and the valves named above is shown in the following figure, due to the courtesy of Wikimedia [6].

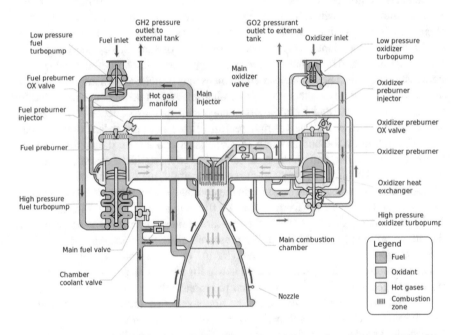

With reference to the preceding figure, the main oxidiser valve, the main fuel valve, and the chamber coolant valve are switched to run schedules during the engine run phase, while the oxidiser pre-burner oxidiser valve and the fuel pre-burner oxidiser valve are switched to closed-loop operations. The run schedule for the main oxidiser valve and for the main fuel valve cause them to simply remain fully open, whereas the run schedule for the chamber coolant valve drives it between half open at 67% thrust (minimum power level) and fully open at 100% thrust and above (in the Space Shuttle main engine, the thrust is variable between the minimum power level or 67% to the full power level or 109%, the rated or 100% value being 2.094×10^6 N in vacuo and 1.667×10^6 N at sea level). This action maintains the appropriate flow relationships among the several parallel fuel flow paths, as the high-pressure fuel turbo-pump output pressure varies with thrust. During engine run, the oxidiser pre-burner oxidiser valve and the fuel pre-burner oxidiser valve are used as control devices for thrust and mixture ratio. Manipulating the valves affects the output of the pre-burners, the speed of the turbo-pumps, and therefore the propellant flow rates. The fuel pre-burner oxidiser valve is driven alone to maintain mixture ratio in the main combustion chamber, while the oxidiser pre-burner oxidiser valve is driven with the fuel pre-burner oxidiser valve to increase or decrease thrust while maintaining the mixture ratio. The control loops include the controller, the valve actuators, and the transducers which sense the flow rates and the pressure in the main combustion chamber, and therefore the thrust. During the engine shutdown phase, all five valves are switched to shutdown schedules. These schedules ensure a smooth and safe shutdown by establishing a fuel lag. In other words, the oxidiser leaves the combustion chambers ahead of the fuel. This lag creates a fuel-rich and cool shutdown environment [7].

The controller of the RS-25 engine provides complete and continuous monitoring and control of engine operation. In addition, it performs maintenance and start preparation checks, and collects data for historical and maintenance purposes. The controller of the RS-25 engine is shown in the following figure, due to the courtesy of Boeing-Rocketdyne [7].

The controller is an electronic package which contains five principal sections:

- power supply section;
- input electronics section;
- output electronics section;
- computer interface section; and
- digital computer unit.

Pressure, temperature, pump speed, flow rate, and position sensors supply the input signals. Output signals operate spark igniters, solenoid valves, and hydraulic actuators. The controller is dual-redundant, which gives it normal, fail-operate, and fail-safe operational mode capability. Fail-operate mode follows a first failure, and is similar to normal mode, but with a loss of some redundancy. Fail-safe mode follows a second failure. In this mode, engine throttling and mixture ratio control are suspended, the main propellant valves are held fixed in their last commanded position, and the engine is subsequently shutdown pneumatically. The controller provides active and continuous control of the engine thrust and of the mixture ratio in the main combustion chamber through closed-loop control. The controller reads the pressure (equivalent to the thrust) in the main combustion chamber, and compares it to the existing thrust reference signal. It uses the error to drive the oxidiser pre-burner oxidiser valve, which adjusts the thrust and eliminates the error. For the mixture ratio in the main combustion chamber, the controller reads the fuel flow-meter and drives the fuel pre-burner oxidiser valve to adjust the fuel flowing to the main combustion chamber, thus maintaining a mixture ratio $o/f = 6$. In addition to these primary functions, the controller performs engine checkout, limit monitoring, start readiness verification, and engine start and shutdown sequencing. The controller instructions to the engine control elements are updated 50 times per second (every 20 ms). The electronics are mounted on modular boards inside a sealed and pressurised chassis, which is cooled by heat convection through pin fins [7].

5.3 Control of Thrust Magnitude

The thrust of a rocket engine is controlled in magnitude by regulating the pressure in the main combustion chamber. Sometimes, a reduction of the thrust magnitude, also known as throttling, is necessary in the last part of the propelled flight of a rocket vehicle. This reduction can be performed by decreasing, either stepwise or continuously, the pressure in the main combustion chamber.

In case of rocket engines fed by turbo-pumps, the pressure in the main combustion chamber can be reduced by regulating either the mass flow rate of the propellants through the gas generator or the mass flow rate of the hot gas through the turbine. When the first method is used, it is also possible to vary the mixture ratio of the propellants.

In case of rocket engines fed by gases under pressure, the pressure in the combustion chamber can be reduced by regulating the pressures in the main tanks of the propellants.

In case of multiple engines arranged in a cluster, thrust control can be performed by shutting off one or more engines of the cluster.

The examples of thrust regulation considered in Sect. 5.1 concern the case of a rocket engine fed by gases under pressure and the case of a rocket engine fed by turbo-pumps. In the second case, the thrust has been controlled in magnitude by using a valve regulating the value of the mass flow rate in the gas generator, and therefore the power delivered by the turbine.

The example given below concerns thrust regulation performed by varying the flow of the propellants. The following, figure, re-drawn from [3], illustrates the scheme of a closed-loop control system based on flow variation.

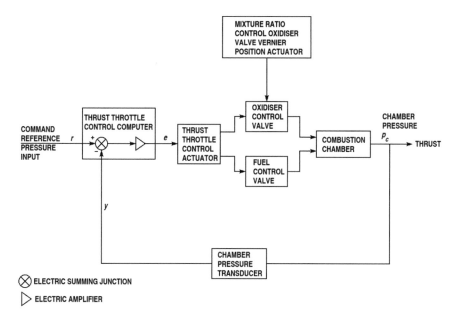

In the present case, the regulation is based on two valves which determine the resistances encountered by the two different propellants in their respective lines towards the combustion chamber. A chamber pressure transducer senses the pressure p_c in the combustion chamber, which is used as an indicator of the thrust magnitude. The output signal y of the chamber pressure transducer is fed back as a secondary input to the thrust throttle control computer, which compares the signal y with its primary input, which is the reference signal r. The error $e = r - y$ is used to drive the thrust throttle control actuator, which in turn regulates the control valves of the propellants in such a way as to reduce the error.

In the RS-25 engine, throttling is accomplished by varying the output of the pre-burners, thus varying the speed of the high-pressure turbo-pumps, and therefore the mass flow rates of the propellants (liquid oxygen and liquid hydrogen). The mixture ratio of the propellants in the main combustion chamber is $o/f = 6.032$. This value is maintained by varying the fuel flow rate around the oxidiser flow rate [7]. In other words, restricting the oxidiser flow to the pre-burners causes the turbine inlet temperature to decrease.

5.4   Control of Propellant Mixture Ratio

An open-loop control system of the mixture ratio of the propellants in a rocket engine can be obtained by installing calibration orifices of proper size in the propellant lines. Further refinements are possible by weighing accurately the propellants loaded in the tanks, by using orifices of adjustable (rather than fixed) size in case of storable fluids, in order to regulate the size of the orifices just before take-off, and by installing valves at the pump inlet of engines fed by turbo-pumps, in order to compensate for effects induced by accelerations acting on the fluid mass. In the last manner, an increase or decrease in pressure in the fluids due to variable accelerations is sensed and fed back to a closed-loop control system, which regulates gradually the valves and also protects the pumps.

In certain cases, it is necessary to perform a continuous control of the mixture ratio by using a closed-loop system. In such cases, Huzel and Huang [3] suggest the control scheme shown in the following figure.

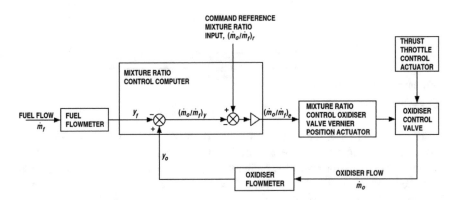

The control system illustrated in the preceding figure has two flow-meters which sense continuously the mass flow rates $\dot{m}_o$ and $\dot{m}_f$ of the propellants. The mixture ratio feedback signal $\left(\dot{m}_o/\dot{m}_f\right)_y$ is compared in a computer with a reference command signal $\left(\dot{m}_o/\dot{m}_f\right)_r$ and the resulting error $\left(\dot{m}_o/\dot{m}_f\right)_e$ is fed to a mixture ratio control oxidiser valve Vernier position actuator. This actuator, in turn, commands the oxidiser control valve, which varies the mass flow rate $\dot{m}_o$ of the oxidiser, in such a way as to correct the error.

A flow-meter is a device which measures either the volume flow rate q (m^3/s) or the mass flow rate $\dot{m}$ (kg/s) of a fluid which passes through it. A volumetric flow-meter (such as a positive displacement flow-meter) measures the volume flow rate $q = Av$ of a fluid stream which passes through a cross-sectional area A (m^2) at a velocity v (m/s). A velocity flow meter (such as a magnetic, turbine, ultrasonic, and vortex shedding and fluidic flow-meter) measures the velocity v of a fluid stream to determine the volume flow rate q. When the volume flow rate q is known, the mass flow rate $\dot{m}$ results from $\dot{m} = \rho q$, where ρ (kg/m³) is the density of the fluid in the given conditions.

A turbine-type flow-meter having a hemispherical hub is illustrated in the following figure, due to the courtesy of NASA [8].

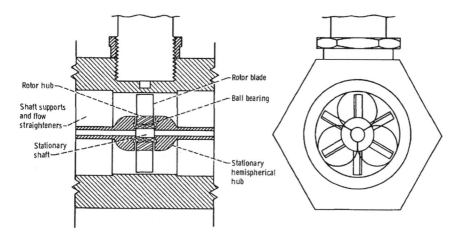

A turbine-type flow-meter is a device in which the entire flow stream turns a bladed rotor at a speed proportional to the volume flow rate of the fluid, and which generates or modulates an output signal, whose frequency is proportional to the angular velocity of the rotor. As shown in the preceding figure, the fluid passing through the flow-meter impinges on the blades of a turbine, which are free to rotate about an axis along the central line of the turbine housing. A permanent magnet placed within the windings of a pickoff coil generates a magnetic field. An electrical cycle is generated by each blade which sweeps through the magnetic field present in the fluid passage.

Another flow-meter, based on the Venturi tube, will be described in Sect. 5.7. Further information on several types of flow-meters can be found in [9].

5.5   Control of Propellant Consumption

This type of control is performed by sensors which measure the amount of propellant remaining in the tanks and the unbalance of one of the two propellants with respect to the other. A closed-loop scheme indicated by Huzel and Huang [3] is shown in the following figure.

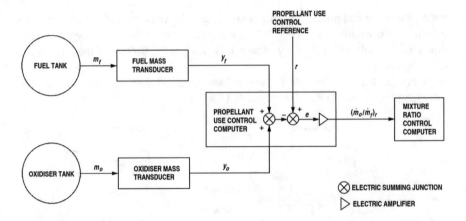

This control system measures continuously the residual masses m_f and m_o of the two propellants by using sensors placed in the respective tanks. The transducer output signals y_f and y_o corresponding to these masses are summed, and the signal resulting from the sum $y_f + y_o$ is compared with a propellant use control reference signal r which is fed to the propellant use control computer.

The error signal $e = r - (y_f + y_o)$ is amplified and then used to modify the command reference mixture ratio signal $(\dot{m}_o/\dot{m}_f)_r$ which is fed to the mixture ratio control computer.

A survey on the sensors used to determine the amount of propellant contained in a tank has been performed by Dodge [10]. A brief account is given below.

Sensors detecting the presence of fluids of a given type at given locations in a tank are usually called wet-dry sensors. Most wet-dry sensors are based on a hot wire or an electrical resistance or an electrical impedance element which carries a small current. The presence of a given type of fluid at the sensor location is detected by a change in the electrical impedance of the sensor, which in turn depends on the type of fluid (gas or liquid) around the sensor. This is because the sensor impedance depends on the amount of heat transfer from the sensor to the surrounding medium, due to the heating caused by the passage of the electric current. The heat transfer is greater when the sensor is surrounded by a liquid. Thus, by measuring a change in current for a constant applied voltage or a change in voltage for a constant applied current, it is possible to detect whether the sensor is in a liquid or in a gas.

Another type of sensor is based on a laser light source incorporated in a prism-like capsule, whose index of refraction is matched to the index of refraction of the liquid contained in a tank. A scheme of this optical sensor is shown in the following figure, due to the courtesy of NASA [10].

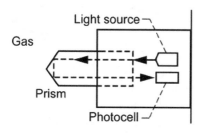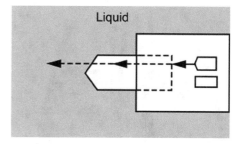

When the sensor is immersed in a gas, as shown on the left-hand side of the preceding figure, the light is reflected off the end of the prism back to a photocell at the light source. By contrast, when the sensor is immersed in a liquid, as shown on the right-hand side of the preceding figure, the light is transmitted through the liquid without reflection off the prism. Thus, the presence or absence of a reflection indicates whether the sensor is dry or wet.

Sensors which detect the location of the surface of a liquid in a tank or the depth of this liquid above the sensor are called level sensors. In this category are pressure sensors, ultrasonic sensors, and electrical capacity sensors.

Pressure sensors cannot be used in weightless conditions, because they measure the pressure head $p/(\rho g_0)$ of the liquid placed above the location of the sensor. A pressure transducer is inserted at the bottom of a gauge line going from the bottom to the top of a tank. A pressure sensor detects the difference of pressure existing between the gas at the top and the liquid at the bottom of the tank. This difference is related to the depth of the liquid when the pressure and the temperature of the substance (liquid or gas) are known.

Ultrasonic sensors generate an ultrasonic pulse which is transmitted through the liquid propellant contained in a tank. A mismatch of acoustic impedance at the liquid-gas interface generates an echo, which is transmitted back through the liquid to an ultrasonic receiver at the transducer location. The time elapsed between the emission of the pulse and the reception of the echo is related to the depth of liquid above the transducer by the sonic velocity in the liquid. A sensor of this type can be used in low-gravity conditions, but the liquid must be maintained in a known configuration above the sensor by a propellant management device based on the principle of capillarity.

An ultrasonic sensor of another type is a torsional wave guide. In this sensor, the speed of a torsional wave propagating along a wave guide, such as a rod, depends on whether the wave guide is immersed in a liquid or in a gas. The impedance mismatch at the liquid-gas interface generates an echo, which is detected to provide information on the length of the wave guide which is immersed in the liquid. However, some tests executed in conditions of zero gravity have shown that the adherence of liquid to portions of the wave guide supposed to be dry degraded seriously the accuracy of the sensor.

A capacity sensor contained in a propellant tank is shown in the following figure, due to the courtesy of NASA [11].

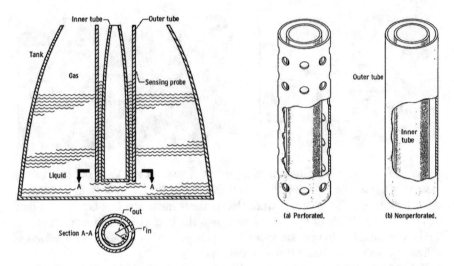

(a) Perforated.   (b) Nonperforated.

A capacity probe senses mass directly. The sensing probe consists essentially of two coaxial tubes contained in a tank. These tubes, when a voltage is applied to them, act as the plates of an electric condenser, whose capacity changes when the proportion between liquid and gas changes. This is because the liquid and the gas contained in the tank have different values of dielectric constant. In other words, the dielectric medium between the two tubes of the probe is the propellant, in liquid and gaseous state, contained in the tank. The total electrical capacity of the probe depends on the level of the liquid which fills it. Thus, the measured electrical capacity of the probe provides information on the mass of liquid in the tank. With a suitable electronic apparatus, the mass of the residual propellant in the tank can be measured continuously. The outer tube of the capacity probe may be perforated or not, as shown in the preceding figure. A non-perforated probe was chosen for the Centaur launch vehicle, because it provided a better measurement [11].

A propellant management system using electrical capacity probes was also chosen for the J-2 engine in the S-IVB stage, as shown in the following figure, due to the courtesy of NASA [12].

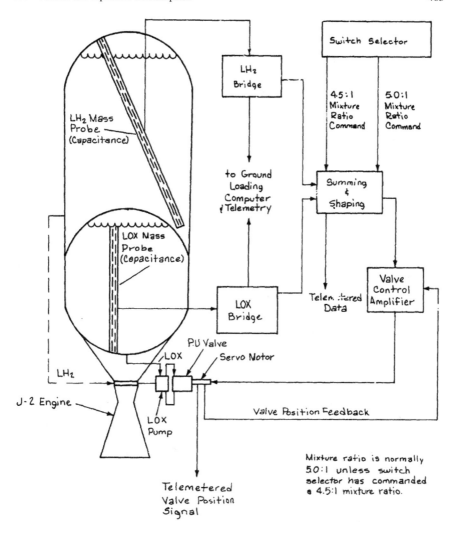

Mixture ratio is normally
5.0:1 unless switch
selector has commanded
a 4.5:1 mixture ratio.

Telemetered
Valve Position
Signal

In the S-IVB stage, the propellant management system, in conjunction with the switch selector, controls the mass propellant loading ratio and the engine mixture ratio (liquid oxygen to liquid hydrogen) to ensure balanced consumption of propellant. The electrical capacity probes, located in the two tanks, monitor the mass of the propellants. During flight, the electrical capacity probes are not used to control the propellant mixture ratio. The mixture ratio is controlled by switch selector outputs, which are used to operate the propellant utilisation (PU) valve. The PU valve is a rotary valve which controls the quantity of liquid oxygen flowing to the engine. The PU valve is commanded to its null position to obtain an engine mixture ratio (EMR) of 5.0:1 prior to engine start. The PU valve remains at the 5.0:1 position during the first burn. Prior to engine restart (first opportunity), the PU valve is commanded by the switch selector to an EMR of 4.5:1 and remain at this position until approximately

2 min of S-IVB burn. Then the PU valve is commanded to its null position (5.0:1) by the switch selector. However, if the S-IVB restart is delayed to the second opportunity, the EMR is shifted from 4.5:1 to 5.0:1 by the switch selector at about the time in which the engine reaches 90% thrust [12].

An electrical capacity probe depends on the liquid being settled, and therefore can only be used for gauging during periods of thrusting [10]. In principle, an electrical capacity probe might be used in low-gravity conditions if the liquid configuration is controlled by propellant management devices. However, in such conditions, capillary forces in the annular gap between the walls of the probe would cause the liquid meniscus location to differ from the liquid level in the tank, and thus cause an inaccurate reading, which may be small and could be compensated for. Depending on the design, the probe can also be sensitive to liquid motions such as occurred with the first landing of the Lunar Module on the Moon, for which non-linear sloshing lowered the effective level of the liquid within the gauge to the point that a premature low liquid level warning was given [10].

In the absence of a specific measurement system, the quantity of propellant remaining in a tank can be estimated by determining the consumption of the propellants which have already been used in comparison with the quantity initially loaded in the tanks. This method is commonly called bookkeeping. Some sensors and devices are used in the bookkeeping method, either separately or in conjunction. They are:

- flow-meters or Venturi-meters in the lines between the tanks and the main combustion chamber or the gas generator to measure the flow rates of the propellants;
- sensors which determine the flow rates of the propellants by measuring their pressures in the pumps in conjunction with the performance curves of the pumps; and
- calibrations of engine thrust versus propellant flow rate and tank pressure.

Various types of flow-meters, turbine-meters, accelerometers, and pressure sensors are required for the bookkeeping method. This method can only be used during periods of engine thrusting, and its accuracy is limited by the accumulation over time of the errors committed in estimating the propellant consumption, in particular near the point of propellant depletion. In addition, the bookkeeping method cannot detect leaks.

Other methods indicated by Dodge [10] to evaluate propellant consumption are:

- measuring the acceleration of the vehicle and the thrust of the engine, in order to determine the mass of the vehicle at any given time in comparison with the mass of the empty vehicle; and
- measuring the decay of pressure in the ullage spaces, in case of tanks which are initially pressurised, as the liquids are drained from their tanks and the ullage spaces increase.

5.6   Control of Thrust Direction

Several methods used to control the direction of the thrust vector in a rocket vehicle have been discussed in Chap. 2, Sect. 2.2, as far as they can affect the design of the thrust chamber of a rocket engine. The present section is meant to describe in further depth the method of controlling the thrust vector by using a gimbal mechanism acting on either the nozzle or the engine assembly. This method is the most used in liquid-propellant rocket engines, due to its reliability and performance. The following figure, re-drawn from [3], shows a control scheme for thrust vector control using hydraulic actuators.

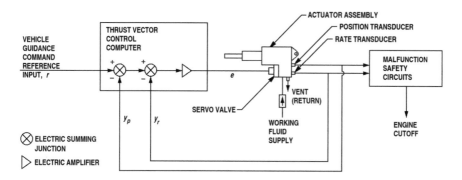

An example of a closed-loop control system using hydraulic actuators for thrust vector control is provided by the S-II, which is the second stage of the Saturn V rocket vehicle. A functional scheme of the control system of the S-II is shown in the following figure, due to the courtesy of NASA [12].

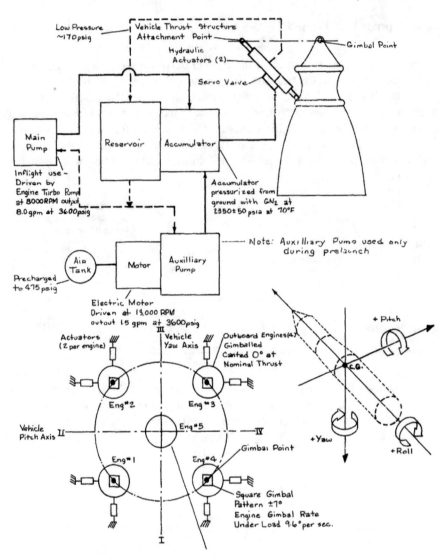

With reference to the preceding figure, the S-II has five J-2 engines arranged in a quincunx pattern (that is, one at the barycentre and four at the vertices of a square). The four outboard engines (No. 1, 2, 3, and 4 in the figure) are gimbal-mounted to provide thrust vector control during powered flight. Attitude control of the S-II stage is maintained by gimballing the four outboard engines in conjunction with electrical control signals from the inertial unit flight control computer. The gimballing system consists of four independent closed-loop hydraulic control subsystems, which provide power for engine gimballing.

The primary components of each control subsystem, also shown in the preceding figure, are an auxiliary pump, a main pump, an accumulator/reservoir manifold assembly, and two servo-actuators. The auxiliary pump is electrically driven from the ground support equipment to provide hydraulic fluid circulation prior to launch. The main pump is mounted to and driven by the engine liquid-oxygen turbo-pump. The accumulator/reservoir manifold assembly consists of a high-pressure accumulator, which receives high-pressure fluid from the pump, and a low-pressure reservoir, which receives return fluid from the servo-actuators. The servo-actuator is a power control unit, which converts electrical signals and hydraulic power into mechanical outputs which gimbal the engine. The components indicated in the preceding functional scheme are also shown in the following figure, due to the courtesy of NASA [13].

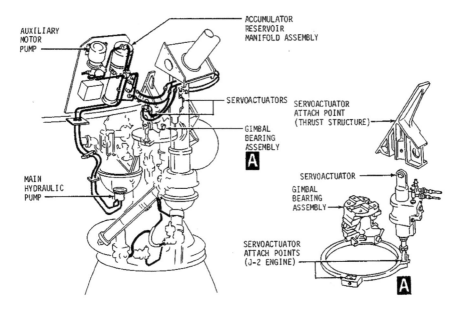

During the pre-launch period, the auxiliary pump circulates the hydraulic fluid to preclude fluid freezing during propellant loading. Circulation is not required during the first stage (S-IC) burn, due to the short duration of the burn. After separation of the second stage (S-II) from the first, a S-II switch selector command unlocks the accumulator lock-up valves, releasing high-pressure fluid to each of the servo-actuators. The accumulators provide gimballing power prior to the main hydraulic pump operation, in the transient of separation of the two stages. During the S-II main-stage operation, the main hydraulic pump supplies high-pressure fluid to each of the servo-actuators. The return fluid from the actuators is routed to the reservoir, which stores hydraulic fluid at sufficient pressure to supply a positive pressure at the main pump inlet [11].

As has been shown in Chap. 2, Sect. 2.2, a gimballed engine is mounted on a spherical joint. In the three main engines (RS-25) of the Space Shuttle, the gimbal bearing is bolted to the vehicle by its upper flange and to the engine by its lower flange. It supports 33,271 N of engine weight and 2,224,000 N of thrust. It is a ball-and-socket universal joint where concave and convex spherical surfaces are interconnected. Sliding contact occurs between these surfaces as the bearing is angulated. Fabroid® inserts located at the sliding contact surfaces reduce friction which occurs during gimbal bearing angulation. The bearing, which is installed during engine assembly, measures approximately 27.9 cm × 36.6 cm, weighs about 467 N, has an angular capability of ±0.218 rad (±12.5 deg) and is made of a titanium alloy (Ti-6Al-6 V-2Sn). It is shown in the following figure, due to the courtesy of Boeing-Rocketdyne [7].

The following figure, adapted from [13], illustrates a typical hydraulic servo-actuator, installed to gimbal the F-1 engine of the S-IC stage of the Saturn V rocket vehicle.

HYDRAULIC SERVOACTUATOR

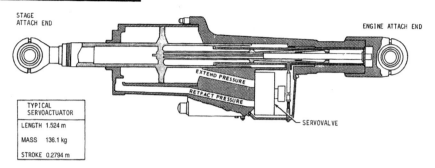

TYPICAL SERVOACTUATOR	
LENGTH	1.524 m
MASS	136.1 kg
STROKE	0.2794 m

A detailed example of design of a control system for a rocket engine mounted on gimbals can be found, for example, in [14].

5.7 Principal Components of Flow Control Systems

Liquid-propellant rocket engines have several components which measure and control the motion of fluids through them. The principal components used for this purpose are valves, pressure regulators, and flow-meters. They are described here and in the following sections.

A simple analysis of flow control devices can be done by considering firstly the motion of an ideal fluid, and then the corrections which are necessary to take account of the behaviour of real fluids. For example, a liquid can be considered inviscid, incompressible, and moving in laminar flow (see below) through a tube. The Bernoulli principle, so called after the Swiss scientist Daniel Bernoulli, applies to this case. According to this principle, the total energy (potential energy plus pressure energy plus kinetic energy) possessed by an ideal fluid in steady motion between any two cross sections 1 and 2 of a tube is constant.

By considering the total energy per unit weight, the Bernoulli principle can be expressed as follows

$$z_1 + \frac{p_1}{\rho g_0} + \frac{v_1^2}{2g_0} = z_2 + \frac{p_2}{\rho g_0} + \frac{v_2^2}{2g_0}$$

where z (m) is the elevation of the chosen point with respect to a reference plane, p (N/m^2) is the absolute pressure in the fluid in the chosen point, ρ (kg/m^3) is the density of the fluid in all points at the given temperature, $g_0 = 9.80665$ m/s^2 is the acceleration of gravity near the surface of the Earth, and v (m/s) is the velocity of the fluid in the chosen point along a stream line.

When the flow takes place at a constant elevation, then $z_1 = z_2$, and therefore the preceding equation may be written as follows

$$\frac{p_1 - p_2}{\rho g_0} = \frac{v_2^2 - v_1^2}{2g_0}$$

In addition, due to the continuity of the flow, the product $\rho v A$ has the same value at any cross section of area A along the tube, and therefore

$$\frac{v_1}{v_2} = \frac{A_2}{A_1}$$

where A_1 (m^2) and A_2 (m^2) are the areas of any two cross sections along the tube.

The preceding equations can be used to measure the volume flow rate q (m^3/s) of a fluid through a fluid control system, as will be shown below.

The following figure, due to the courtesy of NASA [15], shows a longitudinal section of a flow-meter based on the Venturi tube.

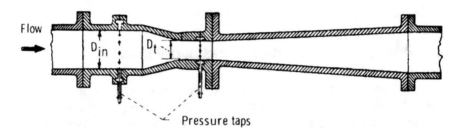

A Venturi-meter, so called after the Italian scientist Giovanni Battista Venturi, is a device which measures the volume flow rate of a liquid moving in a pipe.

It has three main parts, which are

- a convergent cone, whose diameter decreases from the value D_{in} of the pipe to the value D_t of the throat;
- a cylindrical throat, whose diameter D_t is usually one-fourth to one-half of the value D_{in}; and
- a diffuser, whose diameter increases downstream of the throat and reaches again the value D_{in} at the outlet plane.

The included angle of the convergent cone is approximately 0.3665 rad (21°), and the included angle of the diffuser cone ranges from 0.1222 rad (7°) to 0.2618 rad (15°). Two pressure taps are placed one at the inlet of the convergent cone and the other at the throat. A Venturi-meter is based on the Bernoulli principle. When the cross-sectional area of the tube decreases from the inlet plane to the throat, then the pressure head $p/(\rho g_0)$ of the fluid is forced to decrease, and its velocity head $v^2/(2g_0)$ is forced to increase, in order for its total head to be constant. The decrease in static pressure is measurable at the two taps (usually, by reading the difference of head in open vertical tubes inserted through the wall of the tube under pressure), and the volume flow rate q can be expressed as a function of the decrease in static pressure $p_1 - p_2$, as will be shown below. The velocity head acquired by the fluid in the convergent part is then converted back into pressure head (minus a loss due to friction) in the diffuser going from the throat to the outlet plane. The Bernoulli equation, written for $z_1 = z_2$, is

$$p_1 - p_2 = \frac{1}{2}\rho\left(v_2^2 - v_1^2\right)$$

This equation, solved for v_1^2, yields

$$v_1^2 = v_2^2 - \frac{2(p_1 - p_2)}{\rho}$$

By substituting this value of v_1^2 into $v_2 = (A_1/A_2)v_1$, there results

$$v_2^2 = \frac{2(p_1 - p_2)}{\rho}\frac{1}{1 - \left(\frac{A_2}{A_1}\right)^2} = \frac{2g_0\Delta H_p}{1 - \left(\frac{A_2}{A_1}\right)^2}$$

Therefore, the volume flow rate q (m³/s) in ideal conditions results from

$$q = A_2 v_2 = A_2\left[\frac{\frac{2(p_1 - p_2)}{\rho}}{1 - \left(\frac{A_2}{A_1}\right)^2}\right]^{\frac{1}{2}} = A_2\left[\frac{2g_0\Delta H_p}{1 - \left(\frac{A_2}{A_1}\right)^2}\right]^{\frac{1}{2}}$$

where ΔH_p (m) is the difference of pressure head measured between the inlet plane and the throat in the two open vertical tubes inserted through the wall of the Venturi-meter.

In practice, some loss of total head occurs in a Venturi-meter due to friction, and therefore the actual value of the volume flow rate q is slightly lower than the theoretical value resulting from the preceding equation. This loss can be taken into account by means of a discharge coefficient C_d, whose value is determined experimentally and is always less than unity, as follows

$$q = C_d A_2\left[\frac{\frac{2(p_1 - p_2)}{\rho}}{1 - \left(\frac{A_2}{A_1}\right)^2}\right]^{\frac{1}{2}} = C_d A_2\left[\frac{2g_0\Delta H_p}{1 - \left(\frac{A_2}{A_1}\right)^2}\right]^{\frac{1}{2}}$$

When the volume flow rate q (m³/s) of a given fluid has been determined by means of a Venturi-meter, as has been shown above, then the mass flow rate $\dot{m}$ (kg/s) of the same fluid results by multiplying the volume flow rate q by the density ρ (kg/m³) of the fluid at the temperature of interest.

As an example of application of the concepts exposed above, it is required to calculate the mass flow rate of liquid oxygen flowing through an horizontal Venturi-meter which has the following properties: diameter at the inlet plane $D_{in} = 0.1524$ m, diameter at the throat plane $D_t = 0.0762$ m, and discharge coefficient $C_d = 0.92$. The difference of pressure head measured between the inlet and the throat is $\Delta H_p = 13.86$ m, and the density of liquid oxygen at its boiling point is $\rho = 1141$ kg/m^3 [16].

By substituting these data in the preceding equation, the volume flow rate results

$$q = 0.92 \times \frac{3.1416 \times 0.0762^2}{4} \times \left[\frac{2 \times 9.807 \times 13.86}{1 - \left(\frac{0.0762}{0.1524}\right)^4} \right]^{\frac{1}{2}} = 0.07144 \text{ m}^3/\text{s}$$

and the mass flow rate results

$$\dot{m} = \rho q = 1141 \times 0.07144 = 81.52 \text{ kg/s}$$

All real fluids are viscous in various degrees. As has been shown in Chap. 4, Sect. 4.7, viscosity is a physical property which measures the degree of internal resistance opposed by a fluid to motion. The fluids for which the rate of deformation is proportional to the shear stress applied to them are called Newtonian fluids, after Isaac Newton. In case of a one-dimensional motion of a Newtonian fluid along a direction x (planar Couette flow), this linear relationship can be expressed as follows

$$\tau_{xy} = \mu \frac{dv}{dy}$$

where τ_{xy} (N/m^2) is the shear stress, dv/dy (s^{-1}) is the rate of shear strain, that is, the velocity gradient in the direction y perpendicular to the direction x of motion, $v \equiv v(y)$ (m/s) is the velocity of motion in the direction x, and μ (Ns/m^2) is the coefficient of dynamic viscosity of the fluid.

The motion of a viscous fluid in a cylindrical tube may be either laminar or turbulent or transitional. The first two types of flow are shown in the following figure, due to the courtesy of NASA [17].

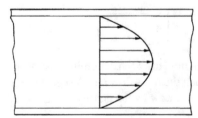 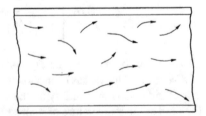

A viscous fluid in laminar flow regime (left) moves smoothly in thin layers, called laminae, which do not mix together. The layers in contact with the walls of the tube are stationary, whereas the internal layers move by sliding one over another. The velocity of the layers increases from the walls to the central line of the tube with a parabolic profile.

By contrast, a viscous fluid in turbulent flow regime (right) moves irregularly with eddies and swirls which mix the layers of fluid together. The mean velocity profile for turbulent flow is approximately elliptic (blunt nose) and is characterised by a much higher shear stress, due to the slope of the velocity profile at the walls [17]. Even in conditions of turbulent flow, a thin layer (called boundary layer) exist near each wall of the tube where the fluid moves in laminar flow regime.

A transitional flow fluctuates between laminar flow and turbulent flow. When this happens, a laminar flow is on the verge of becoming turbulent.

A criterion to indicate whether the motion of a fluid occurs in laminar or turbulent flow is provided by the value of the Reynolds number. As has been shown in Chap. 2, Sect. 2.5, the Reynolds number Re is defined as follows

$$Re = \frac{\rho v d}{\mu}$$

where ρ (kg/m^3) is the density of the fluid at the given temperature, v (m/s) is the mean velocity of the fluid in the tube, d (m) is the hydraulic diameter of the tube, and μ (N s m^{-2}) is the coefficient of dynamic viscosity of the fluid at the given temperature. The Reynolds number is the ratio of the inertial forces to the viscous forces which act on a unit volume of fluid moving in a tube. For low values of the Reynolds number, the viscous forces are sufficiently high to keep the fluid particles in parallel layers, and consequently the flow is laminar. For high values of the Reynolds number, the inertial forces prevail over the viscous forces, and consequently the flow is turbulent. For practical purposes, when the value of the Reynolds number is less than 2000, then the flow is laminar. The transition value for a fluid moving in a tube of circular cross section is $Re = 2300$. When the value of the Reynolds number ranges from about 2300 and 4000, then the flow is unstable, due to an incipient turbulence. When the value of Re is greater than 4000, then the flow is turbulent.

The friction forces acting on the particles of a fluid moving in a tube are due to the rubbing of the particles one against another and also against the walls of the tube. As a result of these forces, part of the kinetic energy possessed by the fluid is converted into heat. This heat may either remain into the fluid or be transferred to the external environment through the walls.

As has also been shown in Chap. 2, Sect. 2.5, the drop of pressure head ΔH_p (m) due to friction in a tube of length L (m) and hydraulic diameter d (m) is expressed as a function of the Darcy friction factor f_D (dimensionless) of the tube by means of the Darcy-Weisbach equation, as follows

$$\Delta H_p = f_D \frac{L}{d} \left(\frac{v^2}{2g_0} \right)$$

where v (m/s) is the average velocity of the fluid, and $g_0 = 9.80665$ m/s^2 is the acceleration of gravity near the surface of the Earth. The Darcy friction factor f_D depends on the Reynolds number Re defined above and also on the shape and smoothness of the tube. This factor can be determined as a function of the Reynolds number Re and of the relative roughness ε/d of the tube not only by means of the Colebrook-White relation (see Chap. 2, Sect. 2.5), but also by means of the Moody diagram (so called after the American scientist Lewis Ferry Moody) shown in the following figure, due to the courtesy of Beck and Collins, through Wikimedia [18].

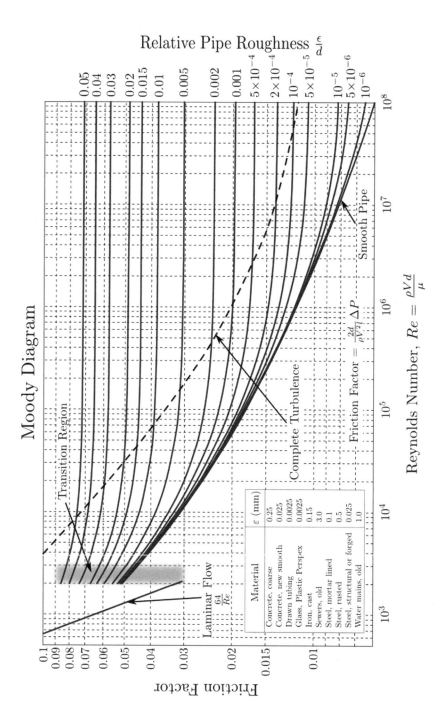

Moody Diagram

Relative Pipe Roughness $\frac{\epsilon}{d}$

Reynolds Number, $Re = \frac{\varrho V d}{\mu}$

Friction Factor

Friction Factor $= \frac{2d}{\varrho V^2 l}\Delta P$

Laminar Flow $\frac{64}{Re}$

Transition Region

Complete Turbulence

Smooth Pipe

Material	ε (mm)
Concrete, coarse	0.25
Concrete, new smooth	0.025
Drawn tubing	0.0025
Glass, Plastic Perspex	0.0025
Iron, cast	0.15
Sewers, old	3.0
Steel, mortar lined	0.1
Steel, rusted	0.5
Steel, structural or forged	0.025
Water mains, old	1.0

The presence of fittings, such as valves, elbows, T's, sudden expansions or contractions, et c., which may be present in a piping system, causes further losses due to friction. These losses can be taken into account by means of a fictitious term L_e (m), called equivalent length, which adds to the actual length L (m) of the pipe. By so doing, the Darcy-Weisbach equation written above becomes

$$\Delta H_p = f_D \left(\frac{L + L_e}{d} \right) \left(\frac{v^2}{2g_0} \right)$$

In other words, the actual length L of a tube and its equivalent length L_e, which takes account of all the fittings placed along the tube, are summed up to form a total length $L + L_e$, which in turn is introduced in the Darcy-Weisbach equation instead of the actual length L. The equivalent lengths L_e for a range of sizes of a given type of fitting have been found experimentally to be in an approximately constant ratio to the diameters d of the fittings. In other words, $L_e/d \approx$ constant. Therefore, a single value is sufficient to cover all sizes of each fitting. A table, which can be found in [19], gives the values of the L_e/d ratio for many valves and other typical fittings.

Another method, called K-method, takes account of the pressure drop due to the fittings by means of a resistance coefficient k assigned to each type of fitting. When this method is used, the total drop of pressure head can be computed by using the Darcy-Weisbach equation, as follows

$$\Delta H_p = \left(\frac{f_D L}{d} + \sum_{i=1}^{N} k_i \right) \left(\frac{v^2}{2g_0} \right)$$

Values of the coefficient k for several fittings can be found in [20]. Other methods are those called the 2-K (Hooper) method and the 3-K (Darby) method. They are refinements of the K method described above. Particulars and values of the coefficients to be used in the 2-K method and in the 3-K method can be found in [21, 22].

As an application of the concepts discussed above, it is required to compute the drop of pressure head for the discharge flexible duct of the oxidiser pump of a rocket engine and also for the main oxidiser valve, which is of the butterfly type, fully open. The following data are known: volume flow rate of liquid oxygen in the flexible duct $q = 0.7836$ m^3/s, inside diameter of the flexible duct $d = 0.2032$ m, actual length of the flexible duct $L = 0.4064$ m, equivalent length of the flexible duct (due to the resistance arising from flow deviation) $L_e = 6d = 6 \times 0.2032 = 1.219$ m, absolute roughness $\varepsilon = 1.524 \times 10^{-5}$ m, and characteristic flow area of the main oxidiser valve $A_2 = 0.78A_1$, where $A_1 = \pi d^2/4$ is the area of the cross section of the flexible duct.

The density and the dynamic viscosity of liquid oxygen at its boiling point are respectively $\rho = 1141$ kg/m^3 and $\mu = 1.95 \times 10^{-4}$ Ns/m^2 [16].

The mean velocity of liquid oxygen in the flexible duct is computed as follows

$$v_1 = \frac{4q}{\pi d^2} = \frac{4 \times 0.7836}{3.1416 \times 0.2032^2} = 24.16 \text{ m/s}$$

The Reynolds number results from

$$Re = \frac{\rho v_1 d}{\mu} = \frac{1141 \times 24.16 \times 0.2032}{0.000195} = 2.873 \times 10^7$$

By inserting $\varepsilon = 1.524 \times 10^{-5}$ m, $d = 0.2032$ m, and $Re = 2.873 \times 10^7$ in the on-line calculator of [23], we find $f_D = 0.01143$, which is the Darcy friction factor.

After substituting $f_D = 0.01143$, $L = 0.4064$ m, $L_e = 1.219$ m, $d = 0.2032$ m, $v_1 = 24.16$ m/s, and $g_0 = 9.807$ m/s^2 in the following equation

$$\Delta H_{p1} = f_D \left(\frac{L + L_e}{d} \right) \left(\frac{v_1^2}{2g_0} \right)$$

we compute the drop of pressure head due to the discharge flexible duct

$$\Delta H_{p1} = 0.01143 \times \left(\frac{0.4064 + 1.219}{0.2032} \right) \times \left(\frac{24.16^2}{2 \times 9.807} \right) = 2.721 \text{ m}$$

This value corresponds to a drop of pressure in the flexible duct

$$\Delta p_1 = \rho g_0 \Delta H_{p1} = 1141 \times 9.807 \times 2.721 = 3.045 \times 10^4 \text{N/m}^2$$

As to the drop of pressure head due only to the butterfly valve, the continuity equation $\rho v_1 A_1 = \rho v_2 A_2$ implies an increase in velocity, from v_1 to a higher value v_2, for the fluid moving through the valve. In the present case, the velocity v_2 is

$$v_2 = \frac{A_1}{A_2} v_1 = \frac{24.16}{0.78} = 30.97 \text{ m/s}$$

The value of the resistance coefficient k for each valve is specified by its manufacturer. In [24], we find the value $k = 0.3$ for a butterfly valve fully open. As a check, we use the following formula (relating to the 2-K method) indicated in [21] to compute the resistance coefficient k for a butterfly valve in a duct of inner diameter d (m):

$$k = \frac{800}{Re} + 0.25 \times \left(1 + \frac{0.0254}{d} \right)$$

After substituting $Re = 2.873 \times 10^7$ and $d = 0.2032$ m in the preceding equation, we find $k = 0.2813$. This confirms the value $k = 0.3$ found in [24].

Therefore, the drop of pressure head, due only to the butterfly valve, is

$$\Delta H_{p2} = k \frac{v_2^2}{2g_0} = 0.3 \times \frac{30.97^2}{2 \times 9.807} = 14.67 \text{ m}$$

This value corresponds to a drop of pressure in the butterfly valve

$$\Delta p_2 = \rho g_0 \Delta H_{p2} = 1141 \times 9.807 \times 14.67 = 1.642 \times 10^5 \text{N/m}^2$$

5.8   Static and Dynamic Seals for Leakage Control in Valves

The dynamic seals considered here are mechanical devices used to prevent or reduce to an acceptable level the leakage of a fluid from one region of a valve in which flow occurs to another, when the surfaces to be protected from leakage are in relative motion. Static seals (such as O-ring seals and gaskets) for stationary surfaces are also considered here. The dynamic seals used for turbo-pumps of rocket engines have been described in Chap. 4, Sect. 4.13.

The leakage from a valve may be either internal leakage, which occurs in the direction of the flow, or external leakage, which occurs from a valve to the external environment, in a direction which differs from the normal direction of the flow. According to Howell and Weathers [25], the factors affecting the leakage requirements for valves are:

- loss of pressure or loss of propellant, to be prevented or kept below an acceptable value in order to avoid system failure due to premature depletion of fluid;
- damage to the system, such as corrosion or fire, which might occur as a result of leakage of propellants;
- damage to personnel, which may occur in case of leakage of fluids which are toxic for inhalation or exposure; and
- interference with experiments, which might occur in case of gases under pressure or propellants enveloping a spacecraft whose mission is to sample the atmosphere of a planet.

The following seals are considered here:

- seals for valves operating at high pressures and temperatures; and
- seals for valves operating at cryogenic temperatures.

Burmeister et al. [26] have made a survey on the matter. A brief account of this survey is given below. Valves for rocket engines operate at pressures ranging from zero to 6.895×10^8 N/m^2, and at temperatures ranging from cryogenic values to over 1366 K. They can be classified into three principal categories, which are

- plug valves;
- gate valves; and
- globe valves.

All of these valves can be used at high pressures and temperatures. Valves of other types, such as butterfly or vane valves, fall into one of these categories or into some combination of them. A tapered plug valve is shown in the following figure, due to the courtesy of NASA [26].

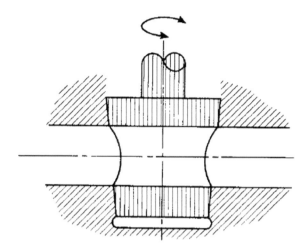

A plug valve has a tapered or cylindrical plug with a hole drilled in the middle. When this plug is rotated, the position of the hole opens, or restricts, or closes the passage of a fluid through the valve. A plug valve can be operated in an intermediate position, in order to throttle the flow. Simple quarter-turn valves, such as the one shown in the preceding figure, are limited in high-pressure application by the rapid increase in operating torque with increasing difference of pressure. When the fluid has access to either the large end or the small end of a tapered plug, then a force unbalance arises into or out of the valve body.

The operating torque depends on the plug taper. The purpose of a tapered plug valve is to shut off flow in a leak-tight manner. When the thickness of the lubricant film is not uniform, then leakage occurs. The lubricant may be washed away by the fluid when unseating the plug, or may also be extruded by the difference of pressure. The three functions fulfilled by the lubricant (lubrication, plug unseating, and viscous sealing) are combined in plug valves of the Nordstrom type, such as the valve shown in the following figure, due to the courtesy of NASA [26].

A valve of this type can bear a higher pressure than the maximum pressure bearable by an ordinary tapered plug valve. An increase in temperature degrades the properties of the lubricant-sealant. Maximum temperatures bearable by this type of valve are about 700 K or 800 K. The high viscosity of the lubricant-sealant at very low temperatures precludes application of grease at cryogenic temperatures.

A gate valve opens or closes the passage of fluid in a pipe by raising or lowering a flat plate across the pipe. This plate slides over sealing surfaces in a direction perpendicular to the fluid stream in the pipe, as shown in the following figure, due to the courtesy of NASA [26].

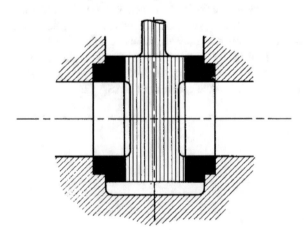

The sealing surfaces of a gate valve must bear the stresses induced by moving the plate from the fully open to the fully closed position. Therefore, the galling properties of the materials used and their ability to withstand stresses are very important. As a general rule, the galling tendency increases and the strength decreases with increasing temperature. To provide good sealing surfaces, the gate and the seat must be flat. Therefore, a gate valve must be thick, especially in large valves subject to high pressures.

Gate valves and tapered plug valves can be protected from leakage by using viscous sealing. For this purpose, a sealant is injected into grooves located around the flow passage in the gate or in the seat, as shown in the following figure, due to the courtesy of NASA [26].

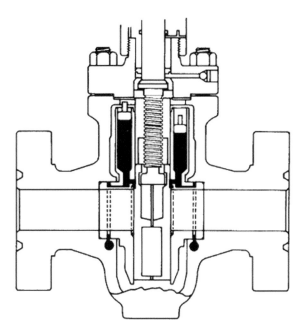

In the valve shown in the preceding figure, the injection pressure of the sealant is proportional to the pressure of the fluid which moves in the pipe. Gate valves using viscous sealants have lower operating torques than conventional parallel slide gate valves. The following figure, due to the courtesy of NASA [26], shows a tapered gate valve, which makes it unnecessary to slide the gate across the entire sealing surface.

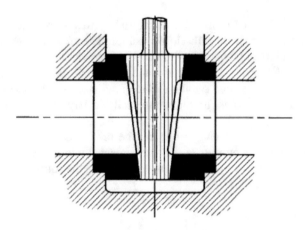

The seat and the gate are not in mutual contact when the gate has been lifted by part of its total travel. Side ribs on a tapered gate valve resist the action exerted by the flow impingement and by the difference of pressure, which tends to move the gate toward the downstream seat, and keep the sealing surfaces from sliding contact.

When a tapered gate valve is closed at a high temperature and then cools down, then the gate tends to bind, because the gate and the body parts have different coefficients of thermal expansion. This can be avoided by using flexible tapered gates for valves operating at high temperatures.

Gate valves not using viscous sealants are subject to leakage, and therefore are not used in case of high pressures. Gate valves using viscous sealants can bear gas and petroleum pressures as high as 1.034×10^8 N/m^2. Gate valves sealed by inserts made of elastomers or fluorocarbons are used when contamination of the fluid is to be avoided. However, the pressure which these valves can bear is limited by plastic flow of these inserts.

Globe valves are those used in water taps to start or stop or regulate the flow in a pipe. They have a moving member which is pushed into the flow passage, as a cork is pushed into the neck of a bottle, and a stationary member, which is a ring seat. The moving member is shaped as a cone frustum, which fits into the mating body seat, as shown in the following figure, due to the courtesy of NASA [26]. The included angle of this cone frustum is much greater than that of a plug used in a tapered plug valve.

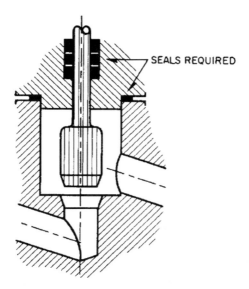

SEALS REQUIRED

A seat shaped as a cone frustum is very resistant to distortions of the valve body due to the pressure exerted by the fluid. In addition, should such distortions occur, the elastic deformation in the seat tends to maintain satisfactory sealing contact. Globe valves lend themselves to be used in cases of high pressures and temperatures. Of course, the materials used in such cases must be carefully selected. The sealing surfaces of these valves are made of cobalt-chromium alloys, such as the Stellites®, or of other materials such as carbides and ceramics. However, these materials have low resistance to impact. Globe valves are widely used for high-pressure and high-temperature applications. Commercial valves of this type are available for pressures up to 1.034×10^8 N/m^2 at atmospheric temperatures, and also for pressures of 0.3447×10^8 N/m^2 at a temperature of 922 K [26].

Since the inner parts of a valve must be contained into the body of the valve, which also contains the fluid under pressure, then it is necessary to provide some means of closure or cover. In addition, an operating mechanism for flow control must be contained into the valve body. Therefore, at least two seals are necessary to a valve. These seals are illustrated in the preceding figure, with reference to a globe valve. However, they are also necessary with valves of other types.

Cover sealing can be obtained in several ways, each of which has its maximum level of pressure and its cost.

A flat gasket for cover sealing, shown on the left-hand side of the following figure, is not suited to either high-pressure or high-temperature applications, because it requires large surfaces. Other types of gaskets which are better suited to these applications are the spiral-wound gaskets and the lens ring gasket. A spiral-wound gasket is shown (right), in comparison with a flat gasket (left), in the following figure, due to the courtesy of NASA [26].

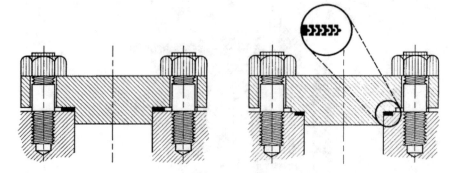

A spiral-wound gasket is made of a V-shaped metallic strip or ribbon mixed with a non-metallic filler material. The metal (usually stainless steel) is wound outwards in a circular spiral, and the filler material (usually graphite or poly-tetra-fluoro-ethylene, whose commercial name is Teflon®) is wound in the same manner, but starting on the opposite side. This results in alternating layers of filler and metal. The filler is the sealing element, and the metal provides structural strength to the gasket. For temperatures above 811 K, ceramic fibre fillers have been used [25]. An inner ring and an outer ring made of steel are used on the gasket for centring and controlling compression.

A lens ring gasket is shown in the following figure, due to the courtesy of NASA [26].

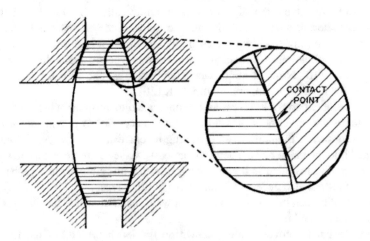

A lens ring gasket is made of a sealing metal of lenticular shape which fits into the recesses or grooves of a flange. Since the metallic ring is designed to be softer than the flange grooves, then the gasket deforms plastically under compressive loads instead of the flange. This deformation spreads the gasket faces, and therefore protects the gasket from overstresses.

In a spiral-wound gasket and in a lens ring gasket, the sealing stress is proportional to the internal pressure. Lens ring gaskets are suitable for pressures of 6.895×10^8 N/m^2 and above [26].

The O-ring is another type of static or dynamic seal, in which the sealing stress is proportional to the internal pressure. An O-ring is placed in a groove which is designed to provide a radial or lateral containment, depending upon the specific application. As pressure increases, an O-ring deforms as shown on the left-hand side of the following figure, due to the courtesy of NASA [26]. The same figure also shows, on the right-hand side, a metallic O-ring having holes to admit the fluid under pressure into the ring, in order to make the sealing stress proportional to the pressure exerted by the fluid contained in the valve.

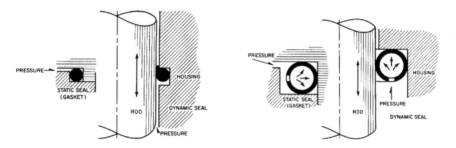

Metallic O-rings may also be coated with elastomers or fluorocarbons to provide better sealing. They are suitable for high pressures, but are also easily damaged, and therefore must be installed in specially designed split grooves [26].

Two further types of seals are the pressure sealing ring and the lapped seal, which are shown on respectively the left-hand side and the right-hand side of the following figure, due to the courtesy of NASA [26].

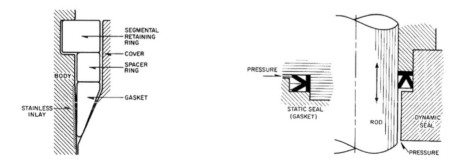

A pressure sealing ring (left) consists of a segmented retaining ring, a spacer ring, and a gasket. A pressure sealing ring uses the displacement of one of its members to develop the sealing stresses. The gasket is initially wedged against the surface of the body by means of bolts. When the internal pressure exerted by the fluid in the body increases, then the gasket is wedged more and more tightly against its constraints. A lapped seal (right) has a very smooth mating surface whose sealing capability is reinforced by lapping. This type of seal is suited to either radial or axial installation, as shown in the preceding figure.

Body-cover joints can also be sealed by using a fillet-weld seal or a canopy seal, which are shown on respectively the left-hand side and the right-hand side of the following figure, due to the courtesy of NASA [26].

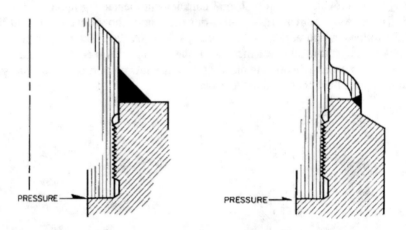

A fillet-weld seal (left) is performed by depositing a fillet weld bead at the interface of the mating parts. A canopy seal (right) is similar to a fillet-weld seal and is used where a slight movement of the mating parts is expected.

Other types of seals for valves used in high-pressure and high-temperature applications are described in [25, 26].

Valves operating at cryogenic temperatures and seals used for them also require a careful design. As has been shown in Chap. 1, Sect. 1.4, by cryogenic fluids we mean gases which can be liquefied at or below 122 K [27].

Permanent gases, such as methane, oxygen, nitrogen, hydrogen, and helium, change from gas to liquid at atmospheric pressure at the temperatures shown in the following table (due to the courtesy of NIST [28]), called the normal boiling point (NBP). Such liquids are known as cryogenic liquids or cryogens. Liquid helium, when cooled further to 2.17 K or below, becomes a superfluid with very unusual properties associated with being in the quantum mechanical ground state.

Cryogen	Temperature (K)
Methane	111.7
Oxygen	90.2
Nitrogen	77.4
Hydrogen	20.3
Helium	4.2

According to Howell and Weathers [25], in the design of valves operating at cryogenic temperatures, the following requirements are to be taken into account:

- reduction of heat transfer from the external environment to the valve, in order to prevent excessive losses of cryogenic fluid by evaporation; and
- reduction of heat transfer between temperature-sensitive parts of valves in cryogenic fluids.

A method often used to thermally insulate temperature-sensitive parts of valves (for example, actuators) from the valve body which contains the cryogenic fluid consists in placing a thermal barrier made of a non-metallic material having a low thermal conductivity (for example, Teflon®) between the valve body and the part to be insulated. Ductile metals of low conductivity which are often used for cryogenic valves are austenitic stainless steels (usually the lower-carbon 304, also known as 18/8, and the ever lower-carbon 304L), aluminium alloys, copper, ASTM B-61 and B-62 bronzes, and nickel [26]. In addition to using materials of low thermal conductivity, other methods are also used for the same purpose. These methods are:

- breaking completely the path of heat conduction, by separating the parts which operate at different temperatures by an insulating space (for example, in a broken stem valve, the valve actuator is thermally insulated from the valve body by means of radial bars on the upper stem which drive against axial pins on the lower stem) and enclosing the entire valve in an evacuated chamber;
- using vacuum-jacketed valves;
- increasing the path of heat conduction, by using extended stems and bonnets including long gaseous columns, in order to keep the stem seal packings exposed only to insulating vapour, and not to the cold liquid;
- providing bonnets with integral black-coated fins, to direct heat away through radiation from the packing to the external environment;
- using plastiform insulating foams, made of polystyrene, polyurethane, rubber, silica, and glass, around the valve body; and
- using low-density materials, such as powders and fibres, with gas at atmospheric pressure in the interstitial spaces.

The following figure, due to the courtesy of NASA [26], shows a vacuum-jacketed valve used for cryogenic fluids, where the inner wall surrounds the primary pressure vessel, and the outer wall is of light construction to reduce to cool-down mass.

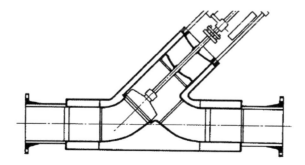

Brightly finished interior surfaces may be used to reduce radiative heat transfer.

In some cases, especially when the cryogenic fluid is liquid helium, a valve is surrounded with a vacuum-jacket, then another cryogenic fluid (usually liquid nitrogen) is circulated in the interstitial space between the two walls to act as a thermal radiation shield, and finally an insulating foam is placed as an outside covering [26]. It is possible to have insulating foam bonded to the surface of the insulation cavity. This type of insulation, usually called "foamed in place" is quite adaptable to valves and other components having irregular surfaces [25]. Trim parts, such as plugs and stems, of valves for cryogenic fluids are often made of stainless steel or Monel®, which is a nickel-copper alloy. Teflon® and Kel-F®/Neoflon® are commonly used for seats. Valves for fluorine have often seats made of copper. Bushings made of Ampco® (aluminium bronze) have been successfully used with stems made of stainless steel. Bushings are also made of Teflon® or glass-impregnated Teflon®. Packing is made of pure or filled Teflon® and Kel-F®/Neoflon®. Welded bellows are sometimes used instead of packings. Gaskets are made of Teflon® or metal-clad Teflon®. Rings made of stainless steel are used in ring-type joint flanges. Fluorine requires soft copper, aluminium 25, or stainless steel for gaskets [26].

The following figure, due to the courtesy of NASA [26], shows a cryogenic fluid valve whose vacuum-insulating jacket encloses the entire body. An expansion bellows for attachment to a piping jacket is provided to prevent the transmission of strains. Air is sealed in the stem cylinder to achieve long, poorly conductive path, and a vapour space is provided between the cylinder and the body extension. Heat transfer is also reduced by the absence of bolts or studs in the valve body [26].

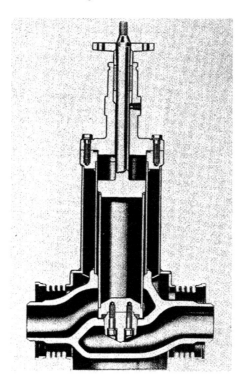

Most of the principal seals considered above for high-pressure and high-temperature applications may also be used for cryogenic fluids. However, special care is necessary because of the different properties of the materials.

The use of soft plastic gaskets is limited to reinforced or laminated Teflon®. Thin gaskets made of hard plastics, such as polyethylene terephthalate (PET), whose trade name is Mylar®, may be used in case of high stress loading. Soft metals, such as copper and aluminium, may be used in case of high shear-stress loading. The following figure, re-drawn from [25], shows two gasket gland designs used for cryogenic service. They are: (left) a rounded-edge seal ring, and (right) a knife-edge shear seal.

Rounded-edge rings for plastics   **Knife-edge shear seal**

The following figure, re-drawn from [25], shows some installation techniques for using elastomeric O-rings in cryogenic applications.

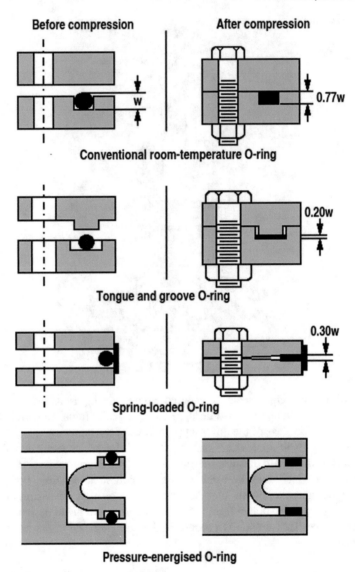

Before compression   **After compression**

Conventional room-temperature O-ring

Tongue and groove O-ring

Spring-loaded O-ring

Pressure-energised O-ring

Metallic O-rings, either solid or hollow, may be used as seals for cryogenic fluids. In particular, solid rings made of soft metals such as copper, indium, or lead have given the best results. When using indium, care must be taken to confine the gland because of the cold-flow tendency of this metal [25].

5.9 Design of Propellant Valves

The valves described in the present section are principally used to initiate and termi-
nate the flow of propellants from the tanks to the main combustion chamber and to
the gas generator of a rocket engine. Therefore, these valves are often of the open-
closed, two-position, normally closed type. However, some valves are also required to
control the flow rate of propellants by means of a restriction of variable area. Valves
of the latter type are used for thrust-throttle and propellant-mixture-ratio control,
because they can continuously vary their operating position, in contrast to shut-off
valves, which are either fully open or fully closed.

Requirements for propellant valves are compatibility with propellants, structural
strength, absence of leakage in closed position, proper actuation time when opening
or closing, and minimum pressure loss.

Apart from the general classification considered in Sect. 5.8, we describe here in
particular the most used types of propellant valves, which are butterfly valves, ball
valves, poppet valves, Venturi valves, gate valves, and needle valves.

An isometric cross section of a butterfly valve is shown in the following figure,
due to the courtesy of NASA [29].

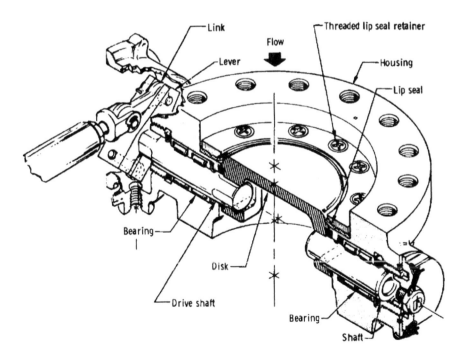

A butterfly valve is used to start, regulate, or stop the motion of a fluid in a duct.
With reference to the preceding figure, the movable part of a butterfly valve is a
flat element, called the disc, which may be rotated to control the flow through the
valve body. The disc can rotate on a single-piece shaft or on a two-piece shaft, which

extends across the diameter of an orifice and supports the disc on both sides. The internal diameter of existing butterfly valves ranges from 50 to 400 mm. These valves are used for propellants at absolute pressures ranging from 1.4×10^5 to 1×10^7 N/m^2 [3]

For greater rigidity, the shaft may be integral with the disc. The centre of rotation of the disc is usually offset, as shown in the preceding figure, in order to allow the disc to rotate off the primary seal. For low leakage, a plastic lip seal (described in Chap. 4, Sect. 4.13) is usually employed in the valve housing.

The discs of butterfly valves have spherical or conical shapes, as shown in the following figure, re-drawn from [29]. A disc of spherical shape (left) has better performance, but is more expensive to fabricate. A disc of conical shape (right) is more subject to leakage because of greater wear [29].

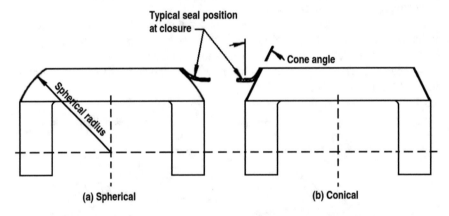

(a) Spherical   (b) Conical

A butterfly valve is operated by an actuator of the piston type, through a connecting link and shaft crank arm. The actuating power is the pressure exerted by either a non-cryogenic liquid propellant or an inert gas, and is controlled by a pilot valve. The shaft and the pins are made of stainless steel, whereas most of the other parts are made of aluminium alloys. Butterfly valves oppose low resistance to the motion of propellants flowing through them.

The characteristic area A_c (m^2) of a butterfly valve results from

$$A_c = \pi \frac{d_s^2}{4} - A_g$$

where d_s (m) is the inside diameter of the valve seat lip seal, and A_g (m^2) is the projected valve disc area at the fully open position. The values of A_c range from $0.65 \times (\pi d_s^2/4)$ for a 50 mm diameter valve to about $0.87 \times (\pi d_s^2/4)$ for a 300 mm diameter valve [3].

A butterfly valve maintains a smooth flow over a wide range of angular positions of the valve disc. The values of the resistance coefficient k as a function of the opening angle are given by the manufacturer of the valve or can be found, for example, in [30, 31].

RP-1 is used sometimes as the actuating fluid for liquid-oxygen valves in rocket engines burning RP-1 with liquid oxygen. In this case, it is necessary to prevent RP-1 from freezing at the actuator by using a heater. A potentiometer is often attached to the drive shaft of a butterfly valve to indicate continuously the position of the valve disc.

The torque required to rotate the shaft and the disc of a butterfly valve depends on the hydraulic torque T_h (Nm) and on the frictional torque T_f (Nm) acting on the valve. The frictional torque opposes always rotation. The hydraulic torque (in spite of the offset of the disc, which is always placed on the side opposite to the direction of the flow, as shown in the preceding figures) acts in the closing direction for most angular positions (from $\pi/20$ to $4\pi/9$ rad) of the disc.

The opening torque T_o (Nm) and the closing torque T_c (Nm) required to operate a butterfly valve can be expressed as follows

$$T_o = T_f + T_h$$
$$T_c = T_f - T_h$$

where the hydraulic torque T_h is supposed to act on the disc of the valve in the closing direction.

The frictional torque T_f depends on the difference of pressure on the two faces of the disc and also on the projected area of the disc, which in turn depends on the angular position of the disc in the valve. The value of the frictional torque can be estimated as follows

$$T_f = k_f \, r_s \, f_m \, d_s^2 \, \Delta p$$

where k_f is the friction factor coefficient whose value is determined experimentally, r_s (m) is the radius of the shaft at the bearing section, f_m is the coefficient of friction between the shaft and the bearing, d_s (m) is the inside diameter of the valve seat lip seal, and Δp (N/m^2) is the difference of pressure across the disc.

The value of the hydraulic torque can be estimated as follows

$$T_h = k_h \, d_s^3 \, \Delta p$$

where k_h is the hydraulic coefficient whose value is determined experimentally and depends on the angular position of the disc in the valve.

In the practical design of a butterfly valve, the actuator is required to provide a torque whose value ranges from two to three times the maximum estimated value necessary to open or close the valve. In addition, at the start of the opening stroke, the actuator must overcome the static friction forces due to all seals. The opening and closing times of butterfly valves range from 0.02 to 0.2 s [3].

As an application of the concepts discussed above, it is required to determine the torques necessary to open and close a butterfly valve having the following data: radius of the valve shaft at the bearing section $r_s = 0.02032$ m, inside diameter of the valve seat lip seal $d_s = 0.1956$ m, coefficient of friction between the shaft and the bearing $f_m = 0.05$. The data found experimentally are given below.

Disc angle (rad)	Δp (N/m²)	k_f	k_h
$\pi/36$	7.295×10^6	0.78	0.00111
$\pi/12$	5.302×10^6	0.78	0.00255
$2\pi/9$	6.033×10^5	1.57	0.0125
$17\pi/36$	1.724×10^5	3.61	−0.01164

By substituting these data in the equation $T_f = k_f\, r_s f_m\, d_s^2\, \Delta p$ we find the following values of the frictional torque T_f at the given angles:

$(\pi/36)$ $T_f = 0.78 \times 0.02032 \times 0.05 \times 0.1956^2 \times 7.295 \times 10^6 = 221.2$ Nm
$(\pi/12)$ $T_f = 0.78 \times 0.02032 \times 0.05 \times 0.1956^2 \times 5.302 \times 10^6 = 160.8$ Nm
$(2\pi/9)$ $T_f = 1.57 \times 0.02032 \times 0.05 \times 0.1956^2 \times 6.033 \times 10^5 = 36.82$ Nm
$(17\pi/36)$ $T_f = 3.61 \times 0.02032 \times 0.05 \times 0.1956^2 \times 1.724 \times 10^5 = 24.19$ Nm

Likewise, by substituting these data in the equation $T_h = k_h\, d_s^3\, \Delta p$ we find the following values of the hydraulic torque T_h at the given angles:

$(\pi/36)$ $T_h = 0.00111 \times 0.1956^3 \times 7.295 \times 10^6 = 60.60$ Nm
$(\pi/12)$ $T_h = 0.00255 \times 0.1956^3 \times 5.302 \times 10^6 = 101.2$ Nm
$(2\pi/9)$ $T_h = 0.0125 \times 0.1956^3 \times 6.033 \times 10^5 = 56.44$ Nm
$(17\pi/36)$ $T_h = -0.01164 \times 0.1956^3 \times 1.724 \times 10^5 = -15.02$ Nm

The torques T_o required to open the butterfly valve at the given angles result from the equation $T_o = T_f + T_h$ as follows

$(\pi/36)$ $T_o = 221.2 + 60.60 = 281.8$ Nm
$(\pi/12)$ $T_o = 160.8 + 101.2 = 262.0$ Nm
$(2\pi/9)$ $T_o = 36.82 + 56.44 = 93.26$ Nm
$(17\pi/36)$ $T_o = 24.19 + (-15.02) = 9.17$ Nm

Likewise, the torques T_c required to close the butterfly valve at the given angles result from the equation $T_c = T_f - T_h$ as follows

$(\pi/36)$ $T_c = 221.2-60.60 = 160.6$ Nm
$(\pi/12)$ $T_c = 160.8-101.2 = 59.6$ Nm
$(2\pi/9)$ $T_c = 36.82-56.44 = -19.62$ Nm
$(17\pi/36)$ $T_c = 24.19-(-15.02) = 39.21$ Nm

An isometric cross section of a ball valve is shown in the following figure, due to the courtesy of NASA [29].

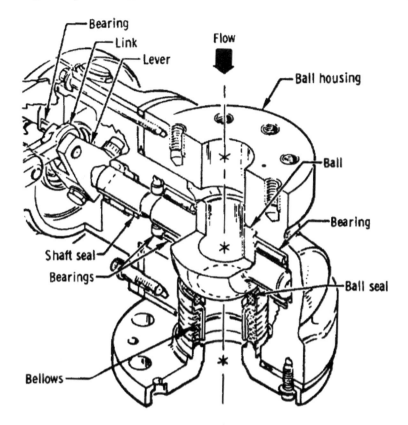

A ball valve is substantially a sphere provided with a port and fitting into a cup-shaped housing, such that a rotation of this sphere through a right angle changes the position of the valve from open to closed. In other words, when the valve is turned to the open position, the ball rotates to a point in which the hole through the ball is aligned with the flow openings (inlet and outlet) of the valve body. When the valve is turned to the closed position, the ball rotates to a point in which the hole through the ball is perpendicular to the flow openings of the valve body, and therefore the flow is stopped. There are two common types of this valve. They are the fixed ball valve and the floating ball valve, which are shown in the following figure, re-drawn from [25].

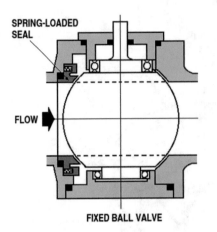
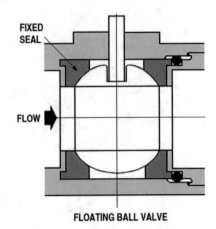

In a fixed ball valve (left), the ball is supported by fixed bearings, and the seal is spring-loaded against the ball. The seal is mounted on the upstream side of the ball, and is usually designed to act in one direction only. However, another spring-loaded seal may also be mounted on the downstream side of the ball, in order to obtain a bi-directional sealing.

In a floating ball valve (right), the ball is supported by fixed seals, and the seating force is provided by the fluid under pressure, which pushes the ball against the seat. In a floating ball valve, the seals are placed both upstream and downstream of the ball, and therefore the valve may be used in either direction. The use of floating ball valves is confined to low-pressure applications.

The materials used for the body of a ball valve may be metals or plastics, such as Unplasticised Polyvinyl Chloride (UPVC). The ball may be made of metal or plastic. Metallic balls have highly polished, hard chrome surfaces. Seal materials may be plastics or elastomers. Teflon® is the most commonly used material for seals, due to its properties of high resistance to corrosion and low friction.

The principal advantage of ball valves is a low pressure drop. They also have a very good leakage control, and can be designed to operate equally well with the flow in either direction. On the other hand, the actuating forces required by them are high, because these valves cannot be pressure-balanced [25].

An isometric cross section of a poppet valve, used for the AJ10-138 rocket engine, is shown in the following figure, due to the courtesy of NASA [32].

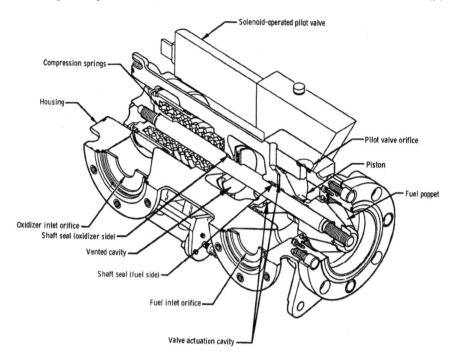

Solenoid-operated pilot valve

Compression springs

Housing

Pilot valve orifice

Piston

Fuel poppet

Oxidizer inlet orifice

Shaft seal (oxidizer side)

Vented cavity

Shaft seal (fuel side)

Fuel inlet orifice

Valve actuation cavity

Poppet valves are those used in cylinders of car engines. They are similar to globe valves (described in Sect. 5.8), because for both of them the movable part of the valve travels perpendicularly to a plane through the seating surface. According to Howell and Weathers [25], the term poppet valve is used synonymously with globe valve.

The designer of a poppet valve chooses the types of sealing surfaces, either hard or soft, which are best suited for the poppet and for the seat. These surfaces are said to be hard or soft depending on the type of material used. A hard sealing surface is made of a material (metal, ceramic, or cermet, the last being a composite material made of ceramic and metallic materials) which does not permanently yield or deform except with wear. By the way, flexible metallic discs are a special type of hard sealing surface. A soft sealing surface is made of plastic or elastomeric materials. Possible configurations for combinations of hard and soft sealing surfaces are indicated below.

Poppet sealing surface	Seat sealing surface	Configuration designation
Hard	Hard	Hard-on-hard
Hard	Soft	Hard-on-soft
Soft	Hard	Hard-on-soft

The following figure, due to the courtesy of NASA [29] shows four examples of hard-on-hard sealing configurations.

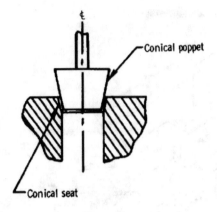

(a) Conical on conical.

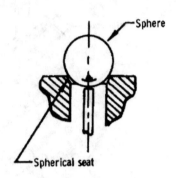

(b) Spherical on spherical.

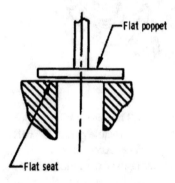

(c) Flat on flat.

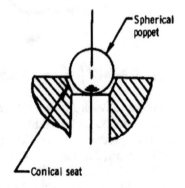

(d) Spherical on conical.

A poppet valve in which the hard sealing surface meets a soft sealing surface may incorporate the soft sealing surface as an insert of elastomer or plastic in the housing. This type of design is known as a soft-seat poppet.

The following figure, due to the courtesy of NASA [29] shows typical configurations for a soft-seat poppet.

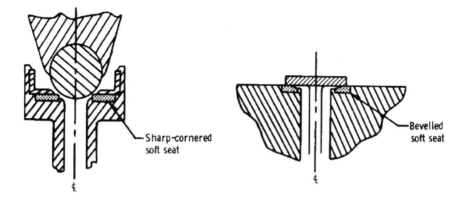

(a) Sphere seating on sharp corner of soft seat   (b) Flat poppet seating on beveled soft seat

It is also possible to incorporate the soft sealing surface either as an integral part of the poppet (for example, of a plastic poppet) or as an insert of elastomer or plastic in the poppet. This type of design is known as a soft poppet. In either location, the soft sealing insert is designated as a seal.

The following figure, due to the courtesy of NASA [29] shows an elastomeric seal insert retained in a poppet by a shrink-fit mechanical retainer.

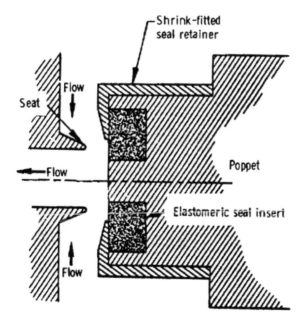

A large flow area in a poppet valve is provided with short travel of the poppet. This makes it possible to use actuators such as solenoids or diaphragms, which are short-stroke devices. A very good leakage control can be achieved by using hard or

soft seals, as has been shown above. When hard seals are used, great care must be taken to eliminate contamination from the fluid and from the duct upstream of the valves.

A poppet valve can be pressure-balanced by using two poppets which are on the same stem and seat on separate seats. By so doing, the pressure acting on the top of one poppet provides the counterbalancing force acting on the bottom of the other poppet. The same result can also be obtained by providing the poppet stem with a pressure-balancing piston area, or by making the poppet stem diameter equal to the seating diameter. However, it is practically impossible to obtain complete pressure balance under both shut-off and flow conditions.

A poppet valve is lighter than valves of other types for many applications, due to the small stroke of its actuator. For example, ball and butterfly valves require a rotation through a right angle for full stroke, and therefore need a larger associated mechanism. A poppet valve is used when the valve is desired to open rapidly from zero to full flow with a short travel of its movable element. Of all types of poppet valve, the in-line valve is the one which has the best flow properties. The body of a poppet valve may be either cast on one piece or split. In the latter case, the body halves are joined by bolts [25].

A cross section of a cavitating Venturi valve is shown in the following figure, re-drawn from [25].

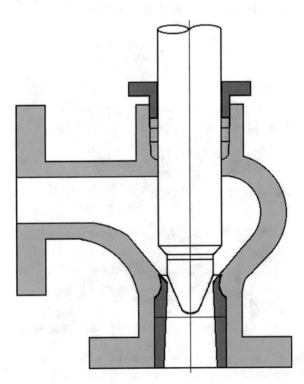

A cavitating Venturi valve is used to control the flow rate of a propellant in a rocket engine as a function of the pressure upstream of the valve and of the throat area. When the minimum pressure of a liquid propellant is made to decrease below its vapour pressure at the throat of a Venturi tube, then cavitation occurs and the propellant evaporates. In case of cavitation, the propellant flow rate is independent of the downstream pressure. Cavitation occurs by decreasing the pressure of a liquid propellant at the throat of a Venturi tube to such an extent that nearly all of the upstream pressure head of the fluid is converted into velocity head. When this occurs, the only static pressure remaining in the fluid at the throat is equal to the vapour pressure of the fluid itself. In such conditions, when the upstream pressure of the fluid is kept constant, a further decrease in downstream pressure does not result in increased flow rate, because cavitation at the throat maintains the pressure equal to the vapour pressure of the fluid.

In other words, the flow rate through the valve remains constant for a given throat area and for a given upstream pressure, independently of the downstream pressure, provided that the downstream pressure does not increase above the level in which cavitation occurs. In practice, the downstream pressure must be less than 85% of the upstream pressure in order for cavitation to occur [25].

Cavitating Venturi valves are used in rocket engines when it is desired to prevent variations of propellant flow rate caused by variable back pressure on the control valve. In the design of a cavitating Venturi valve used to control flow rates, it is necessary to know the total pressure p_T (N/m^2) and the vapour pressure p_v (N/m^2) of the propellant whose flow rate is to be controlled. The ratio

$$\frac{p_T - p_v}{\rho g_0}$$

is the total pressure head available for conversion to velocity head in the Venturi tube. In the preceding equation, ρ (kg/m^3) is the density of the propellant at the given temperature, and $g_0 = 9.80665$ m/s^2 is the acceleration of gravity on the surface of the Earth. The total pressure head $(p_T - p_v)/(\rho g_0)$ is used to determine the theoretical value v_t (m/s) of the propellant velocity at the throat of the Venturi tube, as follows

$$\frac{p_T - p_v}{\rho g_0} = \frac{v_t^2}{2 g_0}$$

The actual volume flow rate q (m^3/s) of propellant through the Venturi tube can be determined by assuming a discharge coefficient of 0.93 [25]. Therefore, the actual volume flow rate is

$$q = 0.93 A_t v_t = 0.93 A_t \left[\frac{2(p_T - p_v)}{\rho}\right]^{\frac{1}{2}}$$

where A_t (m^2) is the area of the throat of the Venturi tube.

It has been found experimentally [25] that the best pressure recovery through a cavitating Venturi valve is obtained when the included angle of the diffuser cone is in the range going from $\pi/36$ rad (5°) to $\pi/30$ rad (6°). Venturi valves have been used successfully in rocket engines for cryogenic and storable propellants [3].

A cross section of a gate valve is shown in the following figure, re-drawn from [25].

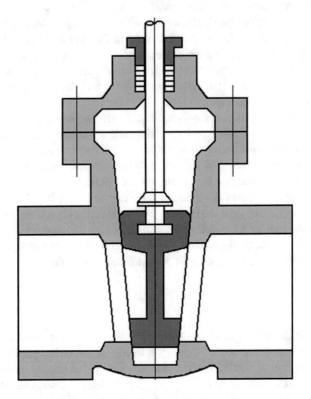

Gate valves have been considered in Sect. 5.8 from the point of view of sealing and leakage control. Their use as propellant valves is discussed below. Gate valves have a flat plate which moves perpendicularly to the fluid stream in a duct in order to open or close the passage of the fluid through them. They may be operated by any of the usual actuators used for this purpose. Quick-return mechanisms have been used when it was desired to open or close such valves at different flow rates. When the plate is completely removed from the fluid stream, a gate valve opposes little or no resistance to the flow. Consequently, the principal advantage of a gate valve is unrestricted fluid stream, and therefore low pressure drop when the valve is in the wide open position. When a gate valve is in the closed position, the contact surface between the plate and the seal extends along the whole circumference of the tube cross section, and therefore this valve provides good sealing, which results in little or no leakage across the plate.

A gate valve has a short distance between the sections of inlet and outlet in the direction of the flow. On the other hand, it has poor throttling properties and is also subject to erosion when they it is in the near closed position. Its response is slow, because of the large travel and the high actuation forces due to friction. Consequently, gate valves are normally used as on-off mechanisms. They have been designed for propellant line pressures up to 2×10^7 N/m^2 [3] and for low propellant flow applications (for example, for gas generator control and ground support services). In ground support systems, they are used as shut-off or block valves in systems subject to moderate pressures. This is because they cannot be balanced. Gate valves have also been used for cryogenic propellants. An example is shown in the following figure, due to the courtesy of NASA [26]. The vent hole on the inlet end of this gate valve is necessary to avoid trapping liquid in the bonnet [26].

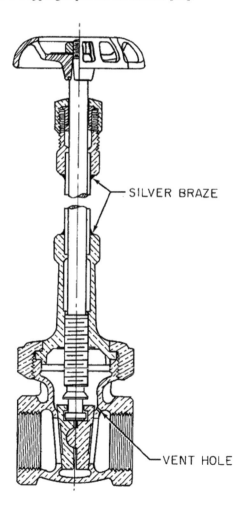

A cross section of a needle valve is shown in the following figure, re-drawn from [25].

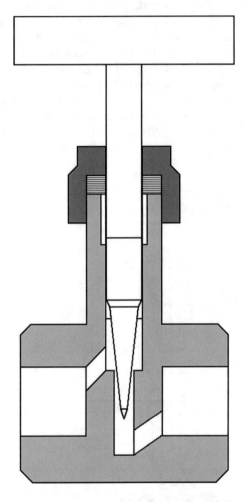

A needle valve is a variation of a globe valve. In comparison with the latter, a needle valve has a control orifice of smaller diameter, and a longer and slimmer movable element, in order to permit throttling. A needle valve is used for regulating the flow of small quantities of fluid, for blending carefully a fluid with another, and for speed control of pumps and actuators. The construction property of a needle valve which makes it apt to be used for a fine flow control is the plug of the valve, which moves inside the seat ring and has long tapered slots milled and ground into its surface. The plug remains inside the seat ring at all times, even when the valve is in the wide-open position, in order for the flow through the valve to be closely controllable by the relative position of the long plug and its gradually tapering grooves. A tight shut-off of a needle valve depends on the concentric position of the plug with the seat at the point of closure. In case of need for a repeated shut-off of a needle valve, a good sealing is obtained by using a disc made of rubber or plastic which closes against a metallic seat. When a non-metallic seal cannot be used due to high temperatures or to the necessity of handling a corrosive fluid, then a non-rotating plug may be designed for closing against the seat, or a device may be used to limit the closing force applied to the stem to the amount strictly necessary for a tight shut-off. Needle valves are used in lines whose diameter is less than 20 mm, a typical value being 6.35 mm. They are used for temperatures ranging from cryogenic values to 811 K and above, and for pressures ranging from vacuum to 2×10^8 N/m^2 and above [25].

5.10   Design of Pilot Valves

Pilot valves are used to control fluids, which in turn control other components of hydraulic circuits (for example, valves for propellants) or a sequence of events occurring in a rocket engine (for example, the admission of fuel for ignition purposes). The control action performed by a pilot valve is of the on-off, open-loop type. The most common types of pilot valves are solenoid valves, pressure-actuated valves, and position-actuated valves in two-way, three-way, and four-way configurations.

   A solenoid valve is an electromagnetic device having a solenoid, which is a helically wound coil of wire, a movable cylindrical core of ferromagnetic material mounted coaxially within the coil, a poppet, and a return spring. This core is called the plunger. When the coil conducts no electric current, the compressed spring closes a small orifice by means of the poppet. When a voltage is applied to the ends of the coil, the electric current flowing through the coil generates a magnetic field, which exerts a force on the plunger. This force pulls the plunger toward the centre of the coil, thereby opening the orifice. The following figure, due to the courtesy of NASA [32], shows a cross section of a solenoid valve having a spherical poppet.

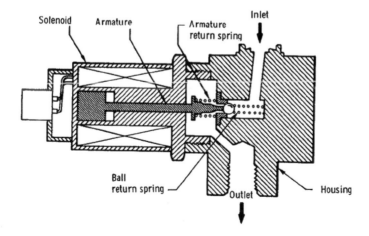

The following figure, re-drawn from [25], shows various shapes of poppets which can be actuated by a solenoid in a pilot valve.

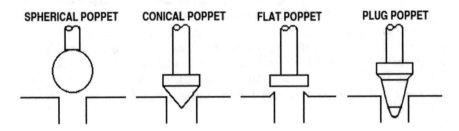

Solenoid actuators can be used not only for simple, two-port, on-off valves, but also for multi-passage valves, in which the flow is directed to three or four ports, as required, as the sequel will show. Solenoid on-off valves are either normally open or normally closed, where the normal position of a valve is the position of its movable element with respect to the upstream pressure when no electric power is applied to the solenoid. In the presence of an electric power, the solenoid places the valve to the desired position by attracting a magnetic core attached to the valve stem. When the electric power is removed, a spring pushes back the magnetic core to its normal position.

Directly acting solenoid valves have response times ranging from 5 to 50 ms [25]. A large opening can be obtained through a short stroke of the movable element. The electric current required to operate these valves depends on their size, on the difference of pressure, and on the response required. Their leakage control is very good, particularly when a soft seat is used. When a hard seat is used, it is advisable to lap the seat parts. Solenoid valves do not require external dynamic seals, and have a very long operating life, of the order of magnitude of hundreds of thousands of cycles without degradation of performance [25].

According to Huzel and Huang [3], the following equations apply to the case of a flat-faced, plunger-type magnetic core

$$F = \frac{B^2 A}{C}$$

$$B = \frac{f P N i}{G}$$

where F (N) is the attracting force which acts on the plunger in its normal position, B (Wb/m^2) is the magnetic flux density in the air gap, A (m^2) is the cross-sectional area of the plunger, C is a factor comprising constants and allowances for stray flux (a value of 2.505×10^{-6} is applicable to round, flat-faced, plunger-type magnets), f is a leakage factor of the magnetic flux, whose value is less than unity and depends on the magnetic circuit, P is a factor comprising constants and the permeability of the fluid in the gap G (m) between the core and the armature (a value $P = 1.255 \times$

10^{-6} applies to an air gap), N is the number of coil turns, and i (A) is the electric current applied to the coil.

The radiating surface of a solenoid valve should be sufficiently large to prevent an overheating of the coil. For this purpose, the resistance of the conductor should be designed according to the maximum allowable temperature. It is necessary to use appropriate seals in order to prevent the propellants from contaminating the coil. When the plunger of a solenoid valve is designed to be in contact with the core in the absence of electric power, then it is also necessary to cover the face of the plunger with a layer of non-magnetic material to avoid sticking [3].

As an application of the concepts discussed above, it is required to determine the electric resistance of the coil for a solenoid valve to be used in a rocket engine.

The following data are known: the required actuating force at the start of the stroke is $F = 120.1$ N, the electric supply has a voltage of 28 V direct current, the maximum current is $i_{max} = 2$ A, the air gap between the solenoid core and the plunger of the valve is $G = 0.00127$ m, the diameter of the plunger is $d = 0.01422$ m, and the leakage factor of the magnetic flux is $f = 0.7$.

The area of the plunger results from

$$A = \frac{1}{4}\pi d^2 = 0.25 \times 3.1416 \times 0.01422^2 = 0.0001588 \text{ m}^2$$

By substituting the value of A found above, $F = 120.1$ N, and $C = 2.505 \times 10^{-6}$ in the equation

$$F = \frac{B^2 A}{C}$$

and solving for B, we find

$$B = \left(\frac{FC}{A}\right)^{\frac{1}{2}} = \left(\frac{120.1 \times 2.505 \times 10^{-6}}{0.0001588}\right)^{\frac{1}{2}} = 1.376 \text{ Wb/m}^2$$

By substituting the value of B found above, $f = 0.7$, $P = 1.255 \times 10^{-6}$, and $G = 0.00127$ m in the following equation

$$B = \frac{fPNi}{G}$$

and solving for Ni, we find

$$Ni = \frac{BG}{fP} = \frac{1.376 \times 0.00127}{0.7 \times 1.255 \times 10^{-6}} = 1989 \text{ turns A}$$

Taking the value $i = 1.4$ A for the current in the coil, the corresponding number of turns is

$$N = \frac{Ni}{i} = \frac{1989}{1.4} = 1421$$

The resistance of the conductor used for the coil is

$$R = \frac{V}{i} = \frac{28}{1.4} = 20\,\Omega$$

We use a copper wire AWG No. 26 for the coil. In [33], we find the following data, at a temperature $T = 293$ K, for this wire: diameter $d = 0.4049$ mm, area $A = \frac{1}{4}\pi d^2 = 0.1288$ mm^2, and resistance per metre $R = 0.1336\,\Omega$/m.

In order for the coil to have a resistance of 20 Ω, the length L of the wire must be

$$L = \frac{20}{0.1336} = 149.7 \text{ m}$$

Since the wire has $N = 1421$ turns, then the average diameter D of each turn results from

$$D = \frac{L}{N\pi} = \frac{149.7}{1421 \times 3.1416} = 0.03353 \text{ m}$$

The pilot valve described above is a two-way valve, which can be either fully open or fully closed. Other pilot valves are multiple-passage valves, which can start, stop, and divert the motion of a fluid between three or more alternate paths. These valves are described below.

Multiple-passage valves are used to control fluids moving to and from actuating cylinders. They are also used to control the direction of flow when it is necessary to switch or direct it between various paths. These valves are known as three-way valves, four-way valves, diversion valves, selection valves, sequence valves, and shuttle valves. Their method of actuation may be manual, mechanical, hydraulic, pneumatic, or electrical. They operate in two or more discrete positions, which change only when these valves are shifted. Their mode of operation is only of the on-off type. Therefore, they cannot be used in proportional or throttling mode. The name three-way valve or four-way valve specifies the number of ports in a valve. The following figure, re-drawn from [25], shows a three-way valve (designed by Valcor Engineering Corporation) of the poppet type, which is actuated by a solenoid.

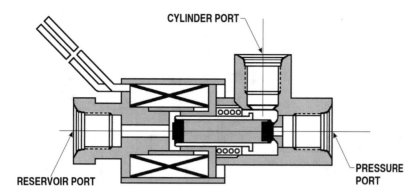

The valve illustrated above has a common port (the cylinder port), which can be connected to either one of two alternate ports (the pressure port and the reservoir port) when the non-connected port is closed. The poppet, when actuated in one direction, opens the cylinder port to the pressure port and closes it to the reservoir port. The same poppet, when actuated in the other direction, opens the cylinder port to the reservoir port and closes it to the pressure port. The cylinder port of a three-way valve is used to control a single-acting cylinder.

A four-way valve is a valve having four external ports arranged so that there are two simultaneous flow paths in the valve. The ports of a four-way valve are identified as pressure port, reservoir port, and two cylinder ports. A four-way valve is used to actuate a double-acting cylinder. For this purpose, the valve is connected so that, when pressure is applied to one of the two cylinder ports, then the other cylinder port is connected to the reservoir, and vice versa. Four-way valves are normally two-position or three-position valves. In a three-position, four-way valve, there is a central position in which all ports are closed.

The following figure, re-drawn from [25], shows how multiple-passage valves can be used to control the position of a pneumatic piston-cylinder.

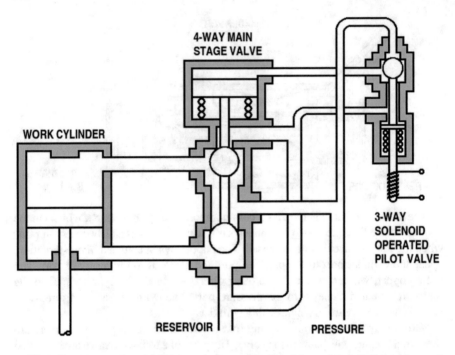

The pilot stage valve (right) is a three-way, solenoid-operated, ball-type poppet valve. The main stage valve (centre) is a four-way, two position, ball-type poppet valve operated by means of a single-acting cylinder, which in turn is actuated by the pilot stage valve. When the solenoid carries no current, then the actuation cylinder of the main stage is not connected with the pressure port, and the work cylinder (left) is in the down position. When the solenoid carries a current, then the actuation cylinder of the main stage is connected with the pressure port and therefore moves down, thereby connecting the upper end of the work cylinder with the reservoir port and the shaft end of the work cylinder with the pressure port, and therefore the work cylinder moves up.

A diversion valve is a three-way valve, such that the common port is the pressure port. A flow can be diverted by a diversion valve to either one of two alternate paths.

A selection valve is similar to a diversion valve, but the number of alternate paths to which the flow from the pressure port can be diverted is unlimited.

A sequence valve is a valve which directs a flow in a pre-determined sequence between two or more paths.

A shuttle valve is a sequence valve which is actuated by pressure, so that, when a desired pressure has been reached, the valve is automatically actuated, directing a flow to two or more paths.

Huzel and Huang [3] describe two examples of multiple-passage valves used for rocket engines. The first example concerns a pressure-actuated, three-way pilot valve, which may be used as an ignition monitor valve. This valve is shown in the following figure, re-drawn from [3].

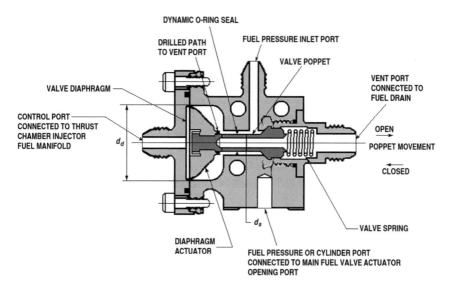

DYNAMIC O-RING SEAL

DRILLED PATH
TO VENT PORT

FUEL PRESSURE INLET PORT

VALVE POPPET

VALVE DIAPHRAGM

VENT PORT
CONNECTED TO
FUEL DRAIN

CONTROL PORT
CONNECTED TO THRUST
CHAMBER INJECTOR
FUEL MANIFOLD

d_d

OPEN

POPPET MOVEMENT

CLOSED

d_s

DIAPHRAGM
ACTUATOR

VALVE SPRING

FUEL PRESSURE OR CYLINDER PORT
CONNECTED TO MAIN FUEL VALVE ACTUATOR
OPENING PORT

The valve illustrated above is held normally closed by a spring, and has a diaphragm at its control port, which is connected to the thrust chamber injector fuel manifold. By the way, a diaphragm is a thin dividing membrane which is used as a seal to prevent fluid leakage and also as a pressure-sensing element having small (less than 1.5 mm) displacements. During engine start and when a satisfactory ignition has been achieved in the main thrust chamber, the increase in pressure sensed at the thrust chamber injector fuel manifold causes the ignition monitor valve to open, because the diaphragm of the valve is put under pressure. This opening, in turn, directs the fuel pressure to the cylinder port connected to the main fuel valve actuator opening port. The valve spring can be calibrated to correspond to the effective area of the diaphragm, so that the valve opens at a preset sensed pressure. During engine cut-off, the decreasing pressure of the fuel causes the ignition monitor valve to close. This closure, in turn, vents the opening side of the main fuel valve actuator, thereby closing the valve.

The poppet of the valve is balanced by the internal pressure of the fluid which acts on a dynamic O-ring seal having the same diameter d_s (m) as the poppet. The valve diaphragm is made of a several layers of thin Mylar® sheets, which are pressure-formed with heat added. The effective area of the diaphragm can be determined experimentally. The required preload of the valve spring may be estimated as follows

$$\frac{1}{4}\pi d_d^2 p_s = F_f + S_p$$

where d_d (m) is the effective diameter of the diaphragm, p_s (N/m²) is the rated sensed threshold pressure to open the valve, F_f (N) is the static friction force of the valve poppet, and S_p (N) is the required preload of the valve spring.

For example, we want to determine the required preload S_p (N) and the output power P (W) for a pressure-actuated, three-way ignition monitor valve of the type described above, which has the following properties: characteristic flow area of the valve at the fully open position $A_c = 1.226 \times 10^{-4}$ m^2, diameter of the diaphragm d_d $= 0.05334$ m, sensed threshold pressure to open the valve $p_s = 1.379 \times 10^5$ N/m^2, static friction force of the poppet $F_f = 62.28$ N, resistance coefficient at the fully open position $k = 3.5$, required volume flow rate $q = 0.003277$ m^3/s, fuel pressure at the inlet port $p_f = 2.413 \times 10^6$ N/m^2. From [34], we also know the density of the fuel (RP-1) to be $\rho = 824$ kg/m^3 at a temperature of 283 K.

By substituting $d_d = 0.05334$ m, $p_s = 1.379 \times 10^5$ N/m^2, $F_f = 62.28$ N into the preceding equation and solving for S_p, we find

$$S_p = \frac{\pi d_d^2 p_s}{4} - F_f = \frac{3.1416 \times 0.05334^2 \times 1.379 \times 10^5}{4} - 62.28 = 245.9 \text{ N}$$

The characteristic velocity of flow in the valve is

$$v = \frac{q}{A_c} = \frac{0.003277}{1.226 \times 10^{-4}} = 26.73 \text{ m/s}$$

The pressure loss through the valve in the design conditions is

$$\Delta p = k\left(\frac{1}{2}\rho v^2\right) = 3.5 \times 0.5 \times 824 \times 26.73^2 = 1.03 \times 10^6 \text{ N/m}^2$$

The pressure of the fuel at the point of discharge of the valve is

$$p_d = 2.413 \times 10^6 - 1.03 \times 10^6 = 1.383 \times 10^6 \text{ N/m}^2$$

The output power of the valve results from the product of the pressure of the fluid at the point of discharge by the volume flow rate, as follows

$$P = p_d q = 1.383 \times 10^6 \times 0.003277 = 4531 \text{ W}$$

The second example concerns a four-way pilot valve of the self-locking type. A scheme of this valve is shown in the following figure, re-drawn from [3].

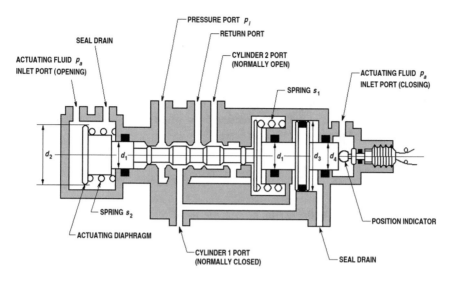

The valve illustrated above is held normally closed at the cylinder 1 port by two springs s_1 and s_2. The actuating pressure p_a is applied to the inlet port (opening) of the valve. The pressure p_a causes the translating shaft to move, and consequently the pressure port is connected to the cylinder 1 port, and the cylinder 2 port is connected to the return port. An unbalanced self-locking force $\frac{1}{4}\pi\,(d_3^2 - d_1^2)p_i$ acts in the opening direction and causes the valve to remain open, even after the actuating pressure p_a is removed from the opening port. The valve can be closed only by applying the actuating pressure p_a to the inlet port (closing) and venting the inlet port (opening).

As an application of the concepts discussed above, it is required to determine the diameters d_2, d_3, and d_4 for the four-way pilot valve shown in the preceding figure, with a contingency factor $c = 1.5$, knowing the following data: pre-load F_1 $= 155.7$ N and elastic constant $k_1 = 3.678 \times 10^4$ N/m of the spring s_1, pre-load F_2 $= 111.2$ N and elastic constant $k_2 = 4.378 \times 10^4$ N/m of the spring s_2, static friction $F_f = 106.8$ N of the poppet, relative pressure $p_i = 2.758 \times 10^6$ N/m^2 at the pressure port, relative pressure $p_a = 1.724 \times 10^6$ N/m^2 of the actuating fluid, ambient pressure (that is, relative pressure equal to zero) at the return port, diameter $d_1 = 0.0127$ m of the poppet guide, and total travel $x = 0.00127$ m of the poppet in the valve.

In order for the valve to be opened, the pressure p_a, which the actuating fluid exerts on the diaphragm area $\frac{1}{4}\pi d_2^2$, must be sufficient to counterbalance the pre-loads F_1 and F_2 of the springs s_1 and s_2 and the static friction force F_f. Therefore, the diameter d_2 of the diaphragm which opens the valve results from the following equation

$$\frac{\pi d_2^2}{4} p_a = c\left(F_1 + F_2 + F_f\right)$$

where $c = 1.5$ is the contingency factor. Solving for d_2, we find

$$d_2 = \left[\frac{4c\left(F_1 + F_2 + F_f\right)}{\pi p_a} \right]^{\frac{1}{2}} = \left[\frac{4 \times 1.5 \times (155.7 + 111.2 + 106.8)}{3.1416 \times 1.724 \times 10^6} \right]^{\frac{1}{2}}$$

$$= 0.02035 \text{ m}$$

The diameter d_3 of the piston depends on the value of the force to be applied in order to lock the valve in the open position even when the actuating pressure p_a is removed from the pressure port. This force is

$$\frac{\pi\left(d_3^2 - d_1^2\right)}{4} p_i = c\left(F_1 + F_2 + k_1 x + k_2 x - F_f\right)$$

By substituting $d_1 = 0.0127$ m, $p_i = 2.758 \times 10^6$ N/m^2, $c = 1.5$, $F_1 = 155.7$ N, $F_2 = 111.2$ N, $k_1 = 3.678 \times 10^4$ N/m, $k_2 = 4.378 \times 10^4$ N/m, $x = 0.00127$ m, and $F_f = 106.8$ N in the preceding equation and solving for d_3, we find

$$d_3 = \left[\frac{4 \times 1.5 \times (155.7 + 111.2 + 36780 \times 0.00127 + 43780 \times 0.00127 - 106.8)}{3.1416 \times 2.758 \times 10^6} \right.$$

$$\left. + 0.0127^2 \right]^{\frac{1}{2}} = 0.01852 \text{ m}$$

The diameter d_4 needed for the actuating pressure p_a to close the valve results from the following equation

$$\frac{\pi d_4^2}{4} p_a = c\left[\frac{\pi\left(d_3^2 - d_1^2\right)}{4} p_i + F_f - F_1 - F_2 - k_1 x - k_2 x \right]$$

By substituting $p_a = 1.724 \times 10^6$ N/m^2, $d_3 = 0.01852$ m, $d_1 = 0.0127$ m, $p_i = 2.758 \times 10^6$ N/m^2, $F_f = 106.8$ N, $F_1 = 155.7$ N, $F_2 = 111.2$ N, $k_1 = 3.678 \times 10^4$ N/m, $k_2 = 4.378 \times 10^4$ N/m, and $x = 0.00127$ m in the preceding equation and solving for d_4, we find

$$d_4 = \left\{ \frac{4 \times 1.5}{3.1416 \times 1.724 \times 10^6} \left[\frac{3.1416 \times (0.01852^2 - 0.0127^2)}{4} \times 2.758 \times 10^6 \right. \right.$$

$$\left. \left. + 106.8 - 155.7 - 36780 \times 0.00127 - 43780 \times 0.00127 \right] \right\}^{\frac{1}{2}} = 0.01205 \text{ m}$$

5.11 Design of Flow Regulating Devices of the Fixed-Area Type

The flow regulating devices described in the present section are nozzles and orifices. Venturi tubes and valves used for the same purpose have been described in Sects. 5.7 and 5.9.

A nozzle is a convergent device inserted coaxially in a conduit and having a curved profile without discontinuities. An orifice is a circular aperture in a thin plate which restricts the cross-sectional area of a conduit in which this plate is inserted coaxially. According to the principle of Bernoulli, a fluid moving through a nozzle or an orifice increases its velocity and decreases its pressure in the direction of flow. These devices can be used in liquid-propellant rocket engines as flow regulators, as will be shown below. Typical shapes used for nozzles and orifices are shown in the following figure, re-drawn from [35].

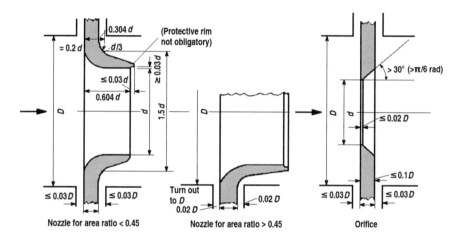

The small turned-out part (protective rim at the discharge of the nozzle) shown in the preceding figure serves to protect the discharge edge, and consequently may be omitted when no damage is to be feared [35].

In Europe, the matter is regulated by the standards ISO 5167 [36]. In the United States of America, it is regulated by the standards of the American Gas Association [37].

The fundamental equation which governs the flow of liquids or gases through nozzles or orifices is the same as that which was shown in Sect. 5.7. This equation is re-written below for convenience of the reader

$$q = C_d A_0 \varepsilon \left[\frac{2\Delta p}{\rho(1 - \beta^4)} \right]^{\frac{1}{2}} = C_d A_0 \varepsilon \left[\frac{2\Delta H}{1 - \beta^4} \right]^{\frac{1}{2}}$$

where q (m³/s) is the volume flow rate of the fluid, C_d is the coefficient of discharge, $A_0 = \pi d^2/4$ (m²) is the area of the throat section, ε is the expansion coefficient (whose value can be taken equal to unity for incompressible fluids) of the fluid, Δp (N/m²) and ΔH (m) are the differences of respectively pressure and pressure head measured upstream and downstream of the device, ρ (kg/m³) is the density of the fluid in the operating conditions, and $\beta = d/D$ is the ratio of the diameter d of the throat to the diameter D of the conduit.

In case of gaseous substances, the value of the expansion coefficient ε can be computed by using the following empirical formula, which holds for $p_2/p_1 \geq 0.75$ [36]:

$$\varepsilon = 1 - \left(0.351 + 0.256\beta^4 + 0.92\beta^8\right)\left[1 - \left(\frac{p_2}{p_1}\right)^{\frac{1}{\gamma}}\right]$$

where $\beta = d/D$ is the ratio of the diameter d of the throat to the diameter D of the conduit, p_1 and p_2 (N/m²) are the static pressures of the gas respectively upstream and downstream of the device, and $\gamma = c_p/c_v$ is the specific heat ratio of the gas.

In addition, [38] gives the following equations for the mass flow rate $\dot{m}$ (kg/s) of a gas through an orifice:

$$\dot{m} = C_d \frac{\pi d^2}{4}\left\{\frac{2\gamma p_1 \rho_1}{\gamma - 1}\left[\left(\frac{p_2}{p_1}\right)^{\frac{2}{\gamma}} - \left(\frac{p_2}{p_1}\right)^{\frac{\gamma+1}{\gamma}}\right]\right\}^{\frac{1}{2}}$$

which holds under non-chocked flow conditions, and

$$\dot{m} = C_d \frac{\pi d^2}{4}\left[\gamma p_1 \rho_1 \left(\frac{2}{\gamma + 1}\right)^{\frac{\gamma+1}{\gamma-1}}\right]^{\frac{1}{2}}$$

which holds under choked (that is, maximum) flow conditions. In the preceding equations, p_1 (N/m²) and ρ_1 (kg/m³) are respectively the static absolute pressure and the density of the gas upstream of the orifice, and p_2 (N/m²) is the static absolute pressure of the gas downstream of the orifice.

The value of the coefficient of discharge C_d depends on the type of device, on the manner in which it is inserted in the conduit, and on the Reynolds number computed upstream of the device

$$Re = \frac{\rho v D}{\mu}$$

where v (m/s) is the stream velocity and μ (Ns/m²) is the coefficient of dynamic viscosity of the fluid.

The coefficient of discharge C_d may be computed by using the Reader-Harris/Gallagher equation, as indicated in [36, 37]. According to [39], in most practical cases, the coefficient of discharge may be computed as follows

$$C_d = C_{d\infty} + \frac{b}{Re^n}$$

where $C_{d\infty}$ (discharge coefficient at infinite Reynolds number), b, and n are three terms whose values can be determined as indicated in [39], depending on the particular case. For example, in case of a nozzle ISA 1932, [39] indicates the following values to be substituted in the preceding equation

$$C_{d\infty} = 0.99 - 0.2262\beta^{4.1}$$
$$b = 1708 - 8936\beta + 19779\beta^{4.7}$$
$$n = 1.15$$

As an application of the concepts discussed above, the discharge duct of the fuel pump of a rocket engine burning liquid oxygen with RP-1 has a diameter $D = 0.1778$ m and a mass flow rate $\dot{m} = 404.6$ kg/s. We want to regulate the pressure in the duct by inserting a plate having an orifice, in order to obtain a pressure drop $\Delta p = 8.895 \times 10^5$ N/m^2 downstream of the plate. From [34], we know the density and the coefficient of dynamic viscosity of RP-1 to be respectively $\rho = 824$ kg/m^3 and $\mu = 2.451 \times 10^{-3}$ Ns/m^2 at 283 K. It is required to estimate the diameter d of the throat of the orifice.

Since RP-1 is liquid at 283 K, then its expansion coefficient ε can be taken equal to unity. The volume flow rate q of RP-1 in the duct results from

$$q = \frac{\dot{m}}{\rho} = \frac{404.6}{824} = 0.491 \text{ m}^3/\text{s}$$

The velocity of the fluid upstream of the orifice is

$$v = \frac{4q}{\pi D^2} = \frac{4 \times 0.491}{3.1416 \times 0.1778^2} = 19.78 \text{ m/s}$$

The Reynolds number upstream of the orifice is

$$Re = \frac{\rho v D}{\mu} = \frac{824 \times 19.78 \times 0.1778}{2.451 \times 10^{-3}} = 1.182 \times 10^6$$

In [39], we find the following expressions

$$C_{d\infty} = 0.5959 + 0.0312\beta^{2.1} - 0.184\beta^6$$
$$b = 91.71\beta^{2.5}$$
$$n = 0.75$$

for the terms of the equation

$$C_d = C_{d\infty} + \frac{b}{Re^n}$$

which expresses the coefficient of discharge C_d for an orifice with corner taps as a function of β. Since $d = \beta D$, we use the following equation

$$q = C_d \frac{\pi \beta^2 D^2}{4} \varepsilon \left[\frac{2\Delta p}{\rho(1 - \beta^4)} \right]^{\frac{1}{2}}$$

and solve this equation iteratively for β. To this end, we define the following function of β

$$f(\beta) = q - C_d \frac{\pi \beta^2 D^2}{4} \varepsilon \left[\frac{2\Delta p}{\rho(1 - \beta^4)} \right]^{\frac{1}{2}}$$

and search the value of β for which the function $f(\beta)$ is equal to zero within a fixed tolerance. Since $\beta = d/D$, then we search the unknown value of β in some interval $0 < \beta < 1$ such that the value of the function $f(\beta)$ changes sign in that interval. We search the value of β in the interval $0.7 \le \beta \le 0.8$.

For $\beta = 0.7$, we find

$$C_{d\infty} = 0.5959 + 0.0312 \times 0.7^{2.1} - 0.184 \times 0.7^6 = 0.5890$$
$$b = 91.71 \times 0.7^{2.5} = 37.60$$
$$n = 0.75$$
$$C_d = 0.5890 + 37.60/(1.182 \times 10^6)^{0.75} = 0.5900$$
$$f(0.7) = 0.491 - 0.25 \times 0.5900 \times 3.1416 \times 0.7^2 \times 0.1778^2 \times 1$$
$$\times \{2 \times 8.895 \times 10^5/[824 \times (1 - 0.7^4)]\}^{\frac{1}{2}} = 0.1084$$

For $\beta = 0.8$, we find

$$C_{d\infty} = 0.5959 + 0.0312 \times 0.8^{2.1} - 0.184 \times 0.8^6 = 0.5672$$
$$b = 91.71 \times 0.8^{2.5} = 52.50$$
$$n = 0.75$$
$$C_d = 0.5672 + 52.50/(1.182 \times 10^6)^{0.75} = 0.5697$$
$$f(0.8) = 0.491 - 0.25 \times 0.5687 \times 3.1416 \times 0.8^2 \times 0.1778^2 \times 1$$
$$\times \{2 \times 8.895 \times 10^5/[824 \times (1 - 0.8^4)]\}^{\frac{1}{2}} = -0.05547$$

Since the value of the function $f(\beta)$ changes sign in the interval $0.7 \le \beta \le 0.8$, then the unknown value of β for which $f(\beta) = 0$ falls within this interval. By using repeatedly the numerical method described in Chap. 1, Sect. 1.2, we find $\beta = 0.7706$ with four significant figures. As is easy to verify, for $\beta = 0.7706$ there results $C_{d\infty} = 0.5754$, $b = 47.81$, $n = 0.75$, $C_d = 0.5768$, and $f(\beta) = -7.743 \times 10^{-5}$. Therefore, the diameter of the orifice necessary to obtain the desired pressure drop $\Delta p = 8.895 \times 10^5$ N/m^2 is

$$d = \beta D = 0.7706 \times 0.1778 = 0.137 \text{ m}$$

The following example concerns the flow of a gas (helium) through an orifice. Knowing the diameter $d = 0.001524$ m and the discharge coefficient $C_d = 0.6$ of the orifice, and the static absolute pressure $p_1 = 3.548 \times 10^6$ N/m^2 and the temperature $T_1 = 311$ K of helium upstream of the orifice, it is required to calculate the mass flow rates of helium through the orifice for the static absolute pressures $p_2 = 0.1013 \times 10^6$ N/m^2 (atmospheric pressure) and $p_2 = 2.514 \times 10^6$ N/m^2 downstream of the orifice.

From [40], we know the specific heat ratio and the density of helium to be respectively $\gamma = 1.663$ and $\rho_1 = 5.408$ kg/m^3 in the given conditions.

The critical pressure ratio of helium is

$$\frac{p_c}{p_1} = \left(\frac{2}{\gamma + 1}\right)^{\frac{\gamma}{\gamma - 1}} = \left(\frac{2}{1.663 + 1}\right)^{\frac{1.663}{1.663 - 1}} = 0.4877$$

When the static absolute pressure downstream of the orifice is $p_2 = 0.1013 \times 10^6$ N/m^2 (atmospheric pressure), then the pressure ratio is

$$\frac{p_2}{p_1} = \frac{0.1013 \times 10^6}{3.548 \times 10^6} = 0.02855$$

and therefore the pressure ratio p_2/p_1 is less than the critical pressure ratio p_c/p_1 (choked flow). By substituting $\rho_1 = 5.408$ kg/m^3, $C_d = 0.6$, $d = 0.001524$ m, $\gamma = 1.663$, and $p_1 = 3.548 \times 10^6$ N/m^2 in the following equation

$$\dot{m} = C_d \frac{\pi d^2}{4} \left[\gamma p_1 \rho_1 \left(\frac{2}{\gamma + 1}\right)^{\frac{\gamma + 1}{\gamma - 1}}\right]^{\frac{1}{2}}$$

we find

$$\dot{m} = 0.6 \times \frac{3.1416 \times 0.001524^2}{4} \times \left[1.663 \times 3.548 \times 10^6 \times 5.408\right.$$

$$\left. \times \left(\frac{2}{1.663 + 1}\right)^{\frac{1.663 + 1}{1.663 - 1}}\right]^{\frac{1}{2}} = 0.003479 \text{ kg/s}$$

When the static absolute pressure downstream of the orifice is $p_2 = 2.514 \times 10^6$ N/m^2, then the pressure ratio is

$$\frac{p_2}{p_1} = \frac{2.514 \times 10^6}{3.548 \times 10^6} = 0.7086$$

and therefore the pressure ratio p_2/p_1 is greater than the critical pressure ratio p_c/p_1 (non-choked flow). By substituting $C_d = 0.6$, $d = 0.001524$ m, $\gamma = 1.663$, $p_1 = 3.548 \times 10^6$ N/m^2, $\rho_1 = 5.408$ kg/m^3, and $p_2/p_1 = 0.7086$ in the following equation

$$\dot{m} = C_d \frac{\pi d^2}{4} \left\{ \frac{2\gamma p_1 \rho_1}{\gamma - 1} \left[\left(\frac{p_2}{p_1} \right)^{\frac{2}{\gamma}} - \left(\frac{p_2}{p_1} \right)^{\frac{\gamma+1}{\gamma}} \right] \right\}^{\frac{1}{2}}$$

we find

$$\dot{m} = 0.6 \times \frac{3.1416 \times 0.001524^2}{4} \times \left[\frac{2 \times 1.663 \times 3.548 \times 10^6 \times 5.408}{1.663 - 1} \right.$$

$$\left. \times \left(0.7086^{\frac{2}{1.663}} - 0.7086^{\frac{1.663+1}{1.663}} \right) \right]^{\frac{1}{2}} = 0.003127 \text{ kg/s}$$

5.12 Design of Servo-Valves

According to the definition given in [41], a servo-valve is a modulating operator that amplifies system signals for variable-displacement, closed-loop control of actuator position. There are two principal types of servo-valves. In a servo-valve of the nozzle-flapper type, the pressure exerted by the fluid on the actuator is controlled directly by a flapper, which restricts the flow through two nozzles of variable size, in response to an electrical input signal applied to a torque motor. In a servo-valve of the spool type, the pressure is controlled indirectly, by means of a sliding spool, whose position in the body of the valve depends on the angular position of the flapper. A spool is a solid cylindrical element having two or more recesses which fits closely in the bore of the valve body, such that the valve opens or closes by translating the spool within the bore. In other words, a servo-valve of the spool type is an electro-hydraulic valve having a spool, whose position in the valve body changes proportionally to an electrical input signal received by the valve. The translational motion of the spool in the valve is obtained by hydraulic pressure, and changes the size of two orifices in order for the valve to control flow. The control exerted by the valve depends on the difference of hydraulic pressure across the orifices, unless some form of compensation is used. A functional scheme of a servo-valve of the spool type is given in the following figure, re-drawn from [42].

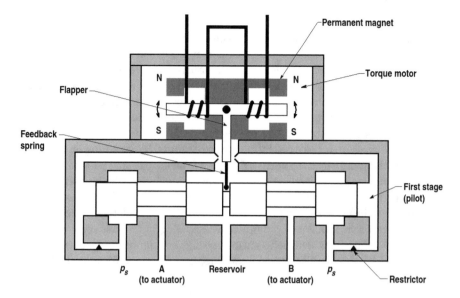

The principal components of the servo-valve shown above are:

- an electromagnetic torque motor, which acts as a transducer to convert an electrical input signal into a mechanical force;
- a flapper driven by the mechanical force generated by the torque motor, which restricts differentially the flow from a pair of nozzles, the flapper stroke being about 0.1 mm;
- a spool, whose translational motion within the valve body is due to the difference of hydraulic pressure when the flapper is off-centre; and
- a feedback spring, which allows the spool to move, the stroke of the spool being about 1 mm, until the restoring force acting on the flapper is in equilibrium with the force generated by the torque motor.

The direction and the magnitude of the displacement of the flapper caused by the torque motor depend on the input signal. The flapper is located between two opposed nozzles, and therefore its motion in one or in the other direction restricts the flow through one or the other of these nozzles. When no electrical signal is applied to the torque motor, then the flapper is located at an equal distance between the two nozzles (neutral position). The presence of an electrical signal causes the flapper to move toward one or the other nozzle, thereby producing an unbalance of hydraulic pressure across the spool. The set of components comprising the torque motor, the nozzles, and the flapper is known as the first stage of a servo-valve. The first stage pilots the second stage, which comprises the spool and the feedback spring. The spring makes the displacement of the spool proportional to the difference of pressure caused by the off-centre position of the flapper in the first stage. The position of the flapper, in turn, depends on the force generated by the torque motor in response to the input signal.

Therefore, the direction and the magnitude of the displacement of the spool are proportional to the direction and to the magnitude of the input signal, and are used to control the direction and the magnitude of flow to the actuator through the pressure at the cylinder ports of the valve. For example, with reference to the preceding figure, an electrical signal resulting in a displacement of the flapper to the right restricts the flow from the nozzle on the right hand side. This restriction increases the pressure upstream of the right nozzle circuit, and decreases the pressure upstream of the left circuit. This unbalance of pressure causes the spool to shift to the left, until the difference of pressure is balanced by the pressure exerted by the spring on the spool. The shift of the spool to the left opens a flow path from the pressure port placed on the left to the cylinder port A, and the returning fluid through the cylinder port B moves through the return line to the reservoir. A servo-valve is not only an electrical-to-hydraulic transducer, but also a power amplifier, because the electrical input power, whose order of magnitude is about 0.1 W, is amplified in the first stage to at least 10 W of hydraulic power, and then used to control, by means of the spool, an amount of about 10 kW of hydraulic power. Therefore, the power amplification factor in a two-stage servo-valve is 10^5 [42].

Another two-stage servo-valve of the spool type is illustrated in the following figure, re-drawn from [25]. In the servo-valve shown below, the spool is spring-loaded, in order for its position in the valve to be proportional to the input received from the first stage. The spool is centred by helical coil springs on each side. Integral filters are used to remove small particles, which would otherwise coalesce and obstruct orifices and nozzles. These filters are also used to protect the spool from contaminants. This servo-valve has also a drain bleed system to protect the torque motor from leakage fluid.

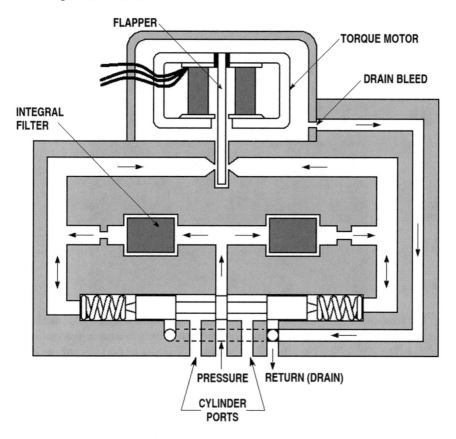

In the main engines (RS-25) of the Space Shuttle, two servo-valves (channel A and channel B) are mounted on each of the five actuators. These servo-valves convert the electrical command signal from the engine controller (described in Sect. 5.2) to hydraulic flow directed to the valve actuator. They convert the polarity and the amplitude of the electrical command signal into respectively the rotation direction and the rotation rate of the shaft. The two servo-valves are redundant, in order for the failure of one of them not to affect the performance of the actuators. An assembly of one of these servo-valves is shown in the following figure, due to the courtesy of Boeing-Rocketdyne [7]. When the torque motor of the servo-valve tilts in response to the polarity and to the amplitude of the input signal, the flow restriction is increased at one nozzle, and decreased at the other. These variable restrictors are paired with constant restrictors at the end of the filter, forming two matched pressure dividers. Therefore, the pressures applied to the ends of the spool in the second stage can be varied, being equal in the null position and not equal otherwise.

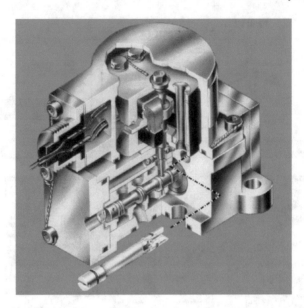

The resulting offset of the spool is opposite to the tilt of the torque motor, and is fed back to the torque motor via the springy connecting rod, thereby assuring proportional control. The spool offset in effect simultaneously moves one port toward the input (higher) pressure, and the other port toward the return (lower) pressure, driving the pistons. Therefore, the polarity of the signal determines the direction of rotation, and the amplitude of the signal determines the rotation rate in the valve [7].

The equations which can be used for the design of servo-valves of the flapper-nozzle type are those shown in Sect. 5.11 for the flow of liquid or gaseous substances through nozzles or orifices. These equations are re-written below for convenience of the reader. In case of liquid substances, the following equation can be used

$$\dot{m} = \rho q = \rho C_d A_0 \varepsilon \left[\frac{2\Delta p}{\rho \left(1 - \beta^4\right)} \right]^{\frac{1}{2}} = \rho C_d A_0 \varepsilon \left[\frac{2\Delta H}{1 - \beta^4} \right]^{\frac{1}{2}}$$

where $\dot{m}$ (kg/s) and q (m^3/s) are respectively the mass flow rate and the volume flow rate of the liquid, C_d is the coefficient of discharge of the valve, $A_0 = \pi d^2/4$ (m^2) is the area of the throat section, ε is the expansion coefficient (whose value can be taken equal to unity) of the liquid, Δp (N/m^2) and ΔH (m) are the differences of respectively pressure and pressure head measured upstream and downstream of the device, ρ (kg/m^3) is the density of the liquid in the operating conditions, and $\beta = d/D$ is the ratio of the diameter d of the throat to the diameter D of the conduit.

In case of gaseous substances under non-chocked flow conditions, the following equation can be used

$$\dot{m} = C_d \frac{\pi d^2}{4} \left\{ \frac{2\gamma p_1 \rho_1}{\gamma - 1} \left[\left(\frac{p_2}{p_1} \right)^{\frac{2}{\gamma}} - \left(\frac{p_2}{p_1} \right)^{\frac{\gamma+1}{\gamma}} \right] \right\}^{\frac{1}{2}}$$

In case of gaseous substances under chocked flow conditions, the following equation can be used

$$\dot{m} = C_d \frac{\pi d^2}{4} \left[\gamma p_1 \rho_1 \left(\frac{2}{\gamma + 1} \right)^{\frac{\gamma+1}{\gamma-1}} \right]^{\frac{1}{2}}$$

In the preceding equations, $\gamma = c_p/c_v$ is the specific heat ratio of the gas, p_1 (N/m^2) and ρ_1 (kg/m^3) are respectively the static pressure and the density of the gas upstream of the device, and p_2 (N/m^2) is the static pressure of the gas downstream of the device.

In addition, Huzel and Huang [3] indicate the following equation to compute the effective, ring-shaped flow area A_n (m^2) of a nozzle of diameter d_n (m) in the presence of a flapper displaced by x (m) from the nozzle

$$A_n = \pi d_n x$$

where the maximum value of the distance x should not exceed $d_n/5$.

The following example of application concerns a pneumatic servo-valve of the flapper-nozzle type. This servo-valve is used as a pilot valve of another valve, which in turn controls the propellant (oxidiser) utilisation in a rocket engine. A scheme of the pneumatic servo-valve, redrawn from [3], is shown below.

The following data are known. The gaseous substance flowing in the circuit is helium. The absolute pressure and the temperature of helium at the supply point are respectively $p_s = 3.447 \times 10^6$ N/m^2 and $T_s = 311$ K. The diameters of the ducts are such that $d_1 = d_2$, and $d_3 = d_4$. The coefficient of discharge for the orifices and for the nozzles is $C_d = 0.7$. The thickness of the flapper is $t = 0.1016$ mm $= 1.016 \times 10^{-4}$ m. The distance between the nozzles is $T = d_3/4 + t$. At the neutral position of the flapper, the absolute pressures in the actuator are $p_c = p_0 = 3.103 \times 10^6$ N/m^2. The mass flow rates through the nozzles are $\dot{m}_3 = \dot{m}_4 = 3.529 \times 10^{-4}$kg/s. The gas is discharged downstream of the nozzles at atmospheric pressure ($p_2 = 0.1013 \times 10^6$ N/m^2). It is required to determine the diameters d_1 and d_2 of the fixed orifices, the diameters d_3 and d_4 of the nozzles, and the distance T between the nozzles.

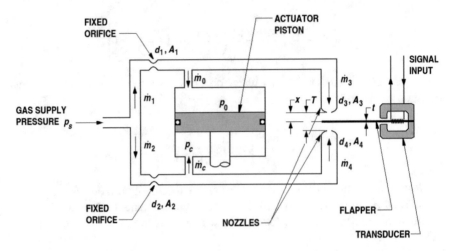

As has been shown in Sect. 5.11, the critical pressure ratio of helium is

$$\frac{p_c}{p_s} = \left(\frac{2}{\gamma + 1}\right)^{\frac{\gamma}{\gamma - 1}} = \left(\frac{2}{1.663 + 1}\right)^{\frac{1.663}{1.663 - 1}} = 0.4877$$

Since the static absolute pressure downstream of the nozzles is $p_2 = 0.1013 \times 10^6$ N/m^2 (atmospheric pressure), then the pressure ratio for the nozzles is

$$\frac{p_2}{p_s} = \frac{0.1013 \times 10^6}{3.447 \times 10^6} = 0.02939$$

and therefore the pressure ratio p_2/p_s is less than the critical pressure ratio $p_c/p_s = 0.4877$ (choked flow). Therefore, the equations to be used to compute the mass flow rates $\dot{m}_3$ and $\dot{m}_4$ through the nozzles are

$$\dot{m}_3 = C_d A_3 \left[\gamma p_0 \rho_s \left(\frac{2}{\gamma + 1}\right)^{\frac{\gamma+1}{\gamma-1}}\right]^{\frac{1}{2}}$$

$$\dot{m}_4 = C_d A_4 \left[\gamma p_c \rho_s \left(\frac{2}{\gamma + 1}\right)^{\frac{\gamma+1}{\gamma-1}}\right]^{\frac{1}{2}}$$

where p_0 and p_c are the static pressures upstream of the nozzles, and $\rho_s = 5.256$ kg/m^3 [40] is the density of helium at $p_s = 3.447 \times 10^6$ N/m^2 and $T_s = 311$ K.

After substituting $\rho_s = 5.256$ kg/m^3, $C_d = 0.7$, and $\gamma = 1.663$ in the two equations which express $\dot{m}_3$ and $\dot{m}_4$ we find

$$\dot{m}_3 = 1.165 A_3 p_0^{\frac{1}{2}}$$

$$\dot{m}_4 = 1.165 A_4 p_c^{\frac{1}{2}}$$

Since the static pressures p_0 and p_c are by far greater than the atmospheric pressure, then the equations to be used to compute the mass flow rates $\dot{m}_1$ and $\dot{m}_2$ through the fixed orifices are

$$\dot{m}_1 = C_d A_1 \left\{ \frac{2\gamma p_s \rho_s}{\gamma - 1} \left[\left(\frac{p_0}{p_s} \right)^{\frac{2}{\gamma}} - \left(\frac{p_0}{p_s} \right)^{\frac{\gamma+1}{\gamma}} \right] \right\}^{\frac{1}{2}}$$

$$\dot{m}_2 = C_d A_2 \left\{ \frac{2\gamma p_s \rho_s}{\gamma - 1} \left[\left(\frac{p_c}{p_s} \right)^{\frac{2}{\gamma}} - \left(\frac{p_c}{p_s} \right)^{\frac{\gamma+1}{\gamma}} \right] \right\}^{\frac{1}{2}}$$

With reference to the preceding figure, the areas A_1, A_2, A_3, and A_4 are

$$A_1 = \tfrac{1}{4}\pi d_1^2$$
$$A_2 = \tfrac{1}{4}\pi d_2^2$$
$$A_3 = \pi d_3 x$$
$$A_4 = \pi d_4 (T - t - x)$$

where x is the displacement of the flapper from the nozzle.

When the flapper is in its neutral position, then the actuator is at rest, and consequently the following equations hold

$$\dot{m}_c = \dot{m}_0 = 0$$

$$\dot{m}_1 = \dot{m}_2 = \dot{m}_3 = \dot{m}_4 = 3.529 \times 10^{-4} \text{ kg/s}$$

$$x = \frac{1}{2}(T - t) = \frac{1}{2}\left(\frac{d_3}{4} + t - t \right) = \frac{d_3}{8}$$

In addition, in the same conditions, the absolute pressures acting on the piston are $p_c = p_0 = 3.103 \times 10^6$ N/m². Therefore, the pressure ratio for both of the fixed orifices is

$$\frac{p_c}{p_s} = \frac{p_0}{p_s} = \frac{3.103 \times 10^6}{3.447 \times 10^6} = 0.9002$$

By introducing this value, $p_s = 3.447 \times 10^6$ N/m², $\rho_s = 5.256$ kg/m³, $C_d = 0.7$, and $\gamma = 1.663$ in the following equations

$$\dot{m}_1 = C_d A_1 \left\{ \frac{2\gamma p_0 \rho_s}{\gamma - 1} \left[\left(\frac{p_0}{p_s} \right)^{\frac{2}{\gamma}} - \left(\frac{p_0}{p_s} \right)^{\frac{\gamma+1}{\gamma}} \right] \right\}^{\frac{1}{2}}$$

$$\dot{m}_2 = C_d A_2 \left\{ \frac{2\gamma p_c \rho_s}{\gamma - 1} \left[\left(\frac{p_c}{p_s} \right)^{\frac{2}{\gamma}} - \left(\frac{p_c}{p_s} \right)^{\frac{\gamma+1}{\gamma}} \right] \right\}^{\frac{1}{2}}$$

we find

$$\dot{m}_1 = 0.6836 \, A_1 p_0^{\frac{1}{2}}$$

$$\dot{m}_2 = 0.6836 \, A_2 p_c^{\frac{1}{2}}$$

which in turn, for $p_0 = p_c = 3.103 \times 10^6$ N/m², yield

$$\dot{m}_1 = 1204 \, A_1$$

$$\dot{m}_2 = 1204 \, A_2$$

By comparing these two equations with

$$\dot{m}_1 = \dot{m}_2 = \dot{m}_3 = \dot{m}_4 = 3.529 \times 10^{-4} \text{ kg/s}$$

there results

$$A_1 = \frac{\pi d_1^2}{4} = \frac{3.529 \times 10^{-4}}{1204} = 2.930 \times 10^{-7} \text{ m}^2$$

$$A_2 = \frac{\pi d_2^2}{4} = \frac{3.529 \times 10^{-4}}{1204} = 2.930 \times 10^{-7} \text{ m}^2$$

hence $d_1 = d_2 = 0.0006108$ m $= 0.6108$ mm.

By comparing the following equations found above

$$\dot{m}_3 = 1.165 \, A_3 p_0^{\frac{1}{2}}$$

$$\dot{m}_4 = 1.165 \, A_4 p_c^{\frac{1}{2}}$$

with

$$\dot{m}_1 = \dot{m}_2 = \dot{m}_3 = \dot{m}_4 = 3.529 \times 10^{-4} \text{ kg/s}$$

where $p_c = p_0 = 3.103 \times 10^6$ N/m², there results

$$A_3 = A_4 = \frac{3.529 \times 10^{-4}}{1.165 \times \left(3.103 \times 10^6\right)^{\frac{1}{2}}} = 1.720 \times 10^{-7}\,\text{m}^2$$

Since

$$A_3 = \pi d_3 x = \pi d_3 \left(\frac{d_3}{8}\right)$$

then

$$d_3 = d_4 = \left(\frac{8A_3}{\pi}\right)^{\frac{1}{2}} = \left(\frac{8 \times 1.720 \times 10^{-7}}{3.1416}\right)^{\frac{1}{2}} = 0.0006617\,\text{m} = 0.6617\,\text{mm}$$

$$T = \frac{d_3}{4} + t = \frac{0.0006617}{4} + 0.0001016 = 0.0002670\,\text{m} = 0.2670\,\text{mm}$$

Now, we suppose the flapper to be deflected $0.03635\,\text{mm} = 3.635 \times 10^{-5}\,\text{m}$ below its neutral position, and the mass flow rates $\dot{m}_0$ and $\dot{m}_c$ respectively to and from the actuator to be $9.525 \times 10^{-5}\,\text{kg/s}$. We want to compute the difference of pressure across the actuator piston.

When the flapper is deflected $3.635 \times 10^{-5}\,\text{m}$ below its neutral position, the distance between the flapper and the upper nozzle is

$$x = \frac{T - t}{2} + 3.635 \times 10^{-5} = \frac{0.0002670 - 0.0001016}{2} + 0.00003635$$
$$= 0.0001191\,\text{m}$$

In this position of the flapper, the flow area of the upper nozzle is

$$A_3 = \pi d_3 x = 3.1416 \times 0.0006617 \times 0.0001191 = 2.475 \times 10^{-7}\,\text{m}^2$$

In these conditions, the mass flow rate $\dot{m}_3$ through the upper nozzle is equal to the mass flow rate $\dot{m}_1$ through the upper orifice plus the mass flow rate $\dot{m}_0 = 9.525 \times 10^{-5}\,\text{kg/s}$ through the upper duct of the actuator, as follows

$$\dot{m}_3 = \dot{m}_1 + \dot{m}_0$$

Since $\dot{m}_3$ has been found above to be

$$\dot{m}_3 = 1.165\,A_3 p_0^{\frac{1}{2}}$$

which, for $A_3 = 2.475 \times 10^{-7}\,\text{m}^2$, yields

$$\dot{m}_3 = 2.884 \times 10^{-7} p_0^{\frac{1}{2}}$$

and since $\dot{m}_1$ depends on the unknown value of p_0 as follows

$$\dot{m}_1 = C_d A_1 \left\{ \frac{2\gamma p_0 \rho_s}{\gamma - 1} \left[\left(\frac{p_0}{p_s} \right)^{\frac{2}{\gamma}} - \left(\frac{p_0}{p_s} \right)^{\frac{\gamma+1}{\gamma}} \right] \right\}^{\frac{1}{2}}$$

$$= 1.053 \times 10^{-6} \times \left\{ p_0 \left[\left(\frac{p_0}{3.447 \times 10^6} \right)^{\frac{2}{1.663}} - \left(\frac{p_0}{3.447 \times 10^6} \right)^{\frac{2.663}{1.663}} \right] \right\}^{\frac{1}{2}}$$

then the equation $\dot{m}_3 = \dot{m}_1 + \dot{m}_0$ becomes

$$2.884 \times 10^{-7} p_0^{\frac{1}{2}}$$

$$= 1.053 \times 10^{-6} \times \left\{ p_0 \left[\left(\frac{p_0}{3.447 \times 10^6} \right)^{\frac{2}{1.663}} - \left(\frac{p_0}{3.447 \times 10^6} \right)^{\frac{2.663}{1.663}} \right] \right\}^{\frac{1}{2}}$$

$$+ 9.525 \times 10^{-5}$$

The preceding equation can be solved numerically for p_0. For this purpose, we define the following function

$$f(p_0) \equiv 2.884 \times 10^{-7} p_0^{\frac{1}{2}} - 1.053 \times 10^{-6} \times \left\{ p_0 \left[\left(\frac{p_0}{3.447 \times 10^6} \right)^{\frac{2}{1.663}} \right. \right.$$

$$\left. \left. - \left(\frac{p_0}{3.447 \times 10^6} \right)^{\frac{2.663}{1.663}} \right] \right\}^{\frac{1}{2}} - 9.525 \times 10^{-5}$$

and search the value of p_0 for which $f(p_0) = 0$ within an interval $p_{01} \le p_0 \le p_{02}$ such that $f(p_{01}) f(p_{02}) < 0$. We choose $p_{01} = \frac{3}{4} p_s = 0.75 \times 3.447 \times 10^6 = 2.585 \times 10^6$ N/m^2 and $p_{02} = \frac{7}{8} p_s = 0.875 \times 3.447 \times 10^6 = 3.016 \times 10^6$ N/m^2.

We find $f(p_{01}) = -0.0001004$ and $f(p_{02}) = 0.00002130$, and therefore the condition $f(p_{01}) f(p_{02}) < 0$ is satisfied. The unknown value of p_0 can be found in the interval indicated above by using a numerical method, for example, the method described in Chap. 1, Sect. 1.2. We find $p_0 = 2.961 \times 10^6$ N/m^2 with four significant figures (for this value, $f(p_0) = 4.471 \times 10^{-8}$).

Likewise, when the flapper is deflected 3.635×10^{-5} m below its neutral position, the flow area A_4 is

$$A_4 = \pi d_4 (T - t - x) = 3.1416 \times 6.617 \times 10^{-4}$$

$$\times \left(2.670 \times 10^{-4} - 1.016 \times 10^{-4} - 1.191 \times 10^{-4} \right)$$

$$= 9.625 \times 10^{-8} \text{m}^2$$

In these conditions, the mass flow rate $\dot{m}_4$ through the lower nozzle is equal to the mass flow rate $\dot{m}_2$ through the lower orifice minus the mass flow rate $\dot{m}_c = 9.525 \times 10^{-5}$ kg/s through the lower duct of the actuator, as follows

$$\dot{m}_4 = \dot{m}_2 - \dot{m}_c$$

Since $\dot{m}_4$ has been found above to be

$$\dot{m}_4 = 1.165 \, A_4 p_c^{\frac{1}{2}}$$

which, for $A_4 = 9.625 \times 10^{-8}$ m^2, yields

$$\dot{m}_4 = 1.121 \times 10^{-7} p_c^{\frac{1}{2}}$$

and since $\dot{m}_2$ depends on the unknown value of p_c as follows

$$\dot{m}_2 = C_d A_2 \left\{ \frac{2\gamma p_c \rho_s}{\gamma - 1} \left[\left(\frac{p_c}{p_s} \right)^{\frac{2}{\gamma}} - \left(\frac{p_c}{p_s} \right)^{\frac{\gamma+1}{\gamma}} \right] \right\}^{\frac{1}{2}}$$

$$= 9.992 \times 10^{-7} \times \left\{ p_c \left[\left(\frac{p_c}{3.447 \times 10^6} \right)^{\frac{2}{1.663}} - \left(\frac{p_c}{3.447 \times 10^6} \right)^{\frac{2.663}{1.663}} \right] \right\}^{\frac{1}{2}}$$

then the equation $\dot{m}_4 = \dot{m}_2 - \dot{m}_c$ becomes

$$1.121 \times 10^{-7} p_c^{\frac{1}{2}}$$

$$= 1.053 \times 10^{-6} \times \left\{ p_c \left[\left(\frac{p_c}{3.447 \times 10^6} \right)^{\frac{2}{1.663}} - \left(\frac{p_c}{3.447 \times 10^6} \right)^{\frac{2.663}{1.663}} \right] \right\}^{\frac{1}{2}}$$

$$- 9.525 \times 10^{-5}$$

The preceding equation can be solved numerically for p_c. For this purpose, we define the following function

$$f(p_c) \equiv 1.121 \times 10^{-7} p_c^{\frac{1}{2}} - 1.053 \times 10^{-6} \times \left\{ p_c \left[\left(\frac{p_c}{3.447 \times 10^6} \right)^{\frac{2}{1.663}} \right. \right.$$

$$\left. \left. - \left(\frac{p_c}{3.447 \times 10^6} \right)^{\frac{2.663}{1.663}} \right] \right\}^{\frac{1}{2}} + 9.525 \times 10^{-5}$$

By operating as has been shown above, we find $p_c = 3.221 \times 10^6$ N/m^2 with four significant figures (for this value, $f(p_c) = 1.243 \times 10^{-7}$).

Therefore, the difference of pressure across the actuator piston is

$$p_c - p_0 = 3.221 \times 10^6 - 2.961 \times 10^6 = 0.26 \times 10^6 \text{ N/m}^2$$

5.13   Design of Pressure-Reducing Regulators

In general terms, a pressure regulator is a control valve which has no auxiliary source of power during operation. The pressure is controlled by varying the flow in the valve as a function of the sensed difference between the actual value and the desired value of pressure. Any unbalanced force resulting from this difference of pressure moves a metering element, which increases or decreases flow to reduce the pressure error to zero [25].

However, in the aerospace field, the name of pressure regulator indicates any device which maintains a desired value of upstream, downstream, or differential pressure by means of a pressure-reducing control element. For example, pressure regulators are used to maintain a constant pneumatic pressure in a tank for the storage of liquid propellants. In a pump-fed rocket engine, this constant pressure is used to protect the structure of the tank and to meet the requirements of pump suction head. A further protection is provided to the tank by a relief valve.

In control systems of large rocket engines, pressure regulators maintain the pressure of gas at a constant value to operate engine control valves, main propellant valves, gas-generator valves, and other control components [43].

In pneumatic control systems, variations of regulated pressure affect valve timing and sequencing in engine start and shutdown operations. In reaction-control systems, the source of energy for propellant feed is supplied by gas stored at high pressure [43].

Pressure regulators can be classified into the following categories:

- pressure-reducing regulators, which are used to reduce the pressure upstream of them to a desired downstream pressure, independently of variations of upstream pressure;
- back pressure regulators, which measure and regulate the pressure upstream of them independently of variations of downstream pressure; and
- differential pressure regulators, which maintain a constant difference of pressure across a restrictor of constant area, for example, across an orifice, to maintain a constant desired value of flow rate.

The pressure regulators described in the present section are only those of the first type. The regulators of the other types (relief valves and differential pressure regulators) will be described in the following sections.

Pressure-reducing regulators may be modulating or non-modulating. A regulator of the modulating type is a device which maintains a constant regulated pressure in a tank. A regulator of the non-modulating type is an on-off device which maintains the pressure in a tank between chosen limits by means of a signal from a pressure switch, which senses variations of pressure in the tank, and consequently opens or closes valves, which control the flow of pressurising gas to the tank. Modulating pressure-reducing regulators include, in turn, directly-operated, dome-loaded, and pilot-operated regulators.

A scheme of a directly-operated regulator is shown in the following figure, due to the courtesy of NASA [43].

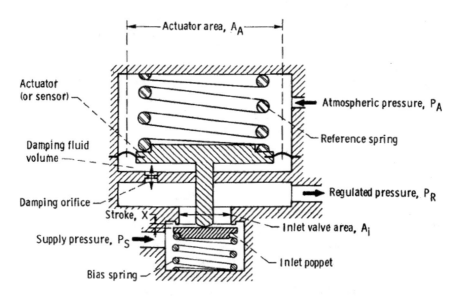

It consists of three principal elements: a sensing element (for example, a diaphragm), an inlet valve (comprising a movable element and a seat), and a source of reference load, which is usually a spring. The reference load due to the spring acts on the sensing element in one direction, and the force due to the regulated pressure acts in the opposite direction. The combination comprising the sensing element, the inlet valve, and the stem, which connects the sensing element with the inlet valve, is called the metering element. The force exerted by the reference spring on the sensing element is called the load. The sensing element is also subject to the actual regulated pressure and compresses the reference spring, which is used to establish the desired pressure setting. The net force acting on the inlet valve controls the flow through the regulator.

The movable element of the inlet valve may, or may not, be pressure-balanced. When the movable element is not pressure-balanced, then the inlet pressure acts on the main seat and changes the force equilibrium within the inlet valve. Consequently, a loss of inlet pressure across the main seat results in an increase in outlet pressure. When the movable element is pressure-balanced, then there is no variation of outlet pressure resulting from changes of inlet pressure. A regulator having a pressure-balanced movable element requires a dynamic seal to pressure-balance the main seat of the inlet valve.

The set point of a direct-acting regulator can be adjusted by means of a screw (not shown in the figure) which lifts or lowers a plate located above the reference spring, in order to extend or compress the spring. To a low value of the constant of the reference spring there corresponds a high value of the stroke of the movable element of the inlet valve which it is possible to achieve.

When a highly accurate regulation with small changes of outlet pressure is desired, then a large area of the sensing element is needed to achieve a high driving force. Therefore, a high force due to the reference spring is needed. Since a low value of the spring constant is desired, then it is necessary to use springs having a high number of coils. This, in turn, increases the size and the mass of the regulator.

The principal disadvantage of a spring-loaded regulator is that any change in the spring length during the regulation cycle changes the load and the set point of the regulator. This undesirable effect can be counteracted by using a spring having low value of spring constant, or by decreasing the stroke of the movable element [25].

A dome-loaded regulator is a regulator in which the source of reference load is a force obtained by pressurising the dome above the sensing element with gas. A scheme of a dome-loaded regulator is shown in the following figure, due to the courtesy of NASA [43].

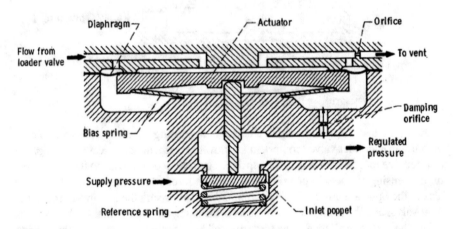

The source of load in a regulator of the dome-loaded type is not a spring. It is the reference pressure which a pilot gas exerts on the dome above the sensing element. The pilot gas, which may be the same as the gas whose pressure is to be regulated, is supplied to the dome through a small loader valve. The loader valve reduces the supply pressure to the desired reference pressure. The excess of pilot gas is directed to a vent side through an orifice. The poppet of the inlet valve is opened or closed by the pressure of the pilot gas and by the stroke of the diaphragm, which depends on the chosen pressure.

A pilot-operated regulator is a small flow control device operated by a small actuation force which controls indirectly a large flow requiring a large actuation force. A scheme of a pilot-operated regulator is shown in the following figure, due to the courtesy of NASA [43].

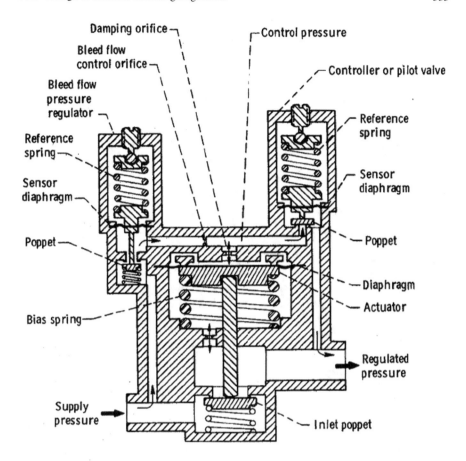

A non-modulating regulator consists principally of a pressure switch and an on-off solenoid valve. The pressure switch senses the pressure in a tank, and the on-off valve controls the flow of a pressurising gas to the tank. A pressure transducer with an amplifying circuit may be used instead of the pressure switch. When the pressure in the tank decreases below the minimum desired value, the electrical contacts of the pressure switch close, and an electrical signal opens the solenoid valve. This causes the pressurising gas to flow to the tank to increase the pressure in the tank. As the pressure in the tank approaches the desired maximum value, the contacts of the pressure switch open, the solenoid valve is not fed by current, and the flow of the pressurising gas stops. Therefore, the pressure in the tank oscillates between a minimum value and a maximum value. Since this regulator requires an external source of energy, then a failure of this source of energy or of any of its components can cause a failure of a system of pressurisation using the regulator [43].

Another example of non-modulating regulator is the device developed by the Frebank Company. This regulator is a single stage, pilot operated valve meant to control the pressure in the tank of a missile. The metering valve is closed until the pressure in the tank falls below the lower set point. When this happens, the metering valve of the regulator opens to full flow until the pressure in the tank reaches the upper set point, at which time the valve closes. A scheme of this valve is shown in the following figure, re-drawn from [25].

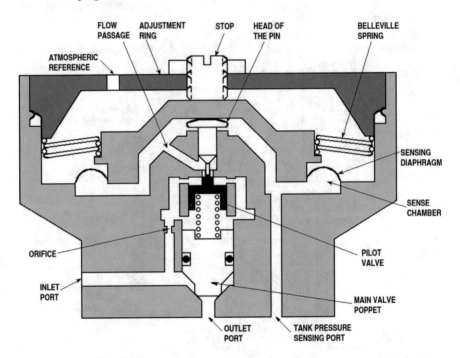

The dead-band of a non-modulating regulator is the pressure range between the upper set point and the lower set point. In the regulator shown in the preceding figure, the two set points are determined by the pre-deflection point and the post-deflection point, which limit the travel of the Belleville springs. The pre-deflection point is determined by the adjustment ring, and the post-deflection point is determined by the stop. The regulated pressure exerts a force on the sensing diaphragm through the sensing port. The force acting on the diaphragm is transmitted by the diaphragm assembly to the Belleville springs. In the absence of pressure, no force acts on the diaphragm. In this condition, the actuator assembly (comprising the diaphragm and the Belleville springs) holds the head of the pin against the pre-deflection travel stop, and causes the pilot valve to be fully open. The adjustment ring provides the pre-deflection adjustment. Gas at high pressure, introduced into the inlet port, flows through an orifice, through the open pilot valve, and into the sense chamber through the flow passage. When the pilot valve is open, the pressure difference across the orifice causes the pressure in the chamber behind the poppet of the main valve to decrease, thereby creating a pressure difference across the poppet. This pressure difference creates a force unbalance, and the poppet opens until it bottoms on the shank of the seat of the pilot valve. Now, the gas flows directly through the main valve into the ullage space. When the pressure in the tank reaches the value corresponding

to the pre-deflection setting of the Belleville springs, the actuator assembly snaps to
the post-deflection stop. Now, the pilot valve, which is no longer held open by the
actuator assembly, is closed by the spring between the poppets. When the pilot valve
is closed, the pressure increases in the chamber between the poppets, and reaches
a value where the poppet of the main valve closes. The pressure in the chamber
continues to increase, until it becomes equal to the inlet pressure, and then both
poppets are held on their seats by the pressure difference between the inlet and the
outlet, and also by the spring force of the poppet. When the actuator assembly is
at the post-deflection stop and both poppets are closed, the regulator is in lock-up.
In this condition, any demand on the system decreases the pressure in the ullage
space. As the pressure drops to a value of the post-deflection setting, the actuator
assembly snaps to the pre-deflection point. The pin moves the poppet of the pilot
valve to the full open position, and the chamber between the poppets evacuates to
create the pressure difference to cause the poppet of the main valve to open. This
cycle is repeated to hold the regulated pressure in the dead-band [25].

For the pressure regulators described above, the characteristic flow area A_c (m^2) of
the regulator, that is, the area of the valve in fully open position, can be determined as
a function of the required mass flow rate $\dot{m}$ (kg/s) and of the regulated outlet pressure
p_r (N/m^2) at the minimum allowable inlet pressure p_i (N/m^2) of the pressurising gas,
by using the equations of Sect. 5.11, which govern the flow of gases through orifices.
These equations are

$$\dot{m} = C_d A_c \left\{ \frac{2\gamma p_i \rho_i}{\gamma - 1} \left[\left(\frac{p_r}{p_i}\right)^{\frac{2}{\gamma}} - \left(\frac{p_r}{p_i}\right)^{\frac{\gamma+1}{\gamma}} \right] \right\}^{\frac{1}{2}}$$

which holds under non-chocked flow conditions, and

$$\dot{m} = C_d A_c \left[\gamma p_i \rho_i \left(\frac{2}{\gamma + 1}\right)^{\frac{\gamma+1}{\gamma-1}} \right]^{\frac{1}{2}}$$

which holds under choked (that is, maximum) flow conditions. In the preceding
equations, C_d is the coefficient of discharge, and ρ_i (kg/m^3) is the density of the gas
upstream of the regulator. The design values of C_d range from 0.6 to 0.7 [3].

As an example of application, the following data are known for the gas (helium)
pressure regulator used in a rocket engine: design mass flow rate $\dot{m} = 0.02177$ kg/s,
minimum allowable inlet pressure $p_i = 1.689 \times 10^6$ N/m^2, inlet temperature $T = 572$ K at the minimum inlet pressure, required regulated outlet pressure $p_r = 1.158 \times 10^6$ N/m^2, and flow discharge coefficient $C_d = 0.65$ of the regulator.

It is required to calculate the characteristic area A_c of the pressure regulator.

From [40] we take the following data for helium in the inlet conditions: specific
heat ratio $\gamma = 1.666$, and density $\rho_i = 1.416$ kg/m^3.

As has been shown in Sect. 5.11, the critical pressure ratio of helium is

$$\frac{p_c}{p_i} = \left(\frac{2}{\gamma + 1}\right)^{\frac{\gamma}{\gamma-1}} = \left(\frac{2}{1.666 + 1}\right)^{\frac{1.666}{1.666-1}} = 0.4873$$

Since

$$\frac{p_r}{p_i} = \frac{1.158 \times 10^6}{1.689 \times 10^6} = 0.6856$$

then the value of the pressure ratio p_r/p_i is greater than the critical value (non-choked flow conditions).

After introducing $\dot{m} = 0.02177$ kg/s, $C_d = 0.65$, $\gamma = 1.666$, $p_i = 1.689 \times 10^6$ N/m², $\rho_i = 1.416$ kg/m³, and $p_r/p_i = 0.6856$ in the equation written above of gas flow through orifices and solving for A_c, we find

$$A_c = \frac{0.02177}{0.65 \times \left[\frac{2 \times 1.666 \times 1.689 \times 10^6 \times 1.416}{1.666-1} \times \left(0.6856^{\frac{2}{1.666}} - 0.6856^{\frac{1.666+1}{1.666}}\right)\right]^{\frac{1}{2}}}$$

$$= 3.245 \times 10^{-5} \text{m}^2$$

5.14 Design of Differential Pressure Regulators

A differential pressure regulator is used to measure the volume flow rate q (m³/s) of a liquid in a line, and to maintain it at a desired constant value.

The measurement can be performed by using a variable head flow-meter (for example, an orifice, or a nozzle, or a Venturi tube) installed in the main line. The flow-meter has static pressure taps placed at two points along the fluid stream, as has been shown in Sect. 5.7, for the Venturi tube. In Sect. 5.7, it has also been shown how the volume flow rate depends on the pressure drop. When the pressure drop across a flow-meter is kept constant, then the volume flow rate will also be constant. The pressure drop is kept constant by using a pressure-reducing regulator. The static pressures are impressed to the fluid on opposite sides of the sensing element of the regulator. These pressures can be adjusted by means of a spring placed on one side of the sensing element. The force provided by the spring can be regulated so as to balance the lower value of static pressure against the upper value. The working principle of a direct-acting differential pressure regulator, designed by the W.A. Kates Company, is illustrated in the following figure, re-drawn from [44].

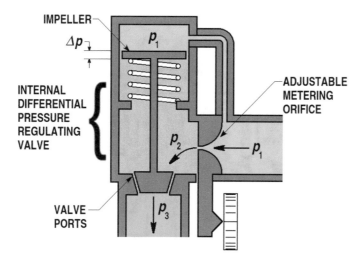

In this regulator, the internal differential pressure regulating valve maintains a constant pressure drop across the adjustable metering orifice to provide the set flow rate, independently of pressure variations which may occur upstream or downstream of the device. The supply pressure p_1 upstream of the valve is balanced by the pressure p_2 within the valve plus the force provided by the spring. When the pressure p_1 upstream of the regulator increases above the desired pressure drop, then the instantaneous imbalance of pressure moves the impeller downward. This movement restricts the valve ports and therefore increases the orifice back-pressure p_2, so as to restore the difference of pressure and the flow rate at the previous settings. The valve works oppositely in case of an increase in the outlet pressure p_3. A direct-acting differential pressure regulator, designed by the W.A. Kates Company, is shown in the following figure, re-drawn from [25].

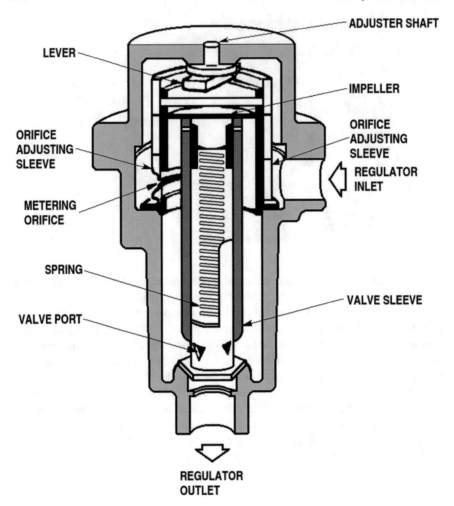

ADJUSTER SHAFT

LEVER

IMPELLER

ORIFICE
ADJUSTING
SLEEVE

ORIFICE
ADJUSTING
SLEEVE

REGULATOR
INLET

METERING
ORIFICE

SPRING

VALVE SLEEVE

VALVE PORT

REGULATOR
OUTLET

5.15 Design of Relief Valves

Relief valves are devices which measure and regulate the pressure of a fluid in a tank
by discharging fluid above the permitted limits to another tank, where the pressure is
lower. They open automatically when a pre-set level of pressure is reached in the tank
under control, and are used primarily in hydraulic systems. Safety valves are particular relief valves which control the pressure in pneumatic systems by discharging
the excessive gas or vapour to the atmosphere.

In aerospace airborne systems, relief valves are used for both liquid and gaseous
substances, and differ one from another depending on the fluids for which they are

used. For example, there are relief valves for liquid oxygen, and relief valves for gaseous helium. The fluid in excess is discharged overboard.

An excess of pressure can occur in a tank due to thermal changes or to leaking valves. Relief valves differ from other valves, because their operation creates a high pressure drop during flow [43].

A relief valve consists of a valve body, a source of reference load (such as a spring), and a control element (such as a poppet) which matches with a seat. The following figure, due to the courtesy of NASA [43], shows a modulating (left) and a non-modulating (right) directly-operated relief valve.

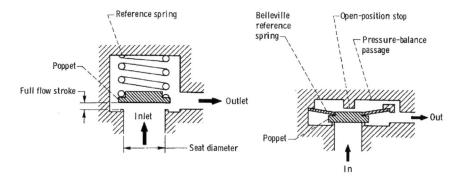

In a direct-operating relief valve, the reference load due to the spring is applied at all times to the control element. The reference load acts on the control element against the pressure in the tank or in the circuit under control. The intensity of the load determines the set point of the relief pressure. When the pressure of the fluid in the tank rises to the level necessary to balance the reference load, then the control element moves from the seat so as to discharge the fluid in excess to a reservoir of lower pressure placed downstream of the valve. Likewise, when the pressure of the fluid in the tank decreases below the set point, then the reference load acting against the pressure closes the valve.

A relief valve has good operating characteristics when the pressure p_f for rated flow and the pressure p_r for reseat closely approach the cracking pressure p_c. The cracking pressure is the relief setting of the valve, that is, the pressure level at which the leakage flow reaches some specified value. The cracking pressure is always less than the allowable working pressure of the tank, and is usually less than 110% of the normal operating pressure. The rated flow is usually established for pressures 10% greater than the pressure setting of the relief valve. The reseat pressure is less than the cracking pressure, by an amount depending on the configuration of the closure (control element and seat). A reseat pressure of 95% of the cracking pressure is a common value [25].

A relief valve may be not only directly-operated, as shown in the preceding figure, but also inversely-operated or pilot-operated. The following figure, due to the courtesy of NASA [43], shows an inversely-operated (left) and a pilot-operated (right) relief valve.

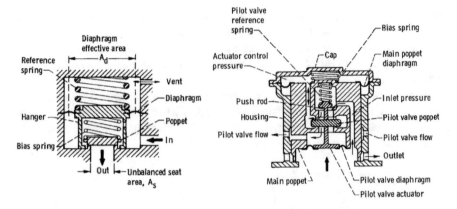

In an inversely-operated relief valve, the reference load holding the poppet on the seat increases with increasing pressure. The diaphragm senses the increasing pressure in the tank and, at some value of pressure, strokes the hanger which comes in contact with the poppet and reduces the closing force applied to the poppet. When the forces applied to the poppet are equal, then the valve is at the cracking pressure. The bias spring applies a load to the poppet to place it in the initial position at low-pressure sealing [43]. It is desirable that the force which holds the poppet on the seat at a specified pressure p_t (where p_t is less than the cracking pressure p_c) should be as great as possible. For this purpose, the effective area A_d of the diaphragm is larger than the unbalanced seat area A_s, as follows

$$A_d \approx \frac{p_c}{p_c - p_t} A_s$$

The effective diameter of the diaphragm becomes increasingly larger than the diameter of the seat when the maximum closing pressure approaches the cracking pressure.

A pilot-operated relief valve is used when it is necessary to control large flows. When this valve is in the non-relieving conditions, the control pressure of the actuator is at the level of the inlet pressure, and the resulting force holds the main poppet in the position of closure. When the inlet pressure increases so as to reach the cracking value, the pilot valve partially strokes, the control pressure of the actuator is ported to the outlet cavity, the pressure in the actuator control-pressure is reduced so much that the resulting force is zero, and the main poppet is ready to open. As the inlet pressure continues to increase, the pilot valve stroke increases, the pressure in the control-pressure cavity decreases, and the resulting force opens the main poppet [43].

In airborne applications, lightweight materials, such as aluminium, are used for the valve body. The source of reference load is a force acting against an increase in pressure on the control element until the set point of pressure relief is reached. The most common element used for this purpose is a compressed spring.

The closure unit comprises a seat and a control element. The seat may have a flat, or spherical, or conical shape. The control element may be a ball, or a poppet (which in turn may be conical or V-shaped), or a piston.

Another type of pilot-operated relief valve, used to protect propellant tanks from overpressure, has been described by Huzel and Huang [3]. A scheme of this valve is illustrated in the following figure, re-drawn from [3]. Both the main valve and the pilot valve are normally held closed by the spring forces F_{sm} and F_{s3} and by the control pressure p_c. The controller of the pilot valve senses the pressure p_t in the tank. When the pressure in the tank reaches or exceeds the preset level, the pilot valve is actuated to crack. This vents the actuator control pressure p_c of the main relief valve and in turn permits the main valve to open. The position of the poppet of the main valve depends on the control pressure p_c, which in turn is controlled by the position of the poppet of the pilot valve and by the pressure p_t in the tank.

With reference to the following figure, it is required to determine the dimensions, in terms of diameters, for the main valve and for the pilot valve. The following data are known: temperature of the pressurising gas (helium) $T = 389$ K, set point for the absolute pressure of relief in the tank $p_t = 1.138 \times 10^6$ N/m², required maximum mass flow rate $\dot{m} = 1.361$ kg/s, discharge coefficient of the main valve $C_d = 0.75$, estimated leakage past the actuator piston seal of the main valve $\dot{m}_s = 0.001361$ kg/s, diameter of the control orifice $d_o = 2.032$ mm $= 0.002032$ m, discharge coefficient for the control orifice and for the pilot valve $C_{do} = 0.60$, and maximum value of the absolute ambient pressure $p_a = 1.031 \times 10^5$ N/m² (atmospheric pressure).

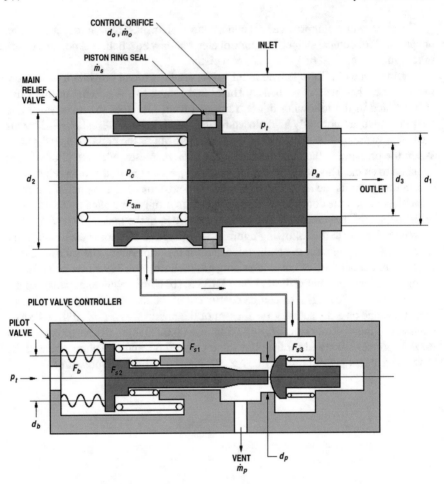

From [40] we take the following data for helium at $T = 389$ K and $p_t = 1.138 \times 10^6$ N/m^2: specific heat ratio $\gamma = 1.666$, and density $\rho_t = 1.403$ kg/m^3.

As has been shown in Sect. 5.11, the critical pressure ratio of helium is

$$\frac{p_c}{p_t} = \left(\frac{2}{\gamma + 1}\right)^{\frac{\gamma}{\gamma - 1}} = \left(\frac{2}{1.666 + 1}\right)^{\frac{1.666}{1.666 - 1}} = 0.4873$$

Since

$$\frac{p_a}{p_t} = \frac{0.1031 \times 10^6}{1.138 \times 10^6} = 0.09060$$

then the value of the pressure ratio p_a/p_t is less than the critical value (choked flow conditions). Therefore, the following equation holds for the mass flow rate $\dot{m}$ through the main valve

$$\dot{m} = C_d \frac{\pi d_3^2}{4} \left[\gamma p_t \rho_t \left(\frac{2}{\gamma + 1} \right)^{\frac{\gamma+1}{\gamma-1}} \right]^{\frac{1}{2}}$$

where C_d is the coefficient of discharge, d_3 (m) is the diameter of the orifice, and p_t (N/m^2) and ρ_t (kg/m^3) are respectively the pressure and the density of the gas upstream of the orifice of the main valve.

After substituting $\dot{m}_o = 1.361$ kg/s, $C_d = 0.75$, $\gamma = 1.666$, $p_t = 1.138 \times 10^6$ N/m^2, and $\rho_t = 1.403$ kg/m^3 in the preceding equation and solving for d_3, we find

$$d_3 = 2 \times \left(\frac{1.361}{3.1416 \times 0.75} \right)^{\frac{1}{2}}$$

$$\times \left[1.666 \times 1.138 \times 10^6 \times 1.403 \times \left(\frac{2}{1.666 + 1} \right)^{\frac{1.666+1}{1.666-1}} \right]^{-\frac{1}{4}}$$

$$= 0.05018 \text{ m} \approx 5 \text{ cm}$$

The displacement x_{mo} (m) of the movable element of the main valve from the closed position to the fully opened position results from the following equation

$$A_3 = \frac{\pi d_3^2}{4} = \pi d_3 x_{mo}$$

Hence

$$x_{mo} = \frac{d_3}{4} = \frac{0.05018}{4} = 0.01255 \text{ m} = 1.255 \text{ cm}$$

The general equation which expresses the equilibrium of the forces acting in the main valve is

$$F_{sm} + p_c A_2 - p_t (A_2 - A_1) - p_a A_3 = F_{seat}$$

where F_{sm} (N) is the pre-load of the spring of the main valve, p_c (N/m^2) is the control pressure, p_t (N/m^2) is the pressure in the tank, $A_2 = \frac{1}{4}\pi d_2^2$ (m^2), $A_1 = \frac{1}{4}\pi d_1^2$ (m^2), $p_a = 1.031 \times 10^5$ N/m^2 is the atmospheric pressure, $A_3 = \frac{1}{4}\pi d_3^2$ (m^2), and F_{seat} (N) is the seating force of the main valve.

In the cracking conditions, the general equation becomes

$$F_{sm} + p_{cc} A_2 - p_{tc} (A_2 - A_1) - p_a A_3 = 0$$

where p_{cc} (N/m^2) is the value of the control pressure p_c in the cracking conditions and p_{tc} (N/m^2) is the value of the tank pressure p_t in the cracking conditions.

For any displacement x_m (m) of the movable element of the main valve from the closed position (such that $x_m < x_{mo}$), the general equation becomes

$$F_{sm} + k_{sm}x_m + p_c A_2 - p_t(A_2 - A_1) - p_a A_3 = 0$$

where k_{sm} (N/m) is the constant of the spring of the main valve.

When the main valve is fully open ($x_m = x_{mo}$), the general equation becomes

$$F_{sm} + k_{sm}x_{mo} + p_{co}A_2 - p_{to}(A_2 - A_1) - p_a A_3 = 0$$

where p_{co} (N/m^2) and p_{to} (N/m^2) are respectively the control pressure p_c and the tank pressure p_t in the fully open conditions.

When the main valve is at the start to reseat ($x_m = x_{mo}$), the general equation becomes

$$F_{sm} + k_{sm}x_{mo} + p_{cr}A_2 - p_{ti}(A_2 - A_1) - p_a A_3 = 0$$

where p_{cr} (N/m^2) and p_{tr} (N/m^2) are respectively the control pressure p_c and the tank pressure p_t in the reseat conditions.

When the main valve is fully reseated ($x_m = 0$), the general equation becomes

$$F_{sm} + p_{cr}A_2 - p_{tr}(A_2 - A_1) - p_a A_3 = 0$$

The mass flow rate $\dot{m}_p$ (kg/s) of the pilot valve must be greater than to the mass flow rate $\dot{m}_o$ (kg/s) through the control orifice of the main valve summed to the leakage $\dot{m}_s$ (kg/s) past the actuator piston seal of the main valve, in order for the control pressure p_c to be sufficiently vented.

The mass flow rate through the control orifice of the main valve results from

$$\dot{m}_o = C_{do}\frac{\pi d_o^2}{4}\left[\gamma p_t \rho_t \left(\frac{2}{\gamma + 1}\right)^{\frac{\gamma+1}{\gamma-1}}\right]^{\frac{1}{2}}$$

After substituting $C_{do} = 0.60$, $d_o = 0.002032$ m, $\gamma = 1.666$, $p_t = 1.138 \times 10^6$ N/m^2, and $\rho_t = 1.403$ kg/m^3 in the preceding equation, we find

$$\dot{m}_o = 0.60 \times \frac{3.1416 \times 0.002032^2}{4} \times \left[1.666 \times 1.138 \times 10^6 \times 1.403\right.$$

$$\left. \times \left(\frac{2}{1.666 + 1}\right)^{\frac{1.666+1}{1.666-1}}\right]^{\frac{1}{2}} = 0.001785 \text{ kg/s}$$

The total mass flow rate through the control cavity of the pilot valve results from

$$\dot{m}_o + \dot{m}_s = 0.001785 + 0.001361 = 0.003146 \text{ kg/s}$$

We set the mass flow rate $\dot{m}_p$ of the pilot valve equal to the preceding value multiplied by 1.5, as follows

$$\dot{m}_p = 1.5 \times 0.003146 = 0.004719 \text{ kg/s}$$

In order for the flow through the restrictions to be maximum, the pressure ratio p_c/p_t of the control pressure to the tank pressure should be at least equal to the critical value

$$\frac{p_c}{p_t} = \left(\frac{2}{\gamma + 1}\right)^{\frac{\gamma}{\gamma - 1}}$$

After substituting $p_t = 1.138 \times 10^6$ N/m^2 and $\gamma = 1.666$ in the preceding equation and solving for p_c, we find the maximum allowable value of the control pressure to be

$$p_c = 1.138 \times 10^6 \times \left(\frac{2}{1.666 + 1}\right)^{\frac{1.666}{1.666-1}} = 5.545 \times 10^5 \text{ N/m}^2$$

We substitute the value found above for p_c, $\dot{m}_p = 0.004719$ kg/s, $C_{do} = 0.60$, $\gamma = 1.666$, and $\rho_t = 1.403$ kg/m^3 in the following equation

$$\dot{m}_p = C_{do} \frac{\pi d_p^2}{4} \left[\gamma p_c \rho_t \left(\frac{2}{\gamma + 1}\right)^{\frac{\gamma+1}{\gamma-1}}\right]^{\frac{1}{2}}$$

and solve this equation for the unknown value of the diameter d_p of the port in the pilot valve. By so doing, we find

$$d_p = 2 \times \left(\frac{0.004719}{3.1416 \times 0.60}\right)^{\frac{1}{2}}$$

$$\times \left[1.666 \times 5.545 \times 10^5 \times 1.403 \times \left(\frac{2}{1.666 + 1}\right)^{\frac{1.666+1}{1.666-1}}\right]^{-\frac{1}{4}} = 0.003954 \text{ m}$$

The displacement x_{po} (m) of the movable element of the pilot valve from the closed position to the fully opened position results from the following equation

$$A_p = \frac{\pi d_p^2}{4} = \pi d_p x_{po}$$

Hence

$$x_{po} = \frac{d_p}{4} = \frac{0.003954}{4} = 0.0009886 \text{ m} \approx 1 \text{ mm}$$

The equation which expresses the equilibrium of the forces acting on the poppet of the pilot valve is

$$(p_c - p_a)A_p + F_{s3} = F_p$$

where $p_c = 5.545 \times 10^5$ N/m^2 is the control pressure, $p_a = 1.031 \times 10^5$ N/m^2 is the ambient (atmospheric) pressure, $A_p = \frac{1}{4}\pi d_p^2$ (m^2) is the area of the port in the pilot valve, $d_p = 0.003954$ m is the diameter of the port in the pilot valve, F_{s3} (N) is the force due to the spring acting on the poppet, and F_p (N) is the seating force of the poppet.

The equation which expresses the equilibrium of the forces acting on the actuator of the pilot valve is

$$(p_t - p_a)A_b - F_b - F_{s1} - F_{s2} = F_a$$

where $p_t = 1.138 \times 10^6$ N/m^2 is the set point for the absolute pressure of relief in the tank, $p_a = 1.031 \times 10^5$ N/m^2 is the ambient (atmospheric) pressure, $A_b = \frac{1}{4}\pi d_b^2$ (m^2) is the area of the cross section of the sensor bellows, F_b (N) is the force due to the sensor bellows, F_{s1} (N) and F_{s2} (N) are the forces due to the springs acting on the sensor bellows, and F_a (N) is the actuating force of the pilot valve.

The pilot valve starts to open when $F_a > F_p$.

5.16   Design of Check Valves

A check valve is a device which prevents a moving fluid from reversing its direction of motion. A check valve allows a fluid to move freely in one direction, and stops the motion in case of a pressure reversal. A pressure reversal may occur in a fluid either normally or as a result of a failure. In the latter case, a flow reversal must be promptly stopped, in order to avoid an overflow or an overpressure in a tank, or an unwanted combination of reactive fluids, or other damages to components of a rocket engine.

A check valve operates automatically, because its movable element is actuated by the forces exerted by the moving fluid. In case of aerospace vehicles, the movable element of a check valve is spring-loaded, in order for the valve to be operated independently of either the gravitational force or the attitude of the vehicle. A check valve requires neither an actuation signal nor a source of power for its operation. Howell and Weathers [25] have identified some criteria to be considered in the design and in the choice of a particular type of check valve. These criteria are briefly indicated below.

The first of them concerns the pressure drop across a check valve. In valves used for aerospace applications, it is desirable to keep the increase in pressure drop with increasing volume flow rate as low as possible. The increase in pressure drop in a valve depends on the type (sphere or cone frustum or poppet having another shape) of its closing element.

Another criterion concerns the type of seal used in a check valve. This is because the sealing properties of a check valve depend on the sealing material, on the initial load of the spring, and on the difference of pressure across the closing element of the valve. In case of a pressure reversal, a check valve leaks until the back pressure acting on it becomes as high as to provide a sufficient seating force. In order for the valve not to leak when the reversal pressure is null or very low, the initial load of the spring must be high, which fact implies a high value of pressure drop when the controlled fluid moves in its normal direction. The use of elastic materials for the seat makes it possible to reduce leakage at low values of the seating load.

Another criterion concerns the cracking and the reseating pressures of a check valve. As has been shown in Sect. 5.15, the cracking pressure of a valve is the minimum pressure which assures a given value of flow rate. The reseating pressure of a valve is the reverse differential pressure needed to keep leakage at or below a given value. A proper choice of the spring loading and of the seating properties can reduce the reseating pressure. The cracking and the reseating pressures of a check valve depend on its pressure drop and on its leakage properties.

Another criterion concerns the type of fluid moving through a check valve. The materials (including metals, seals, and coatings) of which a valve is made must be compatible with the fluid controlled by the valve. In particular, the choice of the materials depends on whether the fluid is a liquid or a gas. In the latter case, special care must be taken with gases of low molar mass, such as hydrogen and helium.

Another criterion concerns the operating pressure of a check valve. The body of a valve must be designed for the maximum pressure to which the valve is subject. In particular, when a valve is rapidly closed, a water hammer can arise in the valve. This phenomenon increases the level of pressure several times the normal operating pressure. When a valve is subject to a large number of pressure cycles, a failure due to fatigue must be taken into account.

Another criterion concerns the temperature at which the fluid and the valve operate. Special care must be taken of the non-metallic materials which are used for the seals. All materials used in a valve must be designed to withstand the worst thermal conditions in their operating environment. Since the mass of the valve acts as a heat sink, short-term excursions of temperature beyond the recommended limits are tolerable. Particular attention should be given to temperature gradients, which can give rise to binding effects due to the different coefficients of thermal expansion of the materials.

Another criterion concerns the sensitivity of either the fluid system or the fluid itself to contamination. Contamination can cause malfunction or internal leakage in a valve. Special care must be taken of the valve body and of the seals.

Another criterion concerns the operations of maintenance required by a valve. The maintenance requirements should be reduced as far as possible, and the need to use special tools should be avoided. Seals requiring periodic replacement should be readily available and easily accessible. Care must be taken to avoid contamination during disassembly and service.

Another criterion concerns the mass and the size of a valve, which are aspects of primary importance in valves used for aerospace applications. For example, in a check valve of the ball type, the mass of the ball is proportional the cube of its radius; by contrast, in a check valve of the flapper type, the mass of the closing element is nearly proportional to the square of its linear size. Therefore, check valves of the ball type are rarely used when the radius of the ball exceeds 1 cm. In this case, flapper-type or swing-type check valves are used.

Another aspect concerns the cost of a check valve. The cost depends on such factors as operating pressure, temperature, flow rate, leakage requirements, type of fluid, mass, reliability, and life-cycle requirements. In case of check valves used for aerospace applications, performance and reliability are factors of primary importance.

Another criterion concerns the operating life of a check valve. The operating life depends on such factors as type of fluid, contamination, operating temperature, stresses, number of cycles, type of seals, materials of construction, and procedures of maintenance. In order to increase the operating life of a valve, loads and stresses acting on the valve components should be reduced as far as possible. However, valves to be used for aerospace applications should not be over-designed for their life-cycle requirements.

According to Tomlinson and Keller [43], check valves can be classified in two categories: poppet valves and flapper valves. Three different valves (ball, cone, and floating ring) of the poppet type are shown in the following figure, due to the courtesy of NASA [43].

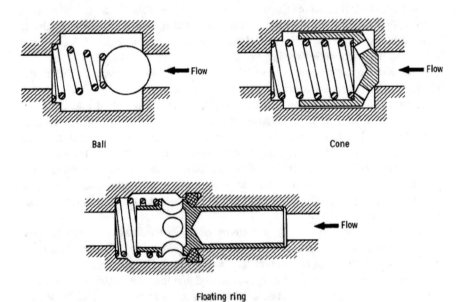

A check valve of the poppet type has a poppet as its closing element. This poppet translates axially in the valve body and a compressed spring of the coil type forces the poppet against the seat. Valves of this type are used in applications requiring low leakage in the direction of the checked flow and low pressure drop in the direction of the free flow. The cracking pressure is low (in the range going from 21000 to 103000 N/m^2, according to [43], and the full-flow pressure is as close to the cracking pressure as possible. The cracking pressure of check valves of the poppet type can be increased by increasing the load of the return spring. In these conditions, a check valve can be used as a relief valve or a back-pressure regulator [43]. A check valve of the ball type is illustrated in the preceding figure. This valve has a hard ball as its closing element. The ball is spring-loaded against a circular, conical, or spherical seat. In order to reduce leakage, a soft seal is sometimes used. This soft seat is made of Teflon®, or of another plastic or elastomeric material. Otherwise, a combination of soft with hard materials is used for the seat. The force due to the fluid stream lifts the ball off the seat against the load provided by the spring, and the flow passes around the ball. Since the flow must surround the ball, then a check valve of the ball type has a tendency toward turbulence and a higher pressure drop than other types of check valves. In normal operating conditions, the ball may rotate slightly on the return spring, thereby causing wear on the ball and on the seat. A check valve of the ball type is prone to chatter when opened by a low flow which does not fully stroke and hold the ball in the full-open position. By chatter we mean an uncontrolled seating and reseating of the closing element of a valve. Check valves of the ball type are simple and cheap to manufacture, since the seat is the only costly detail. They are used in low-flow applications where the stroke is short and the ball will self-guide into the seat [43].

A check valve of the cone type is an improvement of the valve of the ball type described above. The improvement is obtained by guiding and aligning the closing element in a constant direction, in order for its contact with the seat to be continuous and even. For this purpose, the ball is replaced by a piston sliding in a hollow cylinder and having a conical seating surface at one end. This surface is spring-loaded to seat against a spherical or conical surface, which may be made of either a soft material or a hard material. A check valve of the cone type has a lower pressure drop than one of the ball type having the same size, and a lower tendency to chatter, because of the guided movement and the resultant damping of its closing element. The presence of contaminating particles in the seating area and between the piston and the body of the valve can cause sticking or cocking with consequent leakage between the piston and the seating area.

A check valve of the floating-ring type is also a valve having a closing element guided and aligned in a constant direction. It has a mushroom-shaped poppet, whose stem slides in the valve body and whose head seals through an O-ring against a circular flat or tapered seat. The force due to the fluid stream causes the head of the poppet to move off the seat in the flow direction, and the fluid passes through the stem of the poppet, around the head of the poppet, and through the body of the valve. A check valve of this type has a lower pressure drop at an equal flow rate than a ball or a cone check valve. Damping chambers may be incorporated in a floating-ring valve in order to eliminate chatter and hammering. The presence of contaminating particles between the poppet stem and the valve body can cause sticking and leakage. A check valve of the floating-ring type has more components and therefore is more costly than a ball or a cone check valve.

Two check valves of the flapper type, namely, a split-flapper (left) and a swing-flapper (right) valve, are shown in the following figure, due to the courtesy of NASA [43].

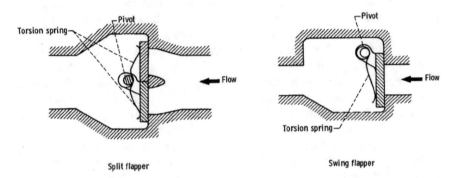

Split flapper   Swing flapper

A check valve of the flapper type has a rotating flapper as its closing element. The loaded spring acting on the flapper is of the torsion type, and the flapper may be either a split element which pivots across the centre of the flow path or a single element which pivots from the side.

Flapper check valves are used in applications requiring low pressure drop with large flow rates. They are often smaller and lighter than those of the poppet type. However, when subject to high pressures, flapper check valves may become excessively heavy [43].

A swing-flapper check valve has a hinged disc which seats against a flat or tapered surface, or a circular sharp edge. A fluid stream moving through the valve in the free-flow direction swings the hinged disc from the seat and out of the flow path, thereby permitting a substantially unrestricted passage with very low pressure drop. When the flow through the valve stops or reverses its direction, the disc swings rapidly to its seat, thereby closing the valve against the reverse flow.

Very low pressure drops con be obtained with swing-flapper check valves, because small opening forces are required, and also because their disc swings out of the flow path. The presence of contaminating particles in the fluid has little effect on their performance. On the other hand, they are particularly subject to water hammer in case of sudden reversal of flow. This is because their disc rotates through a large angle from the open position to the closed position, and consequently the velocity of the reverse flow increases significantly before the complete closure of the disc. For this reason, their performance in eliminating a reverse flow is poor. Check valves of the swing-flapper type are used for applications requiring the minimum value of pressure drop and scarce control of fluid contamination. They are not used for applications having a high tendency toward sudden flow reversal.

Check valves of the split-flapper type have two or more hinged elements instead of one hinged disc. This makes it possible to reduce the moment of inertia of each hinged element. A typical split-flapper valve has two semi-circular elements which are open when the fluid stream moves through the valve in the free-flow direction and closed otherwise. A pressure drop in a split-flapper valve can be reduced to a minimum value, as is also the case with a swing-flapper valve. For two valves of the same size, the flapper is lighter in a split-flapper valve than is in a swing-flapper valve. In addition, a split-flapper valve can bear a heavier life cycle and has a better performance in case of water hammer than a valve of the other type. Split-flapper valves are used principally in pneumatic systems. Their principal advantage is the minimum pressure drop for a given diameter. They are also scarcely sensitive to fluid contamination and have better performance than swing-flapper valves when a sudden flow reversal may occur.

5.17 Design of Burst Discs

A burst disc, also known as a safety disc or a rupture disc or a frangible disc, is a thin metallic diaphragm designed to burst when the pressure in a tank or in a line exceeds a pre-set value. For example, a burst disc is used in a flow-carrying line to initiate flow when ruptured by sufficient pressure in the line. A burst disc commonly used is one of the pre-bulged type, which is shown in the following figure, due to the courtesy of NASA [43].

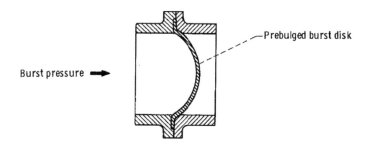

The burst disc shown in the preceding figure bursts due to tension induced by pressure, without the use of local weakening such as coined or machined grooves to initiate and control the bursting.

Another type of burst disc commonly used is the reverse-action knife-blade burst disc, shown in the following figure, due to the courtesy of NASA [43].

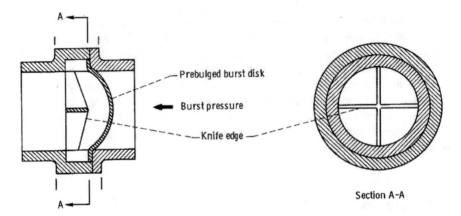

Section A-A

The burst disc illustrated in the preceding figure reverses its form under pressure and buckles onto a knife edge, thus requiring a lower pressure to burst than one of the pre-bulged type.

A burst disc of the shear type is shown in the following figure, due to the courtesy of NASA [43].

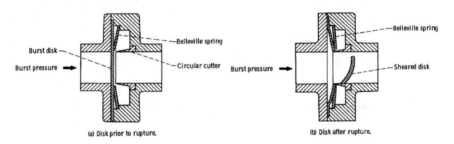

(a) Disk prior to rupture.   (b) Disk after rupture.

The burst disc shown above, before rupture (left) and after rupture (right), is pressure-loaded on a circular cutter by means of a Belleville spring, which is compressed into the negative-spring-rate region of the spring stroke. The Belleville spring supports a portion of the pressure acting on the burst disc up to a pre-set burst pressure. When this value of pressure is exceeded, the spring washer starts to stroke, its supporting force decreases rapidly, and the pressure causes the disc to be sheared by the sharp circular cutter. A catch screen may be required to keep the cut disc from moving downstream. The burst pressure of the disc can be pre-set by replacing the cutter with a flat plug and adjusting the Belleville spring washer until the desired burst pressure is indicated by the motion of the burst disc [43].

Burst discs are used for many components of rocket engines, such as hypergolic-start cartridges, repeat-start turbine spinner assemblies, pump-seal drain lines, and some instrumentation. They are also used to seal a cavity against downstream pressure or temperature, or to contain a liquid until a rise of pressure to a desired level ruptures the disc.

The burst pressure is the primary element to be considered in the design of a burst disc. This is because a burst disc must remain intact in all conditions of storage, handling, and use of storable or cryogenic fluids, until the desired burst pressure is reached. The value of the burst pressure usually ranges from 6.895×10^3 to 6.895×10^7 N/m^2 [43], depending on the application.

Low values of burst pressure are desired for a cartridge containing an ignition aid for the thrust chamber of a rocket engine, in order to burst the cartridge discs, release the ignition aid, and establish ignition as soon as possible. Of course, the burst pressure must not be so low as to cause premature rupture in the discs due to the vapour pressure or to the thermal expansion of the propellants, or to handling loads. High values of burst pressure are desired for a burst disc used for the ignition of a pyrotechnic device, so long as the burning rate of the charge does not become excessive and the pressure level remains below the allowable operating pressure of the container [43].

Burst discs are made of aluminium alloys, mild steel, stainless steel, nickel, Monel®, Inconel®, copper, silver, gold, and platinum. Aluminium is used because of its low physical properties and good forming and machining properties. However, since the properties of this material vary widely with temperature, the upper temperature limit is only about 394 K [43]. Mild steel requires protection from corrosion. This material has low cost and good formability. Stainless steel and Inconel® are difficult to tool. Inconel® can be used over a wide range of temperatures. Silver, gold, and platinum resist corrosion and have good physical properties.

Burst discs can be attached mechanically or by welding. They must not be corroded by the fluid, and the materials of which they are made must not cause decomposition in the fluid.

The thickness of burst discs ranges from 0.05 to 3.18 mm [43], in case of coined-groove discs, for the material remaining under the groove. In this type of disc, the material is weakened locally by grooves which are stamped into the metal. The following figure, due to the courtesy of NASA [43], shows three patterns (radial, single hinge, and double hinge) used for these grooves.

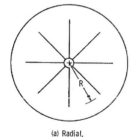

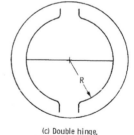

(a) Radial.   (b) Single hinge.   (c) Double hinge.

5.18   Design of Explosive Valves

Explosive valves, also known as squib valves, are shut-off devices actuated by a very compact source of energy only once in a given mission.

The source of energy used by them is an explosive charge which generates high pressure on an actuator. It is possible to actuate repeatedly an explosive valve by using multiple squibs. However, most explosive valves are designed to operate only once, and are either replaced or refurbished after use. Since the explosive charge is expendable, the remaining part of the valve is often designed to be also expendable. This makes it possible to reduce mass and obtain very good sealing.

These valves cannot be tested non-destructively. Therefore, their reliability for a specific mission must be evaluated statistically, by purchasing a large number of them for qualification.

Explosive valves are actuated by mechanical energy resulting from conversion of chemical energy contained in explosive charges. This chemical energy can be released by the charges either by deflagration, which is the rapid combustion of the explosive material, or by detonation, which is the propagation of a shock wave through the explosive material. In case of deflagration, the mechanical energy obtained is the work done by the combusted gases acting on a piston. In case of detonation, the mechanical energy obtained is due to the momentum of the shock wave propagating through the explosive charge. This momentum is transferred to a piston, as if the end of the piston were subject to a sharp blow.

In either case, a small amount of explosive material is necessary to actuate a valve. For example, a typical normally-closed valve, 19.05 mm in diameter, whose working pressure is 2.759×10^7 N/m^2, uses 140 mg of explosive material. The explosive pressure of this valve is about 1.724×10^7 N/m^8 [25].

A deflagrating charge used for explosive valves consists of small flake-like particles of powder similar to shotgun powder. A valve of typical design has a disc 0.254 mm in thickness and 1.27 mm in diameter. A detonating charge uses explosive materials such as PETN (pentaerythritol tetranitrate), or DDNP (diazodinitrophenol), or RDX (1,3,5-trinitroperhydro-1,3,5-triazine).

The explosion of a charge is obtained by a transfer of mechanical or electrical energy to a sensitive primer, such as lead styphnate. For this purpose, the charge is ignited by striking a sharp blow to the primer through a metallic barrier, like the firing pin of a gun, or by piercing abruptly the container of the primer by means of a sharp probe, or by heating electrically the primer by means of a bridge-wire.

The ignition of a detonating or deflagrating charge may also be obtained without the primer, by using an exploding bridge-wire (EBW). In this case, it is necessary to use a firing pulse of several thousand volts, in comparison with a conventional firing pulse of about 28 V or even less. An exploding bridge-wire is extremely insensitive to accidental firing, but requires a high voltage obtained by discharging a condenser contained in a firing circuit.

A typical explosive cartridge (designed by Holex, Inc.), commonly used for explosive valves and having a low-voltage bridge-wire, is shown in the following figure, re-drawn from [25].

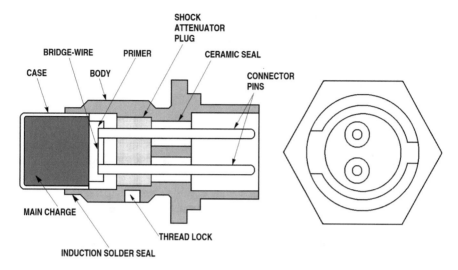

The charge in the cartridge shown above is ignited by means of an electrical connector incorporated in the cartridge. The charge is hermetically sealed to be protected from moisture and damage. The cartridge contains 188 mg of deflagrating charge, and the pressure generated by the ignition can be as high as 4.137×10^8 N/m^8. The shock attenuator plug incorporated in the cartridge weakens the shock waves before they reach the ceramic seals. The electrical conductors are two connector pins made of stainless steel, and the case is made of the same material. The pins are locked into the case by fused ceramic seals. The current necessary to ignite a cartridge like this ranges from 1 to 5 A, and the firing time ranges from 1 to 20 ms [25].

Some explosive valves are illustrated below. The following figure, due to the courtesy of NASA [43], shows a normally closed valve which opens when the explosive charge actuates a pin (ram) which shears the end from the inlet fitting and retains it in a recess of the body.

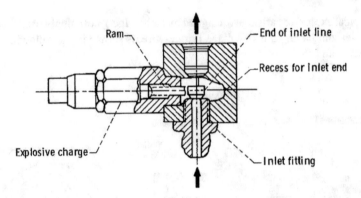

The valve shown above can be made reusable by replacing the explosive charge and the inlet fitting. A reusable valve must be cleaned immediately after firing.

The following figure, due to the courtesy of NASA [43], shows a normally closed non-reusable valve, as it appears after firing. In this valve, a diaphragm is sheared from the body by the ram and is clamped by the ram to be retained within the body.

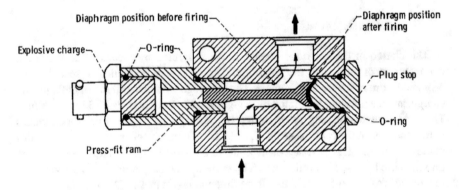

A normally open valve, as it appears before firing, is shown in the following figure, due to the courtesy of NASA [43]. In this valve, the explosive charge actuates the ram, which shears through the flow passage and wedges between the openings to close the flow passage.

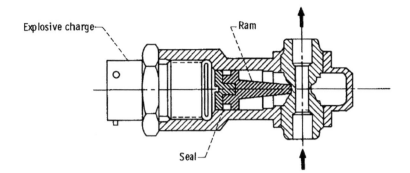

A dual function explosive valve (U.S. Patent No. 3,122,154) is shown in the following figure, due to the courtesy of NASA [43].

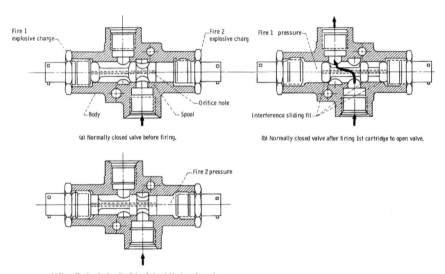

(a) Normally closed valve before firing.

(b) Normally closed valve after firing 1st cartridge to open valve.

(c) Normally closed valve after firing 2nd cartridge to reclose valve.

The valve shown above is normally closed (a), is opened (b) by firing the first explosive charge, and is re-opened (c) by firing the second explosive charge. Sealing is provided by interference sliding fits between the lands on the spool and the body. A hole balances the actuation pressure on each end of the spool after the spool travel.

Explosive valves are usually employed in lines of small diameters (less than 25 mm). However, explosive valves for lines of greater diameters can be obtained by using the energy of the fluid to open and close the valve, in which case the ram is used only as a latch, as shown for the valve (U.S. Patent No. 3,017,894) illustrated in the following figure, due to the courtesy of NASA [43].

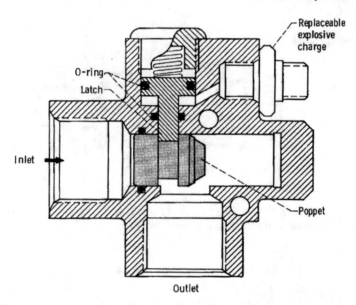

Outlet

Explosive valves are highly reliable components, when used properly. However, redundant explosive valves have also been used for some critical applications.

References

1. Åström KJ, Murray RM (2008) Feedback systems: an introduction for scientists and engineers. Princeton University Press, Princeton, NJ, USA. ISBN 978-0-691-13576-2
2. Lorenzo CF, Musgrave JL (1992) Overview of rocket engine control. NASA Technical Memorandum 105318, Jan 1992, 13pp. https://ntrs.nasa.gov/archive/nasa/casi.ntrs.nasa.gov/199200 04056.pdf
3. Huzel DK, Huang DH (1967) Design of liquid propellant rocket engines. 2nd edn. NASA SP-125, NASA, Washington, D.C., 472pp. https://ntrs.nasa.gov/archive/nasa/casi.ntrs.nasa.gov/ 19710019929.pdf
4. Thomson WT (1960) Laplace transformation, 2nd edn. Prentice-Hall, Englewood Cliffs, NJ, USA
5. Sutton GP, Biblarz O (2001) Rocket propulsion elements, 7th edn. Wiley, New York. ISBN 0-471-32642-9
6. Wikimedia. https://commons.wikimedia.org/wiki/File:Ssme_schematic_(updated).svg. Attribution: Jkwchui with minor adjustments from Chouser [Public domain], via Wikimedia Commons
7. Anonymous (1998) Space Shuttle main engine orientation, Boeing-Rocketdyne Propulsion & Power, Presentation BC98-04, June 1998, 105pp. http://large.stanford.edu/courses/2011/ ph240/nguyen1/docs/SSME_PRESENTATION.pdf
8. Hobart HF, Minkin HL, Warshawsky I (1973) Life tests of small turbine-type flowmeters in liquid hydrogen. NASA TN D-7323, 16 p, June 1973. https://www.ntrs.nasa.gov/archive/nasa/ casi.ntrs.nasa.gov/19730016551.pdf
9. France JT (1972) The measurement of fuel flow. AGARD-AG-160-Vol. 3, Mar 1972. https:// apps.dtic.mil/dtic/tr/fulltext/u2/741610.pdf

10. Dodge FT (2008) Propellant mass gauging: database of vehicle applications and research and development studies. NASA/CR-2008-215281, Aug 2008, 43pp. https://ntrs.nasa.gov/archive/nasa/casi.ntrs.nasa.gov/20080034885.pdf

11. Szabo SV Jr, Berns JA, Stofan AJ (1968) Centaur launch vehicle propellant utilization system. NASA TN D-4848, Oct 1968, 35pp. https://ntrs.nasa.gov/archive/nasa/casi.ntrs.nasa.gov/19680027035.pdf

12. Anonymous (1969) Technical information summary on Apollo-10 (AS-505), Apollo Saturn V space vehicle, NASA, S&E-ASTR-S-69-24, 1 May 1969, 95pp. https://history.nasa.gov/afj/ap10fj/pdf/19700015780_as-505-a10-saturn-v-technical-info-summary.pdf

13. Anonymous (1969) Saturn V flight manual, SA 507, NASA MSFC-MAN-507, Oct 1969. https://www.history.nasa.gov/afj/ap12fj/pdf/a12_sa507-flightmanual.pdf

14. McGlinchey LF (1974) Viking orbiter 1975 thrust vector control system accuracy, NASA-CR-140705, JPL Technical Memorandum 33-703, 15 Oct 1974. https://ntrs.nasa.gov/archive/nasa/casi.ntrs.nasa.gov/19750002097.pdf

15. Glassman AJ (ed) (1994) Turbine design and application, vols 1, 2, and 3, NASA SP-290, NASA Lewis Research Centre, June 1994, 390pp. https://ntrs.nasa.gov/archive/nasa/casi.ntrs.nasa.gov/19950015924.pdf

16. Hands BA. Cryogenic fluids, thermopedia. http://www.thermopedia.com/content/676/

17. Zuk J (1976) Fundamentals of fluid sealing, NASA TN D-8151, Mar 1976. https://ntrs.nasa.gov/archive/nasa/casi.ntrs.nasa.gov/19760012374.pdf

18. Wikimedia, Attribution: Original diagram: S Beck and R Collins, University of Sheffield (Donebythesecondlaw at English Wikipedia) Conversion to SVG: Marc.derumaux [CC BY-SA 4.0 (https://creativecommons.org/licenses/by-sa/4.0)]. https://commons.wikimedia.org/wiki/File:Moody_EN.svg

19. New Jersey Institute of Technology, Newark, NJ, USA. https://web.njit.edu/~barat/ChE396_Spring2011/EquivLengths.pdf

20. Queen's University, Kingston, Ontario, Canada, Faculty of Engineering and Applied Sciences, Mechanical and Materials Engineering, Losses in Pipes. https://me.queensu.ca/People/Sellens/LossesinPipes.html

21. Neutrium. https://neutrium.net/fluid_flow/pressure-loss-from-fittings-2k-method/

22. Neutrium. https://neutrium.net/fluid_flow/pressure-loss-from-fittings-3k-method/

23. AJ Design Software, Colebrook Equations Formulas Calculator. https://www.ajdesigner.com/php_colebrook/colebrook_equation.php

24. Tomoe Valve Co. Ltd., Osaka, Japan. https://www.tomoevalve.com/english/pdf/ValveRelatedData.pdf

25. Howell GW, Weathers TM (eds) (1970) Aerospace fluid component designers' handbook, vol I, Revision D, Report No. RPL-TDR-64-25, TRW Systems Group, One Space Park, Redondo Beach, CA, USA, Feb 1970. https://apps.dtic.mil/dtic/tr/fulltext/u2/874542.pdf

26. Burmeister LC, Loser JB, Sneegas EC (1967) NASA contributions to advanced valve technology, NASA SP-5019, 1967. https://apps.dtic.mil/dtic/tr/fulltext/u2/a306561.pdf

27. Davis ML, Allgeier RK Jr, Rogers ThG, Rysavy G (1970) The development of cryogenic storage systems for space flight, NASA SP-247, 132pp. https://ntrs.nasa.gov/search.jsp?R=19710021434

28. Radebaugh R. Cryogenics, National Institute of Standards and Technology (NIST), US Department of Commerce, Boulder, CO, USA. https://www.nist.gov/sites/default/files/documents/2016/09/08/cryogenicsii.pdf

29. Ellis HJ, Spring TR, Keller RB Jr (eds) (1973) Liquid rocket valve components, NASA SP-8094, Aug 1973. NASA Lewis Research Centre, Cleveland, OH, USA, 150pp. https://ntrs.nasa.gov/archive/nasa/casi.ntrs.nasa.gov/19740019163.pdf

30. Neutrium. https://neutrium.net/fluid_flow/pressure-loss-from-fittings-excess-head-k-method/

31. Suez degremont® water handbook. https://www.suezwaterhandbook.com/formulas-and-tools/formulary/hydraulics/minor-losses-in-the-pipelines-fittings-valves-for-water

32. Spring TR, Keller RB Jr (eds) (1973) Liquid rocket valve assemblies, NASA SP-8097, Nov 1973. NASA Lewis Research Centre, Cleveland, OH, USA, 154pp. https://ntrs.nasa.gov/archive/nasa/casi.ntrs.nasa.gov/19740018866.pdf

33. AWG to Metric Converter. https://technick.net/tools/awg-to-metric-converter/
34. Giovanetti AJ, Spadaccini LJ, Szetela EJ (1983) Deposit formation and heat transfer in hydro-carbon rocket fuels, NASA-CR-168277, 168pp. https://ntrs.nasa.gov/archive/nasa/casi.ntrs.nasa.gov/19840004157.pdf
35. Anonymous (1940) Standards for discharge measurement with standardized nozzles and orifices, German industrial standard 1952, 4th edn. NACA Technical Memorandum No. 952, 78pp. https://ntrs.nasa.gov/archive/nasa/casi.ntrs.nasa.gov/19930094464.pdf
36. EN ISO 5167, Measurement of fluid flow by means of pressure differential devices inserted in circular cross-section conduits running full. https://www.iso.org/obp/ui/#iso:std:iso:5167:-1:ed-2:v1:en
37. AGA 3.1, Orifice metering of natural gas and other related hydrocarbon fluid. https://law.resource.org/pub/us/cfr/ibr/001/aga.3.1.1990.pdf
38. Anonymous, Risk management program guidance for offsite consequence analysis, United States Environmental Protection Agency, Appendix B. https://www.epa.gov/sites/production/files/2017-05/documents/oca-apds.pdf
39. Neutrium. https://neutrium.net/fluid_flow/discharge-coefficient-for-nozzles-and-orifices/
40. National Institute of Standards and Technology (NIST), U.S. Department of Commerce, Thermophysical Properties of Fluid Systems. https://webbook.nist.gov/chemistry/fluid/
41. Absalom JG, Keller RB Jr (eds) (1973) Liquid rocket actuators and operators, NASA SP-8090, May 1973. NASA Lewis Research Centre, Cleveland, OH, USA, 158pp. https://ntrs.nasa.gov/archive/nasa/casi.ntrs.nasa.gov/19740009672.pdf
42. Plummer AR, Electrohydraulic servovalves—past, present, and future. In: Proceedings of the 10th international fluid power conference, IFK2016, Dresden, Germany, 8–16 Mar 2016. University of Bath, UK, pp 405–424. http://tud.qucosa.de/api/qucosa%3A29369/attachment/ATT-0/
43. Tomlinson LE, Keller RB Jr (eds) (1973) Liquid rocket pressure regulators, relief valves, check valves, burst disks, and explosive valves, NASA SP-8080, Mar 1973, 123pp. https://ntrs.nasa.gov/archive/nasa/casi.ntrs.nasa.gov/19740002611.pdf
44. Kates flow control valves, Custom Valves Concepts. https://www.customvalveconcepts.com/products/kates-flow-controllers-2/kates-fc-valve-automatic-flow-rate-controller.html

Chapter 6
Tanks for Propellants

6.1 Fundamental Concepts

Tanks containing liquid propellants for rocket engines can be considered as shells whose thin walls are surfaces of revolution. The following figure, due to the courtesy of NASA [1], shows the external tank of the Space Shuttle.

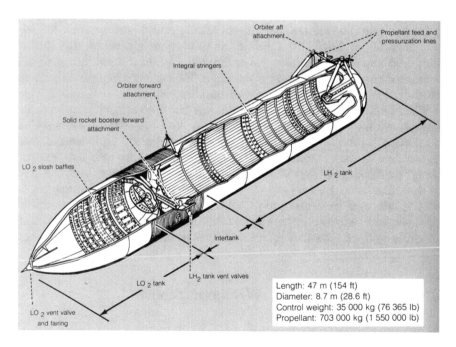

© The Editor(s) (if applicable) and The Author(s), under exclusive license
to Springer Nature Switzerland AG 2021
A. de Iaco Veris, *Fundamental Concepts of Liquid-Propellant Rocket Engines*,
Springer Aerospace Technology,
https://doi.org/10.1007/978-3-030-54704-2_6

The word shell is used here because, at any point of a tank, the thickness of its wall is assumed to be constant and much smaller than the radii of curvature of its middle surface [2].

A surface of revolution results from rotating a plane curve about some straight line (called the axis of revolution) lying in the plane which contains the curve. A line resulting from the intersection of a surface of revolution with a plane containing the axis of revolution is called a meridian. A line resulting from the intersection of a surface of revolution with a plane perpendicular to the axis of revolution is called a circumference. Therefore, the meridian passing through any point of a surface of revolution is perpendicular to the circumference passing through the same point. The structural analysis of tanks which will be presented in this chapter is largely due to Roark's formulas for stress and strain [3].

When a shell is subject to distributed loads resulting from internal or external pressure, the predominant stresses acting on the shell are membrane stresses, which are stresses whose amount is constant through the thickness of the shell. Generally speaking, the stresses acting in a point of a shell are

- a meridian membrane stress σ_1, whose direction is parallel to the local meridian;
- a circumferential membrane stress σ_2, whose direction is parallel to the local circumference; and
- a small radial stress σ_3, which varies through the thickness of the shell.

In addition to these stresses, there may be bending or shear stresses due to loading, or to physical properties of the shell, or to the supporting structure. The stresses considered above cause meridian, circumferential, and radial strains in a shell, and therefore changes in the slopes if its meridians. The circumferences of a shell may also deviate from their circular form when buckling occurs.

When a thin shell (one whose thickness is less than one tenth of its smaller radius of curvature) has no abrupt changes in thickness, slope, or curvature, and is subject to a loading uniformly distributed or smoothly varying and axisymmetric, then the meridian σ_1 and circumferential σ_2 membrane stresses are practically uniform through the thickness of the wall, and are also the most important stresses acting on the shell. The radial stress σ_3 and stresses induced by bending moments are negligible. In this case, the formulae of [3, Table 13.1] can be used to compute stresses and strains such as those described above for shells of cylindrical, conical, spherical, and toroidal shapes. Examples of application will be given in the following sections.

6.2 Tanks Subject Only to Membrane Stresses

Two thin shells can be joined together to form a tank. When it is desired to have no bending stresses at the junction in case of uniformly distributed or smoothly varying loads, then it is necessary to choose shells such that the radial deformations ΔR and the rotations ψ of the meridians be the same for each of them at the point of junction.

For example, a cylindrical shell of radius R (m) and thickness t (m), subject to uniform internal pressure q (N/m²), see [3, Table 13.1, case 1c], has a radial deformation ΔR (m)

$$\Delta R = \frac{q R^2}{Et}\left(1 - \frac{\nu}{2}\right)$$

where E (N/m²) and ν are respectively the Young modulus and the Poisson ratio of the material, whereas a hemispherical shell of equal radius and thickness, subject to the same pressure, see [3, Table 13.1, case 3a], has a radial deformation ΔR (m)

$$\Delta R = \frac{q R^2 (1 - \nu)}{2Et}$$

The rotations ψ of the meridians are the same in both cases. The different radial deformations ΔR cause bending and shear stresses in the vicinity of the junction.

The following figure shows a shell, whose middle surface is a smooth surface of revolution, not necessarily a hemisphere, subject to a uniform internal pressure q [3, Table 13.1, case 4a].

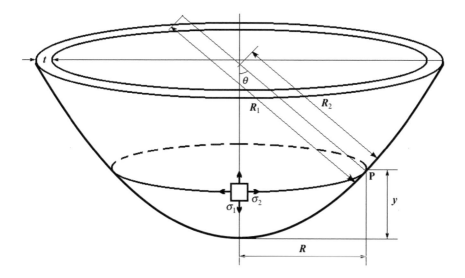

Roark's formulae cited above give the following expressions of respectively ΔR and ψ for this smooth surface of revolution

$$\Delta R = \frac{q R_2^2 \sin\theta}{2Et}\left(2 - \frac{R_2}{R_1} - \nu\right)$$

$$\psi = \frac{q R_2^2}{2Et R_1 \tan\theta}\left[3\frac{R_1}{R_2} - 5 + \frac{R_2}{R_1}\left(2 + \frac{1}{R_1}\frac{dR_1}{d\theta}\tan\theta\right)\right]$$

These formulae show that, when the radius of curvature R_1 of a meridian is infinite at $\theta = \pi/2$, then the radial deformations ΔR and the rotations ψ of the meridians are the same as those of the cylindrical shell of [3, Table 13.1, case 1c]. In the preceding figure, R_1 (m) is the radius of curvature of a meridian in the point P of the shell, R_2 (m) is the length of the normal between the point P of the shell and the axis of rotation, and θ (rad) is the angle between the normal to the surface and the axis of rotation. Flügge [4] has shown that the family of the Cassinian curves (illustrated in Sect. 6.9) has the property indicated above.

As an application (from [3]) of the concepts discussed above, it is required to compute the radial deformations ΔR and the rotations ψ of the meridians at both ends of a segment of toroidal shell used as a transition between a cylinder and a head closure in a tank subject to an internal pressure $q = 1.379 \times 10^6$ N/m^2.

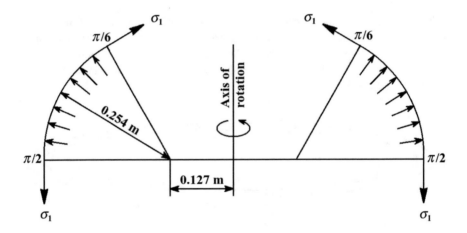

As shown in the preceding figure, the two ends of each toroidal segment are defined by the angles $\theta = \pi/6$ and $\theta = \pi/2$ which the normal to the toroidal surface forms with the axis of rotation. The Young modulus and the Poisson ratio of the material of which the tank is made are respectively $E = 2.068 \times 10^{11}$ N/m^2 and $\nu = 0.3$. The thickness of the wall of the tank is $t = 0.00254$ m $= 2.54$ mm. At the upper end (subscript U) of the toroidal surface, there results

$$\theta_U = \frac{\pi}{6}$$

$$R_1 = 0.254 \, \text{m}$$

$$R_{2U} = 0.254 + \frac{0.127}{\sin(\pi/6)} = 0.508 \, \text{m}$$

where R_1 is the constant radius of curvature of the meridians, and R_{2U} is the length of the normal between the upper end of the toroidal surface and the axis of rotation.

At the upper end, the formulae of [3, Table 13.1, case 4a] give the following results for respectively the radial deformations ΔR_U and the rotations ψ_U of the meridians, taking account that $dR_1/d\theta = 0$, because R_1 is constant

$$
\begin{aligned}
\Delta R_U &= \frac{q R_{2U}^2 \sin \theta_U}{2Et}\left(2 - \frac{R_{2U}}{R_1} - v\right) \\
&= \frac{1.379 \times 10^6 \times 0.508^2 \times \sin(\pi/6)}{2 \times 2.068 \times 10^{11} \times 2.54 \times 10^{-3}} \times \left(2 - \frac{0.508}{0.254} - 0.3\right) \\
&= -5.081 \times 10^{-5}\,\text{m} = -0.05081\,\text{mm}
\end{aligned}
$$

$$
\begin{aligned}
\psi_U &= \frac{q R_{2U}^2}{2Et R_1 \tan \theta_U}\left(3\frac{R_1}{R_{2U}} - 5 + \frac{R_{2U}}{R_1}2\right) \\
&= \frac{1.379 \times 10^6 \times 0.508^2}{2 \times 2.068 \times 10^{11} \times 0.00254 \times 0.254 \times \tan(\pi/6)} \times \left(\frac{3 \times 0.254}{0.508} - 5 + \frac{2 \times 0.508}{0.254}\right) \\
&= 0.001155\,\text{rad}
\end{aligned}
$$

At the lower end (subscript L) of the toroidal surface, there results

$$
\theta_L = \frac{\pi}{2}
$$

$$
R_1 = 0.254\,\text{m}
$$

$$
R_{2L} = 0.254 + 0.127 = 0.381\,\text{m}
$$

At the lower end, the formulae cited above lead to the following results for respectively the radial deformations ΔR_L and the rotations ψ_L of the meridians

$$
\begin{aligned}
\Delta R_L &= \frac{q R_{2L}^2 \sin \theta_L}{2Et}\left(2 - \frac{R_{2L}}{R_1} - v\right) \\
&= \frac{1.379 \times 10^6 \times 0.381^2 \times \sin(\pi/2)}{2 \times 2.068 \times 10^{11} \times 2.54 \times 10^{-3}} \times \left(2 - \frac{0.381}{0.254} - 0.3\right) \\
&= 3.811 \times 10^{-5}\,\text{m} = 0.03811\,\text{mm}
\end{aligned}
$$

$$
\psi_L = 0 \quad \text{because } \tan \theta_L = \tan\left(\frac{\pi}{2}\right) = \infty
$$

Another example of application (also due to [3]) concerns a shell shaped as a cone frustum of semi-aperture angle $\alpha = \pi/12$ rad, supported at its base by the membrane stress σ_1, as shown in the following figure.

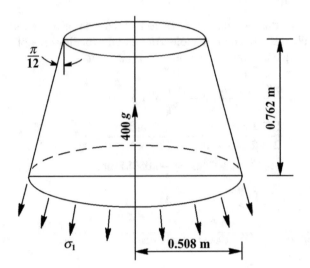

The thickness of the shell is $t = 0.00635$ m $= 6.35$ mm. The density, the Young modulus, and the Poisson ratio of the material (aluminium) of which the shell is made is made are respectively $\rho = 2700$ kg/m^3, $E = 6.9 \times 10^{10}$ N/m^2, and $\nu = 0.3$.

It is required to find the membrane stress σ_1 at the base, the radial deformation ΔR, and the height deformation Δy, supposing the shell to be subject to an acceleration of $400\, g_0 = 400 \times 9.807$ m/s^2 acting in the vertical direction.

The formulae of [3, Table 13.1, case 2c] concern a complete cone loaded by its own weight $\rho V g_0$ N. Since the cone considered here is truncated, then the principle of superposition of effects applies, as will be shown below. For this purpose, we consider firstly a complete cone loaded by its own weight. The specific weight ρg_0 (N/m^3) of this cone is taken with the minus sign, because the vertex of the cone is up instead of down, and this value is multiplied by 400, in order to take account of the acceleration. We use the formulae of [3, Table 13.1, case 2c] with $\alpha = \pi/12$ rad, $R = 0.508$ m, $E = 6.9 \times 10^{10}$ N/m^2, $\nu = 0.3$, and $\rho g_0 = -2700 \times 400 \times 9.807$ N/m^3. By so doing, we find

$$\sigma_1 = \frac{\rho g_0 R}{\sin(2\alpha)} = \frac{-2700 \times 400 \times 9.807 \times 0.508}{\sin(\pi/6)} = -1.076 \times 10^7 \text{ N/m}^2$$

$$\begin{aligned}
\Delta R &= \frac{\rho g_0 R^2}{E \cos \alpha}\left(\sin \alpha - \frac{\nu}{2 \sin \alpha}\right) \\
&= \frac{-400 \times 2700 \times 9.807 \times 0.508^2}{6.9 \times 10^{10} \times \cos(\pi/12)} \times \left[\sin\left(\frac{\pi}{12}\right) - \frac{0.3}{2 \times \sin(\pi/12)}\right] \\
&= 1.315 \times 10^{-5} \text{ m} = 0.01315 \text{ mm}
\end{aligned}$$

$$\Delta y = \frac{\rho g_0 R^2}{E \cos^2 \alpha}\left(\frac{1}{4 \sin^2 \alpha} - \sin^2 \alpha\right)$$

$$= \frac{-400 \times 2700 \times 9.807 \times 0.508^2}{6.9 \times 10^{10} \times \cos^2(\pi/12)} \times \left[\frac{1}{4\sin^2(\pi/12)} - \sin^2\left(\frac{\pi}{12}\right)\right]$$

$$= -1.556 \times 10^{-4}\,\text{m} = -0.1556\,\text{mm}$$

The radius of the circular base at the top of the cone frustum is

$$r = 0.508 - \frac{0.762}{\tan\left(\frac{5\pi}{12}\right)} = 0.3038\,\text{m}$$

The change in length of the upper conical shell to be removed is

$$\Delta y = -1.556 \times 10^{-4} \times \left(\frac{0.3038}{0.508}\right)^2 = -5.565 \times 10^{-5}\,\text{m} = -0.05565\,\text{mm}$$

The slant side of the upper conical shell to be removed is

$$L = \frac{r}{\sin\left(\frac{\pi}{12}\right)}$$

The lateral surface of the upper conical shell to be removed is

$$S = \pi r L = \frac{\pi r^2}{\sin\left(\frac{\pi}{12}\right)}$$

The volume of the upper conical shell to be removed is

$$V = St = \frac{\pi r^2 t}{\sin\left(\frac{\pi}{12}\right)} = \frac{3.1416 \times 0.3038^2 \times 0.00635}{\sin\left(\frac{\pi}{12}\right)} = 0.007114\,\text{m}^3$$

The effective weight of the upper conical shell to be removed is

$$P = 400\,\rho\,g_0\,V = 400 \times 2700 \times 9.807 \times 0.007114 = 7.535 \times 10^4\,\text{N}$$

In order to remove the effective weight P from the truncated conical shell, we substitute $P = 7.535 \times 10^4$ N, $R = 0.508$ m, $r = 0.3038$ m, $h = 0.762$ m, $t = 0.00635$ m, $\alpha = \pi/12$ rad, $E = 6.9 \times 10^{10}$ N/m^2, and $\nu = 0.3$ in the following formulae of [3, Table 13.1, case 2d], which concern a truncated conical shell subject to tangential loading only, with resultant load P, as shown in the following figure.

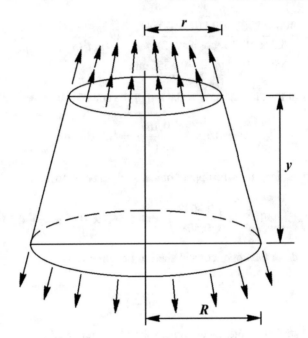

By so doing, we find

$$\sigma_1 = \frac{P}{2\pi Rt \cos\alpha} = \frac{7.535 \times 10^4}{2 \times 3.1416 \times 0.508 \times 0.00635 \times \cos(\pi/12)}$$
$$= 3.849 \times 10^6 \, \text{N/m}^2$$

$$\Delta R = \frac{-\nu P}{2\pi Et \cos\alpha} = \frac{-0.3 \times 7.535 \times 10^4}{2 \times 3.1416 \times 6.9 \times 10^{10} \times 0.00635 \times \cos(\pi/12)}$$
$$= -8.500 \times 10^{-6} \, \text{m} = -0.008500 \, \text{mm}$$

$$\Delta h = \frac{P}{2\pi Et \sin\alpha \cos^2\alpha} \ln\left(\frac{R}{r}\right)$$
$$= \frac{7.535 \times 10^4}{2 \times 3.1416 \times 6.9 \times 10^{10} \times 0.00635 \times \sin(\pi/12) \times \cos^2(\pi/12)} \times \ln\left(\frac{0.508}{0.3038}\right)$$
$$= 5.827 \times 10^{-5} \, \text{m} = 0.05827 \, \text{mm}$$

Therefore, for the original truncated conical shell, there results

$$\sigma_1 = (-10.76 + 3.849) \times 10^6 \, \text{N/m}^2 = -6.911 \times 10^6 \, \text{N/m}^2$$

$$\Delta R = (13.15 - 8.500) \times 10^{-6} \, \text{m} = 4.65 \times 10^{-6} \, \text{m} = 0.00465 \, \text{mm}$$

$$\Delta y = [-15.56 - (-5.565) + 5.827] \times 10^{-5} \, \text{m}$$
$$= -4.168 \times 10^{-5} \, \text{m} = -0.04168 \, \text{mm}$$

6.3 Tanks Subject to Membrane and Bending Stresses

Forces, moments, and displacements for cylindrical tanks can be computed by using the formulae of [3, Table 13.2]. These formulae concern thin-walled cylindrical shells having free ends and subject to axisymmetric loadings. The radial deformations of these shells are assumed to be small in comparison with the thickness of their walls. An example of calculation, from [3], is given below.

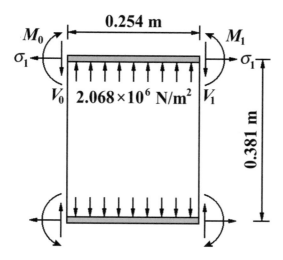

A cylindrical tank of aluminium, shown in the preceding figure, is 0.254 m in length and 0.381 m in diameter. This tank is desired to bear an internal pressure of 2.068×10^6 N/m^2 with a maximum tensile stress of 8.274×10^7 N/m^2. The ends of the tank are capped with flanges, which are sufficiently clamped to the tank to resist any radial or rotational deformations at their ends. Given the Young modulus $E = 6.9 \times 10^{10}$ N/m^2 and the Poisson ratio $v = 0.3$ of aluminium, it is required to determine the thickness t (m) of the wall of the tank.

The value of t may be computed by superposing the effects given in [3, Table 13.1, case 1c, and Table 13.2, cases 8 and 10]. For this purpose, it is necessary to determine the mean radius of the tank

$$R = \frac{0.381}{2} = 0.1905 \, \text{m}$$

the length of the tank

$$l = 0.254 \, \text{m}$$

the bending stiffness of the tank

$$D = \frac{Et^3}{12(1 - v^2)} = \frac{6.9 \times 10^{10}t^3}{12 \times (1 - 0.3^2)} = 6.319 \times 10^9 t^3 \, \text{Nm}$$

and the following quantities

$$\lambda = \left[\frac{3(1 - v^2)}{R^2 t^2}\right]^{\frac{1}{4}} = \left[\frac{3 \times (1 - 0.3^2)}{0.1905^2 t^2}\right]^{\frac{1}{4}} = 2.945 t^{-\frac{1}{2}} \text{m}^{-1}$$

$$\lambda l = 2.945 t^{-\frac{1}{2}} \times 0.254 = 0.748 t^{-\frac{1}{2}}$$

Since the value of t is still unknown, it is necessary to determine whether the loads at one end of the cylindrical shell have (short shell) or have not (long shell) any influence on the deformations at the other end. To this end, we determine approximately the value of t by using the formulae for membrane stresses and deformations given in [3, Table 13.1, case 1c], which concern a cylindrical shell subject to internal pressure q, with ends capped. These formulae are

$$\sigma_1 = \frac{qR}{2t}$$

$$\sigma_2 = \frac{qR}{t}$$

for respectively the meridian stress and the circumferential stress in the shell.

By equating σ_2 to the maximum allowable stress $8.274 \times 10^7 \, \text{N/m}^2$, there results

$$8.274 \times 10^7 = \frac{2.068 \times 10^6 \times 0.1905}{t}$$

The preceding equation, solved for t, yields

$$t = \frac{2.068 \times 0.1905}{82.74} = 0.004769 \, \text{m}$$

Since

$$\lambda l = 0.748 t^{-\frac{1}{2}} = 0.748 \times 0.004769^{-\frac{1}{2}} = 10.83$$

then the cylindrical shell considered here is very long.

A preliminary solution may be found by assuming the deformation at the left end of the cylindrical shell to be independent of the radial load and of the bending moment at the right end. Since the end caps are rigid, the radial deformation and the angular rotation of the left end are set to zero. The formulae of [3, Table 13.1, case 1c], which concern a cylindrical shell subject to internal pressure q, with ends capped, give the following values for respectively the meridian stress, the circumferential stress, the radial deformation, and the angular rotation

$$\sigma_1 = \frac{qR}{2t} = \frac{2.068 \times 10^6 \times 0.1905}{2t} = \frac{0.1970 \times 10^6}{t} \, \text{N/m}^2$$

$$\sigma_2 = \frac{qR}{t} = \frac{2.068 \times 10^6 \times 0.1905}{t} = \frac{0.3940 \times 10^6}{t} \, \text{N/m}^2$$

$$\Delta R = \frac{qR^2}{Et}\left(1 - \frac{\nu}{2}\right) = \frac{2.068 \times 10^6 \times 0.1905^2}{6.9 \times 10^{10}t} \times \left(1 - \frac{0.3}{2}\right) = \frac{9.245 \times 10^{-7}}{t} \, \text{m}$$

$$\psi = 0 \, \text{rad}$$

On the other hand, the formulae of [3, Table 13.2, case 8 and case 10] concern a long shell with the left end free and the right end more than $6/\lambda$ units of length from the closest load, as shown in the following figure.

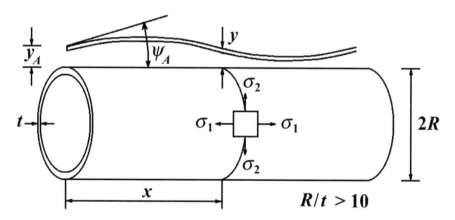

In particular, case 8 concerns the angular rotation ψ_A (rad) and the radial deformation y_A (m) due to the radial load per unit length V_0 (N/m) at the end section A of the cylindrical shell, as follows

$$\psi_A = \frac{V_0}{2D\lambda^2}$$

$$y_A = \frac{-V_0}{2D\lambda^3}$$

After substituting the values of D and λ found above in the preceding equations, we find the following angular rotation and radial deformation for case 8

$$\psi_A = \frac{V_0}{t^2} \times 9.123 \times 10^{-12} \, \text{rad}$$

$$y_A = -\frac{V_0}{t^{\frac{3}{2}}} \times 3.098 \times 10^{-12} \, \text{m}$$

Case 10 concerns the angular rotation ψ_A and the radial deformation y_A due to the end moment per unit length M_0 (Nm/m) at the end section A of the cylindrical shell, as follows

$$\psi_A = \frac{-M_0}{D\lambda}$$

$$y_A = \frac{M_0}{2D\lambda^2}$$

After substituting the values of D and λ found above in the preceding equations, we find the following angular rotations and radial deformations for case 10

$$\psi_A = -\frac{M_0}{t^{\frac{5}{2}}} \times 53.74 \times 10^{-12} \, \text{rad}$$

$$y_A = \frac{M_0}{t^2} \times 9.123 \times 10^{-12} \, \text{m}$$

By summing all the radial deformations found above and equating the sum to zero, we find

$$\frac{9.245 \times 10^{-7}}{t} - \frac{V_0}{t^{\frac{3}{2}}} \times 3.098 \times 10^{-12} + \frac{M_0}{t^2} \times 9.123 \times 10^{-12} = 0$$

By summing all the angular rotations found above and equating the sum to zero, we find

$$0 + \frac{V_0}{t^2} \times 9.123 \times 10^{-12} - \frac{M_0}{t^{\frac{5}{2}}} \times 53.74 \times 10^{-12} = 0$$

By solving the two preceding equations for V_0 and M_0, we find

$$V_0 = t^{\frac{1}{2}} \times 5.968 \times 10^5 \, \text{N/m}$$

$$M_0 = t \times 1.013 \times 10^5 \, \text{Nm/m}$$

Since the maximum bending stress occurs at the ends of the cylindrical shell, then the following stresses must be combined:

(1) the meridian and circumferential membrane stresses computed above

$$\sigma_1 = \frac{qR}{2t} = \frac{2.068 \times 10^6 \times 0.1905}{2t} = \frac{0.1970 \times 10^6}{t} \, \text{N/m}^2$$

$$\sigma_2 = \frac{qR}{t} = \frac{2.068 \times 10^6 \times 0.1905}{t} = \frac{0.3940 \times 10^6}{t} \, \text{N/m}^2$$

(2) the meridian and circumferential membrane and bending stresses given by [3, Table 13.2, case 8]

$$\sigma_1 = 0 \, \text{N/m}^2$$

$$\sigma_2 = \frac{-2V_0\lambda R}{t} = \frac{-2 \times \left(t^{\frac{1}{2}} \times 5.968 \times 10^5\right) \times \left(t^{-\frac{1}{2}} \times 2.945\right) \times 0.1905}{t}$$

$$= \frac{-6.696 \times 10^5}{t} \, \text{N/m}^2$$

$$\sigma_1' = 0 \, \text{N/m}^2$$

$$\sigma_2' = 0 \, \text{N/m}^2$$

(3) the meridian and circumferential membrane and bending stresses given by [3], Table 13.2, case 10

$$\sigma_1 = 0 \, \text{N/m}^2$$

$$\sigma_2 = \frac{2M_0\lambda^2 R}{t} = \frac{2 \times (t \times 1.013 \times 10^5) \times \left(t^{-\frac{1}{2}} \times 2.945\right)^2 \times 0.1905}{t}$$

$$= \frac{3.347 \times 10^5}{t} \, \text{N/m}^2$$

$$\sigma_1' = \frac{6M_0}{t^2} = \frac{6 \times (t \times 1.013 \times 10^5)}{t^2} = \frac{6.078 \times 10^5}{t} \, \text{N/m}^2$$

$$\sigma_2' = \nu\sigma_1' = \frac{0.3 \times 6.078 \times 10^5}{t} = \frac{1.823 \times 10^5}{t} \, \text{N/m}^2$$

At the end of the cylindrical shell, the maximum meridian tensile stress is

$$\frac{0.1970 \times 10^6}{t} + \frac{6.078 \times 10^5}{t} = \frac{8.048 \times 10^5}{t} \, \text{N/m}^2$$

Likewise, at the end of the cylindrical shell, the maximum circumferential tensile stress is

$$\frac{0.3940 \times 10^6}{t} - \frac{6.696 \times 10^5}{t} + \frac{3.347 \times 10^5}{t} + \frac{1.823 \times 10^5}{t}$$
$$= \frac{2.414 \times 10^5}{t} \, \text{N/m}^2$$

Since the maximum allowable stress is 8.274×10^7 N/m^2, then the thickness of the cylindrical shell results from

$$\frac{8.048 \times 10^5}{t} = 8.274 \times 10^7 \, \text{N/m}^2$$

The preceding equation, solved for t, yields

$$t = \frac{8.048}{827.4} = 0.009727 \, \text{m} = 9.727 \, \text{mm}$$

This value of t, substituted in the equation derived above $\lambda l = 0.748 \times t^{-\frac{1}{2}}$, yields

$$\lambda l = 0.748 \times 0.009727^{-\frac{1}{2}} = 7.584$$

This value of λl justifies the assumption made above of having to do with a long cylindrical shell.

Forces, moments, and displacements for tanks of spherical, conical, or toroidal shapes can also be computed by using the formulae of [3], as will be shown below. The following example, from [3], concerns two spherical segments of aluminium ($E = 6.895 \times 10^{10}$ N/m^2, $\nu = 0.33$) welded together to form a symmetrical tank, which is subject to an internal pressure $q = 1.379 \times 10^6$ N/m^2. The angle subtended by each spherical segment is $4\pi/3$ rad. The mean diameter of each spherical segment is 1.219 m, and the thickness of the wall is $t = 0.0127$ m. It is required to compute the stresses at the junction.

Since this tank is symmetrical and symmetrically loaded, then it is possible to consider only one of the two spherical segments. The effects due to the other segment are taken into account by adding the following loads to the internal pressure q:

- a tangential force T, which balances the force due to the internal pressure, and causes only membrane stresses and consequent radial deformations ΔR in the circumferences and no rotations in the meridians;
- a vertical force Q_0, which is added to eliminate the radial component of T; and
- a moment M_0, which is added in order to prevent the edges of the spherical segment from rotating.

The whole tank, the spherical segment considered above, and the loads acting on it are shown in the following figure.

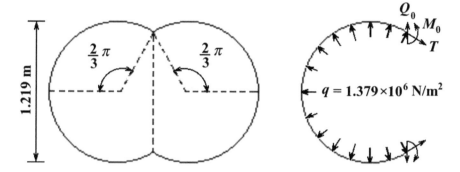

The formulae of [3, Table 13.1, case 3a] concern membrane stresses and deformations in a segment of spherical thin-walled shell of mean radius R_2 subject to an internal or external pressure q with tangential support at the edges, as shown in the following figure.

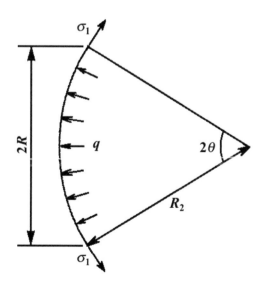

These formulae are

$$\sigma_1 = \sigma_2 = \frac{q\,R_2}{2t} = \frac{1.379 \times 10^6 \times 0.5 \times 1.219}{2 \times 0.0127} = 3.309 \times 10^7 \text{ N/m}^2$$

$$\Delta R = \frac{q\,R_2^2(1-v)\sin\theta}{2Et} = \frac{1.379 \times 10^6 \times (0.5 \times 1.219)^2 \times (1-0.33) \times \sin(2\pi/3)}{2 \times 6.895 \times 10^{10} \times 0.0127}$$

$$= 1.697 \times 10^{-4} \text{ m} = 0.1697 \text{ mm}$$

$$T = \sigma_1 t = 3.309 \times 10^7 \times 0.0127 = 4.203 \times 10^5 \text{ N/m}$$

$$\psi = 0 \text{ rad}$$

Now we apply the formulae of [3, Table 13.3, case 1a], which concern the membrane and bending stresses and strains in a segment of spherical shell with vertical forces Q_0 applied at the edges, as shown in the following figure.

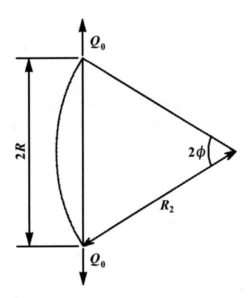

These formulae, applied to the present case, are

$$Q_0 = T \cos\left(\frac{\pi}{3}\right) = 4.203 \times 10^5 \times 0.5 = 2.101 \times 10^5 \text{ N/m}$$

$$\phi = \frac{2}{3}\pi \text{ rad}$$

$$\beta = \left[3(1-v^2)\left(\frac{R_2}{t}\right)^2\right]^{\frac{1}{4}} = \left[3 \times (1-0.33^2) \times \left(\frac{0.5 \times 1.219}{0.0127}\right)^2\right]^{\frac{1}{4}} = 8.858$$

$$K_1 = 1 - \frac{1 - 2\nu}{2\beta \tan\phi} = 1 - \frac{1 - 2 \times 0.33}{2 \times 8.858 \times \tan(2\pi/3)} = 1.011$$

$$K_2 = 1 - \frac{1 + 2\nu}{2\beta \tan\phi} = 1 - \frac{1 + 2 \times 0.33}{2 \times 8.858 \times \tan(2\pi/3)} = 1.054$$

$$\Delta R = \frac{Q_0 R_2 \beta \sin^2\phi}{Et K_1}(1 + K_1 K_2)$$

$$= \frac{2.101 \times 10^5 \times 0.5 \times 1.219 \times 8.858 \times \sin^2(2/3\,\pi)}{6.895 \times 10^{10} \times 0.0127 \times 1.011} \times (1 + 1.011 \times 1.054)$$

$$= 0.001985\,\text{m} = 1.985\,\text{mm}$$

$$\psi = \frac{2Q_0 \beta^2 \sin\phi}{Et K_1} = \frac{2 \times 2.101 \times 10^5 \times 8.858^2 \times \sin(2\pi/3)}{6.895 \times 10^{10} \times 0.0127 \times 1.011} = 0.03225\,\text{rad}$$

$$\sigma_1 = \frac{Q_0 \cos\phi}{t} = \frac{2.101 \times 10^5 \times \cos(2\pi/3)}{0.0127} = -8.272 \times 10^6\,\text{N/m}^2$$

$$\sigma_1' = 0\,\text{N/m}^2$$

$$\sigma_2 = \frac{Q_0 \beta \sin\phi}{2t}\left(\frac{2}{K_1} + K_1 + K_2\right)$$

$$= \frac{2.101 \times 10^5 \times 8.858 \times \sin(2\pi/3)}{2 \times 0.0127} \times \left(\frac{2}{1.011} + 1.011 + 1.054\right)$$

$$= 2.566 \times 10^8\,\text{N/m}^2$$

$$\sigma_2' = \frac{-Q_0 \beta^2 \cos\phi}{K_1 R_2} = \frac{-2.101 \times 10^5 \times 8.858^2 \times \cos(2\pi/3)}{1.011 \times 0.5 \times 1.219}$$

$$= 1.338 \times 10^7\,\text{N/m}^2$$

Now we apply the formulae of [3, Table 13.3. case 1b], which concern the membrane and bending stresses and strains in a segment of spherical shell with moments M_0 applied at the edges, as shown in the following figure.

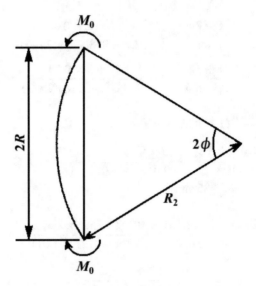

These formulae, applied to the present case, are

$$\Delta R = \frac{2 M_0 \beta^2 \sin \phi}{E t K_1} = \frac{2 \times 8.858^2 \times \sin(2\pi/3)}{6.895 \times 10^{10} \times 0.0127 \times 1.011} M_0 = 1.535 \times 10^{-7} M_0 \text{ m}$$

$$\psi = \frac{4 M_0 \beta^3}{E t R_2 K_1}$$

$$= \frac{4 \times 8.858^3}{6.895 \times 10^{10} \times 0.0127 \times 0.5 \times 1.219 \times 1.011} M_0$$

$$= 5.152 \times 10^{-6} M_0 \text{ rad}$$

Since the total rotation ψ at the edges of the spherical segment must be equal to zero, then

$$0 + 0.03225 + 5.152 \times 10^{-6} M_0 = 0$$

The preceding equation, solved for M_0, yields

$$M_0 = -6.259 \times 10^3 \text{N m/m}$$

and therefore the preceding expression of ΔR becomes

$$\Delta R = 1.535 \times 10^{-7} \times \left(-6.259 \times 10^3\right) = -9.608 \times 10^{-4} \text{ m}$$

The total radial deformation results from

$$\Delta R = (1.697 + 19.85 - 9.608) \times 10^{-4} = 11.94 \times 10^{-4} \text{ m} = 1.194 \text{ mm}$$

Since the value of M_0 is known, we can also use the following formulae of [3, Table 13.3, case 1b]

$$\sigma_1 = 0 \, \text{N/m}^2$$

$$\sigma_1' = \frac{-6M_0}{t^2} = \frac{-6 \times \left(-6.259 \times 10^3\right)}{0.0127^2} = 2.328 \times 10^8 \, \text{N/m}^2$$

$$\sigma_2 = \frac{2M_0\beta^2}{R_2 K_1 t} = \frac{2 \times \left(-6.259 \times 10^3\right) \times 8.858^2}{0.5 \times 1.219 \times 1.011 \times 0.0127} = -1.255 \times 10^8 \, \text{N/m}^2$$

$$M_2 = \frac{M_0}{2\nu K_1}\left[(1 + \nu^2)(K_1 + K_2) - 2K_2\right] = \frac{-6.259 \times 10^3}{2 \times 0.33 \times 1.011}$$
$$\times \left[(1 + 0.33^2) \times (1.011 + 1.054) - 2 \times 1.054\right] = -1.706 \times 10^3 \, \text{Nm/m}$$

$$\sigma_2' = \frac{-6M_2}{t^2} = \frac{-6 \times \left(-1.706 \times 10^3\right)}{0.0127^2} = 0.6347 \times 10^8 \, \text{N/m}^2$$

Therefore, the total stresses at the junction are

$$\sigma_1 = (3.309 - 0.8272 + 0) \times 10^7 = 2.482 \times 10^7 \, \text{N/m}^2$$

$$\sigma_1' = (0 + 0 + 23.28) \times 10^7 = 23.28 \times 10^7 \, \text{N/m}^2$$

$$\sigma_2 = (3.309 + 25.66 - 12.55) \times 10^7 = 16.42 \times 10^7 \, \text{N/m}^2$$

$$\sigma_2' = (0 + 1.338 + 6.347) \times 10^7 = 7.685 \times 10^7 \, \text{N/m}^2$$

The maximum stress at the junction is a tensile meridian stress

$$\sigma_1 + \sigma_1' = (2.482 + 23.28) \times 10^7 = 25.76 \times 10^7 \, \text{N/m}^2$$

This value is greater than the yield stress of aluminium, which is $9.5 \times 10^7 \, \text{N/m}^2$ [5]. In order to reduce the tensile stress at the junction between the two spherical segments, it is possible to add a reinforcing ring of aluminium. We want to compute the cross-sectional area A (m^2) of the reinforcing ring.

If the radial deformation ΔR (m) of the ring were equal to the radial deformation at the edge of each of the two spherical segments due only to membrane stresses, then the bending stresses would be eliminated. The radial deformation at the edge of one of the two spherical segments due only to membrane stresses has been found above to be $\Delta R = 1.697 \times 10^{-4}$ m. Therefore, let the reinforcing ring be subject to a load per unit length $2Q_0$ (N/m) (that is, to a load $F = 2Q_0R$) and have a radial deformation $\Delta R = 1.697 \times 10^{-4}$ m. From Hooke's law, there results

$$\sigma = \varepsilon E$$

By substituting $\sigma = F/A = 2Q_0R/A$ and $\varepsilon = \Delta R/R$ in the preceding equation and solving for A, there results

$$A = \frac{2Q_0R^2}{E\Delta R}$$

In the present case, as has been shown above, $Q_0 = 2.101 \times 10^5$ N/m, $R = R_2 \sin \phi = 0.5 \times 1.129 \times \sin(2\pi/3)$ m, $E = 6.895 \times 10^{10}$ N/m^2, and $\Delta R = 1.697 \times 10^{-4}$ m.

By substituting these values in the preceding equation, there results

$$A = \frac{2 \times 2.101 \times 10^5 \times \left[0.5 \times \sin\left(\frac{2}{3}\pi\right)\right]^2}{6.895 \times 10^{10} \times 1.697 \times 10^{-4}} = 0.01 \text{ m}^2$$

Since this value of A is considerable with respect to $R = 0.5278$ m, then the simple expression of $\Delta R/R$ given above (which is based on a thin ring) is not applicable. In addition, the reinforcing ring is too big to be placed outside the tank. Therefore, we place it inside the tank, as shown in the following figure.

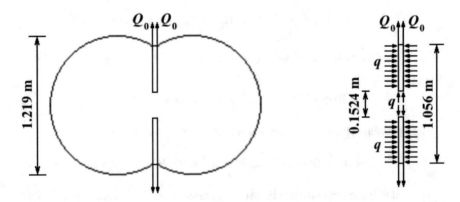

In other words, the two spherical segments of the tank are put together by means of an internal reinforcing disc, which is $2a = 1.219 \times \sin(2\pi/3) = 1.056$ m in diameter. This disc has a coaxial orifice, whose diameter $2b$ is set arbitrarily to 0.1524 m, for the passage of the liquid contained in the tank, and a thickness t_1 (m), whose value is to be determined.

For this purpose, we use the formulae of [3, Table 13.5, case 1a], which concern the stresses and strains in a thick-walled disc subject to a uniform internal radial pressure q (N/m^2) and to zero, or externally balanced, longitudinal pressure. According to the formulae indicated above, the radial deformation Δa (m) of the disc is

$$\Delta a = \frac{q}{E} \frac{2ab^2}{a^2 - b^2}$$

By substituting $q = 1.379 \times 10^6$ N/m^2, $E = 6.895 \times 10^{10}$ N/m^2, $a = 0.5 \times 1.056$ m, and $b = 0.5 \times 0.1524$ m in the preceding equation, we find

$$\Delta a = \frac{1.379 \times 10^6}{6.895 \times 10^{10}} \times \frac{1.056 \times 0.1524^2}{1.056^2 - 0.1524^2} = 4.492 \times 10^{-7} \text{m}$$

The effect of the force per unit length $2Q_0$ on the disc can be evaluated by means of an external negative pressure $-2Q_0/t_1$. We use the formulae of [3, Table 13.5, case 1c], which concern the stresses and strains in a thick-walled disc subject to a uniform external radial pressure q (N/m^2) and to zero, or externally balanced, longitudinal pressure. According to the formulae indicated above, the radial deformation Δa (m) of the disc is

$$\Delta a = \frac{-qa}{E} \left(\frac{a^2 + b^2}{a^2 - b^2} - \nu \right)$$

By substituting the values indicated above, $q = -2 \times 2.101 \times 10^5/t_1$, and $\nu = 0.33$ in the preceding equation, we find

$$\Delta a = \frac{2.101 \times 10^5 \times 1.056}{6.895 \times 10^{10} t_1} \times \left(\frac{1.056^2 + 0.1524^2}{1.056^2 - 0.1524^2} - 0.33 \right) = \frac{2.293 \times 10^{-6}}{t_1} \text{m}$$

In addition, the longitudinal pressure $q = 1.379 \times 10^6$ N/m^2 causes a radial deformation of the disc

$$\Delta a = \frac{q\nu R}{E} = \frac{1.379 \times 10^6 \times 0.33 \times 0.5 \times 1.056}{6.895 \times 10^{10}} = 3.485 \times 10^{-6} \text{m}$$

By summing the three values of the radial deformation Δa found above and equating the result to the desired value 1.697×10^{-4} m, we find

$$0.004492 + \frac{0.02293}{t_1} + 0.03485 = 1.697 \text{ m}$$

The preceding equation, solved for t_1, yields

$$t_1 = \frac{0.02293}{1.697 - 0.004492 - 0.03485} = 0.01383 \text{ m} = 13.83 \text{ mm}$$

Further refinements are possible by varying the diameter $2b$ of the orifice in the reinforcing ring or the thickness t of the wall near the junction.

6.4 Multi-element Tanks

Stresses due to changes in thickness or in shape occur at the junctions of tanks made of shell elements. They are particularly important in case of tanks subject to cyclic or fatigue loads. The following examples show how the tables of [3] can be used to determine the stresses at the junctions of multi-element tanks.

The tank shown in the following figure, in quarter longitudinal section, has a cylindrical shell and two conical shells at its ends.

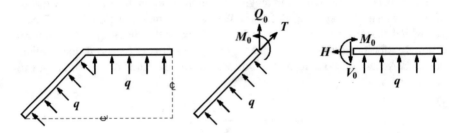

The radius and the thickness of the cylindrical shell are respectively $R = 0.6096$ m and $t = 0.01608$ m. The semi-aperture angle and the thickness of the conical shells are respectively $\alpha = \pi/4$ rad and thickness $t = 0.01918$ m. The cylinder and the cones are welded together. The material of which the tank is made is steel, with Young's modulus $E = 2.068 \times 10^{11}$ N/m^2 and Poisson's ratio $\nu = 0.25$. The tank is subject to an internal pressure $q = 2.068 \times 10^6$ N/m^2. It is required to determine the maximum stress at the junction.

We apply once again the principle of superposition of effects first to the conical shell and then to the cylindrical shell, as will be shown below.

The stresses σ_1 (N/m^2) and σ_2 (N/m^2), the radial deformation ΔR (m), the angular rotation ψ (rad), and the force per unit length T (N/m) at the end of the cone can be determined by using the formulae of [3, Table 13.1, case 2a], which concern the membrane stresses and strains in a conical shell of radius R (m) at the base, thickness t (m) and semi-aperture angle α (rad), subject to a uniform internal or external pressure q (N/m^2), with a tangential support at the edge, as shown in the following figure.

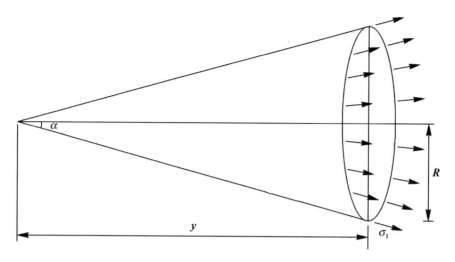

These formulae, applied to the present case, are

$$\sigma_1 = \frac{qR}{2t\cos\alpha} = \frac{2.068 \times 10^6 \times 0.6096}{2 \times 0.01918 \times \cos(\pi/4)} = 4.648 \times 10^7 \,\text{N/m}^2$$

$$T = \sigma_1 t = 4.648 \times 10^7 \times 0.01918 = 8.914 \times 10^5 \,\text{N/m}$$

$$\sigma_2 = \frac{qR}{t\cos\alpha} = 2\sigma_1 = 2 \times 4.648 \times 10^7 = 9.295 \times 10^7 \,\text{N/m}^2$$

$$\sigma_1' = \sigma_2' = 0\,\text{N/m}^2$$

$$\Delta R = \frac{qR^2}{Et\cos\alpha}\left(1 - \frac{\nu}{2}\right) = \frac{2.068 \times 10^6 \times 0.6096^2}{2.068 \times 10^{11} \times 0.01918 \times \cos(\pi/4)} \times \left(1 - \frac{0.25}{2}\right)$$
$$= 0.0002398\,\text{m} = 0.2398\,\text{mm}$$

$$\psi = \frac{3qR\tan\alpha}{2Et\cos\alpha} = \frac{3 \times 2.068 \times 10^6 \times 0.6096 \times \tan(\pi/4)}{2 \times 2.068 \times 10^{11} \times 0.01918 \times \cos(\pi/4)} = 0.0006742\,\text{rad}$$

Now we determine the radial force per unit length Q_0 (N/m) at the edge of the cone by using the formulae of [3, Table 13.3, case 4a]. These formulae concern the membrane and bending stresses and strains for a thin-walled cone of radius R_A (m) at its base and semi-aperture angle α, subject to a uniform radial force per unit length Q_0 at is base. These formulae, applied to the present case ($R_A = 0.6096$ m), are

$$k_A = \frac{2}{\sin\alpha}\left[\frac{12\left(1 - \nu^2\right)R^2\cos^2\alpha}{t^2}\right]^{\frac{1}{4}}$$

$$= \frac{2}{\sin(\pi/4)} \times \left[\frac{12 \times \left(1 - 0.25^2\right) \times 0.6096^2 \times \cos^2(\pi/4)}{0.01918^2} \right]^{\frac{1}{4}}$$

$$= 24.56$$

$$\beta = \left[12\left(1 - \nu^2\right)\right]^{\frac{1}{2}} = \left[12 \times \left(1 - 0.25^2\right)\right]^{\frac{1}{2}} = 3.354$$

$$F_{1A} = F_{3A} = 0$$

$$F_{4A} = 1 - \frac{3.359}{k_A} + \frac{5.641}{k_A^2} - \frac{9.737}{k_A^3} + \frac{14.716}{k_A^4}$$

$$= 1 - \frac{3.359}{24.56} + \frac{5.641}{24.56^2} - \frac{9.737}{24.56^3}$$

$$+ \frac{14.716}{24.56^4} = 0.8720$$

$$F_{10A} = F_{7A} = F_{6A} = 1 - \frac{2.652}{k_A} + \frac{1.641}{k_A^2} - \frac{0.290}{k_A^3} - \frac{2.211}{k_A^4}$$

$$= 1 - \frac{2.652}{24.56} + \frac{1.641}{24.56^2} - \frac{0.290}{24.56^3} - \frac{2.211}{24.56^4} = 0.8947$$

$$F_{2A} = 1 - \frac{2.652}{k_A} + \frac{3.516}{k_A^2} - \frac{2.610}{k_A^3} + \frac{0.038}{k_A^4}$$

$$= 1 - \frac{2.652}{24.56} + \frac{3.516}{24.56^2} - \frac{2.610}{24.56^3} + \frac{0.038}{24.56^4}$$

$$= 0.8977$$

$$F_{8A} = F_{5A} = 1 - \frac{3.359}{k_A} + \frac{7.266}{k_A^2} - \frac{10.068}{k_A^3} + \frac{5.787}{k_A^4}$$

$$= 1 - \frac{3.359}{24.56} + \frac{7.266}{24.56^2} - \frac{10.068}{24.56^3} + \frac{5.787}{24.56^4} = 0.8746$$

$$F_{9A} = C_1 = F_{5A} + \frac{2(2)^{\frac{1}{2}}\nu}{k_A} F_{2A} = 0.8746 + \frac{2 \times 2^{\frac{1}{2}} \times 0.25}{24.56} \times 0.8977 = 0.9005$$

$$\Delta R_A = \frac{Q_0 R_A \sin\alpha}{Et} \frac{k_A}{2^{\frac{1}{2}} C_1} \left(F_{4A} - \frac{4\nu^2}{k_A^2} F_{2A} \right)$$

$$= \frac{Q_0 \times 0.6096 \times \sin(\pi/4)}{2.068 \times 10^{11} \times 0.01918} \times \frac{24.56}{2^{\frac{1}{2}} \times 0.9005} \times \left(0.8720 - \frac{4 \times 0.25^2}{24.56^2} \times 0.8977 \right)$$

$$= 1.827 \times 10^{-9} \times Q_0 \text{ m}$$

$$\psi_A = \frac{Q_0 R_A \beta}{Et^2 C_1} F_{10A} = \frac{Q_0 \times 0.6096 \times 3.354}{2.068 \times 10^{11} \times 0.01918^2 \times 0.9005} \times 0.8947$$
$$= 2.670 \times 10^{-8} \times Q_0 \text{ rad}$$

$$N_{1A} = Q_0 \sin \alpha = Q_0 \times \sin(\pi/4) = 0.7071 \times Q_0 \text{ N/m}$$

$$\sigma_{1A} = \frac{N_{1A}}{t} = \frac{0.7071 \times Q_0}{0.01918} = 36.87 \times Q_0 \text{ N/m}^2$$

$$M_{1A} = 0 \text{ Nm/m}$$

$$\sigma'_{1A} = \frac{-6M_{1A}}{t^2} = 0 \text{ N/m}^2$$

$$N_{2A} = \frac{Q_0 k_A}{2^{\frac{1}{2}} C_1}\left(F_{4A} + \frac{2^{\frac{1}{2}} \nu}{k_A}\right) \sin \alpha$$

$$= \frac{Q_0 \times 24.56}{2^{\frac{1}{2}} \times 0.9005} \times \left(0.8720 + \frac{2^{\frac{1}{2}} \times 0.25}{24.56}\right) \times \sin(\pi/4)$$
$$= 12.09 \times Q_0 \text{ N/m}$$

$$\sigma_{2A} = \frac{N_{2A}}{t} = \frac{12.09 \times Q_0}{0.01918} = 630.3 \times Q_0 \text{ N/m}^2$$

$$M_{2A} = Q_0(1 - \nu^2)\frac{t}{\beta C_1} F_{10A} \sin \alpha$$

$$= Q_0 \times (1 - 0.25^2) \times \frac{0.01918}{3.354 \times 0.9005} \times 0.8947 \times \sin(\pi/4)$$
$$= 0.003767 \times Q_0 \text{ Nm/m}$$

$$\sigma'_{2A} = \frac{-6M_{2A}}{t^2} = \frac{-6 \times 0.003767 \times Q_0}{0.01918^2} = -61.44 \times Q_0$$

Now we determine the moment per unit length M_0 (Nm/m) at the edge of the cone by using the formulae of [3, Table 13.3, case 4b]. These formulae concern the membrane and bending stresses and strains for a thin-walled cone of radius R_A (m) at its base and semi-aperture angle α, subject to a uniform moment per unit length M_0 at is base. These formulae, applied to the present case ($R_A = 0.6096$ m), are

$$\Delta R_A = M_0 \frac{\beta R_A}{Et^2 C_1} F_{7A} = M_0 \frac{3.354 \times 0.6096}{2.068 \times 10^{11} \times 0.01918^2 \times 0.9005} \times 0.8947$$
$$= 2.670 \times 10^{-8} \times M_0 \text{ m}$$

$$\psi_A = M_0 \frac{2(2)^{\frac{1}{2}} \beta^2 R_A}{E t^3 k_A C_1 \sin \alpha} F_{2A}$$

$$= M_0 \times \frac{2 \times 2^{\frac{1}{2}} \times 3.354^2 \times 0.6096}{2.068 \times 10^{11} \times 0.01918^3 \times 24.56 \times 0.9005 \times \sin(\pi/4)} \times 0.8977$$

$$= 7.631 \times 10^{-7} \times M_0 \, \text{rad}$$

$$N_{1A} = 0 \, \text{N/m}$$

$$\sigma_{1A} = \frac{N_{1A}}{t} = 0 \, \text{N/m}^2$$

$$N_{2A} = M_0 \frac{\beta}{t C_1} F_{7A} = M_0 \times \frac{3.354}{0.01918 \times 0.9005} \times 0.8947 = 173.8 \times M_0 \, \text{N/m}$$

$$\sigma_{2A} = \frac{N_{2A}}{t} = \frac{173.8 \times M_0 0}{0.01918} = 9062 \times M_0 \, \text{N/m}^2$$

$$M_{1A} = M_0 \, \text{Nm/m}$$

$$\sigma'_{1A} = \frac{-6 M_{1A}}{t^2} = \frac{-6 \times M_0}{0.01918^2} = -1.631 \times 10^4 \times M_0$$

$$M_{2A} = M_0 \left[v + \frac{2(2)^{\frac{1}{2}} (1 - v^2)}{k_A C_1} F_{2A} \right]$$

$$= M_0 \times \left[0.25 + \frac{2 \times 2^{\frac{1}{2}} \times (1 - 0.25^2)}{24.56 \times 0.9005} \times 0.8977 \right]$$

$$= 0.3576 \times M_0 \, \text{Nm/m}$$

$$\sigma'_{2A} = \frac{-6 M_{2A}}{t^2} = \frac{-6 \times 0.3576 \times M_0}{0.01918^2} = -5833 \times M_0$$

Now, we consider the cylindrical shell. We assume the initial section and the final section of this shell to be at a sufficient distance one from the other, that the stresses and the deformations of the material at one of them do not affect the stresses and the deformations at the other. The radius and the thickness of the cylindrical shell are respectively $R = 0.6096$ m and $t = 0.01608$ m. By using these values and the mechanical properties of the material ($E = 2.068 \times 10^{11}$ N/m^2 and $v = 0.25$), we compute the following quantities

$$\lambda = \left[\frac{3(1 - v^2)}{R^2 t^2} \right]^{\frac{1}{4}} = \left[\frac{3 \times (1 - 0.25^2)}{0.6096^2 \times 0.01608^2} \right]^{\frac{1}{4}} = 13.08 \, \text{m}^{-1}$$

$$D = \frac{Et^3}{12(1-v^2)} = \frac{2.068 \times 10^{11} \times 0.01608^3}{12 \times (1-0.25^2)} = 7.643 \times 10^4 \, \text{Nm}$$

Now we use the formulae of [3, Table 13.1, case 1c], which concern the membrane stresses and strains on a thin-walled cylindrical shell subject to a uniform internal or external pressure q (N/m^2), with ends capped. These formulae, applied to the present case, are

$$\sigma_1 = \frac{qR}{2t} = \frac{2.068 \times 10^6 \times 0.6096}{2 \times 0.01608} = 3.920 \times 10^7 \, \text{N/m}^2$$

$$H = \sigma_1 t = 3.920 \times 10^7 \times 0.01608 = 6.303 \times 10^5 \, \text{N/m}$$

$$\sigma_2 = \frac{qR}{t} = 2\sigma_1 = 2 \times 3.920 \times 10^7 = 7.840 \times 10^7 \, \text{N/m}^2$$

$$\sigma_1' = \sigma_2' = 0 \, \text{N/m}^2$$

$$\Delta R = \frac{qR^2}{Et}\left(1 - \frac{v}{2}\right) = \frac{2.068 \times 10^6 \times 0.6096^2}{2.068 \times 10^{11} \times 0.01608} \times \left(1 - \frac{0.25}{2}\right)$$
$$= 0.0002022 \, \text{m} = 0.2022 \, \text{mm}$$

$$\psi = 0 \, \text{rad}$$

Now we use the formulae of [3, Table 13.2, case 8], which concern the membrane and bending stresses and strains on a long cylindrical shell, with the left end free and the right end more than $6/\lambda$ units of length, subject at its left end to a radial load per unit length V_0 (N/m). These formulae, applied to the present case, are

$$\psi_A = \frac{V_0}{2D\lambda^2} = \frac{V_0}{2 \times 7.643 \times 10^4 \times 13.08^2} = 3.824 \times 10^{-8} \times V_0 \, \text{rad}$$

$$\Delta R_A = y_A = \frac{-V_0}{2D\lambda^3} = \frac{-V_0}{2 \times 7.643 \times 10^4 \times 13.08^3} = -2.923 \times 10^{-9} \times V_0 \, \text{m}$$

$$\sigma_1 = 0 \, \text{N/m}^2$$

$$\sigma_2 = \frac{y_A E}{R} + v\sigma_1 = \frac{-2.923 \times 10^{-9} \times V_0 \times 2.068 \times 10^{11}}{0.6096} = -991.6 \times V_0 \, \text{N/m}^2$$

$$\sigma_1' = 0 \, \text{N/m}^2$$

$$\sigma_2' = \nu\sigma_1' = 0\,\text{N/m}^2$$

Now we use the formulae of [3, Table 13.2, case 10], which concern the membrane and bending stresses and strains on a long cylindrical shell, with the left end free and the right end more than $6/\lambda$ units of length, subject at its left end to a moment per unit length M_0 (Nm/m). These formulae, applied to the present case, are

$$\psi_A = \frac{-M_0}{D\lambda} = \frac{-M_0}{7.643 \times 10^4 \times 13.08} = -1.000 \times 10^{-6} \times M_0\,\text{rad}$$

$$\Delta R_A = y_A = \frac{M_0}{2D\lambda^2} = \frac{M_0}{2 \times 7.643 \times 10^4 \times 13.08^2} = 3.824 \times 10^{-8} \times M_0\,\text{m}$$

$$\sigma_1 = 0\,\text{N/m}^2$$

$$\sigma_2 = \frac{2M_0\lambda^2 R}{t} = \frac{2 \times M_0 \times 13.08^2 \times 0.6096}{0.01608} = 1.297 \times 10^4 \times M_0\,\text{N/m}^2$$

$$\sigma_1' = \frac{-6M_0}{t^2} = \frac{-6 \times M_0}{0.01608^2} = -2.320 \times 10^4 \times M_0\,\text{N/m}^2$$

$$\sigma_2' = \nu\sigma_1' = 0.25 \times \left(-2.320 \times 10^4 \times M_0\right) = -5.801 \times 10^3 \times M_0\,\text{N/m}^2$$

Now, we sum the radial deformations ΔR for the conical shell and equate the result to the sum of the radial deformations ΔR for the cylindrical shell. This yields

$$0.0002398 + 1.827 \times 10^{-9} \times Q_0 + 2.670 \times 10^{-8} \times M_0$$
$$= 0.0002022 - 2.923 \times 10^{-9} \times V_0 + 3.824 \times 10^{-8} \times M_0$$

Then, we do the same operation for the sums of the angular rotations ψ of the two shells. This yields

$$0.0006742 + 2.670 \times 10^{-8} \times Q_0 + 7.631 \times 10^{-7} \times M_0$$
$$= 0 + 3.824 \times 10^{-8} \times V_0 - 1.0 \times 10^{-6} \times M_0$$

Finally, we do the same operation for the radial forces per unit length acting on the two shells. This yields

$$Q_0 + T \cos(\pi/4) = V_0$$

Since $T = 8.914 \times 10^5$ N/m, then

$$Q_0 + 8.914 \times 10^5 \times \cos(\pi/4) = V_0$$

By substituting this value of V_0 in the two preceding equations which express respectively the radial deformations ΔR and the angular rotations ψ, we find

$$11.54\, M_0 - 4.750\, Q_0 = 1.880 \times 10^6$$
$$176.3\, M_0 - 1.154\, Q_0 = 2.343 \times 10^6$$

The preceding system of linear equations, solved for M_0 and Q_0, yields

$$M_0 = 1.087 \times 10^4\, \text{Nm/m}$$
$$Q_0 = -3.694 \times 10^5\, \text{N/m}$$

Substituting $Q_0 = -3.694 \times 10^5$ N/m into $V_0 = Q_0 + 8.914 \times 10^5 \times \cos(\pi/4)$ yields

$$V_0 = -3.694 \times 10^5 + 8.914 \times 10^5 \times \cos(\pi/4) = 2.609 \times 10^5\, \text{N/m}$$

Since M_0, Q_0, and V_0 have known values, we can also evaluate the membrane and bending stresses in the cylindrical shell. They are

$$\sigma_1 = 3.920 \times 10^7 + 0 + 0 = 3.920 \times 10^7\, \text{N/m}^2$$

$$\sigma_2 = 7.840 \times 10^7 - 991.6 \times 2.609 \times 10^5 + 1.297 \times 10^4 \times 1.087 \times 10^4$$
$$= -3.930 \times 10^7\, \text{N/m}^2$$

$$\sigma_1' = 0 + 0 - 2.320 \times 10^4 \times 1.087 \times 10^4 = -2.522 \times 10^8\, \text{N/m}^2$$

$$\sigma_2' = 0 + 0 - 5.801 \times 10^3 \times 1.087 \times 10^4 = -6.306 \times 10^7\, \text{N/m}^2$$

The combined meridian and circumferential stresses in the cylindrical shell are computed as follows:

(1) Combined meridian stress on the outside of the cylindrical shell:

$$0.3920 \times 10^8 - 2.522 \times 10^8 = -2.13 \times 10^8\, \text{N/m}^2$$

(2) Combined meridian stress on the inside of the cylindrical shell:

$$0.3920 \times 10^8 + 2.522 \times 10^8 = 2.914 \times 10^8\, \text{N/m}^2$$

(3) Combined circumferential stress on the outside of the cylindrical shell:

$$-0.393 \times 10^8 - 0.6306 \times 10^8 = -1.024 \times 10^8 \, \text{N/m}^2$$

(4) Combined circumferential stress on the inside of the cylindrical shell:

$$-3.930 \times 10^7 + 6.306 \times 10^7 = 2.376 \times 10^7 \, \text{N/m}^2$$

We evaluate likewise the membrane and bending stresses in the conical shell. They are

$$\sigma_1 = 4.648 \times 10^7 + 36.87 \times \left(-3.694 \times 10^5\right) + 0 = 3.286 \times 10^7 \, \text{N/m}^2$$

$$\sigma_2 = 9.295 \times 10^7 + 630.3 \times \left(-3.694 \times 10^5\right) + 9062 \times 1.087 \times 10^4$$
$$= -4.138 \times 10^7 \, \text{N/m}^2$$

$$\sigma_1' = 0 + 0 - 1.631 \times 10^4 \times 1.087 \times 10^4 = -1.773 \times 10^8 \, \text{N/m}^2$$

$$\sigma_2' = 0 - 61.44 \times \left(-3.694 \times 10^5\right) - 5833 \times 1.087 \times 10^4$$
$$= -4.07 \times 10^7 \, \text{N/m}^2$$

The combined meridian and circumferential stresses in the conical shell are computed as follows:

(1) Combined meridian stress on the outside of the conical shell:

$$0.3286 \times 10^8 - 1.773 \times 10^8 = -1.444 \times 10^8 \, \text{N/m}^2$$

(2) Combined meridian stress on the inside of the conical shell:

$$0.3286 \times 10^8 + 1.773 \times 10^8 = 2.102 \times 10^8 \, \text{N/m}^2$$

(3) Combined circumferential stress on the outside of the conical shell:

$$-4.138 \times 10^7 - 4.07 \times 10^7 = -8.208 \times 10^7 \, \text{N/m}^2$$

(4) Combined circumferential stress on the inside of the conical shell:

$$-4.138 \times 10^7 + 4.07 \times 10^7 = -0.068 \times 10^7 \, \text{N/m}^2$$

6.5 Tanks Subject to Loads Due to Propellant Sloshing

This section considers the loads acting on the tanks of large space vehicles partially filled with liquid propellants. Since the initial masses of these vehicles consist principally of liquid propellants, then the time-varying loads due to the motion of such masses in their tanks are very large. The control systems and the structures of launch vehicles are to be designed in such a way as to respectively counteract and resist such loads.

The dynamical systems which describe the motion of liquid masses into moving containers are very complex, because of couplings of the components of such systems. In particular, the natural frequencies of the control systems, of the elastic bodies, and of the liquid sloshing are to be kept separated as widely as possible, because large forces and moments can be generated by a liquid propellant oscillating at one of its natural frequencies in a partially filled tank. Unfortunately, such is not always the case, as shown in the following table, adapted from [6], which refers to some representative launch vehicles.

Characteristics of Some Representative Launch Vehicles

Vehicle	Length, m	Diameter, m	Thrust, N	Range, km	Control frequency, Hz	Fundamental slosh frequency at liftoff, Hz	Fundamental bending frequency at liftoff, Hz	Important missions
Redstone	21	1.78	1.432×10^8	370.4	0.5	0.8	10–12	Exploration.
Redstone-Mercury	25	1.78	1.432×10^8	370.4	.5	.8	10	Suborbital manned flights.
Jupiter	20	2.65	3.925×10^8	2778	.4	ª.6	9	Reentry, recovery of monkeys Able and Baker.
Juno II	25	2.65	3.925×10^8	--------	.4	.6	8	Moon try, Sun orbit.
Saturn I	60	6.5	3.047×10^9	--------	.3	.45	2	Manned space flight.
Saturn V	130	10	1.512×10^{12}	--------	.16	ᵇ 0.3–0.4	1	Manned space flight.

ª Large slosh masses in unfavorable locations.
ᵇ Exceptionally large slosh masses because of the large tank diameter.

By sloshing we mean the periodic motion of the free surface of a liquid in a partially filled container [7]. This motion results from longitudinal and lateral displacements or angular rotations of the vehicle carrying the container.

In particular, lateral sloshing is the standing wave formed on the free surface of a liquid when a tank partially filled is caused to oscillate horizontally [8], that is, in a plane parallel to the free surface of the quiescent liquid. Lateral sloshing is antisymmetric, and occurs primarily in response to translational or pitching motions of a tank. By contrast, vertical sloshing is symmetric, and occurs primarily in response to motions of a tank in the direction perpendicular to the free surface of the quiescent liquid. Rotational sloshing is a motion exhibiting an apparent swirling of a liquid about a normal axis, and arising as an instability of the antisymmetric lateral sloshing near resonance [6].

The tanks of space vehicles may be subject to oscillations for several causes, which act either separately or in combination. Some examples of these causes, identified by Abramson et al. [9], are

- wind gusts during powered flight;
- programmed changes of attitude of the vehicle;
- control pulses for attitude stabilisation;
- separation impulses; and
- elastic deformations of the vehicle.

The magnitude of the forces and moments due to sloshing depend upon

- shape of the tank;
- properties of the propellants;
- damping;
- height of the propellant in the tank;
- acceleration; and
- perturbing motion of the tank.

The sloshing phenomenon and its consequence on the stability of a space vehicle can be controlled by a proper design of the tank and by the addition of baffles, as will be shown in Sect. 6.6. The shape of the tank has an influence on the natural sloshing frequencies and modes, and on the response to the forced oscillations and to the forces and moments due to the presence of the propellant in the tank. The following figure, due to the courtesy of NASA [6], shows (left) ring baffles mounted on the wall of a circular cylindrical tank wall by means of Z-ring stiffeners, and (right) some typical shapes of tanks.

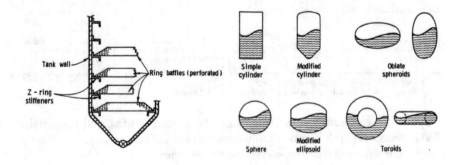

The number and the type of baffles depend on the damping requirements. The strength of the baffles depends on several factors, among which the strength and the rigidity required during manufacturing and handling, the mechanical and thermal stresses due to propellant loading, and the forces and moments due to propellant sloshing [9].

The coupling between a sloshing propellant and the elastic structure of its tank may have a large influence on the vibration frequencies and on the mode shapes of the tank. In addition, the system comprising a sloshing propellant and its tank may have dynamic instabilities either by itself or by coupling with some other components of a rocket engine subject to oscillations, as is the case with the combustion and feed-line systems.

The dynamic response of a space vehicle to sloshing loads can be determined by using an equivalent mechanical model which represents the behaviour of an oscillating tank partially filled with a liquid propellant. This mechanical model consists of fixed masses (m_f) and oscillating masses (m_s) connected to the tank by springs and dashpots or by pendulums and dashpots. The second case is shown in the following figure, re-drawn from [10].

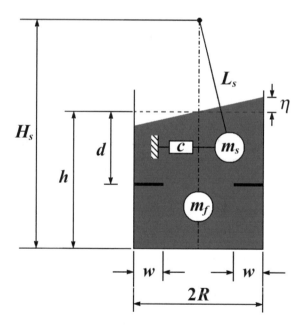

Each mechanical model refers to a particular shape of tank, and is designed so as to have the same resultant force, moment, damping, and natural frequencies as the actual oscillating propellant. This model is then combined with other dynamic elements of the space vehicle being considered, and the dynamic behaviour of the whole system is studied by means of digital or analogue computers.

The number of slosh masses m_s included in an equivalent mechanical model corresponds to the number of slosh modes considered in the analysis. In case of a circular cylindrical tank, the sloshing mass corresponding to the second mode has been found to be about 3% of the mass corresponding to the first node, and therefore the sloshing effects due to the second and higher modes are generally negligible for a tank of this shape. However, for a quarter tank, the sloshing mass corresponding to the second mode is 43% of the mass corresponding to the first mode, and therefore should be included in determining the total sloshing loads acting on the tank [9]. Sloshing effects in tanks having arbitrary shapes have been studied by various authors (see, for example, [6, 11]).

For lateral sloshing of a liquid propellant in a rigid circular cylindrical tank with a flat bottom, the frequencies f_n (Hz) of the oscillating free surface have been found [6, 9] to be

$$f_n = \frac{\omega_n}{2\pi} = \frac{1}{2\pi}\left[\varepsilon_n \frac{g}{R} \tanh\left(\varepsilon_n \frac{h}{R}\right)\right]^{\frac{1}{2}}$$

where ω_n (rad/s) are the angular frequencies, g (m/s^2) is the vertical acceleration of the tank, ε_n are values determined from the roots of the following equation

$$\left(\frac{d J_1(x)}{dx}\right)_{x=\varepsilon_n} = 0$$

$J_1(x)$ is the Bessel function of the first kind and order, R (m) is the radius of the circular cylindrical tank, and h (m) is the height of the free surface of the quiescent liquid propellant. Values of ε_n for the first four slosh modes are given in the following table (from [9]):

Slosh mode (n)	ε_n
1	1.841
2	5.331
3	8.536
4	11.706

Abramson et al. [9] point out that, when the height h of the liquid propellant is greater than the radius R of the circular cylindrical tank, then the equation expressing the frequencies f_n can be approximated as follows

$$f_n = \frac{1}{2\pi}\left(\varepsilon_n \frac{g}{R}\right)^{\frac{1}{2}}$$

A validation of the parameters of the equivalent mechanical model (pendulums and dashpots) illustrated in the preceding figure has been performed by Pérez et al. [10]. This linear (dashpot damping $c = $ constant) model assumes the angles θ of oscillation of the pendulums to be within the interval $\pm\pi/12$ [12].

Since the dominant sloshing frequency f_1 corresponds to $n = 1$ and therefore to the root $\varepsilon_1 = 1.841$ of the equation $J_1' = 0$, then Pérez et al. have considered only one sloshing mass m_s attached to a pendulum of length L_s.

A brief account of the results found by them is given below.

(1) Dominant sloshing frequency

$$f_1 = \frac{1}{2\pi}\left[1.841\frac{g}{R}\tanh\left(1.841\frac{h}{R}\right)\right]^{\frac{1}{2}}$$

(2) Sloshing mass

$$m_s = \frac{mR}{2.199h}\tanh\left(1.841\frac{h}{R}\right)$$

where $m = m_s + m_f$ is the total mass of the propellant.

(3) Height of the suspension point of the pendulum

$$H_s = h - \frac{R}{1.841}\left[\tanh\left(0.9205\frac{h}{R}\right) - \operatorname{csch}\left(1.841\frac{h}{R}\right)\right]^{\frac{1}{2}}$$

(4) Dashpot damping

$$c = 4\pi f_1 m_s \zeta$$

where ζ is the damping ratio, which is defined in terms of the damping factor δ as follows

$$\zeta = \frac{\delta}{\left(4\pi^2 + \delta^2\right)^{\frac{1}{2}}}$$

and the damping factor δ is defined as follows

$$\delta = \frac{1}{k}\ln\left(\frac{A_0}{A_k}\right)$$

where A_0 (m) is the amplitude of the first wave, A_k (m) is the amplitude of the kth wave, and k is the number of cycles over which the decay is measured.

The magnitude of damping in smooth-wall and in baffled-wall tanks has been determined for several shapes. Generally speaking, the amount of damping due to the wiping action of a liquid propellant against the walls of a tank is insufficient, and therefore baffles must be added to provide the damping required to prevent instability. For example, a circular cylindrical tank having ring baffles of width w along its wall is shown in the preceding figure. The damping of liquid propellants in circular cylindrical tanks without baffles has been studied by Stephens et al. [13]. Viscous damping of liquid propellants in tanks of various shapes has been studied by several authors. By viscous damping we mean the damping produced by interaction between the liquid propellant and the wall of the tank. An account of empirical formulae for tanks of various shapes is given in [6, 8]. For example, for a circular cylindrical tank without baffles, the damping ratio ζ is expressed by the following experimental equation due to Mikishev and Dorozhkin [8]:

$$\zeta = 0.79 \left(\frac{\mu}{\rho}\right)^{\frac{1}{2}} R^{-\frac{3}{4}} g^{-\frac{1}{4}} \left\{ 1 + \frac{0.318}{\sinh(1.84h/R)} \left[\frac{1 - h/R}{\cosh(1.84h/R)} + 1 \right] \right\}$$

where μ/ρ is the kinematic viscosity of the liquid, R is the radius of the cross section of the tank, and g is the vertical acceleration. When the depth h of the liquid is greater than the diameter $2R$ of the tank cross section, then the preceding equation can be approximated as follows

$$\zeta = 0.79 \left(\frac{\mu}{\rho}\right)^{\frac{1}{2}} R^{-\frac{3}{4}} g^{-\frac{1}{4}}$$

Effective damping provided by baffles of various types has also been investigated by several authors. An account is given in [6, 8]. With reference to the preceding figure, Dodge [8] indicates the following equation, due to Miles [16], to estimate the damping ratio ζ for a flat ring rigid baffle in a circular cylindrical tank where the liquid depth h is considerably greater than the cross-sectional radius R ($h/R > 2$):

$$\zeta = 2.83 C_1^{\frac{3}{2}} \left(\frac{\eta}{R}\right)^{\frac{1}{2}} \exp\left(-4.6 \frac{d}{R}\right)$$

where C_1 is the ratio of the baffle area A_B to the tank cross-sectional area A_T, and therefore C_1 is for a circular cylindrical tank

$$C_1 = \frac{A_B}{A_T} = \frac{\pi R^2 - \pi (R - w)^2}{\pi R^2} = \frac{w}{R}\left(2 - \frac{w}{R}\right)$$

w is the width of the ring baffles, R is the radius of the cross section of the tank, η is the amplitude of oscillation measured at the tank wall from the free surface of the quiescent liquid, and d is the depth of the baffle below the free surface of the quiescent liquid. For circular cylindrical tanks, the damping provided by a series of ring baffles can be calculated by superposing linearly the contribution of each baffle, when the spacing s between the baffles is greater than the width w of each baffle. For tanks of this shape, ring baffles are usually placed at a distance $s \leq 0.2R$ [10].

Types of baffles studied for circular cylindrical tanks include fixed rings, rings with radial clearance, cruciform baffles, and baffles shaped as conic sections. Baffles have also been studied for spherical tanks and for oblate or prolate spheroidal tanks. Flexible baffles have been compared to rigid baffles, and the former have been found to provide greater damping than the latter under certain conditions [9]. The following figure, due to the courtesy of NASA [6], shows some types of ring baffles used for circular cylindrical tanks.

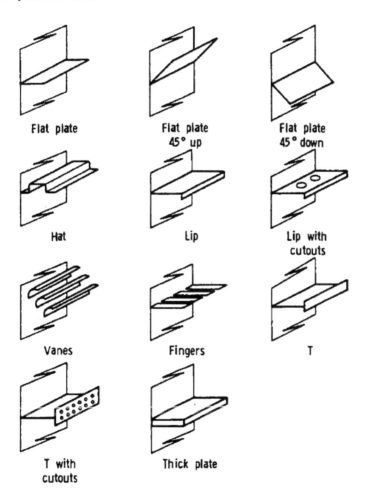

The following figure, also due to the courtesy of NASA [6], shows some types of cruciform baffles used for tanks of various shapes.

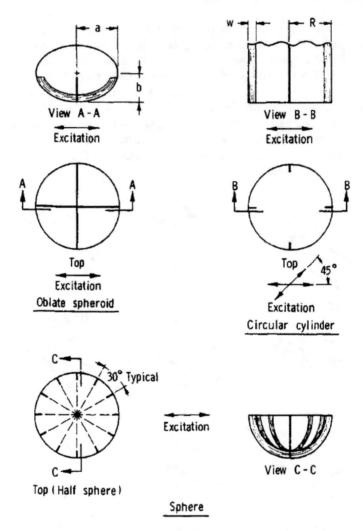

Sloshing loads under low gravity conditions have been found small in comparison with the structural capability of propellant tanks. This topic is dealt with at length in [6, 8].

The lateral sloshing of liquid propellants in their tanks causes a distributed pressure loading on the walls. These loads are important for the structural design of the tanks. In addition, the resultant force and moment due to the distributed pressure are important for the control systems of space vehicles.

The longitudinal sloshing of liquid propellant also causes pressure loads on the tanks. These loads are less important than those due to lateral sloshing. However, the longitudinal pressure modes of a propellant may couple with those of the elastic shell or with those of the feed-line and combustion system. This fact can generate dynamic instability, as is the case with the pogo oscillations.

Loads acting on the top or on the bottom of a circular cylindrical tank can be caused by sudden changes of the net acceleration in the vertical direction. This may occur in such cases as abort just after launch, cut-off of the boost engine, engine start in orbit or in coast flight, et c. These loads are very sensitive to the test conditions and to the shapes of tanks [9].

In case of a circular cylindrical tank subject to lateral sloshing, a simple estimate of the pressure loads can be done by considering the tank pressurisation p_0, the static pressure $\rho g z$ due to the liquid propellant in the tank, and the pressure due to the lateral sloshing at its lowest natural frequency $f_1 = \omega_1/(2\pi)$, corresponding to the first mode of oscillation. In other words, the maximum pressure p_{max} (N/m^2) on a tank subject to lateral sloshing occurs in the plane of oscillation and can be expressed by using the following equation of [9]:

$$p_{max} = p_0 + \rho g z + \rho g \eta \frac{\cosh\left[1.841\left(\frac{h-z}{R}\right)\right]}{\cosh\left[1.841\frac{h}{R}\right]} \sin(\omega_1 t)$$

where ρ (kg/m^3) is the density of the liquid propellant in the operational conditions, g (m/s^2) is the vertical acceleration of the tank, η (m) is the maximum height above the free surface of the quiescent liquid at the wall, R (m) is the radius of the tank, h (m) is the height of the free surface of the quiescent liquid, z (m) is the distance from the free surface of the quiescent liquid to any arbitrary depth (positive downward), and ω_1 (rad/s) is the first natural angular frequency of lateral slosh, whose value has been found above to be

$$\omega_1 = \left[1.841\frac{g}{R}\tanh\left(1.841\frac{h}{R}\right)\right]^{\frac{1}{2}}$$

Abramson et al. [9] point out that the total vertical acceleration acting on a particle of the free surface varies between $g - \eta\omega_1^2$ and $g + \eta\omega_1^2$ at the wall of the tank. When the amplitude of the sloshing propellant becomes large enough for the total vertical acceleration to be instantaneously zero, then the sloshing wave breaks up, and turbulent sloshing begins. Taking this condition as an upper limit yields $\eta = g/\omega_1^2$, and therefore the equation expressing the maximum pressure can be re-written as follows

$$p_{max} = p_0 + \rho g \left\{ z + \frac{R\cosh\left[1.841\left(\frac{h-z}{R}\right)\right]}{1.841 \sinh\left[1.841\frac{h}{R}\right]} \right\}$$

When the height h of the free surface is roughly equal to or greater than the radius R of the tank, the oscillations of the liquid propellant become independent of h, and the equation expressing the maximum pressure can be approximated as follows [9]:

$$p_{max} = p_0 + \rho g \left[z + \frac{R}{1.841} \exp\left(-1.841\frac{z}{R}\right) \right]$$

This equation still holds approximately in cases of tanks whose bottom surfaces are not flat. In case of non-cylindrical tanks, the maximum pressure can be expressed by using other formulae, for example, those of [6, 8], or [11].

Generally speaking, one third of the liquid propellant, the part of it which is near the bottom of the tank, is not affected by sloshing, which occurs near the free surface. The equivalent mechanical models consisting of fixed masses and oscillating masses are based on this fact. The sloshing mass and the resultant force decrease considerably in the higher modes of oscillations. For the lower modes, the sloshing mass can be reduced by dividing a tank into radial or concentric compartments. Some typical configurations used for propellant tanks are shown in the following figure, due to the courtesy of NASA [6].

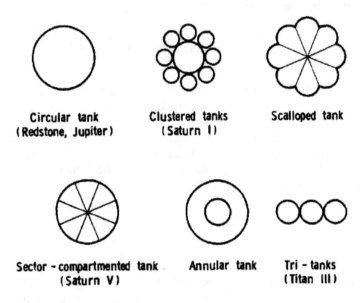

Circular tank
(Redstone, Jupiter)

Clustered tanks
(Saturn I)

Scalloped tank

Sector - compartmented tank
(Saturn V)

Annular tank

Tri - tanks
(Titan III)

On the effectiveness of dividing a tank into compartments for the purpose of reducing the sloshing mass, Dodge [8] notes that, for a circular cylindrical tank, the division into radial sectors raises the first (or fundamental) slosh frequency and lowers the second, so that the two modes are less separated in frequency. In addition, radial compartments are less effective than ring baffles in reducing the amplitude of forces and moments due to sloshing [8].

When a liquid propellant sloshes in a longitudinal mode, the pressures integrated over the bottom and the walls of the tank have a zero resultant force. An estimate of the pressure p (N/m^2) during sloshing in a circular cylindrical tank at the first longitudinal angular frequency ω_1 (rad/s) can be done by using the following formula of [9]:

$$p = p_0 + \rho g\left\{z + \eta \frac{J_0\left(3.83\frac{r}{R}\right)}{J_0(3.83)} \frac{\cosh\left[3.83\left(\frac{h-z}{R}\right)\right]}{\cosh\left[3.83\frac{h}{R}\right]} \sin(\omega_1 t)\right\}$$

where

$$\omega_1 = \left[3.83\frac{g}{R}\tanh\left(3.83\frac{h}{R}\right)\right]^{\frac{1}{2}}$$

Hopfinger and Baumbach [14] is the first of the natural angular frequencies of longitudinal slosh, r (m) is the radial co-ordinate, and J_0 (x) is the zero-order Bessel function of the first kind.

When the height h of the free surface is roughly equal to or greater than the radius R of the tank, assuming again $\eta = g/\omega_1^2$, the maximum pressure becomes

$$p_{max} = p_0 + \rho g\left[z + \frac{R}{3.83}\frac{J_0\left(3.83\frac{r}{R}\right)}{J_0(3.83)}\exp\left(-3.83\frac{z}{R}\right)\right]$$

Abramson et al. [9] also point out that the effect of an elastic tank bottom is to lower the natural sloshing frequencies of the free surface slightly below their values in a rigid tank.

Longitudinal accelerations impressed to a partially filled tank can cause a liquid propellant to impinge on the dome of the tank. The problem of dome impact arises when the resultant acceleration acting on the tank reverses its direction. This may occur during engine shutdown of a launch vehicle flying through the atmosphere, or when an engine is ignited while the propellant is located in the upper part of a tank [6]. The impulsive pressure resulting from one of such events must be taken into account when assessing the structural integrity of a tank. This problem is considered at length in [6]. The impact of a liquid propellant on deflector baffles may also produce considerable loads. There being no general methods of analysis for predicting such loads, empirical methods or tests or both are used for this purpose. Ring baffles and other internal devices can reduce fluid impact loads [9].

The maximum pressure p (N/m^2) acting on a submerged baffle of width w (m), subject to an oscillating velocity $U = U_m \cos(\omega t)$ (m/s), due a liquid propellant sloshing at an angular frequency ω (rad/s), is given by the following equation of [9]:

$$p = K\rho U_m^2$$

where K is a non-dimensional parameter whose values are given in the following figure of [9, 15]

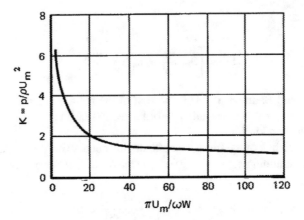

and ρ (kg/m^3) and U_m (m/s) are respectively the density and the maximum velocity of the liquid propellant in the operating conditions.

The maximum vertical velocity U_m (m/s) of the sloshing liquid at any distance z (m), positive downward, from the free surface of the quiescent liquid to any arbitrary depth in a circular cylindrical tank is given by the following equation of [9]:

$$U_m = \frac{1.841g}{\omega_1 R} \eta \frac{\sinh\left[1.841\left(\frac{h-z}{R}\right)\right]}{\cosh\left[1.841\left(\frac{h}{R}\right)\right]}$$

where g (m/s^2) is the vertical acceleration of the tank, R (m) is the radius of the cross section of the tank, h (m) is the height of the free surface of the quiescent liquid, η (m) is the amplitude of oscillation measured at the tank wall from the free surface of the quiescent liquid, and ω_1 (rad/s) is the first natural angular frequency of lateral slosh, whose value has been found above to be

$$\omega_1 = \left[1.841\frac{g}{R}\tanh\left(1.841\frac{h}{R}\right)\right]^{\frac{1}{2}}$$

By using these equation and assuming the height h of the propellant to be equal to or greater than the radius R of the tank, the equation $p = K\rho U_m^2$, which expresses the pressure on the baffle due to the sloshing liquid propellant, can be approximated as follows [9]:

$$p = K\rho\omega_1^2\eta^2 \exp\left(-3.682\frac{z}{R}\right)$$

After setting again $\eta = g/\omega_1^2$, the maximum pressure on a ring baffle placed at the depth z from the free surface can be expressed as follows

$$p = \frac{K\rho g R}{1.841}\exp\left(-3.682\frac{z}{R}\right)$$

The pressure acting on a ring baffle can be reduced by perforating the ring baffle with small holes. This is because the oscillating flow through the holes is an additional source of damping. However, this increased damping is partially offset by a decrease in the effective area of a perforated baffle [8].

A ring baffle, which is just above the free surface of a quiescent liquid propellant, is periodically subject to the slapping action of the sloshing wave. In this case, the pressure acting on the baffle is expressed by the following equation of [9]:

$$p = 2\rho U_m^2$$

where U_m is the velocity of the liquid propellant at impact with the baffle.

By setting $U_m = \omega_1 \eta_1$ and substituting in the preceding equation, there results

$$p = 2\rho \omega_1^2 \eta_1^2$$

By setting again $\eta_1 = g/\omega_1^2$ and assuming $h \geq R$, the maximum pressure on the baffle is [9]:

$$p_{max} = \frac{2\rho g R}{1.841}$$

The methods discussed above have shown how to compute sloshing loads on baffles of the ring type.

When baffles of other types, such as anti-vortex baffles and truss-type baffles, are used, then it is still possible to use the preceding equation $p = K\rho U_m^2$, which expresses the maximum pressure on the baffle, but it is also necessary to determine the value of K and the value of U_m which apply to the case of interest. Keulegan and Carpenter [15] gives values of K for oscillatory flow around cylinders and plates.

Such values depend on the baffle shape and on the non-dimensional parameter $U_m T/(2w)$, where T is the natural period of oscillation. This parameter, in turn, requires the evaluation of U_m. For this purpose, in case of a liquid propellant sloshing laterally in a circular cylindrical tank at its first angular frequency ω_1, the radial component u_r and the tangential component u_θ of the velocity vector $\boldsymbol{u}$ of the liquid propellant can be computed by using the following equations, due to Bauer [17], taken from [9]:

$$u_r = \frac{g\eta}{\omega_1}\left[\frac{1.841 J_0(1.841r/R)}{R J_1(1.841)} - \frac{J_1(1.841r/R)}{r J_1(1.841)}\right](\cos\theta)$$
$$\times \frac{\cosh[1.841(h-z)/R]}{\cosh(1.841h/R)}\cos(\omega_1 t)$$

$$u_\theta = -\frac{g\eta}{\omega_1}\frac{J_1(1.841r/R)}{r J_1(1.841)}(\sin\theta)\frac{\cosh[1.841(h-z)/R]}{\cosh(1.841h/R)}\cos(\omega_1 t)$$

from which the resultant velocity of the liquid propellant can be calculated.

6.6 Slosh-Suppression Devices for Tanks

Section 6.5 has shown how to compute structural loads induced by the periodic motion (sloshing) of liquid propellants in their tanks. The present section describes some devices used for slosh damping. Those of them which have been most frequently used for this purpose are shown in the following figure, due to the courtesy of NASA [7].

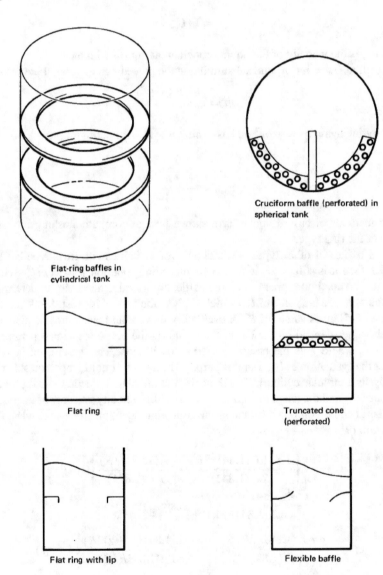

Cruciform baffle (perforated) in
spherical tank

Flat-ring baffles in
cylindrical tank

Flat ring

Truncated cone
(perforated)

Flat ring with lip

Flexible baffle

Such devices are used not only to reduce sloshing loads, but also to protect tank bulkheads from impact loads caused by non-periodic motion of liquid propellants. As has been shown in Sect. 6.5, the sloshing mass m_s of a liquid propellant, that is, the part of the total mass which moves during sloshing, can also be reduced by using tanks of appropriate shapes or divided into compartments. According to [7], slosh-suppression devices include rigid-ring baffles, cruciform baffles, deflectors, flexible flat-ring baffles, floating cans, and positive-expulsion bags or diaphragms. A brief description of them is given below.

The most frequently used manner of providing slosh damping in a tank is to install baffles in the tank. Such baffles are placed in points which are slightly below the level of the free surface of the liquid propellant at the times in which damping is required. The baffles can be installed either for the sole purpose of providing slosh damping in a tank, or also for structural reasons, to be used as wall-stiffener rings. In the latter case, their size is increased to provide the required damping.

Rigid-ring baffles are widely used as slosh-suppression devices. As has been shown in Sect. 6.5, the damping ratio ζ provided by rigid-ring baffles in circular cylindrical tanks is expressed by the Miles formula [16] as follows

$$\zeta = 2.83 \left[\frac{w}{R} \left(2 - \frac{w}{R} \right) \right]^{\frac{3}{2}} \left(\frac{\eta}{R} \right)^{\frac{1}{2}} \exp\left(-4.6 \frac{d}{R} \right)$$

which holds when the depth h of the liquid propellant is considerably greater than the cross-sectional radius R of the tank ($h/R > 2$). In the preceding formula, w is the width of the rigid-ring baffles, η is the amplitude of oscillation measured at the tank wall from the free surface of the quiescent liquid, and d is the depth of the baffle below the free surface of the quiescent liquid. This formula is based on the assumption that the baffle is completely submerged during a slosh cycle.

A series of flat-ring baffles is the most frequently used slosh-suppression device in circular cylindrical tanks. As has been shown in Sect. 6.5, damping ratios for a series of flat-ring baffles can be calculated by using the principle of linear superposition, when the spacing s between the baffles is greater than the width w of each baffle [10].

Damping provided by baffles in tanks of other shapes than circular cylinders has also been studied by several authors. An account of such studies is given in [6, 8] for spherical, oblate and prolate spheroidal, and toroidal tanks. An experimental investigation was conducted by Sumner [18] to determine the slosh-suppression effectiveness of rigid and flexible flat-plate annular-ring baffles in spherical tanks subject to oscillations in the horizontal plane. The baffles caused a variation in the fundamental frequency of oscillation, and were found most effective in reducing the slosh forces and increasing the damping when the free surface of the quiescent liquid was slightly above the baffle, so that the latter remained submerged during the oscillatory cycle of the liquid. The optimum baffle width to tank radius ratio was found to be $w/R = 0.125$ for the spherical tanks considered. A side view of the baffles in the spherical tanks is shown in the following figure, due to the courtesy of NASA [18].

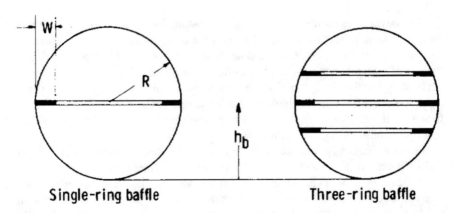

Single-ring baffle   Three-ring baffle

Stephens et al. [19] conducted an experimental investigation of the damping of liquid oscillations in an oblate spheroidal tank. The decay of the fundamental mode was studied for a range of liquid depths in tanks with and without baffles. The results indicate that the addition of ring baffles to the tank results in an increase in the available effective damping when the baffle plane is in a region near the equilibrium liquid surface, and that cruciform baffles are effective in the damping of the fundamental mode in the near-empty tank [19].

As has been anticipated in Sect. 6.5, flexible-ring baffles can offer substantial advantages, upon certain conditions, over rigid-ring baffles. Such advantages concern higher damping effectiveness and lower mass. An experimental study conducted by Stephens and Scholl [20] for large-scale cylindrical tanks fitted with both flexible and rigid annular ring baffles has shown that slosh damping comparable to that provided by rigid baffles can be obtained by using smaller and less massive flexible baffles. Stephens and Scholl have shown that the characteristics of the sloshing liquid, the flexibility of the baffles, and the damping can specified by three non-dimensional parameters. These parameters are: (1) the period parameter P, which describes the velocity of the liquid in the vicinity of the baffle; (2) the flexibility parameter F, which defines the deflection of the baffle per unit loading; and (3) the relative damping parameter δ/δ_r, which is the ratio of the damping factor δ provided by the flexible baffle to the damping factor δ_r provided by a rigid baffle of the same width w and under the same flow conditions as those of the flexible baffle.

The results found by Stephens and Scholl are shown graphically in the following figure, adapted from [20].

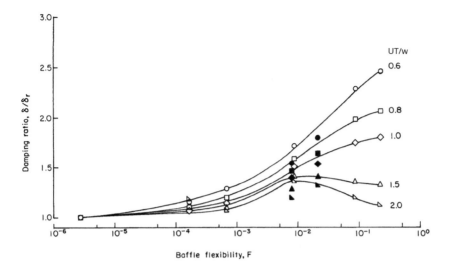

The three non-dimensional parameters P, F, and δ/δ_r are defined as follows:

$$P = \frac{UT}{w}$$

where U (m/s) is the maximum velocity of the liquid propellant at the baffle location, T (s) is the natural period of oscillation, and w (m) is the width of the baffle;

$$F = w_1^3 \left(\frac{1-v^2}{Et^3}\right) \frac{\rho w^2}{T^2} f\left(\frac{w_1}{R}\right)$$

where w_1 (m) is the width of the flexible portion of the baffle, w (m) is the width of the baffle, v is the Poisson ratio of the baffle material, E (N/m^2) is the Young modulus of the baffle material, t (m) is the thickness of the baffle, ρ (kg/m^3) the density of the liquid propellant, $f(w_1/R)$ is a radius correction factor, whose value is close to unity for most applications, and R (m) is the radius of the cylindrical tank;

$$\frac{\delta_r}{2\pi} = 2.83\left[\frac{w}{R}\left(2 - \frac{w}{R}\right)\right]^{\frac{3}{2}} \left(\frac{\eta}{R}\right)^{\frac{1}{2}} \exp\left(-4.6\frac{d}{R}\right)$$

where η (m) is the amplitude of oscillation measured at the tank wall from the free surface of the quiescent liquid, and d (m) is the depth of the baffle below the free surface of the quiescent liquid.

The maximum vertical velocity U of the liquid propellant at the baffle location, due to the lateral oscillation (antisymmetric mode) impressed to the circular cylindrical tank, is given by the following equation of Sect. 6.5:

$$U = \frac{1.841g}{\omega_1 R} \eta \frac{\sinh\left[1.841\left(\frac{h-z}{R}\right)\right]}{\cosh\left[1.841\left(\frac{h}{R}\right)\right]} = \omega_1 \eta \frac{\sinh\left[1.841\left(\frac{h-z}{R}\right)\right]}{\sinh\left[1.841\left(\frac{h}{R}\right)\right]}$$

where h (m) is the height of the free surface of the quiescent liquid propellant, z (m) is the distance from the free surface of the quiescent liquid to any arbitrary depth (positive downward), and ω_1 (rad/s) is the first natural angular frequency of lateral slosh, whose value has been found in Sect. 6.5 to be

$$\omega_1 = \left[1.841\frac{g}{R}\tanh\left(1.841\frac{h}{R}\right)\right]^{\frac{1}{2}}$$

By substituting the preceding expression of U into $P = UT/w$ and remembering that $T = 2\pi/\omega_1$, the period parameter P can be expressed as follows

$$P = \frac{2\pi\eta}{w} \frac{\sinh\left[1.841\left(\frac{h-z}{R}\right)\right]}{\sinh\left[1.841\left(\frac{h}{R}\right)\right]}$$

As shown in the preceding figure, Stephens and Scholl [20] considered values of P ranging from 0.6 to 2, the latter value being the highest attainable and having the appearance of a relatively severe slosh.

The results found by Stephens and Scholl [20] in their experimental investigation on flexible and rigid baffles can be summarised as follows:

- the damping factor δ provided by a flexible baffle is comparable to or greater than the damping factor δ_r provided by a rigid baffle in the same conditions of oscillatory flow;
- the efficiency of damping per unit of weight of a flexible baffle may greatly exceed that of a rigid baffle;
- as a baffle becomes more flexible (that is, for increasing values of the flexibility parameter F), the relative damping parameter δ/δ_r also increases and reaches a maximum value, which depends on the value of the period parameter P; and
- as the flexibility parameter increases further, the relative damping parameter δ/δ_r decreases rapidly to the point at which the baffle opposes no resistance to the flow.

Cruciform baffles are located in tanks in the same manner as stringers. This arrangement makes the damping provided by these baffles in cylindrical tanks independent of the liquid height. Their behaviour has also been investigated in spherical or spheroidal tanks, where of course their damping depends on the liquid height. Cruciform baffles provide a smaller amount of damping than is the case with ring baffles, except when a tank is nearly empty, in which case cruciform baffles suppress rotatory motions and formation of vortices near the tank-drain outlet [7]. An example is given in the following figure, due to the courtesy of NASA [21], which illustrates the liquid-oxygen tank of the S-II stage of the Saturn V launch vehicle.

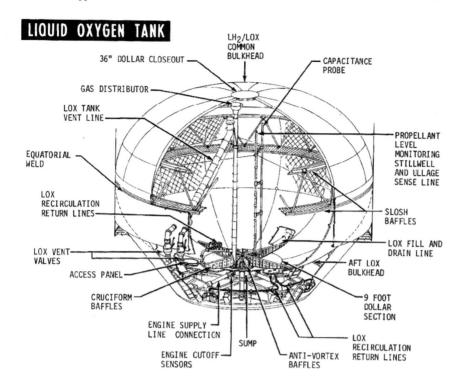

A cutaway view of the same tank is shown in the following figure, due to the courtesy of NASA [22].

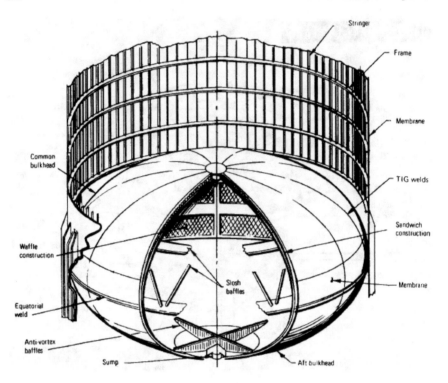

The tank illustrated in the two preceding figures has ring baffles attached to the skin stiffeners, which stabilise the tank wall and reduce sloshing. This tank has also a cruciform baffle and anti-vortex baffles (the latter just above the sump) at its base.

Deflectors have been placed above the surface of a liquid propellant in order to suppress the large-amplitude liquid motions which may be excited by an engine cut-off or by the pulsing of attitude-control engines during orbital coast. Deflectors are shaped as wide, inverted conical-ring baffles. They prevent liquid propellant from reaching the tank vent, and facilitate propellant drainage. They also contribute, when submerged, to damp oscillatory motions [7].

Rigid lids, floating cans or porous mats have also been studied for propellant tanks. Such devices always act at the free surface of a liquid propellant. Rigid lids which cover part of the free surface of a liquid propellant and float up or down as the level changes in the tank have been found effective when they cover 85% or more of the tank diameter [8]. Floating porous mats have also been evaluated for the purpose of increasing viscous effects at the surface. Floating mats (left) and cans (right) are shown in the following figure, due to the courtesy of NASA [23].

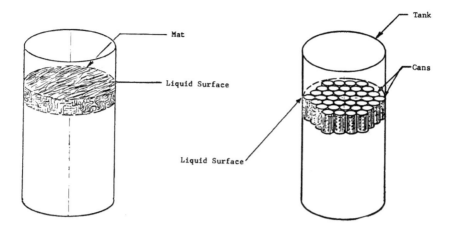

However, lids and covers may "hang up" on internal hardware of a tank, and therefore such devices have not been proved practical for spacecraft [8].

Dividing a tank into radial or concentric compartments has also been used as a means to suppress sloshing motions for the lower frequencies of oscillations. This is shown in the following figure, due to the courtesy of NASA [23].

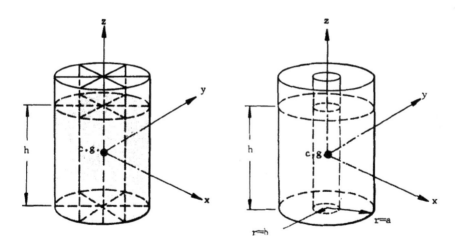

However, this solution has the drawbacks mentioned in Sect. 6.5.

Positive-expulsion bags and diaphragms of elastomeric materials are used in tanks when it is necessary to transfer liquid propellants in low-gravity conditions or when a large flow rate is desired in rocket engines fed without pumps, as has been shown in Chap. 3, Sect. 3.5. These devices provide an impermeable barrier between the liquid and the gas contained in a tank. For this purpose, a membrane of elastomeric material or a flexible metallic diaphragm is attached around the wall of a tank at some section, as shown in the following figure, re-drawn from [8].

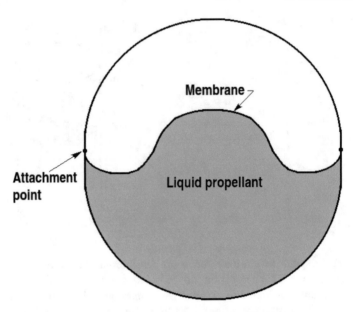

The size of the membrane must be large enough to contain the initial quantity of liquid, and to expel nearly all the liquid which remains near the bottom of the tank.

Stofan and Sumner [24] conducted an experimental investigation to evaluate the slosh-damping effectiveness of positive-expulsion bags and diaphragms in spherical tanks ranging from 0.2413 to 0.8128 m in diameter. The positive-expulsion devices tested were made of butyl rubber and ranged from 0.254 to 1.016 mm in thickness. The excitation was impressed to the tank in the horizontal plane. The maximum slosh forces occurring at the first natural frequency increased with an increase in excitation amplitude and decreased as the thickness of the diaphragm material increased. The damping factor δ was found to be essentially independent of the excitation amplitude and increased with an increase in the thickness of the diaphragm material. The second natural mode force peak and the fluid swirl at the natural mode frequencies, which were observed for the unrestricted liquid sloshing, were completely suppressed [23]. This type of slosh-suppression device has prevalently been used for spherical or oblate spheroidal tanks [8].

6.7 Materials, Processes, and Environmental Conditions of Tanks

The properties of the metallic materials to be considered in the evaluation and selection phases of a given programme have been identified by Wagner and Keller [22] as follows:

- strength/weight efficiency under load/temperature conditions or under other critical failure conditions;
- capability of being fabricated into the desired shapes and sizes without loss of their properties;
- compatibility with all anticipated environments;
- fracture toughness and resistance to subcritical flaw growth;
- availability of shapes and sizes within required schedules; and
- costs of materials and material processing and fabrication.

These properties are discussed below. The principal strength properties to be considered in the design of tanks for rocket vehicles are ultimate tensile strength (F_{tu}), which governs ultimate burst pressure under ductile failure conditions; tensile yield strength (F_{ty}), to comply with the requirement of no yielding at limit load conditions or during proof testing; compressive yield strength (F_{cy}), for compression-critical structures; and the elastic properties (E, G, and ν) of the materials. Further properties are ultimate shear strength (F_{su}), ultimate bearing strength (F_{bru}), and bearing yield strength (F_{bry}), which apply to design details, such as mechanical attachments, and are not usually important factors in the selection of materials. High-frequency, low-stress fatigue data are sometimes required to evaluate the effects of structural vibrations. Low-frequency, high-stress fatigue data are often used to evaluate the effects of multiple pressurisation cycles.

Wagner and Keller [22] have also identified some variable parameters which can affect the mechanical properties of the materials used for tanks. These parameters are temperature, thermal exposure, duration of loading, presence of biaxial and triaxial loads, rate of loading, and unusual environmental conditions such as corrosion and radiation.

The design properties of the materials are to be evaluated for the base metal, for the welds, and sometimes also for the weld zones exposed to heat. It is also necessary to consider the loading direction with respect to the grain orientation of the base metal, and the properties of the materials along and across the direction of the weld. The effects induced on the properties of the materials by processing, forming, and heat treatments are also to be considered.

The metallic materials used for structures of aerospace vehicles form the subject of the military standards MIL-HDBK-5J [25]. The welds form the subject of [26].

A decrease in temperature tends to increase the strength of the material, but often decreases their ductility and toughness. In case of tanks containing cryogenic propellants, an increase in strength at low temperature is desirable, because it results in a mass reduction; however, it is also necessary in this case to ascertain whether the fracture toughness of the material of which the tank is made is adequate to the operational and proof-testing conditions.

Likewise, an increase in temperature tends to reduce the strength of the materials. This fact is to be considered for materials which are particularly sensitive to temperature, as is the case with some titanium alloys. Exposure of such materials to high temperatures for long periods of time can cause a permanent reduction in strength.

Tanks for fluid propellants are required to withstand a high number (sometimes of the order of one hundred) of pressurisation cycles. Such cycles occur in testing and of course in service. Thank failures due to fatigue can occur as a result of these repeated cycles.

Pressurisation stresses in tanks can also cause creep, which is a time-dependent deformation of a material subject to prolonged stresses. Creep usually occurs at high temperatures, but can also occur at moderately high temperatures for some titanium or aluminium alloys.

The multi-axial loading of metals can have significant effects on their mechanical properties. The bi-axial tensile stresses acting in pressurised tanks may in some cases improve, in other cases leave unchanged, and in other cases deteriorate the performance of the materials. Materials which are ductile, homogeneous, and isotropic may show an increase in tensile strength, the amount of this increase depending on the bi-axial stress ratio. The maximum effects usually occurs at a bi-axial tensile-stress ratio of 2:1.

As shown by experience, the magnitude of this effect is often of the order of that predicted by the von Mises criterion [27]. According to this criterion, the effective stress for yielding σ_{eff} in a material element subject to multi-axial loading is

$$\sigma_{eff} = \left[\frac{(\sigma_1 - \sigma_2)^2 + (\sigma_2 - \sigma_3)^2 + (\sigma_3 - \sigma_1)^2}{2} \right]^{\frac{1}{2}}$$

where σ_1, σ_2, and σ_3 are the principal stresses, that is, the stresses acting along three mutually orthogonal planes of zero shear stresses ($\tau_{12} = \tau_{23} = \tau_{31} = 0$).

Anisotropic materials do not conform to the von Mises criterion. Such is the case, for example, with titanium processed in such a way as to obtain preferred orientations of the individual crystals or grains. For such materials, other yield criteria can be used, for example, the Hill 1948 criterion [28]. Further information on the Hill 1948 and other criteria for anisotropic materials can be found in [29]. In addition, Wagner and Keller [22] point out that some homogeneous and isotropic alloys do not appear to behave in full accordance with the von Mises criterion.

The strength of some materials may decrease when they are subject to biaxial tension. This effect may occur with low-elongation materials in a biaxial tensile-stress field. This state of stress is typical of spherical tanks, and limits the ductility of the material to the plane which includes its thickness. In other planes, such materials in this state of stress tend to behave in a brittle manner [22].

The capability possessed by a material of being fabricated into the desired shapes and sizes without loss of its properties is one of the principal qualities considered in the design of tanks for aerospace vehicles. This is because not all high-strength materials can also be manufactured economically to form tanks having the desired characteristics. Some essential fabrication requirements, indicated by Wagner and Keller [22], are indicated below:

- availability in suitable forms, sizes, and levels of quality within the necessary schedules;
- capability of being formed, and machined to the required configurations, on the available equipment, and at the appropriate thicknesses and strength level;
- capability of being welded to suit the common methods of assembly; and
- capability of meeting thermal processing requirements.

Some guidelines to be followed by a designer to choose a method of fabrication for a tank have been suggested by Whitfield and Keller [30]. A brief account is given below.

The method of fabrication chosen for a tank of a rocket engine should be reliable, rapid, and cost-effective for the particular case and needs of the programme to which it applies. A fabrication process should be selected so as to afford the best compromise between fabrication schedule and cost, without reducing reliability below a desired level. An engineering study should include trade-off evaluations of fabrication and welding processes, reliability of various processes based on past experience, schedule effects of material processing, and cost connected with fabrication, tools, and facilities.

A comparison of methods of fabrication used for pressure-vessel components is shown in the following table, taken from [22].

Component	Fabrication method	Advantages	Disadvantages
End domes (complete heads)	**Drawing:** Hydropress (trapped rubber forming)	Moderate production rate Moderate tooling costs Larger sizes than hydroform	Part size and thickness limited Temperature limited Poor control of thickness
	Hydroform (hydraulic fluid forming)	High production rate Better thickness control than hydropress	Limited to small sizes Temperature limited Relatively high tooling costs
	High-Energy-Rate Forming: Explosive	Very large potential sizes (depending on available facility) Good reproducibility Low to moderate tooling costs	Limited to cold forming Low production rate Limited availability of facilities
	Electrical (including spark discharge and magneto-dynamics)	High production rate Good reproducibility	Limited to small sizes Requires specialized equipment and tooling
	Spinning: Shear	Permits integral bosses and skirts Can handle thick material Good thickness control Spinning can be performed hot	Size limited Limited availability of equipment
	Conventional (manual or power)	Moderately large sizes Low tooling costs	Poor thickness control Permits no integral details as formed Temperature limited Thickness limited Low production rate Requires ductile material
	Forging	Not limited to materials with cold- or warm-forming ability Permits complex configurations Permits integral attachments	Size limited High costs Requires considerable machining Low production rate
	Segmenting (formed and welded segments)	Large size capability (starting with smaller individual parts) Reduces difficulty and cost of forming	High total costs – tooling, welding, and inspection Potential for reduced reliability due to increased welding Poor dimensional control Very low production rate
Cylinders	Rolling and Welding	Accommodates large sizes Low cost, simple process	Potential for reduced reliability due to longitudinal weld Permits no integral reinforcements as fabricated
	Shear spinning	Eliminates longitudinal welds Permits integral reinforcements Provides good thickness control Forming can be performed hot	High cost for low production quantity Limited equipment availability Some limitations on size

The size and the shape of a component and the aptitude of its material to be formed and machined are important aspects in choosing a material, a method of fabrication, and a heat treatment. For example, the large size of a component limits the methods of fabrication and the heat treatments which can be chosen.

Specific information on the matter can be found in [31] (titanium and its alloys), [32] (stainless steel), and [33] (precipitation-hardening stainless steel).

The welding properties of a metal are of paramount importance in the choice of a metal for a tank. Desirable properties for a weldable metal are its capability of being fused without the formation of unwanted phases or constituents in or near the fusion zone, its ductility in the range from the melting temperature to room temperature (in order to resist cracking), strength, and fracture-resistance.

Generally speaking, according to NASA [34], the welding processes used for space flight hardware are:

- Gas Metal Arc Welding (GMAW also known as MIG Welding), which is used for quickly fusing mild steel, stainless steel, and aluminium of various thicknesses;
- Tungsten Inert Gas (TIG) Welding, which is used for carrying out work of high quality when a high standard of finish is needed without excessive clean-up by sanding or grinding;
- Shielded Metal Arc Welding (SMAW), which is used for manufacturing, construction, and repair, and is well suited for heavy metal size 4 mm and upward;
- Gas or Oxyacetylene Welding, which is commonly used for brazing soft metals such as copper and bronze, and for welding delicate parts of aluminium; and
- Plasma Cutting, which is used to cut steel and other metals of different thicknesses for metal construction and maintenance.

A further process, used specifically for tanks of liquid-propellant rocket engines, is Friction Stir Welding (FSW), whose working principle is shown in the following figure, due to the courtesy of NASA [35].

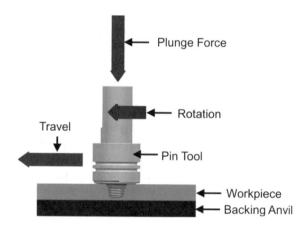

This welding process uses frictional heating combined with forging pressure to produce high-strength bonds virtually free of defects. Friction Stir Welding transforms the metals from a solid state into a plastic-like state, and then mechanically stirs the materials together under pressure to form a welded joint. This process was invented and patented by The Welding Institute (a British research and technology organisation), and has been applied to aerospace, shipbuilding, aircraft, and automotive industries. In particular, it has been used primarily for square butt welds in aluminium alloys of the 2XXX series.

Carter [35] has given some examples of application of Friction Stir Welding to the aerospace industry. They are

- Space Shuttle external tank;
- United Launch Alliance Delta II, Delta IV, and Atlas V;
- Space X Falcon and Falcon 9;
- Japan—JAXA H-IIB; and

- NASA—Space Launch System core stage.

One of the principal benefits of this technology is that it allows welds to be made on aluminium alloys which cannot be readily welded by fusion arc. An example of such alloys is the Al–Li 2195 alloy, which has been used for the Super Light Weight Tank of the Space Shuttle. A further benefit of Friction Stir Welding is that it has fewer elements than other welding techniques to control. In Friction Stir Welding there are only three process variables to control, namely, rotation speed, travel speed, and pressure, all of which are easily controlled. The increase in joint strength combined with the reduction in process variability resulted in an increased margin of safety and in a high degree of reliability for the external tank of the Space Shuttle.

In Friction Stir Welding, a dowel rotates at an angular velocity between 19 and 31 rad/s [36], depending on the thickness of the material. The pin tip of the dowel is forced into the material under a pressure going from 3.4×10^7 to 6.9×10^7 N/m^2 [36]. The pin continues to rotate and moves forward at a speed going from 1.48 to 2.12 mm/s [36]. As the pin rotates, friction heats the surrounding material and rapidly produces a softened plasticised area under the pin. Plasticity starts to occur around 700 K for most materials [37]. The temperature due to friction is carefully controlled not to exceed the point of turning the material to a liquid, which occurs at about 922 K in most cases [37]. As the pin travels forward, the material behind the pin is forged under pressure from the dowel and consolidates to form a bond at a molecular level.

Unlike fusion welding, no actual melting occurs in this process, and the weld is left in the same fine-grained condition as the original metal. For the external tank of the Space Shuttle, a through-spindle retractable pin was developed, which can retract or expand its pin tip within the material. This allows for changes of thickness such as on the longitudinal barrel of the tank [36]. The following figure, due to the courtesy of NASA [37], shows retractable pin tools used for the Friction Stir Welding process.

Retractable pin tools were developed by NASA engineers because the original Friction Stir Welding left a opening, known as a keyhole, which was a point of weakness in the weld. The retractable pin tool retracts automatically when the weld is complete and prevents a keyhole. This improvement makes the weld stronger and eliminates the need to fill the keyhole during manufacturing.

Apart from the choice of a particular welding technique, further processing, such as pre-heat or post-heat or both of them, is generally required for high-carbon, low-alloy steel welds when thickness is higher than 2.5 mm. However, pre-heat and post-heat may be required for welds of thickness equal to or less than 2.5 mm, in order to avoid weld cracking, depending on the material, on the welding process, and on the restraint characteristics of the particular weld. When pre-heat or post-heat is performed by using a torch, then oxidation can reduce the quality of the weld. To avoid this, automatic electric heating is desirable in many cases, because electric heaters controlled by rheostats provide appropriate temperatures and uniform heating. Backup tools made of copper, stainless steel, or refractory-covered metals are used to achieve either high, or low, or negligible heat dissipation, as the case requires, to control the final dimensions and properties of the weld [30]. A reduced strength of a metal in or near the welding zone can be compensated for by increasing its thickness at the joints. Since ductility decreases in the welding zone, then welds are generally located away from high-stress zones. Such zones are those in which a parallel or a meridian of a surface of revolution changes abruptly its radius of curvature.

Residual stresses can still be present in welded zones, unless appropriate steps are taken. For this purpose, such processes as pressure welding and forge welding can reduce residual stresses to minimum values. Residual stresses should be relieved from materials used for tanks to avoid cracking, warping, and reduction of resistance to fatigue and fracture. To this end, such thermal processes as ageing for titanium and heat treatment for steel are applied. For example, the 18%-nickel maraging steel (so called because its strengthening mechanism consists in transforming the alloy to martensite with subsequent age hardening) requires ageing after welding to acquire maximum mechanical properties in welds [22]. Annealing is sometimes required before or during forming of tank components. However, many of the materials commonly used for tanks do not require heat treatments after welding, either to restore mechanical properties or to relieve residual stresses.

The materials which are chosen for tanks of liquid-propellant rocket engines must be compatible with the propellants contained. The following table, due to the courtesy of NASA [22], shows causes and effects of reactions of metals with fluids.

Metal/Fluid reaction	Possible consequences to system	Major sources of metal/fluid reaction			
		A	B	C	D
Metal corrosion (including general corrosion, pitting, intergranular corrosion, and chemical attack)	(1) Metal weakening through loss in cross-sectional area and introduction of stress raisers (2) System contamination with corrosion products	X	X	X	X
Catalytic decomposition of propellants	Loss of efficiency or contamination of system or both			X	
Hydrogen embrittlement of steel	Brittle fracture at low stresses, especially under long-duration loading	X			
Contamination of titanium	Brittle fracture at low stresses	X	X	X	X
Stress corrosion	Metal crack growth or fracture at reduced stress levels	X	X	X	X
Galvanic corrosion	(1) Rapid deterioration of material (2) Stress-corrosion failure	X	X	X	X
Hydrogen-environment embrittlement of metals	Embrittled behavior of metal while exposed to hydrogen gas			X	
Ignition of materials	Catastrophic combustion			X	

Notes:
   A = manufacturing fluids and processes
   B = proof and system testing
   C = service fluid containment
   D = atmospheric exposure

The metallic materials used for tanks should be considered alone or in combination with suitable types of protective finish. Such materials should also be resistant to the effects of exposure to all possible types of external environment.

It is also necessary to prevent the deterioration or the contamination of metals used for tanks during their processing, manufacturing, inspection, test, transportation, and storage [22]. For this purpose, they must be protected from all fluids or processes which might have deleterious effects, as shown in the preceding table.

In particular, many high-strength alloys may be to attacked or contaminated by fluids and treatments commonly used in the phases of manufacturing and process. This is because some fluids or treatments have given rise to undesirable chemical reactions, and must therefore be carefully checked before being used.

Of the alloys commonly used for rocket tanks, those based on titanium have been found to be the most susceptible to contamination. Examples of contamination of such alloys in the manufacturing phase are

- hydrogen contamination at room temperature or at high temperatures;
- oxygen and nitrogen contamination at high temperatures; and
- halogen contamination, resulting from halide-containing materials, before heat treatment or welding.

Titanium contaminated by hydrogen or oxygen or nitrogen becomes brittle. This undesirable effect, when due to heat treatments in air, can sometimes be removed by machining. However, when a titanium alloy is welded in air, then the entire weld becomes brittle.

Steel can also become brittle when contaminated by hydrogen. This happens typically in electrolytic processes. Steel becomes increasingly susceptible to contamination due to hydrogen as its strength increases.

Other type of alloys used for rocket tanks (such as aluminium alloys and alloys based on nickel and cobalt) may, or may not, become brittle in various degrees when contaminated by hydrogen, depending on temperature [22].

Fluids contained in rocket tanks can cause chemical reactions which results in either corrosion or reduced strength of the materials of which the tanks are made. This holds with both testing fluids and propellants. Fluids commonly used for testing purposes are not dangerous by their nature, but may become dangerous as a result of contamination. Examples are tanks made of titanium alloys pressurised with methanol, or made of steel pressurised with water [22].

Some propellants are corrosive or chemically reactive. The containment of these propellants in tanks made of materials sensitive to chemical reactions is possible only when a film or a layer of stable oxide protects the covered material from further reactions. In order to test the susceptibility of a given material to possible chemical attach due to a given propellant, it is necessary to conduct tests in the same conditions of pressure, temperature, and duration of exposure as those which will be experienced by the material in service.

The decomposition of some propellants may be accelerated by the catalysing action exerted by some metals. Examples of propellants subject to catalytic decomposition are hydrogen peroxide and hydrazine.

A tank whose external surface is exposed to atmospheric agents during part of the life of a space vehicle is subject to moisture, salts, and chemical substances of industrial origin. This part includes the times of manufacturing, storage, testing, transportation, and operation. Some metals used for tanks are resistant to atmospheric corrosion, because of the formation of a thin layer of protective oxide. Such is the case with titanium alloys, stainless steels and super-alloys having high percentages of nickel and chromium, and some aluminium alloys. Other alloys and steels must be provided with a protective finish.

The reactions which occur in conditions of service between the tank materials and the fluids contained can be classified as follows

- stress-corrosion cracking;
- galvanic corrosion;
- loss of ductility due to hydrogen; and
- ignition.

A brief description is given below for each type of reaction.

According to the definition given in [38], stress-corrosion cracking is a cracking process caused by the conjoint action of stress and a corrodent. Stress-corrosion cracking is one of the principal causes of failure of high-stressed tanks. Corrosion can be due to either the fluids contained in the tanks or the environments to which the tanks are exposed. The most common of such environments is atmospheric air, which contains moisture, salts, and other chemical substances due to industrial processes. The effects of stress-corrosion on a given material depend on the stress sustained by the material, the degree of corrosion of the environment, and the time of exposure of the material to the environment.

Means commonly used to avoid stress-corrosion cracking in tanks are

- choice of materials resisting to corrosive environments or protection of susceptible materials from corrosive environments;
- reduction of tensile stresses in materials exposed to corrosive environments; and
- reduction of the times of exposure.

The matter of corrosion and protection of metals from corrosion forms the subject of the military standards MIL-HDBK-729 [39]. In particular, Table XIV on p. 127 of [39] indicates the chemical composition of some titanium alloys used for tanks, and the related corrosive substances, for which stress-corrosion cracking has been observed.

In particular, aluminium alloys of the 2XXX and 7XXX series are susceptible to stress-corrosion cracking in atmospheric environments. Titanium alloys are normally resistant to atmospheric environments, but are susceptible to salt and other sources of chlorine which remain in contact with the metals at temperatures above about 561 K. Some titanium alloys have also been found to be susceptible to sea water at room temperatures [22].

A comparison of the resistance of various alloys to stress-corrosion cracking in atmosphere and in other environments is given in Table VII on pp. 35 and 36 of [22].

Galvanic corrosion, so called after the Italian scientist Luigi Galvani, is an electrochemical phenomenon which occurs either on a macro-scale or on a micro-scale.

On a macro-scale, it is the increased corrosion (deterioration) of the more active metal (anode) of a couple of dissimilar metals in an electrolytic solution or medium and the decreased corrosion of the less active metal (cathode) as compared to the corrosion of the individual metals, when not connected, in the same electrolytic environment. By electrolytic solution, we mean a solution capable of conducting an electric current. An example is common sea water.

The dissimilarity which provides the driving force to galvanic corrosion is the difference in the electrode potential of each of the two metals when they are in electrical contact through the electrolyte. Electrode potential is a measure of the tendency of a metal to become more active than another metal when they are immersed in a given electrolyte.

A galvanic series is a list of metals and alloys based on their order and tendency to corrode independently in a particular electrolytic solution or in other environment.

A galvanic series in sea water, due to the courtesy of NASA [40], is given below. This series is arranged in order of increasing activity.

Galvanic Series In Sea Water
Noble (least active)
Platinum
Gold
Graphite
Silver
18-8-3 Stainless steel, type 316 (passive)
18-8 Stainless steel, type 304 (passive)
Titanium
13 percent chromium stainless steel, type 410 (passive)
7Ni-33Cu alloy
75Ni-16Cr-7Fe alloy (passive)
Nickel (passive)
Silver solder
M-Bronze
G-Bronze
70-30 cupro-nickel
Silicon bronze
Copper
Red brass
Aluminium bronze
Admiralty brass
Yellow brass
76Ni-16Cr-7Fe alloy (active)
Nickel (active)
Naval brass
Manganese bronze
Muntz metal
Tin
Lead
18-8-3 Stainless steel, type 316 (active)
18-8 Stainless steel, type 304 (active)
13 percent chromium stainless steel, type 410 (active)
Cast iron
Mild steel
Aluminium 2024
Cadmium
Alclad
Aluminium 6053
Galvanised steel
Zinc
Magnesium alloys

Magnesium
Anodic (most active)

When galvanic corrosion occurs on a macro-scale, the anodic and cathodic areas are easily discerned.

On a micro-scale, galvanic corrosion may occur with one metal having dissimilarities (for example, impurity inclusions, grains of different sizes, difference in composition of grains, differences in mechanical stress), abnormal level of pH, and high temperatures [41]. When galvanic corrosion occurs on a micro-scale, the anodic and cathodic areas can only be discerned by metallographic techniques or deduced by inference or observations of the corroded metal [39].

Galvanic corrosion can occur within a tank containing an electrically conductive fluid, or outside a tank exposed to atmospheric moisture. Generally speaking, in liquid-propellant rocket engines, galvanic corrosion of dissimilar metals is of little or no concern, because most propellants either have little electrical conductivity, or do not develop significant electrode potentials in contact with normal structural metals, or both [22]. However, this problem may arise when new or inadequately tested propellants are used either alone or in the presence of contaminants, or when liquids other than propellants (for example, water) are in contact with dissimilar metals.

Metals can be protected from corrosion by providing them with a coating or a treatment. The coating may be a paint, a sealant, a resin coating, or a metallic coating. Several types of coatings, used to protect specific metals, are described at length in [42]. The treatment may be a surface or bulk treatment, a chemical or mechanical treatment, or a combination of these.

Some metals effectively resist corrosion damages because of their ability to form and maintain, when exposed to an aggressive environment, an adherent and impervious film. These metals are said to have become passivated. However, many of the common or structural metals used for tanks are either scarcely efficient or unable in producing such protective films. Consequently, protective coatings or treatments are applied to many common metals to prevent or reduce corrosion [39].

Metals which are in contact with pure hydrogen should be evaluated for susceptibility to loss of ductility in the presence of hydrogen. This phenomenon, also known as hydrogen-assisted fracture, is due to the easy absorption and subsequent diffusion of hydrogen into metals. On a microscopic scale, hydrogen atoms tend to segregate in certain parts of the crystal lattice of a metal, thereby weakening its chemical bonds. Consequently, the walls of a metallic tank containing hydrogen become brittle, which can lead to failure.

The compatibility of metals with hydrogen has been investigated by several authors. Some of them are Cataldo [43], Caskey [44], San Marchi and Somerday [45], and Chandler [46]. A brief account is given below.

Brittleness of metals due to hydrogen occurs principally in ferrous alloys, such as high-strength and martensitic steels. In most cases, the greater the strength of a metal, the more susceptible it is to brittleness caused by hydrogen. Low-strength metals, such as copper, aluminium, and nickel alloys, have low susceptibility to hydrogen damage. However, in case of these metals having undergone a strain-hardening process, there would be a greater risk of induced brittleness.

In some cases, steels and alloys of low strength can be used to decrease the risk of loss of ductility. However, of course, the metal chosen must withstand the loads applied to it during operation.

In quantitative terms, the following table, adapted from [43], shows the effects induced by gaseous hydrogen, at a pressure of 6.895×10^7 N/m^2, in various alloys.

ARRANGEMENT OF MATERIALS IN ORDER OF PERCENT
REDUCTION OF NOTCH STRENGTH IN 6.895×10^7 N/m^2 HYDROGEN

Alloy	Yield Strength N/m^2	Percent Reduction Notch Strength
18 Ni 250 Maraging	1.710×10^9	88
17-7 PH SS	1.034×10^9	77
H-11	1.682×10^9	76
Rene' 41	1.124×10^9	73
4140	1.234×10^9	60
Inconel 718	1.255×10^9	54
Ti-6A1-4V (STA)	1.076×10^9	45
Nickel 270	0.1568×10^9	30
HY-1Ō0	0.6688×10^9	27
Ti-6A1-4V (Annealed)	0.9101×10^9	25
A302	-	22
HY-80	0.5585×10^9	21
304 ELC SS	0.1655×10^9	13
A-517 (T-1)	0.7515×10^9	11
Be-Cu	0.5447×10^9	7
Ti (C.P.)	0.3654×10^9	5
310 SS	-	3
A-286 SS	0.8481×10^9	3
7075-T73	0.3723×10^9	2
6061-T6	0.2275×10^9	0
1100 A1	-	0
OFHC Copper	0.2689×10^9	0
316 SS	0.4413×10^9	0

As has been anticipated above, the reduction in notch strength resulting from the preceding table is inversely related to the ambient strength of the material, but there are exceptions to this rule. For example, the A-286 stainless steel (an age-hardened iron base superalloy) has a high yield strength, but an apparent low susceptibility to hydrogen. The best alloys are aluminium alloys, copper, and stabilised stainless steels [43].

The materials which are used for tanks containing propellants should not be susceptible to ignition or to violent reactions in the presence of the propellants. Consequently, titanium and titanium alloys should not be used for tanks containing oxidisers such as red fuming nitric acid, liquid oxygen, pressurised gaseous oxygen, mixtures of liquid oxygen and liquid fluorine, and other strong oxidisers. It is also necessary to consider the possibility of ignition due to impact, rupture, friction, electricity, heat, or any other source of highly concentrated energy. This holds in particular with titanium and titanium alloys in contact with oxidisers [22]. For the same reason, copper, lead, zinc, molybdenum, and other alloys which contain free elements should not be used with hydrazine or with other propellants related to hydrazine. As a general rule, it is necessary to ascertain the compatibility of the materials used for tanks with the propellants (in liquid and gaseous phases) to be contained in them, before such materials come in contact with the related propellants.

6.8   Fracture Control of Metals Used for Tanks

Tanks and other pressure vessels containing propellants may have flaws or defects which are either present in the original materials before processing or induced in the materials by processes of fabrication. These defects can reduce the capability of carrying loads and the operational life of the tanks. When such defects are large in comparison with those causing failure, then failure occurs as soon as the tanks are put under pressure for test purposes. When such defects are small, the tanks may withstand several cycles of pressure loading for a considerable number of hours before the size of the defects grows to such an extent as to cause failure.

It is necessary, for reasons of economy, to reduce the possibility of failure of tanks used for a space vehicles during testing. It is mandatory, for reason of both safety of the crew and economy, to reduce the possibility of failure of such tanks during operation.

The present section presents criteria and practices to be used in the design of metallic tanks for propellants, in order to reduce the possibility of failures caused by defects during test, pre-flight, and flight.

For this purpose, it is necessary to consider:

- the initial sizes of the flaws;
- the critical sizes of the flaws (meaning by that, the sizes required to cause fracture at a given level of stress); and
- the subcritical characteristics of the flaws which can cause their growth.

To prevent failures during test, the initial sizes of the flaws must be less than the critical sizes at the level of stress given to the material in test phase. To prevent failures during service, it must be proven that the largest possible size of an initial flaw in a tank cannot grow to the critical size during the required operational life of the tank.

The critical size of a flaw depends on the level of stress, on the toughness of the material to fracture, on the thickness of the wall, and on the location and direction of the flaw. The determination of the initial size of a flaw is limited by the available methods of non-destructive inspection. However, further information may come from the results of a successful proof test. In other words, a proof test in which a tank has not failed provides information on the maximum possible value of the ratio of the initial to critical stress level. This value, in turn, makes it possible to estimate the maximum possible size of the initial flaws.

The growth of subcritical flaws depends on the stress level, the initial size of the flaws, the material, the environment, and the pressure applied as a function of time to a given tank.

Examples of flaws which are rarely detected in metallic tanks or in other pressure vessels are surface flaws and internal flaws. When a tank has an initial flaw whose size exceeds the critical size at the level of stress applied in testing, then the tank fails just in phase of testing. A tank fails in service, when the size of an initial flaw is less than the critical size at the level of stress applied in testing, but this size grows due to the stress applied in service to such an extent as to reach the critical size at the level of stress applied in service. When the size of an initial flaw grows through the thickness of the wall of a tank before reaching the critical size, then leakage occurs.

In the elastic field of stress, the critical sizes of surface and internal flaws depend on the critical value K_{Ic} (Pa m$^{1/2}$, where 1 Pa $= 1$ N/m^2) of the fracture toughness of the material, and on the level of the stress applied. When the critical sizes of the flaws are small in comparison with the thickness of the wall, then a tank is said to be thick-walled. Otherwise, when the critical sizes of the flaws approach or exceed the thickness of the wall, then a tank is said to be thin-walled.

The critical sizes of surface flaws in uniformly stressed thick-walled tanks can be calculated by using the following equation of [47]:

$$\left(\frac{a}{Q}\right)_{cr} = \frac{1}{1.21\pi}\left(\frac{K_{Ic}}{\sigma}\right)^2$$

The critical sizes of small internal flaws in uniformly stressed thick-walled tanks can be calculated by using the following equation of [47]:

$$\left(\frac{a}{Q}\right)_{cr} = \frac{1}{\pi}\left(\frac{K_{Ic}}{\sigma}\right)^2$$

In the two preceding equations, a (m) is the minor semi-axis of the ellipse of equation

$$\frac{x^2}{c^2} + \frac{y^2}{a^2} = 1$$

that is, a (m) is the depth of the semi-elliptic surface flaw, $2c$ (m) is the length of the semi-elliptic surface flaw, Q is the flaw-shape parameter defined as follows

$$Q = \phi^2 - 0.212\left(\frac{\sigma}{\sigma_{ys}}\right)^2$$

ϕ is the complete elliptic integral of the second kind, of modulus $k = (1 - a^2/c^2)^{1/2}$, defined as follows

$$\phi(k) = \int_{0}^{\frac{\pi}{2}} \left(1 - k^2 \sin^2 \theta\right)^{\frac{1}{2}} d\theta = \int_{0}^{\frac{\pi}{2}} \left(1 - \frac{c^2 - a^2}{c^2} \sin^2 \theta\right)^{\frac{1}{2}} d\theta$$

θ (rad) is an angular variable of integration, σ (N/m^2) is the uniform gross stress applied at infinity in a direction perpendicular to the plane of crack, and σ_{ys} (N/m^2) is the uniaxial tensile yield strength of the material. The following figure, due to the courtesy of NASA [47], shows the relationship between the flaw-shape parameter Q and the flaw depth-to-length ratio $a/(2c)$.

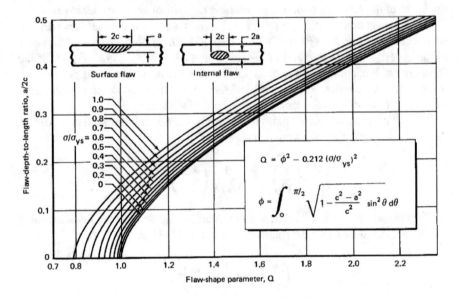

The following figure, also due to the courtesy of NASA [47], is a graphical representation of the preceding equation $(a/Q)_{cr} = (K_{Ic}/\sigma)^2/(1.21\,\pi)$.

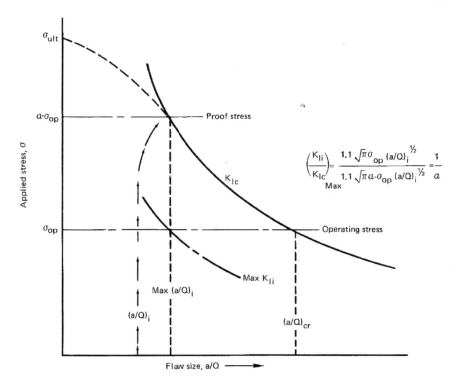

By the way, the complete elliptic integral of the second kind can be evaluated numerically by using the following expansion [48]:

$$\phi(k) = \frac{\pi}{2}\left[1 - \left(\frac{1}{2}\right)^2\frac{k}{1} - \left(\frac{1\cdot 3}{2\cdot 4}\right)^2\frac{k^2}{3} - \left(\frac{1\cdot 3\cdot 5}{2\cdot 4\cdot 6}\right)^2\frac{k^3}{5} - \cdots\right]$$

where $|k| < 1$.

In order to predict critical flaw sizes, failure modes, and operational life of thin-walled tanks or pressure vessels, it is necessary to know the stress intensity for flaws which become very deep with respect to the wall thickness. The solution given by the preceding equation $(a/Q)_{cr} = (K_{Ic}/\sigma)^2/(1.21\pi)$ for a semi-elliptic surface flaw has been found to be sufficiently accurate for flaw depths up to about 50% of the thickness of the material [47]. For flaws of greater depths, the intensity of the stress applied is magnified by the effect of the free surface near the flaw tip. In other words, in thin-walled tanks, the flaw-tip stress intensity can reach the critical value K_{Ic} at a flaw size smaller than that which results from the preceding equation $(a/Q)_{cr} = (K_{Ic}/\sigma)^2/(1.21\pi)$.

The effect of the magnification factor M_K on the critical value K_{Ic} of the fracture toughness of a material is shown in the following figure, due to the courtesy of NASA [47].

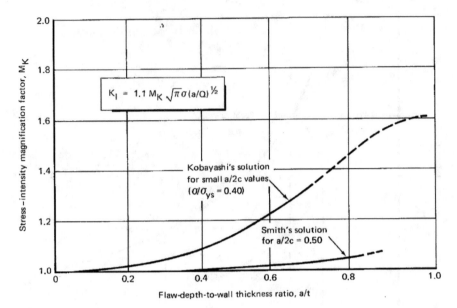

The magnification factor M_K is applied to the equation $(a/Q)_{cr} = (K_{Ic}/\sigma)^2/(1.21\pi)$ as follows

$$K_{Ic} = 1.1 M_K \pi^{\frac{1}{2}} \sigma \left(\frac{a}{Q}\right)^{\frac{1}{2}}_{cr}$$

This application determines the critical value of the fracture toughness in case of deep surface flaws (such that the value of $a/(2c)$ is small).

As shown in the preceding figure, the value of M_K has been found to vary from less than 1.1 for semi-circular flaws ($a = c$, and therefore $a/(2c) = 0.50$) to 1.6 for flaws having smaller values of $a/(2c)$.

In case of tanks having flaws which are long with respect to their depth ($Q \approx 1$), the following equation can be used for a thin-walled tank

$$K_{Ic} = 1.1 M_K \pi^{\frac{1}{2}} \sigma a^{\frac{1}{2}}$$

and the following equation can be used for a thick-walled tank

$$K_{Ic} = 1.1 \pi^{\frac{1}{2}} \sigma a^{\frac{1}{2}}$$

In order to predict the critical sizes for surface and internal flaws by using the preceding equation $(a/Q)_{cr} = (K_{Ic}/\sigma)^2/(1.21\pi)$, it is necessary to know the values of K_{Ic} for the original material and for the welds of a tank. These values can be obtained from laboratory tests.

In order to prevent failure in a tank, either the actual initial flaw sizes or the maximum possible initial flaw sizes must be known. Non-destructive inspection is the only means of determining the actual initial flaw sizes. A successful proof test provides a measure of the maximum possible initial-to-critical stress intensity ratio, and this in turn makes it possible to estimate the maximum possible initial flaw sizes.

Non-destructive inspection techniques commonly used for rocket tanks are radiographic testing, ultrasonic testing, liquid penetrant testing, and magnetic particle testing. Other techniques are eddy current testing and infrared testing [47].

When multiple inspection techniques (for example, radiographic testing, ultrasonic testing, and liquid penetrant testing) are used, then most surface and internal flaws are detected. However, it is not safe to assume that all existing flaws can be detected at any time, because some of them (for example, tight cracks) are particularly difficult to detect. The largest initial flaw sizes which cannot escape detection cannot be established with confidence. The inspection techniques commonly used do not measure precisely the lengths and the depths of the initial flaws to be used in a fracture mechanics analysis. However, it is necessary to rely on non-destructive inspection to prevent proof-test failures of most high-strength tanks [47].

Proof-pressure testing is probably the most reliable non-destructive inspection technique available to ensure the absence of initial flaws of sufficient sizes to cause failure of a tank in operating conditions. Let K_{Ii} (Pa m$^{1/2}$) and K_{Ic} (Pa m$^{1/2}$) be respectively the plane-strain stress-intensity factor at the initial conditions and the fracture toughness of a given material. A successful proof test conducted at a pressure of α times the maximum pressure in operating conditions indicates that the maximum possible value of the ratio K_{Ii}/K_{Ic} is equal to $1/\alpha$. This value can be used with subcritical flaw-growth data to estimate the minimum life of a tank.

From the point of view of the initial design, the minimum required proof-test factor α for a tank is

$$\alpha = \frac{1}{\text{allowable value of the ratio } K_{Ii}/K_{Ic}}$$

The allowable value of the ratio K_{Ii}/K_{Ic} depends on the required service life of the tank and on the subcritical flaw-growth properties of the material [47].

In order to prevent yielding in proof testing, the membrane stresses applied to a tank are limited to a value equal to or less than the yield strength of the material.

In practice, the local level of stress may exceed the yield strength of the material, because there may be discontinuities due to design or manufacturing, and also because of the presence of residual stresses. When the applied stress approaches and exceeds the yield strength of the material, the critical flaw-size curve deviates from the theoretical curve based on a constant value of K_{Ic}, and therefore the critical flaw sizes are smaller than those predicted by the linear-elastic theory of fracture mechanics.

When the stress applied to a tank at proof pressure exceeds the yield strength of the material and the tank also passes the proof test, then the maximum possible value of the ratio K_{Ii}/K_{Ic} resulting from the test is smaller than $1/\alpha$. In this case, the minimum operational life of the tank should exceed the required life, which has previously been used to determine α.

In is to be observed that the equation $\alpha = 1/(\text{allowable value of the ratio } K_{ii}/K_{ic})$, which expresses the required minimum proof-test factor α for a tank, does not take account of the thickness of the tank. However, the value of the proof test in providing assurance against failure in service changes with decreasing wall thickness or with increasing fracture toughness K_{Ic} of the material [47].

When a proof test is performed at a temperature which differs from the temperature in operating conditions, then the required minimum value of the proof-test factor is expressed by the following equation of [47]:

$$\alpha = \frac{1}{\text{allowable } K_{Ii}/K_{Ic} \text{ at operating temperature}} \times \frac{K_{Ic} \text{ at proof} - \text{test temperature}}{K_{Ic} \text{ at operating temperature}}$$

The choice of the proper fluid to be used in a proof test is an important consideration for all alloys. Water may have a detrimental effect in high-strength alloys, because it promotes slow growth of flaws due to the effect of hydrogen cracking. This problem has been solved by using either other fluids, such as oil, or water with corrosion inhibitors, such as sodium bichromate ($Cr_2Na_2O_7$), or distilled water.

Subcritical flaw growth can occur after cyclic loading, sustained-stress loading, and combined cyclic and sustained-stress loading. Data from fracture specimen tests can be used in a fracture mechanics analysis to predict the number of cycles or the time needed for an initial flaw in a tank under sustained pressure to grow to critical size. The time or the number of cycles needed to cause failure has been found to depend primarily on the ratio K_{Ii}/K_{Ic}.

The following figure, adapted from [47], shows the ratio K_{Ii}/K_{Ic} versus cycles to fracture for a heat-treated Ti-6Al-4V alloy in air at room temperature. The quantity R is the ratio of the minimum stress to the maximum stress during a cycle. Both the best-fit, least-square curve and the 96% probability, 99% curve are shown in the plot.

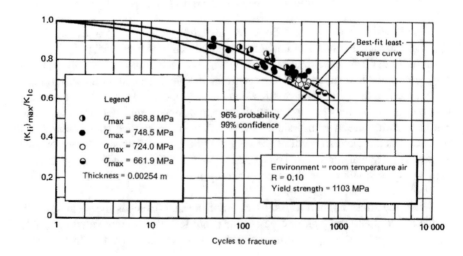

An important property observed in all sustained-stress experiments of flaw growth is the existence of a threshold level of stress intensity for a given material in a given environment.

In other words, below a given value of stress intensity (or of K_{Ii}/K_{Ic} ratio), flaw growth has not been detected. Above this value, flaw growth occurs and can result in fracture. This stress intensity has been designated as K_{TH}.

The discovery of a unique value of K_{TH} for a given material in a given environment makes it possible to design a safe tank subject to a sustained load.

K_{TH} may be 80% of K_{Ic} or higher in an inert environment, but may also be less than 50% of K_{Ic} in a hostile environment.

The threshold value K_{TH} for a Ti-6Al-4V alloy subject to sustained load in two different liquids (distilled water and methanol) is shown in the following figure, due to the courtesy of NASA [47].

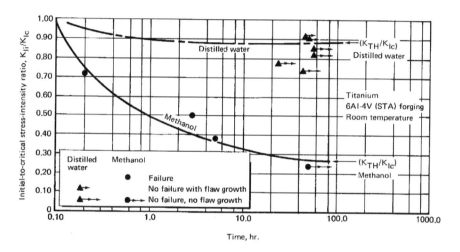

The value of K_{TH} has been found to decrease with increasing yield strength in steel alloys. The flaw growth for sustained load has the highest values in conditions of plane strain. The values of K_{TH}, determined from tests of specimens cracked through the thickness, increase with decrease in specimen thickness.

In chemically inert environments, the crack growth rate initially decreases with increasing stress intensity. When the initial stress intensity is sufficiently low, the crack may halt. At higher stress intensities, the crack growth rate passes through a minimum value and then increases steadily until the crack becomes unstable. This behaviour has been observed for AM 350® steel (a heat-treatable austenitic or martensitic alloy, depending on heat treatment) in a purified argon environment.

For some alloys, such as Ti-5Al-2.5Sn (ELI) and 2219-T87, two thresholds have been observed in environments of room air, liquid nitrogen, and liquid hydrogen. One of these threshold stress intensities was defined as the value of stress above which flaw growth to failure can be expected. The other of them was defined as the value of stress below which there is no flaw growth. In the interval between these two threshold intensities, small amounts of flaw growth can occur, but the growth apparently arrests after a short time at load [47].

Some examples of ratios K_{TH}/K_{Ic}, determined experimentally, are shown in the following table, adapted from [47].

Material	Temp., K	σ_{ys}, MPa	Fluid environment	$\dfrac{K_{TH}}{K_{Ic}}$
6Al-4V (STA) titanium forging	RT [a]	1100	Methanol	0.24
	RT	1100	Freon M.F.	0.58
	RT	1100	N_2O_4 (.30 % NO)	0.74
	RT	1100	N_2O_4 (.60 % NO)	0.83
	RT	1100	H_2O + sodium chromate	0.82
	RT	1100	H_2O	0.86
	RT	1100	Helium, air, or GOX	0.90
	RT	1100	Aerozine 50	0.82
	305	1100	N_2O_4 (.30 % NO)	0.71
	305	1100	N_2O_4 (.60 % NO)	0.75
	314	1100	Monomethyl-hydrazine	0.75
	316	1100	Aerozine 50	0.75
6Al-4V titanium weldments (heat-affected zones)	RT	869	Methanol	0.28
	RT	869	Freon M.F.	0.40
	RT	869	H_2O	0.83
	RT	869	H_2O + sodium chromate	0.82
5A12.5 Sn (ELI) titanium plate	77.6	1240	LN_2 ($\sigma <$ proportional limit)	>0.90
	77.6	1240	LN_2 ($\sigma >$ proportional limit)	0.82
	20.4	1450	LH_2	>0.90
2219-T87 aluminum plate	RT	400	Air	0.90 [b]
	77.6	455	LN_2	0.82 [b]
	20.4	496	LH_2	>0.85 [b]
4330 steel	RT	1410	Water	0.24
4340 steel	RT	> 1380	Salt water	<0.20
GTA welds: 18Ni (200) steel	RT	1380	Salt water spray	>0.70
18Ni (250) steel	RT	1620	Salt water spray	>0.70
12Ni-5Cr-3 Mo steel	RT	1170	Salt water spray	>0.70
9Ni-4Co-2.5C steel	RT	1170	Salt water spray	>0.70
Inconel 718	RT	1140	Gaseous hydrogen at 34.58 MPa (absolute)	<0.25

[a] Room temperature.

[b] No failure K_{TH}, some growth observed at lower values.

Some criteria to be observed in the design of metallic tanks for rocket engines have been identified by Tiffany [47]. They may be summarised as follows. All tanks must be designed to avoid failure in service caused by flaws and to ensure a very low probability of catastrophic failure caused by flaws during proof tests. For this purpose, it is necessary to consider not only the internal pressure to which a tank is subject, but also the pressures, temperatures, environments, and stresses to which the tank is exposed.

The materials used for tanks must have appropriate characteristics of fracture-growth and flaw-growth. These characteristics must come from reliable sources of data, for example, from the military standards MIL-HDBK-5 J [25], and must also be confirmed by tests.

The critical flaw sizes must be determined for the stress levels of interest by either analysis or tests, as appropriate. If possible, the maximum size of initial flaws allowable in tanks should be so great as to be detected by non-destructive inspection, but not so great as to grow to critical size during life in service.

The allowable initial size of flaws must be less than the critical size at the level of stress of proof pressure. The initial stress intensity ratio allowable in a tank must be chosen to ensure that the critical intensity ratio is not attained during the design life of the tank.

Each tank must be proof tested. The level of proof pressure must be chosen to demonstrate that the tank is free of flaws larger than the allowable initial size, or that the actual initial stress-intensity ratio is less than the allowable initial stress-intensity ratio. Account must be taken of the differences between the proof test temperature and the service temperature, and of the time required to apply or remove pressure during the proof test.

In order to prevent failures in proof tests, low levels of stress and materials having high values of fracture toughness must be used during such tests. By so doing, the critical sizes of flaws are large and greater than the thickness of the tank walls. Consequently, the worst case which can occur during proof testing is leakage. In order to obtain maximum assurance of safe performance in service, it is desirable to use large proof-test factors, low levels of operational stresses, and materials having low rates of flaw growth under cyclic loads and high values of K_{TH} in the expected environment of service.

On the other hand, the use of large proof-test factors, low levels of operational stresses, and materials having high values of fracture toughness leads often to high masses of tanks. Therefore, trade-offs can and should be made to arrive to an optimum compromise for a given tank.

6.9 Structural Elements of Tanks

The structural elements considered in the present section are tank shells, weld joints, access openings, support fittings, and accessory attachment provisions.

Not only the tank shells but also the other elements are required to withstand internal and external pressures, the latter acting during propellant servicing or during decontamination. Vacuum is frequently used to remove gas from the liquid side of expulsion devices (described in Sect. 6.6), and also to remove residual propellants from tanks. The tank shells and the other elements are required to withstand an external pressure of one atmosphere (101325 N/m^2).

One of the principal requirements of tanks for rocket engines is low mass. Other requirements are cost, schedule times, easiness of manufacturing, reliability and availability of the materials used for them, technical skills of the personnel, tools, and testing facilities.

In the phase of structural design, the detailed design of a tank includes sidewall, fore and aft bulkheads, access openings, and accessory mounting provisions. The design of these elements depends on the material chosen for the tank. The detailed design of a tank also includes the primary structural junctures.

The design of a tank is based on the relation between the loads imposed and the capability of the tank to withstand such loads. Wagner and Keller [22] have identified the following factors to be taken into account in the analysis of the loads imposed on a tank:

- limit load or stress, which is the maximum stress or pressure which the tank structure is expected to bear, in the specified operating conditions and with allowance for possible variations;
- design safety factor, which is a multiplier, whose value (greater than unity) is chosen by the designer to take account of small variations of the properties of the material, of the quality of manufacturing, and of the magnitude and distribution of the load in the structure;
- design load or stress, which is the limit load multiplied by the design safety factor;
- allowable load or stress, which is the load not to be exceeded to avoid failure, the latter being due to either buckling, or yield, or ultimate loading; and
- margin of safety (m_s), which is the fraction by which the allowable load exceeds the design load, as follows

$$m_s = \frac{\text{allowable load or stress}}{(\text{limit load or stress}) \times (\text{design safety factor})} - 1$$

The magnitude of the design safety factor depends on the degree of confidence in the properties of the materials, on the processes of production, and on the validity of the predicted conditions of use. In practice, a uniform design safety factor is chosen by the designer for the entire structure of a rocket vehicle on the basis of experience and judgement. Values in use for the design safety factor range from 1.0 to 1.1 for yield stress and from 1.25 to 1.5 for ultimate stress, the higher values being used for manned flight vehicles.

The structure of a rocket tank is required to withstand the stresses due to vibration, thermal shock, propellant slosh, and internal pressure. This structure is also required to provide the load path for the loads acting on the body of the rocket vehicle. The structural elements of a rocket tank are subject to different types of stresses, which are briefly described below for each element. The following figure, due to the courtesy of NASA [49], shows schematically the structural elements of the S-IVB (the third stage of the Saturn IV launch vehicle), which has one J-2 engine.

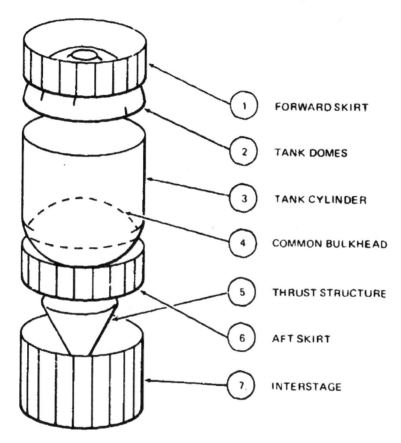

The tank sidewall of a tank is subject to pressure, inertial forces due to the propellant, axial loads, and bending moments. The circumferential stresses are determined by combining the ullage pressure, the load pressure, and the inertial force due to the propellant. The meridian stresses are determined by combining the ullage pressure, the axial loads, and the bending moments.

The stresses acting on the end closures of a tank result from the ullage pressure and the acceleration impressed on the propellant. During the boost time, the aft bulkheads of tanks have a maximum pressure at the apex, and the forward bulkheads have a minimum pressure at the apex. In order to evaluate the stress at a specific location on the end closures, the pressure and the shape of the closure must be considered.

When there are separate bulkheads on two contiguous tanks, the load acting on each of them is to be determined as indicated above for end closures of tanks.

A single common bulkhead is subject to either burst or collapse loads and to temperature gradients through its thickness. The pressure loads acting at any given point of a common bulkhead are determined as described above for end closures of tanks, but a common bulkhead is subject only to the difference between the forward pressure and the aft pressure acting at any given point. The stability under collapsing pressure loads is obtained by designing a bulkhead of large bending stiffness in comparison with its membrane stiffness. A stiffened skin for a tank requires a more complex analysis of internal loads. In this case, a multi-layer shell of revolution is frequently used.

A tank is also subject to local loads at any point in which it is mated to other components of a rocket vehicle. The magnitudes and the directions of these loads depend on the weight of each component multiplied by an amplification factor due to acceleration and vibration. The magnitude of these loads is felt to its full extent by the attachment bolts and by the immediate structural elements. It is felt to a lesser extent by the contiguous structural elements. This lower magnitude is evaluated according to criteria suggested by experience and judgement, taking account of the damping which occurs away from the point of excitation.

In most cases, tanks for liquid-propellant rocket engines are thin-walled surfaces of revolution, meaning by that, structures whose thickness t is assumed constant and much smaller than the radii R of curvature of its middle surface [2]. In practice, the value of R/t is assumed greater than 10. As has been shown in the first four sections of the present chapter, this assumption permits the use of simple formulas for stress and strain, such as those which are given in [3].

The choice of the materials and the evaluation of the corresponding thicknesses and stresses can be done by using only data on mechanical properties with safety factors selected by the designer. Of course, practical considerations concerning manufacturing, handling, and stiffness are also taken into account in this preliminary choice. Before this choice can take place, the level of stress in working condition, the thickness of the skin, and the fracture strength of the materials must be evaluated. The proposed material must possess appropriate values of toughness and resistance to the growth of subcritical flaws, in order to comply with the requirements of life service for a given tank.

After a material has been chosen, the proof-test stress, the operational stress, and the non-destructive inspection requirements are defined according to the desired mission performance. The skin thickness is determined by taking account of the most restrictive conditions, such as safety factor, criteria of fracture control, manufacturing, and handling. The skin thickness for the sidewall of a tank depends on the product of the circumferential tension load under the predicted maximum pressure times the factor of safety [22]. In order to reduce mass, the skin thickness should be as low as possible. There are several methods for constructing thin-walled tanks. Sometimes, a tank can be put under internal pressure to keep its walls from buckling. This is because the net tensional force due to internal pressure is made greater than the compressional force due to the flight loads. Consequently, the tank is only subject to tension, and buckling is avoided. Such was the case with the Atlas and Centaur launch vehicles, in which additional membrane rigidity was obtained through the internal working pressure within the tank [22]. Another method of stabilising thin metallic sheets is used in the construction of conventional aircraft. In this method, stiffening members (stringers) are fastened to the skin in the direction of the compressive loads, as shown in the following figure, due to the courtesy of NASA [50].

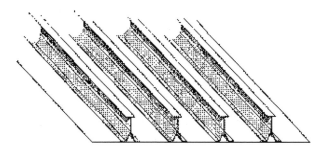

The same result can also be obtained in a single piece of metal by chemically milling or machining a solid sheet to remove all metal except ribs or waffles which act as stringers, as shown in the following figure, due to the courtesy of NASA [50], which illustrates a rectangular waffle structure.

Still another method (developed by McDonnell Douglas) of stiffening a skin is the use of an isogrid structure, in which the ribs are arranged in a repetitive equilateral triangular pattern, as shown in the following figure, due to the courtesy of NASA [51], where an isogrid structure (left) and a square waffle structure (right) are illustrated.

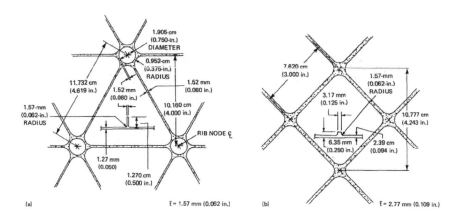

An isogrid structure results in a surface whose stiffness is orthogonally isotropic. Details and methods of calculation for isogrid structures are given in [49]. This method was used for the external skin structure of the Delta launch vehicle [51].

Summarising, the sidewall skin is usually stiffened against buckling by using stringers, frames, or ribs spaced in a grid pattern (either isogrid or rectangular waffle). The sidewall design is particularly important for tanks of large space vehicles. This is because the sidewall in such vehicles not only contains the propellant but also transmits the body loads. The principal types of sidewall designs for tanks under pressure are skin-stringer-frame, grid, and monocoque.

The choice of one or another type of design for a tank depends on the magnitude of the body loads applied externally and on the type of propellant to be contained in the tank. A highly loaded sidewall is usually designed with the skin-stringer-frame structure, whereas a lightly loaded sidewall is designed with the waffle or isogrid structure. A very lightly loaded sidewall can be designed with the monocoque structure, but pressurising is often required for stability.

Integral stiffening is the type of skin-stringer design which is best suited to propellant tanks [22]. This type of design eliminates the many sources of possible leakage associated with the points of mechanical attachment involving the skin, the stringers, and the frames constructed in the conventional manner.

The following figure, due to the courtesy of NASA [22], shows two types of skin-stringer designs. They are: (a) "Blade" stringer; and (b) "T" stringer.

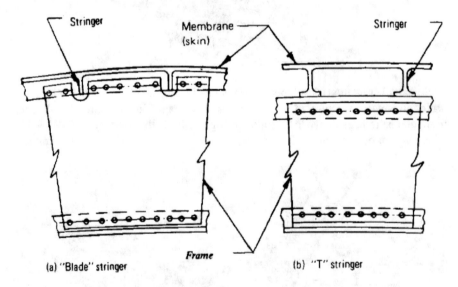

(a) "Blade" stringer   (b) "T" stringer

The type (a), shown on the left-hand side of the preceding figure, consists of panels in which the skin, the "blade" stingers, and the horizontal ribs are machined as an integral unit, where the frames are attached mechanically to the horizontal ribs after the panels are formed.

The type (b), shown on the right-hand side of the preceding figure, consists of panels in which only the skin and the "T" stringers are machined, and the frames are added after forming by mechanical attachment to the inboard flanges of the "T" stringers [22].

The thickness of the skin depends on the circumferential tensional loads under maximum internal pressure. The stringer spacing depends on the local requirements of stability. The stringer configuration and the frame spacing for tanks of minimum mass depend on general requirements of stability under axial compressive loads combined with internal pressures.

The material used for a sidewall of skin-stringer-frame design must be readily machinable and must also have good forming and welding properties. Compensation should be made by increasing the thickness at the weld joints, because the design strength of a weld is generally taken lower than the strength of the original material. The stringer spacing should be designed so as to keep the skin from buckling at limit load, account being taken of the buckling data of the curved panel and also of the stabilising effect of the internal pressure.

The following figure, due to the courtesy of NASA [22], shows a typical sidewall design used successfully for the liquid-hydrogen tank of the Saturn S-II stage.

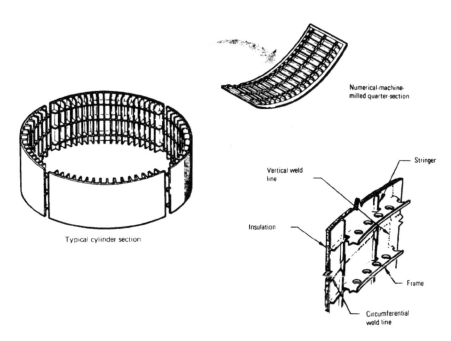

As has been shown above, the skin of a tank can also be stiffened by using integral ribs spaced in a grid pattern (either isogrid or rectangular waffle).

The integral rib stiffeners are usually formed by mechanical or chemical milling of the waffle pattern in a thick plate. Mechanical milling is more efficient than the other method [22]. The waffle structure is usually designed according to criteria of shell stability. Aluminium alloys are frequently used for the waffle plate material.

A waffle design was used for the propellant tank structure of the Saturn S-IVB stage. The waffle pattern of the S-IVB consisted of pockets 241.3 mm on centre, which were oriented $\pm\pi/4$ rad with respect to the longitudinal axis of the vehicle. These pockets were surrounded by ribs which were 15.93 mm high and 3.658 mm wide. The skin or web thickness at the bottom of the pocket was 3.124 mm, and the weld land areas at the edges of these segments were 6.4 mm thick. The following figure, due to the courtesy of NASA [52], shows the propellant tank structure of the Saturn S-IVB stage and, in particular, the machined waffle pattern.

S-IVB PROPELLANT TANK STRUCTURE

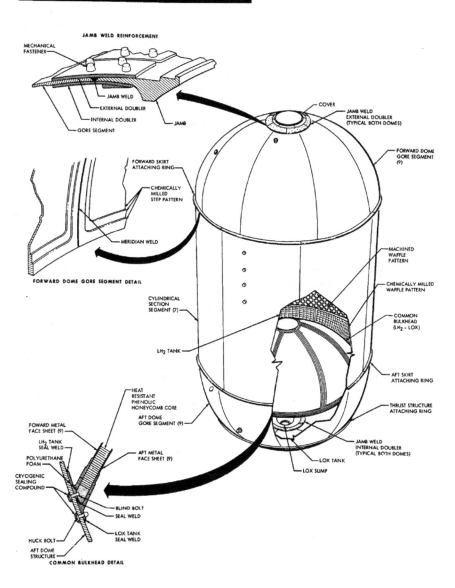

The isogrid pattern has the following advantages over the rectangular or square pattern:

- high twisting rigidity of the structure, which distributes the loading over a wide region;
- isotropic behaviour of the material, which has the same strength in all directions; and

- uncoupling of the bending and the in-plane resultants.

Therefore, the isogrid pattern makes it possible to apply the many available solutions of the classical theory of plates. In other words, a structure stiffened by an isogrid pattern can be analysed as a solid continuous sheet of material having appropriate thickness and elastic modulus [49].

A monocoque structure stabilised by internal pressure has the lowest structural mass. This type of structure requires a material having a high tensile stress. Extreme care must be taken during manufacturing and transportation to avoid handling damage. In order to prevent buckling in a monocoque pressure-stabilised structure, the meridian tensile stress due to internal pressure must be greater than the meridian compressive stress due to external loads. This type of structure has been used for the Atlas and Centaur launch vehicles. It has also been used for the liquid-oxygen tank of the Space Shuttle, which is shown in the following figure, due to the courtesy of NASA [53].

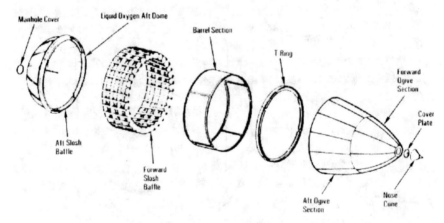

Liquid Oxygen Tank Structure

The shape of the end closures of tanks affects their lengths and also the means used for structural stiffening, to an extent which depends on the materials chosen. A choice should also be made on whether using two separate closures or a common closure for two tanks stacked onto each other. Some surfaces of revolution, which have been analysed by computer subroutines for end closures, are shown in the following figure, due to the courtesy of NASA [22].

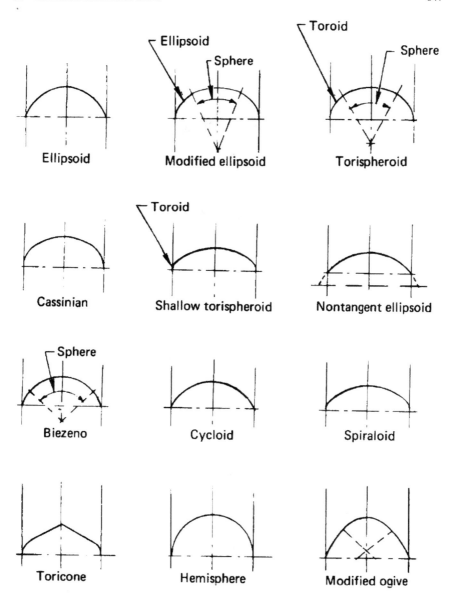

In practice, end closures of hemispherical shape were chosen for the tanks of the Titan and for those of the Saturn S-IV. Oblate spheroids were chosen for the Saturn S-II, and ellipsoids were chosen for the Atlas, the Saturn I-C, and the Centaur [22].

A sandwich structure is used to resist buckling in the presence of compressive loads. Where only tensional loads exist, monocoque structures can be used.

An example of honeycomb sandwich structure is shown in the following figure, due to the courtesy of NASA [22].

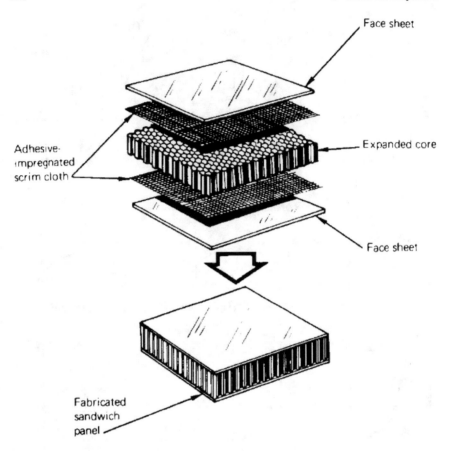

Further information on sandwich structures is given in several textbooks. A short course on the matter can be found in [54].

The forward bulkhead of a tank is usually a surface of revolution, convex on its external side, which is loaded principally by internal pressure. A thin skin subject to membrane stresses is the structural scheme generally used. Compressive stresses should be carefully determined, especially where the radius of curvature of the surface of revolution changes, in order to avoid circumferential buckling, and the results found analytically should be confirmed by tests.

To reduce mass, it is desirable to vary the thickness of the shell so that the bulkhead is subject to the maximum allowable stress along its meridians. On the other end, to reduce cost, it is desirable a shell of constant thickness.

As has been shown in the preceding figures of this section, a bulkhead is generally manufactured by welding together an apical tank skin (also known as the dollar hatch) to a welded sub-assembly consisting of some triangular segments or gores. This method avoids the juncture of the multiple welds where the gore segments meet at a common point.

The aft bulkhead of a tank differs from the forward bulkhead only because some forces can act on it under certain conditions. For example, such forces act during the filling of a tank with a liquid propellant and during the propelled flight. Aft bulkheads are constructed by using sandwich or waffle structures. Engine feed-lines are usually located in the central part of a bulkhead, in case of a single engine.

The fuel and the oxidiser of a bi-propellant rocket engine are separated by either two distinct bulkheads or one common bulkhead, according to the scheme shown in the following figure, due to the courtesy of NASA [22].

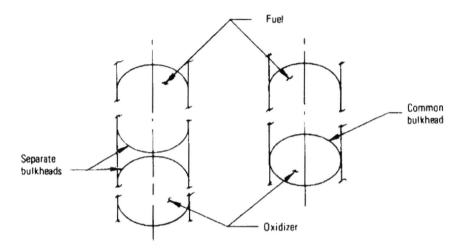

A common bulkhead, if present, may be either self-supporting or pressure-stabilised. A common bulkhead of the self-supporting type must be designed for both bursting and collapsing pressures. Sandwich or waffle structures are generally used for this purpose. The forward surface of a common bulkhead is usually convex, as shown in the preceding figure.

The attachment junctures used in a tank for propellants include:

- weld joints;
- bulkhead/sidewall juncture; and
- bosses and support provisions.

A brief description is given below for each of these elements.

The principal welding processes used in space flight hardware have been described in Sect. 6.7. We describe here the weld joints used specifically in tanks for propellants. Since the strength of the material in the weld area is less than the strength of the original material, then an increase in thickness is necessary at the weld joints. The thicker weld land ($t_{weld} > t_{skin}$) is usually asymmetrical on the two sides (external and internal) of the skin at a weld joint, as shown in the following figure, due to the courtesy of NASA [22].

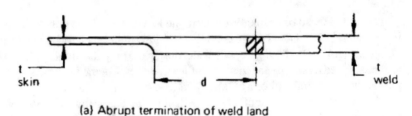

(a) Abrupt termination of weld land

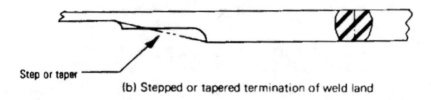

(b) Stepped or tapered termination of weld land

This is because it is desirable to maintain a smooth external surface for aerodynamic reasons, and also because milling on the two sides of a weld land is more expensive than milling on one side only. The preceding figure illustrates two types, (a) and (b), of weld land for aluminium alloys 2014-T6 and 2219-T87, which are materials commonly used for propellant tanks. The thickness t_{weld} of a weld land is greater, by a factor ranging from 2 to 2.25, than the thickness t_{skin} of the skin. The width d of a weld land ranges from 32 to 51 mm. The stepped or tapered termination (type b) of a weld land has the advantage, over the abrupt termination (type a), of avoiding bending stresses and strength reduction where the skin meets the weld land [22].

The following figure, due to the courtesy of NASA [22], shows some pre-weld joint preparations which have proven satisfactory for tank design.

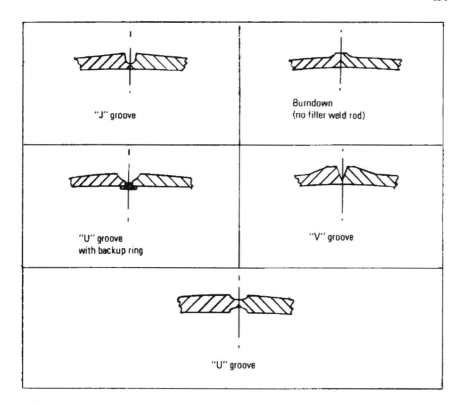

When possible, cylindrical sections of tanks are manufactured by spin forging, in order to eliminate longitudinal welds. Post-weld processing of tanks is limited to ageing, to relieve internal stresses caused by welding. This is because heat treatments after welding could induce distortion and oxidation, and also because complex equipment would be needed for that effect.

Junctures are also needed in a tank between bulkhead and sidewall and between skirt and sidewall. These junctures are usually made at a common location with an appropriate fitting. A fitting commonly used is a Y-ring, so called due to its shape, which has been used for the Saturn I-C and for the Titan II. A Y-ring fitting is shown in the following figure, due to the courtesy of NASA [22].

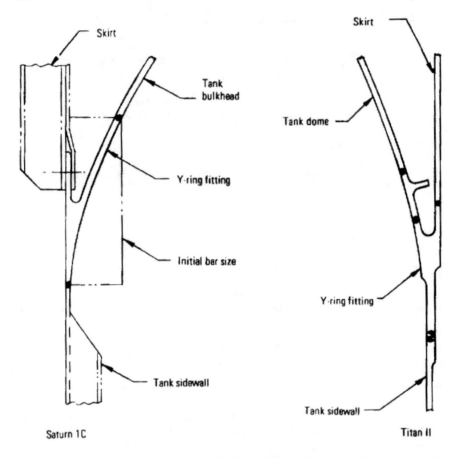

Saturn 1C   Titan II

A different type of fitting for the bulkhead/sidewall juncture is shown in the following figure, due to the courtesy of NASA [22]. It was used for the Centaur rocket stage.

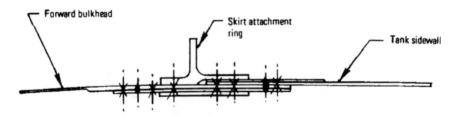

A Y-ring fitting provides a structural path for the loads from the bulkhead to the tank and from the skirt to the tank. It also makes the tank leak-proof at the bulkhead/sidewall juncture.

The system components and the tank mounting structure are usually attached by means of bosses. Such bosses are made deep enough to accommodate mechanical fasteners, in order to avoid tank penetration. When the basic structure is of sufficient thickness, then the bosses are integrally milled with the basic structure. When the basic structure is a thin shell, then the use of integral bosses requires a material of larger initial thickness and a more extensive milling.

Bosses can also be fabricated by welding a circular machined ring, containing the bosses, to the bulkhead. However, this method induces residual stresses in the material as a result of the welding process.

Non-integral support provisions are machined fittings, such as ports, flanges, and support pads, which are welded to the tank skin. For this purpose, it is necessary to provide a material whose thickness is greater than the thickness of the tank skin, in order compensate for the loss of strength due to welding. Otherwise, for the same reason, it is necessary to lower the permissible operating pressure of the tank. In tanks for liquid propellants, a single access opening is frequently used. This is obtained by using a close-out cover which contains the lines for input and output. When multiple tanks are connected in series, the inlet lines are connected to standpipes within the tanks, in order to preclude reverse flow of propellant and to ensure feed-out in series.

Openings and access doors are ports and access points in the skin of a tank. Three examples are shown in the following figure, due to the courtesy of NASA [22].

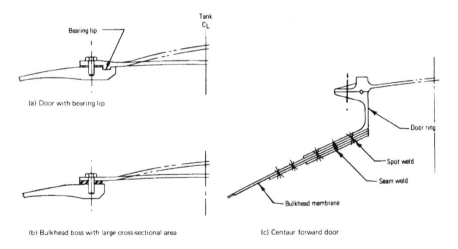

(a) Door with bearing lip

(b) Bulkhead boss with large cross-sectional area

(c) Centaur forward door

Openings and access doors are usually located in or near the apex of a bulkhead. The primary requirements for these access points is the prevention of leakage during the entire service life of a rocket vehicle. The method marked with (a) in the preceding figure uses a door with a bearing lip and oversized holes, in order to reduce rotation at the joint because of eccentric bolt shear loading. The method marked with (b) uses a bulkhead boss of large cross-sectional area, in order to force most of the load to remain in the bolting ring instead of going through the door. The method marked with (c) uses a door ring placed above the apex of the bulkhead, and has been chosen for the forward door of the Centaur rocket stage.

References

1. Anonymous, Space Shuttle, NASA SP-407. https://history.nasa.gov/SP-407/p38.htm
2. Flügge W, Sobel LH (1965) Stability of shells of revolution: general theory and application to the torus. Ph.D. Dissertation, Mar 1965, 138 pp. https://apps.dtic.mil/dtic/tr/fulltext/u2/617197.pdf
3. Young WC, Budynas RG (2002) Roark's formulas for stress and strain, 7th edn. McGraw-Hill, New York. ISBN 0-07-072542-X
4. Flügge W (1960) Stresses in shells. Springer, Berlin
5. The Engineering Toolbox, Young's modulus—tensile and yield strength for common materials. https://www.engineeringtoolbox.com/young-modulus-d_417.html
6. Abramson HN (ed) (1966) The dynamic behaviour of liquids in moving containers with application to space vehicle technology, Southwest Research Institute, NASA SP-106, 464 pp, Jan 1966. https://ntrs.nasa.gov/archive/nasa/casi.ntrs.nasa.gov/19670006555.pdf
7. Abramson HN (1969) Slosh suppression, NASA SP-8031, 36 pp, May 1969. http://everyspec.com/NASA/NASA-SP-PUBS/NASA-SP-8031_20669/
8. Dodge FT (2000) The new "Dynamic behaviour of liquids in moving containers". Southwest Research Institute, 202 pp. ftp://apollo.ssl.berkeley.edu/pub/ME_Archives/Government%20and%20Industry%20Standards/NASA%20Documents/SwRI_Fuel_Slosh_Update.pdf
9. Abramson HN, Bhuta PG, Hutton RE, Stephens DG (1968) Propellant slosh loads, NASA SP-8009, 32 pp, Aug 1968. https://ntrs.nasa.gov/archive/nasa/casi.ntrs.nasa.gov/19690005221.pdf
10. Pérez JG, Parks RA, Lazor DR (2012) Validation of slosh model parameters and anti-slosh baffle designs of propellant tanks by using slosh testing, NASA, 27th Aerospace testing seminar, Oct 2012. https://ntrs.nasa.gov/archive/nasa/casi.ntrs.nasa.gov/20130000590.pdf
11. Lomen DO (1965) Liquid propellant sloshing in mobile tanks of arbitrary shape, NASA CR-222, NASA Langley Research Centre, 72 pp, Apr 1965. https://ntrs.nasa.gov/archive/nasa/casi.ntrs.nasa.gov/19650012759.pdf
12. Fries N, Behruzi P, Arndt T, Winter M, Netter G, Renner U (2012) Modelling of fluid motion in spacecraft propellant tanks—sloshing. In: Space propulsion 2012 conference, 7th–10th May 2012, Bordeaux, France, 11 pp. https://www.flow3d.de/fileadmin/download_publikationen/modelling-fluid-motion-in-spacecraft-propellant-tanks-sloshing-2012.pdf
13. Stephens DG, Leonard HW, Perry TW Jr (1962) Investigation of the damping of liquids in right-circular cylindrical tanks, including the effects of a time-variant liquid depth, NASA TN D-1367, July 1962, 32 pp. https://ntrs.nasa.gov/archive/nasa/casi.ntrs.nasa.gov/19620004069.pdf
14. Hopfinger EJ, Baumbach V (2009) Liquid sloshing in cylindrical fuel tanks. Progr Propul Phys 1:279–292. https://www.eucass-proceedings.eu/articles/eucass/pdf/2009/01/eucass1p279.pdf
15. Keulegan GH, Carpenter LH (1958) Forces on cylinders and plates in an oscillating fluid. J Res Natl Bur Stan 60(5):423–440, Research Paper 2857. https://nvlpubs.nist.gov/nistpubs/jres/60/jresv60n5p423_A1b.pdf
16. Miles JW (1958) Ring damping of free surface oscillations in a circular tank. J Appl Mech 25(2):274–276
17. Bauer HF (1960) Theory of the fluid oscillations in a circular cylindrical ring tank partially filled with liquid, NASA TN D-557, 124 pp
18. Sumner IE (1964) Experimental investigation of slosh-suppression effectiveness of annular-ring baffles in spherical tanks, NASA TN D-2519, 22 pp, Nov 1964. https://ntrs.nasa.gov/archive/nasa/casi.ntrs.nasa.gov/19650001203.pdf
19. Stephens DG, Leonard HW, Silveira MA (1961) An experimental investigation of the damping of liquid oscillations in an oblate spheroidal tank with and without baffles, NASA TN D-808, 28 pp, June 1961. https://ntrs.nasa.gov/archive/nasa/casi.ntrs.nasa.gov/19990017908.pdf
20. Stephens DG, Scholl HF (1967) Effectiveness of flexible and rigid ring baffles for damping liquid oscillations in large-scale cylindrical tanks, NASA TN D-3878, Mar 1967, 34 pp. https://ntrs.nasa.gov/archive/nasa/casi.ntrs.nasa.gov/19670010559.pdf

21. Anonymous, Saturn V Flight Manual SA-503, NASA-TM-X-72151, MSFC-MAN-503, Nov 1968, 243 pp. https://ntrs.nasa.gov/archive/nasa/casi.ntrs.nasa.gov/19750063889.pdf
22. Wagner WA, Keller RB Jr (ed) (1974) Liquid rocket metal tanks and tank components, NASA SP-8088, 165 pp, May 1974. https://ntrs.nasa.gov/archive/nasa/casi.ntrs.nasa.gov/19750004950.pdf
23. Roberts JR, Basurto ER, Chen PY (1966) Slosh design handbook I, NASA CR-406, 328 pp, May 1966. https://ntrs.nasa.gov/archive/nasa/casi.ntrs.nasa.gov/19660014177.pdf
24. Stofan AJ, Sumner IE (1963) Experimental investigation on the slosh damping effectiveness of positive-expulsion bags and diaphragms in spherical tanks, NASA TN D-1712, 22 pp, June 1963. https://ntrs.nasa.gov/archive/nasa/casi.ntrs.nasa.gov/19630006725.pdf
25. MIL-HDBK-5J: Metallic materials and elements for aerospace vehicle structures, 31 Jan 2003. http://everyspec.com/MIL-HDBK/MIL-HDBK-0001-0099/MIL_HDBK_5J_139/
26. Hood DW, Eichenberger TW, Lovell DT (1968) An investigation of the generation and utilisation of engineering data on weldments, The Boeing Company, Technical Report AFML-TR-68-268, Oct 1968. https://apps.dtic.mil/dtic/tr/fulltext/u2/851938.pdf
27. von Mises R (1913) Mechanik der festen Körper im plastisch- deformablen Zustand, Nachrichten von der Gesellschaft der Wissenschaften zu Göttingen, Mathematisch-Physikalische Klasse, vol 1913, pp 582–592. http://eudml.org/doc/58894
28. Hill R (1948) A theory of the yielding and plastic flow of anisotropic metals. Proc R Soc A 193(1033):281–297. https://royalsocietypublishing.org/doi/pdf/10.1098/rspa.1948.0045
29. Meuwissen MHH (1995) Yield criteria for anisotropic elasto-plastic materials. Eindhoven University of Technology, 24 pp. https://pure.tue.nl/ws/portalfiles/portal/4299264/653055.pdf
30. Whitfield HK, Keller RB Jr (eds) (1970) Solid rocket motor metal cases, NASA SP-8025, Apr 1970, 110 pp. https://ntrs.nasa.gov/archive/nasa/casi.ntrs.nasa.gov/19700020430.pdf
31. Gerds AF, Strohecker DE, Byrer TG, Boulger FW (1966) Deformation processing of titanium and its alloys, NASA TM X-53438, Apr 1966. https://ia601209.us.archive.org/13/items/NASA_NTRS_Archive_19660016988/NASA_NTRS_Archive_19660016988.pdf
32. Strohecker DE, Gerds AF, Henning AF, Boulger FW (1966) Deformation processing of stainless steel, NASA TM X-53569. https://ia800308.us.archive.org/22/items/nasa_techdoc_19670022702/19670022702.pdf
33. Slunder CJ, Hoenie AF, Hall AM (1967) Thermal and mechanical treatment for precipitation-hardening stainless steel, NASA SP-5089, Jan 1967, 207 pp. https://ntrs.nasa.gov/archive/nasa/casi.ntrs.nasa.gov/19680010964.pdf
34. Corbett J (2017) Welding, NASA, 7 Aug 2017. https://www.nasa.gov/centers/wstf/supporting_capabilities/machining_and_fabrication/welding.html
35. Carter B (2013) Introduction to friction stir welding (FSW), NASA, Glenn Research Centre, Advanced Metallics Branch, 8 May 2013, 25 pp. https://ntrs.nasa.gov/archive/nasa/casi.ntrs.nasa.gov/20150009520.pdf
36. Anonymous, Space Shuttle Technology Summary, Friction Sir Welding, NASA Marshall Space Flight Centre, FS-2001-03-60-MSFC. https://www.nasa.gov/centers/marshall/pdf/104835main_friction.pdf
37. Anonymous, A Bonding Experience: NASA Strengthens Welds, NASA. https://www.nasa.gov/topics/nasalife/friction_stir.html
38. Brown BF (1970) Stress-corrosion cracking: a perspective review of the problem, NRL Report 7130, Naval Research Laboratory, Washington, D.C., 24 pp, 16 June 1970. https://apps.dtic.mil/dtic/tr/fulltext/u2/711589.pdf
39. MIL-HDBK-729: Corrosion and corrosion prevention metals, 21 Nov 1983, 251 pp. http://everyspec.com/MIL-HDBK/MIL-HDBK-0700-0799/MIL_HDBK_729_1946/
40. McDanels S (2018) Forms of corrosion, galvanic corrosion, NASA, Corrosion Engineering Laboratory, J. F. Kennedy Space Centre, 7 Nov 2018. https://corrosion.ksc.nasa.gov/Corrosion/FormsOf#Galvanic%20Corrosion
41. DOE-HDBK-1015/1-93, DOE Fundamentals handbook—chemistry, vol 1 of 2, FSC-6910, U.S. Department of Energy, Jan 1993. https://www.isibang.ac.in/~library/onlinerz/resources/chem-v1.pdf

42. MIL-STD-889C (2016) Dissimilar metals, 22 Aug 2016. http://everyspec.com/MIL-STD/MIL-STD-0800-0899/MIL-STD-889C_55344/

43. Cataldo CE (1968) Compatibility of metals with hydrogen, NASA TM X-53807, 26 Dec 1968, 26 pp. https://ntrs.nasa.gov/archive/nasa/casi.ntrs.nasa.gov/19690009664.pdf

44. Caskey GR Jr (1983) Hydrogen compatibility handbook for stainless steels, U.S. Department of Energy, Office of Scientific and Technical Information, June 1983. https://www.osti.gov/servlets/purl/5906050

45. San Marchi C, Somerday BP (2012) Technical reference for hydrogen compatibility of materials, Sandia Report SAND2012-7321, Sept 2012. https://www.sandia.gov/matlsTechRef/chapters/SAND2012_7321.pdf

46. Chandler DL (2019) Observing hydrogen effects in metal, MIT News, Massachusetts Institute of Technology, 4 Feb 2019. http://news.mit.edu/2019/observing-hydrogens-effects-metal-0205

47. Tiffany CF (1970) Fracture control of metallic pressure vessels, Technical Report NASA SP-8040, 65 pp, May 1970. https://ntrs.nasa.gov/archive/nasa/casi.ntrs.nasa.gov/19710004655.pdf

48. Abramowitz M, Stegun IA (eds) (1972) Handbook of mathematical functions with formulas, graphs, and mathematical tables, National Bureau of Standards, United States Department of Commerce, Dec 1972. http://people.math.sfu.ca/~cbm/aands/abramowitz_and_stegun.pdf

49. McDonnell Douglas, Isogrid design handbook, NASA-CR-124075, Feb 1973, 222 pp. https://femci.gsfc.nasa.gov/isogrid/NASA-CR-124075_Isogrid_Design.pdf

50. Garber S (1959) Structures and materials. In: Space handbook: Astronautics and its applications, , Staff report of the select committee on astronautics and space exploration, United States, Government printing office, Washington. https://www.history.nasa.gov/conghand/structur.htm

51. Knighton DJ (1972) Delta launch vehicle isogrid structure NASTRAN analysis, NASA, Conference paper, Sept 1972, 23 pp. https://ntrs.nasa.gov/archive/nasa/casi.ntrs.nasa.gov/19720025227.pdf

52. Anonymous, Skylab Saturn IB flight manual, NASA TM-X 70137, NASA, George C. Marshall Space Flight Centre, 30 Sept 1972, 273 pp. https://ntrs.nasa.gov/archive/nasa/casi.ntrs.nasa.gov/19740021163.pdf

53. NASA. https://science.ksc.nasa.gov/shuttle/technology/images/et-lox_1.jpg

54. Joyce P, Sandwich structures, United States Naval Academy. https://www.usna.edu/Users/mecheng/pjoyce/composites/Short_Course_2003/13_PAX_Short_Course_Sandwich-Constructions.pdf

Chapter 7
Interconnecting Components and Structures

7.1 Fundamental Concepts

The present chapter describes interconnecting structures and fluid-carrying ducts which incorporate elements such as bellows, flexible joints, flexible hoses, and flanges. An example is provided by the propellant-supply ducts which convey the fuel and the oxidiser from their tanks in a rocket stage to the respective turbo-pumps. These ducts employ flexible components (hoses, bellows, and joints) to permit freedom of movement during engine gimballing. Some typical assemblies for lines, bellows, and hoses are shown in the following figure, due to the courtesy of NASA [1], which illustrates the J-2 engine, used for the Saturn IB and Saturn V launch vehicles.

© The Editor(s) (if applicable) and The Author(s), under exclusive license to Springer Nature Switzerland AG 2021
A. de Iaco Veris, *Fundamental Concepts of Liquid-Propellant Rocket Engines*, Springer Aerospace Technology, https://doi.org/10.1007/978-3-030-54704-2_7

The ducts of a liquid-propellant rocket engine have often filters, which maintain the fluids at a desired level of cleanliness, by removing contaminant particles which such fluids may contain. A definition of the principal terms used in this chapter is given below. A line or duct is an enclosed leak-proof passageway through which a fluid is conveyed from one component of a rocket engine to another. A bellows is a cylinder, corrugated along its sidewall, which may be enclosed in a line to permit movement by deflection of its corrugations. A bellows joint may be either a bellows with a restraint linkage or a free bellows alone. A flexible hose is a bendable duct used to carry or transfer a fluid from one point to another. A flange is a rib or rim placed at one end of a duct for attachment to another piece or duct. A flange joint is an attachment of ducts, where the connecting pieces have flanges by which the parts are bolted together. A filter is a device which removes a contaminant from a fluid by trapping particles of the contaminant within or on the surface of a porous material. According to Howell and Weathers [2], flexible ducts are considered for use instead of rigid ducts under any of the following conditions:

- existence of angular or lateral misalignment of the points to be connected;
- necessity of removing ducts for repair or replacement;
- occurrence of relative motion caused by thermal expansion, duct movement, or structural bending; and
- presence of severe vibration.

The principal interconnecting ducts and structures in a rocket engine are described in Sect. 7.2. The following sections describe materials, construction, design criteria, and methods of calculation.

7.2 Interconnecting Ducts and Structures in a Rocket Engine

Propellant-supply ducts convey the fuel and the oxidiser from their tanks to the respective turbo-pumps. These ducts, due to their location at the centre of an engine gimbal, are subject to bending and torsional loads. They are also subject to internal pressure, due to the fluids carried by them. The principal requirement for propellant-supply ducts is to keep the pressure losses between the tank outlets and the pump inlets to the minimum possible value. As a result of the many forces acting upon these ducts, restrainers against buckling are frequently required. Such restrainers, if located inside the ducts, contribute to undesired pressure losses. On the other hand, the external location of the restrainers increases the duct size and may create problems of interference of the ducts with other components of the engine. The following figure, due to the courtesy of NASA [1], shows a bellows of the compressed type, externally restrained against buckling.

The following figure, also due to the courtesy of NASA [1], shows an anti-buckling device designed to absorb torsional deflection in a pump inlet duct.

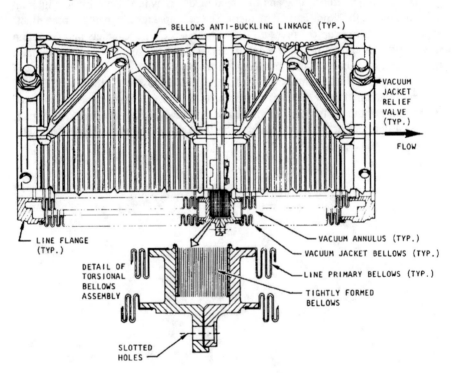

The device shown above is a tightly formed bellows assembly, which was incorporated into the pump inlet ducts of the J-2 engine, when the torsional moment of the ducts was found to be too high for the resistance of the pump casing. The bellows is thin-walled (0.254 mm) and over 2540 mm long. It has 40 deep corrugations stacked in a 31.75 mm height. Flanges encompass the bellows and permit the application of only torsional deflections. The joint is substantially a torsion spring of low elastic constant, which can absorb torsional rotations up to $\pi/60$ rad or 3° [1].

The manufacturer of a rocket vehicle must know the sizes of the connecting flanges, the types of gasket used, and the forces transmitted by the engine ducts to the vehicle when the engine is gimballed. This information is provided by the engine designer. In pump-feed rocket engines, the working relative pressure of propellant-supply ducts ranges from 0.3447×10^6 N/m² to 0.6895×10^6 N/m² and over. In pressure-fed rocket engines, the working absolute pressure of the propellant ducts is less than 3.447×10^6 N/m² [3].

In rocket engines fed by turbo-pumps, the high-pressure propellant ducts for pump discharge connect the outlet sections of the pumps (for respectively the oxidiser and the fuel) to the main fuel valves attached to the thrust chamber. These ducts also contain bellows components. However, some engines have also had rigid (instead of flexible) components, as shown in following figure, due to the courtesy of NASA [1].

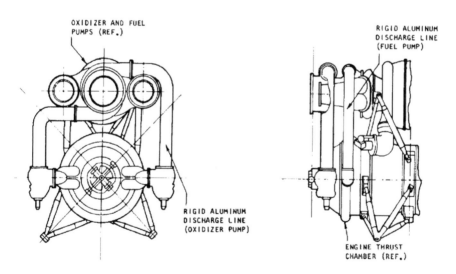

The early F-1 engines incorporated bellows joints in the pump discharge ducts, but these joints were later replaced by hard lines of aluminium with large bends for flexibility. The low modulus of elasticity of aluminium permitted lower end reactions for a given deflection than did a comparable duct of steel or nickel-base alloy. This change was possible because the ducts were not gimballing and were required to absorb only misalignments and thermal effects.

In case of gimballing engines, a classic configuration for pump discharge ducts is the wraparound duct arrangement, which is shown in following figure, due to the courtesy of NASA [1].

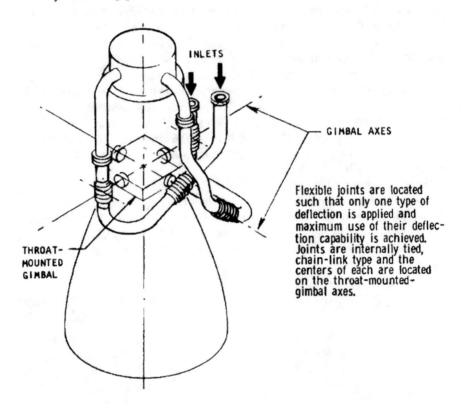

The term wraparound indicates a flexible duct or hose which, as it were, wraps around the thrust-vector-control gimbal of a rocket engine in the plane of the gimbal. This configuration was used for the Navaho, Atlas, Thor, and Jupiter engine pump discharge ducts, for the pump inlet ducts of the H-1 engine, for the gimballing feed lines of the Apollo Service Module engine, and for the descent engine of the Apollo Lunar Module. This configuration was also used for all of the vehicle-to-engine interface lines of the F-1 engine on the S-IC, except the main propellant lines. In addition, the wraparound concept was used extensively for each of the three main engines of the Space Shuttle. Details on the matter can be found in [1, 4].

The following figure, due to the courtesy of NASA [1], shows a typical pump-discharge high-pressure propellant duct with restraining linkage.

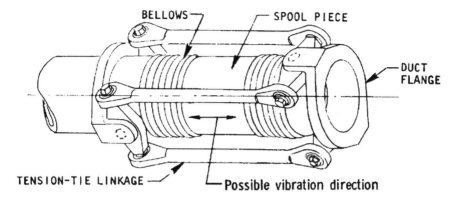

The duct shown above has two bellows separated by a spool piece. This type of duct is subject to vibration of the long unsupported mass.

The following figure, also due to the courtesy of NASA [1], shows a bellows thrust-compensating linkage with external tie bar installed in the oxidiser high-pressure duct of the J-2 rocket engine.

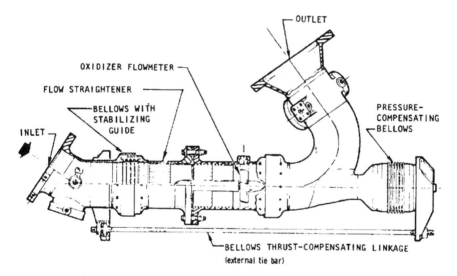

This duct has an external tie bar which acts as a thrust compensator, and two bellows held together by the external tie bar, which limits the movements of the two bellows. One bellows opposes the other in balancing the pressure exerted by the propellant flowing in the duct.

Pressurisation lines for the propellant tanks are used to connect the propellant tanks to sources of pressures, such as gases stored under pressure, gas generators, and heat exchangers for cryogenic propellants. For this purpose, high-pressure hoses and tubing are used. The following figure, due to the courtesy of NASA [5]. shows two 6.35 mm, 34.6 MPa (relative pressure) flexible hoses, which were used for the Saturn I-B rocket stage.

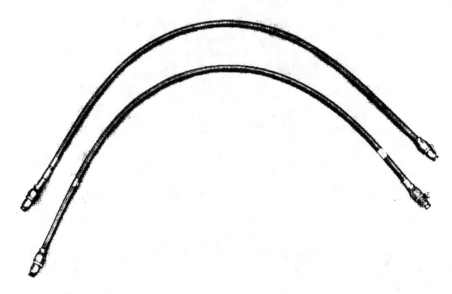

Seal drain lines are often used in case of dynamic seals applied to the shafts or to other moving components of turbo-pumps. Since perfect sealing is difficult to obtain, then drain lines are used between two dynamic seals placed in series. The seal drain lines, which include flexible hoses and tubing, are routed away and overboard. In case of a combination of liquid oxygen with RP-1, the seal drain lines can be routed along the wall of the thrust chamber up to the exit plane of the nozzle. In case of combinations of propellants which can form highly explosive mixtures, these lines are routed to vent ports at the periphery of space vehicles. The seal drain lines are usually routed to the periphery of the vehicles during boost flight, and to the exit of the thrust chamber during stage operation [3].

Seal drains for liquid hydrogen are routed to a safe disposal area, because of the hazard of mixing liquid hydrogen with atmospheric air. Therefore, leaking hydrogen is not allowed to accumulate inside a rocket vehicle. For liquid-hydrogen turbo-pumps, the external drains are usually eliminated, and the seal leakage is allowed to vent into the turbine area, as shown in the following figure, due to the courtesy of NASA [6], which illustrates a particular of the liquid-hydrogen turbo-pump used for the J2-S engine.

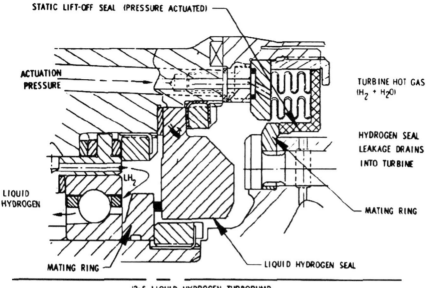

STATIC LIFT-OFF SEAL (PRESSURE ACTUATED)

ACTUATION PRESSURE

TURBINE HOT GAS (H_2 + H_2O)

HYDROGEN SEAL LEAKAGE DRAINS INTO TURBINE

LIQUID HYDROGEN

MATING RING

LIQUID HYDROGEN SEAL

MATING RING

J2-S LIQUID HYDROGEN TURBOPUMP

Pneumatic supply lines are used in rocket engines to provide pneumatic pressure required for several purposes. For example, as described at length in [4], the pneumatic control assembly of the Space Shuttle provide central control of all pneumatic functions, such as:

- engine preparation and shutdown purges;
- bleed valve operation; and
- engine pneumatic shutdown, including pogo post-charge.

The pneumatic pressure is provided by gases stored in vessels, which are charged before flight through high-pressure flexible lines. The design of such lines depends on the mating connexions on the vehicle side, on the type of gas used, and on its temperature and pressure.

Cryogenic-propellant bleed lines are used in cryogenic-propellant engines feed by turbo-pumps. This is because adverse conditions for such engines may occur during start, when the metallic walls containing cryogenic fluids are at insufficiently low temperatures, and also when the cryogenic fluids downstream of the tank outlet are superheated. In addition, the pressure which opens the main valves and starts the turbo-pumps reduces further the static pressure at the inlet of the turbo-pumps, and this accelerates the production of gas. This, in turn, may lead to cavitation in the pumps and to malfunction of the gas generator.

In order to avoid such undesirable events, a continuous bleed from a point farthest downstream of the pump inlet is applied before engine start. By so doing, fresh liquid at the bulk temperature of the tank replaces continuously the warming fluid and cools the metallic walls of the containers. The bleeds are ducted away from the launch site, because they can form combustible or even explosive mixtures. This requires a line to duct away the bleeds. Wire-braided flexible hoses and tubing are generally used for such lines. In case of cryogenic-propellant engines used for upper stages of rockets, a re-circulation system is used, which re-directs the propellants to their tanks instead of dumping them overboard. Flexible lines between the engine and the rocket vehicle are required for this purpose. In order to reduce bleeds and re-circulating flows, it is possible to provide the rocket vehicle with means (such as cooling, thermal insulation, etc.) apt to keep the bulk temperatures of the propellants sufficiently below their boiling temperatures at the operating pressures in the tanks.

Purge lines are required for rocket engines. This is due to the necessity of performing purges during the start and shutdown sequence, and also before and after servicing, to keep the engines dry. These purges are executed by using non-reactive gases, such as nitrogen at ambient temperature, in order to prevent the formation of combustible mixtures, and to expel residual propellants.

As an example, for the F-1 engine, a gaseous nitrogen purge is applied for thermal conditioning and for elimination of explosive mixtures under the engine envelope. Since low temperatures may exist in the space between the engine and its envelope for thermal insulation, then heated nitrogen is applied to this space. This purge is manually operated, whenever there is a prolonged hold of the countdown with liquid oxygen onboard and with an ambient temperature below approximately 286 K. This purge is turned off five minutes prior to ignition command and is continued during the time of umbilical connexion.

A continuous nitrogen purge is also required to expel propellant leakage from the seal housing of the liquid-oxygen turbo-pump and from the liquid-oxygen injector of the gas generator. The pressure of the purging gas improves the properties of the liquid-oxygen seal. This purge is required from the time at which the propellants are loaded and is continuous throughout flight.

A nitrogen purge prevents contaminants from accumulating on the viewing surfaces of the radiation calorimeter. This purge is started 52 s before scheduled lift-off time and is continued during flight.

A gaseous nitrogen purge is required to prevent contaminants from entering the liquid-oxygen system through the liquid-oxygen injector of the main engine or the liquid-oxygen injector of the gas generator. This purge is started prior to engine operation and is continued during the time of umbilical connexion.

At approximately 13 h before scheduled lift-off time, an ethylene glycol solution fills the thrust tubes and the manifolds of the engine. This inert solution serves to smooth out the combustion sequence at engine start. Flow is terminated by a signal from an observer at the engine.

At approximately 5 min before scheduled lift-off time, 186.3 l (0.1893 m^3) are supplied to top off the system to compensate for liquid loss which occurred during engine gimballing [7].

Hydraulic ducts for high-pressure liquids are used in liquid-propellant rocket engines, as shown in the following figure, due to the courtesy of NASA [7], which illustrates the outboard engine fluid system used for the five F-1 engines of the S-IC first stage of the Saturn V rocket.

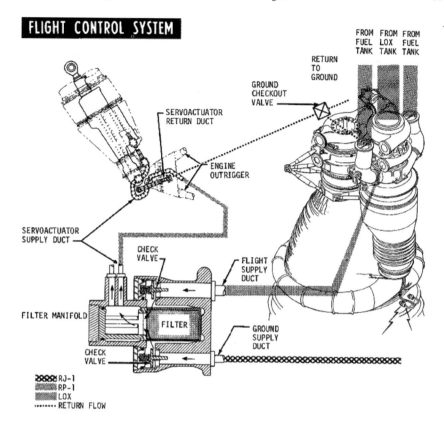

The flight control system of the S-IC gimbals the four outboard engines to provide attitude control during the burn phase of the S-IC. For this purpose, the hydraulic pressure is supplied from a Ground Support Equipment (GSE) pressure source during test, pre-launch checkout, and engine start. When the engine starts, hydraulic pressure is generated by the turbo-pump of the engine.

Pressure from either source is made available to the engine valves, such as the main fuel and liquid-oxygen valves and the igniter fuel valve. These valves are sequenced and controlled by the terminal countdown sequencer, stage switch selector and by mechanical and fluid pressure means. Fluid under pressure also flows through a filter, shown in the preceding figure, and to the two flight control servo-actuators on each outboard engine. The fluid power system illustrated above uses both RJ-1 ramjet fuel and RP-1 rocket propellant as hydraulic fluid.

The RJ-1 is used by the Hydraulic Supply and Checkout Unit (GSE pressure source). RP-1 is the fuel used in the S-IC stage. It is pressurised by the engine turbo-pump. The two pressure sources are separated by check valves while return flow is directed to GSE or stage by the ground checkout valve. Drilled passages in the hydraulic components (valves and servo-actuators) permit a flow of fluid to thermally condition the units and to bleed gases from the fluid power system [7].

In most rocket engines feed by turbo-pumps, the gas generator is connected directly to the turbine inlet, and consequently no special ducts are necessary. In other rocket engines, high-pressure hot-gas ducts are required to connect the gas generator to the turbines. Such is the case with engines having two individual turbo-pumps. Still in other rocket engines, high-pressure hot-gas ducts are required to connect the tap-off ports of the thrust chamber with the turbine. Such is the case with engines (the J-2S, for one) based on the thrust chamber tap-off cycle, which has been described in Chap. 2, Sect. 2.7. High-pressure hot-gas ducts have rigid portions and also flexible portions, which are made of stainless steels and nickel-based alloys apt to resist temperatures up to about 1200 K. These ducts must be capable of absorbing deflections due to thermal expansion and also deflections due to misalignments and dynamic loads [3].

Liquid-propellant rocket engines have ducts for the hot gases exhausted by the turbines. These gases may be ducted either to a region near the exit plane of the nozzle or to a manifold of the thrust chamber. By the way, a manifold is a duct having one or more branches off the main flow stream. Common types are T-shaped and Y-shaped manifolds. The main function of a manifold is to distribute flow from one or more inlet passages to one or more outlet passages. A typical manifold for turbine exhaust gas used in a liquid-propellant rocket engine is shown in the following figure, due to the courtesy of NASA [1].

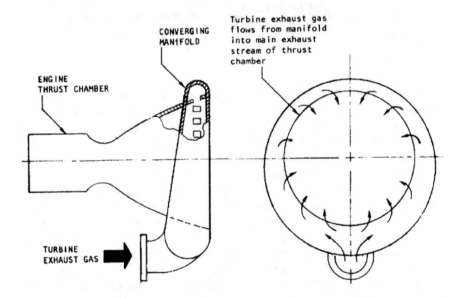

The manifold illustrated in the preceding figure is designed for equal distribution of flow. In other words, the flow areas inside the manifold are designed so as to split equally the flow among the branches. The gas exhausted by the turbine is dumped into the mainstream exhaust of the thrust chamber through an annular manifold. The cross-sectional area of the manifold decreases as it wraps around the thrust chamber, in order for the gas to maintain a constant velocity as it is bled off through openings in the wall of the thrust chamber [1].

In the particular case of the J-2 engine, the turbines of the oxidiser (liquid oxygen) and fuel (liquid hydrogen) turbo-pumps are connected in series by exhaust ducting which directs the discharged exhaust gas from the fuel turbo-pump turbine to the inlet of the oxidiser turbo-pump turbine manifold. One static and two dynamic seals in series prevent the turbo-pump oxidiser fluid and the turbine gas from mixing. Both turbo-pumps are powered in series by a single gas generator, which uses the same propellants as the thrust chamber. During burn periods, the liquid-oxygen tank is pressurised by liquid oxygen flowing through the heat exchanger in the oxidiser turbine exhaust duct. The heat exchanger heats the liquid oxygen, causing it to expand. The liquid-hydrogen tank is pressurised during burn periods by gaseous hydrogen from the thrust chamber fuel manifold [7].

The following figure, due to the courtesy of NASA [8] is an exploded view of the principal interconnecting components of the J-2 engine.

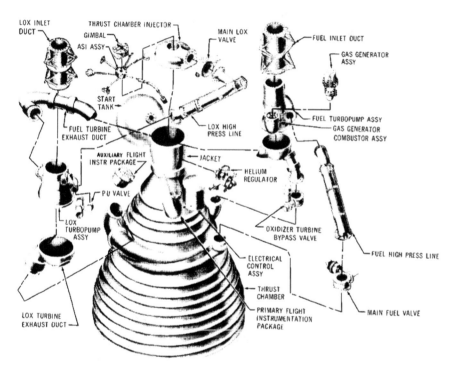

Each stage of a rocket vehicle has a thrust structure assembly, whose principal function is to redistribute the loads applied locally by the engines into a uniform loading about the periphery of a tank (usually, the fuel tank). It also provides support for the engines, engine accessories, base heat shield, engine fairings and fins, propellant lines, retrorockets, and environmental control ducts.

The following figure, adapted from [7], shows the S-IC stage of the Saturn V launch vehicle. This stage is propelled by five F-1 engines.

The lower thrust ring shown in the following figure has four hold-down points, which support the fully loaded Saturn/Apollo (approximately 2.722×10^6 kg) and also, as necessary, restrain the vehicle from lifting off at the full thrust of the F-1 engines. The skin segments are made of 7075-T6 aluminium alloy.

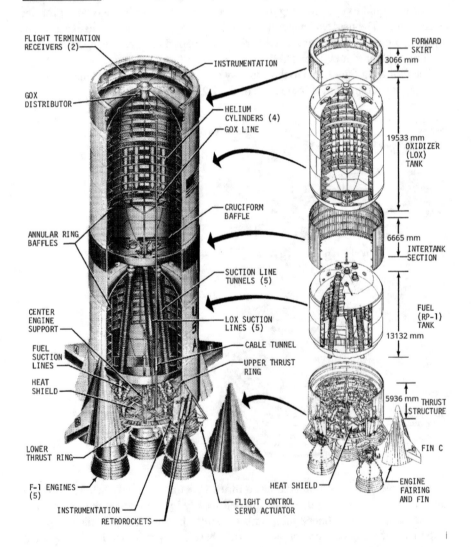

In most liquid-propellant rocket engines fed by turbo-pumps, the complete engine assembly is gimballed by using a gimbal mechanism having a spherical joint, as has been shown in Chap. 5, Sect. 5.6. This joint connects the thrust chamber to the thrust structure (described above) of the rocket vehicle.

A turbo-pump structure is used to fasten the turbo-pumps to either the thrust chamber or other structural elements, such as the thrust structure. The turbo-pump mounts are the connexions between the turbo-pump assembly and the engine. These mounts support the loads due to the weight of the turbo-pump assembly and also react to the loads due to engine inertia, propellant inertia, engine gimballing, differences of pressure in fluids, forces on the flanges, and gyroscopic forces. In addition, the turbo-pump mounts adapt to the differential thermal expansion or contraction of the turbo-pump assembly and of the thrust chamber assembly [9].

The most common types of turbo-pump mounts use struts having at least one ball-joint end connexion to accommodate dimensional tolerances and differential thermal expansion or contraction. Ball-ended struts can be arranged triangularly, to yield the lightest structure by loading its members in pure compression or tension. Another arrangement consists of close-coupled, rigid pads at one end of the turbo-pump, and one or two ball-ended struts at the other end to accommodate the dimensional variations [9].

7.3 Materials Used for Tubing in Rocket Engines

The 18–8 (18% chromium and 8% nickel) corrosion-resistant steels are the most frequently used materials for lines and bellows in rocket engines. Nickel-base alloys, such as those of the Inconel® and Hastelloy® families, are also used for ducts and bellows, because of their higher strengths, greater fatigue lives, and better corrosion resistances. Aluminium alloys are used in some non-critical or low-stress applications. Due to forming requirements, only high-ductility alloys in the annealing condition are used for corrugated sections [1].

Bellows-joint restraining brackets for cryogenic propellants are made of metals having face-centred cubic structure, which have high toughness at low temperatures. Restraining brackets for storable propellants are made of high-strength steels having body-centred cubic or face-centred cubic structure.

Flexible hoses, including wire braid, are made of 321 CRES, which is a stabilised austenitic stainless steel with addition of titanium.

Other alloys, such as Hastelloy® C and Inconel® 718, are used for applications in special environments.

Materials used for tubing in rocket engines are chosen according to criteria of chemical compatibility with fluids, physical and mechanical properties, formability, weldability, and costs. These criteria have been considered at length in Chap. 6, Sect. 6.7 for propellant tanks. A brief account is given here for what concerns specifically fluid-carrying ducts and their flexible components.

As to chemical compatibility of fluids with tubing materials, gaseous hydrogen at high pressure causes a loss of ductility in many metallic alloys. This effects depends on the temperature, pressure, and purity of the gas, and also on the exposure time and level of stress.

Ferritic, martensitic, and bainitic steels, nickel-base alloys, and titanium alloys become brittle when exposed to pure hydrogen at room temperature, this effect being higher at increasing pressure. High-strength alloys are often more susceptible than low-strength alloys to loss of ductility.

Austenitic stainless steels such as 310 and 316, some aluminium alloys such as 6061-T6, 2219-T6, and 7075-T73, pure copper and beryllium copper, and the precipitation-hardened stainless steel A-286 are slightly affected. Inconel® 718, Inconel® X-750, Waspaloy®, and René® 41 are highly susceptible to loss of ductility when exposed to gaseous hydrogen at high pressure [1].

The loss of ductility due to hydrogen is associated with a loss of fracture toughness. The presence of hydrogen, water vapour, and other gases increases the effects of crack initiation and crack growth rate.

In addition to hydrogen, substances which can have undesirable effects on metallic tubing are:

- propellants susceptible to catalytic decomposition, such as anhydrous hydrazine, mono-methyl-hydrazine, and Aerozine 50, which generate hot gas in the presence of molybdenum, iron, copper, or silver;
- nitric acid resulting from absorption of water in nitrogen tetroxide, which attacks aluminium alloys;
- chlorides from cleaning fluids, which can induce stress-corrosion cracking in stainless steel 321 CRES; and
- uninhibited (brown) nitrogen tetroxide, which can induce corrosion cracking in titanium alloys.

It is also necessary to take account of the environment to which metals used for tubing are exposed. For example, titanium exposed to oxygen can ignite and oxidise explosively as a result of an impact.

The principal physical and mechanical properties of metals used for tubing are strength, elongation, and density.

Minimum bend radius and fatigue resistance are desirable properties in materials to be used for corrugated walls of bellows. Cryogenic temperatures reduce the toughness of most materials having body-centred cubic structure. Data of some metals and non-metals exposed to low temperatures (from 20.37 to 533.15 K) are given in [10].

Some materials used for flexible lines are aluminium alloy 6061-T6, stainless steel 321 CRES, Hastelloy® C, Inconel® 625, titanium alloy Ti-6Al-4 V annealed, and Inconel® 718 age-hardened. The principal physical and mechanical properties of these materials are given in Table 3, page 45, of [1].

The materials to be used for flexible parts of interconnecting ducts must be chosen with more demanding requirements than the materials for the remaining parts. Among these requirements, the most important are those of formability, and resistance to corrosion and fatigue. In particular, the materials chosen for bellows corrugations or inner cores of flexible hoses must be ductile. Corrosion resistance is also required for these materials in order to avoid holes and cracks which can give rise to leakage. Fatigue resistance is required in order for a bellows to bear the desired number of flexural cycles. Materials which have shown good properties of strength-to-weight ratio, weldability, and elongation are Inconel® 718, Nitronic® 40 (a high-manganese stainless steel with high strength and excellent resistance to corrosion at high temperatures), and 321 and 347 CRES. In particular, the Nitronic® 40 alloy is scarcely affected by loss of ductility due to hydrogen.

Bolts and nuts for aerospace applications are usually made of corrosion-resistant metals, such as Inconel® 718, Inconel® X-750, Monel® K-500, and René® 41.

The materials indicated above for fluid-carrying ducts have generally good properties of ductility, and therefore deform plastically under stress around the tips of cracks. However, particular conditions of either load or temperature can cause a loss of ductility, in which case a material may fail. Therefore, the choice of a possible material should take account of its fracture toughness and resistance to subcritical flaw growth, as has been shown in Chap. 6, Sect. 6.8.

The formability of a given material depends on its ductility, that is, on its aptitude to form rolls, bends, fittings, and corrugations for bellows. Under this aspect, corrosion-resistant steels and nickel-base alloys have proven to be the best materials for all formed elements of fluid-carrying ducts.

The weldability of a material is important, because welding is the most used method of joining permanently various elements of a line. In case of flexible hoses, brazing is the most common method of connecting permanently the braid of a hose to each of the connecting points. Methods of welding elements of lines include tungsten-inert-gas (TIG) and electron beam, as has been shown in Chap. 6, Sect. 6.7.

Lubricants are sometimes used on the surfaces of bellows or other elements of lines. For this purpose, a corrosion-inhibiting type of molybdenum disulfide (MoS_2) coating can be used for corrosion-resistant steels. Nickel-base alloys resist better than steels to chemical attack. Plating or lubricants are also used on bolts and nuts to prevent thread galling. A thread lubricant used for the Saturn engine systems was a phosphoric-acid-bonded dry-film lubricant. Platings are used at temperatures higher than 422 K [1].

7.4 Coupling Components for Tubing

The components described here are devices which contain and control the flow of fluids in ducts. These devices can be classified into the following categories: disconnects, couplings, fittings, fixed joints, and seals. A brief description is given below for each category.

A disconnect is a type of separable connector consisting of two separable halves, an interface seal, and usually a latch-release locking mechanism. The following figure, due to the courtesy of NASA [11], shows three stages of operation of a typical manually-operated disconnect.

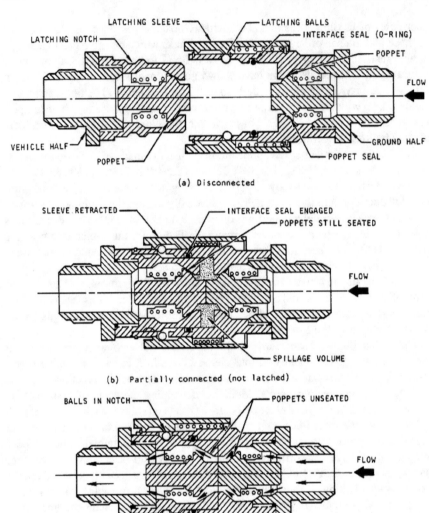

(a) Disconnected

(b) Partially connected (not latched)

(c) Fully connected (latched)

Disconnects are used as interfaces between rocket vehicles and ground systems, or between stages of a rocket vehicle.

A coupling is a mechanically-actuated, separable connector which requires more than a few seconds for engagement or disengagement. Examples of couplings are threaded connectors, bolted flanges, and dynamic swivel couplings. A threaded connector is a line fitting which provides a separable mechanical joint secured by a single threaded nut, whereas a bolted flange connector uses several bolts, a clamp, or a combination of these to secure the joint [2]. The following figure, due to the courtesy of NASA [11], shows four types of threaded connectors.

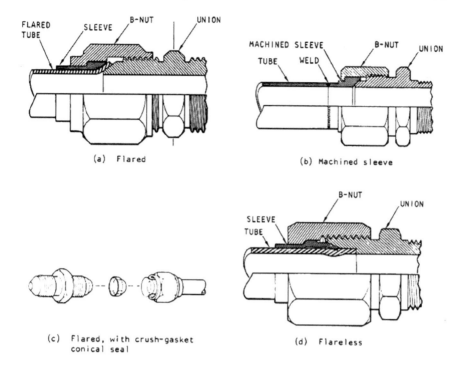

(a) Flared

(b) Machined sleeve

(c) Flared, with crush-gasket
     conical seal

(d) Flareless

The choice between threaded and flanged connectors is determined by the size of the line and by the amount of preloading required to establish a satisfactory seal over the entire range of loads [2]. Flanged connectors are used where loads require the type of restraining force provided by bolts or clamps, or where coupling reliability dictates the use of more than one threaded clamping fastener [11].

The following figure, due to the courtesy of NASA [11], shows eight types of flanged connectors.

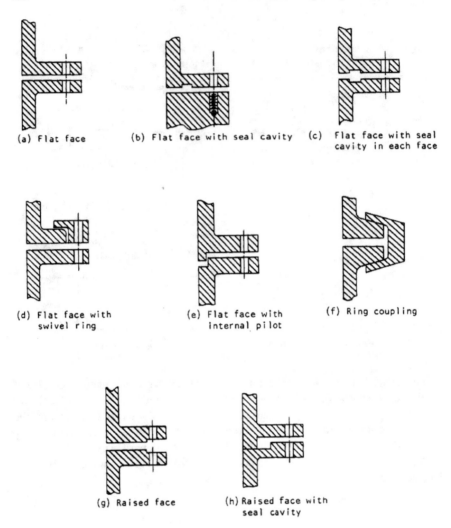

(a) Flat face   (b) Flat face with seal cavity   (c) Flat face with seal
                                                            cavity in each face

(d) Flat face with   (e) Flat face with   (f) Ring coupling
    swivel ring   internal pilot

(g) Raised face   (h) Raised face with
                           seal cavity

A dynamic swivel coupling is a joint designed such that the swivelling or rotary tubular shaft is pressure-balanced, so as to eliminate high sealing and bearing friction forces caused by axial pressure thrust [2]. On the Saturn engines, swivel couplings were used successfully between moving members of mechanical components (for example, valve stems, actuator shafts, and pistons). However, provisions were made to dispose of leakage, which is inherent in this type of coupling [11].

Fittings are devices used to change flow area or direction while connecting two or more straight elements in a tubing, line, or ducting assembly. Such devices are the L-shaped, T-shaped, Y-shaped, et c. tubes used to route fluids to required areas. The following figure, due to the courtesy of NASA [11], shows (above) a low-pressure-loss fitting for joining tubes of different sizes, and (below) two types of tap-off fittings.

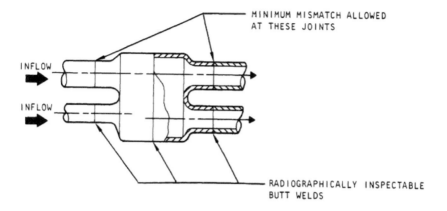

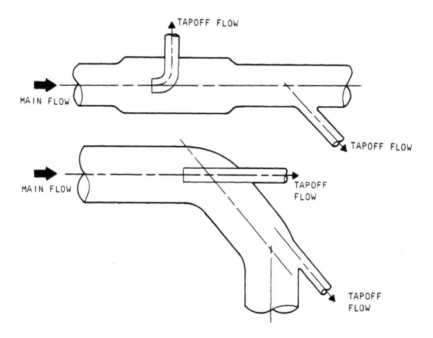

Fixed joints are permanent (that is, non-separable) connexions of fluid-carrying ducts. They are used when low weight and high reliability are more important than ease of separation [2]. The joining methods used for fixed joints are welding, brazing, diffusion bonding, soldering, and interference fit. The following figure, due to the courtesy of NASA [11], shows six types of welded joints used in tubing.

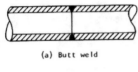

(a) Butt weld

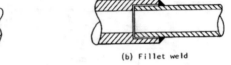

(b) Fillet weld

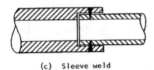

(c) Sleeve weld

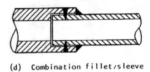

(d)   Combination fillet/sleeve

(e)   Sleeve weld with separate sleeve

(f)   Combination fillet/sleeve with separate sleeve
       (eliminates both internal and external crevices)

With reference to the preceding figure, the butt (a), fillet (b), and sleeve (c) types of welded joints are very common. These three types are sometimes used in combinations, the most common of them being fillet and sleeve (f).

When possible, weld joints are located in areas free from vibrations. In particular, locations in which vibrations are perpendicular to the tube axis are avoided. Tungsten Inert Gas (TIG) Welding is the most widely used of the joining methods for fixed joints. Gas Metal Arc (GMA) Welding is also used to join heavy sections, when it is desirable to reduce the number of weld passes needed to complete a joint. These methods have been described in Chap. 6, Sect. 6.7. Electron Beam (EB) Welding (a vacuum-based process in which a beam of high-velocity electrons is applied to the two metals to be joined) is applied in particular cases, for example, to join metals where a narrow weld bead or minimum heat input to the parts is necessary, or to join parts made of titanium alloys, which must be protected from oxygen.

According to Howell and Weathers [2], brazing is a metal-joining operation performed at temperatures ranging between those of welding and soft soldering, where soft soldering temperatures are considered to be below 723 K. Brazing differs from welding, because in the former: (1) bonding results from wetting rather than melting the base alloy; (2) the brazing filler metal (brazing alloy) is made to flow into the joint by capillary action to create the bond; and (3) the brazing filler metal is an alloy having a composition different from that of the metals to be joined. Brazing filler metals used are often alloys of silver, aluminium, gold, copper, cobalt, or nickel.

According to Kazakov [12], diffusion bonding is a process by which a joint can be made between similar and dissimilar metals, alloys, and non-metals, through the action of diffusion of atoms across the interface, brought about by the bonding pressure and heat applied for a specific length of time.

Diffusion bonding has been used for a titanium-to-stainless-steel tubular transition section. These transition joints were developed to provide a fixed joint between titanium propellant tanks and stainless steel lines. Materials commonly diffusion bonded are Ti-5Al-2.5Sn or Ti-6Al-4 V alloys with 304L, 321, and 347 stainless steel. All titanium alloy-to-stainless-steel combinations have been used successfully [11].

Soldering is a metal-joining operation similar to brazing, but takes place with fillers (also known as solders) which melt at a temperature below 723 K.

Soldered joints have been successful in low-pressure applications. They are light, require simple heating tools, can be assembled in a minimum envelope, do not need inert-gas shielding, and are readily made on in-place hardware [11].

An interference fit joint, also known as tight fit connector, is a fastening between two parts which uses the pressure exerted by one of the parts against the other to obtain the desired result. For this purpose, the inner diameter of enveloping part (also known as the hole or the hub) is smaller than the outer diameter of enveloped part (also known as the shaft). Therefore, in order to fasten the two parts, it is necessary to apply force during assembly. After the parts are joined, the enveloped part exerts a pressure against the enveloping part along the mating surface, with consequent elastic deformation of the whole assembly. Interference fit joints without the use of some other joining method are not used in ducts for propellants, but are used sometimes in ducts for pneumatic or hydraulic fluids [11]. The definitions of the joining methods indicated above (welding, brazing, diffusion bonding, etc.) are rather intuitive than rigorous, for the sake of clarity. Formal definitions of such methods can be found, for example, in [13].

Connecting flanges or glands for tubing used in aerospace vehicles should be rigid enough to maintain the integrity of static seals. Flanges should have the surface finish, the radial clearances, and the rigidity required by the specific seal used. Flanges having insufficient rigidity or inadequately bolted are susceptible to rotation under operational loads. Two types (bowing and rotation) of flange deflection due to lack of rigidity are shown in the following figure, due to the courtesy of NASA [11].

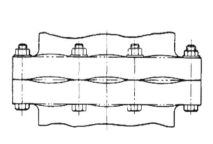

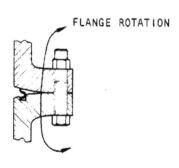
FLANGE ROTATION

(a) Flange bowing   (b) Flange rotation

As a general rule, it is desirable to keep the bolt circumference of a flanged joint as close as possible to the seal, and to use rather many bolts of small diameter than a few bolts of large diameter. NASA [1] recommends the following design for a flanged joint.

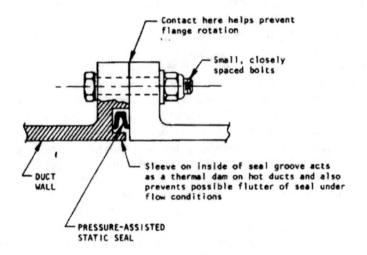

7.5   Control of Pressure Loss in Tubing

Causes of pressure loss in tubing are changes of flow direction, changes of flow area, changes of flow distribution, and friction between a flowing fluid and the walls of a duct.

As to the first cause, when the interfaces between the tanks and the pumps or between the pumps and the combustion chamber of a rocket engine do not permit rectilinear ducts, then elbow-shaped parts of tubing must be used. For a given Reynolds number, the pressure loss coefficient for an elbow decreases, reaches a minimum value, and then increases as the ratio R/D of the bend radius to the inside diameter of the elbow increases.

When the routing of a line requires elbows of small bend radii, the consequent pressure losses can be reduced by either choosing the optimum value of the R/D ratio for each elbow, or adding flow guide vanes to the elbows, as shown in the following figure, due to the courtesy of NASA [1].

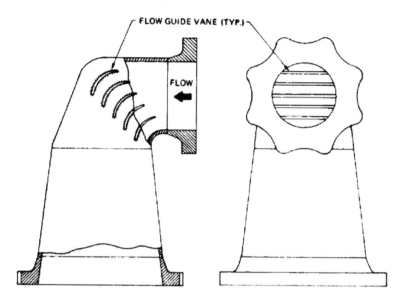

Flow guide lines have been used in sharp elbows at the pump inlet ducts of the engines of the Centaur, Thor, Atlas, and Saturn S-IC. Such vanes have also been used in the pump discharge ducts of the Thor and Jupiter. The reduction of pressure losses improves the performance of a rocket engine. Since oscillations of the fluid can excite the flow guide vanes, then their natural frequencies must be calculated, in order to avoid resonance phenomena.

The following figure, due to the courtesy of NASA [1], shows flow guide vanes used in elbows at the inlet ducts of centrifugal turbo-pumps.

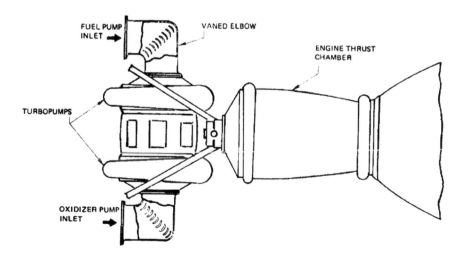

In bellows, flow liners are frequently used to reduce the high losses due to friction at the corrugations.

A change in flow area occurs frequently at the two ends of a duct. The shape of the transition portion near each end is to be determined in such a way as to prevent excessive pressure losses, which would take place in case of an abrupt change in flow area. For the same purpose, the edges of a duct at each end should be rounded instead of sharp. In case of expanding conical ducts, the included angle of the cone frustum should be less than or equal to $\pi/18$ rad ($10°$). For ducts of small diameter having threaded fittings, it is necessary to determine the pressure losses by taking account that the inner diameter of the fitting is usually smaller than the inner diameter of the duct.

The flow distribution at the exit of a duct (for example, at the exit of a feed duct for a rocket engine) is to be determined carefully because of its implications on the performance of the components. For example, when the pumps of a rocket engine have discharge ports placed oppositely, care must be taken in order to have equal pressure losses at the two ports. A flow distribution can be improved by using either flow guide vanes at the elbows, as shown in the preceding figure, or flow straighteners of the egg-crate type placed downstream of the elbows, as shown in the following figure, due to the courtesy of NASA [1].

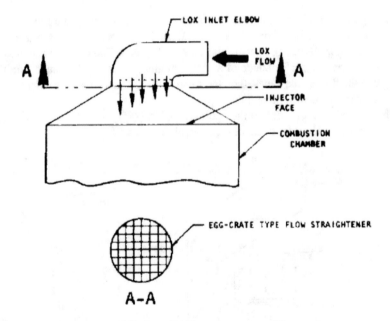

Another method of achieving a balanced distribution of pressure loss at the terminal ends of two ducts is based on a flow splitter. This method was used in the propellant feed lines of the descent engine of the Apollo Lunar Module.

Friction between the walls of a duct and the fluid flowing into it can cause large losses of pressure, in cause of ducts having rough walls. To reduce friction, it necessary to use ducts with smooth walls and to avoid protrusions into the flow stream, such as those which are sometimes caused by welds.

7.6 Control of Vibrations at the Inlet of Pumps

The fundamental concepts on the longitudinal vibrations, called pogo vibrations, of a rocket vehicle have been discussed in Chap. 2, Sect. 2.9. The present paragraph is meant to show in further detail how such vibrations can be suppressed by using devices installed in the feed lines of rocket engines.

The pogo vibrations are due to a feedback interaction between the propulsion system and the structure of a rocket stage. These vibrations are generated by pulsations in the thrust force, which cause a response in the structure. The structural response applies accelerations to the suction part of a propellant feed system. The two feed lines (for respectively the fuel and the oxidiser) respond separately to these accelerations, and cause pressure pulses at the inlet of the two pumps. As a result of these pressure pulses, the pumps and the discharge lines transmit a varying rate of propellant flow to the combustion chamber. The combustion chamber, in turn, generates a pulsating pressure and a pulsating thrust. Instability occurs when the pulsating thrust is fed back to the structure, so as to reinforce the initial perturbation.

The following figure, due to the courtesy of NASA [1], illustrates the vibration suppressing method used for the two pumps of the Titan II.

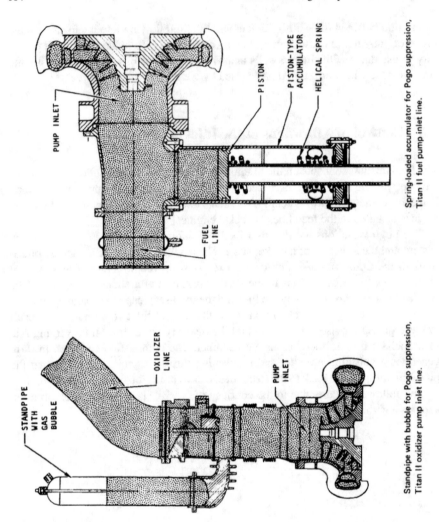

PUMP INLET

PISTON

PISTON-TYPE ACCUMULATOR

HELICAL SPRING

FUEL LINE

Spring-loaded accumulator for Pogo suppression, Titan II fuel pump inlet line.

STANDPIPE WITH GAS BUBBLE

OXIDIZER LINE

PUMP INLET

Standpipe with bubble for Pogo suppression, Titan II oxidizer pump inlet line.

   In the Titan II, a bubble of gas was enclosed in a standpipe connected to the oxidiser feed-line. This bubble provided a cushion or a soft spring, which acted on the mass of the oxidiser in the standpipe. In this manner, the energy due to the pressure oscillations in the oxidiser feed-line was transferred to this spring-and-mass system, by choosing judiciously the volume or the height of the bubble enclosed in the standpipe. The fuel feed-lines had accumulators of the piston type, which used a mechanical arrangement comprising an helical spring and a piston to provide the desired soft-spring action. The fixed mass of this mechanical arrangement and the mass of fuel in the accumulator provided the equivalent mass required for a resonant mechanical system. The oscillation suppression devices described above were designed and tuned so that their frequency responses, coupled with the feed-line properties, would provide the maximum possible attenuation of pressure oscillations in the suction ducts of the pumps, as a result of the tank-structure oscillations.

The pogo suppression system used for the S-IC stage of the Saturn V launch vehicle is shown in the following figure, adapted from [7].

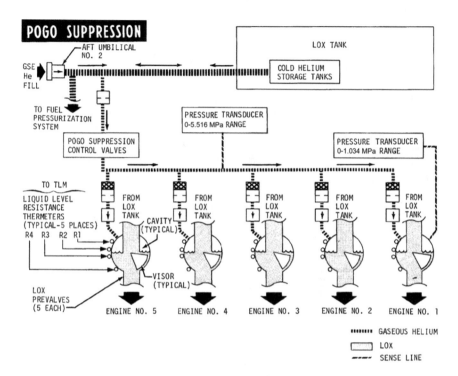

This system shown above uses the liquid oxygen pre-valve cavities as surge chambers to suppress the pogo vibrations. The liquid oxygen pre-valve cavities are pressurised with gaseous helium 11 min before scheduled lift-off time from ground supply by opening the pogo suppression valves. During the initial fill period (from 11 to 9 min before scheduled lift-off time), the filling of the valves is closely monitored by using measurements supplied by the liquid level resistance thermometers R_3 (primary) and R_2 (backup). The gaseous helium ground fill continues to maintain the cavity pressure until umbilical disconnect. After umbilical disconnect, the cavity pressure is maintained by the cold helium spheres located in the liquid oxygen tank. The status on the system operation is monitored through two pressure transducers and four liquid level resistance thermometers. One pressure transducer (0–5.516 MPa absolute pressure) monitors the system input pressure. A second pressure transducer (0–1.034 MPa absolute pressure) monitors the pressure inside the No. 1 engine liquid oxygen pre-valve cavity. The pressure readings are transmitted via telemetry to ground monitors. The liquid level within the pre-valves is monitored by four liquid level resistance thermometers in each pre-valve. These thermometers transmit a "wet" (colder than 108 K) and a "dry" (warmer than 108 K) reading to ground monitors [7].

The following figure, also due to the courtesy of NASA [1], is a scheme of the pogo suppression system used in the liquid oxygen feed-lines on the main engines of the Space Shuttle.

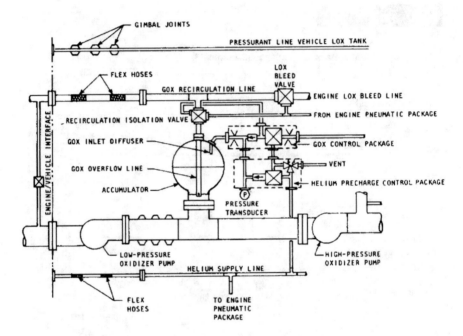

The pogo suppression system shown in the preceding figure is incorporated in the feed system of liquid oxygen at the inlet of the high-pressure oxygen turbo-pump. This system uses an accumulator filled with gas to suppress flow oscillations induced by the vehicle. Gaseous oxygen is tapped off the heat exchanger in the oxidiser-tank pressurisation system. This gas is used as the compliant medium after an initial helium precharge. The system controls the level of liquid oxygen in the accumulator by means of an overflow line, which routes overflowing fluids to the inlet of the low-pressure oxygen turbo-pump.

The accumulator, which is shown in detail in Chap. 2, Sect. 2.9, serves as an attenuator in the flow of liquid oxygen in the circuit, and prevents the transmission of the flow oscillations (at frequencies ranging from 20 to 30 Hz) into the high-pressure oxygen turbo-pump. This system is designed to provide sufficient overflow at the maximum decreasing pressure transient in the discharge duct of the low-pressure oxygen turbo-pump. The engine controller (shown in detail in Chap. 5, Sect. 5.2) provides signals for valve actuation, and monitors the system operation [1].

7.7 Bellows Joints

According to the definition given by Howell and Weathers [2], a bellows joint is an elastic, corrugated, tubular connector used for conducting a fluid between points of relative angular, transverse, lateral, or combined motion.

Bellows joints are used where axial, lateral, angular, or combined deflections exist between interconnected components of lines. A bellows joint in a line consists of the following parts:

- a bellows, which is the flexible corrugated tube carrying the pressure exerted by the fluid conducted in the line;
- restraints on the bellows, which prevent movements caused by internal pressure;
- means of attaching the bellows to the line; and
- a flow liner for the bellows.

A bellows joint in a duct can absorb deflections with much smaller reaction loads than the loads in a hard line. A free bellows can absorb four types of motion applied to its ends: (a) axial motion due to tension or compression forces; (b) offset motion with end planes parallel; (c) angular motion about its centre; and (d) torsional motion about its axis.

Various restrains or linkages can be used in order for a bellows to absorb only certain types of motion. For example, a gimbal ring linkage allows a bellows to absorb only angular motion. An internal (left) and an external (right) mounting of gimbal rings for a bellows joint are shown in the following figure, due to the courtesy of NASA [1]. Other drawings showing bellows joints mounted on gimbals or hinges can be found in [14].

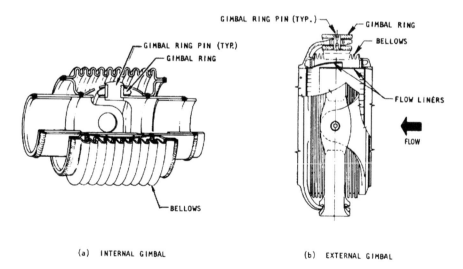

(a) INTERNAL GIMBAL   (b) EXTERNAL GIMBAL

A bellows joint must also absorb vibrations, which are due to motions of either the fluid conducted in the line or mechanical parts of the engine. The frequencies of these vibrations are of the order of magnitude of several thousand hertz. When the stresses resulting from such vibrations exceed the fatigue limits of the material of which a bellows is made, then failure occurs.

Bellows are used for ducts of rocket engines in various sizes and operating conditions. A bellows operates usually in the plastic range of the material of which it is made. This is because the necessity of reducing weight implies small values of thickness for the bellows walls, provided that the magnitudes of the reactions at the bellows ends are acceptable. A formed bellows with straight corrugated sidewall is the type most frequently used for rocket engines.

The types of pressures to be carried by a bellows in operating conditions are:

- normal operating pressure;
- surge pressure;
- proof pressure; and
- burst pressure.

By surge pressure we mean the variation of pressure which occurs in a line conducting fluid as a result of a change in the flow velocity. This variation of pressure may occur in a line, for example, when a pump starts or stops, or a valve is opened or closed rapidly, or the diameter of the line changes abruptly. When a valve is closed rapidly, the velocity head of the fluid moving forward in a line decreases whereas its pressure head increases suddenly, due to the compression which occurs upstream of the valve. By proof pressure we mean:

proof pressure = (normal operating pressure + surge pressure) × safety factor

As has been shown in Chap. 6, Sect. 6.9, the safety factor is a multiplier, whose value (greater than unity) is chosen by the designer to take account of small variations of the properties of the material, of the quality of manufacturing, and of the magnitude and distribution of the load. Finally, by burst pressure we mean:

burst pressure = proof pressure × safety factor

These pressures generate hoop (circumferential) stresses and bulging (meridian) stresses, which must be carried by the corrugated sidewall. A bellows must resist column buckling under application of proof pressure. Under application of burst pressure, a bellows is permitted to deform and take permanent set, but not to leak. The limiting bulging stresses and allowable motion stresses (bending stresses) of materials frequently used for bellows are given in the following table, adapted from [1].

Material	Material condition	Limiting bulging stress[a], MPa		Allowable motion stress, MPa	
		t ≤ 0.3048 mm	t > 0.3048 mm	1000 cycle life	10000 cycle life
321, 347 CRES	Cold worked as formed R_c 10–40	965.3	827.4	827.4	510.2
A-286	Heat treated; R_c 29–40	1379	1103	1379	827.4
Inconel 718	Heat treated, R_c 38–45	1379	1103	1448	930.8
Inconel X-750	Heat treated. R_c 30–37	1379	1103	1138	724.0
Hastelloy C	Cold worked as formed R_c 10–40	1034	896.4	1276	724.0

[a]Reduction in material properties for thicker sheet is due to poorer surface finish and greater variation in thickness than in thinner sheets
R_c—hardness on Rockwell C scale

The principal causes of problems affecting bellows joints used in ducts for liquid-propellant rocket engines are:

- fatigue;
- buckling;
- corrosion;
- manufacturing difficulties; and
- handling damage.

These causes are briefly considered below.

As to fatigue, bellows are subject to large deflections at their ends, due to the offset motion cited above. Bellows have usually thin walls, in order to reduce weight. Consequently, the materials of which bellows are made operate under stresses near or in the plastic range, with resulting low fatigue lives.

Bellows are also subject to vibrations, caused by the fluids or by mechanical parts. They must be designed so that no reduction in fatigue life is caused by vibration. In order to predict vibratory stresses, it is possible to use known vibration inputs relating to other engines in the same conditions, and compare such inputs with the frequencies of resonance of the given ducts, which can be calculated. Otherwise, it is possible to construct a prototype and test it on various engines, while the vibration environment is measured. When a bellows operates at excitation frequencies matching any of its natural frequencies of vibration, then its fatigue life can be exceeded in a short interval of time.

The most frequent failure mode for a bellows is a fatigue crack, which gives rise to leakage through the corrugated wall. The cracks develop circumferentially in areas of maximum bending stress, which are the crowns and the roots of the corrugations.

The primary modes of vibration for a bellows are:

(a)   the axial or accordion mode;
(b)   the lateral mode; and
(c)   the individual convolution mode, in which the individual corrugations rotate back and forth at the inner diameter, and pivot about the outer diameter.

In particular, the individual convolution mode occurs often in vibrations of braided hoses caused by flow oscillations, where the outer diameter of the corrugations is restrained by the braid friction, and consequently the outer point of each corrugation is a fixed pivot point for the corrugation excited by flow forces. The individual convolution mode may be in phase or out of phase, depending on how each corrugation moves with respect to the others.

The primary modes of vibration (a), (b) and (c) described above are shown in the following figure, due to the courtesy of NASA [1].

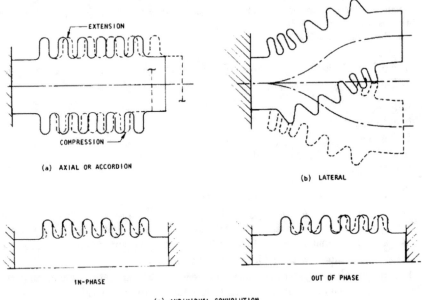

Bellows incorporated in flexible flow lines are frequently subject to vibrations induced by fluids moving at high velocity, with consequent failure due to fatigue stresses. This type of failure is avoided by using internal flow liners, which prevent the flow stream from impinging on the corrugations. However, care must be taken, because internal flow liners, too, may be subject to flow-induced vibrations.

Bellows are also subject to vibrations induced by mechanical parts of engines. When the frequencies of the exciting vibrations are known, for example, through measurements, then bellows can be designed so that their natural frequencies should not match the excitation frequencies.

Bellows may be subject to buckling due to excessive internal or external pressure. Buckling due to high internal pressure results in column instability, whereas buckling due to high external pressure results in crushed corrugations. Buckling can occur only when the pressures in actual operating conditions are beyond the pressures predicted in the original requirements.

A bellows may be subject to corrosion, depending on the material of which it is made. A material commonly used for bellows is stainless steel 321 CRES, which is resistant but still susceptible to corrosion, because of its high percentage of iron. This and other materials have been discussed in paragraph 3.

The principal difficulties in the manufacturing of bellows are material thinning, heat-treatment control, welding, and convolution stack-off.

Bellows are thinned at the corrugation crowns, about 5% or 10% below the nominal thickness of the wall. As a result of this thinning, higher deflection takes place in the thinned section, because its stiffness is reduced, and consequently higher stresses and lower resistance to fatigue can occur.

Heat treatment of bellows is controlled in order to avoid degraded performance caused by improper treatment. In particular, this control requires the maintenance of the proper gaseous environment during heat treatment, to avoid oxidation and consequent strength reduction.

Welding problems are due to insufficient quality control. Such problems can be avoided by inspecting radiographically all welds.

Convolution stack-off is the eccentricity of convolutions or corrugations of the bellows with respect to each other. Since this eccentricity depends on the tools used in the process of fabrication, then all of the eccentricity is accumulated in the same direction, as shown in the following figure, due to the courtesy of NASA [1].

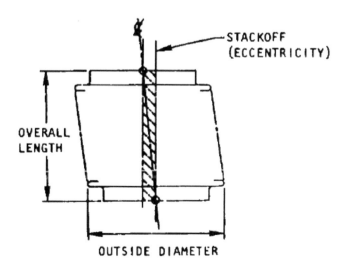

CONVOLUTION STACKOFF

Convolution stack-off makes it difficult to align the bellows for welding into its next assembly. This accumulation can be controlled by rotating the bellows after the forming of each convolution.

It is necessary to use handling-protection devices in the duct assembly and during processing, shipping, and installation of bellows. This is particularly important in case of bellows having thin walls (less than or equal to 0.762 mm).

In order to protect the bellows joints incorporated in a duct from over-deflection, it is necessary to use strong-back fixtures which attach to the duct on either side of each bellows joint and form a protective bridge around it. The following figure, due to the courtesy of NASA [1], shows (right) a strong-back fixture and (left) a shell-type cover for protection of bellows.

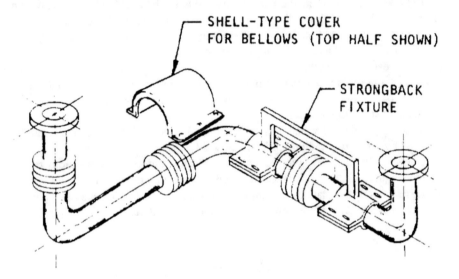

As has been shown above, the most common type of restraint used for bellows joints is the one mounted on gimbal rings. Other types of restraint are the hinged joint, the ball joint, and the braided wire sheath.

The gimbal-ring joint can be designed to withstand high pressures and temperatures. It can also withstand torsional loads and absorb angular deflections in all planes. Excessive angular motion of the bellows can be prevented by means of stops, and the pivot pins which carry the loads can be designed for either single-shear or double-shear support. The same considerations apply to the hinged joint, which allows angular motion in only one plane.

Another type of external, tension-tie restraint is a braided wire sheath, as is the case with a flexible hose. A joint of this type can absorb both angular and shear deflections. It also acts as a vibration damper for the bellows. Friction between the braid and the bellows can be reduced by using an adapter, which provides clearance at the end of the corrugations, and by applying a solid dry-film lubricant on the outer surface of the bellows and on the inner surface of the braid.

The lubricants used for this purpose must be compatible with liquid oxygen, when this is the fluid conducted through the given line.

A link joint with internal tie restraint can be used instead of a gimbal joint when a duct is subject to small or medium pressures. It is used frequently in low-pressure ducts of gas turbines. It has de disadvantage of high friction losses, due to the restriction to flow in the frontal section. In particular, a chain-link joint is a link joint in which the pivot point of the linkage is not fixed, but depends on the plane in which the joint lies. A chain-link joint with internal tie restraint is shown in the following figure die to the courtesy of NASA [1].

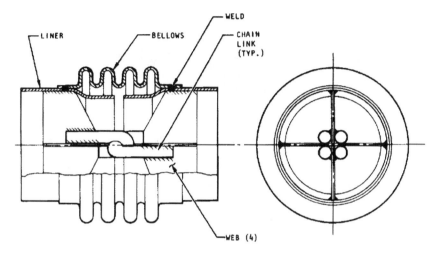

A ball-bearing ball-joint restraint is shown in the following figure due to the courtesy of NASA [1]. This type of joint was used for test in the inlet line to the liquid-oxygen pump of the F-1 engine.

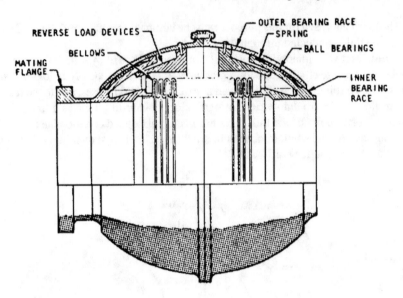

The bellows joint using the restraint illustrated in the preceding figure was located immediately upstream of the pump inlet flange, due to the large pressure separating load. This joint could provide a restraint and a motion of the gimbal type while applying a uniform circumferential load to the pump flange.

The flexible joints used for the RS-25 engine (burning liquid oxygen and liquid hydrogen) are shown in the following figure, due to the courtesy of Boeing-Rocketdyne [4].

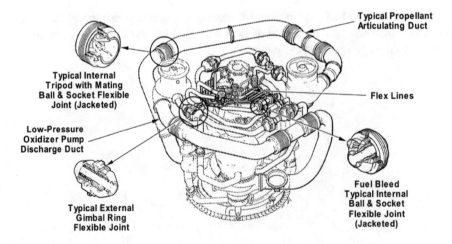

The flexible joints in the lines of the RS-25 engine allow movement for vehicle steering, while maintaining the internal pressure and temperature in the lines. These joints must be flexible, because the lines connect either the gimballed engine with the vehicle or the gimballed engine with non-gimballing components.

A flexible metallic bellows is used as the pressure vessel. This bellows has each of its ends welded into a section of fluid ducting, in order to provide a continuous, leak-proof pathway. The bellows have several thin plies in a sandwich configuration, instead of a single thick sheet of metal. This arrangement makes the bellows more flexible and maintains integrity against pressure.

The ducts which contain flexible joints internally tied have an internal diameter large enough to allow an internal support mounted on gimbals. This internal support arrangement saves weight in comparison with an external support. Were it not for the gimbal joint, which holds both ends of the flexible joint together, the internal operating pressure of the duct would expand the flexible bellows longitudinally like an accordion. The joint of the internal gimballed support is of the ball-and-socket type. Ducts with small internal diameters use an internal gimbal ring to restrain the bellows. The ends of the flexible joint are attached to the gimbal ring (which is centred over the bellows) at two points, so as to form a universal-type joint.

The flexible joints contain an integral flow liner. This liner prevents the propellant flowing through the duct from impinging on the corrugations of the bellows, which could cause turbulence in the flow and vibrations in the materials. The liner is made up of two or three overlapping pieces. The outside ends of the end pieces are welded into the flexible joint, so as to form a continuous, smooth internal diameter when the flexible joint is welded into a duct. The overlapping portions of the liner are shaped as a truncated ball and socket, to allow movement of the flexible joint. The liner allows propellant to fill the space between it and the bellows. Screened ports at the upstream ends of the flow liner provide a path for propellant to exit the space behind the liner after engine shutdown [4].

The flexible joints used in the RS-25 engine are the internal tripod and the external-gimbal ring joints. The internal tripod joint is used for the low-pressure discharge ducts, where the pressure loss can be tolerated and the overall joint envelope must be kept as small as possible. The following figure, due to the courtesy of Boeing-Rocketdyne [4], shows the internal tripod joint non-jacketed (left) and jacketed (right).

According to the data given in [4], these joints are made of Inconel® 718 and ARMCO 21-6-9 stainless steel, and their design life is 200 operational, 1400 non-operational full deflection cycles. The non-jacketed joint is used for the low-pressure discharge duct of the oxidiser. It works at an operating pressure of 2.916 MPa and at an operating temperature of 92.04 K. Its inside diameter is 160 mm, and its angular displacement is ±0.2269 rad (±13°). The jacketed joint is used for the low-pressure discharge duct of the fuel. It works at an operating pressure of 1.924 MPa and at an operating temperature of 22.04 K. Its inside diameter is 132 mm, and its angular displacement is ±0.2007 rad (±11°30′).

The externally-tied gimbal ring joint used in the RS-25 engine is shown in the following figure, due to the courtesy of NASA [1].

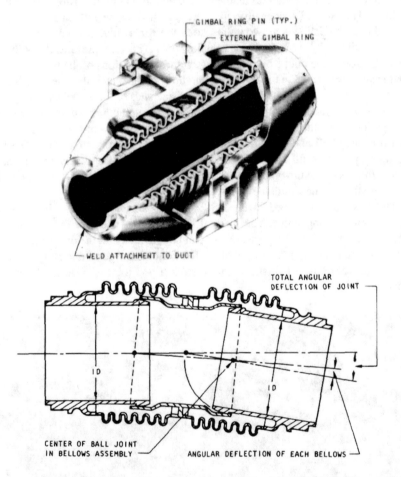

This joint is used in the RS-25 engine on small-diameter (50.80–68.58 mm) high-pressure ducts, where the pressure losses associated with internal ties are not acceptable. The high pressures and gimbal angles of the externally-tied joints make it necessary to use long bellows, which are unstable in column buckling.

This joint is stabilised by a linkage, shown in the preceding figure, which provides lateral support at the mid-span of the bellows live length. By so doing, the angular deflection of the joint assembly is equally distributed between the two bellows.

A flexible joint with linkage restraint known as "Gimbar" (gimbal ring with crossed bars for structural strength) is used in the fuel and oxidiser drain ducts of the Shuttle Orbiter. This joint is shown in the following figure, due to the courtesy of NASA [1].

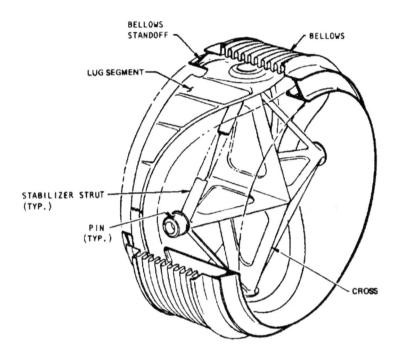

This joint is lighter than a gimbal ring joint having internal or external ties, and can be used in ducts of large diameters carrying fluids moving at low velocities, in which the pressure loss due to the structure across the flow stream can be tolerated. This joint is also capable of carrying torsional loads, by means of its linkage.

Thrust-compensating bellows are sometimes necessary to offset the thrust of primary bellows. An example of thrust-compensating bellows is shown in the following figure, due to the courtesy of NASA [1].

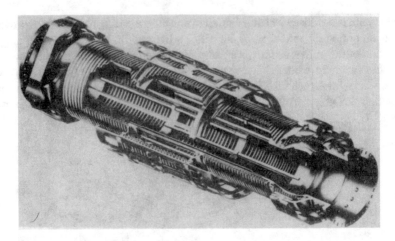

In a thrust-compensating bellows, the axial thrust due to the pressure separating force is balanced by the compensating bellows. Of course, in order to compensate for thrust, the volume of the bellows must also be compensated.

Thrust-compensating bellows also eliminate volumetric changes which occur in straight-run gimballing ducts. Such changes cause pressure perturbations which are detrimental to engine operation.

In the duct at the inlet of the pump of the F-1 engine, a bellows (which is external to and concentric with the main bellows) offsets the separating force of the main bellows, as shown in the following figure, due to the courtesy of NASA [1].

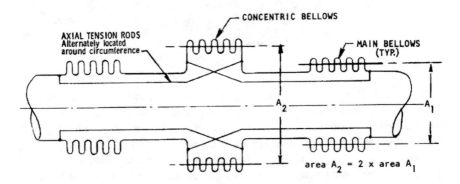

The annular chamber is vented to internal duct pressure. Axial-tension rods tie across the bellows alternately along the circumference. Telescopic flow sleeves can also be provided [1].

Bellows joints are also used in compression systems. A compression system consists of one or more bellows having no tension ties, which are used to absorb the deflections of a duct in operating conditions. In a compression system, the mating structure of the engine reacts to the pressure separating forces and to the elastic forces due to the fluid moving in the duct. The reaction exerted by the mating structure places the duct in compression.

In engines having ducts with compression-restrained bellows (for example, low-pressure ducts, such as those at the outlet of turbines), the support structure of the engine can be overloaded with moments caused by eccentric pressure separating forces acting on the bellows. These moments cause shear stresses in the bellows.

When the deflections induced in a bellows in operational conditions are known, then that bellows con be installed in an engine with a bias which gives rise to opposite deflections. By so doing, the bellows move, when placed in operational conditions, toward a nearly neutral position, corresponding to little or no stress. An installation of a compressed-restrained bellows made according to this criterion is shown in the following figure, due to the courtesy of NASA [1].

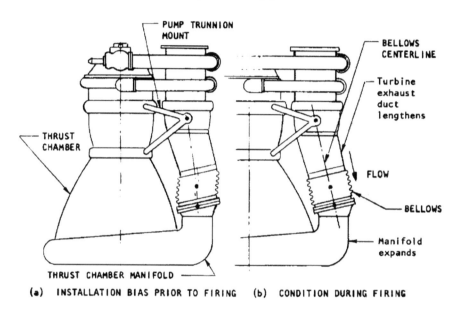

(a)   INSTALLATION BIAS PRIOR TO FIRING   (b)   CONDITION DURING FIRING

As has been shown above, bellows are made of thin, sometimes laminated, and often dissimilar materials. Consequently, junctions between bellows and adjacent parts of higher thickness require the use of careful manufacturing techniques. Electron-beam welding and diffusion bonding (discussed in paragraph 4) offer advantages over other techniques when parts made of dissimilar metals and having different thicknesses are to be joined.

Bellows made of heat-treatable alloys can be welded to ducts made of non-heat-treatable alloys without disturbing the heat treatments in the bellows materials. For this purpose, a transition portion of the duct is welded to the bellows before heat treating the bellows material, so that the final weld to the duct involves identical alloys.

In the RS-25 engine, the welds of the bellows to the ducts are of two principal types:

(a)   welds for thin-wall bellows subject to low pressures; and

(b) welds for thick-wall bellows subject to high pressures.

The necks of the thin-wall bellows are resistance-welded together, then trimmed around the circumference through the centre of the nugget. The bellows is then electron-beam welded to end rings, and the resultant assembly is heat treated.

The two types of welds indicated above are shown in the following figure, adapted from [1].

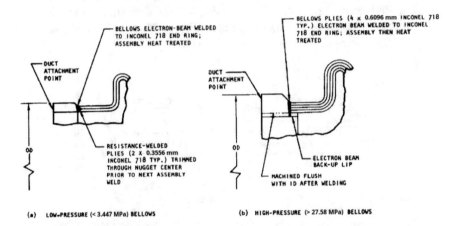

(a) LOW-PRESSURE (< 3.447 MPa) BELLOWS   (b) HIGH-PRESSURE (> 27.58 MPa) BELLOWS

Internal flow liners or sleeves for bellows are used to solve the problem of fatigue failures of bellows. Such failures are due to vibrations induced by the fluid stream. Flow liners are also used to reduce the pressure losses which occur at the joints. Flow liners have been used as structural members supporting tube coils in heat exchangers mounted on the J-2 and J-2S engines.

Flow liners must be designed so that they should not bottom out or bind on either the bellows or the ducts, when the bellows are at the maximum limits of excursion. They must also be designed with drain holes placed as near as possible to the weld attachment, in order to make it possible to remove cleaning fluids and contaminants.

Fatigue failures of flow liners have occurred, in which cracks have appeared in the trailing edge and in the weld joint. These failures have been attributed to vibrations of the cantilevered end of the sleeve. This problem has been solved by an increase in the thickness of the wall, with a consequent increase in the rigidity of the liner [1].

Collapse of liners has also occurred as a result of pressure on the outer diameter of the liner exceeding that of the inner diameter. This has been attributed to sudden changes in flow rates, which cause a different static pressure across the wall of the liner. Vent holes in the liner can reduce some of this difference of pressure, Otherwise, the liner can be strengthened to withstand the applied load.

Some typical configurations for flow liners are shown in the following figure, due to the courtesy of NASA [1].

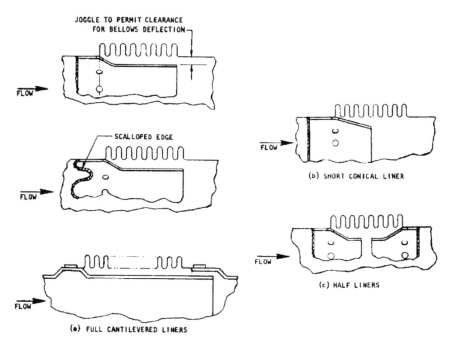

(a) FULL CANTILEVERED LINERS

(b) SHORT CONICAL LINER

(c) HALF LINERS

7.8   Flexible Hoses

A flexible hose consists of a flexible inner liner which conducts a fluid, a reinforcement which braces the inner liner, and two end fittings, which connect the hose ends to other portions of a duct.

Flexible hoses are used more often in large liquid-propellant rocket engines (meant for boosters and upper stages) than in small engines (meant for attitude control or reaction control). This is because little flexibility is required in lines of small engines, which are not mounted on gimbals and use storable propellants.

The J-2 and F-1 engines (both of which have been used in the Saturn V launch vehicle) have many fluid-carrying lines between the engine and the vehicle. These lines are flexible, in order for the engine to be moved on gimbals.

On the J-2 engine, all the lines are flexible hoses and are clamped together to assist in maintaining their relative positions.

On the F-1 engine, the flexible hoses cross the gimbal plane and are attached to non-gimballing flexible hoses on the engine [1].

Coiled tubing is sometimes used in heat exchangers for tank pressurisation. This is done in order to provide the maximum area possible for heat transfer in the minimum space. In the Centaur stage, a coiled tube is used for the pneumatic lines to the main engines, and flexible bends in lines for propellant recirculation and in lines for tank pressurisation, all of which are less than or equal to 25.4 mm in diameter [1].

A typical assembly of flexible hoses is shown in the following figure, due to the courtesy of NASA [1].

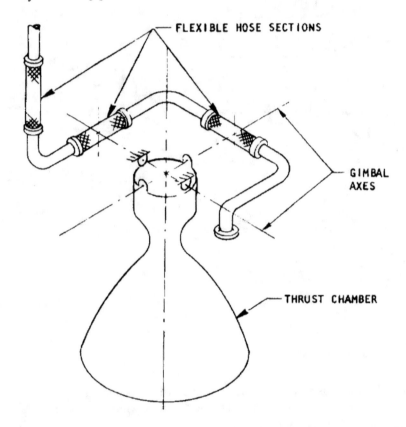

The flexible hose sections shown in the preceding figure are located so as to achieve the maximum motion possible of the line with the minimum motion of the flexible hose sections. In addition, the entire wraparound lies in the gimbal plane rather than passing through it. The preceding figure illustrates a gimbal system mounted on the head of the thrust chamber for thrust vector control. The longitudinal axes of two of the three flexible hose sections are located in the gimbal plane. The midpoint of each of these two flexible hose sections is located on each of the two gimbal axes. The third flexible hose section is located in the vertical direction, to provide universal motion of one end of the line with respect to the other. This method of locating the flexible hose sections limits their motions to the gimbal angles only, in order to reduce bending stresses and consequent fatigue. When the hoses are not stiff enough to resist motions caused by either mechanical vibrations or accelerations load, then clamps and support brackets are used [1].

Other arrangements for flexible hoses than the wraparound arrangement described above have been used, when lack of available space made it necessary to do so. For example, the flexible lines of the J-2 engine at the interface between the engine and the vehicle have been arranged in a U-shaped routing configuration, with a section of braided flexible hose in each leg of the U, as shown in the following figure, due to the courtesy of NASA [1].

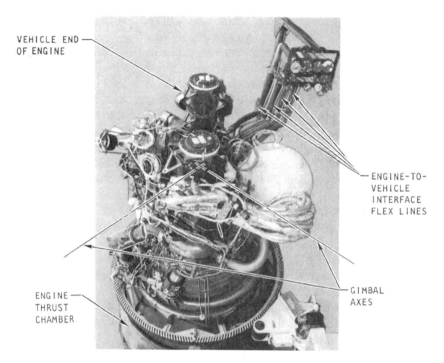

Flexible hoses are sized so that the velocities of gases are kept below the value 0.3 of the Mach number. This avoids pressure losses and vibrations induced by the flow.

The inner core of a flexible hose is subject to vibrations caused by the flow and by mechanical parts, as is the case with a bellows (see paragraph 7), the only difference being that the braid of a flexible hose restrains the corrugations to vibrate individually. Other considerations concerning pressure carried, resistance to corrosion, manufacturing methods, and handling protection for flexible hoses are similar to those which have been discussed in paragraph 7 for bellows.

As a general rule, only metallic materials are used for inner cores of flexible hoses installed in rocket engines, due to the very high or very low temperatures reached in such engines. Some exceptions to this rule are cited below.

The type most frequently used is the metallic, annularly convoluted inner core. The metallic, helically convoluted inner core type is not used in aerospace applications. Rolled and welded tubes or seamless tubes are used, but those of the seamless type are much more expensive. The welds of those of the rolled and welded type require X-ray and leakage verifications before use.

The engines of the Titan III have used successfully inner cores of both normal and carbon-impregnated Teflon® (poly-tetrafluoroethylene) in sizes ranging from 9.525

to 25.4 mm in diameter. On the S-IC stage of the Saturn V launch vehicle, inner cores of Teflon® have been used successfully with RP-1 at a working pressure of 15.17 MPa. Inner cores made of Teflon® have also been used successfully for the airborne hydraulic system on the Centaur. Rubber has been used only for applications involving non-cryogenic or non-high temperatures. In addition, the quality of rubber is subject to deterioration with time.

Bending moments to flexible hoses are to be considered, because the components adjacent to a hose are loaded. Low values of bending moments are desirable, because low bending moments applied to flexible hoses which cross the gimbal plane result in low loads applied to the actuators which gimbal the engines. Bending moments can be reduced by using lubricants applied to braid wires and to the other surfaces of the inner core. An inner core may be compressed axially before installing the braid. This results in a higher number of convolutions per unit length and in a lower bending stiffness, but also in greater weight and cost of the materials. A compressed inner core also improves the ability of a flexible hose to withstand pressure impulses.

Buckling stability is not a concern in the design of flexible hoses. This is because the support provided by the braid for the inner core prevents buckling.

The braid of a flexible hose is a woven tubular cover, which restrains the inner core against elongation and gives it lateral support and protection. The braid absorbs the entire separating load of a flexible hose. In rocket engines, the braid is woven of wire made of stainless steel.

Tubular braid may be woven directly on the inner core of flexible hoses. Sections of braid are cut at a length slightly greater than the length of the convoluted inner core. A tensile load is applied to the ends of the braid section, and the section is attached to end fittings by brazing or welding [1].

The pattern of a braid weave must be such as to prevent the wires from binding or bottoming out, within the design limits of angular deflection for a particular section of flexible hose.

Typical configurations of braids used for flexible hoses are shown in the following figure, due to the courtesy of NASA [1].

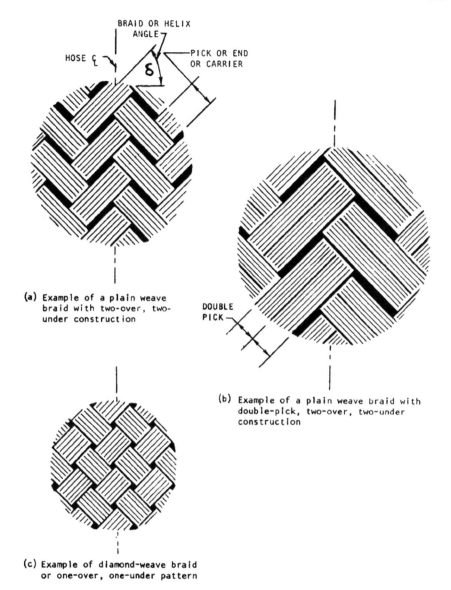

BRAID OR HELIX ANGLE

HOSE ₵

δ

PICK OR END OR CARRIER

(a) Example of a plain weave braid with two-over, two-under construction

DOUBLE PICK

(b) Example of a plain weave braid with double-pick, two-over, two-under construction

(c) Example of diamond-weave braid or one-over, one-under pattern

The braid angle or helix angle δ shown in the preceding figure is the angle whose tangent is the pitch divided by the circumferential length of the braid per pitch. The pitch is the axial distance taken by any given wire to make one complete turn. An angle δ of about $\pi/4$ (45°) is presumed to give the maximum flexibility together with good end strength and resistance to pressure for flexible hoses with metallic inner core [1].

Multiple layers of braid may be used to achieve greater strength. The second layer is assumed to have an efficiency (see below) of about 80%, due to the difficulty of obtaining a perfect distribution of load between two layers.

Tubular braid is available in various metals and sizes up to 457.2 mm in diameter. The ultimate tensile strength of a braid wire made of stainless steel is about 827.4 MPa, but braid wires of higher strength can also be found [1].

The end strength F_B (N) of a braid can be calculated by using the following formula of [1]:

$$F_B = n\, F_w B_e \sin \delta$$

where n is the total number of wires in a braid, F_w (N) is the strength of a single wire in a braid, B_e is the braid efficiency factor (whose values are 0.93 for an annealed wire, 0.85 for a hard-drawn wire, and 0.80 for a second layer), and δ is the braid angle defined above.

The burst pressure is the maximum pressure which a hose can retain without losing pressure or fluid. The burst pressure p_b (N/m^2) of a bellows-type metallic braided hose can be calculated by using the following formula of [1]:

$$p_b = \frac{F_B}{A_{\text{eff}}}$$

where F_B (N) is the end strength of a braid defined above, and A_{eff} (m^2) is the effective area of the bellows, which results from

$$A_{\text{eff}} = \frac{\pi D_m^2}{4}$$

and D_m (m) is the mean diameter of the bellows, such that $D_m = (D_o + D_i)/2$, where D_o (m) and D_i (m) are respectively the outside diameter and the inside diameter of the bellows.

The elongation ξ (m) of the braid, when a flexible hose is under pressure, can be calculated approximately by using the following formula of [1]:

$$\xi = \frac{p_i \ell A_{\text{eff}}}{n E A_w}$$

where p_i (N/m^2) is the internal pressure, ℓ (m) is the total length of one braid wire between its end connexions, E (N/m^2) is the Young modulus of a braid wire, and A_w (m^2) is the cross-sectional area of each wire. The preceding equation includes no allowance for slack in the wires, and holds when the stress is within the elastic limit. Braid wires are subject to corrosion. As a result of the experience gained for the main engines (RS-25) of the Space Shuttle, the braid materials for the flexible hoses used in those engines have been nickel-base alloys.

There are two principal types of end construction used for flexible hoses of small diameter:

(a) a welded-and-brazed type, used at temperatures from cryogenic to 478 K; and
(b) an all-welded type, used at temperatures above 478 K.

These two types are shown in the following figure, adapted from [1].

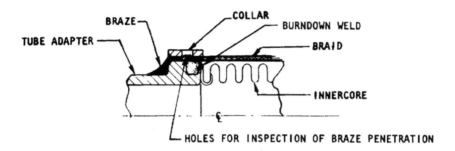

(a) End construction for small-diameter braided metal hose
(temperatures from cryogenic to 478 K)

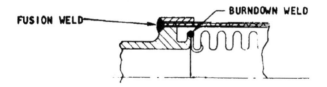

(b) End construction for small-diameter braided metal hose
(temperatures above 478 K)

By end construction we mean a juncture of inner core, tube adapter, and braid. The all-welded type is required at high temperatures, because of the loss of strength of the braze material at such temperatures. These two types have been used for flexible hoses up to 76.2 mm in diameter [1].

In case of tubes whose internal diameter (ID) exceeded 76.2 mm, another type of end construction has been used at temperatures from cryogenic to 478 K. This type (c) is shown in the following figure, adapted from [1].

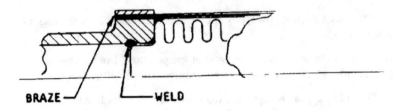

BRAZE   WELD

**(c)   End construction for braided metal hose 76.2 mm ID
and over (temperatures from cryogenic to 478 K)**

The type of construction illustrated in the preceding figure has been used in hoses of large inner diameters (88.9 and 101.6 mm) developed for the gimballing feed systems of the engines of the Atlas and Thor rocket vehicles. In such engines, the internal diameter of the tube was large enough to permit a lap weld of the pressure-carrier neck to the tube adapter [1]. Two further examples of end construction are shown in the following figure, adapted from [1].

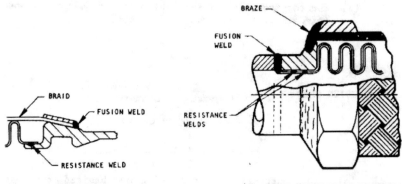

(a) ALL-WELDED CONFIGURATION   (b) WELDED AND BRAZED CONFIGURATION

The moments required to deflect braided metallic hoses depend on numerous parameters, such as inner diameter, outer diameter, thickness of the inner core, number of plies, live length, number of layers of braid, number of wires and their diameters, and operating pressure. Data on the bending moments can be obtained from the manufacturers of flexible hoses or by test. Some typical values can be found in [1].

A guide enables a flexible hose to bend, move, and operate in more than one plane without twisting, in order for the hose to have a longer service life. This guide has proved successful under severe conditions. Two independent bends are curved one

in a horizontal plane, and the other in the vertical plane. The guide provides a neutral
length of hose which separates the bends and prevents interactions between them, as
shown in the following figure, due to the courtesy of NASA [15].

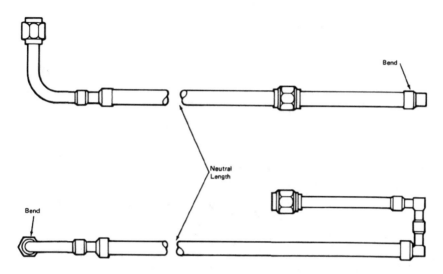

Determination of the minimal neutral distance required for proper operation led to
the identification of a single theoretical point of inflection, where the hose becomes
free of the guide. A roller, for each hose in the assembly, is placed in this point. The
inherent stiffness of the hose makes this point the point of inflection between the
horizontal bend and the vertical bend. Each bend is curved in a single plane with no
components of rotation or twist along the axis of the hose. The arrangement shown
in the following figure, due to the courtesy of NASA [15], can be modified for other
motions of the hose.

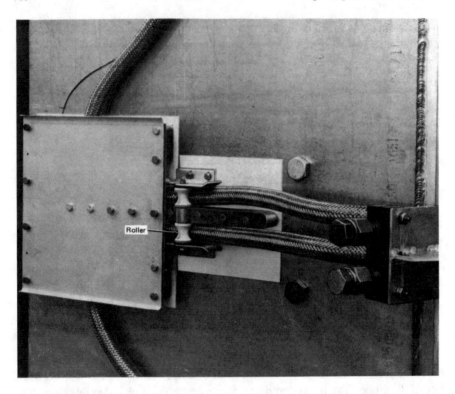

7.9   Filters

As has been shown in paragraph 1, filters are often installed in fluid-carrying ducts
of rocket engines to retain solid particles of contaminants which may be contained in
the fluids. Downstream of filters, the size of the remaining solid particles is reduced
to such an extent as not to affect the performance of components sensitive to contaminants. Examples of such components are valves and actuators. According to Buckingham and Winzel [16], contaminating particles existing in fluid-carrying ducts of
rocket engines are due to three principal causes, which are:

- residual manufacturing debris in the tanks and in other parts of a fluid system;
- contaminating particles in the on-loaded fluids; and
- particles generated by the wear of components in normal operation.

Generally speaking, as has been shown by Howell and Weathers [2], there are two
techniques of filtration, which are briefly described below.

One of them, called surface filtration, is accomplished by impingement and retention of solid particles on a matrix of pores placed in a single planar or curved surface.
Filtration occurs only at that surface, and the particles which are not stopped there

pass through the filtering medium with no further change in direction. This technique of filtration is effective in collecting particles larger than the sizes of the pores, but not effective in collecting fibres and particles smaller than the sizes of the pores. Its capability of retaining solid particles is limited by the area of the surface which can be provided within a given envelope. Examples of surface filtration media are single-layer mesh screens, stacked washers, wound metallic ribbons, and sheets of perforated metals.

The other technique, called depth filtration, is accomplished by impingement and retention of solid particles in a matrix of pores placed in depth. Sand is a typical example of a depth filtering medium, in which the filtering action consists of absorption and entrapment of solid particles at random. Solid particles are retained not only at the surface, but also throughout the thickness of the filtering medium. The particles which pass through a matrix of pores are forced to change direction in a tortuous path. Depth filters are effective in collecting solid particles larger than the maximum size of the pores and fibres. They can also collect a portion of solid particles smaller than the largest size of the pores, depending on the type and on the thickness of the filtering medium. Their capability per unit area of retaining solid particles is large, because the particles are retained not only on the surface but also throughout the depth of the filtering medium. Examples of depth filtration media are multiple layers of mesh, wounded wire cylinders, stacked discs of paper, sintered granulated materials, multiple layers of cloth, compressed or matted organic or inorganic fibres, stacked discs of etched metallic sheet, materials of elastic foam with open pores, and stacked membranes [2].

In the particular case of filters used in fluid-carrying ducts of rocket engines, most surface filters are made of woven wire cloth, and most depth filters are made of stacked etched metallic discs.

In a filter made of woven wire cloth, the filtering medium is woven from strands of metal. A type of wave is plain square wave, in which the strands pass over and under each other in alternating sequence. Another type is twilled square wave, in which the strands pass "over two, under two" in a staggered pattern. In both cases, the resultant openings are square-shaped [16]. Woven wire cloth, also known as woven wire mesh, is woven on looms, by using a process similar to that used to weave clothing. The mesh uses one or another of various crimping pattern for the interlocking segments. Square mesh, described above, is the most common type of mesh. The plain weave is the most common weave for woven wire cloth with a square opening in the plain weave. Materials used are austenitic and ferritic stainless steels. This type of filter is suitable for non-critical hydraulic, pneumatic, and propellant feed applications [2]. The following figure, adapted from [16], shows a wire cloth pleated conical filter.

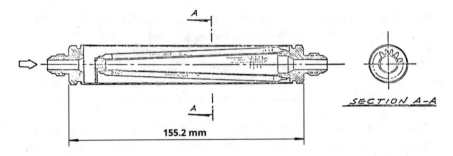

An etched-disc filter is a stack of segments resembling thin washers. Each segment has one face chemically etched to provide a desired flow path, as shown in the following figure, adapted from [16], where all dimensions are expressed in millimetres.

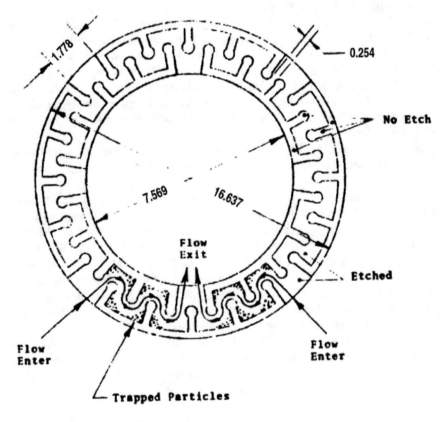

The non-etched face of each segment in the stack is in contact with the etched face of the next adjacent segment, so that the etched areas form minute passages for the flow. The stack of segments is held rigidly by a supporting cage and is tightly compressed. The depth property of an etched disc filter is obtained by compelling the solid particles to cross the surface of each segment through a tortuous path. An etched disc filter can be cleaned by releasing the compression on the segments and back-flushing. This type of filter is suitable for critical hydraulic, pneumatic, and propellant feed applications. Recent advances in wire cloth filter technology have broadened the field of choice for critical applications. However, the etched disc filter remains the only filter made only of metal which can be used for filtering particles below 8×10^{-6} m [2].

A full-flow etched-disc filter, developed by Caltech/JPL [17], has fluid passage-ways in a configuration which allows very low restriction of flow and has also stagna-tion areas for the collection of impurities. A filter housing without a central post has also been developed to improve the flow characteristics. The full-flow etched-disc filter produces a zero reversal of flow and a very low drop of pressure. A perma-nently sealed (welded) filter housing without a central post has improved the flow characteristics. The following figure, due to the courtesy of NASA [17], shows an etched-disc filter element consisting essentially of a thin metallic disc with shallow, radially-disposed, etched, fluid channels incorporating stagnation areas located away from the main flow streams.

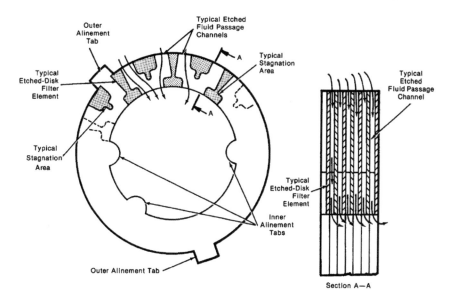

Fluid may flow in either direction. The discs have inner alignment tabs, which ensure proper front-to-back orientation, and outer alignment tabs, which keep the discs aligned within a housing. The following figure, also due to the courtesy of NASA [17], shows: (above) how the discs are oriented, front-to-back, so that the non-etched surfaces are adjacent to the aligned etched channels; and (below) the assembled etched-disc filter.

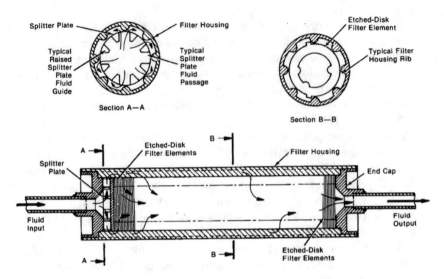

Of various metals used for filters, stainless steel is the most common. Filters of the wire-mesh type have, in comparison with those made of stacked etched discs, the advantages of:

- greater surface area with lower weight; and
- capability of handling higher flow rates.

Filters and filter elements are covered by the military standards MIL-F-5504B [18] and MIL-F-8815D [19].

Some standard filter elements and filter cases used in hydraulic systems for aircraft are illustrated in the following figure, re-drawn from [2]. As shown in this figure, filter cavities can also be machined in the body of a component such as a servo-cylinder, a pump, or a regulator. A filter case provides structural support for a filter element. The design of a filter case takes account of space available, easiness of access and service, pressure loss, fluid compatibility, and pressure and temperature in operating conditions. The size and the shape of a filter case have influence on the capacity and the service life of the filter.

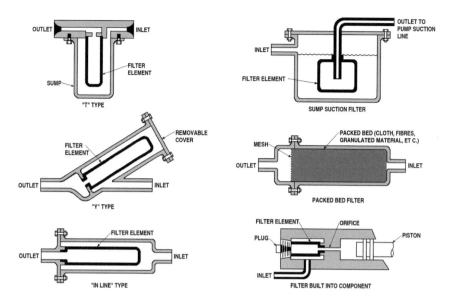

The choice of a type of filter case depends on requirements of accessibility. The in-line filter case is the least accessible type of filter. It requires the removal of the case from the piping system before a filter element can be removed. The T-type, also known as pot type, is designed so that the filter element can be removed by removing the filter pot. The Y-type is designed so that the filter element can be removed through one leg of the Y, by removing a blank flange or a cover plate.

The seal used for a filter is particularly important in case of mechanical vibrations. In critical propellant feed applications, internal seals should be avoided whenever possible, due to compatibility of materials and generation of particles during assembly.

Such devices as supporting cages, bolts, nuts, and O-rings are used to seal the filter element at its interface with the case. The seals or the locking devices used in filters are designed so as to avoid damages due to parts of filters which may migrate downstream.

Some filters have a pressure indicator which senses abnormal high differences in pressure across a filter element. This device may sometimes indicate a high difference in pressure where a liquid and a gas flow intermittently. This may happen in ducts carrying cryogenic fluids.

Some filters used in lubrication and hydraulic ducts have a built-in by-pass valve, which opens and by-passes the filter when the difference in pressure across it exceeds a given value. This device is incorporated in the filter, because it is better in some cases to keep a fluid-carrying duct working with a contaminated fluid than to have a complete failure due to lack of fluid when the filter becomes plugged. This device is rarely used in rocket engines which operate for a very brief period of time, and therefore the level of contamination in these filters cannot grow appreciably during their operating time.

A filter can be cleaned in one or more of the following manners:

- back-flushing;
- ultrasonic cleaning;
- flushing;
- purging; and
- initial cleaning.

A brief description of these methods is given below.

Back-flushing results from reversing the direction of flow through a filtering element, with consequent flushing of solid particles from the inlet side of the filter. This method removes loose particles, but not those which are tightly fixed within the porous medium. Therefore, back-flushing must be followed by other methods to restore a filter element to its original conditions of cleanliness.

Ultrasonic cleanliness consists in immersing a filter element in a tank containing a solvent solution, and applying ultrasonic waves to this solution by means of transducers mounted on or within the tank. The ultrasonic waves produce cavitation on the surface of the filter element immersed in the solution, and this in turn loosens solid particles on the filter element.

Flushing a filter element in the normal direction of flow, after applying back-flushing or ultrasonic cleaning, removes the solid particles which have previously been loosened and improves the degree of cleanliness.

Purging consists in using a pre-filtered dry gas, such as air or nitrogen, which passes through a filter element to dry it and to also remove residual solvent which may be left by previous cleaning processes. Purging a filter element is often less effective than ultrasonic cleaning and back-flushing.

Initial cleaning of a new filter element is necessary, because solid particles may have been introduced during manufacture and assembly of the filter into the filter element and also into the filter case.

Ultrasonic cleaning, flushing, and back-flushing may be combined together to reach a desired level of cleanliness in a filter element.

Cloth elements made of stainless steel can reach a higher degree of initial cleanliness by applying acid passivation after assembly, or by annealing the assembly in an environment of hydrogen [2].

The choice of a particular type of filter to be used for a specific application depends on:

- largest size of solid particles which can be tolerated by the components of a fluid-carrying duct;
- amount and type of solid particles which may be contained in a fluid during the service life of a given rocket engine;
- maximum loss of pressure due to the presence of a given filter; and
- space available within a fluid-carrying duct.

The maximum allowable size of solid particles in a given component (for example, a valve or a servo-actuator) can be determined by evaluating the clearances or the sizes of narrow passages in the given component which are sensitive to contaminants. For this purpose, it is necessary to determine the possible malfunctions of the given component caused by the entrapment of solid particles carried by the fluid.

The amount and the type of solid particles during service life is difficult to determine, because of uncertainties in the prediction of: (a) amount of solid particles which remain after cleaning a given component and its filter; (b) amount of solid particles which are carried by the fluid; and (c) fluids and temperatures in operational conditions, which may differ from those of the tests conducted in laboratories.

The maximum loss of pressure due to a given filter and the space available within the given duct can result from a system analysis and a trade-off study to determine whether it is more advantageous to install a filter mounted externally to the component to be protected or a filter forming an integral part of the component itself.

The largest size of solid particles which can be transmitted through a filter is expressed by a parameter known as filter rating. This parameter measures the degree of protection which a filter provides for downstream components.

Several types of filter rating are specified by filters manufacturers.

One of them is the maximum particle size rating (MPR), which is the longest dimension, in microns, of any solid particle allowed downstream of a filter.

Another type of filter rating is the absolute rating, also known as glass bead rating (GBR), which is the size, in microns, of the largest hard spherical particle (i.e., glass bead) which would be removed by the filter under steady flow conditions [16]. In other words, the definition of absolute filter rating considers all solid particles to have spherical shape, and requires a filter to retain all solid particles whose diameter is greater than or equal to a specified value. Therefore, the absolute rating takes account only of the second largest dimension of any solid particle which can be transmitted through a filter.

Still another type of filter rating is the nominal rating, which assesses the ability of a filter to remove a specified percentage (in either count or weight) of spherical solid particles or graded dust whose size is equal to or greater than the value defined by its absolute rating. Therefore, the nominal rating of a given filter is always less than its absolute rating.

The maximum particle size rating can be determined by test. The test is conducted by using a readily identifiable contaminant with particles of various sizes and shapes. A procedure for determining the maximum particle size rating is described in [16].

The absolute rating of a filter also can be determined by test. A test is conducted by filtration of an artificial contaminant (glass beads) under specified conditions, as described in [19].

The nominal average rating of a filter can be determined by means of a mercury intrusion test, which is described in [20].

Maximum particle size tests are not practical for acceptance testing of production filters. The glass bead test is a destructive type of test, and therefore is never used as an acceptance test. However, correlation of a glass bead test with a non-destructive bubble-point test permits verification of absolute rating for production filters [1]. The bubble-point test is based on the fact that, for a given liquid and for a given size of pores with a constant wetting, the pressure necessary to force a bubble of air through a pore is inversely proportional to the diameter of the pore. In other words, the diameter of the largest pore in a filter can be measured by wetting the filter element with a liquid and then measuring the pressure at which the first stream of bubbles is emitted from the upper surface of the filter element, when air is introduced to the open end. The pressure at which the first stream of bubbles emerges is taken as a measure of the diameter of the largest pore. The value of the constant of inverse proportionality between pressure and diameter is determined experimentally. Further information on the bubble-point test can be found, fir example, in [16, 21]. The bubble-point test is covered by the standards ISO 4003:1977 [22] and ARP 901A [23].

As has been shown above, a filter is necessary to reduce the size of solid particles carried in suspension by an operating fluid to a level which cannot cause damage to critical components of a duct. The determination of the filter area required for this purpose is based on an accurate prediction of the amount of contaminant which can be expected for any given application. The problem of accurately predicting the actual degree of contamination is still open. So far, there appears to be no correlation between the actual and the predicted amounts, sizes, and types of contamination [1]. The reason of this may be the variation in techniques used in manufacturing and building components and assemblies.

In this state of things, for the purpose of demonstrating that a filter of sufficient area has been provided for a given application, a specified amount of contaminant is required to be retained by the filter, without exceeding some maximum difference of pressure between its inlet and outlet. The contaminants most frequently used are AC-Fine and AC-Coarse dusts. These test dusts are Natural Arizona Dusts supplied by General Motors Phoenix Laboratory and classified to specific particle-size distributions by the AC Spark Plug Division of General Motors Corporation [1]. Further information on AC-Fine and AC-Coarse dusts can be found in [24]. The two dusts named above provide a baseline material for evaluating the ability of a filter to retain particles on its upstream side under flow conditions. Buckingham and Winzen [16] gives further information on tests which were conducted to determine the tolerance of various filter materials to contaminants for both gaseous and liquid flow.

Residual contaminants, that is, contaminants remaining in a filter after cleaning, may slough off and pass downstream of the filter. As has been shown above, such contaminants may accumulate in a filter during its manufacturing process, or be introduced there by the operating fluid, or be generated by the wear of mechanical components.

A cleaning of a filter removes a part of contaminants from the filter, the amount of this part depending on the accuracy of the cleaning process. Since this process is expensive, sometimes more expensive than the manufacturing process, then it is desirable to reduce the amount of contaminants which accumulate during fabrication, assembly, and testing.

It is also desirable to reduce media migration, that is, the presence in the operating fluid of contaminants coming from either the filter element or the filter-supporting structure. Filter materials subject to media migration are sintered porous metals, pressed paper, matted fibres, glass fibres, sintered plastic, fired porcelain, bonded carbon, and bonded stone [1, 2].

One of the most common tests for media migration combines thermal shock with vibration. In this test, a filter is exposed to its highest service temperature for a period of time and then to its lowest service temperature for an equal period. This procedure is repeated several times. After this thermal shock, the filter is caused to vibrate at frequencies of service for a given period of time. After testing, the filter is flushed with a given amount of fluid (usually 500 ml of either Freon TF or some ozone-friendly substitute for it), and the fluid is collected and passed through a 0.45 μ membrane pad. The membrane pad is then analysed for particles identifiable as material coming from either the filter element or the filter-supporting structure. When all of the particles are non-metallic and the filter is made entirely of metal, then no media migration has occurred. When some of the particles are metallic, then further analysis is required, and the filter material must be distinguished from residual contaminants [1]. Individual particles taken from a representative sample of the pad can be identified by microscopic examination.

The problem of determining the loss of pressure Δp through a filter as a function of its volumetric flow rate q is difficult to solve. In case of a filter element of the surface type, the pressure loss across the filter medium for either fine or coarse mesh is small in comparison with the pressure loss at filter entrance or exit.

In practice, the problem indicated above can be solved by tests which make it possible to plot curves of pressure loss against volumetric flow rate.

One of the formulae which can be used for determining the properties of a filter is known as the Kozeny-Carman equation [25], which is briefly discussed below.

The Kozeny-Carman equation expresses the pressure loss Δp (N/m^2) of a fluid which passes through a porous material, such as a packed bed of solids. This equation, which holds only in case of laminar flow ($Re \leq 1$), can be expressed as follows

$$\frac{\Delta p}{L} = -\frac{180\mu}{\phi_s^2 D_p^2} \frac{(1-\varepsilon)^2}{\varepsilon^3} u = -\frac{180\mu}{\phi_s^2 D_p^2} \frac{(1-\varepsilon)^2}{\varepsilon^3} \frac{q}{A}$$

where L (m) is the total height of the bed, μ (kg m^{-1} s^{-1}) is the coefficient of dynamic viscosity of the fluid, ϕ_s is the sphericity (see below) of the particles in the packed bed, D_p (m) is the diameter of the volume equivalent spherical particle, ε is the porosity (the fraction of the volume of voids over the total volume) of the bed, μ (m/s) is the mean velocity of the fluid at right angles to the layers of the bed,

q (m^3/s) is the volumetric flow rate of the fluid through the bed, and A (m^2) is the cross-sectional area of a layer (the cross-sectional area of the solids plus the cross-sectional area of the pores) of the bed. The minus sign in front of the right-hand side of the preceding equation takes account of the loss of pressure with increasing mean velocity or volumetric flow rate.

By sphericity ϕ_s of a particle in a packed bed we mean the ratio of the surface area of a sphere having the same volume V_p as the given particle to the surface area A_p of the particle, as follows

$$\phi_s = \frac{\pi^{\frac{1}{3}} \left(6V_p \right)^{\frac{2}{3}}}{A_p}$$

According to this definition, the sphericity of a spherical particle is unity, and the sphericity of any non-spherical particle is less than unity.

A derivation of the Kozeny-Carman equation and worked examples for its application can be found in [26].

Other equations (Darcy-Weisbach and Hagen-Poiseuille) found experimentally to determine the loss of pressure through a filter as a function of its volumetric flow rate are discussed in [16]. The equations cited above provide an excellent means for evaluating filter area requirements. However, they do not consider pressure losses at entrance and exit [1].

Filters are not tested for acceptance with the same fluids as the propellants used by rocket engines, because of difficulties due to testing with cryogenic or toxic or corrosive propellants. Some fluids used for tests are Freon TF, hydraulic oil, trichloroethylene, water, isopropyl alcohol, and methylated ethanol [1].

7.10   Design of a Flange Joint

A flange joint of a type frequently used for a rocket engine is shown in the following figure, re-drawn from [3].

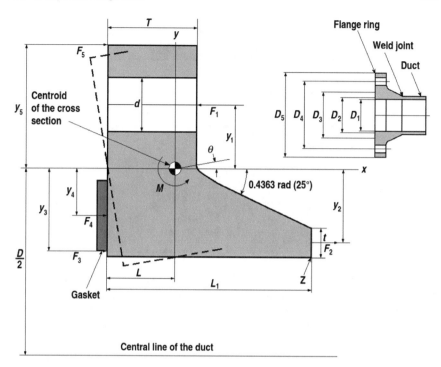

Under various conditions of loads, the flange ring is subject to bending moments *M*, which cause the ring to rotate. The dashed line in the preceding figure indicates exaggeratedly the counter-clockwise direction of rotation of the flange ring illustrated in the preceding figure. Some flange joints of the raised-face type shown in Sect. 7.4 and also in the following figure, re-drawn from [28], have a small gap to eliminate or reduce stresses in the peripheral part of the flange ring. A flange joint of this type is sealed by the bolt force which squeezes the gasket. Seals used for flange joints will be described in Sect. 7.11.

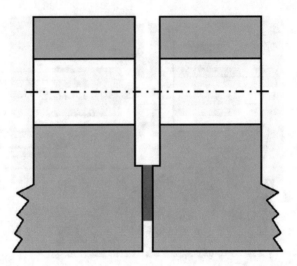

The stresses resulting from the bending moments reach their maximum value at the point Z, where the flange connects with the wall of the inner cylindrical surface of the duct. The bolts used for a flange joint are frequently pre-stressed in tension, in order for a compressive stress to act on the gasket. This compression seals effectively the joint and avoids leakage.

Let us consider an arc of circle of unit length (1 m) passing through the centroid of the flange cross section. According to Huzel and Huang [3], the relations between the forces acting on this arc of circle and the minimum loads acting on the bolts can be written as follows

$$F_1 = F_2 + F_3 + F_4 + F_5$$

$$F_2 = \frac{pD_1^2}{4D} + \frac{W_e}{\pi D}$$

$$F_3 = \frac{p\left(D_2^2 - D_1^2\right)}{4D}$$

$$F_4 = \frac{\sigma_g\left(D_3^2 - D_2^2\right)}{4D}$$

$$\sigma_g = mp$$

$$F_5 = nF_1$$

$$W_b = F_1\pi D$$

where D (m) is the diameter of the circle passing through the centroid of the cross section of the flange ring, D_1 (m) is the inside diameter of the flange and of the duct, D_2 (m) is the inside diameter of the gasket, D_3 (m) is the outside diameter of the gasket, p (N/m^2) is the maximum pressure of the fluid in working conditions, F_1 (N/m) is the force per unit length of the flange ring due to the bolt loading, F_2 (N/m) is the force per unit length of the flange ring due to the longitudinal tension in the duct, F_3 (N/m) is the force per unit length of the flange ring due to the internal pressure, F_4 (N/m) is the force per unit length of the flange ring due to the gasket loading or seal loading, F_5 (N/m) is the force per unit length of the flange ring due to the compressive load at the outside of the flange, σ_g (N/m^2) is the average compressive stress on the gasket required for proper seating against an internal pressure p of the fluid, m is a gasket factor which depends on the gasket design and whose value (to be determined experimentally) ranges from 0.8 to 10, n is a flange factor which depends on the configuration and on the rigidity of the flange and whose value ranges from 0.1 to 0.8, W_e (N) are the end loads on the duct due to inertial and thermal effects (tension or compression), and W_b (N) is the minimum required bolt loading.

As an example of application [3] of the equations given above, the following data are known for a flexible duct at the discharge of the oxidiser pump used in a rocket engine: pressure in normal operating conditions 1.038×10^7 N/m^2, maximum pressure in transient conditions 1.207×10^7 N/m^2, inner diameter of the duct $D_1 = 0.2032$ m, inner diameter of the gasket $D_2 = 0.2032$ m, outer diameter of the gasket $D_3 = 0.2159$ m, end loads in the duct due to thermal contraction $W_e = 1.068 \times 10^4$ N, gasket factor $m = 0.8$, and flange factor $n = 0.3$. It is required to determine the minimum loading needed for the bolts of the flange joint.

By using the maximum value of pressure, we set $p = 1.207 \times 10^7$ N/m^2. After substituting this value and $m = 0.8$ in the equation $\sigma_g = mp$, we find

$$\sigma_g = 0.8 \times 1.207 \times 10^7 = 0.9656 \times 10^7 \text{ N/m}^2$$

By substituting $F_5 = nF_1$ into $F_1 = F_2 + F_3 + F_4 + F_5$ and solving for F_1, there results

$$F_1 = \frac{F_2 + F_3 + F_4}{1 - n}$$

By substituting this expression of F_1 and the following expressions

$$F_2 = \frac{pD_1^2}{4D} + \frac{W_e}{\pi D}$$

$$F_3 = \frac{p\left(D_2^2 - D_1^2\right)}{4D}$$

$$F_4 = \frac{\sigma_g\left(D_3^2 - D_2^2\right)}{4D}$$

into $W_b = F_1 \pi D$, there results

$$W_b = \frac{\pi p D_1^2 + 4W_e + \pi p(D_2^2 - D_1^2) + \pi \sigma_g(D_3^2 - D_2^2)}{4(1-n)}$$

After substituting $p = 1.207 \times 10^7$ N/m^2, $D_1 = D_2 = 0.2032$ m, $W_e = 1.068 \times 10^4$ N, $D_3 = 0.2159$ m, $\sigma_g = 0.9656 \times 10^7$ N/m^2, and $n = 0.3$ in the preceding equation, we find the following value of the minimum loading needed for the bolts

$$W_b = \frac{3.1416 \times 1.207 \times 10^7 \times 0.2032^2 + 4 \times 1.068 \times 10^4}{4 \times (1 - 0.3)}$$
$$+ \frac{3.1416 \times 0.9656 \times 10^7 \times (0.2159^2 - 0.2032^2)}{4 \times (1 - 0.3)} = 6.321 \times 10^5 \text{ N}$$

After the minimum loading needed for the bolts has been determined as has been shown above, the following empirical relation can be used to determine the maximum circumferential spacing P_s (m) between two bolts required for a tight joint

$$P_s = 2d + T$$

where, with reference to the preceding figures, d (m) is the nominal diameter of each bolt, and T (m) is the thickness of the flange.

With reference to the preceding figures, the following empirical relations [3] can be used for the general proportions of a flange joint

$$T = At$$
$$L_1 = Bt$$

where T (m) is the thickness of the flange, t (m) is the thickness of the wall of the duct (depending on the circumferential stress σ_2 acting in the wall, as has been shown in Chap. 6), L_1 (m) is the overall axial length of the flange ring, and A and B are design factors, such that $4 \leq A \leq 8$, and $10 \leq B \leq 14$. The hub portion of the flange ring has a taper angle whose value is usually 0.4363 rad (25°).

With reference to the preceding figures, the following equations of [3] can be used to compute approximately the maximum stresses and strains in a flange ring

$$M = F_1 y_1 + F_2 y_2 + F_3 y_3 + F_4 y_4 - F_5 y_5$$

$$y_1 = \frac{D_4 - D}{2}$$

$$y_2 = \frac{D - D_1 - t}{2}$$

$$y_3 = \frac{2D - D_2 - D_1}{4}$$

$$y_4 = \frac{2D - D_3 - D_2}{4}$$

$$y_5 = \frac{D_5 - D}{2}$$

$$\theta = \frac{MD^2}{4EI}$$

$$\sigma_z = \frac{MD^2(L_1 - L)}{2D_1 I}$$

where, with reference to the preceding figures, M (Nm/m) is the magnitude of the resultant bending moment per unit length of the flange ring, D_4 (m) is the diameter of the bolt circle, D_5 (m) is the outer diameter of the flange ring, y_1 (m), y_2 (m), y_3 (m), y_4 (m), and y_5 (m) are the distances between the centroid of the cross section of the flange ring and the forces per unit length respectively F_1 (N/m), F_2 (N/m), F_3 (N/m), F_4 (N/m), and F_5 (N/m), I (m^4) is the moment of inertia of the cross section of the flange ring about an axis perpendicular to the plane of the sheet, E (N/m^2) is the Young modulus of the material of which the flange is made, θ (rad) is the angle of rotation of the flange ring under maximum working pressure and loads, and σ_z (N/m^2) is the maximum tensile stress which occurs at the point Z of the flange ring in the circumferential direction.

With the same data as those of the preceding example, we want to design the flange for a flexible duct at the discharge of the oxidiser pump used in a rocket engine, with the following materials and related data: Inconel® alloy 718 [27] for flange and duct, minimum yield strength $\sigma_y = 1.172 \times 10^9$ N/m^2, minimum ultimate strength $\sigma_u = 1.379 \times 10^9$ N/m^2, Young modulus $E = 2.041 \times 10^{11}$ N/m^2, duct weld efficiency $e_w = 0.75$, bolt diameter $d = 7.938 \times 10^{-3}$ m, bolt head diameter 1.336×10^{-2} m, A-286 stainless steel, ultimate bolt load 4.565×10^4 N.

In order to determine the design limit pressure, we multiply the value 1.207×10^7 N/m^2 of the maximum transient pressure by the safety factor 1.1, and obtain

$$1.1 \times 1.207 \times 10^7 = 1.328 \times 10^7 \text{ N/m}^2$$

Then, we determine the yield pressure p_y as follows

$$p_y = 1.1 \times 1.328 \times 10^7 = 1.461 \times 10^7 \text{ N/m}^2$$

and use this value to calculate the thickness t of the wall of the duct, as follows

$$t = \frac{p_y D_1}{2\sigma_y e_w} = \frac{1.461 \times 10^7 \times 0.2032}{2 \times 1.172 \times 10^9 \times 0.75} = 0.001689 \text{ m}$$

We determine the ultimate pressure p_u as follows

$$p_u = 1.5 \times 1.328 \times 10^7 = 1.992 \times 10^7 \text{ N/m}^2$$

and use this value to calculate again the thickness t of the wall of the duct, as follows

$$t = \frac{p_u D_1}{2\sigma_u e_w} = \frac{1.992 \times 10^7 \times 0.2032}{2 \times 1.379 \times 10^9 \times 0.75} = 0.001957 \text{ m}$$

We use the higher value (0.001957 m) of t, and round it to 0.002 m (2 mm), which value we choose for the thickness t of the wall.

We assume the values $A = 6$ and $B = 12.2$ for the flange design factors. By using the equation $T = At$, the thickness of the flange can be computed as follows

$$T = At = 6 \times 0.002 = 0.012 \text{ m} = 12 \text{ mm}$$

Likewise, by using the equation $L_1 = Bt$, the overall axial length of the flange ring results

$$L_1 = Bt = 12.2 \times 0.002 = 0.0244 \text{ m} = 24.4 \text{ mm}$$

The following values are given in [3] for the quantities D_4, D_5, L, D, and I: $D_4 = 0.2286$ m, $D_5 = 0.2443$ m, $L = 0.009398$ m, $D = 0.2184$ m, and $I = 1.136 \times 10^{-8}$ m^4. By solving the equation $W_b = F_1 \pi D$ for F_1, we find

$$F_1 = \frac{W_b}{\pi D} = \frac{6.321 \times 10^5}{3.1416 \times 0.2184} = 9.213 \times 10^5 \text{ N/m}$$

Likewise, we compute F_2 as follows

$$F_2 = \frac{p D_1^2}{4D} + \frac{W_e}{\pi D} = \frac{1.207 \times 10^7 \times 0.2032^2}{4 \times 0.2184} + \frac{1.068 \times 10^4}{3.14 \times 0.2184}$$
$$= 5.860 \times 10^5 \text{ N/m}$$

F_3 results from

$$F_3 = \frac{p(D_2^2 - D_1^2)}{4D} = 0 \text{ N/m}$$

because, in the present case, $D_1 = D_2$.

F_4 results from

$$F_4 = \frac{\sigma_g\left(D_3^2 - D_2^2\right)}{4D} = \frac{0.9656 \times 10^7 \times \left(0.2159^2 - 0.2032^2\right)}{4 \times 0.2184}$$

$$= 0.5883 \times 10^5 \text{ N/m}$$

F_5 results from

$$F_5 = n F_5 = 0.3 \times 9.213 \times 10^5 = 2.764 \times 10^5 \text{ N/m}$$

The distance y_3 is of no interest, because $F_3 = 0$. The distances y_1, y_2, y_4, and y_5 result from

$$y_1 = \frac{D_4 - D}{2} = \frac{0.2286 - 0.2184}{2} = 0.0051 \text{ m}$$

$$y_2 = \frac{D - D_1 - t}{2} = \frac{0.2184 - 0.2032 - 0.001689}{2} = 0.006756 \text{ m}$$

$$y_4 = \frac{2D - D_3 - D_2}{4} = \frac{2 \times 0.2184 - 0.2159 - 0.2032}{4} = 0.004425 \text{ m}$$

$$y_5 = \frac{D_5 - D}{2} = \frac{0.2443 - 0.2184}{2} = 0.01295 \text{ m}$$

By using the equation $M = F_1 y_1 + F_2 y_2 + F_3 y_3 + F_4 y_4 - F_5 y_5$ with $F_3 y_3 = 0$, the magnitude of the bending moment per unit length of the flange ring results

$$M = (9.213 \times 0.0051 + 5.86 \times 0.006756 + 0.5883$$

$$\times 0.004425 - 2.764 \times 0.01295) \times 10^5$$

$$= 5339 \text{ Nm/m}$$

The angle of rotation of the flange ring under maximum working pressure and loads results from

$$\theta = \frac{M D^2}{4EI} = \frac{5339 \times 0.2184^2}{4 \times 2.041 \times 10^{11} \times 1.136 \times 10^{-8}} = 0.02746 \text{ rad}$$

The maximum tensile stress, which occurs at the point Z of the flange ring in the circumferential direction, results from

$$\sigma_z = \frac{M D^2 (L_1 - L)}{2D_1 I} = \frac{5339 \times 0.2184^2 \times (0.0244 - 0.009398)}{2 \times 0.2032 \times 1.136 \times 10^{-8}}$$

$$= 8.275 \times 10^8 \text{ N/m}^2$$

The yield load stress at the same point results from

$$1.1 \times 1.1 \times 8.275 \times 10^8 = 1.001 \times 10^9 \text{ N/m}^2$$

The value computed above is less than the minimum yield strength ($\sigma_y = 1.172 \times 10^9$ N/m^2) of the material chosen.

The ultimate load stress at the same point results from

$$1.5 \times 1.1 \times 8.275 \times 10^8 = 1.365 \times 10^9 \text{ N/m}^2$$

The value computed above is less than the minimum ultimate strength ($\sigma_u = 1.379 \times 10^9$ N/m^2) of the material chosen.

Therefore, the design resulting from the preceding calculation is acceptable.

The maximum spacing between the bolts is computed by using the following equation

$$P_s = 2d + T = 2 \times 0.007938 + 0.012 = 0.02788 \text{ m}$$

The required number N of bolts results from the following equation

$$N P_s = \pi D_4$$

which, solved for N, yields

$$N = \frac{\pi D_4}{P_s} = \frac{3.1416 \times 0.2286}{0.02788} = 25.76 \approx 26$$

The minimum required value of the bolt loading ($W_b = 6.321 \times 10^5$ N) has been computed above by using the maximum transient pressure ($p = 1.207 \times 10^7$ N/m^2). The required ultimate bolt loading can be computed as follows

$$1.5 \times 1.1 \times 6.321 \times 10^5 = 1.043 \times 10^6 \text{ N}$$

Consequently, the ultimate loading on each bolt is

$$\frac{1.043 \times 10^6}{26} = 4.012 \times 10^4 \text{ N}$$

This value is smaller than the ultimate bolt load (4.565 $\times$ 10^4 N).

The required pre-load on each bolt is

$$\frac{W_b}{N} = \frac{6.321 \times 10^5}{26} = 2.431 \times 10^4 \text{ N}$$

The preceding description and the related examples of calculation refer to a common type of flange joint, but not to all types. Other types of flange joints have been illustrated in Sect. 7.4. Methods of calculation for such types of joints are described, for example, in [28, 29].

7.11   Gaskets and Other Seals for Flange Joints

Gaskets are used as static seals. The following types of gaskets are generally used for flange joints:

- annular-ring gaskets;
- pressure-actuated gaskets; and
- full-face gaskets.

Gaskets of the annular-ring type are the most widely used for bolted flanges. Pressure-actuated gaskets use the internal pressure of the fluid as an aid to the sealing action. Full-face gaskets generally require higher bolt loads than is the case with annular-ring gaskets. Full-face gaskets are scarcely tolerant to thermal gradients, and have a tendency to concentrate the loads at the bolt holes and at the portion of the gasket outside the bolt circle. These gaskets are largely used for non-critical service conditions, but rarely used for service conditions involving either high (above 505 K) or cryogenic (below 122 K) temperatures, or relative pressures above 2.068 MPa [2].

A gasket used for a flange joint should be thick enough to provide adequate conformity to the surfaces of the metallic parts of the joint, because thicker gaskets are generally better suited to conform to surface scratches and to compensate for sealing waviness. However, the same gasket should be thin enough to provide stability to the joint and to prevent blow-out when residual stresses are low and internal pressures are high [2].

Gaskets of a common type have O-rings made of elastomeric materials, which are natural or synthetic rubbers. They are used over a temperature range going from 211 to 505 K for long periods of time, and at higher temperatures for shorter periods of time [11]. At low temperatures, the elastomeric O-ring is installed between thin foil films, or is coated with indium, which remains ductile at such temperatures, as shown in the following figure, re-drawn from [2].

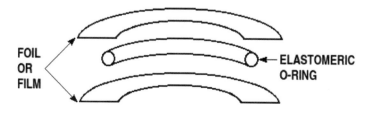

The initial seal is obtained by compressing the elastomer to the desired position of installation. The pressure on the flange ring in service causes the elastomer to conform completely to the flange at the leak path.

Elastomeric O-ring gaskets can be used over a wide range of pressures and are high reliable when installed properly. Incorrect installation can cause damage or improper squeeze or extrusion to the elastomer, with consequent leakage. The squeeze required to prevent damage while at the same time preventing leakage due to insufficient pressure ranges from 8 to 32%. O-ring extrusion and consequent nibbling of the elastomer occur when the clearance between the retaining members permits the elastomer to be forced into the clearance. When the elastomer becomes trapped, it is nibbled or sheared off when the pressure decays. Extrusion can be prevented by combining properly clearances and hard elastomers [11].

Elastomeric materials for flanges can also be used in moulded-in-place seals, which are flat plates having elastomeric inserts moulded into machined grooves, as shown in the following figure, due to the courtesy of NASA [11].

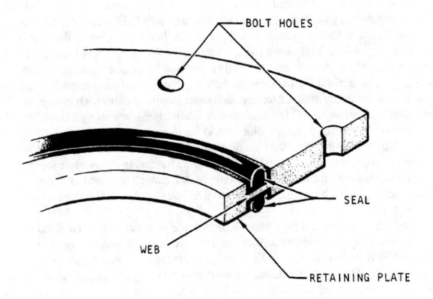

In these devices, the initial sealing action is due to the compression of the elastomer, and the subsequent sealing action is due to the pressure in service, which forces the elastomer against the flange at the leak path.

A moulded-in-place seal has the following advantages over a conventional O-ring:

- the thickness of the metal plate controls the amount of squeeze on the elastomeric portion of the seal, thereby permitting the use of flat-faced flanges on both sides of the seal, and making it unnecessary to machine an O-ring groove in one of the flanges; and
- it is easier to install in large and cumbersome hardware, because it can be installed laterally between two flanges.

Moulded-in-place seals have been used successfully in the fuel system of the F-1 rocket engine [11].

Metallic gaskets can be used at both cryogenic and high temperatures. These gaskets require high seating loads and are scarcely apt to follow flange deflection. Metallic O-rings are used in rocket engines particularly where the flanges or the mating surfaces are very rigid and connected with ample bolting [11]. Metallic O-ring perform satisfactorily in these conditions and offer the advantage of small cross sections where space is limited. They can be either Teflon®-coated for use at cryogenic temperatures or soft-metal plated for use at high or cryogenic temperatures.

Spiral-wound gaskets (see Chap. 5, Sect. 5.8) are made of a V-shaped ribbon of stainless steel wrapped spirally with a soft filler of either graphite or Teflon® between the turns, as shown in the following figure, adapted from [11].

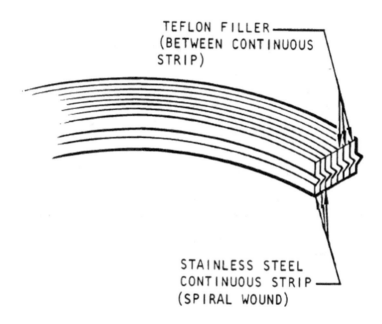

TEFLON FILLER
(BETWEEN CONTINUOUS STRIP)

STAINLESS STEEL
CONTINUOUS STRIP
(SPIRAL WOUND)

These gaskets were used extensively in rocket engines for both cryogenic and high (up to 811 K) temperatures. They need heavy, rigid flanges and high bolt loads (of the order of magnitude of $5 \times 10^5 - 7 \times 10^5$ N per metre of circumference). The high loads at the edges of the steel ribbon caused marring of the mating flanges and reduced the possibility of achieving a good seal when the joint was re-assembled [11]. Spiral-wound gaskets are still used today where leakage is tolerable, which happens in isolate engine locations and in large diameters, where they have a cost advantage over more sophisticated machined seals.

Pressure-actuated seals were originally developed for cryogenic temperatures. Later on, they were modified for hot gases. They have been used extensively in the J-2 and F-1 rocket engines. Their range of use includes cryogenic fluids at temperatures as low as 20 K and pressures as high as 27.56 MPa, and hot gases at temperatures up to 1033 K and pressures up to 10.34 MPa [11].

Pressure-actuated seals are more complex to manufacture and more expensive than those using elastomeric materials, and therefore are used only when those of the latter type cannot meet the requirements. The cross section of a pressure-actuated seal expands due to the internal pressure exerted by the fluid carried in the duct, and this expansion generates the seal. The deflection of the sealing face is larger than the deflection permissible in an elastomeric material. Therefore, a flange using a pressure-actuated seal can be less rigid and heavy than one using an elastomeric seal. A pressure-actuated seal of the cantilever (U-shaped) type is shown in the following figure, due to the courtesy of NASA [28].

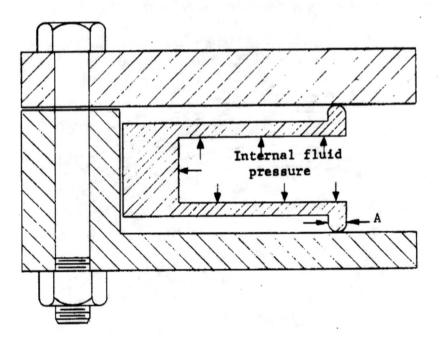

The U-shaped seal sits in a recess of the flange. The cross section of this seal has a web and two legs, each of which has a small tip at its end. The U-shaped seal is supported at the lips and at the outer surface of the web. The internal pressure of the fluid pushes the lips toward the inner surfaces of the flange, thereby exerting the sealing action. The dimensions of the cross-section of the seal are usually much smaller than the inside radius of the seal, and the legs of the seal are much more flexible than the web [28]. The cross section of a pressure-actuated seal may also be C-shaped (open O-ring) instead of U-shaped.

According to [28], pressure-actuated seals have the following advantages over elastomeric or metallic O-rings:

- high resiliency in seal and low clamping pressures, which result in lightweight flanges;
- high localised sealing stress;
- external load taken by other components of the connector;
- use of available pressure to increase sealing; and
- reduction of negative effects of extreme temperatures, due to the ability of the seal to "follow through".

Pressure-actuated seals used for ducts carrying cryogenic fluids are generally coated with Teflon® or with metallic plates. Metallic pressure-actuated seals have been employed in the RS-25 engines of the Space Shuttle. They have been used for cryogenic fluids at temperatures as low as 20 K and pressures up to 62.05 MPa, and also for hot gases at temperatures up to 1255 K and pressures up to 42.75 MPa. The metallic seals have been made of Inconel® 718 plated with silver or gold to provide a soft sealing material at the interface [11]. A groove-type seal, shown in the following figure, adapted from [11], is used to save weight and provide a small envelope.

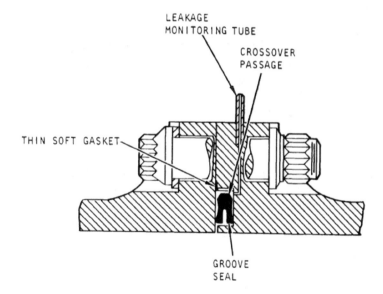

Plastic spring-loaded seals have been developed for fluids at cryogenic temperatures. Such seals are made of a jacket of plastic material, usually Teflon®, which covers a core of metallic spring. The spring provides the force required for the initial seal and also the forces necessary to compensate for dimensional changes resulting from thermal expansions and contractions. These seals are pressure-actuated, and therefore the sealing load increases with pressure.

Plastic spring-loaded seals are less reliable than metallic pressure-actuated seals at cryogenic temperatures, and are therefore used principally in rocket engines fed with storable propellants, when elastomeric materials are not compatible with such propellants.

A seal of the radial type used for a flange is shown in the following figure, due to the courtesy of NASA [11].

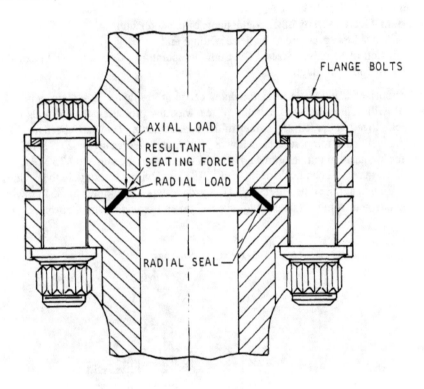

In this seal, the axial load acting on the flange is converted to a radial load acting on the interface through a toggle action within the structure of the seal. The two pieces of the flange joint are designed to confine the outer diameter of the seal, so that radial interference occurs and plastic flow of the seal takes place when an axial load is applied to the joint during installation. The plastic flow makes the seal not re-usable, but the two pieces of the joint are re-usable after installation of new seals. The magnitude of the axial load per unit length required to install a radial seal ranges from 87500 to 105000 N per metre of circumference.

A type of radial seal, the Conoseal® (described in [30]), has been used in some rocket engines (such as the M-1 engine and the Titan engines) and for some rocket vehicles (such as the upper stages S-IV and S-IVB) to seal flange joints for ducts carrying cryogenic propellants, storable propellants, and hot gases [11].

7.12 Design of a Bellows Joint for a Flexible Duct

In the present paragraph, we consider a bellows joint with restraining linkage for a flexible duct at the discharge of the oxidiser pump used in a rocket engine. The bellows joint is shown in the following figure, adapted from [1].

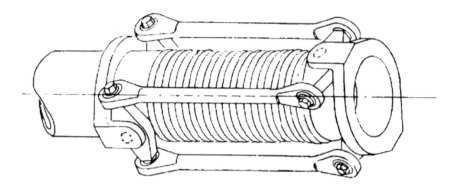

The equations considered below use the following symbols: C_t is the correction factor of the bellows wall-thinning; C_p is the ply inter-reaction factor (whose values are 1.00 for one-ply bellows, 0.90 for two-ply bellows, and 0.85 for three-or-more ply bellows); d_o (m) is the outside diameter of the bellows; d_i (m) is the outside diameter of the convolution root of the bellows; $d_m = \left[(d_i^2 + d_o^2)/2\right]^{\frac{1}{2}}$ (m) is the root-mean-square diameter of the bellows; d_d (m) is the mean diameter of the duct; E (N/m^2) is the Young modulus of the material of which the bellows is made; e_a (m) is the axial deflection of the bellows; e_b (m) is the equivalent axial deflection of the bellows due only to bending; e_p (m) is the equivalent axial deflection of the bellows due to parallel offset; e_s (m) is the equivalent axial deflection of the bellows due only to shear; F_s (N) is the shear load; F_p (N) is the pressure separating load; G (N/m^2) is the shear modulus of elasticity of the material of which the bellows is made; $h = (d_o - d_i)/2$ (m) is the mean height of the convolutions of the bellows; L (m) is the axial length of a convolution of the bellows; $L_a = (N_c - \frac{1}{2})L + N_p t$ (m) is the free axial length of the bellows; N_c is the number of convolutions of the bellows; N_p is the number of plies; t (m) is the thickness of the wall of the bellows; L_b (m) is the axial length of the rigid duct; M (Nm) is the bending moment; p (N/m^2) is the internal or external pressure of the fluid; p_{cr} (N/m^2) is the critical pressure for stability of the bellows; R_a (N/m) is the axial spring constant of the bellows; R_b (N/m) is the bending spring constant of the bellows; R_p (N/m) is the parallel offset spring constant of the bellows; R_s (N/m) is the shear spring constant of the bellows; R_t (Nm/rad) is the torsional spring constant of the bellows; σ_b (N/m^2) is the bulging (meridian) stress of the bellows; σ_h (N/m^2) is the hoop (circumferential) stress of the bellows; σ_m (N/m^2) is the motion stress of the bellows; σ_s (N/m^2) is the shear stress of the bellows; σ_t (N/m^2) is the torsion stress of the bellows; T (Nm) is the torsional moment; T_{cr} (Nm) is the critical stability torsional moment of the bellows; v is the

Poisson ratio of the material of which the bellows is made; y (m) is the transverse deflection of the bellows; θ (rad) is the bending angle of rotation; and ϕ (rad) is the torsional angle of rotation.

Bellows are formed by applying hydraulic pressure to tubes of the same diameter d_i as the bellows diameter at the root of the convolution. The thinning process of the wall of a bellows starts at the original thickness of the material at the root of the convolution, and reduces approximately linearly the thickness to its minimum value at the outside diameter d_o of the convolution. The amounts of thinning range from 10% to 40% [3]. The effects of thinning are taken into account by applying the thinning correction factor to the bellows design.

For the bellows design, we use the following equations indicated by Huzel and Huang [3].

For steel and nickel alloys, the axial spring constant R_a (N/m) is expressed by

$$R_a = \frac{1.49 C_t C_p N_p E d_i t^3}{N_c h^3}$$

For aluminium alloys, the axial spring constant R_a (N/m) is expressed by

$$R_a = \frac{1.23 C_t C_p N_p E d_i t^3}{N_c h^3}$$

The value of the correction factor C_t in the two preceding equations can be determined by substituting the value of the percent thinning x, relating to the bellows of interest, in the following polynomial

$$C_t(x) = -0.0001333x^2 - 0.01133x + 1$$

The percent thinning x is defined as follows

$$x = \left(\frac{t_i - t_o}{t_i} \right) 100$$

where the thicknesses t_i and t_o are shown in the following figure.

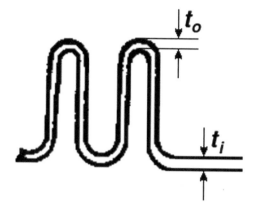

Bulging (meridian) stresses σ_b (N/m^2) are the radial bending stresses induced in the side walls of the bellows by internal or external pressure. They are expressed by

$$\sigma_b = \frac{C_t\ p\ h^2}{2\ N_p\ t^2}$$

The value of the correction factor C_t in the preceding equation can be determined by substituting the value of the percent thinning x, relating to the bellows of interest, in the following polynomial

$$C_t(x) = 0.0004\ x^2 + 0.01\ x + 1$$

The values of σ_b should be kept below those given in the following table, adapted from [3], which indicates mechanical properties of some materials used for bellows.

Material	Yield strength, MPa	Limiting bulging stresses. MPa		Allowable motion stresses, MPa		
		$t \leq 0.3048$ mm	$t \geq 0.3302$ mm	1000 cycles	10,000 cycles	100,000 cycles
321 and 347 stainless steels	269	965	827	1430	1030	634
19-9DL	607	965	827	1430	1030	634
A-2S6	1240	1310	1030	1100	1030	951
Inconel 718	1170	1310	1030	1100	1030	951
Inconel X-750	676	1310	1030	1100	1030	951
6061-T6 aluminum alloy	276	448	448	731	469	193

Bellows are subject to separating loads, because they are loaded by pressure and also by the axial force acting along the duct. The pressure separating loads F_p (N) are expressed by

$$F_p = \frac{p\left(d_m^2 - d_d^2\right)\pi}{4}$$

The hoop (circumferential) stresses σ_h (N/m^2) in a bellows are calculated as follows

$$\sigma_h = \frac{p\,d_m}{2\,N_p\,t\left(\frac{\pi}{2} - 1 + \frac{2h}{L}\right)}$$

The values of these stresses should be kept below the yield and ultimate strengths of the bellows material by a given margin.

Motion stresses σ_m (N/m^2) due to axial deflection of the bellows are caused by bending of the side walls of the bellows. For steel and nickel alloys, motion stresses due to axial deflection are expressed by

$$\sigma_m = \frac{1.40 C_t E t e_a}{N_c h^2}$$

For aluminium alloys, motion stresses due to axial deflection are expressed by

$$\sigma_m = \frac{1.78 C_t E t e_a}{N_c h^2}$$

The value of the correction factor C_t in the preceding equation can be determined by substituting the value of the percent thinning x, relating to the bellows of interest, in the following polynomial

$$C_t(x) = 0.000067 x^2 - 0.0023 x + 1.02$$

Allowable motion stresses for bellows materials, depending on their design cycle life, are given in the preceding table.

Motions of the bellows due to other causes than axial deflection (for example, angular motion and parallel offset motion) can be converted into an equivalent axial deflection of the bellows. By so doing, the two preceding equations expressing σ_m can also be used to calculate the corresponding motion stresses, as will be shown below.

In case of bellows subject to pure bending, the corresponding motion stresses σ_m can be computed by using the following equations

$$R_b = \frac{M}{\theta} = \frac{d_m^2 R_a}{8}$$

$$e_b = \frac{1}{2} d_m \sin\theta$$

$$\sigma_m = \frac{1.40 C_t E t e_b}{N_c h^2} \quad \text{for steel and nickel alloys}$$

$$\sigma_m = \frac{1.78 C_t E t e_b}{N_c h^2} \quad \text{for aluminium alloys}$$

where the value of the correction factor C_t in the preceding equations can be determined by substituting the value of the percent thinning x, relating to the bellows of interest, in the following polynomial

$$C_t(x) = 0.000067 x^2 - 0.0023 x + 1.02$$

In case of bellows subject to pure shear, the corresponding motion stresses σ_m can be computed by using the following equations

$$R_s = \frac{F_s}{y} = \frac{3 d_m^3 R_a}{8 L_a^2}$$

$$\theta = \frac{4 F_s L_a}{R_a d_m^2}$$

$$e_s = \frac{3 d_m y}{2 L_a}$$

$$\sigma_m = \frac{1.40 C_t E t e_s}{N_c h^2} \quad \text{for steel and nickel alloys}$$

$$\sigma_m = \frac{1.78 C_t E t e_s}{N_c h^2} \quad \text{for aluminium alloys}$$

where the value of the correction factor C_t in the preceding equations can be determined by substituting the value of the percent thinning x, relating to the bellows of interest, in the following polynomial

$$C_t(x) = 0.000067 x^2 - 0.0023 x + 1.02$$

In case of bellows subject to parallel offset, the corresponding motion stresses σ_m can be computed by using the following equations

$$R_s = \frac{F_s}{y} = \frac{3 d_m^2 R_a}{2 L_a^2}$$

$$M = \pm \frac{F_s L_a}{2} = \pm \frac{3 d_m^2 y R_a}{4 L_a^2}$$

$$e_p = \frac{3d_m y}{2L_a}$$

$$\sigma_m = \frac{1.40C_t E t e_p}{N_c h^2} \quad \text{for steel and nickel alloys}$$

$$\sigma_m = \frac{1.78C_t E t e_p}{N_c h^2} \quad \text{for aluminium alloys}$$

where the value of the correction factor C_t in the preceding equations can be determined by substituting the value of the percent thinning x, relating to the bellows of interest, in the following polynomial

$$C_t(x) = 0.000067x^2 - 0.0023x + 1.02$$

In case of parallel offset of articulated bellows (two bellows, each of which L_a in free length, separated by a rigid duct L_b in length), the corresponding motion stresses σ_m can be computed by using the following equations

$$R_s = \frac{F_s}{y} = \frac{3d_m^2 R_a}{2(4L_a^2 + 6L_a L_b + 3L_b^2)}$$

$$M = \pm \frac{F_s(2L_a + L_b)}{2} = \pm \frac{3d_m^2(2L_a + L_b)y R_a}{4(4L_a^2 + 6L_a L_b + 3L_b^2)}$$

$$e_p = \frac{3d_m(2L_a + L_b)y}{4L_a^2 + 6L_a L_b + 3L_b^2}$$

$$\sigma_m = \frac{1.40C_t E t e_p}{N_c h^2} \quad \text{for steel and nickel alloys}$$

$$\sigma_m = \frac{1.78C_t E t e_p}{N_c h^2} \quad \text{for aluminium alloys}$$

where the value of the correction factor C_t in the preceding equations can be determined by substituting the value of the percent thinning x, relating to the bellows of interest, in the following polynomial

$$C_t(x) = 0.000067x^2 - 0.0023x + 1.02$$

The torsional stresses σ_t (N/m^2) in a bellows are calculated by using the following equation

$$\sigma_t = \frac{2T}{N_p \pi d_i^2 t}$$

and the torsional spring constant R_t (Nm/rad) of the bellows is

$$R_t = \frac{T}{\phi} = \frac{\pi G d_i t N_p}{4(2h + 0.57L)N_c}$$

A restrained bellows, if pressurised internally beyond a critical value p_{cr} (N/m²), is subject to an instability failure of the same type as a buckling column. The critical value of internal pressure results from

$$p_{cr} = \frac{5.02 R_a}{L_a \frac{d_o}{d_i}}$$

A bellows, if pressurised externally beyond a critical value p_{cr} (N/m²), is subject to buckling in the same manner as a thin cylinder. The critical value of external pressure results from

$$p_{cr} = \frac{4 E t N_p h^3}{(1 - v^2) d_m^3 e_b}$$

A bellows, if loaded by pure torsion, buckles in some manner as one loaded by internal pressure. The critical value T_{cr} (Nm) of the torsional moment results from

$$T_{cr} = \frac{1}{2}\pi d_m^2 R_a$$

The critical pressure p_{cr} and the critical torsional moment T_{cr} have smaller values than those resulting from the preceding equations, in case of bellows under angular and offset deflections. The values of the correction factor to be applied are determined experimentally. These values range from 0.2 to 0.9 [3].

In case of bellows operating at high temperatures due to the flow of hot gases, working stresses used for bellows materials should be adjusted accordingly. An internal liner is generally provided to protect a bellows from high-velocity, high-temperature gases.

A bellows design depends on the forming process. The following geometric limits are generally used for bellows having up to three plies: maximum value of the ratio $d_o/d_i = 1.35$; and maximum value of the axial length of a convolution as a function of thickness $L = (8 + 2N_p)t$.

As an example of application of the concepts discussed above, it is required to design a bellows joint with restraining linkage for a flexible duct at the discharge of the oxidiser pump used in a rocket engine, with the following data: pressure in normal operating conditions 1.038×10^7 N/m², maximum pressure in transient conditions 1.207×10^7 N/m², and inner diameter of the duct $d_i = 0.2032$ m.

We use Inconel 718® for the material of which the bellows is made. This material has the following properties: minimum yield strength $\sigma_y = 1.172 \times 10^9$ N/m², minimum ultimate strength $\sigma_u = 1.379 \times 10^9$ N/m², and Young modulus $E = 2.041 \times 10^{11}$ N/m².

Since the inner diameter of the duct is $d_i = 0.2032$ m, then the outside diameter of the convolution root of the bellows is also $d_i = 0.2032$ m.

We also know that the percent thinning of the wall of the bellows is $x = 20\%$, the maximum value of the free axial length of the bellows is $L_a = 0.1778$ m, the angle of rotation of the bellows is $\theta = \pm\pi/60$ rad ($\pm 3°$), and the life of the bellows is equal to 10000 cycles. In addition to the dimensions, it is also required to determine the axial spring constant R_a (N/m) of the bellows, the magnitude M (Nm) of the bending moment at the angle $\theta = \pm\pi/60$ rad, and the necessary restraining load F (N) due to the linkage at the maximum pressure in transient conditions 1.207×10^7 N/m^2.

As has been shown above, the bulging (meridian) stress σ_b (N/m^2) is expressed by the following equation

$$\sigma_b = \frac{C_t p h^2}{2 N_p t^2}$$

where the value of the correction factor C_t of the bellows wall-thinning can be determined by substituting 20 for x in the following polynomial

$$C_t(x) = 0.0004\,x^2 + 0.01\,x + 1$$

By so doing, we find $C_t = 1.36$.

From the table on the properties of materials given above, we take the value $\sigma_b = 1030$ MPa $= 1.03 \times 10^9$ N/m^2 for the limiting bulging stress relating to Inconel 718® with a life of 10000 cycles and a thickness greater than or equal to 0.3302 mm. From the data of the present example, we also obtain the design limit pressure of the duct as follows

$$1.1 \times 1.207 \times 10^7 = 1.328 \times 10^7 \text{ N/m}^2$$

where 1.1 is the value of the safety factor. We also choose a three-ply bellows, such that $N_p = 3$.

After substituting $C_t = 1.36$, $p = 1.328 \times 10^7$ N/m^2, $\sigma_b = 1.03 \times 10^9$ N/m^2, and $N_p = 3$ in the following equation

$$\sigma_b = \frac{C_t p h^2}{2 N_p t^2}$$

and solving for h/t, we find

$$\frac{h}{t} = \left(\frac{2 N_p \sigma_b}{C_t p}\right)^{\frac{1}{2}} = \left(\frac{2 \times 3 \times 1.03 \times 10^9}{1.36 \times 1.328 \times 10^7}\right)^{\frac{1}{2}} = 18.50$$

We choose a thickness $t = 0.56$ mm $= 5.6 \times 10^{-4}$ m for the wall of the bellows, and consequently the mean height of the convolutions of the bellows results

$$h = 5.6 \times 10^{-4} \times 18.50 = 0.01036\,\text{m} \approx 0.0104\,\text{m}$$

Since $h = (d_o - d_i)/2 = 0.0104$ m, $d_i = 0.2032$ m, and $h = 0.0104$ m, then the outside diameter of the bellows results

$$d_o = d_i + 2h = 0.2032 + 2 \times 0.0104 = 0.2240\,\text{m}$$

and the root-mean-square diameter of the bellows results

$$d_m = \left(\frac{d_i^2 + d_o^2}{2}\right)^{\frac{1}{2}} = \left(\frac{0.2032^2 + 0.2240^2}{2}\right)^{\frac{1}{2}} = 0.2136\,\text{m}$$

The equivalent axial deflection of the bellows due to pure bending results from the following equation

$$e_b = \frac{1}{2} d_m \sin\theta = 0.5 \times 0.2136 \times \sin\left(\frac{\pi}{60}\right) = 0.005589\,\text{m}$$

Again, from the table on the properties of materials given above, we take an allowable motion stress of 1030 MPa $= 1.03 \times 10^9$ N/m^2 for Inconel 718® with a life of 10,000 cycles. However, we use a fraction of this value, that is,

$$0.36 \times 1.03 \times 10^9 = 3.708 \times 10^8\,\text{N/m}^2$$

to improve stability.

The number of convolutions N_c of the bellows can be determined by using the following equation

$$\sigma_m = \frac{1.40 C_t E t e_b}{N_c h^2}$$

where $\sigma_m = 3.708 \times 10^8$ N/m^2, $e_b = 0.005589$ m, and the value of the correction factor C_t of the bellows wall-thinning can be determined by substituting 20 for x in the following polynomial

$$C_t(x) = 0.000067\,x^2 - 0.0023\,x + 1.02$$

By so doing, we find $C_t = 1$. Therefore, N_c results from

$$N_c = \frac{1.40 C_t E t e_b}{h^2 \sigma_m} = \frac{1.40 \times 1 \times 2.041 \times 10^{11} \times 0.00056 \times 0.005589}{0.0104^2 \times 3.708 \times 10^8} = 22.3 \approx 22$$

The axial length of a convolution of the bellows results from

$$L = \left(8 + 2N_p\right)t = (8 + 2 \times 3) \times 5.6 \times 10^{-4} = 0.00784\,\text{m}$$

The free axial length of the bellows results from

$$L_a = (N_c - 1/2)L + N_p t = (22 - 0.5) \times 0.00784 + 3 \times 0.00056 = 0.1702 \text{ m}$$

The axial spring constant of the bellows is expressed by the following equation

$$R_a = \frac{1.49 C_t C_p N_p E d_i t^3}{N_c h^3}$$

where the correction factor C_t results from substituting 20 for x in the following polynomial

$$C_t(x) = -0.0001333\, x^2 - 0.01133\, x + 1$$

Hence, $C_t(20) = 0.72$. Therefore, the axial spring constant of the bellows is

$$R_a = \frac{1.49 \times 0.72 \times 0.85 \times 3 \times 2.041 \times 10^{11} \times 0.2032 \times 0.00056^3}{22 \times 0.0104^3}$$

$$= 8.051 \times 10^5 \text{ N/m}$$

The critical value of internal pressure for the bellows (without angulation) results from

$$p_{cr} = \frac{5.02\, R_a}{L_a \frac{d_o}{d_i}} = \frac{5.02 \times 8.051 \times 10^5}{0.1702 \times \frac{0.2240}{0.2032}} = 2.154 \times 10^7 \text{ N/m}^2$$

This value, divided by the value (1.207×10^7 N/m^2) of the maximum pressure in transient conditions, indicates a safety factor of 1.785, which allows for bellows stability in the presence of angulation.

As has been shown in paragraph 10, the yield pressure p_y and the ultimate pressure p_u result from

$$p_y = 1.1 \times 1.1 \times 1.207 \times 10^7 = 1.461 \times 10^7 \text{ N/m}^2$$
$$p_u = 1.1 \times 1.5 \times 1.207 \times 10^7 = 1.992 \times 10^7 \text{ N/m}^2$$

The yield hoop (circumferential) stress σ_{hy} results from substituting $p_y = 1.461 \times 10^7$ N/m^2 into the following equation

$$\sigma_h = \frac{p\, d_m}{2\, N_p\, t \left(\frac{\pi}{2} - 1 + \frac{2h}{L} \right)}$$

By so doing, we find

$$\sigma_{hy} = \frac{1.461 \times 10^7 \times 0.2136}{2 \times 3 \times 0.00056 \times \left(\frac{3.1416}{2} - 1 + \frac{2 \times 0.0104}{0.00784}\right)} = 2.881 \times 10^8 \,\text{N/m}^2$$

This value is less than the yield strength (1.170×10^9 N/m²) of Inconel® 718, resulting from the preceding table.

The ultimate hoop (circumferential) stress of the bellows results from

$$\sigma_{hu} = 2.881 \times 10^8 \times \frac{1.992 \times 10^7}{1.461 \times 10^7} = 3.928 \times 10^8 \,\text{N/m}^2$$

This value is less than the ultimate strength (1.375×10^9 N/m²) of Inconel® 718.

Summarising, the following results have been found in the preceding calculation: $d_i = 0.2032$ m, $d_o = 0.2240$ m, $d_m = 0.2136$ m, $t = 0.00056$ m, $N_p = 3$, $h = 0.0104$ m, $N_c = 22$, $L = 0.00784$ m, $L_a = 0.1702$ m, and $R_a = 8.051 \times 10^5$ N/m.

The angular spring constant of the bellows results from the following equation

$$R_b = \frac{d_m^2 R_a}{8} = \frac{0.2136^2 \times 8.051 \times 10^5}{8} = 4592 \,\text{Nm/rad}$$

The magnitude of the bending moment on the duct, at an angle $\theta = \pi/60$ rad, is

$$M = R_b\theta = 4592 \times \frac{3.1416}{60} = 240.4 \,\text{Nm}$$

The pressure separating load acting on the bellows results from

$$F_p = \frac{p\left(d_m^2 - d_d^2\right)\pi}{4}$$

Therefore, the necessary restraining load acting on the link at the maximum pressure in transient conditions ($p = 1.207 \times 10^7$ N/m²), considering the axial force, is

$$F = F_p + \frac{pd_d^2\pi}{4} = \frac{pd_m^2\pi}{4} = \frac{1.207 \times 10^7 \times 0.2136^2 \times 3.1416}{4} = 4.325 \times 10^5 \,\text{N}$$

Further information on the structural calculation of bellows can be found in [31, pages 92–94].

References

1. Daniels CM, Keller RB Jr (eds) (1977) Liquid rocket lines, bellows, flexible hoses, and filters, NASA SP-8123, April 1977, 189 p. Web site https://ntrs.nasa.gov/archive/nasa/casi.ntrs.nasa.gov/19780008146.pdf
2. Howell GW, Weathers TM (eds) (1970) Aerospace fluid component designers' handbook, volume I, Revision D, Report No. RPL-TDR-64-25, TRW Systems Group, One Space Park, Redondo Beach, California, USA, February 1970. Web site https://apps.dtic.mil/dtic/tr/fulltext/u2/874542.pdf
3. Huzel DK, Huang DH (1967) Design of liquid propellant rocket engines, 2nd edn. NASA SP-125, NASA, Washington, DC, 472 p. Web site https://ntrs.nasa.gov/archive/nasa/casi.ntrs.nasa.gov/19710019929.pdf
4. Anonymous, Space Shuttle main engine orientation, Boeing-Rocketdyne Propulsion & Power, Presentation BC98-04, June 1998, 105 p. Web site http://large.stanford.edu/courses/2011/ph240/nguyen1/docs/SSME_PRESENTATION.pdf
5. Kolowith R (1966) Saturn IB Program, Test report for flexible hose, 1/4 in, 3000-PSIG, Chrysler Corporation, Space Division, 20 December 1966. Web site https://ntrs.nasa.gov/archive/nasa/casi.ntrs.nasa.gov/19670019497.pdf
6. Burcham RE, Keller RB Jr (eds) (1978) Liquid rocket engine turbopump rotating-shaft seals, NASA SP-8121, February 1978, 168 p. Web site https://ntrs.nasa.gov/archive/nasa/casi.ntrs.nasa.gov/19780022641.pdf
7. Anonymous, Saturn V flight manual SA 503, Technical Manual MSFC-MAN-503, NASA TM X-72151, 243 pages, November 1968. Web site https://ntrs.nasa.gov/archive/nasa/casi.ntrs.nasa.gov/19750063889.pdf
8. NASA, Saturn V news reference, J-2 engine fact sheet, December 1968. Web site https://www.nasa.gov/centers/marshall/pdf/499245main_J2_Engine_fs.pdf
9. Sobin AJ, Bissel WR, Keller RB Jr (eds) (1974) Turbopump systems for liquid rocket engines, NASA SP-8107, August 1974, 168 p. Web site https://ntrs.nasa.gov/archive/nasa/casi.ntrs.nasa.gov/19750012398.pdf
10. Anonymous, Cryogenic materials data handbook, U.S. Department of Commerce, National Bureau of Standards, 1960, 58 p. Web site https://apps.dtic.mil/dtic/tr/fulltext/u2/a286675.pdf
11. Stuck DE, Keller RB Jr (eds) (1976) Liquid rocket disconnects, couplings, fittings, fixed joints, and seals, NASA SP-8119, 164 p. Web site https://ntrs.nasa.gov/archive/nasa/casi.ntrs.nasa.gov/19770017247.pdf
12. Kazakov NF (ed) (1985) Diffusion bonding of materials. Pergamon Press, Oxford. ISBN 0-08-032550-5
13. Anonymous (2009) Standard welding terms and definitions including terms for adhesive bonding, brazing, soldering, and thermal spraying, AWS A3.0 M/A3.0:2010, American Welding Society, 12th edn. ISBN: 978-0-87171-763-4
14. Anderson RV, Jaquay KR (1975) Flexible piping joints for large-scale breeder reactor primary loop applications, Phase I, Technical summary, report, 12 November 1975, United States of America. Web site https://digital.library.unt.edu/ark:/67531/metadc1195050/
15. French EA, George H (1976) Flexible hose guide for multiplane kinematics. In: Fluid handling equipment—a compilation, NASA SP-5976 (03):25. Web site https://ntrs.nasa.gov/archive/nasa/casi.ntrs.nasa.gov/19760013403.pdf
16. Buckingham JR, Winzen J (1974) Shuttle filter study—final report, vol I, NASA technical report CR-140386, June 1974, 158 p. Web site https://ntrs.nasa.gov/archive/nasa/casi.ntrs.nasa.gov/19750005841.pdf
17. Toth LR, Hagler R Jr (1975) Full-flow fluid filter, NASA Tech Brief, B74-10277, January 1975. Web site https://ntrs.nasa.gov/archive/nasa/casi.ntrs.nasa.gov/19740000277.pdf
18. Filters and filter elements, fluid pressure, hydraulic micronic type, Military specification MIL-F-5504B, 17 October 1958. Web site http://everyspec.com/MIL-SPECS/MIL-SPECS-MIL-F/MIL-F-5504B_20463/

19. Filter and filter elements, fluid pressure, hydraulic line, 15 micron absolute and 5 micron absolute, type II systems, Military specification MIL-F-8815D, 27 September 1976. Web site http://everyspec.com/MIL-SPECS/MIL-SPECS-MIL-F/MIL-F-8815D_38187/

20. Howell GW, Weathers TM (eds) (1970) Aerospace fluid component designers' handbook, volume II, Revision D, Report No. RPL-TDR-64-25, TRW Systems Group, One Space Park, Redondo Beach, California, USA, February 1970. Web site https://ntrl.ntis.gov/NTRL/dashboard/searchResults/titleDetail/AD874543.xhtml

21. Scott Laboratories, Bubble point test, Appalachian State University. Web site https://wine.appstate.edu/sites/wine.appstate.edu/files/Bubble%20Point%20Test.pdf

22. ISO 4003:1977, Permeable sintered metal materials—determination of bubble test pore size. Web site https://www.iso.org/standard/9678.html

23. SAE International, Bubble-point test method ARP901A. Web site https://www.sae.org/standards/content/arp901a/

24. Filtration Group, Glossary of filtration terms. Web site https://dm.energy/sites/default/files/Glossary%20of%20Filtration%20Terms_20190619.pdf

25. Kozeny J (1927) Über kapillare Leitung des Wassers im Boden, Sitzungsberichte der Akademie der Wissenschaften, Wien, 136(2a):271–306. Web site https://www.zobodat.at/pdf/SBAWW_136_2a_0271-0306.pdf

26. Anonymous, Freestudy, Free tutorials on engineering and science, Fluid mechanics, Tutorial No. 4, Flow through porous passages. Web site http://www.freestudy.co.uk/fluid%20mechanics/t4203.pdf

27. Special Metals Corporation, Inconel® alloy 718, Pub. No. SMC-045. Web site http://www.specialmetals.com/assets/smc/documents/inconel_alloy_718.pdf

28. Anonymous (1979) Modern flange design, Bulletin 502, Edition VII, Taylor Forge Engineered Systems, Inc., 47 p. Web site https://www.steeltank.com/Portals/0/docs/Modern%20Flange%20Design%20-%20Taylor%20Forge%20Engineered%20Products.pdf

29. Rathbun FO Jr (ed) (1964) Separable connector design handbook (tentative), General Electric Company, NASA CR-64944, December 1964, 269 p. Web site https://ntrs.nasa.gov/archive/nasa/casi.ntrs.nasa.gov/19740079014.pdf

30. Aeroquip Corporation, Aerospace Engineering Bulletin, AEB 197A, 1980. Web site https://www.herberaircraft.com/pdf/117Cat/Clamps/AEB197A.pdf

31. Anderson WF (1964) Analysis of stresses in bellows, Part 1, Design criteria and test results, Report NAA-SR-4527 (Pt. 1), United States of America, Atomic Energy Commission, Division of Technical Information, 15 October 1964. Web site https://www.osti.gov/servlets/purl/4676097

Printed in the United States
by Baker & Taylor Publisher Services